Lecture Notes in Mathematics

Volume 2398

This series reports on new developments in all areas of mathematics and their applications - quickly, informally and at a high level. Mathematical texts analysing new developments in modelling and numerical simulation are welcome. The type of material considered for publication includes:

1. Research monographs
2. Lectures on a new field or presentations of a new angle in a classical field
3. Summer schools and intensive courses on topics of current research.

Texts which are out of print but still in demand may also be considered if they fall within these categories. The timeliness of a manuscript is sometimes more important than its form, which may be preliminary or tentative. Please visit the LNM Editorial Policy (https://drive.google.com/file/d/1MOg4TbwOSokRnFJ3ZR3ciEeKs9hOnNX_/view?usp=sharing)

Titles from this series are indexed by Scopus, Web of Science, Mathematical Reviews, and zbMATH.

Claus Hertling · Khadija Larabi

Triangular Bases of Unimodular Bilinear Lattices and Induced Structures

Automorphism Groups, Vanishing Cycles, Monodromy Groups, Braid Group Actions and Moduli Spaces

Springer

Claus Hertling
School of Business Informatics
and Mathematics
University of Mannheim
Mannheim, Baden-Württemberg, Germany

Khadija Larabi
School of Business Informatics
and Mathematics
University of Mannheim
Mannheim, Baden-Württemberg, Germany

ISSN 0075-8434 ISSN 1617-9692 (electronic)
Lecture Notes in Mathematics
ISBN 978-3-032-25996-7 ISBN 978-3-032-25997-4 (eBook)
https://doi.org/10.1007/978-3-032-25997-4

Mathematics Subject Classification: 57M10, 32S30, 06B15, 14D05, 20F55, 20F36

This work was funded by the Deutsche Forschungsgemeinschaft (DFG, German Research Foundation)—494849004.

This Springer imprint is published by the registered company Springer Nature Switzerland AG
The registered company address is: Gewerbestrasse 11, 6330 Cham, Switzerland

Preface

This monograph starts with an upper triangular matrix with integer entries and 1's on the diagonal. It develops from this a spectrum of structures, which appear in different contexts, in algebraic geometry, representation theory and the theory of irregular meromorphic connections. It provides general tools to study these structures, and it studies systematically the cases of rank 2 and 3. The rank 3 cases lead already to a rich variety of phenomena and give an idea of the general landscape. Their study takes up a large part of the monograph.

Special cases are related to Coxeter groups, generalized Cartan lattices and exceptional sequences, or to isolated hypersurface singularities, their Milnor lattices and their distinguished bases. But these make only a small part of all cases. One case in rank 3 which is beyond them, is related to quantum cohomology of $\mathbb{P}^2$ and to Markov triples.

The first structure associated to the matrix is a $\mathbb{Z}$-lattice with unimodular bilinear form (called Seifert form) and a triangular basis. It leads immediately to an even and an odd intersection form, reflections and transvections, an even and an odd monodromy group, even and odd vanishing cycles. Braid group actions lead to braid group orbits of distinguished bases and of upper triangular matrices.

Finally, complex manifolds, which consist of correctly glued Stokes regions, are associated to these braid group orbits. A report on the case of isolated hypersurface singularities concludes the monograph.

Mannheim, Germany

Claus Hertling
Khadija Larabi

Competing Interests The authors have no competing interests to declare that are relevant to the content of this manuscript.

Contents

Chapter 1
Introduction

This book develops many structures, starting from a single upper triangular $n \times n$ matrix S with integer entries and 1's on the diagonal. The structures are introduced and are called playing characters in Sect. 1.1.

Such a matrix S and induced structures appear in very different mathematical areas. Section 1.2 gives a panorama of these areas. Though these areas are not the subject of the book, except for the area of isolated hypersurface singularities.

Section 1.3 describes the results in this book. The book provides general tools and facts. It treats the cases $n = 2$ and $n = 3$ systematically. In Chap. 10 it overviews the area of isolated hypersurface singularities.

1.1 Playing Characters

$H_{\mathbb{Z}}$ will always be a $\mathbb{Z}$-*lattice*, so a free $\mathbb{Z}$-module of some finite rank $n \in \mathbb{N} = \{1, 2, 3...\}$. Then L will always be a nondegenerate bilinear form $L : H_{\mathbb{Z}} \times H_{\mathbb{Z}} \to \mathbb{Z}$. It is called a *Seifert form*. The pair $(H_{\mathbb{Z}}, L)$ is called a *bilinear lattice*. If for some $\mathbb{Z}$-basis $\underline{e} \in M_{1\times n}(H_{\mathbb{Z}})$ of $H_{\mathbb{Z}}$ the determinant $\det L(\underline{e}^t, \underline{e})$ is 1 the pair $(H_{\mathbb{Z}}, L)$ is called a *unimodular bilinear lattice*. The term *bilinear lattice* is taken from [HK16]. In Chap. 2 we develop the following structures for bilinear lattices, following [HK16]. Though in this introduction and in the Chaps. 3–10 we restrict to unimodular bilinear lattices.

$$T_n(\mathbb{Z}) := \{S \in M_{n\times n}(\mathbb{Z}) \mid S_{ij} = 0 \text{ for } i > j,\ S_{ii} = 1\}$$

denotes the set of all upper triangular matrices with integer entries and 1's on the diagonal.

Let $(H_{\mathbb{Z}}, L)$ be a unimodular bilinear lattice of rank n. A basis $\underline{e}$ of $H_{\mathbb{Z}}$ is called *triangular* if $L(\underline{e}^t, \underline{e})^t \in T_n(\mathbb{Z})$. The transpose in the matrix is motivated by the case

C. Hertling and K. Larabi, *Triangular Bases of Unimodular Bilinear Lattices and Induced Structures*, Lecture Notes in Mathematics 2398,
https://doi.org/10.1007/978-3-032-25997-4_1

of isolated hypersurface singularities. The set of triangular bases is called $\mathcal{B}^{tri}$. By far not every unimodular bilinear lattice has triangular bases. But here we care only about those which have.

For fixed $n \in \mathbb{N}$ there is an obvious 1-1 correspondence between the set of isomorphism classes of unimodular bilinear lattices $(H_\mathbb{Z}, L, \underline{e})$ with triangular bases and the set $T_n(\mathbb{Z})$, given by the map $(H_\mathbb{Z}, L, \underline{e}) \mapsto S := L(\underline{e}^t, \underline{e})^t$. To a given matrix $S \in T_n(\mathbb{Z})$ we always associate the corresponding triple $(H_\mathbb{Z}, L, \underline{e})$.

Let a unimodular bilinear lattice $(H_\mathbb{Z}, L, \underline{e})$ with a triangular basis $\underline{e}$ be given, with matrix $S = L(\underline{e}^t, \underline{e})^t \in T_n(\mathbb{Z})$. The following objects are associated to this triple canonically. The names are motivated by the case of isolated hypersurface singularities.

(i) A symmetric bilinear form $I^{(0)} : H_\mathbb{Z} \times H_\mathbb{Z} \to \mathbb{Z}$ and a skew-symmetric bilinear form $I^{(1)} : H_\mathbb{Z} \times H_\mathbb{Z} \to \mathbb{Z}$ with

$$\begin{aligned} I^{(0)} &= L^t + L, \quad \text{so } I^{(0)}(\underline{e}^t, \underline{e}) = S + S^t, \\ I^{(1)} &= L^t - L, \quad \text{so } I^{(1)}(\underline{e}^t, \underline{e}) = S - S^t, \end{aligned}$$

which are called *even* and *odd intersection form*, respectively.

(ii) An automorphism $M : H_\mathbb{Z} \to H_\mathbb{Z}$ which is defined by

$$L(Ma, b) = L(b, a), \quad \text{so } M(\underline{e}) = \underline{e} \cdot S^{-1} S^t,$$

and called *the monodromy*. It respects L (and $I^{(0)}$ and $I^{(1)}$) because $L(Ma, Mb) = L(Mb, a) = L(a, b)$.

(iii) Six automorphism groups

$$\begin{aligned} O^{(k)} &:= \mathrm{Aut}(H_\mathbb{Z}, I^{(k)}) \qquad \text{for } k \in \{0; 1\}, \\ G_\mathbb{Z}^M &:= \mathrm{Aut}(H_\mathbb{Z}, M) := \{g : H_\mathbb{Z} \to H_\mathbb{Z} \text{ automorphism} \mid gM = Mg\}, \\ G_\mathbb{Z}^{(k)} &:= \mathrm{Aut}(H_\mathbb{Z}, I^{(k)}, M) = O^{(k)} \cap G_\mathbb{Z}^M \qquad \text{for } k \in \{0; 1\}, \\ G_\mathbb{Z} &:= \mathrm{Aut}(H_\mathbb{Z}, L) = \mathrm{Aut}(H_\mathbb{Z}, L, I^{(0)}, I^{(1)}, M). \end{aligned}$$

(iv) The set of *L1-cycles*

$$R^{(0)} := \{a \in H_\mathbb{Z} \mid L(a, a) = 1\},$$

and the set

$$R^{(1)} := H_\mathbb{Z}.$$

(v) For $k \in \{0; 1\}$ and $a \in R^{(k)}$ the *reflection* (if $k = 0$) respectively *transvection* (if $k = 1$) $s_a^{(k)} \in O^{(k)}$ with

$$s_a^{(k)}(b) := b - I^{(k)}(a, b)a \quad \text{for } b \in H_\mathbb{Z}.$$

(vi) For $k \in \{0; 1\}$ the *even* (if $k = 0$) respectively *odd* (if $k = 1$) *monodromy group*

$$\Gamma^{(k)} := \langle s_{e_1}^{(k)}, ..., s_{e_n}^{(k)} \rangle \subset O^{(k)}.$$

(vii) For $k \in \{0; 1\}$ the set of *even* (if $k = 0$) respectively *odd* (if $k = 1$) *vanishing cycles*

$$\Delta^{(k)} := \Gamma^{(k)}\{\pm e_1, ..., \pm e_n\} \subset R^{(k)}.$$

The definitions of all these objects require only $S \in SL_n(\mathbb{Z})$ and $\underline{e} = (e_1, ..., e_n) \in M_{1,n}(R^{(0)})$ a $\mathbb{Z}$-basis of $H_\mathbb{Z}$, not $S \in T_n(\mathbb{Z})$. But the formula (Theorem 2.6)

$$s_{e_1}^{(k)} ... s_{e_n}^{(k)} = (-1)^{k+1} M \quad \text{for } k \in \{0; 1\}$$

depends crucially on $S \in T_n(\mathbb{Z})$.

The even data $I^{(0)}$, $O^{(0)}$, $\Gamma^{(0)}$ and $\Delta^{(0)}$ are in many areas more important and are usually better understood than the odd data $I^{(1)}$, $O^{(1)}$, $\Gamma^{(1)}$ and $\Delta^{(1)}$. But in the area of isolated hypersurface singularities both turn up.

For $k \in \{0; 1\}$ the group $\Gamma^{(k)}$ contains all reflections/transvections $s_a^{(k)}$ with $a \in \Delta^{(k)}$. In the case of a bilinear lattice which is not unimodular this holds for $k = 0$, but not for $k = 1$ (Remark 2.9 (iii)). This is one reason why we restrict in the chapters 3–10 to unimodular bilinear lattices.

Section 3.2 gives an action of a semidirect product $\mathrm{Br}_n \ltimes \{\pm 1\}^n$ of the braid group Br_n of braids with n strings and of a sign group $\{\pm 1\}^n$ on the set $(R^{(k)})^n$ for $k \in \{0; 1\}$. It is compatible with the Hurwitz action of Br_n on $(\Gamma^{(k)})^n$ with connecting map

$$(R^{(k)})^n \to (\Gamma^{(k)})^n, \quad \underline{v} = (v_1, ..., v_n) \mapsto (s_{v_1}^{(k)}, ..., s_{v_n}^{(k)}).$$

The actions on $(R^{(k)})^n$ for $k \in \{0; 1\}$ both restrict to the same action on $\mathcal{B}^{tri}$. Especially, one obtains the orbit

$$\mathcal{B}^{dist} := \mathrm{Br}_n \ltimes \{\pm 1\}^n(\underline{e}) \subset \mathcal{B}^{tri}$$

of *distinguished bases* of $H_\mathbb{Z}$. The triple $(H_\mathbb{Z}, L, \mathcal{B}^{dist})$ (up to isomorphism) is in many cases a canonical object, whereas the choice of a distinguished basis $\underline{e} \in \mathcal{B}^{dist}$ is a true choice. The question whether $\mathcal{B}^{dist} = \mathcal{B}^{tri}$ or $\mathcal{B}^{dist} \subsetneqq \mathcal{B}^{tri}$ is usually a difficult question. The subgroup

$$G_\mathbb{Z}^\mathcal{B} := \{g \in G_\mathbb{Z} \mid g(\mathcal{B}^{dist}) = \mathcal{B}^{dist}\}$$

is in many important cases equal to $G_\mathbb{Z}$. But if $\mathcal{B}^{dist} \subsetneqq \mathcal{B}^{tri}$, then $G_\mathbb{Z}^\mathcal{B} \subsetneqq G_\mathbb{Z}$ is possible.

The action of $\mathrm{Br}_n \ltimes \{\pm 1\}^n$ on $\mathcal{B}^{tri}$ is compatible with an action on $T_n(\mathbb{Z})$. The orbit of S is called

$$\mathcal{S}^{dist} := \mathrm{Br}_n \ltimes \{\pm 1\}^n(S) \subset T_n(\mathbb{Z}),$$

the matrices in it are called *distinguished matrices*. As $\{\pm 1\}^n$ is the normal subgroup in the semidirect product $\mathrm{Br}_n \ltimes \{\pm 1\}^n$, one can first divide out the action of $\{\pm 1\}^n$. One obtains actions of Br_n on $\mathcal{B}^{tri}/\{\pm 1\}^n$ and on $T_n(\mathbb{Z})/\{\pm 1\}^n$. It will be interesting to determine the stabilizers $(\mathrm{Br}_n)_{\underline{e}/\{\pm 1\}^n}$ of $\underline{e}/\{\pm 1\}^n$ and $(\mathrm{Br}_n)_{S/\{\pm 1\}^n}$ of $S/\{\pm 1\}^n$.

Finally, Chap. 8 introduces several complex manifolds which are in fact *semisimple F-manifolds with Euler fields* (Definition 8.8). First, the following manifolds are universal and depend only on $n \in \mathbb{N}$,

$$\begin{aligned} C_n^{pure} &:= \{\underline{u} = (u_1, ..., u_n) \in \mathbb{C}^n \mid u_i \neq u_j \text{ for all } i \neq j\}, \\ C_n^{conf} &:= C_n^{pure}/S_n \subset \mathbb{C}^n/S_n \cong \mathbb{C}^n, \\ C_n^{univ} &:= \text{the universal covering of } C_n^{conf} \text{ and } C_n^{pure}. \end{aligned}$$

C_n^{conf} is the configuration space of Br_n, with $\pi_1(C_n^{conf}) = \mathrm{Br}_n$. The subset of C_n^{pure}

$$F_n := \{\underline{u} \in C_n^{pure} \mid \mathrm{Im}(u_1) < ... < \mathrm{Im}(u_n)\}$$

is a fundamental domain of the action of the symmetric group S_n on C_n^{pure} and maps under the projection $C_n^{pure} \to C_n^{conf}$ almost bijectively to C_n^{conf}.

Second, two manifolds $C_n^{\underline{e}/\{\pm 1\}^n}$ and $C_n^{S/\{\pm 1\}^n}$ between C_n^{conf} and C_n^{univ} are constructed in Sect. 8.1. They are those coverings of C_n^{conf} whose fundamental groups embed into $\mathrm{Br}_n = \pi_1(C_n^{conf})$ as the stabilizer $(\mathrm{Br}_n)_{\underline{e}/\{\pm 1\}^n}$ respectively the stabilizer $(\mathrm{Br}_n)_{S/\{\pm 1\}^n}$. In a more naive way, they are obtained by glueing copies of F_n in an appropriate way, one copy for each element of $\mathcal{B}^{dist}/\{\pm 1\}^n$ respectively $\mathcal{S}^{dist}/\{\pm 1\}^n$. These manifolds, especially $C_n^{\underline{e}/\{\pm 1\}^n}$, are related to manifolds from algebraic geometry which are associated to triples $(H_\mathbb{Z}, L, \underline{e})$, and they have to be compared with these manifolds.

1.2 The Playing Characters in Action: Areas, Literature

Upper triangular integer matrices S with 1's on the diagonal and the induced structures, which are described in Sect. 1.1, arise in many different contexts. Figure 1.1 offers a landscape of mathematical areas where they arise. The lines between the boxes indicate connections. Of course, the chart is very incomplete and subjective and does neither give all areas nor all connections between the mentioned areas.

In the following, for each of the boxes in Fig. 1.1 (and for a few more points), we provide brief explanations and give a few references. One can start with these references and use them to find more references.

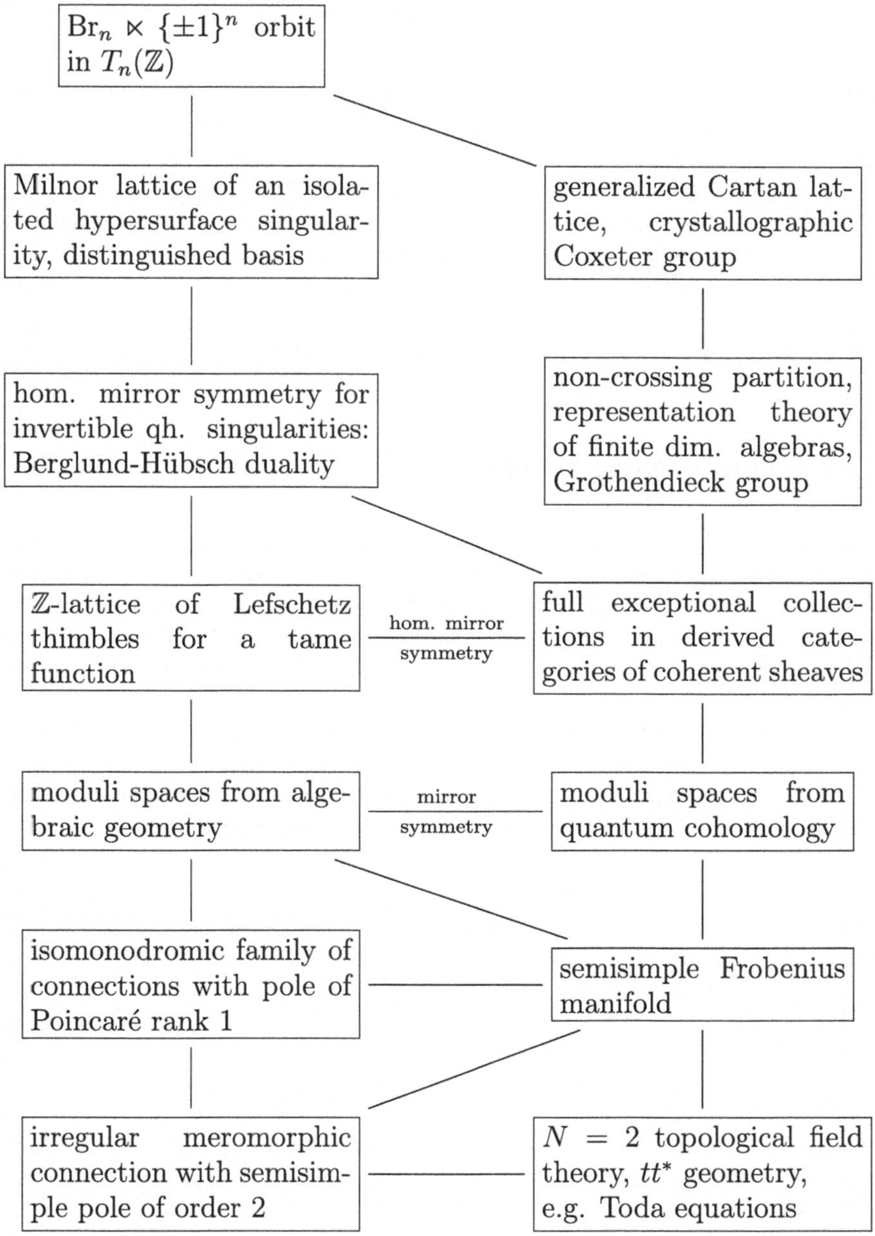

Fig. 1.1 Mathematical areas where upper triangular integer matrices with 1's on the diagonal appear

The mathematical areas in these boxes are not the subject of this book, except for the area of isolated hypersurface singularities, which is treated in Chap. 10. (Therefore a reader could skip the rest of this Sect. 1.2.)

1.2.1 Isolated Hypersurface Singularities

Historically the first appearance of unimodular bilinear lattices $(H_{\mathbb{Z}}, L, \underline{e})$ with triangular bases $\underline{e}$ is in the theory of isolated hypersurfaces singularities. There they arise as Milnor lattices with chosen distinguished bases [Br70, Appendix]. All distinguished bases together form a $\mathrm{Br}_n \ltimes \{\pm 1\}^n$ orbit $\mathcal{B}^{dist}$. A single distinguished basis $\underline{e}$ is a choice. The triple $(H_{\mathbb{Z}}, L, \mathcal{B}^{dist})$ is a canonical object associated to a singularity. These structures were studied a lot ([AGV88][AGLV98][Eb01][Eb20] and references therein). Nevertheless it is not clear how to characterize the triples $(H_{\mathbb{Z}}, L, \underline{e})$ or (equivalently) the matrices $S \in T_n(\mathbb{Z})$ which come from singularities with Milnor number n. See Chap. 10 for more details and more references.

1.2.2 Quiver Representations

A quiver is a directed (finite) multigraph $Q = (Q_0, Q_1)$ where $Q_0 = \{1, ...n\}$ for some $n \in \mathbb{N}$ is the set of vertices and Q_1 is the set of oriented edges. Several edges with the same head and tail are allowed. A quiver gives rise to its *Euler matrix* $E \in M_{n\times n}(\mathbb{Z})$ with $E_{ij} := \delta_{ij} - |\{\text{edges from } i \text{ to } j\}|$ and its *Cartan matrix* $C := E + E^t$. If the quiver is *ordered* (edges only from i to j with $i < j$ [Bo90, 5.]) then $E =: S \in T_n(\mathbb{Z})$.

A representation of a quiver associates to every vertex i a finite dimensional k-vector space V_i (for a fixed field k) and to an edge from i to j a linear map $V_i \to V_j$. An important problem is to classify all indecomposable representations. Each indecomposable representation leads to a dimension vector $(\dim V_i)_{i\in Q_0}$.

Consider the case $E = S \in T_n(\mathbb{Z})$ and the associated triple $(H_{\mathbb{Z}}, L, \underline{e})$. Then L is the *Euler form* and $e_1, ..., e_n$ are the *simple roots*. The dimension vectors give rise to *real roots*, *imaginary roots* and *Schur roots* in $H_{\mathbb{Z}}$ [Ka90][DW05][Ki16][DW17]. The set of real roots is the set $\Delta^{(0)}$. In [HK16] the elements of $R^{(0)}$ are called *pseudo-real roots* or just *roots*. We took the name *L1-cycles*.

The quivers and their representations motivated Hubery and Krause [HK16] to consider bilinear lattices $(H_{\mathbb{Z}}, L)$ with triangular bases $\underline{e}$. Chapter 2 in this book takes up their definition. Though the Chaps. 3–10 in this book restrict to *unimodular* bilinear lattices with triangular bases. One reason is that we consider, contrary to [HK16], also the odd monodromy group $\Gamma^{(1)}$ and the set $\Delta^{(1)}$ of odd vanishing cycles. They do not seem to behave well in the case of a not unimodular bilinear lattice with triangular basis (Remark 2.9 (iii)).

[HK16] is interested especially in the case of a *generalized Cartan lattice* (Remark 2.4 (v)). In the case of $E = S \in T_n(\mathbb{Z})$ this means $S_{ij} \leq 0$ for $i < j$, which holds by definition of $E = S$. The intersection with the case of isolated hypersurface singularities consists only of the ADE-singularities.

Every generalized Cartan lattice arises as the Grothendieck group of a hereditary Artin algebra [HK16, Proposition 4.6]. A crucial result here (cited in Theorem 3.6) is the surjectivity of the braid group action on the set of those n-tuples of reflections in $\Gamma^{(0)}$, whose product is $-M$ [IS10][BDSW14]. Using this and additional work, [HK16] can reprove results of Ingalls & Thomas, Igusa & Schiffler, Crawley-Boevey and Ringel on exceptional sequences and non-crossing partitions related to finite dimensional modules of finite dimensional hereditary algebras. [BBGKK19] gives a survey on non-crossing partitions.

1.2.3 Full Exceptional Collections in Derived categories of Coherent Sheaves

Not only the people mentioned above, but also the Moscow school studied full exceptional collections in triangulated categories, with special emphasis on the derived categories of coherent sheaves on algebraic manifolds. These include Rudakov, Gorodentsev, Bondal, Kapranov [Ru90], Orlov, Kuznetsov and others. Such a derived category is (under sufficiently good conditions) equivalent to the derived category of modules of the path algebra of a quiver with relations [Bo90, Theorem 6.2]. Not the semidirect product $\mathrm{Br}_n \ltimes \{\pm 1\}^n$, but a semidirect product $\mathrm{Br}_n \ltimes \mathbb{Z}^n$ acts on the set of full exceptional collections [Bo90][BP94]. The ICM talks [BO02], [Ku14] and [Ku23] are devoted to these structures. The introduction and §1 in [Bo04] offer a bridge to mirror symmetry and functions with isolated singularities.

The Grothendieck group with Euler form is a unimodular bilinear lattice $(H_{\mathbb{Z}}, L)$ with triangular bases if full exceptional collections exist. Then the monodromy M is induced by the *Serre operator*. In the case of a projective manifold of dimension d, $(-1)^d M$ is unipotent [BP94, Lemma 3.1].

The Chaps. 3 and 4 of the book [CDG24] give a survey on helices in triangulated categories, the Grothendieck group of a small triangulated category and the derived category of coherent sheaves of an algebraic manifold. It also recalls results of Gorodentsev [Go94-1][Go94-2] on pairs $(H_{\mathbb{C}}, L)$ over $\mathbb{C}$. They split into indecomposable objects, which can be classified. A finer classification of pairs $(H_{\mathbb{R}}, L)$ over $\mathbb{R}$ was done by Nemethi [Ne98] and in a more explicit way in [BH19]. Also pairs $(H_{\mathbb{R}}, L)$ split into indecomposable objects, which can be classified. Pairs $(H_{\mathbb{R}}, L)$ up to isomorphism are in 1-1 correspondence with tuples of *spectral pairs* where the first entry is modulo $2\mathbb{Z}$.

A unimodular bilinear lattice with monodromy $(H_{\mathbb{Z}}, L, M)$ as an object in its own right is named and studied besides [Go94-1][Go94-2] only in more recent papers. It is called *unimodular Mukai lattice with canonical operator* in [CDG24], *Serre lattice* in [dTVdB16], *unimodular pseudolattice with Serre operator* in [Ku17]. The last two references study cases related to algebraic surfaces. There M is unipotent by [BP94, Lemma 3.1], and some more conditions hold. The condition to have triangular bases comes as an additional condition.

1.2.4 Meromorphic Connections with Poles Of Poincaré Rank 1

Upper triangular matrices $S \in T_n(\mathbb{C})$ arise also as the carrier of the Stokes structure of a holomorphic vector bundle on the unit disk $\mathbb{D}$ with a (flat) meromorphic connection with a semisimple order 2 pole at $0 \in \mathbb{D}$ and pairwise different eigenvalues of the pole part. After some choice this structure is given by a tuple $\underline{u} \in C_n^{pure} \subset \mathbb{C}^n$ of eigenvalues of the pole part, two upper triangular matrices in $T_n(\mathbb{C})$, which are called *Stokes matrices*, and a diagonal matrix carrying the exponents of the formal connection [Ma83][Sa02][HS11].

Generic deformations of isolated hypersurface singularities as well as tame functions on affine manifolds give rise to such a connection via oscillating integrals of a distinguished basis of Lefschetz thimbles [Ph83][Ph85]. Then the two Stokes matrices coincide and are in $T_n(\mathbb{Z})$, and $\underline{u}$ is the tuple of critical values.

Varying $\underline{u}$ in a suitable covering of C_n^{conf} (the best is the covering $C_n^{\underline{e}/\{\pm 1\}^n}$ in the notation of Chap. 8) gives rise to a holomorphic bundle on $\mathbb{D} \times C_n^{\underline{e}/\{\pm 1\}^n}$ with a meromorphic connection with a pole of Poincaré rank 1 along $\{0\} \times C_n^{\underline{e}/\{\pm 1\}^n}$ [Sa02].

Some additional choices and enrichments allow to construct from this isomonodromic family of connections the structure of a Frobenius manifold on the covering $C_n^{\underline{e}/\{\pm 1\}^n}$ [Sa02][He03][Du99]. Chapter 8 describes the underlying F-manifold.

1.2.5 Mirror Symmetry

Also quantum cohomology gives rise to Frobenius manifolds. The construction is described e.g. in [Ma99]. One version of mirror symmetry says that in good cases the Frobenius manifold from the quantum cohomology of some toric Fano manifold is isomorphic to the Frobenius manifold from oscillating integrals of some family of functions with isolated singularities. This builds on a mirror symmetry result of Givental [Gi98] and was established in [RS15]. These Frobenius manifolds are generically semisimple and also in other aspects rather special, as they come from algebraic geometry.

Iritani considered and compared the integral structures on both sides [Ir09]. The integral structure on the A-side uses some *Gamma classes*.

A homological mirror symmetry for polynomials with singularities was proposed by Takahashi. It compares structures above pairs of invertible polynomials. One result for chain type singularities is in [AT20].

Mirror symmetry comes from mathematical physics. The paper [CV93] of the physicist Cecotti and Vafa is not on mirror symmetry, but it starts with matrices in $T_n(\mathbb{Z})$ and builds from them rich structures: first certain meromorphic connection which are formalized as tt^* geometry in [He03], then in good cases $N = 2$ topological field theories behind them. As concrete cases, [CV93] classifies and interprets those

$\mathrm{Br}_3 \ltimes \{\pm 1\}^3$ orbits of matrices $S \in T_3(\mathbb{Z})$ whose monodromy matrices $S^{-1}S^t$ have roots of unity as eigenvalues. We refine this in Theorem 4.6. [GL12] studies certain higher rank cases, which are related to the Toda equations. Also [GL12] considers especially the cases related to matrices in $T_n(\mathbb{Z})$.

1.2.6 Some Problems

Finally we list some problems, smaller and bigger ones.

(P1) This book studies systematically the rank 2 and rank 3 cases. The next cases are the rank 4 cases. van Straten [vS09] has some unpublished work on the $\mathrm{Br}_4 \ltimes \{\pm 1\}^4$ orbits of matrices $S \in T_4(\mathbb{Z})$ whose monodromy $S^{-1}S^t$ has roots of unity as eigenvalues. [dTVdB16] gives results on the cases when $S^{-1}S^t$ is unipotent. In view of this book, much more can be asked about.

(P2) An important question for each unimodular bilinear lattice $(H_\mathbb{Z}, L)$ with triangular bases is whether all $\mathbb{Z}$-bases $\underline{v}$ of $H_\mathbb{Z}$ with $s^{(k)}_{v_1} \dots s^{(k)}_{v_n} = (-1)^{k+1}M$ (for a chosen $k \in \{0; 1\}$) are in a single $\mathrm{Br}_n \ltimes \{\pm 1\}^n$ orbit. Ten other natural questions for $(H_\mathbb{Z}, L)$ are posed at the beginning of the next Sect. 1.3.

(P3) In arbitrary rank, one can try to understand special families of matrices in $T_n(\mathbb{Z})$. [Ku17] gives a beautiful example.

(P4) [CHS25] give counterexamples to Conjecture 2.2 in [BP94] on the transitivity of the action of $\mathrm{Br}_n \ltimes \mathbb{Z}^n$ on full exceptional collections in triangulated categories which have full exceptional collections. The proof in [CHS25] is geometric. [NT25] gives a simpler algebraic proof. Can this be understood in a still simpler way on the level of $(H_\mathbb{Z}, L)$? This would relate it to (P2) and to the question when the inclusion $\mathcal{B}^{dist} \subset \mathcal{B}^{tri}$ is an equality.

(P5) The results in this book on the pair $(H_\mathbb{Z}, L)$ which is associated to the matrix $S(\mathcal{H}_{1,2}) \in T_3(\mathbb{Z})$ in Example 1.1 should find their place in relation to the survey on algebraic geometric structures (three constructions of a triangulated category, a Frobenius manifold) which are associated to $S(\mathcal{H}_{1,2})$ in [NST25].

(P6) This book contains quite some results on the pair $(H_\mathbb{Z}, L)$ which is associated to the matrix $S(\mathbb{P}^2) \in T_3(\mathbb{Z})$ in Example 1.1: $\Delta^{(0)} = R^{(0)}$, $\Gamma^{(0)}$, $\Gamma^{(1)}$, some control on $\Delta^{(1)}$, $\mathcal{B}^{dist} = \mathcal{B}^{tri} = \{\underline{v} \in (\Delta^{(k)})^3 \mid s^{(k)}_{v_1} s^{(k)}_{v_2} s^{(k)}_{v_3} = (-1)^{k+1}M\}$, the space $C_3^{\underline{e}/\{\pm 1\}^3} \cong C_3^{univ} \cong \mathbb{C}^2 \times \mathbb{H}$. Theorem 4.6 (e) recovers the (well known) surjectivity of the action of Br_3 on the set $\{\underline{x} \in \mathbb{N}^3 \mid x_1^2 + x_2^2 + x_3^2 = x_1 x_2 x_3\}$ of *Markov triples*. One could try to attack the uniqueness conjecture for Markov triples [Fr13] (see also [Ai13]), which says that a Markov triple is (up to ordering) uniquely determined by its biggest number. One should also compare $C_3^{\underline{e}/\{\pm 1\}^3}$ with the Frobenius manifold from quantum cohomology of $\mathbb{P}^2$.

(P7) How directly can the real roots, imaginary root and Schur roots in $H_\mathbb{Z}$ which come from a ordered quiver be derived from the associated unimodular bilinear lattice $(H_\mathbb{Z}, L, \underline{e})$ with triangular basis? Which are good cases, which are bad cases? [DW05] gives a bad rank 3 case.

(P8) How can the matrices in $T_n(\mathbb{Z})$ which come from distinguished bases of isolated hypersurface singularities with Milnor number n be separated from the other matrices in $T_n(\mathbb{Z})$? The paper [BH25] answers this for those matrices $S \in T_n(\mathbb{Z})$ for which $S + S^t$ is positive definite or positive semidefinite. For the cases when $S + S^t$ is indefinite it only offers some ideas. Other ideas are very welcome.

(P9) One should also mention Dubrovin's conjecture, first because there matrices in $T_n(\mathbb{Z})$ are central objects, and second because it relates several of the structures mentioned above. It relates the monodromy data of a meromorphic connection of a Fano manifold with generically semisimple quantum cohomology to full exceptional collections in the derived category of coherent sheaves on the same Fano manifold [Du98]. It was refined independently by Galkin & Golyshev & Iritani [GGI16] and by Cotti & Dubrovin & Guzzetti [CDG24]. It was proved in a number of cases. It was extended in [SS20] to cases where the quantum cohomology is not generically semisimple.

1.3 Results

Section 1.1 associated to each matrix $S \in T_n(\mathbb{Z})$ an impressive list of algebraic-combinatorial data and also a few complex manifolds. For a given matrix S there are many natural questions which all aim at controlling parts of these data, for example:

(i) What can one say about the $\mathbb{Z}$-lattice $(H_\mathbb{Z}, I^{(0)})$ with the even intersection form, e.g. its signature?
(ii) What can one say about the $\mathbb{Z}$-lattice $(H_\mathbb{Z}, I^{(1)})$ with the odd intersection form?
(iii) What are the eigenvalues and the Jordan block structure of the monodromy M?
(iv) How big are the groups $G_\mathbb{Z}$, $G_\mathbb{Z}^{(0)}$, $G_\mathbb{Z}^{(1)}$ and $G_\mathbb{Z}^M$?
(v) How good can one understand the even monodromy group $\Gamma^{(0)}$? Is it determined by the pair $(H_\mathbb{Z}, I^{(0)})$ alone?
(vi) How good can one understand the odd monodromy group $\Gamma^{(1)}$? Is it determined by the pair $(H_\mathbb{Z}, I^{(1)})$ alone?
(vii) Is $\Delta^{(0)} = R^{(0)}$ or $\Delta^{(0)} \subsetneqq R^{(0)}$? How explicitly can one control these two sets? How explicitly can one control $\Delta^{(1)}$?
(viii) Is there an easy description of the set $\mathcal{B}^{dist}$ of distinguished bases? Is $\mathcal{B}^{dist} = \mathcal{B}^{tri}$ or $\mathcal{B}^{dist} \subsetneqq \mathcal{B}^{tri}$?
(ix) Is $G_\mathbb{Z}^\mathcal{B} = G_\mathbb{Z}$ or $G_\mathbb{Z}^\mathcal{B} \subsetneqq G_\mathbb{Z}$?
(x) How do the manifolds $C_n^{\underline{e}/\{\pm 1\}^n}$ and $C_n^{S/\{\pm 1\}^n}$ look like? Do they have natural partial compactifications?

In the cases from isolated hypersurface singularities, S is quite special. There some of the questions have beautiful answers ((i), (ii), (iii) and (v), partially (vi) and (vii)), others are wide open. The even case is better understood than the odd case. Chapter 10 overviews this.

In the case of a generalized Cartan lattice, some S is easy to give. The even case is well understood ((i), (iii) and (v), partially (vii) and (viii)), the odd case is wide open and has probably much less beautiful structure.

In this book we concentrate on general tools and on the cases of rank 2 and rank 3. The cases of rank 2 are already interesting, but still very special. The cases of rank 3 are still in some sense small, but they show already a wide variety of different types and phenomena. We consider them as sufficiently general to give an idea of the landscape for arbitrary rank $n \in \mathbb{N}$. The large number of pages of this book is due to the systematic study of all cases of rank 3.

Here the singularity cases form just two cases (A_3, A_2A_1), and also the cases from generalized Cartan lattices form a subset which one can roughly estimate as one third of all cases, not containing some of the most interesting cases ($\mathcal{H}_{1,2}$ and $\mathbb{P}^2$).

Examples 1.1 In the following examples, some matrices in $T_2(\mathbb{Z})$ and $T_3(\mathbb{Z})$ are distinguished. They cover the most important cases in $T_2(\mathbb{Z})$ and $T_3(\mathbb{Z})$. This will be made precise in Theorem 1.2, which gives results on the braid group action on $T_3(\mathbb{Z})$.

$$S(A_1^2)\ \begin{pmatrix}1&0\\0&1\end{pmatrix} \quad S(A_2)\ \begin{pmatrix}1&-1\\0&1\end{pmatrix} \quad S(\mathbb{P}^1)\ \begin{pmatrix}1&-2\\0&1\end{pmatrix} \quad S(x) \text{ for } x\in\mathbb{Z}\ \begin{pmatrix}1&x\\0&1\end{pmatrix} \quad S(A_1^3)\ \begin{pmatrix}1&0&0\\0&1&0\\0&0&1\end{pmatrix}$$

$$S(\mathbb{P}^2)\ \begin{pmatrix}1&-3&3\\0&1&-3\\0&0&1\end{pmatrix} \quad S(A_2A_1)\ \begin{pmatrix}1&-1&0\\0&1&0\\0&0&1\end{pmatrix} \quad S(A_3)\ \begin{pmatrix}1&-1&0\\0&1&-1\\0&0&1\end{pmatrix} \quad S(\mathbb{P}^1A_1)\ \begin{pmatrix}1&-2&0\\0&1&0\\0&0&1\end{pmatrix}$$

$$S(\widehat{A}_2)\ \begin{pmatrix}1&-1&-1\\0&1&-1\\0&0&1\end{pmatrix} \quad S(\mathcal{H}_{1,2})\ \begin{pmatrix}1&-2&2\\0&1&-2\\0&0&1\end{pmatrix} \quad S(-l,2,-l) \text{ for } l\geq 3\ \begin{pmatrix}1&-l&2\\0&1&-l\\0&0&1\end{pmatrix} \quad S(x_1,x_2,x_3) \text{ for } x_1,x_2,x_3\in\mathbb{Z}\ \begin{pmatrix}1&x_1&x_2\\0&1&x_3\\0&0&1\end{pmatrix}$$

The notations A_1^2 , A_2, A_1^3, A_2A_1, A_3 and $\widehat{A}_2$ are due to the facts that $(H_\mathbb{Z}, I^{(0)})$ is in these cases the corresponding root lattice respectively in the case $\widehat{A}_2$ the affine root lattice of type $\widehat{A}_2$. The notations $\mathbb{P}^1$ and $\mathbb{P}^2$ come from the quantum cohomology of $\mathbb{P}^1$ and $\mathbb{P}^2$: The matrix S is in these cases one Stokes matrix of the associated meromorphic connection with a semisimple pole of order 2 [Gu99][Du99, Example 4.4 and (4.97)]. The notation $\mathcal{H}_{1,2}$ is related to a Hurwitz space: Here $(H_\mathbb{Z}, L)$ comes from a $\mathbb{Z}$-lattice of Lefschetz thimbles for a branched covering of degree 2 from an elliptic curve to $\mathbb{P}^1$ with 4 simple branch points, one of them above ∞ [Du96, Lecture 5]. A different point of view is that here $(H_\mathbb{Z}, I^{(0)})$ is an extended affine root lattice of type $A_1^{(1,1)*}$. We prefer the notation $\mathcal{H}_{1,2}$.

A large part of this book is devoted to answering the questions above for the cases of rank 2 and 3. Though the Chaps. 2, 3, 8 and the Sects. 5.1 and 6.1 offer also a lot of background material and tools. Chapter 10 overviews the state of the art in the case of isolated hypersurface singularities. In the following, we present some key results from the Chaps. 4 to 9.

The action of $\mathrm{Br}_3 \ltimes \{\pm 1\}^3$ on $T_3(\mathbb{Z})$ boils down to an action of $PSL_2(\mathbb{Z}) \ltimes G^{sign}$ on $T_3(\mathbb{Z})$ where $G^{sign} \cong \{\pm 1\}^2$ comes from the action of the sign group $\{\pm 1\}^3$. As the action of $PSL_2(\mathbb{Z})$ is partially nonlinear, it is good to write it as a semidirect product $PSL_2(\mathbb{Z}) \cong G^{phi} \rtimes \langle \gamma \rangle$ where γ acts cyclically and linearly of order 3 and G^{phi} is a free Coxeter group with 3 generators which act nonlinearly.

The Sects. 4.2–4.4 analyze the action on $T_3(\mathbb{Z})$ carefully. The first result Theorem 4.6 builds on coarser classifications of Krüger [Kr90, §12] and Cecotti-Vafa [CV93, Ch. 6.2]. The following theorem gives a part of Theorem 4.6.

Theorem 1.2 *(Part of Theorem 4.6)*

(a) The characteristic polynomial of $S^{-1}S^t$ and of the monodromy M of $(H_{\mathbb{Z}}, L, \underline{e})$ for $S = S(\underline{x}) \in T_3(\mathbb{Z})$ with $\underline{x} \in \mathbb{Z}^3$ is

$$p_{ch,M} = (t-1)(t^2 - (2 - r(\underline{x}))t + 1),$$
$$\text{where} \quad r : \mathbb{Z}^3 \to \mathbb{Z}, \quad \underline{x} \mapsto x_1^2 + x_2^2 + x_3^2 - x_1x_2x_3.$$

The characteristic polynomial and $r(\underline{x})$ are invariants of the $\mathrm{Br}_3 \ltimes \{\pm 1\}^3$ *orbit of $S(\underline{x})$. All eigenvalues of $p_{ch,M}$ are roots of unity if and only if $r(\underline{x}) \in \{0, 1, 2, 3, 4\}$.*

(b) For $\rho \in \mathbb{Z} - \{4\}$ the fiber $r^{-1}(\rho) \subset \mathbb{Z}^3$ consists only of finitely many $\mathrm{Br}_3 \ltimes \{\pm 1\}^3$ *orbits. The following table gives the symbols in Example 1.1 for the fibers over $r \in \{0, 1, 2, 3, 4\}$, so there are only seven orbits plus one series of orbits over $r \in \{0, 1, 2, 3, 4\}$.*

$r(\underline{x})$	0	1	2	3	4
	$A_1^3, \mathbb{P}^2$	A_2A_1	A_3	–	$\mathbb{P}^1 A_1, \widehat{A}_2, \mathcal{H}_{1,2}, S(-l, 2, -l)$ with $l \geq 3$

With the help of certain (beautiful) graphs, in Theorem 4.13 the stabilizers $(\mathrm{Br}_3)_{S/\{\pm 1\}^3}$ are calculated for certain representatives of all $\mathrm{Br}_3 \ltimes \{\pm 1\}^3$-orbits in $T_3(\mathbb{Z})$. We work with 14 graphs and 24 sets of representatives.

Lemma 5.8 gives informations on the characteristic polynomial and the signature of $I^{(0)}$ in all rank 3 cases.

Lemma 1.3 *(Lemma 5.8 (b))*

Consider $\underline{x} \in \mathbb{Z}^3$ with $r = r(\underline{x}) < 0$ or > 4 or with $S(\underline{x})$ one of the cases in the table in Theorem 1.2. Then $p_{ch,M} = (t - \lambda_1)(t - \lambda_2)\Phi_1$ and sign $I^{(0)}$ *are as follows (Φ_m = the cyclotomic polynomial of m-th primitive roots of unity).*

$r(\underline{x})$	$p_{ch,M}$	sign $I^{(0)}$	$S(\underline{x})$
$r<0$	$\lambda_1,\lambda_2>0$	$(+--)$	$S(\underline{x})$
$r=0$	Φ_1^3	$(+++)$	$S(A_1^3)$
$r=0$	Φ_1^3	$(+--)$	$S(\mathbb{P}^2)$
$r=1$	$\Phi_6\Phi_1$	$(+++)$	$S(A_2A_1)$
$r=2$	$\Phi_4\Phi_1$	$(+++)$	$S(A_3)$
$r=4$	$\Phi_2^2\Phi_1$	$(++\,0)$	$S(\mathbb{P}^1A_1)$
$r=4$	$\Phi_2^2\Phi_1$	$(++\,0)$	$S(\widehat{A}_2)$
$r=4$	$\Phi_2^2\Phi_1$	$(+\,0\,0)$	$S(\mathcal{H}_{1,2})$
$r=4$	$\Phi_2^2\Phi_1$	$(+\,0\,-)$	$S(-l,2,-l)$ with $l\geq 3$
$r>4$	$\lambda_1,\lambda_2<0$	$(++-)$	$S(\underline{x})$

Chapter 5 analyzes the groups $G_{\mathbb{Z}}$, $G_{\mathbb{Z}}^{(0)}$, $G_{\mathbb{Z}}^{(1)}$ and $G_{\mathbb{Z}}^{M}$ in all rank 3 cases. This leads into an intricate case discussion. The case $\mathcal{H}_{1,2}$ is different from all other cases as it is the only case where $G_{\mathbb{Z}}$ is not abelian and where the subgroup $\{\pm M^m \mid m\in\mathbb{Z}\}$ does not have finite index in $G_{\mathbb{Z}}$. The automorphism $Q\in G_{\mathbb{Q}}:=\text{Aut}(H_{\mathbb{Q}},L)$ is defined for $r(\underline{x})\neq 0$. It is id on $\ker(M-\text{id})$ and $-\text{id}$ on $\ker(M^2-(2-r)M+\text{id})$ (Definition 5.9). It is only in a few cases in $G_{\mathbb{Z}}$ (Theorem 5.11).

Theorem 1.4 *(Part of the Theorems 5.11, 5.13, 5.14, 5.16, 5.18, 3.28)*
(a) For S in the $\text{Br}_3\ltimes\{\pm 1\}^3$ *orbit of $S(\mathcal{H}_{1,2})$*

$$G_{\mathbb{Z}}\cong SL_2(\mathbb{Z})\times\{\pm 1\},\quad M=Q,$$

and the subgroup $\{\pm M^m \mid m\in\mathbb{Z}\}=\{\pm\text{id},\pm Q\}$ has infinite index in $G_{\mathbb{Z}}$.
(b) In all other rank 3 cases the subgroup $\{\pm M^m \mid m\in\mathbb{Z}\}$ has finite index in $G_{\mathbb{Z}}$ and $G_{\mathbb{Z}}$ is abelian. Then one of the five possibilities holds,

$$G_{\mathbb{Z}}=O_3(\mathbb{Z}), \tag{1.1}$$
$$G_{\mathbb{Z}}=\{\text{id},Q\}\times\{\pm(M^{root})^m \mid m\in\mathbb{Z}\}, \tag{1.2}$$
$$G_{\mathbb{Z}}=\{\pm(M^{root})^m \mid m\in\mathbb{Z}\}, \tag{1.3}$$
$$G_{\mathbb{Z}}=\{\text{id},Q\}\times\{\pm M^m \mid m\in\mathbb{Z}\}, \tag{1.4}$$
$$G_{\mathbb{Z}}=\{\pm M^m \mid m\in\mathbb{Z}\}, \tag{1.5}$$

where M^{root} is a root of $\pm M$ or of MQ. The following table gives the index $[G_{\mathbb{Z}}:\{\pm M^m \mid m\in\mathbb{Z}\}]\in\mathbb{N}$ and informations on M^{root}.

matrix	index	M^{root}
(1.1) $S(A_1^3)$	24	
(1.2) $S(x,0,0)$ with $x<0$	4	$(M^{root})^2 = MQ$
(1.2) $S(-l,2,-l)$ with l even	l^2-4	$(M^{root})^{l^2/2-2} = MQ$
(1.2) $S(4,4,4)$ and $S(5,5,5)$	6	$(M^{root})^3 = -M$
(1.2) $S(4,4,8)$	4	$(M^{root})^2 = M$
(1.3) $S(\mathbb{P}^2)$	3	$(M^{root})^3 = M$
(1.3) $S(\widehat{A}_2)$ and $S(x,x,x)$ with $x\in\mathbb{Z}-\{-1,0,...,5\}$	3	$(M^{root})^3 = -M$
(1.3) $S(-l,2,-l)$ with l odd	l^2-4	$(M^{root})^{l^2-4} = -M$
(1.3) $S(2y,2y,2y^2)$ with $y\in\mathbb{Z}_{\geq 3}$	2	$(M^{root})^2 = M$
(1.4) $S(3,3,4)$ and $S(x,x,0)$ with $x\in\mathbb{Z}_{\geq 2}$	2	
(1.5) $S(A_3)$ and $S(\underline{x})$ in other $\mathrm{Br}_3\ltimes\{\pm 1\}^3$ orbits	1	

(c) $G_{\mathbb{Z}} = G_{\mathbb{Z}}^{\mathcal{B}}$ holds for all rank 3 cases except four cases, the $\mathrm{Br}_3\ltimes\{\pm 1\}^3$ *orbits of $S(\underline{x})$ with*

$$\underline{x}\in\{(3,3,4),(4,4,4),(5,5,5),(4,4,8)\}.$$

In these four cases $Q\in G_{\mathbb{Z}}-G_{\mathbb{Z}}^{\mathcal{B}}$ and $G_{\mathbb{Z}} = \langle Q, G_{\mathbb{Z}}^{\mathcal{B}}\rangle$.

Though in higher rank it is easier to construct matrices S with $G_{\mathbb{Z}}^{\mathcal{B}}\subsetneqq G_{\mathbb{Z}}$ (Remarks 3.29).

Chapter 6 studies the even and odd monodromy groups and the sets of even and of odd vanishing cycles in the rank 2 and rank 3 cases. The following theorem catches some of the results on the even monodromy group $\Gamma^{(0)}$ and the set $\Delta^{(0)}$ of even vanishing cycles for the rank 3 cases. The group $O^{(0),*}$ (Definition 6.4) is a certain subgroup of $O^{(0)}$ which is determined only by $(H_{\mathbb{Z}}, I^{(0)})$ (so independently of $\mathcal{B}$). Part (b) of Theorem 1.5 discusses only the (in general difficult) problem whether $\Delta^{(0)} = R^{(0)}$ or $\Delta^{(0)}\subsetneqq R^{(0)}$. Theorem 6.14 contains many more informations on $\Delta^{(0)}$. Theorem 6.11 contains much more information on $\Gamma^{(0)}$ than part (a) of Theorem 1.5. Remarkably, $\Gamma^{(0)}\cong G^{fCox,3}$ (the free Coxeter group with three generators) holds not only for the Coxeter cases $\underline{x}\in\mathbb{Z}^3_{\leq -2}$ (which all satisfy $r(\underline{x})>4$), but also in all cases $\underline{x}\in\mathbb{Z}^3$ with $r(\underline{x})<0$ and in the case $\mathbb{P}^2$.

Theorem 1.5 *(a) (Part of Lemma 2.11 and Theorem 6.11)*

(i) (Part of Lemma 2.11) The case A_1^n, $n\in\mathbb{N}$:

$$\Gamma^{(0)}\cong\{\pm 1\}^n,\quad \Gamma^{(1)} = \{\mathrm{id}\},\quad \Delta^{(0)} = R^{(0)} = \Delta^{(1)} = \{\pm e_1,...,\pm e_n\}.$$

(ii) The cases with $r(\underline{x})>4$ and the cases A_3, $\widehat{A}_2$, A_2A_1, $\mathbb{P}^1A_1$: They contain all reducible rank 3 cases except A_1^3. Then $\Gamma^{(0)}$ is a Coxeter group. If $\underline{x}\in\mathbb{Z}^3_{\leq -2}$ then $\Gamma^{(0)}\cong G^{fCox,3}$.

(iii) The cases A_3, $\widehat{A}_2$, $\mathcal{H}_{1,2}$: Then $\Gamma^{(0)} = O^{(0),*}$.

(iv) The cases $S(-l, 2, -l)$ with $l \geq 3$: Then $\Gamma^{(0)} \overset{1:l}{\subset} O^{(0),*}$.

(v) The cases $\mathbb{P}^2$ and $\underline{x} \in \mathbb{Z}^3$ with $r(\underline{x}) < 0$: Then $\Gamma^{(0)} \cong G^{fCox,3}$.

(b) (Part of Theorem 6.14)

(i) $\Delta^{(0)} = R^{(0)}$ holds in the following cases: $A_3, \widehat{A}_2, \mathbb{P}^2$, all $S(\underline{x})$ with $\underline{x} \in \{0, -1, -2\}^3$, all reducible cases.

(ii) $\Delta^{(0)} \subsetneqq R^{(0)}$ holds in the following cases: $\mathcal{H}_{1,2}$, all $S(-l, 2, -l)$ with $l \geq 3$, $S(3, 3, 4)$, $S(4, 4, 4)$, $S(5, 5, 5)$, $S(4, 4, 8)$.

(iii) In the cases of the other $\mathrm{Br}_3 \ltimes \{\pm 1\}^3$ *orbits in $T_3(\mathbb{Z})$, we do not know whether $\Delta^{(0)} = R^{(0)}$ or $\Delta^{(0)} \subsetneqq R^{(0)}$ holds.*

Let E_n denote the $n \times n$ unit matrix. Given $S \in T_n(\mathbb{Z})$ with associated triple $(H_\mathbb{Z}, L, \underline{e})$, consider the matrix $\widetilde{S} := 2E_n - S \in T_n(\mathbb{Z})$ with the associated triple $(H_\mathbb{Z}, \widetilde{L}, \underline{e})$. Then $\widetilde{L}, \widetilde{I}^{(0)}$ and $\widetilde{M}$ are far from $L, I^{(0)}$ and M, but $\widetilde{I}^{(1)} = -I^{(1)}$, $\widetilde{\Gamma}^{(1)} = \Gamma^{(1)}$ and $\widetilde{\Delta}^{(1)} = \Delta^{(1)}$ (see the Remarks 4.17). For example the cases A_3 and $\widehat{A}_2$ are related in this way, and also the Coxeter case $(-2, -2, -2)$ and the case $\mathcal{H}_{1,2}$ are related in this way (in both cases after an action of $\mathrm{Br}_3 \ltimes \{\pm 1\}^3$).

This motivates in the rank 3 cases to consider the action of the bigger group $(G^{phi} \ltimes \widetilde{G}^{sign}) \rtimes \langle \gamma \rangle$ on $\mathbb{Z}^3$ where $\widetilde{G}^{sign}$ is generated by G^{sign} and the total sign change $\delta^\mathbb{R} : \underline{x} \mapsto -\underline{x}$. Lemma 4.18 gives representatives for all orbits of this action on $\mathbb{Z}^3$ (respectively $T_3(\mathbb{Z})$). Still it is difficult to see for a given triple $\underline{x} \in \mathbb{Z}^3$ in which orbit it is.

We had for some time the hope that the beautiful facts on the even monodromy group $\Gamma^{(0)}$ for the Coxeter cases $\underline{x} \in \mathbb{Z}^3_{\leq 0}$ would have analogs for the odd monodromy group $\Gamma^{(1)}$, but this does not hold in general. In the case $(-2, -2, -2)$ $\Gamma^{(0)} \cong G^{fCox,3}$, but in the case $\mathcal{H}_{1,2}$ not, and in both cases together $\Gamma^{(1)} \not\cong G^{free,3}$. On the other hand, $\Gamma^{(1)} \cong G^{free,3}$ for $\underline{x} \in B_1$, where $B_1 \subset \mathbb{Z}^3$ is as follows.

$$B_1 := (G^{phi} \ltimes \widetilde{G}^{sign}) \rtimes \langle \gamma \rangle (\{\underline{x} \in \mathbb{Z}^3 - \{(0,0,0)\} \mid r(\underline{x}) \leq 0\}),$$
$$B_2 := \{\underline{x} \in \mathbb{Z}^3 - \{(0,0,0)\} \mid S(\underline{x}) \text{ is reducible}\},$$
$$B_3 := \{(0,0,0)\}.$$

Though the set B_1 is difficult to understand (see the Examples 4.20). It contains $(3, 3, 3)$, so the orbit of $\mathbb{P}^2$. $B_2 \cup B_3$ consists of the triples $\underline{x}$ with reducible $S(\underline{x})$, so with two or three zero entries.

Consider $\underline{x} \in \mathbb{Z}^3 - B_3$. The radical $\mathrm{Rad}\, I^{(1)}$ has rank 1, so the quotient lattice $\overline{H_\mathbb{Z}}^{(1)} := H_\mathbb{Z} / \mathrm{Rad}\, I^{(1)}$ has rank 2. Denote by $\Gamma_s^{(1)}$ the image of $\Gamma^{(1)}$ under the natural homomorphism $\Gamma^{(1)} \to \mathrm{Aut}(\overline{H_\mathbb{Z}}^{(1)})$ and by $\Gamma_u^{(1)}$ the kernel of it. There is an exact sequence

$$\{1\} \to \Gamma_u^{(1)} \to \Gamma^{(1)} \to \Gamma_s^{(1)} \to \{1\}.$$

Denote by $\overline{\Delta^{(1)}} \subset \overline{H_\mathbb{Z}}^{(1)}$ the image of $\Delta^{(1)}$ in $\overline{H_\mathbb{Z}}^{(1)}$. Often $\overline{\Delta^{(1)}}$ is easier to describe than $\Delta^{(1)}$.

The long Theorems 6.18 and 6.21 offer detailed results about $\Gamma^{(1)}$ and $\Delta^{(1)}$ for the representatives in Lemma 4.18 of the $(G^{phi} \ltimes \widetilde{G}^{sign}) \rtimes \langle\gamma\rangle$ orbits in $\mathbb{Z}^3$. The next theorem gives only a rough impression.

Theorem 1.6 *Consider $S = S(\underline{x}) \in T_3(\mathbb{Z})$ and the associated triple $(H_{\mathbb{Z}}, L, \underline{e})$.*
(a) (Part of Theorem 6.18) Consider $\underline{x} \neq (0, 0, 0)$.

$$\begin{aligned} \Gamma^{(1)} \cong G^{free,3} &\iff \underline{x} \in B_1, \\ \Gamma_u^{(1)} = \{\mathrm{id}\} &\iff \underline{x} \in B_1 \cup B_2, \\ \Gamma_u^{(1)} \cong \mathbb{Z}^2 &\iff \underline{x} \in \mathbb{Z}^3 - (B_1 \cup B_2 \cup B_3), \end{aligned}$$

$$\Gamma_s^{(1)} \cong \text{one of the groups } SL_2(\mathbb{Z}),\ G^{free,2},\ G^{free,2} \times \{\pm 1\} \\ \text{for } \underline{x} \in \mathbb{Z}^3 - (B_1 \cup B_3).$$

(b) (Part of Theorem 6.21)
(i) In the cases of A_3 and $\widehat{A}_2$ $\overline{\Delta^{(1)}} = \overline{H_{\mathbb{Z}}}^{(1),prim}$, so $\overline{\Delta^{(1)}}$ is the set of primitive vectors in $\overline{H_{\mathbb{Z}}}^{(1)}$, and $\Delta^{(1)}$ is the full preimage in $H_{\mathbb{Z}}$ of $\overline{\Delta^{(1)}}$.
(ii) Though in many other cases $\overline{\Delta^{(1)}} \not\subset \overline{H_{\mathbb{Z}}}^{(1),prim}$, and $\Delta^{(1)}$ is not the full preimage in $H_{\mathbb{Z}}$ of $\overline{\Delta^{(1)}}$, but each fiber has infinitely many elements.
(iii) But for $\underline{x} \in B_1$ the map $\Delta^{(1)} \to \overline{\Delta^{(1)}}$ is a bijection. Especially for $\mathbb{P}^2$ $\overline{\Delta^{(1)}}$ is easy to describe (Theorem 6.21 (h)), but $\Delta^{(1)}$ not.

Chapter 7 studies the set $\mathcal{B}^{dist} = \mathrm{Br}_n \ltimes \{\pm 1\}^n(\underline{e})$ of distinguished bases for a given triple $(H_{\mathbb{Z}}, L, \underline{e})$. In general, it is difficult to characterize this orbit in easy terms. We know that the following inclusions hold,

$$\mathcal{B}^{dist} \subset \{\underline{v} \in (\Delta^{(0)})^n \mid s_{v_1}^{(0)} \ldots s_{v_n}^{(0)} = -M\}, \tag{3.3}$$
$$\mathcal{B}^{dist} \subset \{\underline{v} \in (\Delta^{(1)})^n \mid s_{v_1}^{(1)} \ldots s_{v_n}^{(1)} = M\}. \tag{3.4}$$

We are interested when they are equalities. In general, this is a difficult question. In the rank 3 cases Theorem 7.3 and Theorem 7.7 give our results for (3.3) and (3.4).

Theorem 1.7 *Consider $S(\underline{x}) \in T_3(\mathbb{Z})$ and the associated triple $(H_{\mathbb{Z}}, L, \underline{e})$.*
(a) (Part of Theorem 7.3)
(3.3) is an equality for all cases except for $\underline{x}$ in the $\mathrm{Br}_3 \ltimes \{\pm 1\}^3$ orbit of $\mathcal{H}_{1,2}$. There the right hand side of (3.3) consists of countably many $\mathrm{Br}_3 \ltimes \{\pm 1\}^3$ orbits.
(b) (Part of Theorem 7.7)
(i) The inclusion in (3.4) is an equality $\iff \underline{x} \in B_1 \cup B_2$.
(ii) The cases A_3, $\widehat{A}_2$, $\mathcal{H}_{1,2}$ and $S(-l, 2, -l)$ with $l \geq 3$ are not in B_1. But there the inclusion in (3.4) becomes an equality if one adds on the right hand side of (3.4) the condition $\sum_{i=1}^n \mathbb{Z}v_i = H_{\mathbb{Z}}$.

The last Sect. 7.4 of Chap. 7 builds on Theorem 4.16 which determines for a representative $S \in T_3(\mathbb{Z})$ of each $\mathrm{Br}_3 \ltimes \{\pm 1\}$ orbit in $T_3(\mathbb{Z})$ the stabilizer $(Br_3)_{S/\{\pm 1\}^3}$. Theorem 7.11 determines in each of these cases the stabilizer $(\mathrm{Br}_3)_{\underline{e}/\{\pm 1\}^3}$. The graphs $\mathcal{G}_1, ..., \mathcal{G}_{14}$ in Sect. 4.4 used the groups $G^{phi} \rtimes \langle\gamma\rangle$. At the end of Sect. 7.4 different graphs, which use the group Br_3, are introduced for the orbits of matrices as well as the orbits of triangular bases. For the cases of finite orbits and for the case $\widehat{A}_2$ the graphs are given explicitly. In the case of A_3 the orbit $\mathcal{S}^{dist}/\{\pm 1\}^3$ has four elements and the orbit $\mathcal{B}^{dist}/\{\pm 1\}^3$ has 16 elements.

The stabilizers $(\mathrm{Br}_3)_{\underline{e}/\{\pm 1\}^3}$ and $(\mathrm{Br}_3)_{S/\{\pm 1\}^3}$ are used in Sect. 9 for the construction of the manifolds $C_3^{\underline{e}/\{\pm 1\}^3}$ and $C_3^{S/\{\pm 1\}^3}$ in the rank 3 cases. Recall that they are obtained by glueing Stokes regions, one for each element of an orbit $\mathcal{B}^{dist}/\{\pm 1\}^3$ or $\mathcal{S}^{dist}/\{\pm 1\}^3$. Theorem 9.3 gives complete results, it covers all cases. Remarkably, in all cases where $\Gamma^{(0)} \cong G^{fCox,3}$ or $\Gamma^{(1)} \cong G^{free,3}$, the manifold $C^{\underline{e}/\{\pm 1\}^3}$ is just $C_3^{univ} \cong \mathbb{C}^2 \times \mathbb{H}$. This holds especially for the case $\mathbb{P}^2$. In this aspect, the case $\mathbb{P}^2$ is easier than the case A_3, where $C_3^{\underline{e}/\{\pm 1\}^3}$ consists of 16 Stokes regions and is isomorphic to $\mathbb{C}^3$ minus two hypersurfaces (a caustic and a Maxwell stratum).

Corollary 9.8 discusses natural partial compactifications. Section 9.1 prepares this by giving explicitly deck transformations of C_3^{univ} which generate the group Br_3 of deck transformations of the covering $\mathrm{pr}_3^{u,c} : C_3^{univ} \to C_3^{conf}$. The formulas for the deck transformations use a Schwarzian triangle function which is equivalent to the λ-function and a certain lift of the logarithm.

Chapter 8 gives general background for these manifolds: the configuration space C_n^{conf} of Br_n and the related manifolds C_n^{pure} and C_n^{univ}, the notion of a (semisimple) F-manifold with Euler field, distinguished systems of paths, two a priori different, but closely related braid group actions on (homotopy classes) of distinguished systems of paths, and a construction of $\mathbb{Z}$-lattice bundles on C_n^{univ}, $C_n^{\underline{e}/\{\pm 1\}^n}$ and $C_n^{S/\{\pm 1\}^n}$. They are the door to more transcendent structures on these manifolds, like Dubrovin-Frobenius manifolds.

We refer to the beginning of Sect. 10 for an introduction to the material of Sect. 10, which presents known results in the case of isolated hypersurface singularities.

Appendix A recalls properties of the hyperbolic plane and subgroups of its group of isometries. The upper half plane, Möbius transformations from matrices in $SL_2(\mathbb{R})$ and hyperbolic polygons are mentioned. Three special cases of the Poincaré-Maskit theorem are made explicit. Later also a model of the hyperbolic plane from a cone of positive vectors in $\mathbb{R}^3$ with a metric with signature $(+--)$ is explained. This connects groups of 3×3 matrices with groups of isometries. Appendix A is mainly used in Chap. 6 for the determination of some groups $\Gamma^{(0)}$ and $\Gamma^{(1)}$, but also in Chap. 4 for the group G^{phi}.

Appendix B is used for the explicit formulas for some generating deck transformations of the covering $\mathrm{pr}_3^{u,c} : C_3^{univ} \to C_3^{conf}$ in Sect. 9.1. It recalls properties of the first congruence subgroups $\Gamma(n) \subset SL_2(\mathbb{Z})$, those with $n \in \{2, 3, 4, 5\}$, with special emphasis on the most important case $n = 2$. In this case, the Schwarzian triangle function $T : \mathbb{H} \to \mathbb{C} - \{0; 1\} \cong \mathbb{H}/\Gamma(2)$, which we use, is obtained by composing the λ-function with some automorphism of $\mathbb{C} - \{0; 1\}$. Later we introduce a lift $\kappa : \mathbb{H} \to \mathbb{C}$

of the restriction $\ln : \mathbb{C} - \{0; 1\} \dashrightarrow \mathbb{C}$ of the logarithm $\ln : \mathbb{C} - \{0\} \dashrightarrow \mathbb{C}$, lifting it with T.

Appendix C contains Lemma C.1 which determines the unit groups in two families of rings of quadratic algebraic integers. The lemma is used at many places, especially in Chap. 5 for the determination of the groups $G_{\mathbb{Z}}$, $G_{\mathbb{Z}}^{(0)}$, $G_{\mathbb{Z}}^{(1)}$, $G_{\mathbb{Z}}^{M}$. In order to present an elegant proof of Lemma C.1, we use the theory of continued fractions. The results are certainly known, but we did not find a completely satisfying reference.

Also Appendix D is used in Chap. 5. It studies powers of units in rings of quadratic algebraic integers.

Acknowledgements We thank Martin Guest and Atsushi Takahashi for discussions on topics related to this book.

Chapter 2
Bilinear Lattices and Induced Structures

This chapter fixes the basic notions, a bilinear lattice and its associated data, namely a Seifert form, an even and an odd intersection form, a monodromy, the L1-cycles, the triangular bases, an even and an odd monodromy group, the even and the odd vanishing cycles. The notion of a bilinear lattice and the even part of the associated data are considered in [HK16]. The more special case of a unimodular bilinear lattice and even and odd data are considered since long time in singularity theory [AGV88, Eb01]. In this book we are mainly interested in unimodular bilinear lattices. Only this chapter treats the general case, partially following [HK16].

Notation 2.1 In these notations, R will be either the ring $\mathbb{Z}$ or one of the fields $\mathbb{Q}$, $\mathbb{R}$ or $\mathbb{C}$. Later we will work mainly with $\mathbb{Z}$. If $R = \mathbb{Z}$ write $\widetilde{R} := \mathbb{Q}$, else write $\widetilde{R} := R$.

In the whole paper, $H_R \supsetneqq \{0\}$ is a finitely generated free R-module, so a $\mathbb{Z}$-lattice if $R = \mathbb{Z}$, and a finite dimensional R-vector space if R is $\mathbb{Q}$, $\mathbb{R}$ or $\mathbb{C}$. Its rank will usually be called $n \in \mathbb{N} = \{1, 2, 3, ...\}$ (it is its dimension if R is $\mathbb{Q}$, $\mathbb{R}$ or $\mathbb{C}$). If R_1 and R_2 are both in the list $\mathbb{Z}, \mathbb{Q}, \mathbb{R}, \mathbb{C}$ and R_1 is left of R_2 and H_{R_1} is given, then $H_{R_2} := H_{R_1} \otimes_{R_1} R_2$.

In the whole paper, $L : H_R \times H_R \to R$ will be a nondegenerate R-bilinear form. If $U \subset H_R$ is an R-submodule, then $U^\perp := \{b \in H_R \mid L(U, b) = 0\}$ and ${}^\perp U := \{a \in H_R \mid L(a, U) = 0\}$. In the case $R = \mathbb{Z}$, $U^\perp$ and ${}^\perp U$ are obviously primitive $\mathbb{Z}$-submodules of $H_\mathbb{Z}$.

In Lemma 2.2 we will start with H_R and a symmetric R-bilinear form $I^{[0]} : H_R \times H_R \to R$ or a skew-symmetric R-bilinear form $I^{[1]} : H_R \times H_R \to R$. With the square brackets in the index we distinguish them from the bilinear forms $I^{(0)}$ and $I^{(1)}$, which are induced in Definition 2.3 by a given bilinear form L. Though later they will be identified.

Suppose that $M : H_R \to H_R$ is an automorphism. Then $M_s, M_u, N : H_{\widetilde{R}} \to H_{\widetilde{R}}$ denote the semisimple part, the unipotent part and the nilpotent part of M with $M = M_s M_u = M_u M_s$ and $N = \log M_u$, $e^N = M_u$. Denote $H_\lambda := \ker(M_s - \lambda \cdot \mathrm{id}) : H_\mathbb{C} \to H_\mathbb{C}$.

C. Hertling and K. Larabi, *Triangular Bases of Unimodular Bilinear Lattices and Induced Structures*, Lecture Notes in Mathematics 2398,
https://doi.org/10.1007/978-3-032-25997-4_2

For $m \in \mathbb{N}$ denote by $\Phi_m \in \mathbb{Z}[t]$ the cyclotomic polynomial whose zeros are the primitive m-th roots of unity.

The following lemma is elementary and classical. We skip the proof.

Lemma 2.2 *Let $R \in \{\mathbb{Q}, \mathbb{R}, \mathbb{C}\}$ and let H_R be an R-vector space of dimension $n \in \mathbb{N}$.*

(a) Let $I^{[0]} : H_R \times H_R \to R$ be a symmetric bilinear form. Consider $a \in H_R$ with $I^{[0]}(a, a) \neq 0$. The map

$$s_a^{[0]} : H_R \to H_R, \quad s_a^{[0]}(b) := b - \frac{2I^{[0]}(a, b)}{I^{[0]}(a, a)} a,$$

is a reflection, *so it is in* $\mathrm{Aut}(H_r, I^{[0]})$*, it fixes the codimension 1 subspace $\{b \in H_R \mid I^{[0]}(a, b) = 0\}$ and it maps a to $-a$. Especially* $(s_a^{[0]})^2 = \mathrm{id}$.

(b) Let $I^{[1]} : H_R \times H_R \to R$ be a skew-symmetric bilinear form. Consider $a \in H_R$. The map

$$s_a^{[1]} : H_R \to H_R, \quad s_a^{[1]}(b) := b - I^{[1]}(a, b)a,$$

is in $\mathrm{Aut}(H_R, I^{[1]})$ *with*

$$(s_a^{[1]})^{-1}(b) = b + I^{[1]}(a, b)a.$$

It is id *if* $a \in \mathrm{Rad}(I^{[1]})$*. If* $a \notin \mathrm{Rad}(I^{[1]})$ *then it fixes the codimension 1 subspace $\{b \in H_R \mid I^{[1]}(a, b) = 0\}$, and* $s_a^{[1]} - \mathrm{id}$ *is nilpotent with a single* 2×2 *Jordan block. Then it is called a* transvection.

(c) Fix $k \in \{0; 1\}$ and consider $I^{[k]}$ as in (a) or (b). An element $g \in \mathrm{Aut}(H_R, I^{[k]})$ *and an element $a \in H_R$ with $I^{[0]}(a, a) \neq 0$ if $k = 0$ satisfy*

$$g\, s_a^{[k]}\, g^{-1} = s_{g(a)}^{[k]}.$$

Definition 2.3 (a) [HK16, Chap. 2] A *bilinear lattice* is a pair $(H_\mathbb{Z}, L)$ with $H_\mathbb{Z}$ a $\mathbb{Z}$-lattice of some rank $n \in \mathbb{N}$ together with a nondegenerate bilinear form $L : H_\mathbb{Z} \times H_\mathbb{Z} \to \mathbb{Z}$. If $\det L(\underline{e}^t, \underline{e}) = 1$ for some $\mathbb{Z}$-basis $\underline{e} = (e_1, ..., e_n)$ of $H_\mathbb{Z}$ then L and the pair $(H_\mathbb{Z}, L)$ are called *unimodular*. The bilinear form is called *Seifert form* in this paper.

(b) A bilinear lattice induces several structures:

(i) [HK16, Chap. 2] A symmetric bilinear form

$$I^{(0)} = L^t + L : H_\mathbb{Z} \times H_\mathbb{Z} \to \mathbb{Z}, \quad \text{so } I^{(0)}(a, b) = L(b, a) + L(a, b),$$

which is called *even intersection form*.

(ii) A skew-symmetric bilinear form

$$I^{(1)} = L^t - L : H_\mathbb{Z} \times H_\mathbb{Z} \to \mathbb{Z}, \quad \text{so } I^{(1)}(a, b) = L(b, a) - L(a, b),$$

which is called *odd intersection form.*

(iii) [HK16, Chap. 2] An automorphism $M : H_{\mathbb{Q}} \to H_{\mathbb{Q}}$ which is defined by

$$L(Ma, b) = L(b, a) \quad \text{for } a, b \in H_{\mathbb{Q}},$$

which is called *monodromy*.

(iv) Six automorphism groups

$$\begin{aligned} O^{(k)} &:= \mathrm{Aut}(H_{\mathbb{Z}}, I^{(k)}) \qquad \text{for } k \in \{0; 1\}, \\ G_{\mathbb{Z}}^{M} &:= \mathrm{Aut}(H_{\mathbb{Z}}, M) := \{g : H_{\mathbb{Z}} \to H_{\mathbb{Z}} \text{ automorphism} \mid gM = Mg\}, \\ G_{\mathbb{Z}}^{(k)} &:= \mathrm{Aut}(H_{\mathbb{Z}}, I^{(0)}, M) = O^{(k)} \cap G_{\mathbb{Z}}^{M} \qquad \text{for } k \in \{0; 1\}, \\ G_{\mathbb{Z}} &:= \mathrm{Aut}(H_{\mathbb{Z}}, L). \end{aligned}$$

(v) [HK16, Chap. 2] The set $R^{(0)} \subset H_{\mathbb{Z}}$ of *L1-cycles* (*L1* is motivated by Lemma 2.6 (c))

$$R^{(0)} := \{a \in H_{\mathbb{Z}} \mid L(a, a) > 0; \frac{L(a, b)}{L(a, a)}, \frac{L(b, a)}{L(a, a)} \in \mathbb{Z} \text{ for all } b \in H_{\mathbb{Z}}\}.$$

(vi) [HK16, Chap. 2] The set $\mathcal{B}^{tri}$ of *triangular bases*,

$$\mathcal{B}^{tri} := \{\underline{e} = (e_1, ..., e_n) \in (R^{(0)})^n \mid \bigoplus_{i=1}^{n} \mathbb{Z}e_i = H_{\mathbb{Z}}, L(e_i, e_j) = 0 \text{ for } i < j\}.$$

(c) Let $n \in \mathbb{N}$ and $R \in \{\mathbb{Z}, \mathbb{Q}, \mathbb{R}, \mathbb{C}\}$. The sets T_n^{tri} and $T_n(R)$ of upper triangular matrices are defined by

$$\begin{aligned} T_n(R) &:= \{S = (s_{ij}) \in M_{n\times n}(R) \mid s_{ii} = 1, s_{ij} = 0 \text{ for } i > j\}, \\ T_n^{tri} &:= \{S = (s_{ij}) \in M_{n\times n}(\mathbb{Z}) \mid s_{ii} \in \mathbb{N}, s_{ij} = 0 \text{ for } i > j, \\ &\qquad \frac{s_{ij}}{s_{ii}}, \frac{s_{ji}}{s_{ii}} \in \mathbb{Z} \text{ for } i \neq j\}. \end{aligned}$$

Obviously $T_n(\mathbb{Z}) \subset T_n^{tri}$.

Remark 2.4 (i) There are bilinear lattices with $\mathcal{B}^{tri} = \emptyset$. We are interested only in bilinear lattices with $\mathcal{B}^{tri} \neq \emptyset$.

(ii) A triangular basis $\underline{e} \in \mathcal{B}^{tri}$ is called in [HK16] a *complete exceptional sequence*.

(iii) In the case $\mathcal{B}^{tri} \neq \emptyset$, [HK16] considers the bilinear form L^t (with $L^t(a, b) = L(b, a)$). Our choice L is motivated by singularity theory. Also the names for L, $I^{(0)}$, $I^{(1)}$ and M, namely *Seifert form, even intersection form, odd intersection form* and *monodromy* are motivated by singularity theory. The L1-cycles in $R^{(0)}$ are in [HK16] called *pseudo-real roots* or simply *roots*.

(iv) In this paper we are mainly interested in the cases of unimodular bilinear lattices with $\mathcal{B}^{tri} \neq \emptyset$. Singularity theory leads to such cases.

(v) [HK16] is mainly interested in the cases of *generalized Cartan lattices*. A generalized Cartan lattice is a triple $(H_{\mathbb{Z}}, L, \underline{e})$ with $(H_{\mathbb{Z}}, L)$ a bilinear lattice and $\underline{e} \in \mathcal{B}^{tri}$ with $L(e_i, e_j) \leq 0$ for $i > j$.

Remark 2.5 (i) The classification of pairs $(H_{\mathbb{R}}, L)$ and pairs $(H_{\mathbb{C}}, L)$ with L a nondegenerate bilinear form on $H_{\mathbb{R}}$ respectively $H_{\mathbb{C}}$ is well understood. Such a pair decomposes into an orthogonal sum of irreducible pairs. This and the classification of the irreducible pairs over $\mathbb{R}$ is carried out in [Ne98] and, more explicitly, in [BH19].

In both references it is also proved that a pair $(H_{\mathbb{R}}, L)$ of rank $n \in \mathbb{N}$ up to isomorphism is uniquely determined by an unordered tuple of n spectral pairs modulo $2\mathbb{Z}$, i.e. by n pairs $([\alpha_1], l_1), ..., ([\alpha_n], l_n) \in \mathbb{R}/2\mathbb{Z} \times \mathbb{Z}$. Here $\alpha_1, ..., \alpha_n \in \mathbb{R}$. The eigenvalues of the monodromy M are the numbers $e^{-2\pi i\alpha_1}, ..., e^{-2\pi i\alpha_n}$. The numbers $l_1, ..., l_n$ determine the Jordan block structure, see [BH19] for details.

The classification over $\mathbb{C}$ follows easily. Though it was carried out before in [Go94-1, Go94-2], and it is formulated also in [CDG24, Theorem 4.22].

(ii) A unimodular bilinear lattice $(H_{\mathbb{Z}}, L)$ is called in [CDG24] a *Mukai pair*. In [CDG24, 4.1–4.4] basic results of Gorodentsev for $R = \mathbb{Z}$ or $R = \mathbb{C}$ are rewritten. The monodromy is there called *canonical operator*. A triangular basis is there called *exceptional*.

(iii) The classification over $\mathbb{Z}$, so of unimodular bilinear lattices $(H_{\mathbb{Z}}, L)$, is wide open for larger n. The case $n = 3$ is treated in great detail in this book.

Lemma 2.6 *(a) Let $(H_{\mathbb{Z}}, L)$ be a bilinear lattice of rank $n \in \mathbb{N}$.*

(i) Let $\underline{e} = (e_1, ..., e_n)$ be a $\mathbb{Z}$-basis of $H_{\mathbb{Z}}$. Define $S := L^t(\underline{e}^t, \underline{e}) = L(\underline{e}^t, \underline{e})^t \in M_{n\times n}(\mathbb{Z}) \cap GL_n(\mathbb{Q})$. Then

$$I^{(0)}(\underline{e}^t, \underline{e}) = S + S^t,\ I^{(1)}(\underline{e}^t, \underline{e}) = S - S^t,\ M(\underline{e}) = \underline{e}S^{-1}S^t.$$

(ii)

$$\begin{aligned} I^{(0)}(a, b) &= L((M + \mathrm{id})a, b), \quad \mathrm{Rad}\, I^{(0)} = \ker((M + \mathrm{id}) : H_{\mathbb{Z}} \to H_{\mathbb{Z}}), \\ I^{(1)}(a, b) &= L((M - \mathrm{id})a, b), \quad \mathrm{Rad}\, I^{(1)} = \ker((M - \mathrm{id}) : H_{\mathbb{Z}} \to H_{\mathbb{Z}}). \end{aligned}$$

(iii)

$$G_{\mathbb{Z}} = \mathrm{Aut}(H_{\mathbb{Z}}, L, I^{(0)}, I^{(1)}, M) \subset \left\{ \begin{matrix} G_{\mathbb{Z}}^{(0)} \\ G_{\mathbb{Z}}^{(1)} \end{matrix} \right\} \subset G_{\mathbb{Z}}^{M}.$$

(iv) $M \in G_{\mathbb{Z}}$ if $(H_{\mathbb{Z}}, L)$ is unimodular or if $\mathcal{B}^{tri} \neq \emptyset$.

(v) If $a \in R^{(0)}$ then

$$s_a^{(0)} := s_a^{[0]} \text{ and } s_a^{(1)} := s_{a/\sqrt{L(a,a)}}^{[1]}, \quad \text{so } s_a^{(1)}(b) = b - \frac{I^{(1)}(a, b)}{L(a, a)}a,$$

are in $O^{(0)}$ *respectively* $O^{(1)}$.

(vi) [HK16, Lemma 2.1] If $a, b \in R^{(0)}$ *then* $s_a^{(0)}(b) \in R^{(0)}$ *(but not necessarily* $s_a^{(1)}(b) \in R^{(0)}$*).*

(vii) If $a, b \in R^{(0)}$ *with* $L(a, b) = 0$ *then*

$$L(s_a^{(1)}b, s_a^{(1)}b) = L(b, b), \quad s_a^{(1)}(b) \in R^{(0)}, \quad s^{(1)}_{s_a^{(1)}(b)} = s_a^{(1)} s_b^{(1)} (s_a^{(1)})^{-1}.$$

(b) The map

$$\{(H_{\mathbb{Z}}, L, \underline{e}) \mid \begin{matrix} (H_{\mathbb{Z}}, L) \text{ is a bilinear} \\ \text{lattice of rank } n, \ \underline{e} \in \mathcal{B}^{tri} \end{matrix}\}/\text{isomorphism} \to T_n^{tri}$$

is a bijection and restricts to a bijection

$$\{(H_{\mathbb{Z}}, L, \underline{e}) \mid \begin{matrix} (H_{\mathbb{Z}}, L) \text{ is a unimodular} \\ \text{bilinear lattice of rank } n, \ \underline{e} \in \mathcal{B}^{tri} \end{matrix}\}/\text{isom.} \to T_n(\mathbb{Z}).$$

(c) [HK16, Lemma 3.10] Let $(H_{\mathbb{Z}}, L)$ *be a unimodular bilinear lattice with* $\mathcal{B}^{tri} \neq \emptyset$. *Then*

$$R^{(0)} = \{a \in H_{\mathbb{Z}} \mid L(a, a) = 1\}.$$

(d) Let $(H_{\mathbb{Z}}, L)$ *be a unimodular bilinear lattice with* $\mathcal{B}^{tri} \neq \emptyset$. *Define for* $a \in H_{\mathbb{Z}}$ $s_a^{(1)} := s_a^{[1]} \in O^{(1)}$. *This definition is compatible with the definition of* $s_a^{(1)}$ *for* $a \in R^{(0)}$ *in part (a) (v). Furthermore now for* $a, b \in H_{\mathbb{Z}}$

$$s^{(1)}_{s_a^{(1)}(b)} = s_a^{(1)} s_b^{(1)} (s_a^{(1)})^{-1}.$$

Proof (a) (i) The defining equation for M can be written as $L((M\underline{e})^t, \underline{e}) = L(\underline{e}, \underline{e})^t$, which implies $M\underline{e} = \underline{e}S^{-1}S^t$. The rest is trivial.

(ii) Trivial.

(iii) $g \in G_{\mathbb{Z}}$ commutes with M because

$$L(gMa, gb) = L(Ma, b) = L(b, a) = L(gb, ga) = L(Mga, gb).$$

Of course it respects $I^{(0)}$ and $I^{(1)}$. Therefore $G_{\mathbb{Z}} = \text{Aut}(H_{\mathbb{Z}}, L, I^{(0)}, I^{(1)}, M)$. The rest is trivial.

(iv) The calculation $L(Ma, Mb) = L(Mb, a) = L(a, b)$ shows that M respects L. It remains to show $M \in \text{Aut}(H_{\mathbb{Z}})$.

This is clear if $(H_{\mathbb{Z}}, L)$ is unimodular. Suppose $\mathcal{B}^{tri} \neq \emptyset$ and $(H_{\mathbb{Z}}, L)$ not unimodular. Consider $\underline{e} \in \mathcal{B}^{tri}$, $S := L(\underline{e}^t, \underline{e})^t \in T_n^{tri}$ and $D := \text{diag}(s_{11}, ..., s_{nn}) \in M_{n\times n}(\mathbb{Z})$. Then $D^{-1}S, SD^{-1} \in T_n(\mathbb{Z})$ and

$$S^{-1}S^t = S^{-1}DD^{-1}S^t = (D^{-1}S)^{-1}(SD^{-1})^t \in GL_n(\mathbb{Z}),$$

so $M \in \mathrm{Aut}(H_{\mathbb{Z}})$.

(v) If $a \in R^{(0)}$ and $b \in H_{\mathbb{Z}}$ then $\frac{L(a,b)}{L(a,a)}, \frac{L(b,a)}{L(a,a)} \in \mathbb{Z}$, so

$$\frac{2I^{(0)}(a,b)}{I^{(0)}(a,a)}, \frac{I^{(1)}(a,b)}{L(a,a)} \in \mathbb{Z} \text{ and } s_a^{(0)}(b), s_a^{(1)}(b), (s_a^{(1)})^{-1}(b) \in H_{\mathbb{Z}}.$$

(vi) $L(s_a^{(0)}(b), s_a^{(0)}(b)) = L(b,b)$ because $s_a^{(0)} \in G_{\mathbb{Z}}^{(0)}$ and $I^{(0)} = L^t + L$ (in general $L(s_a^{(1)}(b), s_a^{(1)}(b)) \neq L(b,b)$). For $c \in H_{\mathbb{Z}}$

$$\begin{aligned}\frac{L(s_a^{(0)}(b), c)}{L(s_a^{(0)}(b), s_a^{(0)}(b))} &= \frac{L(b - \frac{L(a,b)+L(b,a)}{L(a,a)}a, c)}{L(b,b)} \\ &= \frac{L(b,c)}{L(b,b)} - \frac{L(a,b)+L(b,a)}{L(b,b)}\frac{L(a,c)}{L(a,a)} \in \mathbb{Z},\end{aligned}$$

and analogously $\frac{L(c, s_a^{(0)}(b))}{L(s_a^{(0)}(b), s_a^{(0)}(b))} \in \mathbb{Z}$.

(vii) $I^{(1)}(a,b) = L(b,a)$ because of $L(a,b) = 0$.

$$\begin{aligned}L(s_a^{(1)}(b), s_a^{(1)}(b)) &= L(b - \frac{L(b,a)}{L(a,a)}a, b - \frac{L(b,a)}{L(a,a)}a) \\ &= L(b,b) - \frac{L(b,a)}{L(a,a)}L(b,a) - 0 + (-\frac{L(b,a)}{L(a,a)})^2 L(a,a) \\ &= L(b,b).\end{aligned}$$

For $c \in H_{\mathbb{Z}}$

$$\begin{aligned}\frac{L(s_a^{(1)}(b), c)}{L(s_a^{(1)}(b), s_a^{(1)}(b))} &= \frac{L(b - \frac{L(b,a)}{L(a,a)}a, c)}{L(b,b)} \\ &= \frac{L(b,c)}{L(b,b)} - \frac{L(b,a)}{L(b,b)}\frac{L(a,c)}{L(a,a)} \in \mathbb{Z},\end{aligned}$$

so $s_a^{(1)}(b) \in R^{(0)}$. Finally

$$\begin{aligned}s_a^{(1)} s_b^{(1)} (s_a^{(1)})^{-1} &= s_a^{(1)} s^{[1]}_{b/\sqrt{L(b,b)}} (s_a^{(1)})^{-1} \\ &\overset{\text{Lemma 2.2 (c)}}{=} s^{[1]}_{s_a^{(1)}(b/\sqrt{L(b,b)})} = s^{[1]}_{s_a^{(1)}(b)/\sqrt{L(b,b)}} \\ &= s^{[1]}_{s_a^{(1)}(b)/\sqrt{L(s_a^{(1)}(b), s_a^{(1)}(b))}} = s^{(1)}_{s_a^{(1)}(b)}.\end{aligned}$$

(b) Starting with $S \in T_n^{tri}$, one can define $H_{\mathbb{Z}} := M_{n\times 1}(\mathbb{Z})$ with standard $\mathbb{Z}$-basis $\underline{e} = (e_1, ..., e_n)$, and one can define $L : H_{\mathbb{Z}} \times H_{\mathbb{Z}} \to \mathbb{Z}$ by $L(\underline{e}^t, \underline{e}) = S^t$. Then $\underline{e} \in \mathcal{B}^{tri}$.

If $(H_{\mathbb{Z}}, L)$ is unimodular then $\pm 1 = \det L(\underline{e}^t, \underline{e}) = L(e_1, e_1)...L(e_n, e_n)$ and $L(e_i, e_i) \in \mathbb{N}$, so $L(e_i, e_i) = 1$ and $L(\underline{e}^t, \underline{e})^t \in T_n(\mathbb{Z})$. The rest is trivial.

(c) The inclusion $R^{(0)} \supset \{a \in H_{\mathbb{Z}} \mid L(a, a) = 1\}$ is obvious. Consider $\underline{e} \in \mathcal{B}^{tri}$. By part (b), the matrix $S := L(\underline{e}^t, \underline{e})^t$ is in $T_n(\mathbb{Z})$. Consider an L1-cycle $a = \sum_{i=1}^n \alpha_i e_i \in R^{(0)}$. Then

$$(L(a, e_1), ..., L(a, e_n)) = (\alpha_1, ..., \alpha_n)S,$$
$$\text{so } \gcd(L(a, e_1), ..., L(a, e_n)) = \gcd(\alpha_1, ..., \alpha_n).$$

But $L(a, a)$ divides $\gcd(L(a, e_1), ..., L(a, e_n))$ because a is an L1-cycle. Therefore $L(a, a)$ divides each α_i. Thus $L(a, a)^2$ divides $\sum_{i=1}^n \sum_{j=1}^n \alpha_i s_{ij} \alpha_j = L(a, a)$, so $L(a, a) = 1$.

(d) By part (c) $L(a, a) = 1$ for $a \in R^{(0)}$. □

Up to now, the triangular shape of the matrix $L(\underline{e}^t, \underline{e})^t \in T_n^{tri}$ has not been used. It leads to the result in Theorem 2.7. In algebraic geometry and the theory of meromorphic differential equations, this result is well known, it is a piece of Picard-Lefschetz theory. In the frame of singularity theory, it is treated in [AGV88, Eb01]. An elementary direct proof for a unimodular lattice is given in [BH20]. The case $k = 0$ is proved in [HK16].

Theorem 2.7 *Let $(H_{\mathbb{Z}}, L, \underline{e})$ be a bilinear lattice with a triangular basis $\underline{e}$. Let $k \in \{0, 1\}$. Then*

$$s_{e_1}^{(k)}...s_{e_n}^{(k)} = (-1)^{k+1}M.$$

Proof The case $k = 0$ is a special case of Proposition 2.4 in [HK16]. The case $k = 1$ can be proved by an easy modification of Lemma 2.3 (5) and Proposition 2.4 in [HK16]. Both cases are proved for a unimodular bilinear lattice in [BH20, Theorem 4.1]. □

In Picard-Lefschetz theory and singularity theory also the following notions are standard.

Definition 2.8 Let $(H_{\mathbb{Z}}, L, \underline{e})$ be a bilinear lattice with a triangular basis $\underline{e}$. It induces several structures:

(a) The *even monodromy group* $\Gamma^{(0)} := \langle s_{e_1}^{(0)}, ..., s_{e_n}^{(0)} \rangle \subset O^{(0)}$.

(b) The *odd monodromy group* $\Gamma^{(1)} := \langle s_{e_1}^{(1)}, ..., s_{e_n}^{(1)} \rangle \subset O^{(1)}$.

(c) The set $\Delta^{(0)} := \Gamma^{(0)}\{\pm e_1, ..., \pm e_n\} \subset H_{\mathbb{Z}}$ of *even vanishing cycles*.

(d) The set $\Delta^{(1)} := \Gamma^{(1)}\{\pm e_1, ..., \pm e_n\} \subset H_{\mathbb{Z}}$ of *odd vanishing cycles*.

Remark 2.9 (i) The even vanishing cycles are L1-cycles, i.e. $\Delta^{(0)} \subset R^{(0)}$, because of Lemma 2.6 (a) (vi). In general $\Delta^{(1)} \not\subset R^{(0)}$. The name *vanishing cycles* for the elements of $\Delta^{(0)}$ and $\Delta^{(1)}$ and the name *monodromy group* stem from singularity theory. In [HK16] the elements of $\Delta^{(0)}$ are called *real roots*. $\Gamma^{(1)}$ and $\Delta^{(1)}$ are not considered in [HK16].

(ii) A matrix $S \in T_n^{tri}$ or $T_n(\mathbb{Z})$ determines by Lemma 2.6 (b) a bilinear lattice $(H_\mathbb{Z}, L, \underline{e})$ with a triangular basis (up to isomorphism). This leads to the program to determine for a given matrix S the data $I^{(0)}, I^{(1)}, G_\mathbb{Z}, G_\mathbb{Z}^{(0)}, G_\mathbb{Z}^{(1)}, G_\mathbb{Z}^M, \Gamma^{(0)}, \Gamma^{(1)}, \Delta^{(0)}$ and $\Delta^{(1)}$. One should start with relevant invariants like sign $I^{(0)}$, Rad $I^{(0)}$, Rad $I^{(1)}$, the characteristic polynomial and the Jordan normal form of M.

(iii) The odd monodromy group $\Gamma^{(1)}$ arises naturally in many cases where $(H_\mathbb{Z}, L)$ is a unimodular bilinear lattice, for example in cases from isolated hypersurface singularities. But it is not clear whether it is natural in cases where $(H_\mathbb{Z}, L)$ is a bilinear lattice which is not unimodular. Theorem 2.7 is positive evidence. But the following is negative evidence. The monodromy group

$$\Gamma^{(1)} = \langle s^{[1]}_{e_i/\sqrt{L(e_i,e_i)}} \mid i \in \{1, ..., n\}\rangle$$

contains because of Lemma 2.2 (c) all transvections $s^{[1]}_{g(e_i)/\sqrt{L(e_i,e_i)}}$ for $g \in \Gamma^{(1)}$. Only in the unimodular cases these coincide with the transvections $s^{[1]}_a$ for $a \in \Delta^{(1)}$. We will only consider the unimodular cases.

(iv) We will work on this program rather systematically in the Chaps. 5 and 6 for $S \in T_2(\mathbb{Z})$ and $S \in T_3(\mathbb{Z})$. We will discuss in Chap. 9 known results for matrices $S \in T_n(\mathbb{Z})$ from singularity theory, with emphasis on the simple singularities (ADE-type, i.e. the ADE root lattices) and the simple elliptic singularities (i.e. $\widetilde{E}_6, \widetilde{E}_7, \widetilde{E}_8$).

Definition 2.10 and Lemma 2.11 discuss the case when a unimodular bilinear lattice $(H_\mathbb{Z}, L, \underline{e})$ with triangular basis is *reducible*. Then also the monodromy groups, the set of L1-cycles and the sets of vanishing cycles split. But beware that here reducibility involves not only $(H_\mathbb{Z}, L)$, but also $\underline{e}$.

Definition 2.10 (a) Let $(H_\mathbb{Z}, L, \underline{e})$ be a unimodular bilinear lattice of rank $n \in \mathbb{N}$ with a triangular basis $\underline{e}$. Let $\{1, ..., n\} = I_1 \dot{\cup} I_2$ be a decomposition into disjoint subsets such that

$$L(e_i, e_j) = L(e_j, e_i) = 0 \quad \text{for } i \in I_1,\ j \in I_2.$$

Then the triple $(H_\mathbb{Z}, L, \underline{e})$ is called *reducible*. If such a decomposition does not exist the triple is called *irreducible*.

(b) A matrix $S \in T_n(\mathbb{Z})$ is called reducible if the triple $(H_\mathbb{Z}, L, \underline{e})$ (which is unique up to isomorphism) is reducible, where $(H_\mathbb{Z}, L)$ is a unimodular bilinear lattice and $\underline{e}$ is a triangular basis with $S = L(\underline{e}^t, \underline{e})^t$.

Lemma 2.11 *Keep the situation of Definition 2.10. For $l \in \{1, 2\}$ let $\sigma_l : \{1, 2, ..., |I_l|\} \to I_l$ be the unique bijection with $\sigma_l(i) < \sigma_l(j)$ for $i < j$. Define*

$$\underline{e}_l := (e_{1,l}, e_{2,l}, ..., e_{|I_l|,l}) := (e_{\sigma_l(1)}, e_{\sigma_l(2)}, ..., e_{\sigma_l(|I_l|)}),$$
$$H_{\mathbb{Z},l} := \bigoplus_{i=1}^{|I_l|} \mathbb{Z} \cdot e_{i,l}, \quad L_l := L|_{H_{\mathbb{Z},l}}.$$

Then $(H_{\mathbb{Z},l}, L_l, \underline{e}_l)$ *is a unimodular bilinear lattice with triangular basis. The decomposition* $H_{\mathbb{Z}} = H_{\mathbb{Z},1} \oplus H_{\mathbb{Z},2}$ *is left and right L-orthogonal. Denote by* $\Gamma_l^{(0)}, \Gamma_l^{(1)}, \Delta_l^{(0)}, \Delta_l^{(1)}$ *and* $R_l^{(0)}$ *the monodromy groups and sets of vanishing cycles and L1-cycles of* $(H_{\mathbb{Z},l}, L_l, \underline{e}_l)$. *Denote by* $\widetilde{M}_l$ *the automorphism of* $H_{\mathbb{Z}}$ *which extends the monodromy* M_l *on* $H_{\mathbb{Z},l}$ *by the identity on* $H_{\mathbb{Z},m}$*, where* $\{l, m\} = \{1, 2\}$. *Then*

$$\begin{aligned} \Gamma^{(k)} &= \Gamma_1^{(k)} \times \Gamma_2^{(k)}, \\ R^{(0)} &= R_1^{(0)} \,\dot{\cup}\, R_2^{(0)}, \\ \Delta^{(k)} &= \Delta_1^{(k)} \,\dot{\cup}\, \Delta_2^{(k)}, \\ M &= \widetilde{M}_1\widetilde{M}_2 = \widetilde{M}_2\widetilde{M}_1. \end{aligned}$$

The proof is trivial. Because of this lemma, we will study the monodromy groups and the sets of vanishing cycles only for irreducible triples. In the Example 1.1 this excludes the cases $S(A_1^2)$, $S(A_1^3)$, $S(A_2A_1)$, $S(\mathbb{P}^1 A_1)$ and all cases $S(x_1, x_2, x_3)$ where two of the three numbers x_1, x_2, x_3 are zero.

The following lemma treats the cases $S(A_1^n) := E_n$ for $n \in \mathbb{N}$. It is a trivial consequence of the special case $S(A_1) = (1) \in M_{1\times 1}(\mathbb{Z})$ and Lemma 2.11, but worth to be stated.

Lemma 2.12 *The case* A_1^n *for any* $n \in \mathbb{N}$:

$$H_{\mathbb{Z}} = \bigoplus_{i=1}^n \mathbb{Z}\cdot e_i, \quad S = S(A_1^n) := E_n, \quad I^{(0)} = 2L, \quad I^{(1)} = 0,$$

the reflections $s_{e_i}^{(0)}$ with $s_{e_i}^{(0)}(e_j) = \begin{Bmatrix} e_j & \text{if } j \neq i, \\ -e_i & \text{if } j = i, \end{Bmatrix}$ *commute, the transvections* $s_{e_i}^{(1)}$ *are* $s_{e_i}^{(1)} = \mathrm{id}$,

$$\begin{aligned} \Gamma^{(0)} &= \{\prod_{i=1}^n (s_{e_i}^{(0)})^{l_i} \mid (l_1, ..., l_n) \in \{0; 1\}^n\} \cong \{\pm 1\}^n, \\ \Gamma^{(1)} &= \{\mathrm{id}\}, \\ \Delta^{(0)} &= R^{(0)} = \{\pm e_1, ..., \pm e_n\} = \Delta^{(1)}. \end{aligned}$$

Chapter 3
Braid Group Actions

In the Sects. 3.2–3.4 a unimodular bilinear lattice $(H_{\mathbb{Z}}, L)$ of some rank $n \geq 2$ is considered. The braid group Br_n is introduced in Sect. 3.1. It acts on several sets of n-tuples and of matrices associated to $(H_{\mathbb{Z}}, L)$.

Section 3.1 starts with the Hurwitz action on G^n where G is a group. Results of Artin, Birman-Hilden and Igusa-Schiffler for G a free group, a free Coxeter group or any Coxeter group are cited and applied. This is relevant as many of the monodromy groups $\Gamma^{(1)}$ and $\Gamma^{(0)}$ of rank 2 or rank 3 unimodular bilinear lattices with triangular bases are such groups.

It turns out that the Hurwitz action of Br_n on $(O^{(k)})^n$ lifts to an action of a semidirect product $\mathrm{Br}_n \ltimes \{\pm 1\}^n$ on sets of certain n-tuples of cycles in $H_{\mathbb{Z}}$. This is discussed in Sect. 3.2.

Most important is the set $\mathcal{B}^{tri}$ of triangular bases of $(H_{\mathbb{Z}}, L)$ (if this set is not empty) and the subset $\mathcal{B}^{dist} = \mathrm{Br}_n \ltimes \{\pm 1\}^n(\underline{e})$ of a chosen triangular basis $\underline{e}$. Section 3.3 poses questions on the characterization of such a set $\mathcal{B}^{dist}$ of *distinguished bases* which will guide our work in Chap. 7. It also offers several examples with quite different properties.

Section 3.4 connects the group $\mathrm{Br}_n \ltimes \{\pm 1\}^n$ via its action on the orbit of a triangular basis $\underline{e}$ with the group $G_{\mathbb{Z}}$. There is a group antihomomorphism $Z : (\mathrm{Br}_n \ltimes \{\pm 1\}^n)_S \to G_{\mathbb{Z}}$, where $(\mathrm{Br}_n \ltimes \{\pm 1\}^n)_S$ denotes the stabilizer of a matrix $S \in T_n(\mathbb{Z})$. In this way certain braids induce automorphisms in $G_{\mathbb{Z}}$, and in many cases these automorphisms generate $G_{\mathbb{Z}}$, i.e. Z is surjective. Theorem 3.26 (b) states the well known fact $Z((\delta^{1-k}\sigma^{root})^n) = (-1)^{k+1}M$ for $k \in \{0; 1\}$. Theorem 3.26 (c) gives a condition when $Z(\delta^{1-k}\sigma^{root})$ is in $G_{\mathbb{Z}}$ and thus an n-th root of $(-1)^{k+1}M$. Theorem 3.28 states that for almost all cases with rank ≤ 3 the map Z is surjective. The exceptions are only four cases.

C. Hertling and K. Larabi, *Triangular Bases of Unimodular Bilinear Lattices and Induced Structures*, Lecture Notes in Mathematics 2398,
https://doi.org/10.1007/978-3-032-25997-4_3

3.1 The Braid Group and the Hurwitz Action, Some Classical Results

Choose $n \in \mathbb{Z}_{\geq 2}$. The braid group Br_n of braids with n strings was introduced by Artin [Ar25]. It is the fundamental group of a configuration space. We will come to this geometric point of view in Chap. 8. Here we take a purely algebraic point of view. Artin [Ar25, Satz 1] showed that Br_n is generated by $n-1$ elementary braids $\sigma_1, ..., \sigma_{n-1}$ and that all relations come from the relations

$$\begin{aligned} \sigma_i\sigma_j &= \sigma_j\sigma_i \quad \text{for } i, j \in \{1, ..., n-1\} \text{ with } |i-j| \geq 2, \\ \sigma_i\sigma_{i+1}\sigma_i &= \sigma_{i+1}\sigma_i\sigma_{i+1} \quad \text{for } i \in \{1, ..., n-2\}. \end{aligned}$$

He also showed [Ar47, Theorem 19] that the center of Br_n is

$$\mathrm{Center}(\mathrm{Br}_n) = \langle \sigma^{mon} \rangle,$$

where

$$\sigma^{root} := \sigma_{n-1}\sigma_{n-2}...\sigma_2\sigma_1, \quad \sigma^{mon} := (\sigma^{root})^n.$$

An important action of Br_n is the *Hurwitz action* on the n-th power G^n for any group G. The braid group Br_n acts via

$$\begin{aligned} \sigma_i(g_1, ..., g_n) &:= (g_1, ..., g_{i-1}, g_i g_{i+1} g_i^{-1}, g_i, g_{i+2}, ..., g_n), \\ \sigma_i^{-1}(g_1, ..., g_n) &:= (g_1, ..., g_{i-1}, g_{i+1}, g_{i+1}^{-1} g_i g_{i+1}, g_{i+2}, ..., g_n). \end{aligned}$$

The fibers of the map

$$\pi_n : G^n \to G, \quad \underline{g} = (g_1, ..., g_n) \mapsto g_1...g_n,$$

are invariant under this action,

$$\pi_n(\underline{g}) = \pi_n(\sigma_i \underline{g}) = \pi_n(\sigma_i^{-1} \underline{g}).$$

We will study this action for $n = 3$ in the cases of the monodromy groups for the rank 3 unimodular bilinear lattices. The following results in Theorem 3.2 of Artin [Ar25] and Birman-Hilden [BH73] will be relevant.

Definition 3.1 (a) Let $G^{free,n}$ be the free group with n generators $x_1, ..., x_n$. Let

$$\Delta(G^{free,n}) := \bigcup_{i=1}^{n} \{w x_i w^{-1} \mid w \in G^{free,n}\}$$

be the set of elements conjugate to $x_1, ..., x_n$. Obviously $\mathrm{Br}_n((x_1, ..., x_n)) \subset \Delta(G^{free,n})^n$.

(b) Let $G^{fCox,n}$ be the free Coxeter group with n generators $x_1, ..., x_n$, so all relations are generated by the relations $x_1^2 = ... = x_n^2 = e$. Let

$$\Delta(G^{fCox,n}) := \bigcup_{i=1}^{n} \{w x_i w^{-1} \mid w \in G^{fCox,n}\}$$

be the set of elements conjugate to $x_1, ..., x_n$. Obviously $\mathrm{Br}_n((x_1, ..., x_n)) \subset \Delta(G^{fCox,n})^n$.

Theorem 3.2 *(a) [Ar25, Satz 7 and Satz 9]* Br_n *acts simply transitively on the set of tuples*

$$\{(w_1, ..., w_n) \in \Delta(G^{free,n})^n \mid w_1 ... w_n = x_1 ... x_n\}.$$

(b) [BH73, Theorem 7] Br_n *acts simply transitively on the set of tuples*

$$\{(w_1, ..., w_n) \in \Delta(G^{fCox,n})^n \mid w_1 ... w_n = x_1 ... x_n\}.$$

Remark 3.3 (i) Both results were reproved by Krüger in [Kr90, Satz 7.6].

(ii) Theorem 1.31 in [KT08] gives a weaker version of Artin's result Theorem 3.2 (a). Theorem 1.31 in [KT08] is equivalent to the statement that Br_n acts simply transitively on the set of tuples

$$\begin{aligned}\{(w_1, ..., w_n) \in \Delta(G^{free,n})^n \mid\ & w_1 ... w_n = x_1 ... x_n, \\ & w_1, ..., w_n \text{ generate } G^{free,n}, \text{ a permutation} \\ & \sigma \in S_n \text{ exists with } w_i \text{ conjugate to } x_{\sigma(i)}\}.\end{aligned}$$

(iii) The formulation of Theorem 1.31 in [KT08] is different. There a group automorphism φ of $G^{free,n}$ is called a *braid automorphism* if $\varphi(x_1 ... x_n) = x_1 ... x_n$ and if a permutation $\sigma \in S_n$ with $\varphi(x_i)$ conjugate to $x_{\sigma(i)}$ exists. The group of all braid automorphisms is called $\widetilde{\mathrm{Br}}_n$. Theorem 1.31 in [KT08] states that the map

$$\begin{aligned} \widetilde{Z} : \{\sigma_1^{\pm 1}, ..., \sigma_{n-1}^{\pm 1}\} \to \widetilde{Br}_n \quad \text{with} \\ (\widetilde{Z}(\sigma_i)(x_1), ..., \widetilde{Z}(\sigma_i)(x_n)) = \sigma_i^{-1}(x_1, ..., x_n), \\ (\widetilde{Z}(\sigma_i^{-1})(x_1), ..., \widetilde{Z}(\sigma_i^{-1})(x_n)) = \sigma_i(x_1, ..., x_n), \end{aligned}$$

extends to a group isomorphism $\mathrm{Br}_n \to \widetilde{\mathrm{Br}}_n$.

(iv) In order to understand the equivalence of the statements in (ii) and (iii), it is crucial to see that the extension $\widetilde{Z} : \mathrm{Br}_n \to \widetilde{\mathrm{Br}}_n$ which is defined by

$$(\widetilde{Z}(\beta)(x_1), ..., \widetilde{Z}(\beta)(x_n)) = \beta^{-1}(x_1, ..., x_n) \quad \text{for } \beta \in \mathrm{Br}_n$$

is a group homomorphism. This follows from the equations

$$\widetilde{Z}(\beta\sigma_i) = \widetilde{Z}(\beta)\widetilde{Z}(\sigma_i) \quad \text{and} \quad \widetilde{Z}(\beta\sigma_i^{-1}) = \widetilde{Z}(\beta)\widetilde{Z}(\sigma_i^{-1})$$

for $\beta \in \mathrm{Br}_n$ and $i \in \{1, ...n-1\}$. The first equation holds because of

$$\begin{aligned}
&(\widetilde{Z}(\beta\sigma_i)(x_1), ..., \widetilde{Z}(\beta\sigma_i)(x_n)) \\
&= (\beta\sigma_i)^{-1}(x_1, ..., x_n) \\
&= \sigma_i^{-1}(\beta^{-1}(x_1, ..., x_n)) \\
&= \sigma_i^{-1}(\widetilde{Z}(\beta)(x_1), ..., \widetilde{Z}(\beta)(x_n)) \\
&= (\widetilde{Z}(\beta)(x_1), ..., \widetilde{Z}(\beta)(x_{i-1}), \widetilde{Z}(\beta)(x_{i+1}), \\
&\quad (\widetilde{Z}(\beta)(x_{i+1}))^{-1}\widetilde{Z}(\beta)(x_i)\widetilde{Z}(\beta)(x_{i+1}), \widetilde{Z}(\beta)(x_{i+2}), ..., \widetilde{Z}(\beta)(x_n)) \\
&= (\widetilde{Z}(\beta)(x_1), ..., \widetilde{Z}(\beta)(x_{i-1}), \widetilde{Z}(\beta)(x_{i+1}), \\
&\quad \widetilde{Z}(\beta)(x_{i+1}^{-1}x_i x_{i+1}), \widetilde{Z}(\beta)(x_{i+2}), ..., \widetilde{Z}(\beta)(x_n)) \\
&= (\widetilde{Z}(\beta)\widetilde{Z}(\sigma_i)(x_1), ..., \widetilde{Z}(\beta)\widetilde{Z}(\sigma_i)(x_n)).
\end{aligned}$$

The second equation is proved by a similar calculation.

Example 3.4 Theorem 3.2 will be applied in the following situation, which in fact arises quite often.

Consider a unimodular bilinear lattice $(H_{\mathbb{Z}}, L, \underline{e})$ of rank $n \geq 2$ with triangular basis $\underline{e}$ such that for some $k \in \{0, 1\}$ the following holds:

$$\Gamma^{(k)} = \begin{cases} G^{fCox,n} & \text{with generators } s_{e_1}^{(0)}, ..., s_{e_n}^{(0)} \quad \text{if } k = 0, \\ G^{free,n} & \text{with generators } s_{e_1}^{(1)}, ..., s_{e_n}^{(1)} \quad \text{if } k = 1. \end{cases}$$

Then in the notation of Definition 3.1

$$\begin{aligned}
\Delta(G^{fCox,n}) &= \{s_v^{(0)} \mid v \in \Delta^{(0)}\} \quad \text{if } k = 0, \\
\Delta(G^{free,n}) &= \{s_v^{(1)} \mid v \in \Delta^{(1)}\} \quad \text{if } k = 1.
\end{aligned}$$

By Theorem 3.2, two statements hold:

(1) The set

$$\{(s_{v_1}^{(k)}, ..., s_{v_n}^{(k)}) \mid v_1, ..., v_n \in \Delta^{(k)}, s_{v_1}^{(k)}...s_{v_n}^{(k)} = (-1)^{k+1}M\}$$

is a single orbit under the Hurwitz action of Br_n.

(2) The stabilizer of any such tuple $(s_{v_1}^{(k)}, ..., s_{v_n}^{(k)})$ under the Hurwitz action of Br_n is $\{\mathrm{id}\}$.

Theorem 3.2 (b) concerns a free Coxeter group with n generators. The transitivity of the action generalizes to arbitrary Coxeter groups and can be applied to generalize the statement (1) in Example 3.4, as is explained in the following.

Definition 3.5 (Classical, e.g. [Hu90, 5.1]) A *Coxeter system* (W, S^{gen}) consists of a group W and a finite set $S^{gen} = \{s_1, ..., s_n\} \subset W$ for some $n \in \mathbb{N}$ of generators of the group such that there are generating relations as follows. There is a subset $I \subset \{(i, j) \in \{1, ..., n\}^2 \mid i < j\}$ and a map $a : I \to \mathbb{Z}_{\geq 2}$ such that the generating relations are

$$s_1^2 = ... = s_n^2 = 1, \quad 1 = (s_i s_j)^{c(i,j)} \text{ for } (i, j) \in I.$$

The group W is then called a *Coxeter group*.

The following theorem was proved by Deligne [De74] for the ADE Weyl groups and in general by Igusa and Schiffler [IS10]. A short proof was given by Baumeister, Dyer, Stump and Wegener [BDSW14].

Theorem 3.6 *[De74][IS10][BDSW14] Let (W, S^{gen}) with $S^{gen} = \{s_1, ..., s_n\}$ be a Coxeter system with $n \geq 2$. Define $\Delta(W, S^{gen}) := \bigcup_{i=1}^n \{w s_i w^{-1} \mid w \in W\}$. The set*

$$\{(w_1, ..., w_n) \in \Delta(W, S^{gen})^n \mid w_1 ... w_n = s_1 ... s_n\}$$

is a single orbit under the Hurwitz action of Br_n.

Part (a) of the following theorem is classical if $S_{ij} \in \{0, -1, -2\}$ for $i < j$ and due to Vinberg [Vi71] in the general case.

Theorem 3.7 *Let $(H_{\mathbb{Z}}, L)$ be a unimodular bilinear lattice of rank $n \geq 2$ and let $\underline{e}$ be a triangular basis such that the matrix $S = L(\underline{e}^t, \underline{e})^t \in T_n(\mathbb{Z})$ satisfies $S_{ij} \leq 0$ for $i < j$.*

(a) (Classical for $S_{ij} \in \{0, -1, -2\}$, [Vi71, Proposition 6, Theorem 1, Theorem 2, Proposition 17] for $S_{ij} \leq 0$) The pair $(\Gamma^{(0)}, \{s_{e_1}^{(0)}, ..., s_{e_n}^{(0)}\})$ is a Coxeter system with

$$I = \{(i, j) \in \{1, ..., n\}^2 \mid i < j, S_{ij} \in \{0, -1\}\} \quad \textit{and}$$

$$a(i, j) = \begin{cases} 2 & \textit{if } S_{ij} = 0, \\ 3 & \textit{if } S_{ij} = -1. \end{cases}$$

(b) The set

$$\{(g_1, ..., g_n) \in \left(\{s_v^{(0)} \mid v \in \Delta^{(0)}\}\right)^n \mid g_1 ... g_n = -M\}$$

is a single orbit under the Hurwitz action of Br_n.

Proof of part (b): Observe

$$\Delta(\Gamma^{(0)}, \{s_{e_1}^{(0)}, ..., s_{e_n}^{(0)}\}) = \{s_v^{(0)} \mid v \in \Delta^{(0)}\}.$$

Apply Theorem 3.6. □

Remark 3.8 (i) The transitivity result in part (b) holds also for a bilinear lattice $(H_{\mathbb{Z}}, L)$ which is not necessarily unimodular, if it comes equipped with a triangular basis $\underline{e}$ with $L(e_i, e_j) \le 0$ for $i > j$. This is the case of a generalized Cartan lattice (Remark 2.4 (v)). This is crucial in [HK16].

(ii) Especially in the case of a unimodular bilinear lattice $(H_{\mathbb{Z}}, L, \underline{e})$ with triangular basis $\underline{e}$ and matrix $S = L(\underline{e}^t, \underline{e})^t \in T_n(\mathbb{Z})$ with $S_{ij} \le -2$ for all $i < j$ we have $\Gamma^{(0)} = G^{fCox,n}$ with generators $s^{(0)}_{e_1}, ..., s^{(0)}_{e_n}$.

(iii) Theorem 6.11 (g) gives in the case $n = 3$ $\Gamma^{(0)} = G^{fCox,3}$ with generators $s^{(0)}_{e_1}, s^{(0)}_{e_2}, s^{(0)}_{e_3}$ also in the following cases: if $S_{ij} \ge 3$ for $i < j$ and if additionally

$$2S_{12} \le S_{13}S_{23}, \quad 2S_{13} \le S_{12}S_{23}, \quad 2S_{23} \le S_{12}S_{13}.$$

(iv) Theorem 6.18 (g) gives in the situation of part (iii) also $\Gamma^{(1)} = G^{free,3}$ with generators $s^{(1)}_{e_1}, s^{(1)}_{e_2}, s^{(1)}_{e_3}$.

(v) Though in the situation of part (ii) there are cases with $\Gamma^{(1)} = G^{free,n}$, and there are cases with $\Gamma^{(1)} \neq G^{free,n}$. The odd cases are more complicated than the even cases. For the cases with $n = 3$ see the Remarks 4.17, Lemma 4.18 and Theorem 6.18.

3.2 Braid Group Action on Tuples of Cycles

Consider a unimodular bilinear lattice $(H_{\mathbb{Z}}, L)$ of some rank $n \ge 2$ and the groups $O^{(k)} = \text{Aut}(H_{\mathbb{Z}}, I^{(k)})$ for $k \in \{0; 1\}$. Recall that here the set of L1-cycles $R^{(0)}$ is

$$R^{(0)} = \{\delta \in H_{\mathbb{Z}} \mid L(\delta, \delta) = 1\}.$$

In order to treat the even case $k = 0$ and the odd case $k = 1$ uniformly, we define

$$R^{(1)} := H_{\mathbb{Z}}.$$

The Hurwitz action of Br_n on $(O^{(k)})^n$ restricts because of

$$s^{(k)}_a s^{(k)}_b (s^{(k)}_a)^{-1} = s^{(k)}_{s^{(k)}_a(b)} \quad \text{for} \quad a, b \in R^{(k)} \tag{3.1}$$

(Lemma 2.2 (c)) to an action on the subset $(\{s^{(k)}_v \mid v \in R^{(k)}\})^n$.

It turns out that this action has a natural lift to an action of a certain semidirect product $\mathrm{Br}_n \ltimes \{\pm 1\}^n$ on the set $(R^{(k)})^n$. Here the sets $(R^{(k)})^n$ and $(\{s_v^{(k)} \mid v \in R^{(k)}\})^n$ are related by the map

$$\pi_n^{(k)} : (R^{(k)})^n \to (\{s_v^{(k)} \mid v \in R^{(k)}\})^n \subset (O^{(k)})^n,$$
$$\underline{v} = (v_1, ..., v_n) \mapsto (s_{v_1}^{(k)}, ..., s_{v_n}^{(k)}).$$

Recall also the map

$$\pi_n : (O^{(k)})^n \to O^{(k)}, \quad (g_1, ..., g_n) \mapsto g_1 ... g_n$$

which was defined for an arbitrary group before Definition 3.1.

Furthermore it turns out that both actions, for $k = 0$ and for $k = 1$, restrict to the same action on the set $\mathcal{B}^{tri}$ of triangular bases if this set is not empty. This is the action in which we are interested most. In the case of a unimodular bilinear lattice from singularity theory, it is well known [Eb01, 5.7] [AGV88, §1.9] and has been studied by A'Campo, Brieskorn, Ebeling, Gabrielov, Gusein-Zade, Krüger and others. In fact it works also for bilinear lattices with $\mathcal{B}^{tri} \neq \emptyset$ which are not necessarily unimodular, see Remark 3.14.

Finally, the actions induce actions of $\mathrm{Br}_n \ltimes \{\pm 1\}^n$ on several spaces of matrices. The purpose of this section is to fix all these well known actions.

Lemma 3.9 presents the semidirect product $\mathrm{Br}_n \ltimes \{\pm 1\}^n$. Lemma 3.10 gives its action on $(R^{(k)})^n$. Lemma 3.11 gives its action on $\mathcal{B}^{tri}$ if this set is not empty.

Lemma 3.9 *Fix* $n \in \mathbb{Z}_{\geq 2}$.

(a) The multiplicative group $\{\pm 1\}^n$ *is called sign group. It is generated by the elements* $\delta_j = ((-1)^{\delta_{ij}})_{i=1,...,n} \in \{\pm 1\}^n$ *(here* δ_{ij} *is the Kronecker symbol) for* $j \in \{1, ..., n\}$.

(b) The following relations define a semidirect product $\mathrm{Br}_n \ltimes \{\pm 1\}^n$ *of* Br_n *and* $\{\pm 1\}^n$ *with* $\{\pm 1\}^n$ *as normal subgroup,*

$$\sigma_j \delta_i \sigma_j^{-1} = \delta_i \quad \textit{for } i \in \{1, ..., n\} - \{j, j+1\},$$
$$\sigma_j \delta_j \sigma_j^{-1} = \delta_{j+1}, \quad \sigma_j \delta_{j+1} \sigma_j^{-1} = \delta_j.$$

In the following $\mathrm{Br}_n \ltimes \{\pm 1\}^n$ *always means this semidirect product.*

Proof Part (a) is a notation. Part (b) requires a proof. We have the exact sequence

$$\{1\} \to \mathrm{Br}_n^{pure} \to \mathrm{Br}_n \to S_n \to \{1\}$$

where $\mathrm{Br}_n^{pure} \subset \mathrm{Br}_n$ is the normal subgroup of pure braids (and $\sigma_i \in \mathrm{Br}_n$ maps to the transposition $(i\ i+1) \in S_n$). See Remark 8.4 (vii) for this group and this exact sequence. The natural action of S_n on $\{\pm 1\}^n$,

$$S_n \ni \alpha : \underline{\varepsilon} = (\varepsilon_1, ..., \varepsilon_n) \mapsto (\varepsilon_{\alpha^{-1}(1)}, ..., \varepsilon_{\alpha^{-1}(n)}) =: \alpha.\underline{\varepsilon}$$

lifts to an action of Br_n on $\{\pm 1\}^n$, $\sigma : \underline{\varepsilon} \mapsto \sigma.\underline{\varepsilon}$. This action can be used to define a semidirect product of Br_n and $\{\pm 1\}^n$ by $\sigma\underline{\varepsilon}\sigma^{-1} := \sigma.\underline{\varepsilon}$. It is the semidirect product in part (b). □

Lemma 3.10 *Let $(H_{\mathbb{Z}}, L)$ be a unimodular bilinear lattice of rank $n \geq 2$. Fix $k \in \{0, 1\}$.*

(a) The following formulas define an action of the semidirect product $\mathrm{Br}_n \ltimes \{\pm 1\}^n$ *from Definition 3.9 (b) on the set $(R^{(k)})^n$,*

$$\begin{aligned}\sigma_j(\underline{v}) &= (v_1, ..., v_{j-1}, s^{(k)}_{v_j}(v_{j+1}), v_j, v_{j+2}, ..., v_n),\\ \sigma_j^{-1}(\underline{v}) &= (v_1, ..., v_{j-1}, v_{j+1}, (s^{(k)}_{v_{j+1}})^{-1}(v_j), v_{j+2}, ..., v_n),\\ \delta_j(\underline{v}) &= (v_1, ..., v_{j-1}, -v_j, v_{j+1}, ..., v_n),\end{aligned}$$

for $\underline{v} = (v_1, ..., v_n) \in (R^{(k)})^n$.

(b) The map $\pi_n^{(k)} : (R^{(k)})^n \to (O^{(k)})^n$ is compatible with the action of $\mathrm{Br}_n \ltimes \{\pm 1\}^n$ *on $(R^{(k)})^n$ from part (a) and the Hurwitz action of* Br_n *on $(O^{(k)})^n$, so the diagram*

$$\begin{array}{ccc} (R^{(k)})^n & \xrightarrow{\ \sigma_j\ } & (R^{(k)})^n \\ {\scriptstyle \pi_n^{(k)}}\downarrow & & \downarrow{\scriptstyle \pi_n^{(k)}} \\ (O^{(k)})^n & \xrightarrow{\ \sigma_j\ } & (O^{(k)})^n \end{array}$$

commutes. Here the sign group $\{\pm 1\}^n$ acts trivially on $(O^{(k)})^n$. Especially, each orbit in $(R^{(k)})^n$ is contained in one fiber of the projection $\pi_n \circ \pi_n^{(k)} : (R^{(k)})^n \to O^{(k)}$.

Proof (a) We denote the actions in part (a) by $\sigma_j^{(k)}$, $(\sigma_j^{-1})^{(k)}$ and $\delta_j^{(k)}$ (of course $\delta_j^{(0)} = \delta_j^{(1)}$). The identities

$$\begin{aligned}\sigma_j^{(k)}(\sigma_j^{-1})^{(k)} = (\sigma_j^{-1})^{(k)}\sigma_j^{(k)} &= \mathrm{id} \quad \text{for } j \in \{1, ..., n-1\}\\ \sigma_i^{(k)}\sigma_j^{(k)} &= \sigma_j^{(k)}\sigma_i^{(k)} \quad \text{for } |i-j| \geq 2,\\ \sigma_j^{(k)}\delta_i^{(k)}(\sigma_j^{-1})^{(k)} &= \delta_i^{(i)} \quad \text{for } i \in \{1, ..., n\} - \{j, j+1\},\\ \sigma_j^{(k)}\delta_j^{(k)}(\sigma_j^{-1})^{(k)} &= \delta_{j+1}^{(k)} \quad \text{and}\\ \sigma_j^{(k)}\delta_{j+1}^{(k)}(\sigma_j^{-1})^{(k)} &= \delta_j^{(k)} \quad \text{for } j \in \{1, ..., n-1\}\end{aligned}$$

are obvious or easy to see. The identities

$$\sigma_i^{(k)}\sigma_{i+1}^{(k)}\sigma_i^{(k)} = \sigma_{i+1}^{(k)}\sigma_i^{(k)}\sigma_{i+1}^{(k)} \quad \text{for } i \in \{1, ..., n-2\}$$

are proved by the following calculation with $\underline{v} \in (R^{(k)})^n$,

$$\sigma_i^{(k)}\sigma_{i+1}^{(k)}\sigma_i^{(k)}(\underline{v})$$

$$
\begin{aligned}
&= \sigma_i^{(k)}\sigma_{i+1}^{(k)}(\dots, v_{i-1}, s_{v_i}^{(k)}(v_{i+1}), v_i, v_{i+2}, v_{i+3}\dots) \\
&= \sigma_i^{(k)}(\dots, v_{i-1}, s_{v_i}^{(k)}(v_{i+1}), s_{v_i}^{(k)}(v_{i+2}), v_i, v_{i+3}, \dots) \\
&= (\dots, v_{i-1}, s_{s_{v_i}^{(k)}(v_{i+1})}^{(k)}(s_{v_i}^{(k)}(v_{i+2})), s_{v_i}^{(k)}(v_{i+1}), v_i, v_{i+3}, \dots) \\
&\overset{(3.1)}{=} (\dots, v_{i-1}, s_{v_i}^{(k)}s_{v_{i+1}}^{(k)}(s_{v_i}^{(k)})^{-1}(s_{v_i}^{(k)}(v_{i+2})), s_{v_i}^{(k)}(v_{i+1}), v_i, v_{i+3}, \dots) \\
&= (\dots, v_{i-1}, s_{v_i}^{(k)}(s_{v_{i+1}}^{(k)}(v_{i+2})), s_{v_i}^{(k)}(v_{i+1}), v_i, v_{i+3}, \dots) \\
&= \sigma_{i+1}^{(k)}(\dots, v_{i-1}, s_{v_i}^{(k)}(s_{v_{i+1}}^{(k)}(v_{i+2})), v_i, v_{i+1}, v_{i+3}, \dots) \\
&= \sigma_{i+1}^{(k)}\sigma_i^{(k)}(\dots, v_{i-1}, v_i, s_{v_{i+1}}^{(k)}(v_{i+2}), v_{i+1}, v_{i+3}, \dots) \\
&= \sigma_{i+1}^{(k)}\sigma_i^{(k)}\sigma_{i+1}^{(k)}(\underline{v}).
\end{aligned}
$$

The maps $\sigma_j^{(k)}$ and $\delta_i^{(k)}$ satisfy all relations between the generators σ_j and δ_i of the group $\mathrm{Br}_n \ltimes \{\pm 1\}^n$. Therefore the formulas in part (a) define an action of this group on the set $(R^{(k)})^n$.

(b) The actions are compatible because of (3.1). The sign group acts trivially on $(O^{(k)})^n$ because $s_v^{(k)} = s_{-v}^{(k)}$ for $v \in R^{(k)}$. Each orbit of the Hurwitz action on $(O^{(k)})^n$ is contained in one fiber of the map π_n, as was remarked in Sect. 3.1. □

Lemma 3.11 *Let $(H_{\mathbb{Z}}, L)$ be a unimodular bilinear lattice with nonempty set $\mathcal{B}^{tri}$ of triangular bases. The actions in Lemma 3.10 of* $\mathrm{Br}_n \ltimes \{\pm 1\}^n$ *on $(R^{(0)})^n$ and on $(R^{(1)})^n$ both restrict to the same action on $\mathcal{B}^{tri}$. This action can also be written as follows,*

$$
\begin{aligned}
\sigma_j(\underline{v}) &= (v_1, \dots, v_{j-1}, v_{j+1} - L(v_{j+1}, v_j)v_j, v_j, v_{j+2}, \dots, v_n), \\
\sigma_j^{-1}(\underline{v}) &= (v_1, \dots, v_{j-1}, v_{j+1}, v_j - L(v_{j+1}, v_j)v_{j+1}, v_{j+2}, \dots, v_n), \\
\delta_j(\underline{v}) &= (v_1, \dots, v_{j-1}, -v_j, v_{j+1}, \dots, v_n),
\end{aligned}
$$

for $\underline{v} = (v_1, \dots, v_n) \in \mathcal{B}^{tri}$.

Proof $\underline{v} \in \mathcal{B}^{tri}$ implies $L(v_j, v_{j+1}) = 0$ and $2L(v_j, v_j) = 2 = I^{(0)}(v_j, v_j)$. Recall $I^{(0)} = L + L^t$ and $I^{(1)} = L^t - L$. Therefore

$$
\begin{aligned}
s_{v_j}^{(k)}(v_{j+1}) &= v_{j+1} - I^{(k)}(v_j, v_{j+1})v_j = v_{j+1} - L(v_{j+1}, v_j)v_j, \\
(s_{v_{j+1}}^{(k)})^{-1}(v_j) &= v_j - (-1)^k I^{(k)}(v_{j+1}, v_j)v_{j+1} = v_j - L(v_{j+1}, v_j)v_{j+1}.
\end{aligned}
$$

So $\sigma_j(\underline{v})$ and $\sigma_j^{-1}(\underline{v})$ are given by the formulas in Lemma 3.11. It remains to see that the images are again in $\mathcal{B}^{tri}$. They are in $(R^{(0)})^n$ because of the even case $k = 0$. They form $\mathbb{Z}$-bases of $H_{\mathbb{Z}}$ because $\underline{v}$ is a $\mathbb{Z}$-basis of $H_{\mathbb{Z}}$. They are triangular bases because

$$
\begin{aligned}
L(\sigma_j(\underline{v})_j, \sigma_j(\underline{v})_{j+1}) &= L(v_{j+1} - L(v_{j+1}, v_j)v_j, v_j) = 0, \\
L(\sigma_j^{-1}(\underline{v})_j, \sigma_j^{-1}(\underline{v})_{j+1}) &= L(v_{j+1}, v_j - L(v_{j+1}, v_j)v_{j+1}) = 0.
\end{aligned}
$$

Of course $\delta_j(\underline{v}) \in \mathcal{B}^{tri}$. □

Definition 3.12 Fix $n \in \mathbb{N}$ and $R \in \{\mathbb{Z}, \mathbb{Q}, \mathbb{R}, \mathbb{C}\}$.

(a) Recall the definition of the set $T_n(R)$ of upper triangular $n \times n$ matrices with entries in R and 1's on the diagonal in Definition 2.3 (c). Additionally we define the sets of symmetric and skewsymmetric matrices

$$T_n^{(0)}(R) := \{A \in M_{n\times n}(R) \mid A^t = A, A_{ii} = 2\},$$
$$T_n^{(1)}(R) := \{A \in M_{n\times n}(R) \mid A^t = -A\}.$$

(b) For $a \in \mathbb{Z}$ define the $n \times n$ matrix

$$C_{n,j}(a) = \begin{pmatrix} 1 & & & & \\ & \ddots & & & \\ & & a\ 1 & & \\ & & 1\ 0 & & \\ & & & \ddots & \\ & & & & 1 \end{pmatrix}$$

which differs from the unit matrix only in the positions $(j, j), (j, j+1), (j+1, j), (j+1, j+1)$. Its inverse is

$$C_{n,j}^{-1}(a) = \begin{pmatrix} 1 & & & & \\ & \ddots & & & \\ & & 0\ \ 1 & & \\ & & 1\ -a & & \\ & & & \ddots & \\ & & & & 1 \end{pmatrix}$$

with $-a$ at the position $(j+1, j+1)$.

Lemma 3.13 *Fix $n \in \mathbb{Z}_{\geq 2}$ and $R \in \{\mathbb{Z}, \mathbb{Q}, \mathbb{R}, \mathbb{C}\}$.*

(a) The following formulas define an action of the semidirect product $\mathrm{Br}_n \ltimes \{\pm 1\}^n$ *on each of the sets of matrices $T_n^{(0)}(R)$, $T_n^{(1)}(R)$ and $T_n(R)$,*

$$\sigma_j(A) = C_{n,j}(-A_{j,j+1}) \cdot A \cdot C_{n,j}(-A_{j,j+1}) \quad \textit{for } j \in \{1, ..., n-1\},$$
$$\sigma_j^{-1}(A) = C_{n,j}^{-1}(A_{j,j+1}) \cdot A \cdot C_{n,j}^{-1}(A_{j,j+1}) \quad \textit{for } j \in \{1, ..., n-1\},$$
$$\delta_j(A) = \mathrm{diag}(((-1)^{\delta_{ij}})_{i=1,...,n}) \cdot A \cdot \mathrm{diag}(((-1)^{\delta_{ij}})_{i=1,...,n}) \quad \textit{for } j \in \{1, ..., n\}.$$

(b) Let $(H_{\mathbb{Z}}, L)$ be a unimodular bilinear lattice of rank n.

(i) Fix $k \in \{0; 1\}$. The map

$$(R^{(k)})^n \to T_n^{(k)}(\mathbb{Z}), \quad \underline{v} \mapsto I^{(k)}(\underline{v}^t, \underline{v}),$$

is compatible with the actions of $\mathrm{Br}_n \ltimes \{\pm 1\}^n$ *on* $(R^{(k)})^n$ *and on* $T_n^{(k)}(\mathbb{Z})$.

(ii) Suppose that $\mathcal{B}^{tri}$ *is not empty. The map*

$$\mathcal{B}^{tri} \to T_n(\mathbb{Z}), \quad \underline{v} \mapsto L(\underline{v}^t, \underline{v})^t,$$

is compatible with the actions of $\mathrm{Br}_n \ltimes \{\pm 1\}^n$ *on* $\mathcal{B}^{tri}$ *and on* $T_n(\mathbb{Z})$.

Proof (a) For $A \in T_n^{(k)}(\mathbb{Z})$ there is a unique matrix $S \in T_n(\mathbb{Z})$ with $A = S + (-1)^k S^t$. Consider a unimodular bilinear lattice $(H_\mathbb{Z}, L, \underline{e})$ with a triangular basis $\underline{e}$ with matrix $S = L(\underline{e}^t, \underline{e})^t$. Then

$$\begin{aligned}
\sigma_j(\underline{e}) &= (..., e_{j-1}, e_{j+1} - S_{j,j+1}e_j, e_j, e_{j+2}, ...) \\
&= \underline{e} \cdot C_{n,j}(-S_{j,j+1}), \\
\sigma_j^{-1}(\underline{e}) &= (..., e_{j-1}, e_{j+1}, e_j - S_{j,j+1}e_{j+1}, e_{j+2}, ...) \\
&= \underline{e} \cdot C_{n,j}^{-1}(S_{j,j+1}), \\
\delta_j(\underline{e}) &= (..., e_{i-1}, -e_i, e_{i+1}, ...) \\
&= \underline{e} \cdot \mathrm{diag}(((-1)^{\delta_{ij}})_{i=1,...,n}).
\end{aligned}$$

Observe that $C_{n,j}(a)$ for $a \in \mathbb{Z}$ is symmetric. Therefore for $k \in \{0; 1\}$

$$\begin{aligned}
I^{(k)}(\sigma_j(\underline{e})^t, \sigma_j(\underline{e})) &= C_{n,j}(-S_{j,j+1}) \cdot I^{(k)}(\underline{e}^t, \underline{e}) \cdot C_{n,j}(-S_{j,j+1}), \\
I^{(k)}(\sigma_j^{-1}(\underline{e})^t, \sigma_j^{-1}(\underline{e})) &= C_{n,j}^{-1}(S_{j,j+1}) \cdot I^{(k)}(\underline{e}^t, \underline{e}) \cdot C_{n,j}^{-1}(S_{j,j+1}),
\end{aligned}$$

similarly for L instead of $I^{(k)}$, and also similarly for the action of δ_j. This shows part (a) for $R = \mathbb{Z}$. Changing the set of scalars does not change the matrix identities which say that the group $\mathrm{Br}_n \ltimes \{\pm 1\}^n$ acts.

(b) This follows from the proof of part (a). □

Remark 3.14 Let $(H_\mathbb{Z}, L)$ be a bilinear lattice, not necessarily unimodular.

(i) The action of $\mathrm{Br}_n \ltimes \{\pm 1\}^n$ on $(R^{(0)})^n$ in Lemma 3.10 (a) works also in this case. It restricts as in Lemma 3.11 to an action on $\mathcal{B}^{tri}$, if this set is not empty, though here for $\underline{v} \in \mathcal{B}^{tri}$

$$\begin{aligned}
\sigma_j(\underline{v}) &= (v_1, ..., v_{j-1}, v_{j+1} - \frac{L(v_{j+1}, v_j)}{L(v_j, v_j)} v_j, v_j, v_{j+2}, ..., v_n), \\
\sigma_j^{-1}(\underline{v}) &= (v_1, ..., v_{j-1}, v_{j+1}, v_j - \frac{L(v_{j+1}, v_j)}{L(v_{j+1}, v_{j+1})} v_{j+1}, v_{j+2}, ..., v_n).
\end{aligned}$$

(ii) The action of $\mathrm{Br}_n \ltimes \{\pm 1\}^n$ in Lemma 3.10 on $(R^{(1)})^n$ does not generalize. In Lemma 2.6 (a) (v) we defined in the case of a general bilinear lattice $s_a^{(1)}$ only for $a \in R^{(0)}$. We defined $s_a^{(1)}$ for any $a \in R^{(1)}$ in Lemma 2.6 (d) only in the case of a unimodular bilinear lattice.

(iii) On the other hand, part (a) (vii) of Lemma 2.6 says that the action in Lemma 3.10 for $k = 1$ works for $\underline{v} \in \mathcal{B}^{tri}$. Though at the end this is just the action in (i) above.

(iv) The action in (i) on $\mathcal{B}^{tri}$ is compatible with an action on the set T_n^{tri} of matrices in Lemma 2.3 (c), which generalizes the action in Lemma 3.13 (a). Here $C_{n,j}(-S_{j,j+1})$ and $C_{n,j}^{-1}(S_{j,j+1})$ in Lemma 3.13 (a) have to be replaced by $C_{n,j}(-\frac{S_{j,j+1}}{S_{j,j}})$ and $C_{n,j}^{-1}(\frac{S_{j,j+1}}{S_{j+1,j+1}})$.

$\{\pm 1\}^n$ is the normal subgroup in the semidirect product $\mathrm{Br}_n \ltimes \{\pm 1\}^n$. Therefore, if $\mathrm{Br}_n \ltimes \{\pm 1\}^n$ acts on some set Σ, the group Br_n acts on the quotient $\Sigma/\{\pm 1\}^n$. Often it is good to consider this quotient and the action of Br_n on it.

Lemma 3.15 *Let $(H_{\mathbb{Z}}, L)$ be a unimodular bilinear lattice of some rank $n \geq 2$. Fix $k \in \{0; 1\}$.*

(a) The map

$$\pi_n^{(k)} : (R^{(k)})^n \to (\{s_v^{(k)} \mid v \in R^{(k)}\})^n \subset (O^{(k)})^n, \quad \underline{v} \mapsto (s_{v_1}^{(k)}, ..., s_{v_n}^{(k)}),$$

factors into maps

$$(R^{(k)})^n \longrightarrow (R^{(k)})^n/\{\pm 1\}^n \xrightarrow{\pi_n^{(k)}/\{\pm 1\}^n} (\{s_v^{(k)} \mid v \in R^{(k)}\})^n.$$

Br_n acts on the quotient $(R^{(k)})^n/\{\pm 1\}^n$, and the second map $\pi_n^{(k)}/\{\pm 1\}^n$ is Br_n equivariant. The image of $\underline{v}$ in $(R^{(k)})^n/\{\pm 1\}^n$ is denoted by $\underline{v}/\{\pm 1\}^n$.

(b) The second map

$$\pi_n^{(k)}/\{\pm 1\}^n : (R^{(k)})^n/\{\pm 1\}^n \to (\{s_v^{(k)} \mid v \in R^{(k)}\})^n$$

in part (a) is a bijection if $k = 0$ or if $k = 1$ and $\mathrm{Rad}\, I^{(1)} = \{0\}$.

(c) Consider the case $k = 1$ and $\mathrm{Rad}\, I^{(1)} \supsetneqq \{0\}$. Consider a triangular basis $\underline{e} \in \mathcal{B}^{tri}$ and the induced set $\Delta^{(1)}$ of odd vanishing cycles. The second map restricts to a Br_n equivariant bijection

$$(\Delta^{(1)})^n/\{\pm 1\}^n \to (\{s_v^{(1)} \mid v \in \Delta^{(1)}\})^n, \quad \underline{v} \mapsto (s_{v_1}^{(1)}, ..., s_{v_n}^{(1)}),$$

if $(H_{\mathbb{Z}}, L, \underline{e})$ is either irreducible or reducible with at most one summand of type A_1.

Proof Part (a) is trivial. (b) Suppose $k = 0$ or ($k = 1$ and $\mathrm{Rad}\, I^{(1)} = \{0\}$). If $k = 1$ and $v = 0$ then $s_v^{(1)} = \mathrm{id}$. If $k = 0$ and $v \in R^{(0)}$ or if $k = 1$ and $v \in R^{(1)} - \{0\}$ then $v \notin \mathrm{Rad}\, I^{(k)}$ and $s_v^{(k)} \neq \mathrm{id}$ for any $v \in R^{(k)}$. Then one can recover $\pm v$ from $s_v^{(k)}$, essentially because of

$$\{0\} \subsetneqq (s_v^{(k)} - \mathrm{id})(H_{\mathbb{Z}}) \subset \mathbb{Z}v.$$

(c) If $(H_{\mathbb{Z}}, L, \underline{e})$ is irreducible then $\Delta^{(1)} \cap \mathrm{Rad}\, I^{(1)} = \emptyset$, and the argument of part (b) holds. If $(H_{\mathbb{Z}}, L, \underline{e})$ is reducible with only one summand of type A_1 then for a unique $j \in \{1, ..., n\}$ $e_j \in \mathrm{Rad}\, I^{(1)}$, and then $\Delta^{(1)} \cap \mathrm{Rad}\, I^{(1)} = \{\pm e_j\}$. Then $v \in \Delta^{(1)}$ satisfies $s_v^{(1)} = \mathrm{id}$ if and only if $v = \pm e_j$. So also then one can recover $\pm v$ from $s_v^{(1)}$ for any $v \in \Delta^{(1)}$. □

Remark 3.16 (i) Consider the action of Br_n on $T_n(\mathbb{Z})$. The elementary braid σ_j maps $S = (S_{ij}) \in T_n(\mathbb{Z})$ to

$$\sigma_j(S) = C_{n,j}(-S_{j,j+1}) \cdot S \cdot C_{n,j}(-S_{j,j+1})$$
$$\text{with} \quad \begin{pmatrix} \sigma_j(S)_{jj} & \sigma_j(S)_{j,j+1} \\ \sigma_j(S)_{j+1,j} & \sigma_j(S)_{j+1,j+1} \end{pmatrix} = \begin{pmatrix} 1 & -S_{j,j+1} \\ 0 & 1 \end{pmatrix}.$$

(ii) Especially, in the case $n = 2$ δ_1, δ_2 and σ_1 all map $S = \begin{pmatrix} 1 & x \\ 0 & 1 \end{pmatrix}$ to $\begin{pmatrix} 1 & -x \\ 0 & 1 \end{pmatrix}$, so the $\mathrm{Br}_2 \ltimes \{\pm 1\}^2$ orbit equals the Br_2 orbit and the $\langle \delta_1 \rangle$ orbit and consists only of $S = \begin{pmatrix} 1 & x \\ 0 & 1 \end{pmatrix}$ and $\begin{pmatrix} 1 & -x \\ 0 & 1 \end{pmatrix}$.

(iii) Under rather special circumstances, also in higher rank n the sign group action is eaten up by the braid group action. Ebeling proved the following lemma.

Lemma 3.17 *Let $(H_\mathbb{Z}, L)$ be a unimodular bilinear lattice and $\underline{e} \in \mathcal{B}^{tri}$ a triangular basis with $S = L(\underline{e}^t, \underline{e})^t \in T_n(\mathbb{Z})$.*

(a) [Eb83, proof of Prop. 2.2] Suppose $S_{j,j+1} = \varepsilon \in \{\pm 1\}$ for some $j \in \{1, ..., n-1\}$. Then

$$\sigma_j^{3\varepsilon}(\underline{e}) = \delta_j(\underline{e}), \quad \sigma_j^{-3\varepsilon}(\underline{e}) = \delta_{j+1}(\underline{e}).$$

(b) [Eb83, Prop. 2.2] Suppose $S_{ij} \in \{0, 1, -1\}$ for all i, j and that $(H_\mathbb{Z}, L)$ is irreducible. Then the orbit of $\underline{e}$ under Br_n *coincides with the orbit of $\underline{e}$ under* $\mathrm{Br}_n \ltimes \{\pm 1\}^n$.

3.3 Distinguished Bases

In this section we continue the discussion of the braid group action on $\mathcal{B}^{tri}$. Now we fix one triangular basis $\underline{e}$. Definition 3.18 gives notations. The Remarks 3.19 pose questions on the orbit of $\underline{e}$ under $\mathrm{Br}_n \ltimes \{\pm 1\}^n$. The questions will guide the work which will be done in Chap. 7.

Definition 3.18 Let $(H_\mathbb{Z}, L)$ be a unimodular lattice of rank $n \geq 2$ with nonempty set $\mathcal{B}^{tri}$ of triangular bases.

Given a triangular basis $\underline{e} \in \mathcal{B}^{tri}$ and $k \in \{0, 1\}$, we are interested in the following orbits:

the set $\mathcal{B}^{dist} := \mathrm{Br}_n \ltimes \{\pm 1\}^n(\underline{e})$ of *distinguished bases*, the set $\mathcal{R}^{(k),dist} := \mathrm{Br}_n(\pi_n^{(k)}(\underline{e}))$ of *distinguished tuples of reflections or transvections*,

the set $\mathcal{S}^{dist} := \mathrm{Br}_n \ltimes \{\pm 1\}^n(S)$ of *distinguished matrices*, where $S = L(\underline{e}^t, \underline{e})^t \in T_n(\mathbb{Z})$,

the quotient sets $\mathcal{B}^{dist}/\{\pm 1\}^n$ and $\mathcal{S}^{dist}/\{\pm 1\}^n$, which are Br_n orbits. We are also interested in the stabilizers in Br_n of the points $\underline{e}/\{\pm 1\}^n \in \mathcal{B}^{dist}/\{\pm 1\}^n$ and $S/\{\pm 1\}^n \in \mathcal{S}^{dist}/\{\pm 1\}^n$, namely the groups

$$(\mathrm{Br}_n)_{\underline{e}/\{\pm 1\}^n} \subset (\mathrm{Br}_n)_{S/\{\pm 1\}^n} \subset \mathrm{Br}_n.$$

Remark 3.19 In the situation of Definition 3.18 the following constraints on the set $\mathcal{B}^{dist}$ of distinguished bases are clear from what has been said,

$$\begin{aligned}\mathcal{B}^{dist} &\subset \mathcal{B}^{tri} \cap (\Delta^{(0)})^n \cap (\Delta^{(1)})^n \cap \{\underline{v} \in (H_\mathbb{Z})^n \mid \sum_{i=1}^n \mathbb{Z}v_i = H_\mathbb{Z}\} \\ &\cap (\pi_n \circ \pi_n^{(0)})^{-1}(-M) \cap (\pi_n \circ \pi_n^{(1)})^{-1}(M),\end{aligned} \tag{3.2}$$

where $\pi_n \circ \pi_n^{(k)} : (R^{(k)})^n \to O^{(k)}, \quad \underline{v} \mapsto s_{v_1}^{(k)} \dots s_{v_n}^{(k)}$.

An interesting problem is which–if any–of these constraints are sufficient to characterize the orbit $\mathcal{B}^{dist}$. We are most interested in the questions whether the inclusions

$$\mathcal{B}^{dist} \subset \{\underline{v} \in (\Delta^{(0)})^n \mid \pi_n \circ \pi_n^{(0)}(\underline{v}) = -M\}, \tag{3.3}$$

$$\mathcal{B}^{dist} \subset \{\underline{v} \in (\Delta^{(1)})^n \mid \pi_n \circ \pi_n^{(1)}(\underline{v}) = M\}, \tag{3.4}$$

are equalities. We will study this problem systematically in Chap. 7 for $n = 2$ and $n = 3$. In this Sect. 3.3 we give some examples.

Remark 3.20 Consider a unimodular bilinear lattice $(H_\mathbb{Z}, L)$ of rank $n \geq 2$ and $k \in \{0; 1\}$.

(i) Two basic invariants of the $\mathrm{Br}_n \ltimes \{\pm 1\}^n$ orbit of a tuple $\underline{v} \in (R^{(k)})^n$ are the product $(\pi_n \circ \pi_n^{(k)})(\underline{v})) = s_{v_1}^{(k)} \dots s_{v_n}^{(k)} \in O^{(k)}$ and the sublattice $\sum_{i=1}^n \mathbb{Z}v_i \subset H_\mathbb{Z}$, namely

$$(\pi_n \circ \pi_n^{(k)})(\sigma_j(\underline{v})) = (\pi_n \circ \pi_n^{(k)})(\underline{v}) \quad \text{and} \quad \sum_{i=1}^n \mathbb{Z}\sigma_j(\underline{v})_i = \sum_{i=1}^n \mathbb{Z}v_i.$$

(ii) A triangular basis $\underline{e} \in \mathcal{B}^{tri}$ induces the even and odd monodromy groups $\Gamma^{(0)}$ and $\Gamma^{(1)}$ and the sets $\Delta^{(0)}$ and $\Delta^{(1)}$ of even and odd vanishing cycles. Each distinguished basis $\underline{v} \in \mathcal{B}^{dist}$ induces the same even and odd monodromy groups $\Gamma^{(0)}$ and $\Gamma^{(1)}$ and the same sets $\Delta^{(0)}$ and $\Delta^{(1)}$ of even and odd vanishing cycles. This is obvious from the action of $\mathrm{Br}_n \ltimes \{\pm 1\}^n$ on $\mathcal{B}^{dist}$, the Hurwitz action of Br_n on

$\mathcal{R}^{(k),dist}$ and the definition of $\Gamma^{(k)}$ and $\Delta^{(k)}$. So they are invariants of the set $\mathcal{B}^{dist}$ of distinguished bases.

We did not mention the monodromy M, because it is by Theorem 2.7 an invariant of $(H_{\mathbb{Z}}, L)$ if $\mathcal{B}^{tri} \neq \emptyset$, so it does not depend on the choice of a $\mathrm{Br}_n \ltimes \{\pm 1\}^n$ orbit in $\mathcal{B}^{tri}$.

(iii) A matrix $S \in T_n(\mathbb{Z})$ determines a triple $(H_{\mathbb{Z}}, L, \underline{e})$ with $(H_{\mathbb{Z}}, L)$ a unimodular bilinear lattice and $\underline{e}$ a triangular basis with $S = L(\underline{e}^t, \underline{e})^t$ up to isomorphism. For a second matrix $\widetilde{S}$ in the $\mathrm{Br}_n \ltimes \{\pm 1\}^n$ orbit of S then a triangular basis $\widetilde{\underline{e}}$ of $(H_{\mathbb{Z}}, L)$ with $\widetilde{S} = L(\widetilde{\underline{e}}^t, \widetilde{\underline{e}})^t$ exists (but it is not unique in general). Therefore the triple $(H_{\mathbb{Z}}, L, \mathcal{B}^{dist})$ and all induced data depend only on the $\mathrm{Br}_n \ltimes \{\pm 1\}^n$ orbit $\mathcal{S}^{dist}$ of S. These induced data comprise $R^{(0)}$, $\Gamma^{(0)}$, $\Gamma^{(1)}$, $\Delta^{(0)}$ and $\Delta^{(1)}$.

(iv) Choose a triangular basis $\underline{e}$. Then $s_\delta^{(k)} \in \Gamma^{(k)}$ for $\delta \in \Delta^{(k)}$. This follows from the definition of a vanishing cycle and from formula 3.2. It also implies that the set $(\Delta^{(k)})^n$ is invariant under the action of $\mathrm{Br}_n \ltimes \{\pm 1\}^n$ on $(R^{(k)})^n$.

Remark 3.21 (i) Consider a unimodular bilinear lattice which is not a lattice of type A_1^n and a fixed triangular basis $\underline{e}$. Then $\Delta^{(0)} \subset R^{(0)}$, but $\Delta^{(1)} \not\subset R^{(0)}$ by Corollary 6.22 (a). Nevertheless $\mathcal{B}^{dist} \subset (\Delta^{(0)} \cap \Delta^{(1)})^n$, so many odd vanishing cycles do not turn up in bases in the braid group orbit of $\underline{e}$, i.e. in distinguished bases.

(ii) In all cases except A_1^n $\Delta^{(1)} \not\subset \Delta^{(0)}$ because $\Delta^{(1)} \not\subset R^{(0)}$. In some cases $\Delta^{(0)} \subset \Delta^{(1)}$, in many cases $\Delta^{(0)} \not\subset \Delta^{(1)}$. See Corollary 6.22.

Given a unimodular bilinear lattice $(H_{\mathbb{Z}}, L)$, any element $g \in O^{(k)}$ acts on $(R^{(k)})^n$ by $g(\underline{v}) := (g(v_1), ..., g(v_n))$. Part (a) of the next Lemma 3.22 says especially that this action commutes with the action of $\mathrm{Br}_n \ltimes \{\pm 1\}^n$ on $(R^{(k)})^n$. Part (b) gives implications, which will be used to construct the interesting Examples 3.23 (i) and (ii).

Lemma 3.22 *Let $(H_{\mathbb{Z}}, L)$ be a unimodular bilinear lattice. Fix $k \in \{0, 1\}$.*

(a) If $g \in O^{(k)}$ and $(\alpha, \varepsilon) \in \mathrm{Br}_n \ltimes \{\pm 1\}^n$ then for $\underline{v} \in (R^{(k)})^n$

$$g((\alpha, \varepsilon)(\underline{v})) = (\alpha, \varepsilon)(g(\underline{v})),$$
$$(\pi_n \circ \pi_n^{(k)})(g(\underline{v})) = g \circ (\pi_n \circ \pi_n^{(k)})(\underline{v}) \circ g^{-1}.$$

Here $g(\underline{v})$ means $(g(v_1), ..., g(v_n))$, and similarly for $g((\alpha, \varepsilon)(\underline{v}))$.

(b) If $g \in G_{\mathbb{Z}}^{(k)} - G_{\mathbb{Z}}$ and $\underline{e} \in \mathcal{B}^{tri}$ then

$$g(\underline{e}) \notin \mathcal{B}^{tri},$$
$$\textit{so especially} \quad g(\underline{e}) \notin \mathrm{Br}_n \ltimes \{\pm 1\}^n(\underline{e}),$$
$$\textit{but} \quad (\pi_n \circ \pi_n^{(k)})(g(\underline{e})) = (\pi_n \circ \pi_n^{(k)})(\underline{e}) = (-1)^{k+1} M.$$

Proof (a) $g((\mathrm{id}, \varepsilon)(\underline{v})) = (\mathrm{id}, \varepsilon)(g(\underline{v}))$ is trivial. Consider $(\alpha, \varepsilon) = (\sigma_j, (1, ..., 1)) = \sigma_j$.

$$g(\sigma_j(\underline{v})) = (g(v_1), ..., g(v_{j-1}), g s_{v_j}^{(k)}(v_{j+1}), g(v_j), g(v_{j+2}), ..., g(v_n))$$

$$
\begin{aligned}
&= (g(v_1), ..., g(v_{j-1}), s^{(k)}_{g(v_j)}(gv_{j+1}), g(v_j), g(v_{j+2}), ..., g(v_n)) \\
&= \sigma_j(g(\underline{v})),
\end{aligned}
$$

because of $gs^{(k)}_{v_j}g^{-1} = s^{(k)}_{g(v_j)}$ (Lemma 2.2 (c)).

$$
\begin{aligned}
(\pi_n \circ \pi_n^{(k)})(g(\underline{v})) &= s^{(k)}_{g(v_1)}...s^{(k)}_{g(v_n)} = \big(gs^{(k)}_{v_1}g^{-1}\big)...\big(gs^{(k)}_{v_n}g^{-1}\big) \\
&= gs^{(k)}_{v_1}...s^{(k)}_{v_n}g^{-1} = g \circ (\pi_n \circ \pi_n^{(k)})(\underline{v}) \circ g^{-1}.
\end{aligned}
$$

(b) Suppose $g \in G^{(k)}_{\mathbb{Z}} - G_{\mathbb{Z}}$ and $\underline{e} \in \mathcal{B}^{tri}$. Then $gMg^{-1} = M$ and

$$
\begin{aligned}
(\pi_n \circ \pi_n^{(k)})(g(\underline{e})) &\overset{(a)}{=} g \circ (\pi_n \circ \pi_n^{(k)})(\underline{e}) \circ g^{-1} \\
&= g\big((-1)^{k+1}M\big)g^{-1} = (-1)^{k+1}M = (\pi_n \circ \pi_n^{(k)})(\underline{e}).
\end{aligned}
$$

Furthermore $S = L(\underline{e}^t, \underline{e})^t \in T_n(\mathbb{Z})$, $I^{(k)} = L^t + (-1)^k L$ and

$$
\begin{aligned}
I^{(k)}(g(\underline{e})^t, g(\underline{e})) &= I^{(k)}(\underline{e}^t, \underline{e}) = S + (-1)^k S^t \quad \text{because } g \in G^{(k)}_{\mathbb{Z}}, \\
L(g(\underline{e})^t, g(\underline{e}))^t &\neq L(\underline{e}^t, \underline{e})^t = S \quad \text{because } g \notin G_{\mathbb{Z}}, \\
\text{so } L(g(\underline{e})^t, g(\underline{e})) &\notin T_n(\mathbb{Z}),
\end{aligned}
$$

so $g(\underline{e}) \notin \mathcal{B}^{tri}$, so $g(\underline{e}) \notin \mathrm{Br}_n \ltimes \{\pm 1\}^n(\underline{e})$. □

Example 3.23 Let $(H_{\mathbb{Z}}, L)$ be a unimodular bilinear lattice of rank $n \geq 2$, $\underline{e} \in \mathcal{B}^{tri}$ a triangular basis, $S = L(\underline{e}^t, \underline{e})^t \in T_n(\mathbb{Z})$ and $k \in \{0, 1\}$.

(i) $k = 1$, $n = 3$, $S = S(-3, 3, -3) = S(\mathbb{P}^2)$, the odd case $\mathbb{P}^2$. To carry out this example we need two results which will be proved later, Theorem 5.14 (b) and Theorem 6.21 (h).

By Theorem 5.14 (b) (i) there is a root $M^{root} \in G_{\mathbb{Z}}$ of the monodromy with $(M^{root})^3 = M$ and

$$
\begin{aligned}
G^{(1)}_{\mathbb{Z}} &= \{\pm(M^{root})^l(\mathrm{id} + a(M^{root} - \mathrm{id})^2) \mid l, a \in \mathbb{Z}\} \\
&\supsetneqq G_{\mathbb{Z}} = \{\pm(M^{root})^l \mid l \in \mathbb{Z}\}.
\end{aligned}
$$

Also, here $\mathrm{Rad}\, I^{(1)} = \mathbb{Z} f_3$ with $f_3 = e_1 + e_2 + e_3$. The shape of M^{root} in Theorem 5.14 (b) shows

$$
(M^{root} - \mathrm{id})^2(\underline{e}) = \underline{e}(\begin{pmatrix} 3 & -3 & 1 \\ 1 & 0 & 0 \\ 0 & 1 & 0 \end{pmatrix} - E_3)^2 = f_3(1, -2, 1).
$$

For example $g := \mathrm{id} + (M^{root} - \mathrm{id})^2 \in G^{(1)}_{\mathbb{Z}} - G_{\mathbb{Z}}$ satisfies

$$
g(\underline{e}) = \underline{e} + f_3(1, -2, 1),
$$

$$
\begin{aligned}
I^{(1)}(g(\underline{e})^t, g(\underline{e})) &= I^{(1)}(\underline{e}^t, \underline{e}) = S - S^t = \begin{pmatrix} 0 & -3 & 3 \\ 3 & 0 & -3 \\ -3 & 3 & 0 \end{pmatrix}, \\
L(g(\underline{e})^t, g(\underline{e}))^t &= \begin{pmatrix} 3 & -7 & 5 \\ -4 & 9 & -7 \\ 2 & -4 & 3 \end{pmatrix}, \\
s^{(1)}_{g(e_1)} s^{(1)}_{g(e_2)} s^{(1)}_{g(e_3)} &= M, \\
g(e_1), g(e_2), g(e_3) &\notin R^{(0)}, \quad \text{because } 3 \neq 1 \text{ and } 9 \neq 1, \\
g(e_1), g(e_2), g(e_3) &\notin \Delta^{(1)}.
\end{aligned}
$$

The last claim $g(e_j) \notin \Delta^{(1)}$ holds because of $g(e_j) \neq e_j$, but $g(e_j) \in e_j + \mathbb{Z} f_3$, and because the projection $H_\mathbb{Z} \to H_\mathbb{Z}/\mathbb{Z} f_3$ restricts to an injective map $\Delta^{(1)} \to H_\mathbb{Z}/\mathbb{Z} f_3$ by Theorem 6.21 (h).

(ii) $k = 0, n = 3, S = S(-2, 2, -2) = S(\mathcal{H}_{1,2})$, the even case $\mathcal{H}_{1,2}$. To carry out this example we need two results which will be proved later, Theorem 5.14 (a) and Theorem 6.14 (e).

Recall Theorem 5.14 (b) (i),

$$
\begin{aligned}
&(H_\mathbb{Z}, L) = (H_{\mathbb{Z},1}, L_1) \oplus (H_{\mathbb{Z},2}, L_2) \\
&\text{with } H_{\mathbb{Z},1} = \mathbb{Z} f_1 \oplus \mathbb{Z} f_2 = \ker \Phi_2(M), \quad H_{\mathbb{Z},2} = \mathbb{Z} f_3 = \ker \Phi_1(M), \\
&(f_1, f_2, f_3) = \underline{e} \begin{pmatrix} 1 & 0 & 1 \\ 1 & 1 & 1 \\ 0 & 1 & 1 \end{pmatrix}, \\
&G^{(0)}_\mathbb{Z} = G^{(0)}_{\mathbb{Z},1} \times G^{(0)}_{\mathbb{Z},2} \supsetneqq G_\mathbb{Z} = G_{\mathbb{Z},1} \times G_{\mathbb{Z},2} \\
&\text{with } G^{(0)}_{\mathbb{Z},1} = \mathrm{Aut}(H_{\mathbb{Z},1}) \supsetneqq G_{\mathbb{Z},1} = \{g \in \mathrm{Aut}(H_{\mathbb{Z},1}) \mid \det g = 1\}, \\
&G^{(0)}_{\mathbb{Z},2} = G_{\mathbb{Z},2} = \{\pm \mathrm{id}|_{\mathbb{Z} f_3}\}.
\end{aligned}
$$

For example $g := ((f_1, f_2, f_3) \mapsto (f_1, -f_2, f_3)) \in G^{(0)}_\mathbb{Z} - G_\mathbb{Z}$ satisfies

$$
\begin{aligned}
g(\underline{e}) &= g(f_3 - f_2, -f_3 + f_1 + f_2, f_3 - f_1) \\
&= (f_3 + f_2, -f_3 + f_1 - f_2, f_3 - f_1) \\
&= \underline{e} + f_2(2, -2, 0), \\
s^{(0)}_{g(e_1)} s^{(0)}_{g(e_2)} s^{(0)}_{g(e_3)} &= -M,
\end{aligned}
$$

$$
\begin{aligned}
I^{(0)}(g(\underline{e})^t, g(\underline{e})) &= I^{(0)}(\underline{e}^t, \underline{e}) = S + S^t, \\
L(g(\underline{e})^t, g(\underline{e}))^t &= \begin{pmatrix} 1 & 0 & 0 \\ -2 & 1 & 0 \\ 2 & -2 & 1 \end{pmatrix} = S^t \neq S = L(\underline{e}^t, \underline{e})^t,
\end{aligned}
$$

$$g(\underline{e}) \notin \mathcal{B}^{tri}, \text{ so } g(\underline{e}) \notin \mathcal{B}^{dist},$$
$$g(e_1), g(e_2), g(e_3) \in \Delta^{(0)}.$$

The last claim $g(e_j) \in \Delta^{(0)}$ holds because of Theorem 6.14 (e). Therefore

$$g(\underline{e}) \in \{\underline{v} \in (\Delta^{(0)})^3 \mid (\pi_3 \circ \pi_3^{(0)})(\underline{v}) = -M, \sum_{i=1}^{3} \mathbb{Z}v_i = H_\mathbb{Z}\}.$$

Especially, here the inclusion in (3.3) is not an equality. In a certain sense, this example is the worst case within all cases $S(\underline{x})$ with $k = 0$, $n = 3$ and eigenvalues of M roots of unity. See Theorem 7.3 (b).

(iii) $n \in \mathbb{N}$, $S = E_n$, the even and odd case A_1^n. Compare Lemma 2.12.

$$\begin{aligned}
\Delta^{(0)} &= R^{(0)} = \Delta^{(1)} = \{\pm e_1, ..., \pm e_n\}, \\
M &= \mathrm{id}, \\
\mathcal{B}^{dist} &= \{(\varepsilon_1 e_{\sigma(1)}, ..., \varepsilon_n e_{\sigma(n)}) \mid \varepsilon_1, ..., \varepsilon_n \in \{\pm 1\}, \sigma \in S_n\}, \\
&= \{\underline{v} \in (\Delta^{(0)})^n \mid (\pi_n \circ \pi_n^{(0)})(\underline{e}) = -M = -\mathrm{id}\}
\end{aligned}$$

The last equality follows from $-M = -\mathrm{id}$ and

$$s_{e_i}^{(0)}|_{\mathbb{Z}e_i} = -\mathrm{id}, \quad s_{e_i}^{(0)}|_{\sum_{j\neq i}\mathbb{Z}e_j} = \mathrm{id}\,.$$

Here the inclusion in (3.3) is an equality. On the contrary, the inclusion (3.4) is not an equality, the set

$$\{\underline{v} \in (\Delta^{(1)})^n \mid (\pi_n \circ \pi_n^{(1)})(\underline{v}) = M = \mathrm{id}\}$$

is much bigger than $\mathcal{B}^{dist}$ if $n \geq 2$, it consists of many $\mathrm{Br}_n \ltimes \{\pm 1\}^n$ orbits. Each of these orbits contains a unique one of the following tuples,

$$(e_{l(1)}, e_{l(2)}, ..., e_{l(n)}) \quad \text{with} \quad 1 \leq l(1) \leq l(2) \leq ... \leq l(n) \leq n.$$

This follows from $s_{e_i}^{(1)} = \mathrm{id}$.

The braids act just by permutations on the Br_n orbit $\mathcal{B}^{dist}/\{\pm 1\}^n$. Therefore the stabilizer of $\underline{e}/\{\pm 1\}^n$ is the group Br_n^{pure} of pure braids (see Remark 8.2 (vii) for this group). The stabilizer of $S/\{\pm 1\}^n$ is the whole group Br_n.

(iv) Reconsider Example 3.4, so a case where

$$\Gamma^{(k)} = \begin{cases} G^{fCox,n} & \text{with generators } s_{e_1}^{(0)}, ..., s_{e_n}^{(0)} \text{ if } k = 0, \\ G^{free,n} & \text{with generators } s_{e_1}^{(1)}, ..., s_{e_n}^{(1)} \text{ if } k = 1. \end{cases}$$

In the notation of Definition 3.1 $\Delta(G^{fCox,n}) = \{s_\delta^{(0)} \mid \delta \in \Delta^{(0)}\}$ if $k = 0$ and $\Delta(G^{free,n}) = \{s_\delta^{(1)} \mid \delta \in \Delta^{(1)}\}$ if $k = 1$. By Theorem 3.2

$$\mathcal{R}^{dist,(k)} = \{(s_{v_1}^{(k)}, ..., s_{v_n}^{(k)}) \mid \underline{v} \in (\Delta^{(k)})^n, s_{v_1}^{(k)}...s_{v_n}^{(k)} = (-1)^{k+1}M\}.$$

Furthermore, the shape of $\Gamma^{(k)}$ shows that $(H_{\mathbb{Z}}, L, \underline{e})$ is irreducible. Lemma 3.15 (b) or (c) applies. Therefore

$$\mathcal{B}^{dist} = \{\underline{v} \in (\Delta^{(k)})^n \mid s_{v_1}^{(k)}...s_{v_n}^{(k)} = (-1)^{k+1}M\}.$$

So in this case only the constraints $\underline{v} \in (\Delta^{(k)})^n$ and $s_{v_1}^{(k)}...s_{v_n}^{(k)} = (-1)^{k+1}M$ in the Remarks 3.19 are needed in order to characterize the orbit $\mathcal{B}^{dist}$. The inclusions in (3.3) and (3.4) are here equalities.

By Theorem 3.2 the stabilizer of $\pi_n^{(k)}(\underline{e})$ and of $\underline{e}/\{\pm 1\}^n$ is $\{\mathrm{id}\} \subset \mathrm{Br}_n$. The size of the stabilizer $(\mathrm{Br}_n)_{S/\{\pm 1\}^n}$ depends on the case. Theorem 7.11 gives cases with $n = 3$ where it is $\langle \sigma_2\sigma_1 \rangle$ or $\langle \sigma_2\sigma_1^2 \rangle$ or $\langle \sigma^{mon} \rangle$.

(v) Suppose $S_{ij} \leq 0$ for $i < j$, so $(H_{\mathbb{Z}}, L, \underline{e})$ is a generalized Cartan lattice as in Theorem 3.7. By Theorem 3.7 (b)

$$\mathcal{R}^{dist,(0)} = \{(g_1, ..., g_n) \in \left(\{s_\delta^{(0)} \mid \delta \in \Delta^{(0)}\}\right)^n \mid g_1...g_n = -M\}.$$

With Lemma 3.15 (b) this implies

$$\mathcal{B}^{dist} = \{\underline{v} \in (\Delta^{(0)})^n \mid s_{v_1}^{(k)}...s_{v_n}^{(k)} = -M\}.$$

Also here only the two constraints $\underline{v} \in (\Delta^{(0)})^n$ and $s_{v_1}^{(k)}...s_{v_n}^{(k)} = -M$ in the Remarks 3.19 are needed in order to characterize the orbit $\mathcal{B}^{dist}$. The inclusion in (3.3) is here an equality.

3.4 From $\mathrm{Br}_n \ltimes \{\pm 1\}^n$ to $G_{\mathbb{Z}}$

Definition 3.24 gives a map $Z : \mathrm{Br}_n \ltimes \{\pm 1\}^n \to \mathrm{Aut}(H_{\mathbb{Z}})$, which restricts to a group antihomomorphism $Z : (\mathrm{Br}_n \ltimes \{\pm 1\}^n)_S \to G_{\mathbb{Z}}$. The definition and the restriction to a group antihomomorphism are classical. Lemma 3.25 provides basic facts around this map. Also Theorem 3.26 (b) is classical. It states that $Z((\delta_n^{1-k}\sigma^{root})^n) = (-1)^{k+1}M$.

Theorem 3.26 (c) gives a condition when $Z(\delta_n^{1-k}\sigma^{root}) \in G_{\mathbb{Z}}$. Then this is an n-th root of $(-1)^{k+1}M$. Theorem 3.26 (c) embraces Theorem 4.5 (a)+(b) in [BH20]. It gives more because in [BH20] no braids are considered. The braids allow a new and more elegant proof than the one in [BH20]. Theorem 3.26 (c) will be used in the discussion of the groups $G_{\mathbb{Z}}$ in many cases in Chap. 5.

Definition 3.24 Let $(H_{\mathbb{Z}}, L, \underline{e})$ be a unimodular bilinear lattice of rank $n \geq 2$ with a triangular basis $\underline{e}$. For $(\alpha, \varepsilon) \in \mathrm{Br}_n \ltimes \{\pm 1\}^n$ define an automorphism $Z((\alpha, \varepsilon)) \in Aut(H_{\mathbb{Z}})$ by the following action on the $\mathbb{Z}$-basis $\underline{e}$ of $H_{\mathbb{Z}}$,

$$\begin{aligned} Z : \mathrm{Br}_n \ltimes \{\pm 1\}^n &\to \mathrm{Aut}(H_{\mathbb{Z}}), \\ (\alpha, \varepsilon) &\mapsto Z((\alpha, \varepsilon)) = (\underline{e} \to (\alpha, \varepsilon)(\underline{e})). \end{aligned}$$

Lemma 3.25 *Let $(H_{\mathbb{Z}}, L, \underline{e})$ be a unimodular bilinear lattice of rank $n \geq 2$ with a triangular basis $\underline{e}$.*

(a) For $(\alpha, \varepsilon) \in \mathrm{Br}_n \ltimes \{\pm 1\}^n$

$$Z((\alpha, \varepsilon)) \in G_{\mathbb{Z}} \iff Z((\alpha, \varepsilon)) \in G_{\mathbb{Z}}^{(0)} \iff Z((\alpha, \varepsilon)) \in G_{\mathbb{Z}}^{(1)}.$$

(b) The stabilizer of S in $\mathrm{Br}_n \ltimes \{\pm 1\}^n$ is

$$(\mathrm{Br}_n \ltimes \{\pm 1\}^n)_S = \{(\alpha, \varepsilon) \in \mathrm{Br}_n \ltimes \{\pm 1\}^n \mid Z((\alpha, \varepsilon)) \in G_{\mathbb{Z}}\}.$$

(c) The restriction of the map Z to the stabilizer $(\mathrm{Br}_n \ltimes \{\pm 1\}^n)_S$ is also denoted Z,

$$Z : (\mathrm{Br}_n \ltimes \{\pm 1\}^n)_S \to G_{\mathbb{Z}}.$$

It is a group antihomomorphism with kernel the stabilizer $(\mathrm{Br}_n \ltimes \{\pm 1\}^n)_{\underline{e}}$ of $\underline{e}$.

(d) The triple $(H_{\mathbb{Z}}, L, \mathcal{B}^{dist})$ with $\mathcal{B}^{dist} = \mathrm{Br}_n \ltimes \{\pm 1\}^n(\underline{e})$ the set of distinguished bases (Definition 3.18) gives rise to the subgroup $G_{\mathbb{Z}}^{\mathcal{B}}$ of $G_{\mathbb{Z}}$,

$$G_{\mathbb{Z}}^{\mathcal{B}} := \mathrm{Aut}(H_{\mathbb{Z}}, L, \mathcal{B}^{dist}) := \{g \in G_{\mathbb{Z}} \mid g(\mathcal{B}^{dist}) = \mathcal{B}^{dist}\} \subset G_{\mathbb{Z}}.$$

It does not depend on $\underline{e}$, but only on the triple $(H_{\mathbb{Z}}, L, \mathcal{B}^{dist})$. Then $G_{\mathbb{Z}}^{\mathcal{B}}$ is the image of $(\mathrm{Br}_n \ltimes \{\pm 1\}^n)_S$ under Z in $G_{\mathbb{Z}}$,

$$G_{\mathbb{Z}}^{\mathcal{B}} = Z((\mathrm{Br}_n \ltimes \{\pm 1\}^n)_S) \subset G_{\mathbb{Z}}.$$

(e) The subgroup $Z((\{\pm 1\}^n)_S)$ of $G_{\mathbb{Z}}^{\mathcal{B}}$ is a normal subgroup of $G_{\mathbb{Z}}^{\mathcal{B}}$, and the group antihomomorphism Z in part (c) induces a group antihomomorphism

$$\overline{Z} : (\mathrm{Br}_n)_{S/\{\pm 1\}^n} \to G_{\mathbb{Z}}^{\mathcal{B}} / Z((\{\pm 1\}^n)_S)$$

with kernel $(\mathrm{Br}_n)_{\underline{e}/\{\pm 1\}^n}$, which is isomorphic to $(\mathrm{Br}_n \ltimes \{\pm 1\}^n)_{\underline{e}}$.

(f) Suppose that $(H_{\mathbb{Z}}, L, \underline{e})$ is irreducible (Definition 2.10 (a)). Then

$$\begin{aligned} (\{\pm 1\}^n)_S &= \{(1, ..., 1), (-1, ..., -1)\}, \\ Z((-1, ..., -1)) &= -\mathrm{id} \in G_{\mathbb{Z}}, \\ Z((\{\pm 1\}^n)_S) &= \{\pm \mathrm{id}\}. \end{aligned}$$

$\{\pm \mathrm{id}\}$ *is a normal subgroup of* $G_{\mathbb{Z}}$*. The group antihomorphism* $\overline{Z}$ *in part (d) becomes*

$$\overline{Z} : (\mathrm{Br}_n)_{S/\{\pm 1\}^n} \to G_{\mathbb{Z}}/\{\pm \mathrm{id}\}$$

with kernel $(\mathrm{Br}_n)_{\underline{e}/\{\pm 1\}^n}$ *and image* $G_{\mathbb{Z}}^{\mathcal{B}}/\{\pm \mathrm{id}\}$.

Proof (a) Fix $k \in \{0, 1\}$ and $(\alpha, \varepsilon) \in \mathrm{Br}_n \ltimes \{\pm 1\}^n$. Then

$$Z((\alpha, \varepsilon)) \in G_{\mathbb{Z}}^{(k)} \iff I^{(k)}((\alpha, \varepsilon)(\underline{e})^t, (\alpha, \varepsilon)(\underline{e})) = S + (-1)^k S^t.$$

If this equality holds then $I^{(k)} = L^t + (-1)^k L$ and $L((\alpha, \varepsilon)(\underline{e})^t, (\alpha, \varepsilon)(\underline{e}))^t \in T_n(\mathbb{Z})$ imply $L((\alpha, \varepsilon)(\underline{e})^t, (\alpha, \varepsilon)(\underline{e}))^t = S$, so $Z((\alpha, \varepsilon)) \in G_{\mathbb{Z}}$.

(b) Trivial with the compatibility of the actions of $\mathrm{Br}_n \ltimes \{\pm 1\}^n$ on $\mathcal{B}^{dist}$ and on $T_n(\mathbb{Z})$ in Theorem 3.4 (d).

(c) The following calculation shows that the map $Z : (\mathrm{Br}_n \ltimes \{\pm 1\}^n)_S \to G_{\mathbb{Z}}$ is a group antihomomorphism,

$$\begin{aligned} Z((\alpha, \varepsilon)(\beta, \widetilde{\varepsilon}))(\underline{e}) &= (\alpha, \varepsilon)(\beta, \widetilde{\varepsilon})(\underline{e}) \\ &= (\alpha, \varepsilon)(Z((\beta, \widetilde{\varepsilon}))(e_1), ..., Z((\beta, \widetilde{\varepsilon}))(e_n)) \\ &\stackrel{3.22\,(a)}{=} Z((\beta, \widetilde{\varepsilon}))(\alpha, \varepsilon)(\underline{e}) = Z((\beta, \widetilde{\varepsilon}))Z((\alpha, \varepsilon))(\underline{e}). \end{aligned}$$

It is trivial that the kernel of this map Z is $(\mathrm{Br}_n \ltimes \{\pm 1\}^n)_{\underline{e}}$.

(d) $G_{\mathbb{Z}} \subset Z((\mathrm{Br}_n \ltimes \{\pm 1\}^n)_S)$: Consider $g \in G_{\mathbb{Z}}^{\mathcal{B}}$. Then $g(\underline{e}) \in \mathcal{B}^{dist}$ comes with the same matrix S as $\underline{e}$ because g respects L. There is a pair $(\alpha, \varepsilon) \in \mathrm{Br}_n \ltimes \{\pm 1\}^n$ with $Z((\alpha, \varepsilon))(\underline{e}) = (\alpha, \varepsilon)(\underline{e}) = g(\underline{e})$. Therefore $g = Z((\alpha, \varepsilon))$.

$G_{\mathbb{Z}} \supset Z((\mathrm{Br}_n \ltimes \{\pm 1\}^n)_S)$: Consider $(\alpha, \varepsilon) \in (\mathrm{Br}_n \ltimes \{\pm 1\}^n)_S$, $(\beta, \widetilde{\varepsilon}) \in \mathrm{Br}_n \ltimes \{\pm 1\}^n$ and $\underline{v} := (\beta, \widetilde{\varepsilon})(\underline{e})$. We have to show $Z((\alpha, \varepsilon))(\underline{v}) \in \mathcal{B}^{dist}$. This is rather obvious with the commutativity of the actions of $O^{(k)}$ and $\mathrm{Br}_n \ltimes \{\pm 1\}^n$ in Lemma 3.22 (a),

$$\begin{aligned} Z((\alpha, \varepsilon))(\underline{v}) &= Z((\alpha, \varepsilon))(\beta, \widetilde{\varepsilon})(\underline{e}) \\ &\stackrel{3.22(a)}{=} (\beta, \widetilde{\varepsilon})Z((\alpha, \varepsilon))(\underline{e}) \\ &= (\beta, \widetilde{\varepsilon})(\alpha, \varepsilon)(\underline{e}). \end{aligned}$$

(e) Elementary group theory gives the group isomorphisms

$$\begin{aligned} (\mathrm{Br}_n)_{S/\{\pm 1\}^n} &\cong \frac{(\mathrm{Br}_n \ltimes \{\pm 1\}^n)_S}{(\{\pm 1\}^n)_S}, \\ (\mathrm{Br}_n)_{\underline{e}/\{\pm 1\}^n} &\cong \frac{(\mathrm{Br}_n \ltimes \{\pm 1\}^n)_{\underline{e}}}{(\{\pm 1\}^n)_{\underline{e}}} \cong (\mathrm{Br}_n \ltimes \{\pm 1\}^n)_{\underline{e}} \end{aligned}$$

Here use $(\{\pm 1\}^n)_{\underline{e}} = \{(1, ..., 1)\}$. $\{\pm 1\}^n$ is a normal subgroup of $\mathrm{Br}_n \ltimes \{\pm 1\}^n$. Therefore $(\{\pm 1\}^n)_S$ is a normal subgroup of $(\mathrm{Br}_n \ltimes \{\pm 1\}^n)_S$. Therefore $Z((\{\pm 1\}^n)_S)$ is a normal subgroup of $G_{\mathbb{Z}}^{\mathcal{B}}$. Therefore $\overline{Z}$ is well defined. Its kernel is still $(\mathrm{Br}_n)_{\underline{e}/\{\pm 1\}^n} \cong (\mathrm{Br}_n \ltimes \{\pm 1\}^n)_{\underline{e}}$ because $(\mathrm{Br}_n \ltimes \{\pm 1\}^n)_{\underline{e}} \cap (\{\pm 1\}^n)_S = (\{\pm 1\}^n)_{\underline{e}} = \{(1, ..., 1)\}$.

(f) If $(H_{\mathbb{Z}}, L, \underline{e})$ is irreducible, then the following graph is connected: its vertices are $e_1, ..., e_n$, and it has an edge between e_i and e_j for $i < j$ if $S_{ij} (= L(e_j, e_i)) \neq 0$. Therefore then $(\{\pm 1\}^n)_S = \{(1, ..., 1), (-1, ..., -1)\}$. Everything else follows from this and from part (d). □

The antihomomorphism $Z : (\mathrm{Br}_n \ltimes \{\pm 1\}^n)_S \to G_{\mathbb{Z}}$ is not always surjective, but in many cases. See Theorem 3.28, the Remarks 3.29 and the Remarks 9.1. Theorem 3.26 (b) writes $(-1)^{k+1}M$ as an image of a braid by Z. Theorem 3.26 (c) gives conditions when it has an n-th root which is also an image of a braid by Z.

Theorem 3.26 *Let $(H_{\mathbb{Z}}, L, \underline{e})$ be a unimodular bilinear lattice of rank $n \geq 2$ with a triangular basis $\underline{e}$ and matrix $S = L(\underline{e}^t, \underline{e})^t \in T_n(\mathbb{Z})$. Fix $k \in \{0, 1\}$. Recall from Chap. 3.1*

$$\begin{aligned} \sigma^{root} &:= \sigma_{n-1}\sigma_{n-2}...\sigma_2\sigma_1 \in \mathrm{Br}_n, \\ \sigma^{mon} &:= (\sigma^{root})^n, \\ center(\mathrm{Br}_n) &= \langle \sigma^{mon} \rangle. \end{aligned}$$

(a)

$$\begin{aligned} \delta_n^{1-k}\sigma^{root}(\underline{e}) &= Z(\delta_n^{1-k}\sigma^{root})(\underline{e}) \\ &= (s_{e_1}^{(k)}(e_2), s_{e_1}^{(k)}(e_3), ..., s_{e_1}^{(k)}(e_n), s_{e_1}^{(k)}(e_1)) \\ &= \underline{e} \cdot R \end{aligned}$$

$$\text{with } R := \left(\begin{array}{ccc|c} -q_{n-1} & \dots & -q_1 & -q_0 \\ \hline & & & \\ & E_{n-1} & & \\ & & & \end{array} \right)$$

$$\text{so } R_{ij} = \begin{cases} -q_{n-j} & \text{if } i = 1, \\ \delta_{i-1,j} & \text{if } i \geq 2, \end{cases}$$

where $q_0 = (-1)^k$, $q_{n+1-j} = S_{1j}$ for $j \in \{2, ..., n\}$.

(b)

$$\begin{aligned} Z((\delta_n^{1-k}\sigma^{root})^n) &= (-1)^{k+1}M, \\ \text{so especially } (\delta_n^{1-k}\sigma^{root})^n &\in (\mathrm{Br}_n \ltimes \{\pm 1\}^n)_S. \end{aligned}$$

(c) Write $q_0 = (-1)^k$, $q_{n+1-j} = S_{1j}$ *for* $j \in \{2, ..., n\}$ *as in part (a) and additionally* $q_n := 1$. *Suppose* $q_{n-j} = q_0 q_j$ *for* $j \in \{1, ..., n-1\}$, *and suppose that S has the following shape,*

$$S := \begin{pmatrix} 1 & q_{n-1} & \cdots & q_1 \\ & \ddots & \ddots & \vdots \\ & & \ddots & q_{n-1} \\ & & & 1 \end{pmatrix},$$

$$\text{so } S_{ij} = \begin{cases} 0 & \text{if } i > j, \\ q_{n-(j-i)} & \text{if } i \le j, \end{cases}$$

Then

$$\begin{aligned} M^{root} &:= Z(\delta_n^{1-k}\sigma^{root}) \in G_{\mathbb{Z}} \\ \text{with } (M^{root})^n &= (-1)^{k+1} M. \end{aligned}$$

M^{root} *is regular and cyclic and has the characteristic polynomial* $q(t) := \sum_{i=0}^n q_i t^i \in \mathbb{Z}[t]$.

Proof (a) The second line follows from the definition of the action of $\delta_n^{1-k}\sigma^{root}$ on $\underline{e}$. For the third line observe $s_{e_1}^{(k)}(e_j) = e_j - S_{1j}e_1$ for $j \ge 2$ and $s_{e_1}^{(k)}(e_1) = -q_0 e_1$.

(b) Use part (a) and

$$s^{(k)}_{s^{(k)}_{e_1}(e_2)}(s^{(k)}_{e_1}(e_j)) = s^{(k)}_{e_1} s^{(k)}_{e_2} (s^{(k)}_{e_1})^{-1} s^{(k)}_{e_1}(e_j) = s^{(k)}_{e_1} s^{(k)}_{e_2}(e_j)$$

to find

$$(\delta_n^{1-k}\sigma^{root})^2(\underline{e}) = (s^{(k)}_{e_1}s^{(k)}_{e_2}(e_3), ..., s^{(k)}_{e_1}s^{(k)}_{e_2}(e_n), s^{(k)}_{e_1}s^{(k)}_{e_2}(e_1), s^{(k)}_{e_1}s^{(k)}_{e_2}(e_2)).$$

One continues inductively and finds

$$(\delta_n^{1-k}\sigma^{root})^n(\underline{e}) = (s^{(k)}_{e_1}...s^{(k)}_{e_n}(e_1), ..., s^{(k)}_{e_1}...s^{(k)}_{e_n}(e_n)) = (-1)^{k+1}M(\underline{e}),$$

so $Z((\delta_n^{1-k}\sigma^{root})^n) = (-1)^{k+1}M$.

(c) If S is as in part (c) then

$$\begin{aligned} I^{(k)}(\delta_n^{1-k}\sigma^{root}(\underline{e})^t, \delta_n^{1-k}\sigma^{root}(\underline{e})) &\overset{(1)}{=} I^{(k)}((e_2\, e_3\, ...\, e_n\, e_1)^t, (e_2\, e_3\, ...\, e_n\, e_1)) \\ &\overset{(2)}{=} S + (-1)^k S^t = I^{(k)}(\underline{e}^t, \underline{e}). \end{aligned}$$

Here $\overset{(1)}{=}$ uses $s^{(k)}_{e_1} \in O^{(k)}$, and $\overset{(2)}{=}$ uses that $I^{(k)}(\underline{e}^t, \underline{e}) = S + (-1)^k S^t$ and that S is as in part (c).

Therefore $M^{root} := Z(\delta_n^{1-k}\sigma^{root}) \in G_{\mathbb{Z}}^{(k)}$, so by Lemma 3.25 (a)

$$M^{root} \in G_{\mathbb{Z}} \quad \text{and} \quad \delta_n^{1-k}\sigma^{root} \in (\mathrm{Br}_n \ltimes \{\pm 1\}^n)_S.$$

Also

$$(M^{root})^n = (Z(\delta_n^{1-k}\sigma^{root}))^n = Z((\delta_n^{1-k}\sigma^{root})^n) = (-1)^{k+1}M.$$

Let $\underline{e}^*$ be the $\mathbb{Z}$-basis of $H_{\mathbb{Z}}$ which is left L-dual to the $\mathbb{Z}$-basis $\underline{e}$, so with $L((\underline{e}^*)^t, \underline{e}) = E_n$. Remark 4.8 in [BH20] says

$$M^{root}\underline{e}^* = \underline{e}^* R^{-t} = \underline{e}^* \cdot \begin{pmatrix} & -q_0 \\ \hline & -q_1 \\ E_{n-1} & \vdots \\ & -q_{n-1} \end{pmatrix}.$$

The matrix R^{-t} is the companion matrix of the polynomial $q(t)$. Therefore M^{root} is regular, cyclic with generating vector $c = e_1^*$ and has the characteristic polynomial $q(t)$. □

Remark 3.27 The main part of part (c) of Theorem 3.26 has also the following matrix version: For $q(t) = \sum_{i=0}^n q_i t^i \in \mathbb{Z}[t]$ with $q_n = 1$, $q_0 = (-1)^k$ for some $k \in \{0; 1\}$ and $q_{n-j} = q_0 q_j$ the matrix R in part (a) and the matrix S in part (c) of Theorem 3.26 satisfy

$$R^n = (-1)^{k+1} S^{-1} S^t.$$

A proof using matrices of this version of part (c) of Theorem 3.26 was given in [BH20, Theorem 4.5 (a)+(b)]. The proof here with the braid group action is more elegant.

The antihomomorphism $Z : (\mathrm{Br}_n \ltimes \{\pm 1\}^n)_S \to G_{\mathbb{Z}}$ is not surjective in general. A simple example with $n = 4$ is given in the Remarks 3.29. But it is surjective in the case $n = 1$, in all cases with $n = 2$ and in almost all cases with $n = 3$. Theorem 3.28 gives precise statements. Its proof requires first a good control of the braid group action on $T_3(\mathbb{Z})$, which is the subject of Chap. 4 and second complete knowledge of the group $G_{\mathbb{Z}}$ for all cases with $n \leq 3$, which is the subject of Chap. 5. Theorem 3.28 is proved within the theorems in Chap. 5 which treat the different cases with $n \in \{1, 2, 3\}$, namely Lemma 5.4 (the cases A_1^n), Theorem 5.5 (the rank 2 cases), Theorem 5.13 (the reducible rank 3 cases), Theorem 5.14 (the irreducible rank 3 cases with all eigenvalues in S^1), Theorem 5.16 (some special other rank 3 cases), Theorem 5.18 (the rest of the rank 3 cases).

Theorem 3.28 *Let $(H_{\mathbb{Z}}, L, \underline{e})$ be a unimodular bilinear lattice of rank $n \leq 3$ with triangular basis $\underline{e}$ and matrix $S = L(\underline{e}^t, \underline{e})^t \in T_n(\mathbb{Z})$. The group antihomomorphism $Z : (\mathrm{Br}_n \ltimes \{\pm 1\}^n)_S \to G_{\mathbb{Z}}$ is not surjective in the four cases with $n = 3$ where S is in the $\mathrm{Br}_3 \ltimes \{\pm 1\}^3$ orbit of $S(\underline{x})$ with*

$$\underline{x} \in \{(3, 3, 4), (4, 4, 4), (5, 5, 5), (4, 4, 8)\},$$

so then $G_{\mathbb{Z}} \supsetneqq G_{\mathbb{Z}}^{\mathcal{B}}$. It is surjective in all other cases with $n \leq 3$, so then $G_{\mathbb{Z}} = G_{\mathbb{Z}}^{\mathcal{B}}$.

Remark 3.29 (i) Contrary to Theorem 3.28 for the cases $n = 1, 2, 3$, it is in the cases $n \geq 4$ easy to find matrices $S \in T_n(\mathbb{Z})$ such that the group antihomomorphism $Z : (\mathrm{Br}_n \ltimes \{\pm 1\}^n)_S \to G_{\mathbb{Z}}$ is not surjective. Though the construction which we propose in part (ii) and carry out in one example in part (iii) leads to matrices which are rather particular. For a given matrix S it is in general not easy to see whether Z is surjective or not.

(ii) Consider a reducible unimodular bilinear lattice $(H_{\mathbb{Z}}, L, \underline{e})$ of rank n with triangular basis $\underline{e}$ and matrix $S = L(\underline{e}^t, \underline{e})^t \in T_n(\mathbb{Z})$. There are an L-orthogonal decomposition $H_{\mathbb{Z}} = \bigoplus_{j=1}^{l} H_{\mathbb{Z},j}$ with $l \geq 2$ and rank $H_{\mathbb{Z},j} \geq 1$ and a surjective map $\alpha : \{1, ..., n\} \to \{1, ..., l\}$ with $e_i \in H_{\mathbb{Z},\alpha(i)}$. Then for $k \in \{0, 1\}$

$$\Gamma^{(k)}\{e_i\} \subset H_{\mathbb{Z},\alpha(i)},$$
$$\text{so} \quad \Delta^{(k)} \subset \bigcup_{j=1}^{l} H_{\mathbb{Z},j}.$$

Especially any $g \in G_{\mathbb{Z}}^{\mathcal{B}} \subset G_{\mathbb{Z}}$ maps each e_i to an element of $\bigcup_{j=1}^{l} H_{\mathbb{Z},j}$. This does not necessarily hold for any $g \in G_{\mathbb{Z}}$. Part (iii) gives an example.

(iii) Consider the unimodular bilinear lattice $(H_{\mathbb{Z}}, L, \underline{e})$ of rank 4 with triangular basis $\underline{e}$ and matrix

$$S = L(\underline{e}^t, \underline{e})^t = \begin{pmatrix} 1 & 2 & 0 & 0 \\ 0 & 1 & 0 & 0 \\ 0 & 0 & 1 & 2 \\ 0 & 0 & 0 & 1 \end{pmatrix} \in T^{uni,4}(\mathbb{Z}).$$

Then $H_{\mathbb{Z}} = H_{\mathbb{Z},1} \oplus H_{\mathbb{Z},2}$ with $H_{\mathbb{Z},1} = \mathbb{Z}e_1 \oplus \mathbb{Z}e_2$ and $H_{\mathbb{Z},2} = \mathbb{Z}e_3 \oplus \mathbb{Z}e_4$. The $\mathbb{Z}$-linear map $g : H_{\mathbb{Z}} \to H_{\mathbb{Z}}$ with

$$(g(e_1), g(e_2), g(e_3), g(e_4)) = (e_1 + (e_3 - e_4), e_2 + (e_3 - e_4), \\ e_3 + (e_1 - e_2), e_4 + (e_1 - e_2))$$

is not in $G_{\mathbb{Z}}^{\mathcal{B}}$ because $g(e_1), g(e_2), g(e_3), g(e_4) \notin H_{\mathbb{Z},1} \cup H_{\mathbb{Z},2}$. But $g \in G_{\mathbb{Z}}$ because

$$L(e_1 - e_2, e_1 - e_2) = 0 = L(e_3 - e_4, e_3 - e_4),$$
$$\text{so} \quad L(g(e_i), g(e_j)) = L(e_i, e_j) \quad \text{for } \{i, j\} \subset \{1, 2\} \text{ or } \{i, j\} \subset \{3, 4\},$$

and also

$$L(g(e_i), g(e_j)) = L(e_i, e_j) \quad \text{for } (i, j) \in (\{1, 2\} \times \{3, 4\}) \cup (\{3, 4\} \times \{1, 2\}).$$

So here $g \in G_{\mathbb{Z}} - G_{\mathbb{Z}}^{\mathcal{B}}$, so $G_{\mathbb{Z}}^{\mathcal{B}} \subsetneqq G_{\mathbb{Z}}$.

Chapter 4
Braid Group Action on Upper Triangular 3 × 3 Matrices

The subject of this chapter is the case $n = 3$ of the action in Lemma 3.13 of $\mathrm{Br}_n \ltimes \{\pm 1\}^n$ on the matrices in $T_n(\mathbb{Z})$.

In Sect. 4.1 the action on $T_3(\mathbb{R})$ is made concrete. The (quotient) group of actions is given in new generators. It is

$$(G^{phi} \ltimes G^{sign}) \rtimes \langle\gamma\rangle \cong (G^{phi} \rtimes \langle\gamma\rangle) \ltimes G^{sign},$$

where G^{phi} is a free Coxeter group with three generators, G^{sign} is the group of actions in $T_3(\mathbb{R})$ which the sign group $\{\pm 1\}^3$ induces, and γ acts cyclically of order 3. In fact, $G^{phi} \rtimes \langle\gamma\rangle \cong PSL_2(\mathbb{Z})$, so we have a nonlinear action of $PSL_2(\mathbb{Z})$, but this way to look at it is less useful than the presentation as $G^{phi} \rtimes \langle\gamma\rangle$.

The action on $T_3(\mathbb{Z})$ had been studied already by Krüger [Kr90, §12] and by Cecotti-Vafa [CV93, Ch. 6.2]. Section 4.2 recovers and refines their results. Like them, it puts emphasis on the cases where the monodromy of a corresponding unimodular bilinear lattice has eigenvalues in S^1. Section 4.2 follows largely Krüger [Kr90, §12].

Section 4.3 uses *pseudo-graphs* to systematically study all cases, not only those where the monodromy has eigenvalues in S^1. This goes far beyond Krüger and Cecotti-Vafa.

The results of Sect. 4.3 and the pseudo-graphs are used in Sect. 4.4 to determine in all cases the stabilizer $(\mathrm{Br}_3 \ltimes \{\pm 1\}^3)_S$ respectively the stabilizer $(\mathrm{Br}_3)_{S/\{\pm 1\}^3}$. Section 7.4 will build on this and determine the stabilizer $(\mathrm{Br}_3)_{\underline{e}/\{\pm 1\}^3}$ of a distingushed basis $\underline{e} \in \mathcal{B}^{dist}$ for any unimodular bilinear lattice of rank 3 with a fixed triangular basis.

Section 4.5 starts with the observation that a matrix $S \in T_n(\mathbb{Z})$ and the matrix $\widetilde{S} \in T_n(\mathbb{Z})$ with $\widetilde{S}_{ij} = -S_{ij}$ for $i < j$ lead to unimodular bilinear lattices with the same odd monodromy groups and the same odd vanishing cycles. This motivates to study the action on $T_n(\mathbb{Z})$ which extends the action of $\mathrm{Br}_n \ltimes \{\pm 1\}^n$ by this global sign

C. Hertling and K. Larabi, *Triangular Bases of Unimodular Bilinear Lattices and Induced Structures*, Lecture Notes in Mathematics 2398,
https://doi.org/10.1007/978-3-032-25997-4_4

change. Section 4.5 carries this out in the case $n = 3$ and gives standard representatives for each orbit. Though examples show that the action is rather wild. Similar looking triples in $\mathbb{Z}^3$ are in the orbits of very different standard representatives.

4.1 Braid Group Action on Real Upper Triangular 3 x 3 Matrices

The action of $\mathrm{Br}_3 \ltimes \{\pm 1\}^3$ on $T_3(\mathbb{Z})$ will be studied in the next sections. It extends to an action on $T_3(\mathbb{R}) \cong \mathbb{R}^3$ which will be studied here. By Theorem 3.4 (d), σ_1 acts on $T_3(\mathbb{Z})$ by

$$\sigma_1 : \begin{pmatrix} 1 & x_1 & x_2 \\ 0 & 1 & x_3 \\ 0 & 0 & 1 \end{pmatrix} \mapsto \begin{pmatrix} -x_1 & 1 & 0 \\ 1 & 0 & 0 \\ 0 & 0 & 1 \end{pmatrix} \begin{pmatrix} 1 & x_1 & x_2 \\ 0 & 1 & x_3 \\ 0 & 0 & 1 \end{pmatrix} \begin{pmatrix} -x_1 & 1 & 0 \\ 1 & 0 & 0 \\ 0 & 0 & 1 \end{pmatrix}$$
$$= \begin{pmatrix} 1 & -x_1 & x_3 - x_1 x_2 \\ 0 & 1 & x_2 \\ 0 & 0 & 1 \end{pmatrix}.$$

It extends to an action on $T_3(\mathbb{R})$. With the isomorphism

$$T_3(R) \xrightarrow{\cong} R^3, \quad \begin{pmatrix} 1 & x_1 & x_2 \\ 0 & 1 & x_3 \\ 0 & 0 & 1 \end{pmatrix} \mapsto (x_1, x_2, x_3) \quad \text{for } R \in \{\mathbb{Z}, \mathbb{Q}, \mathbb{R}, \mathbb{C}\}$$

this gives the action

$$\sigma_1^{\mathbb{R}} : \mathbb{R}^3 \to \mathbb{R}^3, \quad (x_1, x_2, x_3) \mapsto (-x_1, x_3 - x_1 x_2, x_2).$$

Analogously

$$\begin{aligned}
(\sigma_1^{\mathbb{R}})^{-1} &: \mathbb{R}^3 \to \mathbb{R}^3, \quad (x_1, x_2, x_3) \mapsto (-x_1, x_3, x_2 - x_1 x_3), \\
\sigma_2^{\mathbb{R}} &: \mathbb{R}^3 \to \mathbb{R}^3, \quad (x_1, x_2, x_3) \mapsto (x_2 - x_1 x_3, x_1, -x_3), \\
(\sigma_2^{\mathbb{R}})^{-1} &: \mathbb{R}^3 \to \mathbb{R}^3, \quad (x_1, x_2, x_3) \mapsto (x_2, x_1 - x_2 x_3, -x_3), \\
\delta_1^{\mathbb{R}} &: \mathbb{R}^3 \to \mathbb{R}^3, \quad (x_1, x_2, x_3) \mapsto (-x_1, -x_2, x_3), \\
\delta_2^{\mathbb{R}} &: \mathbb{R}^3 \to \mathbb{R}^3, \quad (x_1, x_2, x_3) \mapsto (-x_1, x_2, -x_3), \\
\delta_3^{\mathbb{R}} &: \mathbb{R}^3 \to \mathbb{R}^3, \quad (x_1, x_2, x_3) \mapsto (x_1, -x_2, -x_3).
\end{aligned}$$

One sees

$$\delta_3^{\mathbb{R}} = \delta_1^{\mathbb{R}}\delta_2^{\mathbb{R}} \quad \text{and} \quad G^{sign} := \langle \delta_1^{\mathbb{R}}, \delta_2^{\mathbb{R}} \rangle \cong \{\pm 1\}^2.$$

The group $\langle \sigma_1^{\mathbb{R}}, \sigma_2^{\mathbb{R}} \rangle \ltimes G^{sign} \subset \mathrm{Aut}_{pol}(\mathbb{R}^3)$ of polynomial automorphisms of $\mathbb{R}^3$ will be more transparent in other generators.

Definition 4.1 Define the polynomial automorphisms of $\mathbb{R}^3$

$$\begin{aligned}
\varphi_1 &: \mathbb{R}^3 \to \mathbb{R}^3, \quad (x_1, x_2, x_3) \mapsto (x_2x_3 - x_1, x_3, x_2), \\
\varphi_2 &: \mathbb{R}^3 \to \mathbb{R}^3, \quad (x_1, x_2, x_3) \mapsto (x_3, x_1x_3 - x_2, x_1), \\
\varphi_3 &: \mathbb{R}^3 \to \mathbb{R}^3, \quad (x_1, x_2, x_3) \mapsto (x_2, x_1, x_1x_2 - x_3), \\
\gamma &: \mathbb{R}^3 \to \mathbb{R}^3, \quad (x_1, x_2, x_3) \mapsto (x_3, x_1, x_2),
\end{aligned}$$

and the group $G^{phi} := \langle \varphi_1, \varphi_2, \varphi_3 \rangle \subset \mathrm{Aut}_{pol}(\mathbb{R}^3)$.

Theorem 4.2 *(a) The group G^{phi} is a free Coxeter group with the three generators $\varphi_1, \varphi_2, \varphi_3$, so $G^{phi} \cong G^{fCox,3}$.*

(b) $\langle \gamma \rangle \cong \mathbb{Z}/3\mathbb{Z} \cong A_3 \subset S_3$.

(c) $\langle \sigma_1^{\mathbb{R}}, \sigma_2^{\mathbb{R}} \rangle \ltimes G^{sign} = (G^{phi} \ltimes G^{sign}) \rtimes \langle \gamma \rangle$.

Proof (a) $\varphi_1^2 = \varphi_2^2 = \varphi_3^2 = \mathrm{id}$ is obvious, and also that $(2, 2, 2) \in \mathbb{R}^3$ is a fixed point of G^{phi}. We will show that the group $\langle d_{(2,2,2)}\varphi_1, d_{(2,2,2)}\varphi_2, d_{(2,2,2)}\varphi_3 \rangle$ of induced actions on the tangent space $T_{(2,2,2)}\mathbb{R}^3$ is a free Coxeter group with three generators. This will imply $G^{phi} \cong G^{fCox,3}$.

Affine linear coordinates $(\widetilde{x}_1, \widetilde{x}_2, \widetilde{x}_3)$ on $\mathbb{R}^3$ which vanish at $(2, 2, 2)$ with

$$(x_1, x_2, x_3) = (2 + \widetilde{x}_1, 2 + \widetilde{x}_2, 2 + \widetilde{x}_3) = (2, 2, 2) + (\widetilde{x}_1, \widetilde{x}_2, \widetilde{x}_3)$$

are also linear coordinates on $T_{(2,2,2)}\mathbb{R}^3$. We have

$$\begin{aligned}
\varphi_1((2, 2, 2) + (\widetilde{x}_1, \widetilde{x}_2, \widetilde{x}_3)) &= (2, 2, 2) + (\widetilde{x}_2\widetilde{x}_3 + 2\widetilde{x}_2 + 2\widetilde{x}_3 - \widetilde{x}_1, \widetilde{x}_3, \widetilde{x}_2), \\
d_{(2,2,2)}\varphi_1(\widetilde{x}_1, \widetilde{x}_2, \widetilde{x}_3) &= (\widetilde{x}_1, \widetilde{x}_2, \widetilde{x}_3) \begin{pmatrix} -1 & 0 & 0 \\ 2 & 0 & 1 \\ 2 & 1 & 0 \end{pmatrix},
\end{aligned}$$

and analogously

$$d_{(2,2,2)}\varphi_2(\widetilde{x}_1, \widetilde{x}_2, \widetilde{x}_3) = (\widetilde{x}_1, \widetilde{x}_2, \widetilde{x}_3) \begin{pmatrix} 0 & 2 & 1 \\ 0 & -1 & 0 \\ 1 & 2 & 0 \end{pmatrix},$$

$$d_{(2,2,2)}\varphi_3(\widetilde{x}_1, \widetilde{x}_2, \widetilde{x}_3) = (\widetilde{x}_1, \widetilde{x}_2, \widetilde{x}_3) \begin{pmatrix} 0 & 1 & 2 \\ 1 & 0 & 2 \\ 0 & 0 & -1 \end{pmatrix}.$$

The group G^{phi} respects the fibers of the map

$$r_{\mathbb{R}} : \mathbb{R}^3 \to \mathbb{R}, \quad (x_1, x_2, x_3) \mapsto x_1^2 + x_2^2 + x_3^2 - x_1 x_2 x_3.$$

The group $\langle d_{(2,2,2)}\varphi_1, d_{(2,2,2)}\varphi_2, d_{(2,2,2)}\varphi_3 \rangle$ respects the tangent cone at $(2, 2, 2)$ of the fiber $r_{\mathbb{R}}^{-1}(4)$. This tangent cone is the zero set of the quadratic form

$$\begin{aligned} &q_{\mathbb{R}^3,(2,2,2)} : \mathbb{R}^3 \to \mathbb{R}, \\ &(\widetilde{x}_1, \widetilde{x}_2, \widetilde{x}_3) \mapsto -\widetilde{x}_1^2 - \widetilde{x}_2^2 - \widetilde{x}_3^2 + 2\widetilde{x}_1\widetilde{x}_2 + 2\widetilde{x}_1\widetilde{x}_3 + 2\widetilde{x}_2\widetilde{x}_3. \end{aligned}$$

This quadratic form is indefinite with signature $(+, -, -)$. As in Theorem A.4, its cone of positive vectors is called $\mathcal{K}$. Consider the six vectors

$$\begin{aligned} v_1 &= (1, 1, 0), \ v_2 = (1, 0, 1), \ v_3 = (0, 1, 1), \\ w_1 &= v_1 + v_2, \ w_2 = v_1 + v_3, \ w_3 = v_2 + v_3. \end{aligned}$$

Then

$$\begin{aligned} v_1, v_2, v_3 \in \partial\mathcal{K}, &\quad w_1, w_2, w_3 \in \mathcal{K}, \\ d_{(2,2,2)}\varphi_1 &: v_1 \leftrightarrow v_2, \ w_1 \mapsto w_1, \\ d_{(2,2,2)}\varphi_2 &: v_1 \leftrightarrow v_3, \ w_2 \mapsto w_2, \\ d_{(2,2,2)}\varphi_3 &: v_2 \leftrightarrow v_3, \ w_3 \mapsto w_3. \end{aligned}$$

Compare Theorem A.4. In the model $\mathcal{K}/\mathbb{R}^*$ of the hyperbolic plane, $d_{(2,2,2)}\varphi_i$ for $i \in \{1, 2, 3\}$ gives a rotation with angle π and elliptic fixed point $\mathbb{R}^* w_i$, which maps the hyperbolic line with euclidean boundary points $\mathbb{R}^* v_1 \& \mathbb{R}^* v_2$ respectively $\mathbb{R}^* v_1 \& \mathbb{R}^* v_3$ respectively $\mathbb{R}^* v_2 \& \mathbb{R}^* v_3$ to itself.

Theorem A.2 (b) applies and shows $\langle d_{(2,2,2)}\varphi_i \,|\, i \in \{1, 2, 3\} \rangle \cong G^{fCox,3}$. Figure 4.1 illustrates this.

(b) Trivial.

(c) The equality of groups

$$\langle \sigma_1^{\mathbb{R}}, \sigma_2^{\mathbb{R}} \rangle \ltimes G^{sign} = \langle \varphi_1, \varphi_2, \varphi_3, \gamma, \delta_1^{\mathbb{R}}, \delta_2^{\mathbb{R}} \rangle$$

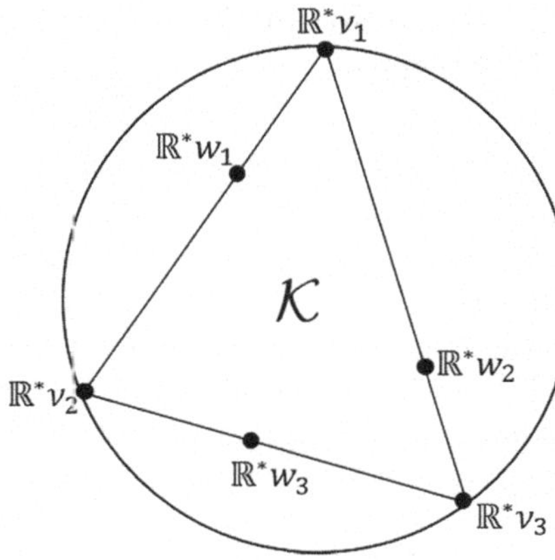

Fig. 4.1 $G^{fCox,3}$ generated by 3 elliptic Möbius transformations, an application of Theorem A.2 (b)

follows from

$$\gamma = \delta_3^{\mathbb{R}} \sigma_2^{\mathbb{R}} \sigma_1^{\mathbb{R}}, \tag{4.1}$$
$$\varphi_1 = \delta_1^{\mathbb{R}} \gamma^{-1} (\sigma_2^{\mathbb{R}})^{-1}, \tag{4.2}$$
$$\varphi_2 = \delta_1^{\mathbb{R}} \gamma \sigma_2^{\mathbb{R}} = \delta_3^{\mathbb{R}} \gamma^{-1} (\sigma_1^{\mathbb{R}})^{-1}, \tag{4.3}$$
$$\varphi_3 = \delta_3^{\mathbb{R}} \gamma \sigma_1^{\mathbb{R}}. \tag{4.4}$$

and

$$\sigma_1^{\mathbb{R}} = \gamma^{-1} \delta_3^{\mathbb{R}} \varphi_3, \quad \sigma_2^{\mathbb{R}} = \gamma^{-1} \delta_1^{\mathbb{R}} \varphi_2. \tag{4.5}$$

G^{phi} fixes $(2, 2, 2)$, and therefore $G^{phi} \cap G^{sign} = \{\mathrm{id}\}$. As G^{sign} is a normal subgroup of $\langle \sigma_1^{\mathbb{R}}, \sigma_2^{\mathbb{R}} \rangle \ltimes G^{sign}$, it is also a normal subgroup of $\langle \varphi_1, \varphi_2, \varphi_3, \delta_1^{\mathbb{R}}, \delta_2^{\mathbb{R}} \rangle$, so $\langle \varphi_1, \varphi_2, \varphi_3, \delta_1^{\mathbb{R}}, \delta_2^{\mathbb{R}} \rangle = G^{phi} \ltimes G^{sign}$. More precisely

$$\varphi_i \delta_j^{\mathbb{R}} \varphi_i^{-1} = \delta_k^{\mathbb{R}} \quad \text{for } (i, j, k) \in \{(1, 3, 3), (2, 2, 2), (3, 1, 1), (1, 1, 2), (1, 2, 1), (2, 1, 3), (2, 3, 1), (3, 2, 3), (3, 3, 2)\}.$$

We claim $\gamma \notin G^{phi} \ltimes G^{sign}$. If γ were in $G^{phi} \ltimes G^{sign}$ then $\gamma \in G^{phi}$ as γ fixes $(2, 2, 2)$. But all elements of finite order in $G^{phi} \cong G^{fCox,3}$ have order two, though γ has order three. Hence $\gamma \notin G^{phi}$ and $\gamma \notin G^{phi} \ltimes G^{sign}$.

The claim and

$$\gamma \varphi_1 \gamma^{-1} = \varphi_2, \ \gamma \varphi_2 \gamma^{-1} = \varphi_3, \ \gamma \varphi_3 \gamma^{-1} = \varphi_1, \tag{4.6}$$
$$\gamma \delta_1^{\mathbb{R}} \gamma^{-1} = \delta_3^{\mathbb{R}}, \ \gamma \delta_2^{\mathbb{R}} \gamma^{-1} = \delta_1^{\mathbb{R}}, \ \gamma \delta_3^{\mathbb{R}} \gamma^{-1} = \delta_2^{\mathbb{R}}, \tag{4.7}$$

show

$$\langle \varphi_1, \varphi_2, \varphi_3, \delta_1^{\mathbb{R}}, \delta_2^{\mathbb{R}}, \gamma \rangle = (G^{phi} \ltimes G^{sign}) \rtimes \langle \gamma \rangle. \qquad \square$$

4.2 Braid Group Action on Integer Upper Triangular 3 x 3 Matrices

In this section we will give a partial classification of the orbits of the action of $\mathrm{Br}_3 \ltimes \{\pm 1\}^3$ on $T_3(\mathbb{Z})$. This refines results which were obtained independently by Krüger [Kr90, §12] and Cecotti-Vafa [CV93, Ch. 6.2] (building on Mordell [Mo69, p 106 ff]).

The refinement consists in the following. By Theorem 4.2 the action of $\mathrm{Br}_3 \ltimes \{\pm 1\}^3$ on $T_3(\mathbb{Z})$ coincides with the action of $(G^{phi} \ltimes G^{sign}) \rtimes \langle\gamma\rangle$. For reasons unknown to us, Krüger and Cecotti-Vafa considered the action of the slightly larger group $(G^{phi} \ltimes G^{sign}) \rtimes \langle\gamma, \gamma_2\rangle$ with

$$\gamma_2 : \mathbb{R}^3 \to \mathbb{R}^3, \quad (x_1, x_2, x_3) \mapsto (x_2, x_1, x_3), \quad \text{so } \langle\gamma, \gamma_2\rangle \cong S_3.$$

Thus they obtained a slightly coarser classification. Nevertheless Theorem 4.6 is essentially due to them (and Mordell [Mo69, p 106 ff]). The following definition and lemma prepare it. They are due to Krüger [Kr90, §12].

Definition 4.3 [Kr90, Def. 12.2] For $\underline{x} = (x_1, x_2, x_3) \in \mathbb{R}^3$ we set as usual $\|\underline{x}\| := \sqrt{x_1^2 + x_2^2 + x_3^2}$. A tuple $\underline{x} \in \mathbb{R}^3$ is called a *local minimum* if

$$\|\underline{x}\| \leq \min(\|\sigma_1^{\mathbb{R}}(\underline{x})\|, \|(\sigma_1^{\mathbb{R}})^{-1}(\underline{x})\|, \|\sigma_2^{\mathbb{R}}(\underline{x})\|, \|(\sigma_2^{\mathbb{R}})^{-1}(\underline{x})\|).$$

This is obviously equivalent to

$$\|\underline{x}\| \leq \min(\|\varphi_1(\underline{x})\|, \|\varphi_2(\underline{x})\|, \|\varphi_3(\underline{x})\|).$$

Lemma 4.4 *[Kr90, Lemma 12.3] $\underline{x} \in \mathbb{R}^3$ is a local minimum if and only if it satisfies (i) or (ii),*

(i) $x_1x_2x_3 \leq 0$,
(ii) $x_1x_2x_3 > 0,\ 2|x_1| \leq |x_2x_3|,\ 2|x_2| \leq |x_1x_3|,\ 2|x_3| \leq |x_1x_2|$.

In the case (ii) also $|x_1| \geq 2$, $|x_2| \geq 2$ and $|x_3| \geq 2$ hold.

Proof $\underline{x} \in \mathbb{R}^3$ is a local minimum if for all i, j, k with $\{i, j, k\} = \{1, 2, 3\}$

$$x_i^2 + x_j^2 + x_k^2 \leq x_i^2 + x_j^2 + (x_k - x_ix_j)^2$$

holds, which is equivalent to

$$2x_ix_jx_k \leq x_i^2x_j^2.$$

1st case, $x_1x_2x_3 \leq 0$: Then $\underline{x}$ is a local minimum.
2nd case, $x_1x_2x_3 > 0$: Then the condition $2x_ix_jx_k \leq x_i^2x_j^2$ is equivalent to

$$2|x_k| \leq |x_i x_j|.$$

These three conditions together imply

$$4|x_k| \leq 2|x_i||x_j| \leq |x_i||x_i||x_k|, \quad \text{so } 4 \leq |x_i|^2, \quad \text{so } 2 \leq |x_i|. \qquad \square$$

The square $\|.\|^2 : \mathbb{Z}^3 \to \mathbb{Z}_{\geq 0}$ of the norm has on $\mathbb{Z}^3$ values in $\mathbb{Z}_{\geq 0}$. Therefore each $\mathrm{Br}_3 \ltimes \{\pm 1\}^3$ orbit in $\mathbb{Z}^3$ has local minima. Krüger showed that the only $\mathrm{Br}_3 \ltimes \{\pm 1\}^3$ orbits in $\mathbb{R}^3$ without local minimal are of the following shape. We will not use this result, but we find it interesting enough to cite it.

Theorem 4.5 *[Kr90, Theorem 12.6] Let $\underline{x} \in \mathbb{R}^3$ whose $\mathrm{Br}_3 \ltimes \{\pm 1\}^3$ orbit does not contain a local minimum. Then*

$$x_1 x_2 x_3 > 0, \quad 2 < \min(|x_1|, |x_2|, |x_3|),$$
$$4 = r_{\mathbb{R}}(\underline{x})(= x_1^2 + x_2^2 + x_3^2 - x_1 x_2 x_3).$$

Furthermore, there is a sequence $(\psi_n)_{n \in \mathbb{N}}$ with $\psi_n \in \{\varphi_1, \varphi_2, \varphi_3\}$ with $\psi_n \neq \psi_{n+1}$ such that the sequence $(\underline{x}^{(n)})_{n \in \mathbb{N} \cup \{0\}}$ with $\underline{x}^{(0)} = \underline{x}$ and $\underline{x}^{(n+1)} = \psi_n(\underline{x}^{(n)})$ satisfies

$$\|\underline{x}^{(n+1)}\| < \|\underline{x}^{(n)}\| \quad \text{for all } n \in \mathbb{N} \cup \{0\},$$
$$\lim_{n \to \infty} (|x_1^{(n)}|, |x_2^{(n)}|, |x_3^{(n)}|) = (2, 2, 2).$$

Now we come to the classification of $\mathrm{Br}_3 \ltimes \{\pm 1\}^3$ orbits in $\mathbb{Z}^3$. The following result is except for its part (f) a refinement of [Kr90, Theorem 12.7] and of [CV93, Ch 6.2]. The proof below follows (except for the part (f)) the proof in [Kr90]. Recall that each $\mathrm{Br}_3 \ltimes \{\pm 1\}^3$ orbit in $\mathbb{R}^3$ is contained in one fiber of the map $r_{\mathbb{R}} : \mathbb{R}^3 \to \mathbb{R}$, $\underline{x} \mapsto x_1^2 + x_2^2 + x_3^2 - x_1 x_2 x_3$.

Theorem 4.6 *(a) Each fiber of*

$$r : \mathbb{Z}^3 \to \mathbb{Z}, \ r(\underline{x}) = x_1^2 + x_2^2 + x_3^2 - x_1 x_2 x_3$$

except the fiber $r^{-1}(4)$ contains only finitely many local minima.

(b) Each fiber of $r : \mathbb{Z}^3 \to \mathbb{Z}$ except the fiber $r^{-1}(4)$ consists of only finitely many $\mathrm{Br}_3 \ltimes \{\pm 1\}^3$ orbits.

(c) For $\rho \in \mathbb{Z}_{<0}$, each local minimum $\underline{x} \in r^{-1}(\rho)$ satisfies $x_1 x_2 x_3 > 0$ and $|x_1| \geq 3$, $|x_2| \geq 3$, $|x_3| \geq 3$.

(d) For $\rho \in \mathbb{N} - \{4\}$, each local minimum $\underline{x} \in r^{-1}(\rho)$ satisfies $x_1 x_2 x_3 \leq 0$.

(e) The following table gives all local minima in $r^{-1}(\{0, 1, 2, 3, 4\})$. The local minima in one $\mathrm{Br}_3 \ltimes \{\pm 1\}^3$ orbit are in one line. The last entry in each line is one matrix in the corresponding orbit in $T_3(\mathbb{Z})$.

$r=3$	–	–
$r=0$	$(0,0,0)$	$S(A_1^3)$
$r=0$	$(3,3,3),(-3,-3,3),(-3,3,-3),(3,-3,-3)$	$S(\mathbb{P}^2)$
$r=1$	$(\pm 1,0,0),(0,\pm 1,0),(0,0,\pm 1)$	$S(A_2A_1)$
$r=2$	$(\pm 1,\pm 1,0),(\pm 1,0,\pm 1),(0,\pm 1,\pm 1)$	$S(A_3)$
$r=4$	$(\pm 2,0,0),(0,\pm 2,0),(0,0,\pm 2)$	$S(\mathbb{P}^1A_1)$
$r=4$	$(-1,-1,-1),(1,1,-1),(1,-1,1),(-1,1,1)$	$S(\widehat{A}_2)$
$r=4$	$(2,2,2),(-2,-2,2),(-2,2,-2),(2,-2,-2)$	$S(\mathcal{H}_{1,2})$
$\begin{Bmatrix} r=4 \\ l\in\mathbb{Z}_{\geq 3}\end{Bmatrix}$	$\begin{Bmatrix} (\varepsilon_1 2,\varepsilon_2 l,\varepsilon_3 l),(\varepsilon_1 l,\varepsilon_2 2,\varepsilon_3 l),(\varepsilon_1 l,\varepsilon_2 l,\varepsilon_3 2) \\ \text{for } \varepsilon_1,\varepsilon_2,\varepsilon_3\in\{\pm 1\} \text{ with } \varepsilon_1\varepsilon_2\varepsilon_3=1\end{Bmatrix}$	$S(-l,2,-l)$

So there are seven single $\mathrm{Br}_3\ltimes\{\pm 1\}^3$ *orbits and one series with parameter* $l\in\mathbb{Z}_{\geq 3}$ *of* $\mathrm{Br}_3\ltimes\{\pm 1\}^3$ *orbits with* $r\in\{0,1,2,3,4\}$. *These are the most interesting orbits as the monodromy matrix* $S(\underline{x})^{-1}S(\underline{x})^t$ *for* $\underline{x}\in\mathbb{R}^3$ *has eigenvalues in* S^1 *if and only if* $r(\underline{x})\in[0,4]$.

(f) For a given local minimum $\underline{x}\in\mathbb{Z}^3$ *the set of all local minima in the* $\mathrm{Br}_3\ltimes\{\pm 1\}^3$ *orbit of* $\underline{x}$ *is either the set* $G^{sign}\rtimes\langle\gamma\rangle(\underline{x})$ *or the set* $G^{sign}\rtimes\langle\gamma,\gamma_2\rangle(\underline{x})$ *(see Theorem 4.13 (b) for details).*

Proof (a) Fix $\rho\in\mathbb{Z}-\{4\}$. Let $\underline{x}\in\mathbb{Z}^3$ be a local minimum with $r(\underline{x})=\rho$.

1st case, $x_1x_2x_3\leq 0$: Then $\rho=r(\underline{x})=x_1^2+x_2^2+x_3^2-x_1x_2x_3\geq\|\underline{x}\|^2$, so $\rho\geq 0$. The closed ball of radius $\sqrt{\rho}$ around 0 in $\mathbb{R}^3$ intersects $\mathbb{Z}^3$ only in finitely many points.

2nd case, $x_1x_2x_3>0$: We can suppose $x_i>0$ for $i\in\{1,2,3\}$ because of the action of G^{sign} on $\mathbb{R}^3$ and $\mathbb{Z}^3$. Lemma 4.4 says $2x_1\leq x_2x_3$, $2x_2\leq x_1x_3$, $2x_3\leq x_1x_2$, $x_i\geq 2$ for $i\in\{1,2,3\}$.

We can suppose $x_1=\min(x_1,x_2,x_3)$ (the other cases are analogous). If $x_1=2$ then $4\neq\rho=r(\underline{x})=4+x_2^2+x_3^2-2x_2x_3=4+(x_2-x_3)^2$, so $x_2\neq x_3$, which is a contradiction to $2x_2\leq x_1x_3=2x_3$, $2x_3\leq x_1x_2=2x_2$. Therefore $x_1\geq 3$.

We can suppose $x_1\leq x_2\leq x_3$ (the other cases are analogous).

$$\begin{aligned}
\rho=r(\underline{x}) &= x_1^2+x_2^2+(x_3-\frac{1}{2}x_1x_2)^2-\frac{1}{4}x_1^2x_2^2\\
&\leq x_1^2+x_2^2+(x_2-\frac{1}{2}x_1x_2)^2-\frac{1}{4}x_1^2x_2^2 \quad (\text{because } x_2\leq x_3\leq\frac{1}{2}x_1x_2)\\
&= x_1^2+2x_2^2-x_1x_2^2=(x_1-2)(x_1+2-x_2^2)+4\\
&\leq (x_1-2)(x_1+2-x_1^2)+4=-(x_1-2)(x_1-2)(x_1+1)+4\\
&\leq \begin{cases} -(3-2)(3-2)(3+1)+4=0,\\ -(x_1-2)^3+4.\end{cases}
\end{aligned}$$

This implies $\rho\leq 0$ and $x_1\leq 2+\sqrt[3]{4-\rho}$, so x_1 is one of the finitely many values in $\mathbb{Z}\cap[3,2+\sqrt[3]{4-\rho}]$.

The inequality $\rho \leq (x_1 - 2)(x_1 + 2 - x_2^2) + 4$ implies

$$x_2^2 \leq \frac{4-\rho}{x_1 - 2} + x_1 + 2,$$

so x_2 is one of the finitely many values in $\mathbb{Z} \cap [x_1, \sqrt{\frac{4-\rho}{x_1-2} + x_1 + 2}]$.

Because of $x_3 \leq \frac{1}{2}x_1x_2$ also x_3 can take only finitely many values.

(b) Each $\mathrm{Br}_3 \ltimes \{\pm 1\}^3$ orbit in $\mathbb{Z}^3$ is mapped by $\|.\|^2$ to a subset of $\mathbb{Z}_{\geq 0}$. A preimage in this orbit of the minimum of this subset is a local minimum. Therefore (a) implies (b).

(c) Suppose $\rho < 0$ and $\underline{x} \in r^{-1}(\rho)$ is a local minimum. $0 > \rho = \|\underline{x}\|^2 - x_1x_2x_3$ implies $x_1x_2x_3 > 0$. Lemma 4.4 gives $|x_1| \geq 2$.

If $x_1 = 2\varepsilon$ with $\varepsilon \in \{\pm 1\}$ then $\rho = r(\underline{x}) = 4 + (x_2 - \varepsilon x_3)^2 \geq 4$, a contradiction. So $|x_1| \geq 3$. Analogously $|x_2| \geq 3$ and $|x_3| \geq 3$.

(d) Suppose $\rho \in \mathbb{N} - \{4\}$ and $\underline{x} \in r^{-1}(\rho)$ is a local minimum. In the second case $x_1x_2x_3 > 0$ in the proof of part (a) we concluded $\rho \leq 0$. Therefore we are in the first case in the proof of part (a), so $x_1x_2x_3 \leq 0$.

(e) Suppose $\rho \in \{0, 1, 2, 3, 4\}$, and $\underline{x} \in r^{-1}(\rho)$ is a local minimum.

In the cases $\rho \in \{1, 2, 3\}$ by part (d) $x_1x_2x_3 \leq 0$ and $\rho = r(\underline{x}) = x_1^2 + x_2^2 + x_3^2 + |x_1x_2x_3|$, so in these cases all $x_i \neq 0$ is impossible, so some $x_i = 0$, so $\rho = r(\underline{x}) = x_j^2 + x_k^2$ where $\{i, j, k\} = \{1, 2, 3\}$.

The case $\rho = 3$**:** $3 = x_j^2 + x_k^2$ is impossible, the case $\rho = 3$ is impossible, $r^{-1}(3) = \emptyset$.

The case $\rho = 1$**:** $1 = x_j^2 + x_k^2$ is solved only by $(x_j, x_k) \in \{(\pm 1, 0), (0, \pm 1)\}$. The six local minima $(\pm 1, 0, 0), (0, \pm 1, 0), (0, 0, \pm 1)$ are in one orbit of $\mathrm{Br}_3 \ltimes \{\pm 1\}^3$ because $\gamma(1, 0, 0) = (0, 1, 0), \gamma(0, 1, 0) = (0, 0, 1)$.

The case $\rho = 2$**:** $x_j^2 + x_k^2 = 2$ is solved only by $(x_j, x_k) \in \{(\pm 1, \pm 1)\}$. The twelve local minimal $(\pm 1, \pm 1, 0), (\pm 1, 0, \pm 1), (0, \pm 1, \pm 1)$ are in one orbit of $\mathrm{Br}_3 \ltimes \{\pm 1\}^3$ because $\gamma(1, 1, 0) = (0, 1, 1), \gamma(0, 1, 1) = (1, 0, 1)$.

The case $\rho = 0$**:** We use the proof of part (a).

1st case, $x_1x_2x_3 \leq 0$**:** $0 = \rho = \|x\|^2 - x_1x_2x_3 \geq \|x\|^2$, so $\underline{x} = (0, 0, 0)$. Its $\mathrm{Br}_3 \ltimes \{\pm 1\}^3$ orbit consists only of $(0, 0, 0)$.

2nd case, $x_1x_2x_3 > 0$: We can suppose $x_i > 0$ for each $i \in \{1, 2, 3\}$. Suppose $x_i \leq x_j \leq x_k$ for $\{i, j, k\} = \{1, 2, 3\}$. The proof of part (a) gives

$$3 \leq x_i \leq 2 + \sqrt[3]{4 - \rho} = 2 + \sqrt[3]{4}, \text{ so } x_i = 3$$

and

$$3 = x_i \leq x_j \leq \sqrt{\frac{4-\rho}{x_i - 2} + x_i + 2} = 3, \text{ so } x_j = 3.$$

$0 = r(\underline{x}) = 9 + 9 + x_k^2 - 9x_k = (x_k - 3)(x_k - 6)$ and $x_k \le \frac{1}{2}x_ix_j = \frac{9}{2}$ show $x_k = 3$. The four local minima $(3, 3, 3)$, $(-3, -3, 3)$, $(-3, 3, -3)$ and $(3, -3, -3)$ are in one $\mathrm{Br}_3 \ltimes \{\pm 1\}^3$ orbit because of the action of G^{sign}.

The case $\rho = 4$:

1st case, some $x_i = 0$: Then with $\{i, j, k\} = \{1, 2, 3\}$ $4 = r(\underline{x}) = x_j^2 + x_k^2$. This is solved only by $(x_j, x_k) \in \{(\pm 2, 0), (0, \pm 2)\}$. The six local minima $(\pm 2, 0, 0)$, $(0, \pm 2, 0)$, $(0, 0, \pm 2)\}$ are in one $\mathrm{Br}_3 \ltimes \{\pm 1\}^3$ orbit because $\gamma((2, 0, 0)) = (0, 2, 0)$, $\gamma((0, 2, 0)) = (0, 0, 2)$.

2nd case, all $x_i \neq 0$ and $x_1x_2x_3 < 0$: $4 = r(\underline{x}) = x_1^2 + x_2^2 + x_3^2 + |x_1x_2x_3|$, so $(x_1, x_2, x_3) \in \{(-1, -1, -1), (1, 1, -1), (1, -1, 1), (-1, 1, 1)\}$. These four local minima are in one $\mathrm{Br}_3 \ltimes \{\pm 1\}^3$ orbit because of the action of G^{sign}.

3rd case, all $x_i \neq 0$ and $x_1x_2x_3 > 0$: We can suppose $x_i > 0$ for each $i \in \{1, 2, 3\}$ and $x_i \le x_j \le x_k$ for some i, j, k with $\{i, j, k\} = \{1, 2, 3\}$. As in the proof of part (a) we obtain the estimate

$$4 = \rho = r(\underline{x}) \le -(x_i - 2)^3 + 4, \quad \text{so } x_i = 2,$$

and

$$4 = \rho = r(\underline{x}) = 4 + (x_j - x_k)^2, \quad \text{so } l := x_j = x_k \ge 2.$$

For $l = 2$ the four local minima

$$(2, 2, 2), (-2, -2, 2), (-2, 2, -2), (2, -2, -2)$$

and for $l \ge 3$ the 24 local minima

$$(\varepsilon_1 2, \varepsilon_2 l, \varepsilon_3 l), (\varepsilon_1 l, \varepsilon_2 2, \varepsilon_3 l), (\varepsilon_1 l, \varepsilon_2 l, \varepsilon_3 2)$$
$$\text{with } \varepsilon_1, \varepsilon_2, \varepsilon_3 \in \{\pm 1\}, \varepsilon_1\varepsilon_2\varepsilon_3 = 1,$$

are in one $\mathrm{Br}_3 \ltimes \{\pm 1\}^3$ orbit because of the action of G^{sign} and γ.

It remains to see that local minima in different lines in the list in part (e) are in different $\mathrm{Br}_3 \ltimes \{\pm 1\}^3$ orbits. One reason is part (f). Another way to argue is given in the Remarks 4.7.

(f) See Lemma 4.10 (e). □

Remarks 4.7 Part (f) of Theorem 4.6 is strong and allows easily to see when $\mathrm{Br}_3 \ltimes \{\pm 1\}^3$ orbits are separate. Nevertheless it is also interesting to find invariants of the orbits which separate them.

Now we discuss several invariants which help to prove the claim that local minima in different lines in the list in part (e) are in different $\mathrm{Br}_3 \ltimes \{\pm 1\}^3$ orbits.

The number $r(\underline{x}) \in \mathbb{Z}$ is such an invariant. Furthermore the set $\{(0, 0, 0)\}$ is a single orbit and thus different from the orbit of $S(\mathbb{P}^2)$. Therefore the claim is true for the lines with $r \in \{0, 1, 2, 3\}$.

It remains to consider the $3+\infty$ lines with $r=4$. Certainly the reducible case $S(\mathbb{P}^1 A_1)$ is separate from the other cases, which are all irreducible. The signature of $I^{(0)}$ is an invariant. It is given in Lemma 5.7. It allows to see that the orbits of $S(\widehat{A}_2)$, $S(\mathcal{H}_{1,2})$ are different from one another and from the orbits of $S(-l,2,-l)$ for $l \geq 3$. In order to see that the orbits of $S(-l,2,-l)$ for $l \geq 3$ are pairwise different, we can offer Lemma 7.10, which in fact allows to separate all the lines with $r=4$. It considers the induced monodromy on the quotient lattice $H_{\mathbb{Z}}/\operatorname{Rad} I^{(1)}$.

Remarks 4.8 Krüger [Kr90, §12] and Cecotti-Vafa [CV93, Ch. 6.2] considered the action of the group $(G^{phi} \ltimes G^{sign}) \rtimes \langle \gamma, \gamma_2 \rangle$ with $\gamma_2 : \mathbb{R}^3 \to \mathbb{R}^3$, $(x_1,x_2,x_3) \mapsto (x_2,x_1,x_3)$, so $\langle \gamma, \gamma_2 \rangle \cong S_3$ which is slightly larger than $(G^{phi} \ltimes G^{sign}) \rtimes \langle \gamma \rangle$. Because of

$$\gamma_2\varphi_1\gamma_2^{-1} = \varphi_2,\ \gamma_2\varphi_2\gamma_2^{-1} = \varphi_1,\ \gamma_2\varphi_3\gamma_2^{-1} = \varphi_3,\ \gamma_2\gamma\gamma_2^{-1} = \gamma^{-1},$$

we have

$$\mathrm{Br}_3 \ltimes \{\pm 1\}^3\big(\gamma_2(\underline{x})\big) = \gamma_2\big(\mathrm{Br}_3 \ltimes \{\pm 1\}^3(\underline{x})\big).$$

Especially, the $\mathrm{Br}_3 \ltimes \{\pm 1\}^3$ orbit of $\underline{x}$ coincides with the $(G^{phi} \ltimes G^{sign}) \rtimes \langle \gamma, \gamma_2 \rangle$ orbit of $\underline{x}$ in the following cases:

(i) if $x_i = x_j$ for some $i \neq j$,
(ii) if $x_i = 0$ for some i (observe $\delta_3^{\mathbb{R}}\gamma\varphi_1(x_1,x_2,0) = (x_2,x_1,0)$),
(iii) if $\underline{x} = (x_1,x_2,\frac{1}{2}x_1x_2)$ with $|x_i| \geq 3$ and $|x_1| \neq |x_2|$ (observe $\varphi_3(\underline{x}) = (x_2,x_1,\frac{1}{2}x_1x_2)$).

In Lemma 4.12 24 sets C_1, ..., C_{24} of local minima are considered. The only local minima $\underline{x} \in \bigcup_{i=1}^{24} C_i$ which satisfy none of the conditions (i), (ii) and (iii) are those in $C_{16} \cup C_{22} \cup C_{24}$. Theorem 4.13 (b) shows that in these cases the $\mathrm{Br}_3 \ltimes \{\pm 1\}^3$ orbits of $\underline{x}$ and of $\gamma_2(\underline{x})$ are indeed disjoint.

Especially all orbits in the fibers $r^{-1}(\rho)$ with $\rho \in \{0,1,2,3,4\}$ contain local minima which satify (i), (ii) or (iii), so there the classifications in Theorem 4.6 and the classification by Krüger and Cecotti-Vafa coincide.

We do not know whether for $\underline{x} \in C_{16} \cup C_{22} \cup C_{24}$ and $\gamma_2(\underline{x})$ the corresponding unimodular bilinear lattices with sets of distinguished bases are isomorphic or not.

For $\underline{x}$ and $\gamma_2(\underline{x})$ in one $\mathrm{Br}_3 \ltimes \{\pm 1\}^3$ orbit they are isomorphic, see Remark 3.20 (iii).

4.3 A Classification of the $\mathrm{Br}_3 \ltimes \{\pm 1\}^3$ Orbits in $\mathbb{Z}^3$

This section refines the results of Sect. 4.2 on the braid group action on integer upper triangular 3×3 matrices. Using pseudo-graphs, it gives a classification of all orbits of Br_3 on $\mathbb{Z}^3/\{\pm 1\}^3$. Definition 4.9 makes precise what is meant here by a pseudo-graph, and it defines a pseudo-graph $\mathcal{G}(\underline{x})$ for any local minimum $\underline{x} \in \mathbb{Z}^3$.

As $\mathrm{Br}_3 \ltimes \{\pm 1\}^3$ and $(G^{phi} \ltimes \{\pm 1\}^3) \rtimes \langle\gamma\rangle$ are semidirect products with normal subgroups $\{\pm 1\}^3$, the groups Br_3 and $G^{phi} \rtimes \langle\gamma\rangle$ act on $\mathbb{Z}^3/\{\pm 1\}^3$.

Definition 4.9 (a) For any set $\mathcal{V}$, $\mathcal{P}(\mathcal{V})$ denotes its power set, so the set of all its subsets, and $\mathcal{P}_k(\mathcal{V})$ for some $k \in \mathbb{N}$ denotes the set of all subsets with k elements. We will use only $\mathcal{P}_1(\mathcal{V})$ and $\mathcal{P}_2(\mathcal{V})$.

(b) A *pseudo-graph* is here a tuple $\mathcal{G} = (\mathcal{V}, \mathcal{V}_0, \mathcal{V}_1, \mathcal{V}_2, v_0, \mathcal{E}_1, \mathcal{E}_2, \mathcal{E}_3, \mathcal{E}_\gamma)$ with the following ingredients:

$\mathcal{V}$ is a non-empty finite or countably infinite set of vertices.

$\mathcal{V}_0, \mathcal{V}_1, \mathcal{V}_2 \subset \mathcal{V}$ are pairwise disjoint subsets, $\mathcal{V}_0$ is not empty (the sets $\mathcal{V}_1$ and $\mathcal{V}_2$ may be empty, the union $\mathcal{V}_0 \cup \mathcal{V}_1 \cup \mathcal{V}_2$ can be equal to $\mathcal{V}$ or a proper subset of $\mathcal{V}$).

$v_0 \in \mathcal{V}_0$ is a distinguished vertex in $\mathcal{V}_0$.

$\mathcal{E}_1, \mathcal{E}_2, \mathcal{E}_3 \subset \mathcal{P}_1(\mathcal{V}) \cup \mathcal{P}_2(\mathcal{V})$ are sets of undirected edges. A subset of $\mathcal{V}$ with two elements means an edge between the two vertices. A subset of $\mathcal{V}$ with one element means a loop from the vertex to itself.

$\mathcal{E}_\gamma = \{(v_0, v_1), (v_2, v_0)\}$ for some $v_1, v_2 \in \mathcal{V}_0$ is a set of two or one directed edges, one only if $v_1 = v_2 = v_0$, and then it is a directed loop.

(c) An isomorphism between two pseudo-graphs $\mathcal{G}$ and $\widetilde{\mathcal{G}}$ is a bijection $\phi : \mathcal{V} \to \widetilde{\mathcal{V}}$ with $\phi(v_0) = \widetilde{v_0}$ which induces bijections $\phi : \mathcal{V}_i \to \widetilde{\mathcal{V}_i}$, $\phi : \mathcal{E}_j \to \widetilde{\mathcal{E}_j}$ and $\phi : \mathcal{E}_\gamma \to \widetilde{\mathcal{E}_\gamma}$.

(d) $\mathcal{G}|_{\mathcal{V}_0 \cup \mathcal{V}_1}$ denotes the restriction of a pseudo-graph $\mathcal{G}$ to the vertex set $\mathcal{V}_0 \cup \mathcal{V}_1$, so one deletes all vertices in $\mathcal{V} - (\mathcal{V}_0 \cup \mathcal{V}_1)$ and all edges with at least one end in $\mathcal{V} - (\mathcal{V}_0 \cup \mathcal{V}_1)$. Analogously, $\mathcal{G}|_{\mathcal{V}_0}$ denotes the restriction of a pseudo-graph $\mathcal{G}$ to the vertex set $\mathcal{V}_0$.

(e) Define

$$\begin{aligned}
\mathcal{L}_0 &:= \{\underline{x}/\{\pm 1\}^3 \mid \underline{x} \in \mathbb{Z}^3 \text{ is a local minimum}\}, \\
\mathcal{L}_1 &:= \{\underline{y}/\{\pm 1\}^3 \mid \underline{y} \in \mathbb{Z}^3 \text{ with } |y_i| = 1 \text{ for some } i\} - \mathcal{L}_0, \\
\mathcal{L}_2 &:= \mathbb{Z}^3/\{\pm 1\}^3 - (\mathcal{L}_0 \cup \mathcal{L}_1).
\end{aligned}$$

(f) A pseudo-graph $\mathcal{G}(\underline{x})$ is associated to a local minimum $\underline{x} \in \mathbb{Z}^3$ in the following way:

$$\begin{aligned}
\mathcal{V} &:= \mathrm{Br}_3(\underline{x}/\{\pm 1\}^3) \subset \mathbb{Z}^3/\{\pm 1\}^3, \\
\mathcal{V}_0 &:= \mathcal{V} \cap \mathcal{L}_0 \text{ is the set of sign classes in } \mathcal{V} \text{ of local minima}, \\
\mathcal{V}_1 &:= \mathcal{V} \cap \mathcal{L}_1, \\
\mathcal{V}_2 &:= \{w \in \mathcal{V} \cap \mathcal{L}_2 \mid \text{an } i \in \{1, 2, 3\} \text{ with } \varphi_i(w) \in \mathcal{V}_0 \cup \mathcal{V}_1 \text{ exists}\}, \\
v_0 &:= \underline{x}/\{\pm 1\}^3, \\
\mathcal{E}_i &:= \{\{w, \varphi_i(w)\} \mid w \in \mathcal{V}\} \quad \text{for } i \in \{1, 2, 3\}, \\
\mathcal{E}_\gamma &:= \{(v_0, \gamma(v_0)), (\gamma^{-1}(v_0), v_0)\}.
\end{aligned}$$

(g) An *infinite tree* $(\mathcal{W}, \mathcal{F})$ consists of a countably infinite set $\mathcal{W}$ of vertices and a set $\mathcal{F} \subset \mathcal{P}_2(\mathcal{W})$ of undirected edges such that the graph is connected and has no cycles. A $(2, \infty \times 3)$-tree is an infinite tree with a distinguished vertex with two neighbours such that any other vertex has three neighbours.

The next lemma gives already structural results about the pseudo-graphs $\mathcal{G}(\underline{x})$ for the local minima $\underline{x} \in \mathbb{Z}^3$. Theorem 4.13 and the Remarks 4.14 will give a complete classification of all isomorphism classes of pseudo-graphs $\mathcal{G}(\underline{x})$ for the local minima $\underline{x} \in \mathbb{Z}^3$.

Lemma 4.10 *Let $\underline{x} \in \mathbb{Z}^3$ be a local minimum with pseudo-graph $\mathcal{G}(\underline{x}) = (\mathcal{V}, \mathcal{V}_0, \mathcal{V}_1, \mathcal{V}_2, v_0, \mathcal{E}_1, \mathcal{E}_2, \mathcal{E}_3, \mathcal{E}_\gamma)$.*

(a) ($\underline{x}$ and $\mathcal{G}(\underline{x})$ are not used in part (a)) For $w \in \mathcal{L}_2$, there are i, j, k with $\{i, j, k\} = \{1, 2, 3\}$ and $\|\varphi_i(w)\| < \|w\|$, $\|\varphi_j(w)\| > \|w\|$, $\|\varphi_k(w)\| > \|w\|$, $\varphi_j(w) \in \mathcal{L}_2$, $\varphi_k(w) \in \mathcal{L}_2$ and $\varphi_j(w) \neq \varphi_k(w)$.

(b) Let $w \in \mathcal{V}_2$, so especially $w \in \mathcal{L}_2$. Choose i, j, k as in part (a). The edge which connects $w \in \mathcal{V}_2$ with $\mathcal{V}_0 \cup \mathcal{V}_1$ is in $\mathcal{E}_i$. After deleting this edge, the component of the remaining pseudo-graph which contains w is a $(2, \infty \times 3)$-tree with distinguished vertex w and all vertices in $\mathcal{L}_2$.

(c) The pseudo-graph $\mathcal{G}(\underline{x})$ is connected.

(d) The pseudo-graph $\mathcal{G}(\underline{x})|_{\mathcal{V}_0 \cup \mathcal{V}_1}$ is connected.

(e) The pseudo-graph $\mathcal{G}(\underline{x})|_{\mathcal{V}_0 \cup \mathcal{V}_1}$ is finite, and

$$\mathcal{V}_0 = \langle \gamma \rangle (v_0) \quad \text{or} \quad \mathcal{V}_0 = \langle \gamma, \gamma_2 \rangle (v_0).$$

Proof (a) Consider $w = \underline{y}/\{\pm 1\}^3 \in \mathcal{L}_2$. Then $|y_i| \geq 2$ for $i \in \{1, 2, 3\}$ because $w \notin \mathcal{L}_0 \cup \mathcal{L}_1$. $w \notin \mathcal{L}_0$ implies $y_1 y_2 y_3 > 0$. We can suppose $y_1, y_2, y_3 \in \mathbb{Z}_{\geq 2}$. Observe for i, j, k with $\{i, j, k\} = \{1, 2, 3\}$

$$\begin{aligned} &\|\varphi_j(\underline{y})\|^2 - \|\varphi_i(\underline{y})\|^2 \\ &= (y_i^2 + (y_i y_k - y_j)^2 + y_k^2) - ((y_j y_k - y_i)^2 + y_j^2 + y_k^2) \\ &= (y_i^2 - y_j^2) y_k^2. \end{aligned}$$

Consider the case $2 \leq y_1 \leq y_2 \leq y_3$. The other cases are analogous. Then

$$\|\varphi_3(\underline{y})\| \leq \|\varphi_2(\underline{y})\| \leq \|\varphi_1(\underline{y})\|.$$

Because $w \notin \mathcal{L}_0$ $\|\varphi_3(\underline{y})\| < \|\underline{y}\|$. Also $\varphi_2(\underline{y}) = (y_3, y_1 y_3 - y_2, y_1)$ with

$$\varphi_2(\underline{y})_2 = y_1 y_3 - y_2 \geq (y_1 - 1) y_3 \begin{cases} \geq 2y_3 > y_2 & \text{if } y_1 > 2, \\ = y_3 > y_2 & \text{if } y_1 = 2. \end{cases}$$

Here $y_3 > y_2$ if $y_1 = 2$ because $\underline{y} = (2, y_2, y_3)$ is not a local minimum.

Therefore $\|\varphi_2(\underline{y})\| > \|\underline{y}\|$, and also $\|\varphi_1(\underline{y})\| \geq \|\varphi_2(\underline{y})\| > \|\underline{y}\|$. Especially $\varphi_2(\underline{y})$ and $\varphi_1(\underline{y})$ are not local minima, so $\varphi_2(w) \notin \mathcal{L}_0$ and $\varphi_1(w) \notin \mathcal{L}_0$.

Obviously $\varphi_2(\underline{y})_i \geq 2$ and $\varphi_1(\underline{y})_i \geq 2$ for $i \in \{1, 2, 3\}$, so $\varphi_2(w) \in \mathcal{L}_2$ and $\varphi_1(w) \in \mathcal{L}_2$. The inequality $\varphi_2(w) \neq \varphi_1(w)$ follows from

$$\varphi_2(\underline{y})_2 \geq 2y_3 > y_3 = \varphi_1(\underline{y})_2 \text{ if } y_1 > 2,$$
$$\varphi_2(\underline{y})_2 = y_1 y_3 - y_2 > (y_1 - 1)y_3 = y_3 = \varphi_1(\underline{y})_2 \text{ if } y_1 = 2.$$

(b) Because $\varphi_j(w), \varphi_k(w) \in \mathcal{L}_2$, the edge which connects w to $\mathcal{V}_0 \cup \mathcal{V}_1$ cannot be in $\mathcal{E}_j$ or $\mathcal{E}_k$, so it is in $\mathcal{E}_i$.

Using part (a) again and again, one sees that the component of $\mathcal{G}(\underline{x})$ – (this edge) which contains w is a $(2, \infty \times 3)$ tree.

(c) Any vertex $w \in \mathcal{V} = (G^{phi} \rtimes \langle\gamma\rangle)(v_0)$ is obtained from v_0 by applying an element $\psi\gamma^\xi$ with $\psi \in G^{phi}$ and $\xi \in \{0, \pm 1\}$. As G^{phi} is a free Coxeter group with generators $\varphi_1, \varphi_2, \varphi_3$, applying $\psi\gamma^\xi$ to v_0 yields a path in $\mathcal{G}(\underline{x})$ from v_0 to w.

(d) This follows from (b) and (c).

(e) Consider $w = \underline{y}/\{\pm 1\}^3 \in \mathcal{V}_1$. Because $w \notin \mathcal{V}_0$, $y_1 y_2 y_3 > 0$. We can suppose $y_1, y_2, y_3 \in \mathbb{N}$, and one of them is equal to 1. Suppose $1 = y_1 \leq y_2 \leq y_3$. Then

$$\begin{aligned} \varphi_3(\underline{y}) &= (y_2, 1, y_2 - y_3), \text{ so } y_2 \cdot 1 \cdot (y_2 - y_3) \leq 0, \\ &\quad \text{so } \varphi_3(w) \in \mathcal{V}_0, \end{aligned} \tag{4.8}$$
$$\begin{aligned} \varphi_2(\underline{y}) &= (y_3, y_3 - y_2, 1), \\ &\quad \text{so } \varphi_2(w) \in \mathcal{V}_0 \cup \mathcal{V}_1 \text{ (in } \mathcal{V}_0 \text{ only if } y_3 = y_2), \end{aligned} \tag{4.9}$$
$$\begin{aligned} \varphi_1\varphi_2(\underline{y}) &= (-y_2, 1, y_3 - y_2), \\ &\quad \text{so } \varphi_1\varphi_2(w) = \varphi_3(w) \in \mathcal{V}_0. \end{aligned} \tag{4.10}$$

Especially, each vertex in $\mathcal{V}_1$ is connected by an edge to a vertex in $\mathcal{V}_0$.

Therefore the main point is to show $\mathcal{V}_0 = \langle\gamma\rangle(v_0)$ or $\mathcal{V}_0 = \langle\gamma, \gamma_2\rangle(v_0)$. Then $\mathcal{V}_0$ and $\mathcal{V}_1$ are finite.

First case, the restricted pseudo-graph $\mathcal{G}(\underline{x})|_{\mathcal{V}_0}$ is connected: Then $\|w\| = \|v_0\|$ for each $w \in \mathcal{V}_0$. This easily implies $\mathcal{V}_0 = \langle\gamma\rangle(v_0)$ or $\mathcal{V}_0 = \langle\gamma, \gamma_2\rangle$.

Second case, the restricted pseudo-graph $\mathcal{G}(\underline{x})|_{\mathcal{V}_0}$ is not connected: We will lead this to a contradiction. Within all paths in $\mathcal{G}(\underline{x})|_{\mathcal{V}_0 \cup \mathcal{V}_1}$ which connect vertices in different components of $\mathcal{G}(\underline{x})|_{\mathcal{V}_0}$, consider a shortest path. It does not contain an edge in $\mathcal{E}_\gamma$, because else one could go over to a path of the same length with an edge in $\mathcal{E}_\gamma$ at one end and between the same vertices, but dropping that edge would lead to a shorter path. Because each vertex in $\mathcal{V}_1$ is connected by an edge to a vertex in $\mathcal{V}_0$, a shortest path contains either one or two vertices in $\mathcal{V}_1$. The observations (4.8)–(4.10) lead in both cases to the vertices at the end of the path being in the same component of $\mathcal{G}(\underline{x})|_{\mathcal{V}_0}$, so to a contradiction. □

Examples 4.11 The following 14 figures show the pseudo-graphs $\mathcal{G}_1, \ldots, \mathcal{G}_{14}$ for 14 values $v_0 = \underline{x}/\{\pm 1\}^3$ with $\underline{x} \in \mathbb{Z}^3$ a local minimum. The ingredients of the figures have the following meaning.

- $\bullet$ a vertex in $\mathcal{V}_0$,
- $\otimes$ a vertex in $\mathcal{V}_1$,
- ♉ a vertex in $\mathcal{V}_2$ together with the $(2, \infty \times 3)$ tree (compare Lemma 4.10 (b)),
- $\overset{i}{\text{---}}$ an edge in $\mathcal{E}_i$,
- $\overset{\gamma}{\rightarrow}$ an edge in $\mathcal{E}_\gamma$.

The pseudo-graphs are enriched in the following way. Each vertex is labeled with a value $\underline{y}$ of its sign class $\underline{y}/\{\pm 1\}^3$. We have chosen $\underline{y} \in \mathbb{N}^3$ if $y_1 y_2 y_3 > 0$ (this holds for all $\underline{y}/\{\pm 1\}^3 \in \mathcal{V}_1 \cup \mathcal{V}_2$ and some $\underline{y}/\{\pm 1\}^3 \in \mathcal{V}_0$) and $\underline{y} \in \mathbb{Z}^3_{\leq 0}$ if $y_1 y_2 y_3 \leq 0$ (this holds for some $\underline{y}/\{\pm 1\}^3 \in \mathcal{V}_0$). The vertex v_0 can be recognized by the edges in $\mathcal{E}_\gamma$ leading to and from it. The sets C_i are defined in Lemma 4.12. The relations $\mathcal{G}_j : C_i$ are explained in Theorem 4.13 (Figs. 4.2, 4.3, 4.4, 4.5, 4.6, 4.7, 4.8, 4.9, 4.10, 4.11, 4.12, 4.13, 4.14, 4.15).

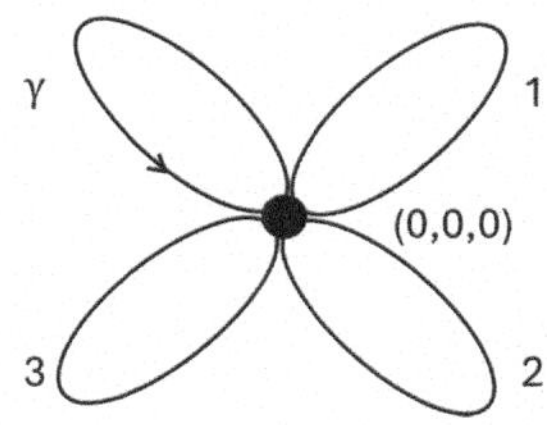

Fig. 4.2 G_1 : C_1 (A_1^3), C_2 $(\mathcal{H}_{1,2})$. Here $\underline{x} = (0, 0, 0) \in C_1$

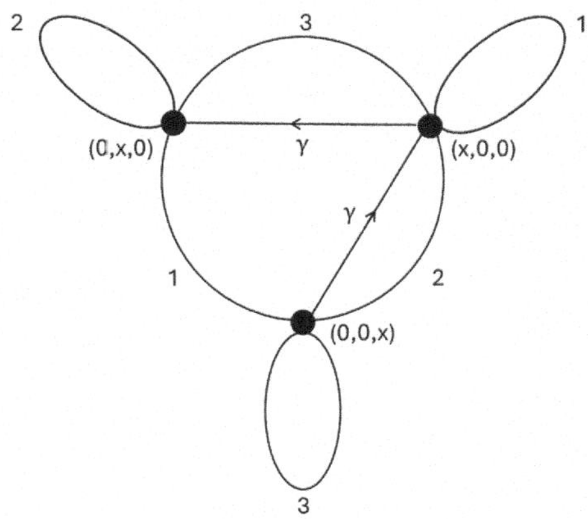

Fig. 4.3 G_2 : C_3, C_4, C_5, so the reducible cases without A_1^3. Here $\underline{x} = (x, 0, 0) \in C_3 \cup C_4 \cup C_5$, $x < 0$

Fig. 4.4 $G_3 : C_6\ (A_3)$. Here $\underline{x} = (-1, 0, -1)$

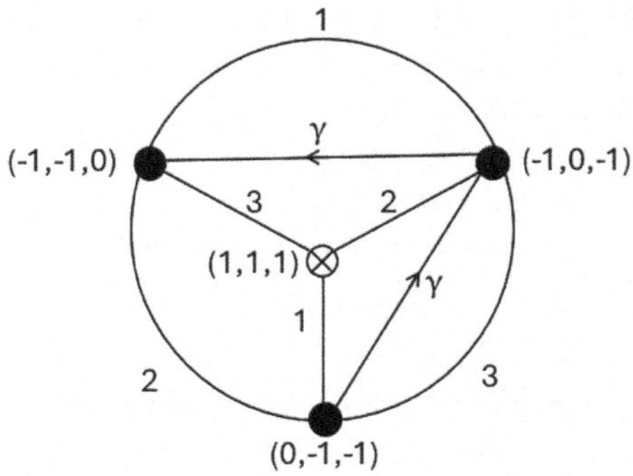

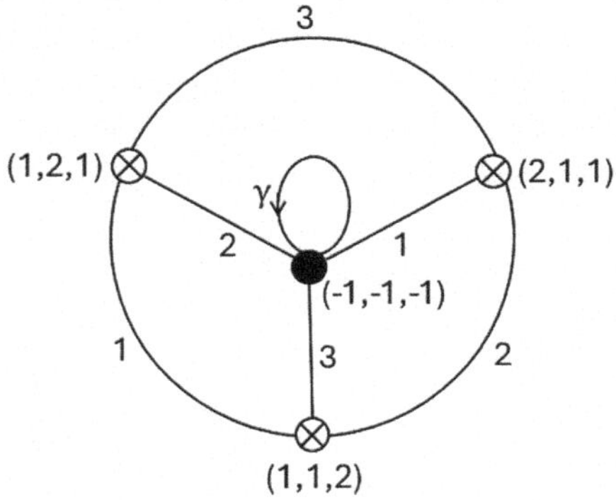

Fig. 4.5 $G_4 : C_7\ (\widehat{A}_2)$. Here $\underline{x} = (-1, -1, -1)$

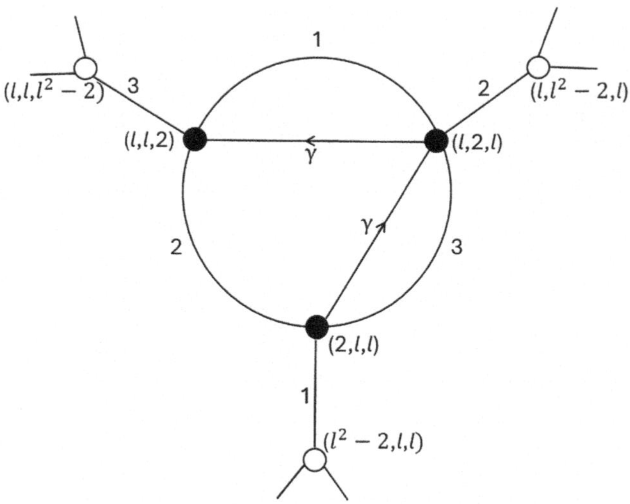

Fig. 4.6 $G_5 : C_8, C_9$. Here $\underline{x} = (l, 2, l) \sim (-l, 2, -l)$ with $l \geq 3$

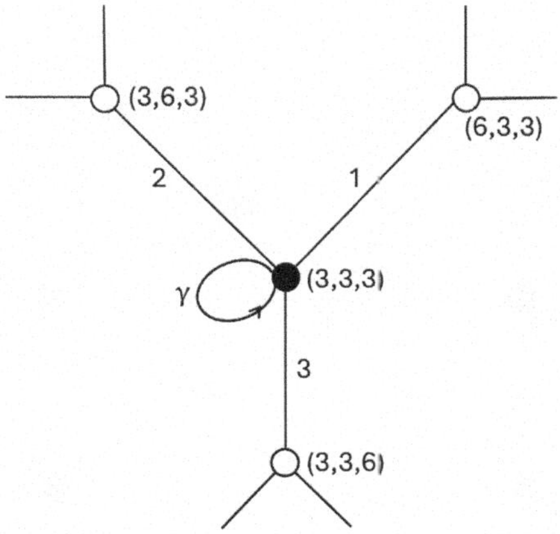

Fig. 4.7 $G_6 : C_{10}, C_{11}, C_{12}$. Here $\underline{x} = (3, 3, 3) \in C_{10}$

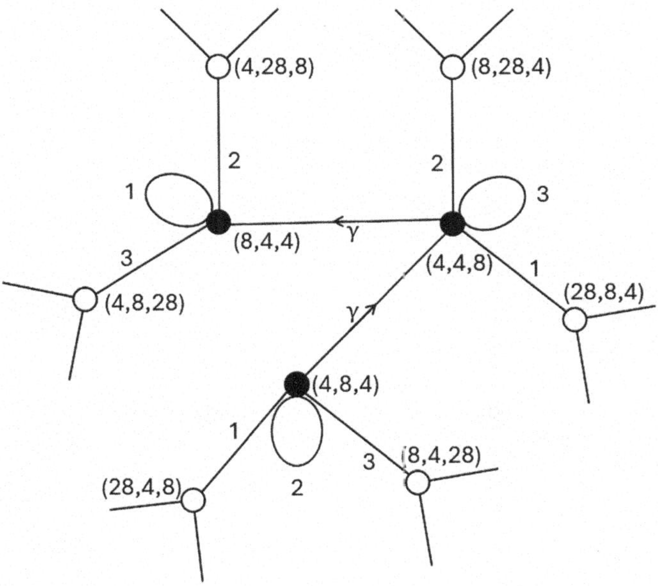

Fig. 4.8 $G_7 : C_{13}$. Here $\underline{x} = (4, 4, 8)$

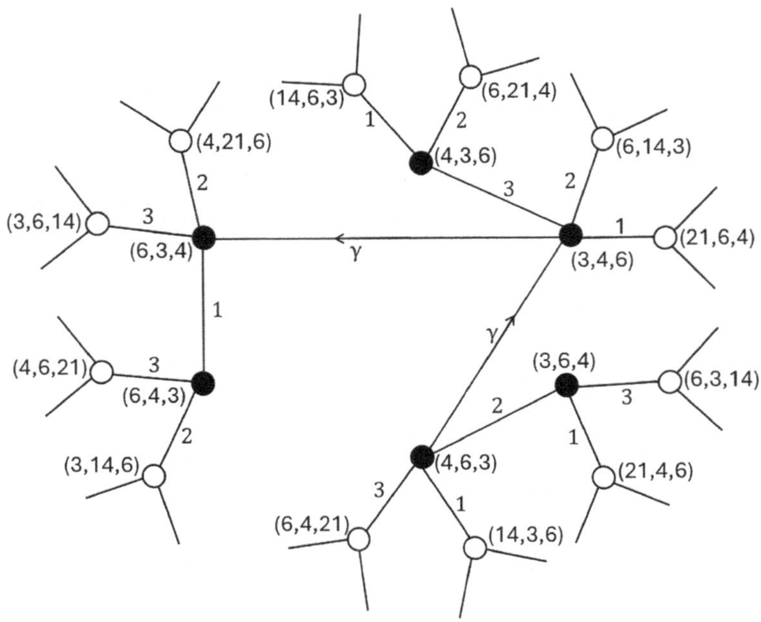

Fig. 4.9 $G_8 : C_{14}$. Here $\underline{x} = (3, 4, 6)$

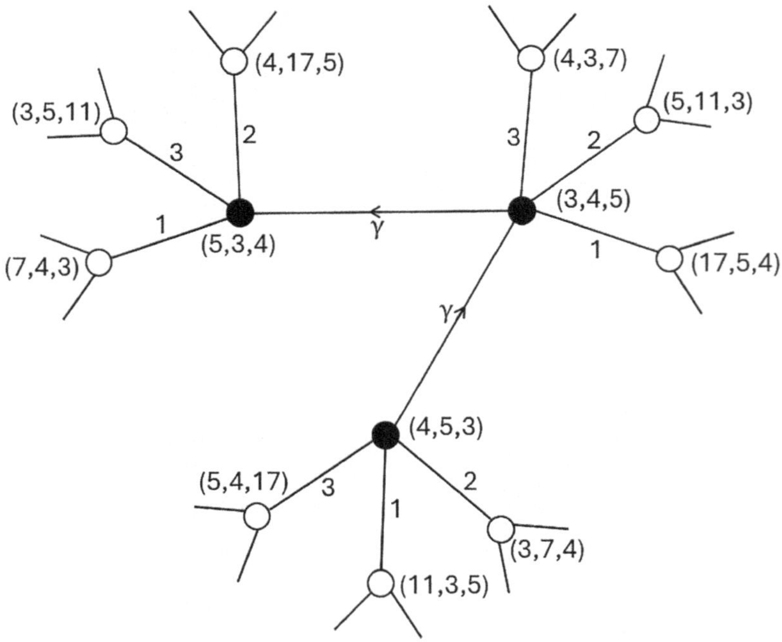

Fig. 4.10 $G_9 : C_{15}, C_{16}, C_{23}, C_{24}$. Here $\underline{x} = (3, 4, 5) \in C_{16}$

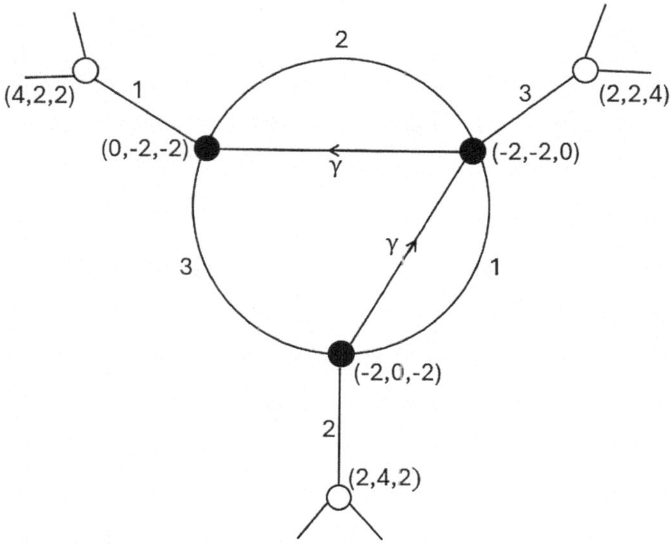

Fig. 4.11 G_{10} : C_{17}. Here $\underline{x} = (-2, -2, 0)$

Fig. 4.12 G_{11} : C_{18}. Here $\underline{x} = (-3, -2, 0)$

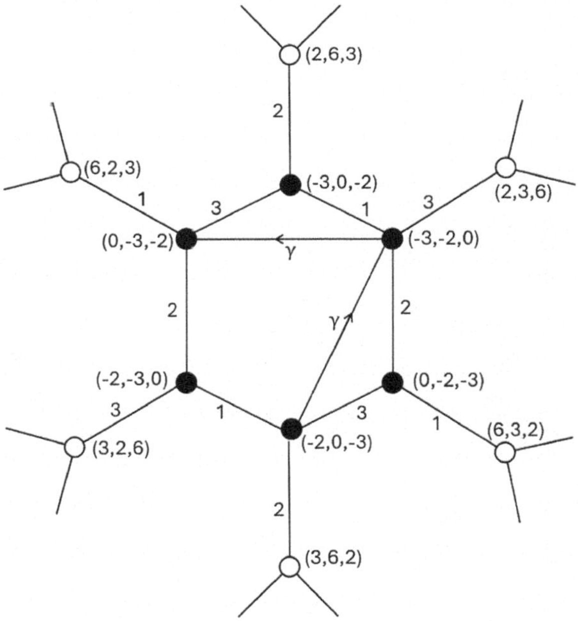

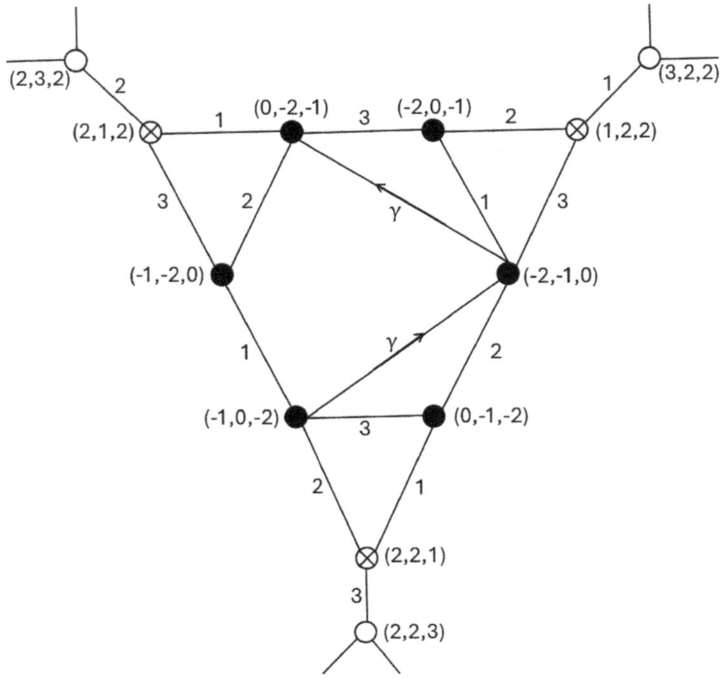

Fig. 4.13 $G_{12} : C_{19}$. Here $\underline{x} = (-2, -1, 0)$

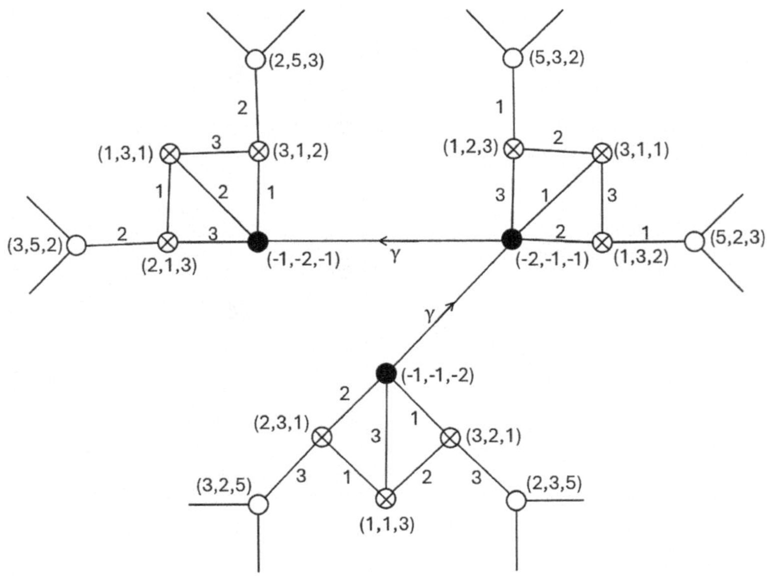

Fig. 4.14 $G_{13} : C_{20}$. Here $\underline{x} = (-2, -1, -1)$

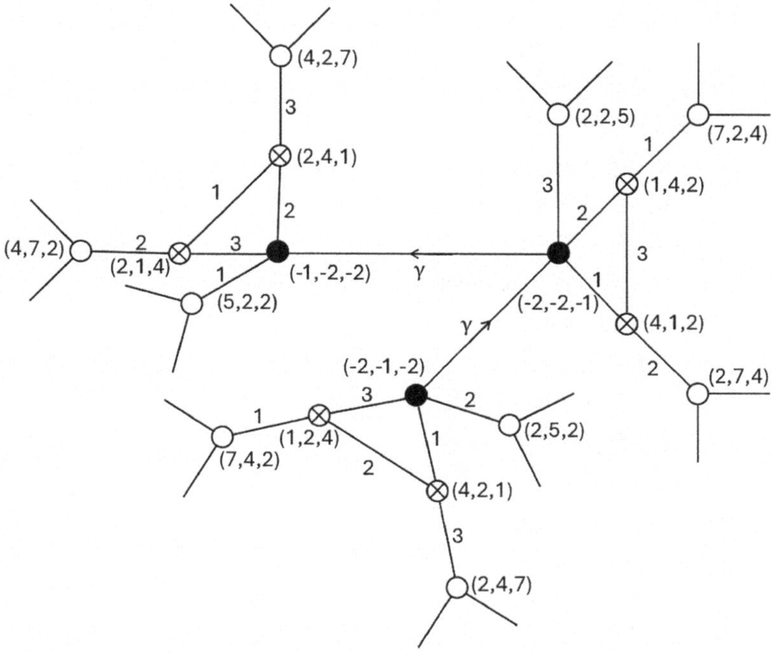

Fig. 4.15 G_{14} : C_{21}, C_{22}. Here $\underline{x} = (-2, -2, -1) \in C_{21}$

Lemma 4.12 *Consider the following 24 sets C_i, $i \in \{1, 2, ..., 24\}$, of triples in $\mathbb{Z}^3$.*

$$
\begin{aligned}
C_1 &= \{(0,0,0)\} \quad (A_1^3),\\
C_2 &= \{(2,2,2)\} \quad (\mathcal{H}_{1,2}),\\
C_3 &= \{(-1,0,0)\} \quad (A_2A_1),\\
C_4 &= \{(-2,0,0)\} \quad (\mathbb{P}^1 A_1),\\
C_5 &= \{(x,0,0) \mid x \in \mathbb{Z}_{\le -3}\}\\
&\qquad \text{(the reducible cases without } A_1^3, A_2A_1, \mathbb{P}^1A_1),\\
C_6 &= \{(-1,0,-1)\} \quad (A_3),\\
C_7 &= \{(-1,-1,-1)\} \quad (\widehat{A}_2),\\
C_8 &= \{(-l,2,-l) \mid l \in \mathbb{Z}_{\ge 3} \text{ odd}\},\\
C_9 &= \{(-l,2,-l) \mid l \in \mathbb{Z}_{\ge 4} \text{ even}\},\\
C_{10} &= \{(3,3,3)\} \quad (\mathbb{P}^2),\\
C_{11} &= \{(x,x,x) \mid x \in \mathbb{Z}_{\ge 4}\},\\
C_{12} &= \{(x,x,x) \mid x \in \mathbb{Z}_{\le -2}\},\\
C_{13} &= \{(2y,2y,2y^2) \mid y \in \mathbb{Z}_{\ge 2}\},\\
C_{14} &= \{(x_1,x_2,\tfrac{1}{2}x_1x_2 \mid 3 \le x_1 < x_2, x_1x_2 \text{ even}\},\\
C_{15} &= \{(x_1,x_1,x_3) \mid 3 \le x_1 < x_3 < \tfrac{1}{2}x_1^2\}\\
&\quad \cup \{(x_1,x_2,x_2) \mid 3 \le x_1 < x_2\},\\
C_{16} &= \{(x_1,x_2,x_3) \mid 3 \le x_1 < x_2 < x_3 < \tfrac{1}{2}x_1x_2\}\\
&\quad \cup \{(x_1,x_2,x_3) \mid 3 \le x_2 < x_1 < x_3 < \tfrac{1}{2}x_1x_2\},\\
C_{17} &= \{(x,x,0) \mid x \in \mathbb{Z}_{\le -2}\},\\
C_{18} &= \{(x_1,x_2,0) \mid x_1 < x_2 \le -2\},\\
C_{19} &= \{(x,-1,0) \mid x \in \mathbb{Z}_{\le -2}\},\\
C_{20} &= \{(x,-1,-1) \mid x \in \mathbb{Z}_{\le -2}\},\\
C_{21} &= \{(x,x,-1) \mid x \in \mathbb{Z}_{\le -2}\},\\
C_{22} &= \{(x_1,x_2,-1) \mid x_1 < x_2 \le -2\}\\
&\quad \cup \{(x_1,x_2,-1) \mid x_2 < x_1 \le -2\},\\
C_{23} &= \{(x_1,x_1,x_3) \mid x_1 < x_3 \le -2\}\\
&\quad \cup \{(x_1,x_2,x_2) \mid x_1 < x_2 \le -2\},\\
C_{24} &= \{(x_1,x_2,x_3) \mid x_1 < x_2 < x_3 \le -2\}\\
&\quad \cup \{(x_1,x_2,x_3) \mid x_2 < x_1 < x_3 \le -2\}.
\end{aligned}
$$

(a) Each triple in $\bigcup_{i=1}^{24} C_i$ is a local minimum. All $\underline{x}$ in one set C_i have the value in the following table or satisfy the inequality in the following table,

ρ	i with $r(\underline{x}) = \rho$ for $\underline{x} \in C_i$
0	1, 10
1	3
2	6
4	2, 4, 7, 8, 9
< 0	11, 13, 14, 15, 16
> 4	5, 12, 17, 18, 19, 20, 21, 22, 23, 24

(b) The following table makes statements about the $\langle\gamma\rangle$ orbits and the $\langle\gamma, \gamma_2\rangle$ orbits of $v_0 = \underline{x}/\{\pm 1\}^3$ with $\underline{x} \in \bigcup_{i=1}^{24} C_i$,

$i \in \{1, 2, 7, 10, 11, 12\}$	$\langle\gamma\rangle(v_0)$ is γ_2-invariant and has size 1
$i \in \{3, 4, 5, 6, 8, 9, 13, 15, 17, 20, 21, 23\}$	$\langle\gamma\rangle(v_0)$ is γ_2-invariant and has size 3
$i \in \{14, 16, 18, 19, 22, 24\}$	$\langle\gamma\rangle(v_0)$ is not γ_2-invariant and has size 3, $\langle\gamma, \gamma_2\rangle(v_0)$ has size 6

(c) The set of all local minima in $\mathbb{Z}^3$ is the following disjoint union,

$$\Big(\dot{\bigcup}_{i\in\{1,\dots,24\}-\{14,18,19\}} \dot{\bigcup}_{\underline{x}\in C_i} (G^{sign} \rtimes \langle\gamma\rangle)\{\underline{x}\}\Big)$$
$$\dot{\cup} \Big(\dot{\bigcup}_{i\in\{14,18,19\}} \dot{\bigcup}_{\underline{x}\in C_i} (G^{sign} \rtimes \langle\gamma, \gamma_2\rangle)\{\underline{x}\}\Big).$$

Proof Part (b) is trivial. The parts (a) and (c) follow with the characterization of local minima in Lemma 4.4 and Theorem 4.6 (c)–(e). □

Theorem 4.13 *(a)* $\mathbb{Z}^3$ *is the disjoint union*

$$\dot{\bigcup}_{i\in\{1,\dots,24\}}\dot{\bigcup}_{\underline{x}\in C_i}(\mathrm{Br}_3\ltimes\{\pm 1\}^3)(\underline{x}).$$

(b) For $v_0=\underline{x}/\{\pm 1\}^3$ *with* $\underline{x}\in\bigcup_{i=1}^{24}C_i$*, the set* $\mathcal{V}_0=\mathrm{Br}_3(v_0)\cap\mathcal{L}_0$ *of sign classes of local minima in the* $\mathrm{Br}_3\ltimes\{\pm 1\}^3$ *orbit of* $\underline{x}$ *is as follows,*

$$\mathcal{V}_0=\langle\gamma\rangle(v_0)\quad\text{if } i\in\{1,\dots,24\}-\{14,18,19\},$$
$$\mathcal{V}_0=\langle\gamma,\gamma_2\rangle(v_0)\qquad\text{if } i\in\{14,18,19\}.$$

(c) The set $\{\mathcal{G}(\underline{x})\,|\,\underline{x}\in\bigcup_{i=1}^{24}C_i\}$ *of pseudo-graphs* $\mathcal{G}(\underline{x})$ *for* $\underline{x}\in\bigcup_{i=1}^{24}C_i$ *consists of the 14 isomorphism classes* $\mathcal{G}_1,\dots,\mathcal{G}_{14}$ *in the Examples 4.11. All* $\underline{x}$ *in one set* C_i *have the same pseudo-graph. The first and second column in the following table give for each of the 14 pseudo-graphs* $\mathcal{G}_j$ *the set or the sets* C_i *with* $\mathcal{G}(\underline{x})=\mathcal{G}_j$ *for* $\underline{x}\in C_i$*. The third and fourth column in the following table are subject of Theorem 4.16.*

	sets	$(G^{phi}\rtimes\langle\gamma\rangle)_{\underline{x}/\{\pm 1\}^3}$	$(\mathrm{Br}_3)_{\underline{x}/\{\pm 1\}^3}$
$\mathcal{G}_1$	$C_1\ (A_1^3),\ C_2\ (\mathcal{H}_{1,2})$	$G^{phi}\rtimes\langle\gamma\rangle$	Br_3
$\mathcal{G}_2$	C_3, C_4, C_5 (red. cases)	$\langle\varphi_1,\gamma^{-1}\varphi_3\rangle$	$\langle\sigma_1,\sigma_2^2\rangle$
$\mathcal{G}_3$	$C_6\ (A_3)$	$\langle\gamma\varphi_3,\varphi_2\varphi_1\varphi_3\rangle$	$\langle\sigma_1\sigma_2,\sigma_1^3\rangle$
$\mathcal{G}_4$	$C_7\ (\widehat{A}_2)$	$\langle\gamma,\varphi_2\varphi_1\varphi_3\rangle$	$\langle\sigma_2\sigma_1,\sigma_1^3\rangle$
$\mathcal{G}_5$	$C_8,\ C_9\ ((-l,2,-l))$	$\langle\gamma^{-1}\varphi_1\rangle$	$\langle\sigma^{mon},\sigma_1^{-1}\sigma_2^{-1}\sigma_1\rangle$
$\mathcal{G}_6$	$C_{10}(\mathbb{P}^2),\ C_{11},\ C_{12}$	$\langle\gamma\rangle$	$\langle\sigma_2\sigma_1\rangle$
$\mathcal{G}_7$	C_{13} (e.g. $(4,4,8)$)	$\langle\varphi_3\rangle$	$\langle\sigma_2\sigma_1^2\rangle$
$\mathcal{G}_8$	C_{14}(e.g. $(3,4,6)$)	$\langle\mathrm{id}\rangle$	$\langle\sigma^{mon}\rangle$
$\mathcal{G}_9$	$C_{15},\ C_{16},\ C_{23},\ C_{24}$	$\langle\mathrm{id}\rangle$	$\langle\sigma^{mon}\rangle$
$\mathcal{G}_{10}$	C_{17} (e.g. $(-2,-2,0)$)	$\langle\gamma^{-1}\varphi_2\rangle$	$\langle\sigma^{mon},\sigma_2\rangle$
$\mathcal{G}_{11}$	C_{18} (e.g. $(-3,-2,0)$)	$\langle\gamma^{-1}\varphi_3\varphi_1\rangle$	$\langle\sigma^{mon},\sigma_2^2\rangle$
$\mathcal{G}_{12}$	C_{19} (e.g. $(-2,-1,0)$)	$\langle\gamma^{-1}\varphi_3\varphi_1,\varphi_3\varphi_2\varphi_1\rangle$	$\langle\sigma^{mon},\sigma_2^2,\sigma_2\sigma_1^3\sigma_2^{-1}\rangle$
$\mathcal{G}_{13}$	C_{20} (e.g. $(-2,-1,-1)$)	$\langle\varphi_2\varphi_3\varphi_1,\varphi_3\varphi_2\varphi_1\rangle$	$\langle\sigma^{mon},\sigma_2^3,\sigma_2\sigma_1^3\sigma_2^{-1}\rangle$
$\mathcal{G}_{14}$	$C_{21},\ C_{22}$ (e.g. $(-2,-2,-1)$)	$\langle\varphi_2\varphi_3\varphi_1\rangle$	$\langle\sigma^{mon},\sigma_2^3\rangle$

Proof We start with part (c). It can be seen rather easily for all $\underline{x}$ in one family C_i simultaneously. We do not give more details.

(b) The pseudo-graphs G_8, G_{11} and G_{12} are the only of the 14 pseudo-graphs with $|\mathcal{V}_0|=6$. By inspection of them or by Lemma 4.10 (e), for them $\mathcal{V}_0=\langle\gamma,\gamma_2\rangle(v_0)$. The table in part (c) gives the correspondence $G_8\leftrightarrow C_{14}$, $G_{11}\leftrightarrow C_{18}$, $G_{12}\leftrightarrow C_{19}$. The other 11 pseudo-graphs satisfy $|\mathcal{V}_0|=1$ or $|\mathcal{V}_0|=3$, so in any case $\mathcal{V}_0=\langle\gamma\rangle(v_0)$.

(a) Part (c) of Lemma 4.12 alone shows already that $\mathbb{Z}^3$ is the union given. Part (b) of Theorem 4.13 adds only the fact that this is a disjoint union. □

Remarks 4.14 (i) We have 14 pseudo-graphs $\mathcal{G}_1,\dots,\mathcal{G}_{14}$, but 24 sets $C_1,\dots,C_{24}$ because the separation into the sets shall be fine enough for the table in Theorem 4.13 (c) and the tables in Lemma 4.12 (a) and (b).

(ii) In the pseudo-graphs $\mathcal{G}_j$ with $|\mathcal{V}_0| = 3$ or $|\mathcal{V}_0| = 6$ one can choose another distinguished vertex $\widetilde{v_0} \in \mathcal{V}_0$ and change the set $\mathcal{E}_\gamma$ to a set $\widetilde{\mathcal{E}_\gamma}$ accordingly. This gives a pseudo-graph $\widetilde{\mathcal{G}_j}$ which is not equal to $\mathcal{G}_j$, but closely related.

The graphs $\mathcal{G}_5$ and $\mathcal{G}_{10}$ are related by such a change.

The following table shows for each $\mathcal{G}_j$ except $\mathcal{G}_{10}$ the number of isomorphism classes of pseudo-graphs obtained in this way (including the original pseudo-graphs). In the cases $\mathcal{G}_8$ and $\mathcal{G}_{11}$ there is $|\mathcal{V}_0| = 6$, but because of some symmetry of the pseudo-graph without $\mathcal{E}_\gamma$, there are only 3 related pseudo-graphs.

$\mathcal{G}_i$, i	1	2	3	4	5	6	7	8	9	10	11	12	13	14	$\sum$
related pseudo-graphs	1	3	3	1	3	1	3	3	3	$\mathcal{G}_5$	3	6	3	3	36

The total number 36 is the number of isomorphism classes of pseudo-graphs $\mathcal{G}(\underline{x})$ for $\underline{x} \in \mathbb{Z}^3$ a local minimum.

4.4 The Stabilizers of Upper Triangular 3 x 3 Matrices

The groups $G^{phi} \ltimes \langle\gamma\rangle$ and Br_3 act on $\mathbb{Z}^3/\{\pm 1\}^3$. The pseudo-graphs in the Examples 4.11 and Theorem 4.13 offer a convenient way to determine the stabilizers $(G^{phi} \ltimes \langle\gamma\rangle)_{v_0}$ and $(\mathrm{Br}_3)_{v_0}$ for $v_0 = \underline{x}/\{\pm 1\}^3$ with $\underline{x} \in \bigcup_{i=1}^{24} C_i$ a local minimum.

The stabilizers of v_0 depend only on the pseudo-graph G_j with $G_j = \mathcal{G}(\underline{x})$. The results are presented in Theorem 4.16. The Remarks 4.15 prepare this.

Remarks 4.15 (i) First we recall some well known facts about the groups $SL_2(\mathbb{Z})$ and $PSL_2(\mathbb{Z})$ and their relation to Br_3. The group $SL_2(\mathbb{Z})$ is generated by the matrices

$$A_1 := \begin{pmatrix} 1 & -1 \\ 0 & 1 \end{pmatrix} \quad \text{and} \quad A_2 := \begin{pmatrix} 1 & 0 \\ 1 & 1 \end{pmatrix}.$$

Generating relations are

$$A_1A_2A_1 = A_2A_1A_2 \quad \text{and} \quad (A_2A_1)^6 = E_2.$$

The group Br_3 is generated by the elementary braids σ_1 and σ_2. The only generating relation is $\sigma_1\sigma_2\sigma_1 = \sigma_2\sigma_1\sigma_2$. Therefore there is a surjective group homomorphism

$$\mathrm{Br}_3 \to SL_2(\mathbb{Z}), \quad \sigma_1 \mapsto A_1, \quad \sigma_2 \mapsto A_2,$$

with kernel $\langle(\sigma_2\sigma_1)^6\rangle = \langle(\sigma^{mon})^2\rangle$. It induces a surjective group homomorphism

$$\mathrm{Br}_3 \to PSL_2(\mathbb{Z}), \quad \sigma_1 \mapsto [A_1], \quad \sigma_2 \mapsto [A_2],$$

with kernel $\langle\sigma^{mon}\rangle$ because $(A_2A_1)^3 = -E_2$.

(ii) The action of $\mathrm{Br}_3 \ltimes \{\pm 1\}^3$ on $T_3(\mathbb{Z})$ and on $\mathbb{Z}^3$ is fixed in the beginning of Sect. 4.1. One sees that $\sigma^{mon} = (\sigma_2\sigma_1)^3$ acts trivially on $T_3(\mathbb{Z})$ and $\mathbb{Z}^3$. This can be checked directly. Or it can be seen as a consequence of the following two facts.

(1) The action of $\mathrm{Br}_3 \ltimes \{\pm 1\}^3$ on $(\mathrm{Br}_3 \ltimes \{\pm 1\}^3)(\underline{x})$ for some $\underline{x} \in \mathbb{Z}^3$ is induced by the action of $\mathrm{Br}_3 \ltimes \{\pm 1\}^3$ on the set $\mathcal{B}^{dist}$ of distinguished bases of a triple $(H_\mathbb{Z}, L, \underline{e})$ with $L(\underline{e}^t, \underline{e})^t = S(\underline{x})$ by $S((\alpha,\varepsilon)(\underline{x})) = L((\alpha,\varepsilon)(\underline{e})^t, (\alpha,\varepsilon)(\underline{e}))^t$ for $(\alpha,\varepsilon) \in \mathrm{Br}_3 \ltimes \{\pm 1\}^3$.
(2) For $(\alpha,\varepsilon) = (\sigma^{mon}, (1,1,1))$ $(\alpha,\varepsilon)(\underline{e}) = Z((\alpha,\varepsilon))(\underline{e}) = M(\underline{e})$ by Theorem 3.10, and $L(M(\underline{e})^t, M(\underline{e})) = L(\underline{e}^t, \underline{e})$.

In any case the action of $\mathrm{Br}_3 \ltimes \{\pm 1\}^3$ on $\mathbb{Z}^3$ boils down to a nonlinear action of $PSL_2(\mathbb{Z}) \ltimes G^{sign}$ where $G^{sign} = \langle\delta_1^\mathbb{R}, \delta_2^\mathbb{R}\rangle$ catches the action of $\{\pm 1\}^3$ on $\mathbb{Z}^3$, see Sect. 4.1.

(iii) The shape of this nonlinear action led us in Definition 4.1 and Theorem 4.2 to the group $(G^{phi} \ltimes G^{sign}) \rtimes \langle\gamma\rangle = (G^{phi} \rtimes \langle\gamma\rangle) \ltimes G^{sign}$. In fact, $G^{phi} \rtimes \langle\gamma\rangle \cong PSL_2(\mathbb{Z})$. This can be seen as follows.

The formulas (4.1)–(4.4) in the proof of Theorem 4.2 (c) give lifts to $\mathrm{Br}_3 \ltimes \{\pm 1\}^3$ of the generators $\varphi_1, \varphi_2, \varphi_3$ and γ of $G^{phi} \ltimes \langle\gamma\rangle$. Dropping the generators of the sign action in these lifts, we obtain the following lifts to Br_3,

$$\left.\begin{aligned} &l(\gamma) = \sigma_2\sigma_1, \quad l(\gamma^{-1}) = \sigma_1^{-1}\sigma_2^{-1}, \\ &l(\varphi_1) = l(\gamma)^{-1}\sigma_2^{-1} = \sigma_1^{-1}\sigma_2^{-2}, \\ &l(\varphi_2) = l(\gamma)\sigma_2 = \sigma_2\sigma_1\sigma_2 = \sigma_1\sigma_2\sigma_1, \\ &l(\varphi_3) = l(\gamma)\sigma_1 = \sigma_2\sigma_1^2. \end{aligned}\right\} \tag{4.11}$$

The equality of groups in Theorem 4.2 (c) boils down after dropping the sign action to an equality of groups

$$\langle\sigma_1^\mathbb{R}, \sigma_2^\mathbb{R}\rangle \cong G^{phi} \rtimes \langle\gamma\rangle. \tag{4.12}$$

As $(\sigma_2^\mathbb{R}\sigma_1^\mathbb{R})^3 = \mathrm{id}$, we obtain a surjective group homomorphism $PSL_2(\mathbb{Z}) \to G^{phi} \rtimes \langle\gamma\rangle$ with $[A_i] \mapsto \sigma_i^\mathbb{R}$. The subgroup $\langle [A_1]^{-1}[A_2]^{-2}, [A_2][A_1][A_2], [A_2][A_1]^2\rangle$ of $PSL_2(\mathbb{Z})$ is mapped to G^{phi}. One easily calculates that this subgroup is the free Coxeter group with three generators which was considered in Remark 6.12 (iv) and which has index three in $PSL_2(\mathbb{Z})$. As G^{phi} is also a free Coxeter group with three generators and has index 3 in $G^{phi} \rtimes \langle\gamma\rangle$, the map $PSL_2(\mathbb{Z}) \to G^{phi} \rtimes \langle\gamma\rangle$ is a group isomorphism.

(iv) For use in the proof of Theorem 4.16 we recall the formulas (4.6)

$$\left.\begin{aligned} \gamma\varphi_1 &= \varphi_2\gamma, & \gamma\varphi_2 &= \varphi_3\gamma, & \gamma\varphi_3 &= \varphi_1\gamma, \\ \varphi_1\gamma^{-1} &= \gamma^{-1}\varphi_2, & \varphi_2\gamma^{-1} &= \gamma^{-1}\varphi_3, & \varphi_3\gamma^{-1} &= \gamma^{-1}\varphi_1, \end{aligned}\right\} \tag{4.13}$$

from the proof of Theorem 4.2.

(v) The relation $\sigma_1\sigma_2\sigma_1 = \sigma_2\sigma_1\sigma_2$ is equivalent to each of the two relations

$$\sigma_1\sigma_2\sigma_1^{-1} = \sigma_2^{-1}\sigma_1\sigma_2 \quad \text{and} \quad \sigma_1^{-1}\sigma_2\sigma_1 = \sigma_2\sigma_1\sigma_2^{-1}$$

and induces for any $m \in \mathbb{Z}$ the relations

$$\sigma_1\sigma_2^m\sigma_1^{-1} = \sigma_2^{-1}\sigma_1^m\sigma_2 \quad \text{and} \quad \sigma_1^{-1}\sigma_2^m\sigma_1 = \sigma_2\sigma_1^m\sigma_2^{-1}. \tag{4.14}$$

Also this will be useful in the proof of Theorem 4.16.

Theorem 4.16 *Consider $v_0 := \underline{x}/\{\pm 1\}^3$ with $\underline{x} \in \bigcup_{i=1}^{24} C_i \subset \mathbb{Z}^3$ a local minimum, and consider the pseudo-graph $\mathcal{G}_j$ with $\mathcal{G}_j = \mathcal{G}(\underline{x})$. The entries in the third and fourth column in the table in Theorem 4.13, which are in the line of $\mathcal{G}_j$, give the stabilizers $(G^{phi} \rtimes \langle\gamma\rangle)_{v_0}$ and $(\mathrm{Br}_3)_{v_0}$ of v_0.*

Proof First we treat the stabilizer $(G^{phi} \rtimes \langle\gamma\rangle)_{v_0}$.

The *total set* $|\mathcal{G}_j|$ of the pseudo-graph $\mathcal{G}_j$ means the union of vertices and edges in an embedding of the pseudo-graph in the real plane $\mathbb{R}^2$ as in the figures in the Examples 4.11. The fundamental group $\pi_1(|\mathcal{G}_j|, v_0)$ is a free group with 0, 1, 3, 4, 5 or 6 generators. The number of generators is the number of compact components in $\mathbb{R}^2 - |\mathcal{G}_j|$. A generator is the class of a closed path which starts and ends at v_0 and turns once around one of these compact components. Any such closed path induces a word in $\varphi_1, \varphi_2, \varphi_3, \gamma$ and γ^{-1}. This word gives an element of $(G^{phi} \rtimes \langle\gamma\rangle)_{v_0}$. We obtain a group homomorphism

$$\pi_1(|\mathcal{G}_j|, v_0) \to (G^{phi} \rtimes \langle\gamma\rangle)_{v_0}.$$

It is surjective because any element of $(G^{phi} \times \langle\gamma\rangle)_{v_0}$ can be written as $\psi\gamma^\xi$ with $\psi \in G^{phi}$ and $\xi \in \{0, \pm 1\}$. The element $\psi\gamma^\xi$ leads to and comes from a closed path in $|\mathcal{G}_j|$ which starts and ends at v_0.

In fact, this shows that we could restrict to closed paths which run never or only once at the beginning through an edge in $\mathcal{E}_\gamma$. But we will not use this fact.

The following list gives for each of the 14 pseudo-graphs $\mathcal{G}_1, ..., \mathcal{G}_{14}$ in the first line one word in $\varphi_1, \varphi_2, \varphi_3, \gamma$ and γ^{-1} for each compact component of $\mathbb{R}^2 - |\mathcal{G}_j|$. In the following lines the relations (4.13) are used to show that all these words are generated in $G^{phi} \rtimes \langle\gamma\rangle$ by the generators in the third column in the table in Theorem 4.13. The generators are underlined.

$\mathcal{G}_1$: $\underline{\varphi_1}, \underline{\varphi_2}, \underline{\varphi_3}, \underline{\gamma}$.
$\mathcal{G}_2$: $\underline{\varphi_1}, \varphi_3\gamma, \gamma\varphi_2, \gamma^{-1}\varphi_2\gamma, \gamma\varphi_3\gamma^{-1}, \gamma\varphi_1\gamma$.

$$\begin{aligned} \gamma\varphi_1\gamma &= \gamma^2\varphi_3 = \underline{\gamma^{-1}\varphi_3}, \\ \gamma\varphi_2 &= \varphi_3\gamma = (\gamma^{-1}\varphi_3)^{-1}, \\ \gamma^{-1}\varphi_2\gamma &= \varphi_1, \\ \gamma\varphi_3\gamma^{-1} &= \varphi_1. \end{aligned}$$

$\mathcal{G}_3$: $\varphi_1\gamma$, $\varphi_2\varphi_3\gamma$, $\gamma\varphi_2\gamma$, $\gamma\varphi_1\varphi_2$, $\underline{\gamma\varphi_3}$.

$$\begin{aligned} \varphi_1\gamma &= \gamma\varphi_3, \\ \varphi_2\varphi_3\gamma &= \varphi_2\gamma\varphi_2 = \gamma\varphi_1\varphi_2 = (\gamma\varphi_3)(\underline{\varphi_2\varphi_1\varphi_3})^{-1}, \\ \gamma\varphi_2\gamma &= \gamma^2\varphi_1 = \gamma^{-1}\varphi_1 = \varphi_3\gamma^{-1} = (\gamma\varphi_3)^{-1}. \end{aligned}$$

$\mathcal{G}_4$: $\underline{\gamma}$, $\underline{\varphi_2\varphi_1\varphi_3}$, $\varphi_3\varphi_2\varphi_1$, $\varphi_1\varphi_3\varphi_2$.

$$\begin{aligned} \gamma\varphi_2\varphi_1\varphi_3\gamma^{-1} &= \gamma\varphi_2\varphi_1\gamma^{-1}\varphi_1 = \gamma\varphi_2\gamma^{-1}\varphi_2\varphi_1 = \varphi_3\varphi_2\varphi_1, \\ \gamma^{-1}\varphi_2\varphi_1\varphi_3\gamma &= \gamma^{-1}\varphi_2\varphi_1\gamma\varphi_2 = \gamma^{-1}\varphi_2\gamma\varphi_3\varphi_2 = \varphi_1\varphi_3\varphi_2. \end{aligned}$$

$\mathcal{G}_5$: $\underline{\gamma^{-1}\varphi_1}$, $\gamma\varphi_2\gamma$, $\gamma\varphi_3$.

$$\begin{aligned} \gamma\varphi_2\gamma &= \gamma^2\varphi_1 = \gamma^{-1}\varphi_1, \\ \gamma\varphi_3 &= \varphi_1\gamma = (\gamma^{-1}\varphi_1)^{-1}. \end{aligned}$$

$\mathcal{G}_6$: $\underline{\gamma}$.
$\mathcal{G}_7$: $\underline{\varphi_3}$, $\gamma^{-1}\varphi_1\gamma$, $\gamma\varphi_2\gamma^{-1}$.

$$\begin{aligned} \gamma^{-1}\varphi_1\gamma &= \varphi_3, \\ \gamma\varphi_2\gamma^{-1} &= \varphi_3. \end{aligned}$$

$\mathcal{G}_8$: $\underline{\mathrm{id}}$.
$\mathcal{G}_9$: $\underline{\mathrm{id}}$.
$\mathcal{G}_{10}$: $\underline{\gamma^{-1}\varphi_2}$, $\gamma\varphi_3\gamma$, $\gamma\varphi_1$.

$$\begin{aligned} \gamma\varphi_3\gamma &= \gamma^2\varphi_2 = \gamma^{-1}\varphi_2, \\ \gamma\varphi_1 &= \varphi_2\gamma = (\gamma^{-1}\varphi_2)^{-1}. \end{aligned}$$

$\mathcal{G}_{11}$: $\underline{\gamma^{-1}\varphi_3\varphi_1}$, $\gamma\varphi_1\varphi_2\gamma$, $\varphi_2\varphi_3\gamma^{-1}$.

$$\begin{aligned} \gamma\varphi_1\varphi_2\gamma &= \gamma\varphi_1\gamma\varphi_1 = \gamma^2\varphi_3\varphi_1 = \gamma^{-1}\varphi_3\varphi_1, \\ \varphi_2\varphi_3\gamma^{-1} &= \varphi_2\gamma^{-1}\varphi_1 = \gamma^{-1}\varphi_3\varphi_1. \end{aligned}$$

$\mathcal{G}_{12}$: $\underline{\varphi_3\varphi_2\varphi_1}$, $\underline{\gamma^{-1}\varphi_3\varphi_1}$, $\gamma^{-1}\varphi_1\varphi_3\varphi_2\gamma$, $\gamma\varphi_1\varphi_2\gamma$, $\varphi_2\varphi_3\gamma^{-1}$, $\gamma\varphi_2\varphi_1\varphi_3\gamma^{-1}$.

$$\begin{aligned} \gamma^{-1}\varphi_1\varphi_3\varphi_2\gamma &= \gamma^{-1}\varphi_1\varphi_3\gamma\varphi_1 = \gamma^{-1}\varphi_1\gamma\varphi_2\varphi_1 = \varphi_3\varphi_2\varphi_1, \\ \gamma\varphi_1\varphi_2\gamma &= \gamma\varphi_1\gamma\varphi_1 = \gamma^2\varphi_3\varphi_1 = \gamma^{-1}\varphi_3\varphi_1, \\ \varphi_2\varphi_3\gamma^{-1} &= \varphi_2\gamma^{-1}\varphi_1 = \gamma^{-1}\varphi_3\varphi_1, \\ \gamma\varphi_2\varphi_1\varphi_3\gamma^{-1} &= \gamma\varphi_2\varphi_1\gamma^{-1}\varphi_1 = \gamma\varphi_2\gamma^{-1}\varphi_2\varphi_1 = \varphi_3\varphi_2\varphi_1. \end{aligned}$$

$\mathcal{G}_{13}$: $\underline{\varphi_2\varphi_3\varphi_1}$, $\underline{\varphi_3\varphi_2\varphi_1}$, $\gamma^{-1}\varphi_3\varphi_1\varphi_2\gamma$, $\gamma^{-1}\varphi_1\varphi_3\varphi_2\gamma$, $\gamma\varphi_1\varphi_2\varphi_3\gamma^{-1}$, $\gamma\varphi_2\varphi_1\varphi_3\gamma^{-1}$.

$$\begin{aligned}
\gamma^{-1}\varphi_3\varphi_1\varphi_2\gamma &= \gamma^{-1}\varphi_3\varphi_1\gamma\varphi_1 = \gamma^{-1}\varphi_3\gamma\varphi_3\varphi_1 = \varphi_2\varphi_3\varphi_1,\\
\gamma^{-1}\varphi_1\varphi_3\varphi_2\gamma &= \gamma^{-1}\varphi_1\varphi_3\gamma\varphi_1 = \gamma^{-1}\varphi_1\gamma\varphi_2\varphi_1 = \varphi_3\varphi_2\varphi_1,\\
\gamma\varphi_1\varphi_2\varphi_3\gamma^{-1} &= \gamma\varphi_1\varphi_2\gamma^{-1}\varphi_1 = \gamma\varphi_1\gamma^{-1}\varphi_3\varphi_1 = \varphi_2\varphi_3\varphi_1,\\
\gamma\varphi_2\varphi_1\varphi_3\gamma^{-1} &= \gamma\varphi_2\varphi_1\gamma^{-1}\varphi_1 = \gamma\varphi_2\gamma^{-1}\varphi_2\varphi_1 = \varphi_3\varphi_2\varphi_1.
\end{aligned}$$

$\mathcal{G}_{14}$: $\underline{\varphi_2\varphi_3\varphi_1}$, $\gamma^{-1}\varphi_3\varphi_1\varphi_2\gamma$, $\gamma\varphi_1\varphi_2\varphi_3\gamma^{-1}$.

$$\begin{aligned}
\gamma^{-1}\varphi_3\varphi_1\varphi_2\gamma &= \varphi_2\varphi_3\varphi_1 \quad (\text{see } \mathcal{G}_{13}),\\
\gamma\varphi_1\varphi_2\varphi_3\gamma^{-1} &= \varphi_2\varphi_3\varphi_1 \quad (\text{see } \mathcal{G}_{13}).
\end{aligned}$$

Therefore the stabilizer $(G^{phi} \rtimes \langle\gamma\rangle)_{v_0}$ is as claimed in the third column of the table in Theorem 4.13.

Now we treat the stabilizer $(\mathrm{Br}_3)_{v_0}$. It is the preimage in Br_3 of $(G^{phi} \rtimes \langle\gamma\rangle)_{v_0}$ under the surjective group homomorphism $\mathrm{Br}_3 \to G^{phi} \rtimes \langle\gamma\rangle$ with kernel $\langle\sigma^{mon}\rangle$. So if $(G^{phi} \rtimes \langle\gamma\rangle)_{v_0} = \langle g_1, ..., g_m\rangle$ and $h_1, ..., h_m$ are any lifts to Br_3 of $g_1, ..., g_m$ then $(\mathrm{Br}_3)_{v_0} = \langle\sigma^{mon}, h_1, ..., h_m\rangle$.

For any word in $\varphi_1, \varphi_2, \varphi_3, \gamma$ and γ^{-1} we use the lifts in (4.11) to construct a lift of this word.

The following list gives for each of the 14 pseudo-graphs $\mathcal{G}_1, ..., \mathcal{G}_{14}$ for each generator of $(G^{phi} \times \langle\gamma\rangle)_{v_0}$ in the third column of the table in Theorem 4.13 this lift and rewrites it using the relations (4.14). The generators in the fourth column of the table in Theorem 4.13 are underlined.

$$\begin{aligned}
\mathcal{G}_1 :& G^{phi} \rightsquigarrow \mathrm{Br}_3,\\
\mathcal{G}_2 :& \varphi_1 \rightsquigarrow \sigma_1^{-1}\sigma_2^{-2} = \sigma_2(\sigma_2^{-1}\sigma_1^{-1}\sigma_2^{-1})\sigma_2^{-1} = \sigma_2(\sigma_1^{-1}\sigma_2^{-1}\sigma_1^{-1})\sigma_2^{-1}\\
&= (\sigma_2^2\sigma_1)(\sigma_1^{-1}\sigma_2^{-1})^3 = \underline{\sigma_2^2}\sigma_1(\sigma^{mon})^{-1},\\
& \gamma^{-1}\varphi_3 \rightsquigarrow (\sigma_1^{-1}\sigma_2^{-1})(\sigma_2\sigma_1^2) = \underline{\sigma_1},\\
\mathcal{G}_3 :& \gamma\varphi_3 \rightsquigarrow (\sigma_2\sigma_1)(\sigma_2\sigma_1^2) = (\sigma_2\sigma_1\sigma_2)\sigma_1^2 = (\underline{\sigma_1\sigma_2})\sigma_1^3,\\
& \varphi_2\varphi_1\varphi_3 \rightsquigarrow (\sigma_1\sigma_2\sigma_1)(\sigma_1^{-1}\sigma_2^{-2})(\sigma_2\sigma_1^2) = \underline{\sigma_1^3},\\
\mathcal{G}_4 :& \gamma \rightsquigarrow \underline{\sigma_2\sigma_1},\\
& \varphi_2\varphi_1\varphi_3 \rightsquigarrow \underline{\sigma_1^3},\\
\mathcal{G}_5 :& \gamma^{-1}\varphi_1 \rightsquigarrow (\sigma_1^{-1}\sigma_2^{-1})(\sigma_1^{-1}\sigma_2^{-2}) = (\sigma_1^{-1}\sigma_2^{-1})^3\sigma_2\sigma_1\sigma_2^{-1}\\
& \stackrel{(4.14)}{=} (\sigma^{mon})^{-1}\underline{(\sigma_1^{-1}\sigma_2^{-1}\sigma_1)^{-1}},\\
\mathcal{G}_6 :& \gamma \rightsquigarrow \underline{\sigma_2\sigma_1},
\end{aligned}$$

$$\mathcal{G}_7 : \varphi_3 \rightsquigarrow \underline{\sigma_2\sigma_1^2},$$

$$\begin{aligned}
\mathcal{G}_8 &: \mathrm{id} \rightsquigarrow \underline{\mathrm{id}},\\
\mathcal{G}_9 &: \mathrm{id} \rightsquigarrow \underline{\mathrm{id}},\\
\mathcal{G}_{10} &: \gamma^{-1}\varphi_2 \rightsquigarrow (\sigma_1^{-1}\sigma_2^{-1})(\sigma_2\sigma_1\sigma_2) = \underline{\sigma_2},\\
\mathcal{G}_{11} &: \gamma^{-1}\varphi_3\varphi_1 \rightsquigarrow (\sigma_1^{-1}\sigma_2^{-1})(\sigma_2\sigma_1^2)(\sigma_1^{-1}\sigma_2^{-2}) = \sigma_2^{-2} = (\underline{\sigma_2^2})^{-1},\\
\mathcal{G}_{12} &: \gamma^{-1}\varphi_3\varphi_1 \rightsquigarrow (\underline{\sigma_2^2})^{-1},\\
&\quad \varphi_3\varphi_2\varphi_1 \rightsquigarrow (\sigma_2\sigma_1^2)(\sigma_1\sigma_2\sigma_1)(\sigma_1^{-1}\sigma_2^{-2}) = \underline{\sigma_2\sigma_1^3\sigma_2^{-1}},\\
\mathcal{G}_{13} &: \varphi_2\varphi_3\varphi_1 \rightsquigarrow (\sigma_1\sigma_2\sigma_1)(\sigma_2\sigma_1^2)(\sigma_1^{-1}\sigma_2^{-2}) = (\sigma_1\sigma_2)^3\sigma_2^{-3}\\
&= \sigma^{mon}(\underline{\sigma_2^3})^{-1},\\
&\quad \varphi_3\varphi_2\varphi_1 \rightsquigarrow \underline{\sigma_2\sigma_1^3\sigma_2^{-1}},\\
\mathcal{G}_{14} &: \varphi_2\varphi_3\varphi_1 \rightsquigarrow \sigma^{mon}(\underline{\sigma_2^3})^{-1}.
\end{aligned}$$

Observe

$$(\sigma_2\sigma_1)^3 = \sigma^{mon}, \quad (\sigma_2\sigma_1^2)^2 = \sigma^{mon}.$$

Therefore the stabilizer $(\mathrm{Br}_3)_{v_0}$ is as claimed in the fourth column of the table in Theorem 4.13. □

4.5 A Global Sign Change, Relevant for the Odd Case

Remarks 4.17 (i) For $S \in T_n(\mathbb{Z})$ consider a unimodular bilinear lattice $(H_\mathbb{Z}, L)$ with a triangular basis $\underline{e} = (e_1, ..., e_n)$ with $L(\underline{e}^t, \underline{e})^t = S$. Consider also the matrix $\widetilde{S} \in T_n(\mathbb{Z})$ with $\widetilde{S}_{ij} = -S_{ij}$ for $i < j$.

On the same lattice $H_\mathbb{Z}$ and the same basis $\underline{e}$ we define a second unimodular bilinear form $\widetilde{L}$ by $\widetilde{L}(\underline{e}^t, \underline{e})^t = \widetilde{S}$ and denote all objects associated to $(H_\mathbb{Z}, \widetilde{L}, \underline{e})$ with a tilde, $\widetilde{I}^{(k)}$, $\widetilde{\mathcal{B}}^{tri}$, $\widetilde{\Gamma}^{(k)}$, $\widetilde{\Delta}^{(k)}$.

Most of them differ a lot from the objects associated to $(H_\mathbb{Z}, L, \underline{e})$. But the odd intersection forms differ only by the sign. Therefore the monodromies M and $\widetilde{M}$ are different (in general), but the odd monodromy groups and the sets of odd vanishing cycles coincide:

$$\begin{aligned}
&\widetilde{I}^{(1)} = -I^{(1)},\\
&\text{so } \widetilde{s}_{e_i}^{(1)} = (s_{e_i}^{(1)})^{-1},\\
&\widetilde{M} = \widetilde{s}_{e_1}^{(1)} \circ ... \circ \widetilde{s}_{e_n}^{(1)} \overset{\text{in general}}{\neq} M = s_{e_1}^{(1)} \circ ... \circ s_{e_n}^{(1)},\\
&\text{but } \widetilde{\Gamma}^{(1)} = \langle \widetilde{s}_{e_1}^{(1)}, ..., \widetilde{s}_{e_n}^{(1)} \rangle = \langle s_{e_1}^{(1)}, ..., s_{e_n}^{(1)} \rangle = \Gamma^{(1)},
\end{aligned}$$

$$\widetilde{\Delta}^{(1)} = \widetilde{\Gamma}^{(1)}\{\pm e_1, ..., \pm e_n\} = \Gamma^{(1)}\{\pm e_1, ..., \pm e_n\} = \Delta^{(1)}.$$

Because of $\widetilde{\Gamma}^{(1)} = \Gamma^{(1)}$ and $\widetilde{\Delta}^{(1)} = \Delta^{(1)}$ the global sign change from S to $\widetilde{S}$ is interesting.

(ii) In this section we will study the action on $T_3(\mathbb{Z})$ of the extension of the action of $\mathrm{Br}_3 \ltimes \{\pm 1\}^3$ by this global sign change. Define

$$\delta^{\mathbb{R}} : \mathbb{R}^3 \to \mathbb{R}^3,\ (x_1, x_2, x_3) \mapsto (-x_1, -x_2, -x_3)$$

and

$$\widetilde{G}^{sign} := \langle \delta_1^{\mathbb{R}}, \delta_2^{\mathbb{R}}, \delta^{\mathbb{R}} \rangle \cong \{\pm 1\}^3.$$

It is easy to see that the double semidirect product $(G^{phi} \ltimes G^{sign}) \rtimes \langle \gamma \rangle$ extends to the double semidirect product $(G^{phi} \ltimes \widetilde{G}^{sign}) \rtimes \langle \gamma \rangle$.

Lemma 4.18 *Each $(G^{phi} \ltimes \widetilde{G}^{sign}) \rtimes \langle \gamma \rangle$ orbit in $\mathbb{Z}^3$ contains at least one local minimum of one of the following types:*

(a) $\underline{x} \in \mathbb{Z}^3_{\geq 3}$ *with* $2x_i \leq x_j x_k$ *for* $\{i, j, k\} = \{1, 2, 3\}$.
(b) $(-l, 2, -l)$ *for some* $l \in \mathbb{Z}_{\geq 2}$.
(c) $(x_1, x_2, 0)$ *for some* $x_1, x_2 \in \mathbb{Z}_{\geq 0}$ *with* $x_1 \geq x_2$.

Proof In (c) we can restrict to $x_1 \geq x_2$ because $\delta_3^{\mathbb{R}} \gamma \varphi_1(x_1, x_2, 0) = (x_2, x_1, 0)$.

Each $(G^{phi} \ltimes \widetilde{G}^{sign}) \rtimes \langle \gamma \rangle$ orbit consists of one or several $(G^{phi} \ltimes G^{sign}) \rtimes \langle \gamma \rangle$ orbits and thus contains local minima.

Suppose that $\underline{x} \in \mathbb{Z}^3$ is such a local minimum and is not obtained with G^{sign} from a local minimum in (a), (b) or (c). Then either $\underline{x}$ is a local minimum associated to $S(\widehat{A}_2)$, so $\underline{x} \in \{(-1, -1, -1), (1, 1, -1), (1, -1, 1), (-1, 1, 1)\}$ or $x_1 x_2 x_3 < 0$ and $r(\underline{x}) > 4$.

In the first case $\delta^{\mathbb{R}}(-1, -1, -1) = (1, 1, 1) = \varphi_3(1, 1, 0)$, so the orbit contains a local minimum in (c).

In the second case $r(\delta^{\mathbb{R}}(\underline{x})) = r(-\underline{x}) = r(\underline{x}) + 2x_1 x_2 x_3 < r(\underline{x})$. We consider a local minimum $\underline{x}^{(1)}$ in the $\mathrm{Br}_3 \ltimes \{\pm 1\}^3$ orbit of $-\underline{x}$. If it is not obtained with G^{sign} from a local minimum in (a), (b) or (c), then $\underline{x}^{(1)}$ is a local minimum associated to $S(\widehat{A}_2)$ or $x_1^{(1)} x_2^{(1)} x_3^{(1)} < 0$ and $r(\underline{x}^{(1)}) > 4$.

We repeat this procedure until we arrive at a local minimum obtained with G^{sign} from one in (a), (b) or (c). This stops after finitely many steps because $r(\underline{x}^{(1)}) = r(-\underline{x}) < r(\underline{x})$. □

Remark 4.19 Corollary 6.23 will say that the $(G^{phi} \ltimes \widetilde{G}^{sign}) \rtimes \langle \gamma \rangle$ orbits of the local minima in the parts (b) and (c) of Lemma 4.18 are pairwise different and also different from the orbits of the local minima in part (a).

Examples 4.20 Given an element $\underline{x} \in \mathbb{Z}^3$, it is not obvious which local minimum of a type in (a), (b) or (c) is contained in the $(G^{phi} \ltimes \widetilde{G}^{sign}) \rtimes \langle \gamma \rangle$ orbit of $\underline{x}$. We give four families of examples. An arrow $\mapsto$ between two elements of $\mathbb{Z}^3$ means that these two elements are in the same orbit.

(i) Start with $(x_1, x_2, -1)$ with $x_1 \geq x_2 > 0$.

$$
\begin{aligned}
(x_1, x_2, -1) &\overset{\widetilde{G}^{sign}}{\mapsto} (x_1, x_2, 1) \overset{\varphi_1}{\mapsto} (x_2 - x_1, 1, x_2) \overset{\widetilde{G}^{sign}}{\mapsto} (x_1 - x_2, 1, x_2) \\
&\mapsto \dots \mapsto (\gcd(x_1, x_2), 1, 0).
\end{aligned}
$$

(ii) Start with $(x_1, x_2, -2)$ with $x_1 \geq x_2 \geq 2$.

$$
\begin{aligned}
(x_1, x_2, -2) &\overset{\widetilde{G}^{sign}}{\mapsto} (x_1, x_2, 2) \overset{\varphi_1}{\mapsto} (2x_2 - x_1, 2, x_2) \mapsto \dots \\
&\mapsto \begin{cases} (\gcd(x_1, x_2), 2, 0) & \text{if } x_1 \text{ and } x_2 \\ & \quad \text{contain different powers of 2,} \\ (\gcd(x_1, x_2), 2, \gcd(x_1, x_2)) & \text{if } x_1 \text{ and } x_2 \\ & \quad \text{contain the same power of 2.} \end{cases}
\end{aligned}
$$

In order to understand the case discussion, observe that $2x_2 - x_1$ and x_2 contain the same power of 2 if and only if x_1 and x_2 contain the same power of 2. Furthermore 0 is divisible by an arbitrarily large power of 2.

(iii) The examples in part (ii) lead in the special case $\gcd(x_1, x_2) = 1$ to the following,

$$
\begin{aligned}
(x_1, x_2, -2) \mapsto \dots \mapsto &\mapsto (1, 2, 0) \text{ or } (-1, 2, -1) \\
&\mapsto (2, 1, 0) \text{ or } (1, 1, 0),
\end{aligned}
$$

again depending on whether x_1 and x_2 contain different powers of 2 or the same power of 2. In the second case x_1 and x_2 are odd with $\gcd(x_1, x_2) = 1$, and then $(x_1, x_2, -2) \mapsto (1, 1, 0) \mapsto (x_1, x_2, -1)$.

(iv) Start with $(-3, -3, -l)$ for some $l \in \mathbb{Z}_{\geq 2}$.

$$
\begin{aligned}
(-3, -3, -l) &\overset{\delta^{\mathbb{R}}}{\mapsto} (3, 3, l) \mapsto (3, 3, \pm(l - 9)) \\
&\mapsto (3, 3, \widetilde{l}) \text{ for some } \widetilde{l} \in \{0, 1, 2, 3, 4\} \\
&\mapsto (3, 3, 0),\ (3, 1, 0),\ (-3, 2, -3),\ (3, 3, 3) \text{ or } (4, 3, 3).
\end{aligned}
$$

$(3, 3, 0)$ and $(3, 1, 0)$ are in part (c), $(-3, 2, -3)$ is in part (b), and $(3, 3, 3)$ and $(4, 3, 3)$ are in part (a) of Lemma 4.18.

Chapter 5
Automorphism Groups

A unimodular bilinear lattice $(H_{\mathbb{Z}}, L)$ comes equipped with four automorphism groups $G_{\mathbb{Z}}$, $G_{\mathbb{Z}}^{(0)}$, $G_{\mathbb{Z}}^{(1)}$ and $G_{\mathbb{Z}}^{M}$ of $H_{\mathbb{Z}}$, which all respect the monodromy M and possibly some bilinear form. They are the subject of this chapter.

Section 5.1 gives two basic observations which serve as general tools to control these groups under reasonable conditions. Another very useful tool is Theorem 3.26 (c), which gives in favourable situations an n-th root of $(-1)^{k+1}M$ in $G_{\mathbb{Z}}$. Section 5.1 treats also the cases A_1^n.

Section 5.2 takes care of the rank 2 cases. It makes use of some statements on quadratic units in Lemma C.1 (a) in Appendix C.

All further Sects. 5.3–5.7 are devoted to the rank 3 cases. Section 5.3 discusses the setting and the basic data. It also introduces the special automorphism $Q \in \mathrm{Aut}(H_{\mathbb{Q}}, L)$ which is id on $\ker(M - \mathrm{id})$ and $-\mathrm{id}$ on $\ker(M^2 - (2-r)M + \mathrm{id})$ and determines the (rather few) cases where Q is in $G_{\mathbb{Z}}$.

The treatment of the reducible rank 3 cases in Sect. 5.4 builds on the rank 2 cases and is easy.

The irreducible rank 3 cases with all eigenvalues in S^1 form the four single cases $A_3, \widehat{A}_2, \mathbb{P}^2, \mathcal{H}_{1,2}$ and the series $S(-l, 2, -l)$ with $l \in \mathbb{Z}_{\geq 3}$. Here in Sect. 5.5 third roots of the monodromy and in the series $S(-l, 2, -l)$ even higher roots of the monodromy turn up.

Sections 5.6 and 5.7 treat all irreducible rank 3 cases with eigenvalues not all in S^1 (1 is always one eigenvalue). Section 5.6 takes care of those families of cases where $G_{\mathbb{Z}} \supsetneqq \{\pm M^m \mid m \in \mathbb{Z}\}$, Sect. 5.7 of all others.

Section 5.6 is rather long. Here again roots of the monodromy turn up, and statements on quadratic units in Lemma C.1 are used.

Section 5.7 is very long. The main result $G_{\mathbb{Z}} = \{\pm M^m \mid m \in \mathbb{Z}\}$ in these cases requires an extensive case discussion.

This chapter determines $G_{\mathbb{Z}}$ in all cases with rank $n \leq 3$. An application is a proof of Theorem 3.28, which says that in almost all cases with rank $n \leq 3$ the map $Z : (\mathrm{Br}_n \ltimes \{\pm 1\}^n)_S \to G_{\mathbb{Z}}$ is surjective, the exception being four cases in Sect. 5.6.

C. Hertling and K. Larabi, *Triangular Bases of Unimodular Bilinear Lattices and Induced Structures*, Lecture Notes in Mathematics 2398,
https://doi.org/10.1007/978-3-032-25997-4_5

5.1 Basic Observations

Given a bilinear lattice $(H_{\mathbb{Z}}, L)$, the most important of the four automorphisms groups $G_{\mathbb{Z}}^{M}$, $G_{\mathbb{Z}}^{(0)}$, $G_{\mathbb{Z}}^{(1)}$ and $G_{\mathbb{Z}}$ in Definition 2.3 (b) (iv) and Lemma 2.6 (a) (iii) is the smallest group $G_{\mathbb{Z}}$. But the key to it is often the largest group $G_{\mathbb{Z}}^{M}$. We collect some elementary observations on these groups.

Lemma 5.1 *Let $H_{\mathbb{Z}}$ be a $\mathbb{Z}$-lattice of some rank $n \in \mathbb{N}$ and let $M : H_{\mathbb{Z}} \to H_{\mathbb{Z}}$ be an automorphism of it.*

(a) The characteristic polynomial $p_{ch,M}(t) \in \mathbb{Z}[t]$ of the automorphism M is unitary. Each eigenvalue $\lambda \in \mathbb{C}$ of M is an algebraic integer and a unit in the ring $\mathcal{O}_{\mathbb{Q}[\lambda]} \subset \mathbb{Q}[\lambda]$ of algebraic integers in $\mathbb{Q}[\lambda]$, so in $\mathcal{O}^{*}_{\mathbb{Q}[\lambda]}$. Also $\lambda^{-1} \in \mathcal{O}^{*}_{\mathbb{Q}[\lambda]}$.

(b) Suppose that M is regular, *that means, $M : H_{\mathbb{C}} \to H_{\mathbb{C}}$ has for each eigenvalue only one Jordan block.*

(i) Then

$$\mathbb{Q}[M] = \bigoplus_{i=0}^{n-1} \mathbb{Q}M^{i} \stackrel{!}{=} \mathrm{End}(H_{\mathbb{Q}}, M) := \{g : H_{\mathbb{Q}} \to H_{\mathbb{Q}} \mid gM = Mg\}.$$

(ii) Consider a polynomial $p(t) = \sum_{i=0}^{n-1} p_i t^i \in \mathbb{Q}[t]$ and the endomorphism $g = p(M) \in \mathrm{End}(H_{\mathbb{Q}}, M)$ of $H_{\mathbb{Q}}$ and $H_{\mathbb{C}}$. Then g has the eigenvalue $p(\lambda)$ on the generalized eigenspace H_{λ} of M with eigenvalue λ. If $g \in \mathrm{End}(H_{\mathbb{Z}}, M)$ then $p(\lambda) \in \mathcal{O}_{\mathbb{Q}[\lambda]}$ for each eigenvalue λ of M. If $g \in G_{\mathbb{Z}}^{M}$ then $p(\lambda) \in \mathcal{O}^{*}_{\mathbb{Q}[\lambda]}$ for each eigenvalue λ of M.

(c) Suppose that M is the monodromy of a bilinear lattice $(H_{\mathbb{Z}}, L)$. Suppose that M is regular.

(i) Then

$$G_{\mathbb{Z}}^{(0)} \cup G_{\mathbb{Z}}^{(1)} \subset \{p(M) \mid p(t) = \sum_{i=0}^{n-1} p_i t^i \in \mathbb{Q}[t],\ p(M) \in \mathrm{End}(H_{\mathbb{Z}}),$$

$$p(\lambda)p(\lambda^{-1}) = 1 \text{ for each eigenvalue } \lambda \text{ of } M\}.$$

(ii) If M is semisimple then $G_{\mathbb{Z}} = G_{\mathbb{Z}}^{(0)} = G_{\mathbb{Z}}^{(1)}$, and this set is equal to the set on the right hand side of (i).

Proof (a) Trivial.

(b) (i) As M is regular, one can choose a vector $c \in H_{\mathbb{Q}}$ with $H_{\mathbb{Q}} = \bigoplus_{i=0}^{n-1} \mathbb{Q}M^{i}c$. The inclusion $\mathbb{Q}[M] \subset \mathrm{End}(H_{\mathbb{Q}}, M)$ is clear. Suppose $g \in \mathrm{End}(H_{\mathbb{Q}}, M)$. Write $gc = p(M)c$ for some polynomial $p(t) = \sum_{i=0}^{n-1} p_i t^i \in \mathbb{Q}[t]$. As

$$gM^{k}c = M^{k}gc = M^{k}p(M)c = p(M)M^{k}c$$

for each $k \in \{0, 1, ..., n-1\}$, $g = p(M)$. Thus $\mathbb{Q}[M] = \mathrm{End}(H_{\mathbb{Q}}, M)$.

(ii) Similar to part (a).

(c) (i) Suppose $g = p(M) \in G_{\mathbb{Z}}^{(k)}$ for some $k \in \{0; 1\}$ with $p(t) = \sum_{i=0}^{n-1} p_i t^i \in \mathbb{Q}[t]$. Recall $\mathrm{Rad}\, I^{(k)} = \ker(M - (-1)^{k+1}\mathrm{id} : H_{\mathbb{Z}} \to H_{\mathbb{Z}})$.

For $\lambda \neq (-1)^{k+1}$ the generalized eigenspaces H_λ and $H_{\lambda^{-1}}$ are dual to one another with respect to $I^{(k)}$, and g has eigenvalue $p(\lambda)$ on H_λ and eigenvalue $p(\lambda^{-1})$ on $H_{\lambda^{-1}}$. That g respects $I^{(k)}$ implies $p(\lambda)p(\lambda^{-1}) = 1$.

For $\lambda = (-1)^{k+1}$, g restricts to an automorphism of the sublattice $H_\lambda \cap H_{\mathbb{Z}}$ of $H_{\mathbb{Z}}$ with determinant $\pm 1 = \det(g|_{H_\lambda \cap H_{\mathbb{Z}}}) = p(\lambda)^{\dim H_\lambda}$, so $p(\lambda) = \pm 1$.

(ii) Suppose additionally that M is semisimple and that $g = p(M)$ with $p(t) = \sum_{i=0}^{n-1} p_i t^i \in \mathbb{Q}[t]$ satifies $g \in \mathrm{End}(H_{\mathbb{Z}})$ and $p(\lambda)p(\lambda^{-1}) = 1$ for each eigenvalue λ of M. As M is regular and semisimple, each eigenvalue has multiplicity 1. The (1-dimensional) eigenspaces H_λ and $H_{\lambda^{-1}}$ are dual to one another with respect to L. The conditions $p(\lambda)p(\lambda^{-1}) = 1$ imply that g respects L and also that $\det g = \prod_{\lambda \text{ eigenvalue}} p(\lambda) = \pm 1$. Together with $g \in \mathrm{End}(H_{\mathbb{Z}})$ this shows $g \in G_{\mathbb{Z}}$. □

The situation in the following Lemma 5.2 arises surprisingly often. One reason is Theorem 3.26 (c). See the Remarks 5.3 below.

Lemma 5.2 *Let $H_{\mathbb{Z}}$ be a $\mathbb{Z}$-lattice of some rank $n \in \mathbb{N}$, let $M : H_{\mathbb{Z}} \to H_{\mathbb{Z}}$ and $M^{root} : H_{\mathbb{Z}} \to H_{\mathbb{Z}}$ be automorphisms of $H_{\mathbb{Z}}$, and let $l \in \mathbb{N}$ and $\varepsilon \in \{\pm 1\}$ be such that the following holds: M is regular,*

$$(M^{root})^l = \varepsilon M,$$

and M^{root} is cyclic, *that means, a vector $c \in H_{\mathbb{Z}}$ with $\bigoplus_{i=0}^{n-1} \mathbb{Z}(M^{root})^i c = H_{\mathbb{Z}}$ exists.*

(a) Then M^{root} is regular, and

$$\mathrm{End}(H_{\mathbb{Z}}, M) = \mathrm{End}(H_{\mathbb{Z}}, M^{root}) = \mathbb{Z}[M^{root}],$$

$$\mathrm{Aut}(H_{\mathbb{Z}}, M) = \mathrm{Aut}(H_{\mathbb{Z}}, M^{root}) = \{p(M^{root}) \mid p(t) = \sum_{i=0}^{n-1} p_i t^i \in \mathbb{Z}[t], \\ p(\kappa) \in (\mathbb{Z}[\kappa])^* \text{ for each eigenvalue } \kappa \text{ of } M^{root}\}.$$

(b) Suppose that M is the monodromy of a bilinear lattice $(H_{\mathbb{Z}}, L)$ and that the set of eigenvalues of M^{root} is invariant under inversion, that means, with κ an eigenvalue of M^{root} also κ^{-1} is an eigenvalue of M^{root}.

(i) Then $M^{root} \in G_{\mathbb{Z}}$ and

$$G_{\mathbb{Z}}^{(0)} \cup G_{\mathbb{Z}}^{(1)} \subset \{p(M^{root}) \mid p(t) = \sum_{i=0}^{n-1} p_i t^i \in \mathbb{Z}[t], \\ p(\kappa)p(\kappa^{-1}) = 1 \text{ for each eigenvalue } \kappa \text{ of } M^{root}\}.$$

(ii) If M is semisimple then $G_{\mathbb{Z}} = G_{\mathbb{Z}}^{(0)} = G_{\mathbb{Z}}^{(1)}$, and this set is equal to the set on the right hand side if (i).

Proof (a) M^{root} is regular as M is regular and $(M^{root})^l = \varepsilon M$. This equation also implies $\mathbb{Q}[M] \subset \mathbb{Q}[M^{root}]$. As these $\mathbb{Q}$-vector spaces have both dimension n,

$$\mathrm{End}(H_{\mathbb{Q}}, M) = \mathbb{Q}[M] = \mathbb{Q}[M^{root}] = \mathrm{End}(H_{\mathbb{Q}}, M^{root}).$$

Then $\mathrm{End}(H_{\mathbb{Z}}, M) = \mathrm{End}(H_{\mathbb{Z}}) \cap \mathbb{Q}[M^{root}]$. Consider $g \in \mathrm{End}(H_{\mathbb{Z}}, M)$. Choose a cyclic generator $c \in H_{\mathbb{Z}}$ with $\bigoplus_{i=0}^{n-1} \mathbb{Z}(M^{root})^i c = H_{\mathbb{Z}}$. Write $g(c) = p(M^{root})c$ for some polynomial $p(t) = \sum_{i=0}^{n-1} p_i t^i \in \mathbb{Z}[t]$. As in the proof of Lemma 5.1, one finds $g = p(M^{root})$, so $\mathrm{End}(H_{\mathbb{Z}}, M) = \mathbb{Z}[M^{root}]$. The element $g = p(M^{root})$ above is in $\mathrm{Aut}(H_{\mathbb{Z}}, M)$ if and only if $\det g = \pm 1$, and this holds if and only if the algebraic integer $p(\kappa)$ is a unit in $\mathbb{Z}[\kappa]$ for each eigenvalue κ of M^{root}.

(b) (i) The main point is to show $M^{root} \in G_{\mathbb{Z}}$. As M and M^{root} are both regular, the map $\kappa \mapsto \varepsilon\kappa^l$ is a bijection from the set of eigenvalues of M^{root} to the set of eigenvalues of M. For $\lambda = \varepsilon\kappa^l$, the generalized eigenspaces H_λ and $H_{\lambda^{-1}}$ of M are the generalized eigenspaces of M^{root} with eigenvalues κ and κ^{-1}. These two spaces are dual to one another with respect to $I^{(0)}$ (if $\lambda \neq -1$), $I^{(1)}$ (if $\lambda \neq 1$) and L.

Consider the decomposition of M^{root} into the commuting semisimple part M_s^{root} and unipotent part M_u^{root} with nilpotent part N^{root} with $\exp(N^{root}) = M_u^{root}$, and also the decomposition $M = M_s M_u$ with nilpotent part N with $\exp N = M_u$.

M_s^{root} and M_s respect L because they have eigenvalue κ and λ on H_λ and eigenvalue κ^{-1} and λ^{-1} on $H_{\lambda^{-1}}$.

As M and M_s respect L, also M_u respects L. Therefore N is an infinitesimal isometry. Because $N = lN^{root}$, also N^{root} is an infinitesimal isometry. Therefore M_u^{root} respects L. Thus also $M^{root} = M_s^{root} M_u^{root}$ respects L, so $M^{root} \in G_{\mathbb{Z}}$.

Part (ii) and the rest of part (i) are proved as part (b) in Lemma 5.1. □

Remarks 5.3 (i) The pair $(H_{\mathbb{Z}}, M^{root})$ in Lemma 5.2 is an *Orlik block* if M^{root} is of finite order. Orlik blocks will be defined and discussed in the beginning of Sect. 10.3. They are important building blocks in the unimodular bilinear lattices $(H_{\mathbb{Z}}, L)$ for many isolated hypersurface singularities.

(ii) If the matrix $S = L(\underline{e}^t, \underline{e})^t$ of a unimodular bilinear lattice $(H_{\mathbb{Z}}, L)$ with a triangular basis $\underline{e}$ has the special shape in Theorem 3.26 (c), then the monodromy $(-1)^{k+1}M$ has by Theorem 3.26 (c) a specific n-th root $M^{root} \in G_{\mathbb{Z}}$. This situation is special, but it arises surprisingly often, in singularity theory and in the cases in part (ii).

(iii) Theorem 3.26 (c) applies to all matrices $S(x) = \begin{pmatrix} 1 & x \\ 0 & 1 \end{pmatrix}$ with $x \in \mathbb{Z}$ and to all matrices $S = \begin{pmatrix} 1 & x & \varepsilon x \\ 0 & 1 & x \\ 0 & 0 & 1 \end{pmatrix}$ with $x \in \mathbb{Z}$ and $\varepsilon \in \{\pm 1\}$.

It applies especially to the matrices $S(A_1^3)$, $S(\widehat{A}_2)$, $S(\mathcal{H}_{1,2})$ and $S(\mathbb{P}^2)$ in the Examples 1.1 and to the matrix $S(-1, 1, -1)$ in the $\mathrm{Br}_3 \ltimes \{\pm 1\}^3$ orbit of $S(A_3)$.

Though it is not useful in the cases $S(A_1^3)$ and $S(\mathcal{H}_{1,2})$ because their monodromies are not regular. In the case $S(A_3)$ we will not use it as there the monodromy itself is cyclic.

The completely reducible cases A_1^n for $n \in \mathbb{N}$ can be treated easily, building on Lemma 2.12.

Lemma 5.4 *Fix $n \in \mathbb{N}$ and consider the case A_1^n with $S = S(A_1^n) = E_n$. Then*

$$\begin{aligned} G_{\mathbb{Z}} = G_{\mathbb{Z}}^{(0)} &\cong O_n(\mathbb{Z}) = \{A \in GL_n(\{0; \pm 1\}) \mid \exists\, \sigma \in S_n \\ &\qquad \exists\, \varepsilon_1, ..., \varepsilon_n \in \{\pm 1\} \text{ such that } A_{ij} = \varepsilon_i \delta_{i\sigma(j)}\}, \\ G_{\mathbb{Z}}^{(1)} = G_{\mathbb{Z}}^{M} &= \mathrm{Aut}(H_{\mathbb{Z}}) \cong GL_n(\mathbb{Z}). \end{aligned}$$

The map $Z : (\mathrm{Br}_n \ltimes \{\pm 1\}^n)_S = \mathrm{Br}_n \ltimes \{\pm 1\}^n \to G_{\mathbb{Z}}$ is surjective.

Proof The groups $G_{\mathbb{Z}}^{(1)}$ and $G_{\mathbb{Z}}^{M}$ are as claimed because $M = \mathrm{id}$ and $I^{(1)} = 0$. The groups $G_{\mathbb{Z}}$ and $G_{\mathbb{Z}}^{(0)}$ map the set $R^{(0)} = \{\pm e_1, ..., \pm e_n\}$ to itself and are therefore also as claimed.

The stabilizer of $S = E_n$ is the whole group $\mathrm{Br}_n \ltimes \{\pm 1\}^n$. The subgroup $\{\pm 1\}^n$ gives all sign changes of the basis $\underline{e} = (e_1, ..., e_n)$. The subgroup Br_n gives under Z all permutations of the elements of the tuple $(e_1, ..., e_n)$. Therefore Z is surjective. □

5.2 The Rank 2 Cases

For $x \in \mathbb{Z}$ consider the matrix $S = S(x) = \begin{pmatrix} 1 & x \\ 0 & 1 \end{pmatrix} \in T_2(\mathbb{Z})$, and consider a unimodular bilinear lattice $(H_{\mathbb{Z}}, L)$ with a triangular basis $\underline{e} = (e_1, e_2)$ with $L(\underline{e}^t, \underline{e})^t = S$. Then

$$\begin{aligned} M\underline{e} &= \underline{e} S^{-1} S^t = \underline{e} \begin{pmatrix} 1 - x^2 & -x \\ x & 1 \end{pmatrix}, \\ p_{ch,M}(t) &= t^2 - (2 - x^2)t + 1 \\ \text{with zeros } \lambda_{1/2} &= \frac{2 - x^2}{2} \pm \frac{1}{2} x \sqrt{x^2 - 4}. \end{aligned}$$

Theorem 3.26 (c) applies with $(n, k, q_0, q_1) = (2, 0, 1, x)$, namely

$$\begin{aligned} \delta_2 \sigma^{root} &= \delta_2 \sigma_1 \in (\mathrm{Br}_2 \ltimes \{\pm 1\}^2)_S, \\ M^{root} &:= Z(\delta_2 \sigma_1) \in G_{\mathbb{Z}} \\ \text{with } M^{root}\underline{e} &= \underline{e} \begin{pmatrix} -x & -1 \\ 1 & 0 \end{pmatrix} \end{aligned}$$

$$\text{and } (M^{root})^2 = -M.$$

M^{root} is regular and cyclic,

$$\begin{aligned} p_{ch,M^{root}}(t) &= t^2 + xt + 1 \\ \text{with zeros } \kappa_{1/2} &= -\frac{x}{2} \pm \frac{1}{2}\sqrt{x^2-4} \\ \text{with } \kappa_i^2 &= -\lambda_i = -x\kappa_i - 1,\ \kappa_1+\kappa_2 = -x,\ \kappa_1\kappa_2 = 1. \end{aligned}$$

M is regular if $x \neq 0$. M and M^{root} are semisimple if $x \neq \pm 2$. If $x = \pm 1$, M has eigenvalues $e^{\pm 2\pi i/6}$ and M^{root} has eigenvalues $e^{\pm 2\pi i/3}$ respectively $e^{\pm 2\pi i/6}$. If $|x| \geq 3$, M and M^{root} have real eigenvalues and infinite order. If $x = \pm 2$, they have a 2×2 Jordan block with eigenvalue -1 respectively $-\frac{x}{2}$.

Theorem 5.5 *(a) If $x \neq 0$ then*

$$\begin{aligned} G_{\mathbb{Z}} &= G_{\mathbb{Z}}^{(0)} = G_{\mathbb{Z}}^{(1)} = \{\pm (M^{root})^l \mid l \in \mathbb{Z}\}, \\ G_{\mathbb{Z}}^M &= \begin{cases} G_{\mathbb{Z}} & \text{if } x \neq \pm 3, \\ \{\pm(M^{root} + \frac{x}{|x|}\operatorname{id})^l \mid l \in \mathbb{Z}\} & \text{if } x = \pm 3. \end{cases} \end{aligned}$$

If $x = 0$ then

$$\begin{aligned} G_{\mathbb{Z}} &= G_{\mathbb{Z}}^{(0)} \cong O_2(\mathbb{Z}) = \{\begin{pmatrix} \varepsilon_1 & 0 \\ 0 & \varepsilon_2 \end{pmatrix}, \begin{pmatrix} 0 & \varepsilon_1 \\ \varepsilon_2 & 0 \end{pmatrix} \mid \varepsilon_1, \varepsilon_2 \in \{\pm 1\}\}, \\ G_{\mathbb{Z}}^{(1)} &= G_{\mathbb{Z}}^M = \operatorname{Aut}(H_{\mathbb{Z}}) \cong GL_2(\mathbb{Z}). \end{aligned}$$

In all cases $G_{\mathbb{Z}} = G_{\mathbb{Z}}^{\mathcal{B}}$, so $Z : (\mathrm{Br}_2 \ltimes \{\pm 1\}^2)_S \to G_{\mathbb{Z}}$ is surjective.

(b) Properties of $I^{(0)}$ and $I^{(1)}$:

$$\begin{aligned} x = 0 &: I^{(1)} = 0,\ \operatorname{Rad} I^{(1)} = H_{\mathbb{Z}},\ L(\underline{e}^t, \underline{e})^t = E_2,\ I^{(0)}(\underline{e}^t, \underline{e}) = 2E_2. \\ x \neq 0 &: \operatorname{Rad} I^{(1)} = \{0\}. \\ |x| \leq 1 &: I^{(0)} \text{ is positive definite.} \\ |x| = 2 &: I^{(0)} \text{ is positive semi-definite, } \operatorname{Rad} I^{(0)} = \mathbb{Z}(e_1 - \frac{x}{|x|}e_2). \\ |x| > 2 &: I^{(0)} \text{ is indefinite, } \operatorname{Rad} I^{(0)} = \{0\}. \end{aligned}$$

Proof Part (b) is obvious.

The case $x = 0$ is the case A_1^2. It is covered by Lemma 5.4.

Consider the cases $x \neq 0$ in part (a). We can restrict to $x < 0$ because of $L((e_1, -e_2)^t, (e_1, -e_2))^t = \begin{pmatrix} 1 & -x \\ 0 & 1 \end{pmatrix}$. So suppose $x < 0$.

We know $\{\pm(M^{root})^l \mid l \in \mathbb{Z}\} \subset G_{\mathbb{Z}} \subset G_{\mathbb{Z}}^M$. By Lemma 5.2 (a)

$$G_{\mathbb{Z}}^M = \{p(M^{root}) \mid p(t) = p_1 t + p_0 \in \mathbb{Z}[t],\ p(\kappa_1)p(\kappa_2) \in \{\pm 1\}\}.$$

The map

$$Q_2 : \mathbb{Z}^2 \to \mathbb{Z}, \quad (p_1, p_0) \mapsto p(\kappa_1)p(\kappa_2) = p_1^2 - p_1 p_0 x + p_0^2,$$

is a quadratic form. Lemma 5.2 (a) shows

$$G_{\mathbb{Z}}^M = \{p_1 M^{root} + p_0 \,\mathrm{id} \mid (p_1, p_0, \varepsilon_1) \in \mathbb{Z}^2 \times \{\pm 1\}, Q_2(p_1, p_0) = \varepsilon\}.$$

Lemma 5.2 (b) (ii) shows for $x \neq -2$

$$G_{\mathbb{Z}} = G_{\mathbb{Z}}^{(0)} = G_{\mathbb{Z}}^{(1)} = \{p_1 M^{root} + p_0 \,\mathrm{id} \mid (p_1, p_0) \in \mathbb{Z}^2, Q_2(p_1, p_0) = 1\}.$$

For $x = -2$ it shows only

$$G_{\mathbb{Z}}, G_{\mathbb{Z}}^{(0)}, G_{\mathbb{Z}}^{(1)} \subset \{p_1 M^{root} + p_0 \,\mathrm{id} \mid (p_1, p_0) \in \mathbb{Z}^2, Q_2(p_1, p_0) = 1\}.$$

We discuss the cases $x = -1$, $x = -2$ and $x \leq -3$ separately.

The case $x = -1$**:** Q_2 is positive definite, so $Q_2(p_1, p_0) = -1$ is impossible, and

$$\begin{aligned}\{(p_1, p_0) \mid Q_2(p_1, p_0) = 1\} &= \{\pm(0, 1), \pm(1, 0), \pm(1, -1)\}, \\ G_{\mathbb{Z}} = G_{\mathbb{Z}}^{(0)} = G_{\mathbb{Z}}^{(1)} = G_{\mathbb{Z}}^M &= \{\pm \,\mathrm{id}, \pm M^{root}, \pm (M^{root})^2\}.\end{aligned}$$

Because of $M^2 = -M^{root}$ this equals $\{\pm \,\mathrm{id}, \pm M, \pm M^2\}$.

The case $x = -2$**:** This follows from Lemma 5.6.

Remark: Here Q_2 is positive semidefinite with $Q_2(p_1, p_0) = (p_1 + p_0)^2$. The solution $(p_1, p_0) = (p_1, -p_1 + \varepsilon_2)$ with $\varepsilon_2 \in \{\pm 1\}$ of $Q_2(p_1, p_0) = 1$ corresponds to

$$\begin{aligned}p_1 M^{root} + (-p_1 + \varepsilon_2)\,\mathrm{id} &= \varepsilon_2(\mathrm{id} + \varepsilon_2 p_1 (M^{root} - \mathrm{id})) \\ &= \varepsilon_2(\mathrm{id} + (M^{root} - \mathrm{id}))^{\varepsilon_2 p_1} \\ &= \varepsilon_2 (M^{root})^{\varepsilon_2 p_1}.\end{aligned}$$

The cases $x \leq -3$**:** The arguments above show

$$\begin{aligned}&G_{\mathbb{Z}}^M \cong (\mathbb{Z}[\kappa_1])^* \\ &\supset G_{\mathbb{Z}} \cong \{p_1\kappa_1 + p_0 \in \mathbb{Z}[\kappa_1] \mid (p_1\kappa_1 + p_0)(p_1\kappa_2 + p_0) = 1\}.\end{aligned}$$

So we need to understand the unit group of $\mathbb{Z}[\kappa_1]$ and the subgroup of elements with norm 1. Both are treated in Lemma C.1 (a) in Appendix C.

It remains to show $G_{\mathbb{Z}} = G_{\mathbb{Z}}^{\mathcal{B}}$ in all cases $x \in \mathbb{Z}_{\leq -1}$. This follows from $G_{\mathbb{Z}} = \{\pm (M^{root})^l \mid l \in \mathbb{Z}\}$ and

$$Z(\delta_1\delta_2) = -\,\mathrm{id}, \quad Z(\delta_2\sigma_1) = M^{root}. \qquad \square$$

Lemma 5.6 *Let $H_{\mathbb{Z}}$ be a $\mathbb{Z}$-lattice of rank 2, and let $\widetilde{M} : H_{\mathbb{Z}} \to H_{\mathbb{Z}}$ be an automorphism with a 2×2 Jordan block and eigenvalue $\lambda \in \{\pm 1\}$.*

(a) Then a cyclic automorphism $\widetilde{M}^{root} : H_{\mathbb{Z}} \to H_{\mathbb{Z}}$ with eigenvalue 1 and a number $l \in \mathbb{N}$ with $(\widetilde{M}^{root})^l = \lambda M$ exist. They are unique.

$$\mathrm{Aut}(H_{\mathbb{Z}}, \widetilde{M}) = \{\pm(\widetilde{M}^{root})^l \mid l \in \mathbb{Z}\}.$$

(b) If $\widetilde{I} : H_{\mathbb{Z}} \times H_{\mathbb{Z}} \to \mathbb{Z}$ is an $\widetilde{M}$-invariant bilinear form then it is also $\widetilde{M}^{root}$-invariant and

$$\mathrm{Aut}(H_{\mathbb{Z}}, \widetilde{M}, \widetilde{I}) = \mathrm{Aut}(H_{\mathbb{Z}}, \widetilde{M}) = \{\pm(\widetilde{M}^{root})^l \mid l \in \mathbb{Z}\}.$$

Proof (a) There is a $\mathbb{Z}$-basis $\underline{f} = (f_1, f_2)$ of $H_{\mathbb{Z}}$ and an $l \in \mathbb{N}$ with

$$\widetilde{M}\underline{f} = \underline{f}\lambda \begin{pmatrix} 1 & l \\ 0 & 1 \end{pmatrix}.$$

Here f_1 is a generator of the rank 1 $\mathbb{Z}$-lattice $\ker(\widetilde{M} - \lambda\,\mathrm{id}) \subset H_{\mathbb{Z}}$. It is unique up to the sign. It is a primitive element of $H_{\mathbb{Z}}$. An element f_2 with $H_{\mathbb{Z}} = \mathbb{Z}f_1 \oplus \mathbb{Z}f_2$ exists. It is unique up to sign and up to adding a multiple of f_1. The sign is fixed by l in the matrix above being positive. l is unique.

Define $\widetilde{M}^{root} : H_{\mathbb{Z}} \to H_{\mathbb{Z}}$ by

$$\widetilde{M}^{root}\underline{f} = \underline{f} \begin{pmatrix} 1 & 1 \\ 0 & 1 \end{pmatrix}.$$

Obviously $(\widetilde{M}^{root})^l = \lambda\widetilde{M}$.

Any $g \in \mathrm{Aut}(H_{\mathbb{Z}}, \widetilde{M})$ must fix $\mathbb{Z}f_1 = \ker(\widetilde{M} - \lambda\,\mathrm{id})$. Therefore it must be up to the sign a power of $\widetilde{M}^{root}$.

(b) That $\widetilde{M}^{root}$ respects $\widetilde{I}$ follows by the same arguments as $M^{root} \in G_{\mathbb{Z}}$ in the proof of Lemma 5.2 (b) (i) (but now the situation is simpler, as $\widetilde{M}$ and $\widetilde{M}^{root}$ have a single 2×2 Jordan block). The rest follows with part (a). □

5.3 Generalities on the Rank 3 Cases

For $\underline{x} = (x_1, x_2, x_3) \in \mathbb{Z}^3$ consider the matrix $S = S(\underline{x}) = \begin{pmatrix} 1 & x_1 & x_2 \\ 0 & 1 & x_3 \\ 0 & 0 & 1 \end{pmatrix} \in T_3(\mathbb{Z})$, and consider a unimodular bilinear lattice $(H_{\mathbb{Z}}, L)$ with a triangular basis $\underline{e} = (e_1, e_2, e_3)$ with $L(\underline{e}^t, \underline{e})^t = S$. Then

$$S^{-1} = \begin{pmatrix} 1 & -x_1 & x_1x_3 - x_2 \\ 0 & 1 & -x_3 \\ 0 & 0 & 1 \end{pmatrix},$$

$$S^{-1}S^t = \begin{pmatrix} 1 - x_1^2 - x_2^2 + x_1x_2x_3 & -x_1 - x_2x_3 + x_1x_3^2 & x_1x_3 - x_2 \\ x_1 - x_2x_3 & 1 - x_3^2 & -x_3 \\ x_2 & x_3 & 1 \end{pmatrix},$$

$$M\underline{e} = \underline{e}S^{-1}S^t,$$

$$I^{(0)}(\underline{e}^t, \underline{e}) = S + S^t = \begin{pmatrix} 2 & x_1 & x_2 \\ x_1 & 2 & x_3 \\ x_2 & x_3 & 2 \end{pmatrix},$$

$$I^{(1)}(\underline{e}^t, \underline{e}) = S - S^t = \begin{pmatrix} 0 & x_1 & x_2 \\ -x_1 & 0 & x_3 \\ -x_2 & -x_3 & 0 \end{pmatrix},$$

$$p_{ch,M}(t) = (t-1)(t^2 - (2 - r(\underline{x}))t + 1),$$

where

$$r : \mathbb{Z}^3 \to \mathbb{Z}, \quad \underline{x} = (x_1, x_2, x_3) \mapsto x_1^2 + x_2^2 + x_3^2 - x_1x_2x_3.$$

For $(x_1, x_2, x_3) \neq (0, 0, 0)$ define

$$f_3 := \underline{e}\, \frac{1}{\gcd(x_1, x_2, x_3)} \begin{pmatrix} -x_3 \\ x_2 \\ -x_1 \end{pmatrix}.$$

This is a primitive vector in $H_{\mathbb{Z}}$.

$$\operatorname{Rad} I^{(1)} \overset{2.5\,(a)(ii)}{=} \ker(M - \mathrm{id}) = \begin{cases} \mathbb{Z}f_3 & \text{if } (x_1, x_2, x_3) \neq (0, 0, 0), \\ H_{\mathbb{Z}} & \text{if } (x_1, x_2, x_3) = (0, 0, 0). \end{cases}$$

Also

$$p_{ch,S+S^t}(t) = t^3 - 6t^2 + (12 - x_1^2 - x_2^2 - x_3^2)t - 2(4 - r),$$

$$L(f_3, f_3) = \frac{r(\underline{x})}{\gcd(x_1, x_2, x_3)^2}, \quad I^{(0)}(f_3, f_3) = 2L(f_3, f_3).$$

The eigenvalues of M are called

$$\lambda_{1/2} = \frac{2 - r}{2} \pm \frac{1}{2}\sqrt{r(r-4)}, \quad \lambda_3 = 1$$

with $\lambda_1 + \lambda_2 = 2 - r$, $\lambda_1\lambda_2 = 1$.

The following Lemma gives implicitly precise information on $p_{ch,M}$ and sign $I^{(0)}$ for all $\underline{x} \in \mathbb{Z}^3$. *Implicitly*, because one has to determine with the tools from section 4.2 in the cases $r(\underline{x}) \in \{0, 1, 2, 4\}$ in which $\mathrm{Br}_3 \ltimes \{\pm 1\}^3$ orbit in Theorem 4.6 (e) the matrix $S(\underline{x})$ is.

Lemma 5.7 *(a)* $r^{-1}(3l) = \emptyset$ *for* $l \in \mathbb{Z} - 3\mathbb{Z}$.

(b) Consider $\underline{x} \in \mathbb{Z}^3$ *with* $r = r(\underline{x}) < 0$ *or* > 4 *or with* $S(\underline{x})$ *one of the cases in Theorem 4.6 (e). Then* $p_{ch,M}$ *and* sign $I^{(0)}$ *are as follows.*

	$p_{ch,M}$	sign $I^{(0)}$	$S(\underline{x})$
$r < 0$	$\lambda_1, \lambda_2 > 0$	$(+--)$	$S(\underline{x})$
$r = 0$	Φ_1^3	$(+++)$	$S(A_1^3)$
$r = 0$	Φ_1^3	$(+--)$	$S(\mathbb{P}^2)$
$r = 1$	$\Phi_6\Phi_1$	$(+++)$	$S(A_2A_1)$
$r = 2$	$\Phi_4\Phi_1$	$(+++)$	$S(A_3)$
$r = 4$	$\Phi_2^2\Phi_1$	$(++\,0)$	$S(\mathbb{P}^1A_1)$
$r = 4$	$\Phi_2^2\Phi_1$	$(++\,0)$	$S(\widehat{A}_2)$
$r = 4$	$\Phi_2^2\Phi_1$	$(+\,0\,0)$	$S(\mathcal{H}_{1,2})$
$r = 4$	$\Phi_2^2\Phi_1$	$(+\,0\,-)$	$S(-l, 2, -l)$ with $l \geq 3$
$r > 4$	$\lambda_1, \lambda_2 < 0$	$(++-)$	$S(\underline{x})$

Proof (a) If $(3|x_1, 3|x_2, 3|x_3)$ then $9|r$.
If $(3|x_1, 3|x_2, 3\nmid x_3)$ then $3|(r-1), 3\nmid r$.
If $(3|x_1, 3\nmid x_2, 3\nmid x_3)$ then $3|(r-2), 3\nmid r$.
If $(3\nmid x_1, 3\nmid x_2, 3\nmid x_3)$ then $3|(x_1^2 + x_2^2 + x_3^2), 3\nmid r$.
(b) The statements on $p_{ch,M}$ are obvious.

$$\mathrm{Rad}\, I^{(0)} \overset{2.5\,(a)(ii)}{=} \ker(M + \mathrm{id}) \supsetneqq \{0\} \iff \Phi_2 | p_{ch,M} \iff r = 4.$$

In the cases with $r = 4$, one calculates $p_{ch,S+S^t}(t)$ and reads off sign $I^{(0)}$ from the zeros of $p_{ch,S+S^t}(t)$. The case $S(A_1^3) = S(0, 0, 0)$ is trivial.

Consider the cases with $r \neq 4$ and $\underline{x} \neq (0, 0, 0)$. Then $I^{(0)}$ is nondegenerate. The product of the signs in the signature of $I^{(0)}$ is the sign of $\det(S + S^t) = 2(4 - r)$. Because of the 2's on the diagonal of $S + S^t$, $I^{(0)}$ cannot be negative definite. Also recall $I^{(0)}(f_3, f_3) = 2r(\gcd(x_1, x_2, x_3))^{-2}$. This shows sign $I^{(0)} = (+--)$ for $r < 0$, sign $I^{(0)} = (++-)$ for $r > 4$ and sign $I^{(0)} = (+++)$ or $(+--)$ for $r \in \{0, 1, 2\}$.

The classification of $\mathrm{Br}_3 \ltimes \{\pm 1\}$ orbits in $T_3(\mathbb{Z})$ in Theorem 4.6 (e) says that for each of the cases $r \in \{0, 1, 2\}$ there is only one orbit (with $\underline{x} \neq (0, 0, 0)$ in the case $r = 0$), namely $S(\mathbb{P}^2)$, $S(A_2A_1)$ and $S(A_3)$. One checks the claims on sign $I^{(0)} = \mathrm{sign}(S + S^t)$ immediately. □

Remarks 5.8 (i) It is very remarkable that the fibers $r^{-1}(1)$ and $r^{-1}(2) \subset \mathbb{Z}^3$ of $r : \mathbb{Z}^3 \to \mathbb{Z}$ consist each of only one orbit. If one looks at the fibers of the real map

$$r_{\mathbb{R}} : \mathbb{R}^3 \to \mathbb{R}, \quad \underline{x} \mapsto x_1^2 + x_2^2 + x_3^2 - x_1x_2x_3,$$

this does not hold. Each real fiber $r_{\mathbb{R}}^{-1}(\rho)$ with $\rho \in (0, 4)$ has five components, one compact (homeomorphic to a 2-sphere), four non-compact (homeomorphic to $\mathbb{R}^2$). The four non-compact components are related by the action of G^{sign}. It is remarkable that the fibers $r_{\mathbb{R}}^{-1}(1)$ and $r_{\mathbb{R}}^{-1}(2) \subset \mathbb{R}^3$ intersect $\mathbb{Z}^3$ only in the central piece.

(ii) By $p_{ch,M} = (t-1)(t^2 - (2 - r(\underline{x}))t + 1)$, the monodromy matrix $S^{-1}S^t$ for $\underline{x} \in \mathbb{R}^3$ and $S = S(\underline{x})$ has all eigenvalues in S^1 if and only if $r_{\mathbb{R}}(\underline{x}) \in [0, 4]$.

(iii) The semialgebraic subvariety $r_{\mathbb{R}}^{-1}([0, 4]) \subset \mathbb{R}^3$ was studied in [BH20, 5.2]. It has a central piece which is G^{sign} invariant and which looks like a tetrahedron with smoothened edges and four other pieces which are permuted by G^{sign}. Each other piece is homeomorphic to $[0, 1] \times \mathbb{R}^2$ and is glued in one point (its only singular point) to one of the vertices of the central piece. The four vertices are $(2, 2, 2), (2, -2, -2), (-2, 2, -2), (-2, -2, 2)$, so the elements of the $\mathrm{Br}_3 \ltimes \{\pm 1\}^3$ orbit of $(2, 2, 2)$.

For $\underline{x}$ inside the central piece $S + S^t$ is positive definite, on its boundary except the vertices $\mathrm{sign}(S + S^t) = (++0)$, at a vertex $\mathrm{sign}(S + S^t) = (+00)$. On those boundary components of the other four pieces which contain one of the vertices $\mathrm{sign}(S + S^t) = (+0-)$ (except at the vertex). On the interior of $r_{\mathbb{R}}^{-1}(-\infty, 4]$ except the central piece $\mathrm{sign}(S + S^t) = (+--)$. On the exterior of $r_{\mathbb{R}}^{-1}(-\infty, 4]$ $\mathrm{sign}(S + S^t) = (++-)$.

(iv) Due to Lemma 5.7 (b) and Theorem 4.6, the seven cases $S(A_1^3)$, $S(\mathbb{P}^2)$, $S(A_2A_1)$, $S(A_3)$, $S(\mathbb{P}^1A_1)$, $S(\widehat{A}_2)$, $S(\mathcal{H}_{1,2})$ and the series $S(-l, 2, -l)$ for $l \geq 3$ give the only rank 3 unimodular bilinear lattices where all eigenvalues of the monodromy are roots of unity. In the Sects. 5.4 and 5.5 we will focus on the reducible cases and these cases. In the Sects. 5.6 and 5.7 we will treat the other cases.

The following definition presents a special automorphism Q in $\mathrm{Aut}(H_{\mathbb{Q}}, L)$. Theorem 5.11 will say in which cases Q is in $G_{\mathbb{Z}}$ and in which cases not. The determination of the group $G_{\mathbb{Z}}$ in all irreducible rank 3 cases in the Sects. 5.5–5.7 will build on this result. It is preceded by Lemma 5.10 which provides notations and estimates which will be used in the proof of Theorem 5.11 and also later.

Definition 5.9 Consider $\underline{x} \in \mathbb{Z}^3$ with $r(\underline{x}) \neq 0$. Then $H_{\mathbb{Q}} = H_{\mathbb{Q},1} \oplus H_{\mathbb{Q},2}$ with $H_{\mathbb{Q},1} := \ker(M^2 - (2 - r(\underline{x}))M + \mathrm{id} : H_{\mathbb{Q}} \to H_{\mathbb{Q}})$ and $H_{\mathbb{Q},2} := \ker(M - \mathrm{id} : H_{\mathbb{Q}} \to H_{\mathbb{Q}})$. This decomposition is left and right L-orthogonal. Then $Q : H_{\mathbb{Q}} \to H_{\mathbb{Q}}$ denotes the automorphism with $Q|_{H_{\mathbb{Q},1}} = -\mathrm{id}$ and $Q|_{H_{\mathbb{Q},2}} = \mathrm{id}$. It is in $\mathrm{Aut}(H_{\mathbb{Q}}, L)$.

Lemma 5.10 *(a) For $\underline{x} \in \mathbb{Z}^3 - \{(0, 0, 0)\}$ write $r := r(\underline{x})$ and define*

$$g := g(\underline{x}) := \gcd(x_1, x_2, x_3) \in \mathbb{N},$$
$$\widetilde{\underline{x}} := (\widetilde{x}_1, \widetilde{x}_2, \widetilde{x}_3) := g^{-1}\underline{x} \in \mathbb{Z}^3.$$

Then $f_3 = -\widetilde{x}_3e_1 + \widetilde{x}_2e_2 - \widetilde{x}_1e_3$, $\gcd(\widetilde{x}_1, \widetilde{x}_2, \widetilde{x}_3) = 1$ and

$$g^2 \,|\, r, \quad \frac{r}{g^2} = \widetilde{x}_1^2 + \widetilde{x}_2^2 + \widetilde{x}_3^2 - g\widetilde{x}_1\widetilde{x}_2\widetilde{x}_3. \tag{5.1}$$

(b) Consider a local minimum (Definition 4.3) $\underline{x} \in \mathbb{Z}^3_{\geq 3}$ with $x_i \leq x_j \leq x_k$ for some i, j, k with $\{i, j, k\} = \{1, 2, 3\}$. Then

$$\widetilde{x}_i \leq \frac{2 + (4 - r)^{1/3}}{g}, \tag{5.2}$$

$$x_j^2 \leq \frac{4 - r}{x_i - 2} + x_i + 2, \tag{5.3}$$

$$x_k \leq \frac{1}{2}x_i x_j \quad \text{and} \quad \widetilde{x}_k \leq \frac{g}{2}\widetilde{x}_i\widetilde{x}_j. \tag{5.4}$$

Proof (a) Trivial.

(b) Lemma 4.4 shows $x_k \leq \frac{1}{2}x_i x_j$, which is (5.4). This is equivalent to $\widetilde{x}_k \leq \frac{g}{2}\widetilde{x}_i\widetilde{x}_j$. We also have $\widetilde{x}_i \leq \widetilde{x}_j \leq \widetilde{x}_k$, and we know $r \leq 0$ from Theorem 4.6.

The proof of (5.2) is similar to the second case in the proof of Theorem 4.6 (a).

$$\begin{aligned}
\frac{r}{g^2} &= \widetilde{x}_i^2 + \widetilde{x}_j^2 + (\widetilde{x}_k - \frac{g}{2}\widetilde{x}_i\widetilde{x}_j)^2 - \frac{g^2}{4}\widetilde{x}_i^2\widetilde{x}_j^2 \\
&\leq \widetilde{x}_i^2 + \widetilde{x}_j^2 + (\widetilde{x}_j - \frac{g}{2}\widetilde{x}_i\widetilde{x}_j)^2 - \frac{g^2}{4}\widetilde{x}_i^2\widetilde{x}_j^2 \quad (\text{because } \widetilde{x}_j \leq \widetilde{x}_k \leq \frac{g}{2}\widetilde{x}_i\widetilde{x}_j) \\
&= \widetilde{x}_i^2 + 2\widetilde{x}_j^2 - g\widetilde{x}_i\widetilde{x}_j^2 \\
&= (\widetilde{x}_i - g\widetilde{x}_j^2 + \frac{2}{g})(\widetilde{x}_i - \frac{2}{g}) + \frac{4}{g^2}.
\end{aligned}$$

If $\widetilde{x}_i < \frac{2}{g}$, then (5.2) holds anyway. If $\widetilde{x}_i \geq \frac{2}{g}$ then we can further estimate the last formula using $-g\widetilde{x}_j^2 \leq -g\widetilde{x}_i^2$. Then we obtain

$$\begin{aligned}
\frac{r}{g^2} &\leq (\widetilde{x}_i - g\widetilde{x}_i^2 + \frac{2}{g})(\widetilde{x}_i - \frac{2}{g}) + \frac{4}{g^2} \\
&= -g(\widetilde{x}_i - \frac{2}{g})^2(\widetilde{x}_i + \frac{1}{g}) + \frac{4}{g^2} \\
&\leq -g(\widetilde{x}_i - \frac{2}{g})^3 + \frac{4}{g^2},
\end{aligned}$$

$$\text{so } (\widetilde{x}_i - \frac{2}{g})^3 \leq \frac{4 - r}{g^3}.$$

This shows (5.2). The inequality (5.3) was proved within the second case in the proof of Theorem 4.6 (a). □

Theorem 5.11 *Consider $\underline{x} \in \mathbb{Z}^3$ with $r \neq 0$. The automorphism $Q \in \mathrm{Aut}(H_{\mathbb{Q}}, L)$ which was defined in Definition 5.9 can be written in two interesting ways,*

$$Q = (\underline{e} \mapsto -\underline{e} + \frac{2g^2}{r} f_3(-\widetilde{x}_3, \widetilde{x}_2 - g\widetilde{x}_1\widetilde{x}_3, -\widetilde{x}_1)), \tag{5.5}$$

$$Q = \mathrm{id} + 2(M - \mathrm{id}) + \frac{2}{r}(M - \mathrm{id})^2. \tag{5.6}$$

We have

$$Q \in G_{\mathbb{Z}} \iff \frac{2g^2}{r} \in \mathbb{Z} \iff \frac{r}{g^2} \in \{\pm 1, \pm 2\}. \tag{5.7}$$

This holds if and only if $\underline{x}$ *is in the* $\mathrm{Br}_3 \ltimes \{\pm 1\}^3$ *orbit of a triple in the following set:*

$$\begin{aligned} &\{(x, 0, 0) \mid x \in \mathbb{N}\} \quad \text{(these are the reducible cases except } A_1^3\text{)},\\ \cup\, &\{(x, x, 0) \mid x \in \mathbb{N}\} \quad \text{(these cases include } A_3\text{)},\\ \cup\, &\{(-l, 2, -l) \mid l \geq 2 \text{ even}\} \quad \text{(these cases include } \mathcal{H}_{1,2}\text{)},\\ \cup\, &\{(3, 3, 4), (4, 4, 4), (5, 5, 5), (4, 4, 8)\}. \end{aligned}$$

So, within the cases with $r \in \{0, 1, 2, 3, 4\}$, Q *is not defined for* A_1^3 *and* $\mathbb{P}^2$, *and* $Q \notin G_{\mathbb{Z}}$ *for* $\widehat{A}_2$ *and* $S(-l, 2, -l)$ *with* $l \geq 3$ *odd.*

Proof First we prove (5.5). The 2-dimensional subspace $H_{\mathbb{Q},1} = \ker(M^2 - (2 - r)M + \mathrm{id}) \subset H_{\mathbb{Q}}$, on which Q is $-\mathrm{id}$, can also be characterized as the right L-orthogonal subspace $H_{\mathbb{Q},1} = (\mathbb{Q}f_3)^{\perp}$ to $\mathbb{Q}f_3$. For $b = \underline{e} \cdot \underline{y}^t \in H_{\mathbb{Q}}$ with $\underline{y} \in \mathbb{Q}^3$

$$\begin{aligned} L(f_3, b) &= (-\widetilde{x}_3, \widetilde{x}_2, -\widetilde{x}_1) \begin{pmatrix} 1 & 0 & 0 \\ x_1 & 1 & 0 \\ x_2 & x_3 & 1 \end{pmatrix} \begin{pmatrix} y_1 \\ y_2 \\ y_3 \end{pmatrix} \\ &= (-\widetilde{x}_3, \widetilde{x}_2 - g\widetilde{x}_1\widetilde{x}_3, -\widetilde{x}_1) \begin{pmatrix} y_1 \\ y_2 \\ y_3 \end{pmatrix}, \end{aligned}$$

so

$$H_{\mathbb{Q},1} = \{\underline{e} \cdot \underline{y}^t \mid \underline{y} \in \mathbb{Q}^3, 0 = (-\widetilde{x}_3, \widetilde{x}_2 - g\widetilde{x}_1\widetilde{x}_3, -\widetilde{x}_1) \begin{pmatrix} y_1 \\ y_2 \\ y_3 \end{pmatrix}\}. \tag{5.8}$$

Denote the endomorphisms on the right hand sides of (5.5) and (5.6) by $Q^{(5.5)}$ respectively $Q^{(5.6)}$. The formulas (5.5) and (5.8) show $Q^{(5.5)}|_{H_{\mathbb{Q},1}} = -\mathrm{id}$. Also

$$Q^{(5.5)}(f_3) = -f_3 + f_3 \frac{2g^2}{r}(\widetilde{x}_3^2 + (\widetilde{x}_2 - g\widetilde{x}_1\widetilde{x}_3)\widetilde{x}_2 + \widetilde{x}_1^2) = f_3.$$

Therefore $Q^{(5.5)} = Q$, so (5.5) holds.

Now we prove (5.6). Because $(M - \mathrm{id})(f_3) = 0$, we have $Q^{(5.6)}(f_3) = f_3$. Consider $b \in H_{\mathbb{Q},1}$. Then

$$\begin{aligned} 0 &= (M^2 - (2-r)M + \mathrm{id})(b), \quad \text{so} \\ (M - \mathrm{id})^2(b) &= -rM(b), \quad \text{so} \\ Q^{(5.6)}(b) &= (\mathrm{id} + 2(M - \mathrm{id}) + \frac{2}{r}(-rM))(b) = (-id)(b) = -b. \end{aligned}$$

Therefore $Q^{(5.6)} = Q$, so (5.6) holds.

(5.7) can be proved either with (5.6) and Lemma 5.17 below or with (5.5), which is easier and which we do now. Observe $\gcd(-\widetilde{x}_3, \widetilde{x}_2 - g\widetilde{x}_1\widetilde{x}_3, -\widetilde{x}_1) = \gcd(\widetilde{x}_1, \widetilde{x}_2, \widetilde{x}_3) = 1$. Also, f_3 is a primitive vector in $H_{\mathbb{Z}}$. This shows that $Q(e_1), Q(e_2), Q(e_3)$ are all in $H_{\mathbb{Z}}$ if and only if $\frac{2g^2}{r} \in \mathbb{Z}$. This shows (5.7).

It is easy to see that all triples in the set in Theorem 5.11 satisfy $\frac{r}{g^2} \in \{\pm 1, \pm 2\}$:

$$\frac{r}{g^2}((x,0,0)) = \frac{x^2}{x^2} = 1, \quad \frac{r}{g^2}((x,x,0)) = \frac{2x^2}{x^2} = 2,$$
$$\text{for even } l \quad \frac{r}{g^2}((-l,2,-l)) = \frac{4}{4} = 1,$$

$$\frac{r}{g^2}((3,3,4)) = \frac{-2}{1} = -2, \quad \frac{r}{g^2}((4,4,4)) = \frac{-16}{16} = -1,$$
$$\frac{r}{g^2}((5,5,5)) = \frac{-50}{25} = -2, \quad \frac{r}{g^2}((4,4,8)) = \frac{-32}{16} = -2.$$

The difficult part is to see that there are no other $\underline{x} \in \mathbb{Z}^3$ with $r \neq 0$ and $\frac{r}{g^2} \in \{\pm 1, \pm 2\}$. It is sufficient to consider local minima (Definition 4.3). The calculations

$$\frac{r}{g^2}((-l,2,-l)) = \frac{4}{1} = 4 \quad \text{for odd } l \geq 3$$
$$\text{and} \quad \frac{r}{g^2}((-1,-1,-1)) = \frac{4}{1} = 4$$

deal with the other cases with $r \in \{1,2,3,4\}$, see Theorem 4.6 (e).

Consider $\underline{x} \in \mathbb{Z}^3_{\leq 0}$ with $x_i \leq x_j \leq x_k$ for some i, j, k with $\{i, j, k\} = \{1, 2, 3\}$. Then $\frac{r}{g^2} = \widetilde{x}_1^2 + \widetilde{x}_2^2 + \widetilde{x}_3^2 + g|\widetilde{x}_1\widetilde{x}_2\widetilde{x}_3|$ can be 1 or 2 only if $\widetilde{x}_k = 0$ and $\widetilde{x}_i = \widetilde{x}_j = -1$. Then $\underline{x} = (-g, -g, 0)$. This is in the $\mathrm{Br}_3 \ltimes \{\pm 1\}^3$ orbit of $(g, g, 0)$.

Consider a local minimum $\underline{x} \in \mathbb{Z}^3_{\geq 3}$ with $x_i \leq x_j \leq x_k$ for some i, j, k with $\{i, j, k\} = \{1, 2, 3\}$. Suppose $\frac{r}{g^2} \in \{-1, -2\}$. We have to show $\underline{x} \in \{(3,3,4), (4,4,4), (5,5,5), (4,4,8)\}$. Of course $\underline{x} \neq (3,3,3)$ because $r \neq 0$.

If $g = 1$ then $r \in \{-1, -2\}$. One sees easily that $r = -1$ is impossible and that $r = -2$ is only satisfied for $\underline{x} = (3,3,4)$.

From now on suppose $g \geq 2$. Write $\rho := |\frac{r}{g^2}| \in \{1, 2\}$. (5.2) takes the shape

$$\widetilde{x}_i \leq \frac{2}{g} + \left(\frac{\rho}{g} + \frac{4}{g^3}\right)^{1/3}.$$

The only pairs $(\widetilde{x}_i, g) \in \mathbb{N} \times \mathbb{Z}_{\geq 2}$ which satisfy this and $x_i = \widetilde{x}_i g \geq 3$ are in the following two tables,

$\rho = 1$:

g	2	3	4	5
$\widetilde{x}_i$	2	1	1	1

$\rho = 2$:

g	2	3	4	5	6
$\widetilde{x}_i$	2	1	1	1	1

The following table discusses these nine cases. Three of them lead to $(4, 4, 4)$, $(5, 5, 5)$ and $(4, 4, 8)$, six of them are impossible. The symbol $\circledast$ denotes *impossible*. The inequalities $x_i \leq x_j \leq x_k$ and (5.3) and (5.4) are used, and also $x_i = g\widetilde{x}_i$ and $x_j, x_k \in g\mathbb{N}$.

ρ	g	$-r$	$\widetilde{x}_i$	x_i	$x_j^2 \leq \frac{4-r}{x_i-2} + x_i + 2$	x_j	$x_k \leq \frac{1}{2}x_i x_j$	x_k
1	2	4	2	4	$x_j^2 \leq 10$	$\circledast$		
1	3	9	1	3	$x_j^2 \leq 18$	3	$x_k \leq \frac{9}{2}$	3 $\circledast$
1	4	16	1	4	$x_j^2 \leq 16$	4	$x_k \leq 8$	4
1	5	25	1	5	$x_j^2 \leq 16 + \frac{2}{3}$	$\circledast$		
2	2	8	2	4	$x_j^2 \leq 12$	$\circledast$		
2	3	18	1	3	$x_j^2 \leq 27$	3	$x_k \leq \frac{9}{2}$	3 $\circledast$
2	4	32	1	4	$x_j^2 \leq 24$	4	$x_k \leq 8$	8
2	5	50	1	5	$x_j^2 \leq 25$	5	$x_k \leq 12 + \frac{1}{2}$	5
2	6	72	1	6	$x_j^2 \leq 27$	$\circledast$		

This finishes the proof of Theorem 5.11. □

5.4 The Reducible Rank 3 Cases

Definition 2.10 proposed the notion of reducible triple $(H_{\mathbb{Z}}, L, \underline{e})$ where $(H_{\mathbb{Z}}, L)$ is a unimodular bilinear lattice and $\underline{e}$ is a triangular basis. The following remarks propose the weaker notion when a unimodular bilinear lattice $(H_{\mathbb{Z}}, L)$ (without a triangular basis) is reducible. Then the groups $G_{\mathbb{Z}}$, $G_{\mathbb{Z}}^{(0)}$, $G_{\mathbb{Z}}^{(1)}$ and $G_{\mathbb{Z}}^{M}$ split accordingly if also the eigenvalues of M split in a suitable sense.

Remarks 5.12 (i) Suppose that a unimodular bilinear lattice $(H_{\mathbb{Z}}, L)$ splits into a direct sum $H_{\mathbb{Z},1} \oplus H_{\mathbb{Z},2}$ which is left and right L-orthogonal. Then the restrictions of L, $I^{(0)}$, $I^{(1)}$ and M to $H_{\mathbb{Z},i}$ are called L_i, $I_i^{(0)}$, $I_i^{(1)}$ and M_i for $i \in \{1; 2\}$. We say that $(H_{\mathbb{Z}}, L)$ is *reducible* and that it splits into the direct sum $(H_{\mathbb{Z},1}, L_1) \oplus (H_{\mathbb{Z},2}, L_2)$.

(ii) In the situation of (i), suppose that the eigenvalues of M_1 are pairwise different from the eigenvalues of M_2. Then any element of $G_{\mathbb{Z}}^M$ respects the splitting. For $\{i, j\} = \{1, 2\}$ write $G_{\mathbb{Z},i}^M := G_{\mathbb{Z}}^M(H_{\mathbb{Z},i}, L_i)$. Then

$$G_{\mathbb{Z}}^M = G_{\mathbb{Z},1}^M \times G_{\mathbb{Z},2}^M,$$

and with analogous notations

$$G_{\mathbb{Z}} = G_{\mathbb{Z},1} \times G_{\mathbb{Z},2},\ G_{\mathbb{Z}}^{(0)} = G_{\mathbb{Z},1}^{(0)} \times G_{\mathbb{Z},2}^{(0)},\ G_{\mathbb{Z}}^{(1)} = G_{\mathbb{Z},1}^{(1)} \times G_{\mathbb{Z},2}^{(1)}.$$

(iii) There is only one unimodular bilinear lattice of rank 1. We call it A_1-lattice and denote the matrix $S = S(A_1) = (1) \in M_{1\times 1}(\mathbb{Z})$. Here $G_{\mathbb{Z}} = G_{\mathbb{Z}}^{(0)} = G_{\mathbb{Z}}^{(1)} = G_{\mathbb{Z}}^M = \{\pm \mathrm{id}\}$.

(iv) Suppose that the characteristic polynomial $p_{ch,M}(t) \in \mathbb{Z}[t]$ of the monodromy M of a unimodular bilinear lattice $(H_{\mathbb{Z}}, L)$ splits into a product $p_{ch,M} = p_1 p_2$ of non-constant polynomials p_1 and p_2 with $\gcd(p_1, p_2) = 1$. Then $\ker p_1(M) \oplus \ker p_2(M)$ is a sublattice of finite index in $H_{\mathbb{Z}}$, and the summands are left and right L-orthogonal to one another. If the index is 1, we are in the situation of (ii). Theorem 5.14 will show that this applies to the cases $S(\mathcal{H}_{1,2})$ and $S(-l, 2, -l)$ with $l \geq 4$ even, but not to the cases $S(A_3)$, $S(\widehat{A}_2)$ and $S(-l, 2, -l)$ with $l \geq 3$ odd.

These remarks apply especially to the reducible 3×3 cases except A_1^3 (which is part of Lemma 5.4). This includes the two reducible cases A_2A_1 and $\mathbb{P}^1A_1$ with eigenvalues in S^1.

Theorem 5.13 *Consider $\underline{x} = (x, 0, 0) \in \mathbb{Z}^3$ with $x \neq 0$ and the unimodular bilinear lattice $(H_{\mathbb{Z}}, L, \underline{e})$ with triangular basis $\underline{e}$ with $L(\underline{e}^t, \underline{e})^t = S(\underline{x}) \in T_3(\mathbb{Z})$.*

Then $(H_{\mathbb{Z}}, L, \underline{e})$ is reducible with the summands $(H_{\mathbb{Z},1}, L_1, (e_1, e_2))$ and $(H_{\mathbb{Z},2}, L_2, e_3)$ with $H_{\mathbb{Z},1} = \mathbb{Z}e_1 \oplus \mathbb{Z}e_2$ and $H_{\mathbb{Z},3} = \mathbb{Z}e_3$. The first summand is an irreducible rank two unimodular bilinear lattice with triangular basis. Its groups $G_{\mathbb{Z},1}, G_{\mathbb{Z},1}^{(0)}, G_{\mathbb{Z},1}^{(1)}$ and $G_{\mathbb{Z},1}^M$ are treated in Theorem 5.5. The second summand is of type A_1. See Remark 5.12 (iii) for its groups.

The decompositions in Remark 5.12 (ii) hold for the groups $G_{\mathbb{Z}}, G_{\mathbb{Z}}^{(0)}, G_{\mathbb{Z}}^{(1)}$ and $G_{\mathbb{Z}}^M$. Here $G_{\mathbb{Z}}^M = G_{\mathbb{Z}}$ if $x \neq \pm 3$ and $G_{\mathbb{Z}} = G_{\mathbb{Z}}^{(0)} = G_{\mathbb{Z}}^{(1)}$ always.

The map $Z : (\mathrm{Br}_3 \ltimes \{\pm 1\}^3)_S \to G_{\mathbb{Z}}$ is surjective.

Proof The first point to see is that Remark 5.122 (ii) applies. It does because the characteristic polynomials of the monodromies M_1 and M_2 of the two summands are $t^2 - (2 - x^2)t + 1$ and $t - 1$, and here $x \neq 0$, so that the eigenvalues of M_1 are not equal to the eigenvalue 1 of M_2.

The second point to see is the surjectivity of the map Z. This follows from the surjectivity of the map Z in the irreducible rank 2 cases in Theorem 5.5 and in the case A_1 in Lemma 5.4. □

5.5 The Irreducible Rank 3 Cases with All Eigenvalues in S^1

Theorem 5.14 is the only point in this section. It treats the irreducible rank 3 cases with all eigenvalues in S^1.

Theorem 5.14 *Consider for each of the matrices $S(\mathbb{P}^2)$, $S(A_3)$, $S(\widehat{A}_2)$, $S(\mathcal{H}_{1,2})$ and $S(-l,2,-l)$ for $l \geq 3$ in the Examples 1.1 a unimodular bilinear lattice $(H_\mathbb{Z}, L)$ with a triangular basis $\underline{e}$ with $L(\underline{e}^t, \underline{e})^t = S$.*

(a) The cases $S(\mathcal{H}_{1,2})$ and $S(-l,2,-l)$ for $l \geq 4$ even: Then $(H_\mathbb{Z}, L)$ is reducible (in the sense of Remark 5.12 (i)), $H_\mathbb{Z} = H_{\mathbb{Z},1} \oplus H_{\mathbb{Z},2}$ with

$$\begin{aligned} H_{\mathbb{Z},1} &:= \ker(M + \mathrm{id})^2 \text{ of rank 2} \\ \text{and } H_{\mathbb{Z},2} &:= \ker(M - \mathrm{id}) \text{ of rank 1.} \end{aligned}$$

$(H_{\mathbb{Z},2}, L_2)$ is an A_1-lattice. In all cases the decompositions in Remark 5.12 (ii) hold for the groups $G_\mathbb{Z}^M$, $G_\mathbb{Z}^{(0)}$, $G_\mathbb{Z}^{(1)}$ and $G_\mathbb{Z}$. The groups $G_{\mathbb{Z},1}^M$, $G_{\mathbb{Z},1}^{(0)}$, $G_{\mathbb{Z},1}^{(1)}$ and $G_{\mathbb{Z},1}$ are as follows.

(i) $S(\mathcal{H}_{1,2})$: $H_{\mathbb{Z},1}$ has a $\mathbb{Z}$-basis $\underline{f} = (f_1, f_2)$ with

$$\begin{aligned} L(\underline{f}^t, \underline{f})^t &= \begin{pmatrix} 0 & -1 \\ 1 & 0 \end{pmatrix}, \quad I^{(0)}(\underline{f}^t, \underline{f}) = \begin{pmatrix} 0 & 0 \\ 0 & 0 \end{pmatrix}, \\ I^{(1)}(\underline{f}^t, \underline{f}) &= \begin{pmatrix} 0 & -2 \\ 2 & 0 \end{pmatrix}, \quad M\underline{f} = \underline{f}\begin{pmatrix} -1 & 0 \\ 0 & -1 \end{pmatrix}, \end{aligned}$$

$$\begin{aligned} G_{\mathbb{Z},1}^M &= G_{\mathbb{Z},1}^{(0)} = \mathrm{Aut}(H_{\mathbb{Z},1}) \cong GL_2(\mathbb{Z}), \\ G_{\mathbb{Z},1} &= G_{\mathbb{Z},1}^{(1)} = \{g \in \mathrm{Aut}(H_{\mathbb{Z},1}) \mid \det g = 1\} \cong SL_2(\mathbb{Z}). \end{aligned}$$

(ii) $S(-l,2,-l)$ for $l \geq 4$ even: $H_{\mathbb{Z},1}$ has a $\mathbb{Z}$-basis $\underline{f} = (f_1, f_2)$ with

$$\begin{aligned} L(\underline{f}^t, \underline{f})^t &= \begin{pmatrix} 0 & -1 \\ 1 & 1 - \frac{l^2}{4} \end{pmatrix}, \quad I^{(0)}(\underline{f}^t, \underline{f}) = \begin{pmatrix} 0 & 0 \\ 0 & 2 - \frac{l^2}{2} \end{pmatrix}, \\ I^{(1)}(\underline{f}^t, \underline{f}) &= \begin{pmatrix} 0 & -2 \\ 2 & 0 \end{pmatrix}, \quad M\underline{f} = \underline{f}\begin{pmatrix} -1 & 2 - \frac{l^2}{2} \\ 0 & -1 \end{pmatrix}, \end{aligned}$$

Define $M_1^{root} \in \mathrm{Aut}(H_{\mathbb{Z},1})$ by $M_1^{root}\underline{f} = \underline{f}\begin{pmatrix} 1 & 1 \\ 0 & 1 \end{pmatrix}$. Then

$$\begin{aligned} (M_1^{root})^{l^2/2-2} &= -M_1 \qquad \text{and} \\ G_{\mathbb{Z},1} = G_{\mathbb{Z},1}^{(0)} = G_{\mathbb{Z},1}^{(1)} = G_{\mathbb{Z},1}^M &= \{\pm(M_1^{root})^m \mid m \in \mathbb{Z}\}. \end{aligned}$$

(b) The cases $S(\mathbb{P}^2)$, $S(A_3)$, $S(\widehat{A}_2)$, $S(-l,2,-l)$ with $l \geq 3$ odd: Then $(H_\mathbb{Z}, L)$ is irreducible.

$$G_\mathbb{Z} = G_\mathbb{Z}^{(0)} = \{\pm (M^{root})^m \,|\, m \in \mathbb{Z}\}.$$

Here M^{root} is defined by

$$M^{root} := Z(\sigma^{root}) \quad \text{in the case } S(\mathbb{P}^2), \tag{5.9}$$

$$M^{root} := M = Z(\sigma^{mon}) \quad \text{in the case } S(A_3), \tag{5.10}$$

$$M^{root} := Z(\delta_3 \sigma^{root}) \quad \text{in the case } S(\widehat{A}_2), \tag{5.11}$$

$$M^{root} := (-M) \circ Z(\delta_3 \sigma_1^{-1} \sigma_2^{-1} \sigma_1)^{(5-l^2)/2} \quad \text{in the case } S(-l,2,-l) \text{ for odd } l \geq 3. \tag{5.12}$$

It satisfies

$$M^{root} \underline{e} = \underline{e} M^{root,mat} \quad \text{and} \quad (M^{root})^m = \varepsilon M$$

where $M^{root,mat}$, m and ε are as follows:

	$S(\mathbb{P}^2)$	$S(A_3)$
$M^{root,mat}$	$\begin{pmatrix} 3 & -3 & 1 \\ 1 & 0 & 0 \\ 0 & 1 & 0 \end{pmatrix}$	$\begin{pmatrix} 0 & 0 & 1 \\ -1 & 0 & 1 \\ 0 & -1 & 1 \end{pmatrix}$
(m,ε)	$(3,1)$	$(1,1)$

	$S(\widehat{A}_2)$	$S(-l,2,-l)$ with $l \geq 3$ odd
$M^{root,mat}$	$\begin{pmatrix} 1 & 1 & -1 \\ 1 & 0 & 0 \\ 0 & 1 & 0 \end{pmatrix}$	$\frac{1}{2}\begin{pmatrix} 1-l^2 & l^3-l & -1-l^2 \\ -2l & 2l^2-2 & -2l \\ 1-l^2 & l^3-3l & 3-l^2 \end{pmatrix}$
(m,ε)	$(3,-1)$	$(l^2-4,-1)$

M and M^{root} are regular. M^{root} is cyclic. In the cases $S(A_3)$, $S(\widehat{A}_2)$ and $S(-l,2,-l)$ for $l \geq 3$ odd

$$G_\mathbb{Z} = G_\mathbb{Z}^{(0)} = G_\mathbb{Z}^{(1)} = G_\mathbb{Z}^M = \{\pm (M^{root})^m \,|\, m \in \mathbb{Z}\}.$$

Some additional information:

(i) $S(\mathbb{P}^2)$*:* sign $I^{(0)} = (+\,-\,-)$, $p_{ch,M} = p_{ch,M^{root}} = \Phi_1^3$, *$M$ and M^{root} have a 3×3 Jordan block,*

$$G_\mathbb{Z}^{(1)} = G_\mathbb{Z}^M = \{\pm (M^{root})^m (\mathrm{id} + a(M^{root} - \mathrm{id})^2) \,|\, m, a \in \mathbb{Z}\} \supsetneqq G_\mathbb{Z}.$$

(ii) $S(A_3)$*:* $\text{sign}\, I^{(0)} = (+++)$*,* $p_{ch,M} = \Phi_4\Phi_1$*,* $M = M^{root}$*,* $|G_{\mathbb{Z}}| = 8$.

(iii) $S(\widehat{A}_2)$*:* $\text{sign}\, I^{(0)} = (++\,0)$*,* $p_{ch,M} = \Phi_2^2\Phi_1$*,* $p_{ch,M^{root}} = \Phi_1^2\Phi_2$*, M and M^{root} have a 2×2 Jordan block with eigenvalue -1 respectively* 1.

(iv) $S(-l,2,-l)$ *with $l \geq 3$ odd:* $\text{sign}\, I^{(0)} = (+0-)$*,* $p_{ch,M} = \Phi_2^2\Phi_1$*,* $p_{ch,M^{root}} = \Phi_1^2\Phi_2$*, M and M^{root} have a 2×2 Jordan block with eigenvalue -1 respectively* 1.

(c) In all cases in this theorem the map $Z : (\text{Br}_3 \ltimes \{\pm 1\}^3)_S \to G_{\mathbb{Z}}$ is surjective, so $G_{\mathbb{Z}} = G_{\mathbb{Z}}^{\mathcal{B}}$.

Proof (a) Recall

$$H_{\mathbb{Z},2} = \ker(M - \text{id}) = \text{Rad}\, I^{(1)} = \mathbb{Z}f_3,\ f_3 = -\widetilde{x}_3e_1 + \widetilde{x}_2e_2 - \widetilde{x}_1e_3.$$

We will choose a $\mathbb{Z}$-basis $\underline{f} = (f_1, f_2)$ of $H_{\mathbb{Z},1} := \ker(M+\text{id})^2$. Denote $\underline{\widetilde{f}} := (f_1, f_2, f_3)$. In all cases it will be easy to see that $\underline{\widetilde{f}}$ is a $\mathbb{Z}$-basis of $H_{\mathbb{Z}}$. Therefore in all cases $H_{\mathbb{Z}} = H_{\mathbb{Z},1} \oplus H_{\mathbb{Z},2}$.

(i) $S(\mathcal{H}_{1,2})$: Recall $p_{ch,M} = \Phi_2^2\Phi_1$,

$$S = \begin{pmatrix} 1 & -2 & 2 \\ 0 & 1 & -2 \\ 0 & 0 & 1 \end{pmatrix},\ S^{-1}S^t = \begin{pmatrix} 1 & -2 & 2 \\ 2 & -3 & 2 \\ 2 & -2 & 1 \end{pmatrix},\ f_3 = \underline{e}\begin{pmatrix} 1 \\ 1 \\ 1 \end{pmatrix}.$$

Define

$$f_1 := \underline{e}\begin{pmatrix} 1 \\ 1 \\ 0 \end{pmatrix},\ f_2 := \underline{e}\begin{pmatrix} 0 \\ 1 \\ 1 \end{pmatrix}.$$

Then

$$L(\underline{\widetilde{f}}^t, \underline{\widetilde{f}})^t = \begin{pmatrix} 0 & -1 & 0 \\ 1 & 0 & 0 \\ 0 & 0 & 1 \end{pmatrix},\ M\underline{\widetilde{f}} = \underline{\widetilde{f}}\begin{pmatrix} -1 & 0 & 0 \\ 0 & -1 & 0 \\ 0 & 0 & 1 \end{pmatrix},\ H_{\mathbb{Z},1} = \mathbb{Z}f_1 \oplus \mathbb{Z}f_2.$$

The claims on the groups $G_{\mathbb{Z},1}^M, G_{\mathbb{Z},1}^{(1)}, G_{\mathbb{Z},1}^{(0)}$ and $G_{\mathbb{Z},1}$ follow from the shape of the matrices of $M, I^{(1)}, I^{(0)}$ and L with respect to the basis $\underline{f}$ of $H_{\mathbb{Z},1}$.

(ii) $S(-l,2,-l)$ with $l \geq 4$ even: Recall $p_{ch,M} = \Phi_2^2\Phi_1$,

$$S = \begin{pmatrix} 1 & -l & 2 \\ 0 & 1 & -l \\ 0 & 0 & 1 \end{pmatrix},\ S^{-1}S^t = \begin{pmatrix} l^2-3 & -l^3+3l & l^2-2 \\ l & -l^2+1 & l \\ 2 & -l & 1 \end{pmatrix},\ f_3 = \underline{e}\begin{pmatrix} \frac{l}{2} \\ 1 \\ \frac{l}{2} \end{pmatrix}.$$

Define

$$f_1 := \underline{e}\begin{pmatrix}1\\0\\-1\end{pmatrix},\ f_2 := \underline{e}\begin{pmatrix}\frac{l^2-2}{2}\\ \frac{l}{2}\\ 0\end{pmatrix}.$$

Then

$$L(\underline{\widetilde{f}}^t, \underline{\widetilde{f}})^t = \begin{pmatrix}0 & -1 & 0\\ 1 & 1-\frac{l^2}{4} & 0\\ 0 & 0 & 1\end{pmatrix},\ M\underline{\widetilde{f}} = \underline{\widetilde{f}}\begin{pmatrix}-1 & 2-\frac{l^2}{2} & 0\\ 0 & -1 & 0\\ 0 & 0 & 1\end{pmatrix},$$

$$H_{\mathbb{Z},1} = \mathbb{Z}f_1 \oplus \mathbb{Z}f_2,\ H_{\mathbb{Z}} = H_{\mathbb{Z},1} \oplus H_{\mathbb{Z},2},$$

$M_1 = M|_{H_{\mathbb{Z},1}}$ and M_1^{root} have each a 2×2 Jordan block, M_1^{root} is cyclic and $(M_1^{root})^{l^2/2-2} = -M_1$. Lemma 5.6 shows

$$G_{\mathbb{Z},1}^M = G_{\mathbb{Z},1}^{(0)} = G_{\mathbb{Z},1}^{(1)} = G_{\mathbb{Z},1} = \{\pm(M_1^{root})^m \mid m \in \mathbb{Z}\}.$$

(b) Recall

	$S(\mathbb{P}^2)$	$S(A_3)$
$S^{-1}S^t$	$\begin{pmatrix}10 & -15 & 6\\ 6 & -8 & 3\\ 3 & -3 & 1\end{pmatrix}$	$\begin{pmatrix}0 & 0 & 1\\ -1 & 0 & 1\\ 0 & -1 & 1\end{pmatrix}$
$p_{ch,M}$	Φ_1^3	$\Phi_4\Phi_1$

	$S(\widehat{A}_2)$	$S(-l, 2, -l)$ with $l \geq 3$ odd
$S^{-1}S^t$	$\begin{pmatrix}-2 & -1 & 2\\ -2 & 0 & 1\\ -1 & -1 & 1\end{pmatrix}$	$\begin{pmatrix}l^2-3 & -l^3+3l & l^2-2\\ l & -l^2+1 & l\\ 2 & -l & 1\end{pmatrix}$
$p_{ch,M}$	$\Phi_2^2\Phi_1$	$\Phi_2^2\Phi_1$

The case $S(-l, 2, -l)$ with $l \geq 3$ odd will be treated separately below. Theorem 3.26 (c) applies in the case $S(\mathbb{P}^2)$ with $k = 1$ and in the case $S(\widehat{A}_2)$ with $k = 0$. It shows in these cases $(M^{root})^3 = \varepsilon M$. By Theorem 3.26 (c) the matrices $M^{root,mat}$ are as claimed in the cases $S(\mathbb{P}^2)$ and $S(\widehat{A}_2)$. In the case A_3 by definition $M^{root} = M$.

In all three cases $S(\mathbb{P}^2)$, $S(A_3)$ and $S(\widehat{A}_2)$ M^{root} is cyclic with cyclic generator e_1. In the cases $S(\mathbb{P}^2)$ and $S(A_3)$ $p_{ch,M^{root}} = p_{ch,M}$. In the case $S(\widehat{A}_2)$ $p_{ch,M^{root}} = \phi_1^2\phi_2$ and $p_{ch,M} = \phi_2^2\phi_1$. Lemma 5.2 (a) shows in all three cases

$$G_{\mathbb{Z}}^M = \{p(M^{root}) \mid p(t) = \sum_{i=0}^{2} p_i t^i \in \mathbb{Z}[t],$$
$$p(\kappa) \in (\mathbb{Z}[\kappa])^* \text{ for each eigenvalue } \kappa \text{ of } M^{root}\}.$$

(i) $S(\mathbb{P}^2)$: $M^{root} - \mathrm{id}$ is nilpotent with $(M^{root} - \mathrm{id})^2 \neq 0$, $(M^{root} - \mathrm{id})^3 = 0$. An element of $\mathbb{Z}[M^{root}]$ can be written in the form

$$q_0\,\mathrm{id} + q_1(M^{root} - \mathrm{id}) + q_2(M^{root} - \mathrm{id})^2 \text{ with } q_0, q_1, q_2 \in \mathbb{Z}.$$

It is in $\mathrm{Aut}(H_\mathbb{Z})$ if and only if $q_0 \in \{\pm 1\}$. Then it can be written as

$$q_0(M^{root})^{q_0q_1}(\mathrm{id} + \widetilde{q}_2(M^{root} - \mathrm{id})^2) \text{ for some } \widetilde{q}_2 \in \mathbb{Z}.$$

Therefore $G_\mathbb{Z}^M$ is as claimed.

Because $(M^{root} - \mathrm{id})^3 = 0$, $(M^{root} - \mathrm{id})^2(H_\mathbb{Z}) \subset \ker(M^{root} - \mathrm{id}) = \mathrm{Rad}\, I^{(1)}$. Therefore $\mathrm{id} + \widetilde{q}_2(M^{root} - \mathrm{id})^2$ and thus any element of $G_\mathbb{Z}^M$ respects $I^{(1)}$, so $G_\mathbb{Z}^M = G_\mathbb{Z}^{(1)}$.

On the other hand, one easily checks that $\mathrm{id} + \widetilde{q}_2(M^{root} - \mathrm{id})^2$ respects $I^{(0)}$ only if $\widetilde{q}_2 = 0$. Therefore

$$G_\mathbb{Z} = G_\mathbb{Z}^{(0)} = \{\pm(M^{root})^m \mid m \in \mathbb{Z}\} \subsetneqq G_\mathbb{Z}^{(1)} = G_\mathbb{Z}^M.$$

(ii) $S(A_3)$: For $p(t) = \sum_{i=0}^2 p_i t^i \in \mathbb{Z}[t]$ write $\mu_j := p(\lambda_j)$ for $j \in \{1, 2, 3\}$ for the eigenvalues of the element $p(M) \in \mathbb{Z}[M] = \mathrm{End}(H_\mathbb{Z}, M)$ where $\lambda_1 = i, \lambda_2 = -i, \lambda_3 = 1$ are the eigenvalues of M. Because of $(\mathbb{Z}[i])^* = \{\pm 1, \pm i\}$, one can multiply a given element of $G_\mathbb{Z}^M$ with a suitable power of M and obtain an element with $\mu_1 = \mu_2 = 1$. Therefore

$$\begin{aligned} G_\mathbb{Z}^M = \{M^{m_1}(\mathrm{id} + m_2\Phi_4(M)) | m_1 \in \{0, 1, 2, 3\}, m_2 \in \mathbb{Z}, \\ 1 + m_2\Phi_4(1) \in \{\pm 1\}\}. \end{aligned}$$

This forces $m_2 \in \{0, -1\}$. The case $m_2 = -1$ gives $\mathrm{id} + (-1)(M^2 + \mathrm{id}) = -M^2$. Therefore

$$\{\pm M^m \mid m \in \{0, 1, 2, 3\}\} = \{\pm M^m \mid m \in \mathbb{Z}\} = G_\mathbb{Z}^M = G_\mathbb{Z}^{(0)} = G_\mathbb{Z}^{(1)} = G_\mathbb{Z}.$$

(iii) $S(\widehat{A}_2)$: Write $\underline{f} = (f_1, f_2) := \underline{e}\begin{pmatrix} 1 & 2 \\ 1 & 1 \\ 1 & 0 \end{pmatrix}$. Then

$$\begin{aligned} M^{root}\underline{f} &= \underline{f}\begin{pmatrix} 1 & 1 \\ 0 & 1 \end{pmatrix}, \\ H_{\mathbb{Z},1} &= \ker \Phi_2^2(M) = \ker \Phi_1^2(M^{root}) = \mathbb{Z}f_1 \oplus \mathbb{Z}f_2. \end{aligned}$$

Lemma 5.6 implies

$$\begin{aligned} \{g|_{H_{\mathbb{Z},1}} \mid g \in G_\mathbb{Z}^M\} &= \{\pm(M^{root}|_{H_{\mathbb{Z},1}})^m \mid m \in \mathbb{Z}\}, \\ G_\mathbb{Z}^M &= \{\pm(M^{root})^m \mid m \in \mathbb{Z}\} \times \{g \in G_\mathbb{Z}^M \mid g|_{H_{\mathbb{Z},1}} = \mathrm{id}\}. \end{aligned}$$

But $p(t) = 1 + q\Phi_1^2(t)$ with $q \in \mathbb{Z}$ satisfies $p(-1) = 1 + q \cdot 2^2 \in \{\pm 1\}$ only if $q = 0$. Therefore

$$\{\pm (M^{root})^m \mid m \in \mathbb{Z}\} = G_{\mathbb{Z}}^M = G_{\mathbb{Z}}^{(0)} = G_{\mathbb{Z}}^{(1)} = G_{\mathbb{Z}}.$$

(iv) $S(-l, 2, -l)$ with $l \geq 3$ odd: Recall f_3 and define $(f_1, \widetilde{f_2})$ and $\underline{d} = (d_1, d_2, d_3)$ with

$$(f_1, \widetilde{f_2}, f_3) = \underline{e} \begin{pmatrix} 1 & l^2 - 2 & l \\ 0 & l & 2 \\ -1 & 0 & l \end{pmatrix}, \quad \underline{d} = \underline{e} \begin{pmatrix} \frac{l^2-l-2}{2} & \frac{l^2-1}{2} & \frac{l^2-l}{2} \\ \frac{l-1}{2} & \frac{l+1}{2} & \frac{l-1}{2} \\ 0 & \frac{l-1}{2} & -1 \end{pmatrix}.$$

The matrix which expresses $(f_1, \widetilde{f_2}, f_3)$ with $\underline{e}$ has determinant 4, the matrix which expresses $\underline{d}$ with $\underline{e}$ has determinant 1. Therefore $\mathbb{Z}f_1 \oplus \mathbb{Z}\widetilde{f_2} \oplus \mathbb{Z}f_3$ is a sublattice of index 4 in $H_{\mathbb{Z}}$, and $\underline{d}$ is a $\mathbb{Z}$-basis of $H_{\mathbb{Z}}$. One calculates

$$M(f_1, \widetilde{f_2}, f_3) = (f_1, \widetilde{f_2}, f_3) \begin{pmatrix} -1 & -l^2 + 4 & 0 \\ 0 & -1 & 0 \\ 0 & 0 & 1 \end{pmatrix}.$$

Especially

$$\mathbb{Z}f_1 \oplus \mathbb{Z}\widetilde{f_2} = \ker(M^2 + \mathrm{id})^2 \supset \mathbb{Z}f_1 = \ker(M + \mathrm{id}) = \mathrm{Rad}\, I^{(0)}.$$

Observe

$$\delta_3 \sigma_1^{-1} \sigma_2^{-1} \sigma_1 \in (\mathrm{Br}_3 \ltimes \{\pm 1\}^3)_S. \tag{5.13}$$

Define

$$\widetilde{M} := Z(\delta_3 \sigma_1^{-1} \sigma_2^{-1} \sigma_1) \in G_{\mathbb{Z}}. \tag{5.14}$$

Then $M^{root} = (-M) \circ \widetilde{M}^{(5-l^2)/2} \in G_{\mathbb{Z}}$. One calculates

$$\widetilde{M}\underline{e} = \underline{e} + f_1(-1, l, -1), \tag{5.15}$$

$$\widetilde{M}(f_1, \widetilde{f_2}, f_3) = (f_1, \widetilde{f_2}, f_3) \begin{pmatrix} 1 & 2 & 0 \\ 0 & 1 & 0 \\ 0 & 0 & 1 \end{pmatrix}, \tag{5.16}$$

$$M^{root}(f_1, \widetilde{f_2}, f_3) = (f_1, \widetilde{f_2}, f_3) \begin{pmatrix} 1 & 1 & 0 \\ 0 & 1 & 0 \\ 0 & 0 & -1 \end{pmatrix},$$

$$(M^{root})^{l^2-4} = -M,$$

$$(M^{root})^2 = \widetilde{M}.$$

Finally, one calculates

$$M^{root}\underline{d} = \underline{d}\begin{pmatrix} 0 & 0 & -1 \\ 1 & 0 & 1 \\ 0 & 1 & 1 \end{pmatrix}.$$

Therefore M^{root} is cyclic with cyclic generator d_1 and regular. Lemma 5.2 (a) shows

$$G_{\mathbb{Z}}^M = \{p(M^{root}) \mid p(t) = \sum_{i=0}^{2} p_i t^i \in \mathbb{Z}[t],\ p(1), p(-1) \in \{\pm 1\}\}.$$

As in the case $S(\widehat{A}_2)$ one finds with Lemma 5.6

$$\begin{aligned} \{g|_{H_{\mathbb{Z},1}} \mid g \in G_{\mathbb{Z}}^M\} &= \{\pm(M^{root}|_{H_{\mathbb{Z},1}})^m \mid m \in \mathbb{Z}\}, \\ G_{\mathbb{Z}}^M &= \{\pm(M^{root})^m \mid m \in \mathbb{Z}\} \times \{g \in G_{\mathbb{Z}}^M \mid g|_{H_{\mathbb{Z},1}} = \mathrm{id}\}. \end{aligned}$$

But $p(t) = 1 + q\Phi_1^2(t)$ with $q \in \mathbb{Z}$ satisfies $p(-1) = 1 + q \cdot 2^2 \in \{\pm 1\}$ only if $q = 0$. Therefore

$$\{\pm(M^{root})^m \mid m \in \mathbb{Z}\} = G_{\mathbb{Z}}^M = G_{\mathbb{Z}}^{(0)} = G_{\mathbb{Z}}^{(1)} = G_{\mathbb{Z}}.$$

(c) Of course $Z(\delta_1\delta_2\delta_3) = -\mathrm{id}$. In the cases in part (b) $G_{\mathbb{Z}} = \{\pm(M^{root})^m \mid m \in \mathbb{Z}\}$. The definitions (5.9)–(5.12) show that Z is surjective.

The case $\mathcal{H}_{1,2}$: $G_{\mathbb{Z}} \cong SL_2(\mathbb{Z}) \times \{\pm 1\}$. The group $G_{\mathbb{Z}}$ is generated by $-\mathrm{id}$, h_1 and h_2 with

$$\begin{aligned} h_1 &:= (\underline{\widetilde{f}} \mapsto \underline{\widetilde{f}}\begin{pmatrix} 1 & -1 & 0 \\ 0 & 1 & 0 \\ 0 & 0 & 1 \end{pmatrix}) = (\underline{e} \mapsto \underline{e}\begin{pmatrix} 2 & -1 & 0 \\ 1 & 0 & 0 \\ 0 & 0 & 1 \end{pmatrix}) = Z(\delta_2\sigma_1), \\ h_2 &:= (\underline{\widetilde{f}} \mapsto \underline{\widetilde{f}}\begin{pmatrix} 1 & 0 & 0 \\ 1 & 1 & 0 \\ 0 & 0 & 1 \end{pmatrix}) = (\underline{e} \mapsto \underline{e}\begin{pmatrix} 1 & 0 & 0 \\ 0 & 2 & -1 \\ 0 & 1 & 0 \end{pmatrix}) = Z(\delta_3\sigma_2). \end{aligned}$$

The cases $S(-l, 2, -l)$ with $l \geq 4$ even: (5.13)–(5.16) hold also for even l. With respect to the $\mathbb{Z}$-basis $\underline{\widetilde{f}} = (f_1, \widetilde{f}_2, f_3) = (f_1, \frac{1}{2}\widetilde{f}_2, f_3)$ of $H_{\mathbb{Z}}$

$$\widetilde{M}\underline{\widetilde{f}} = \underline{\widetilde{f}}\begin{pmatrix} 1 & 1 & 0 \\ 0 & 1 & 0 \\ 0 & 0 & 1 \end{pmatrix}.$$

One sees

$$G_{\mathbb{Z}} = \langle -\mathrm{id}, \widetilde{M}, Q \rangle$$

with

$$Q = (\underline{\widetilde{f}} \mapsto \underline{\widetilde{f}} \begin{pmatrix} -1 & 0 & 0 \\ 0 & -1 & 0 \\ 0 & 0 & 1 \end{pmatrix}) = M \circ \widetilde{M}^{2-l^2/2}$$
$$= Z(\sigma^{mon}) \circ Z(\delta_3\sigma_1^{-1}\sigma_2^{-1}\sigma_1)^{2-l^2/2}.$$

□

5.6 Special Rank 3 Cases with Eigenvalues Not All in S^1

This section starts with a general lemma for all rank 3 cases with eigenvalues not all in S^1. It gives coarse information on the four groups $G_{\mathbb{Z}}^M, G_{\mathbb{Z}}^{(0)}, G_{\mathbb{Z}}^{(1)}$ and $G_{\mathbb{Z}}$. Afterwards Theorem 5.16 determines these groups precisely for three series of cases and one exceptional case. Theorem 5.18 in Sect. 5.7 will treat the other irreducible cases with eigenvalues not all in S^1.

Lemma 5.15 *Fix $\underline{x} \in \mathbb{Z}^3$ with $r(\underline{x}) < 0$ or $r(\underline{x}) > 4$. Then*

$$G_{\mathbb{Z}}^M \overset{(2:1) \text{ or } (1:1)}{\supset} G_{\mathbb{Z}} = G_{\mathbb{Z}}^{(0)} = G_{\mathbb{Z}}^{(1)} \overset{(\text{finite}:1)}{\supset} \{\pm M^l \mid l \in \mathbb{Z}\}.$$

Proof For $p(t) = \sum_{i=0}^2 p_i t^i \in \mathbb{Q}[t]$ write $\mu_j := p(\lambda_j)$ for $j \in \{1, 2, 3\}$. Then μ_1, μ_2 and μ_3 are the eigenvalues of $p(M) \in \mathrm{End}(H_{\mathbb{Q}})$. The monodromy is because of $r(\underline{x}) \in \mathbb{Z} - \{0, 1, 2, 3, 4\}$ and Lemma 5.7 semisimple and regular. Lemma 5.1 (c) (i) and (ii) applies and gives

$$G_{\mathbb{Z}}^M = \{p(M) \mid p(t) = \sum_{i=0}^{2} p_i t^i \in \mathbb{Q}[t], \tag{5.17}$$
$$p(M) \in \mathrm{End}(H_{\mathbb{Z}}), \mu_1\mu_2 \in \{\pm 1\}, \mu_3 \in \{\pm 1\}\}$$
$$\supset G_{\mathbb{Z}} = G_{\mathbb{Z}}^{(0)} = G_{\mathbb{Z}}^{(1)} = \{p(M) \in G_{\mathbb{Z}}^M \mid \mu_1\mu_2 = 1\} \tag{5.18}$$

Especially $h \in G_{\mathbb{Z}}^M \Rightarrow h^2 \in G_{\mathbb{Z}}$. Also, $\mathrm{End}(H_{\mathbb{Q}}, M) = \mathbb{Q}[M]$, and the map

$$\mathrm{End}(H_{\mathbb{Q}}, M) = \mathbb{Q}[M] \to \mathbb{Q}[\lambda_1] \times \mathbb{Q}, \quad p(M) \mapsto (\mu_1, \mu_3)$$

is an isomorphism of $\mathbb{Q}$-algebras. This is a special case of the chinese remainder theorem. Q is mapped to $(-1, 1)$.

Observe $(-\mathrm{id}) \in G_{\mathbb{Z}} \subset G_{\mathbb{Z}}^M$. Therefore the subgroup $\{h \in G_{\mathbb{Z}} \mid \mu_3 = 1\}$ has index 2 in $G_{\mathbb{Z}}$, and the subgroup $\{h \in G_{\mathbb{Z}}^M \mid \mu_3 = 1\}$ has index 2 in $G_{\mathbb{Z}}^M$. The map

$$\{h \in G_{\mathbb{Z}}^M \mid \mu_3 = 1\} \to \mathcal{O}_{\mathbb{Q}[\lambda_1]}^*, \quad h \mapsto \mu_1, \tag{5.19}$$

is injective. The element $-1 \in \mathcal{O}^*_{\mathbb{Q}[\lambda_1]}$ is in the image of the map in (5.19) if and only if $Q \in G_{\mathbb{Z}}^M$, and then it is the image of Q. By Dirichlet's unit theorem [BSh73, Ch. 2 4.3 Theorem 5] the group $\mathcal{O}^*_{\mathbb{Q}[\lambda_1]}$ is isomorphic to the group $\{\pm 1\} \times \mathbb{Z}$. Therefore $\{h \in G_{\mathbb{Z}}^M \mid \mu_3 = 1\}$ is isomorphic to $\{\pm 1\} \times \mathbb{Z}$ if $Q \in G_{\mathbb{Z}}^M$ and to $\mathbb{Z}$ if $Q \notin G_{\mathbb{Z}}^M$. If $Q \in G_{\mathbb{Z}}^M$ then because of $Q \in \mathrm{Aut}(H_{\mathbb{Q}}, L)$ also $Q \in G_{\mathbb{Z}}$. This and the implication $h \in G_{\mathbb{Z}}^M \Rightarrow h^2 \in G_{\mathbb{Z}}$ show

$$[G_{\mathbb{Z}}^M : G_{\mathbb{Z}}] = [\{h \in G_{\mathbb{Z}}^M \mid \mu_3 = 1\} : \{h \in G_{\mathbb{Z}} \mid \mu_3 = 1\}] \in \{1, 2\}.$$

The group $\{\pm M^l \mid l \in \mathbb{Z}\} \subset G_{\mathbb{Z}} \subset G_{\mathbb{Z}}^M$ is isomorphic to $\{\pm 1\} \times \mathbb{Z}$, so it has a free part of rank 1, just as $G_{\mathbb{Z}}$ and $G_{\mathbb{Z}}^M$. Therefore $[G_{\mathbb{Z}} : \{\pm M^l \mid l \in \mathbb{Z}\}] < \infty$. □

Later we will see precisely how much bigger $G_{\mathbb{Z}}^M$ and $G_{\mathbb{Z}}$ are than $\{\pm M^l \mid l \in \mathbb{Z}\}$. In the majority of the cases they are not bigger, but $G_{\mathbb{Z}}^M = G_{\mathbb{Z}} = \{\pm M^l \mid l \in \mathbb{Z}\}$.

The following theorem determines the groups $G_{\mathbb{Z}}$ and $G_{\mathbb{Z}}^M \supset G_{\mathbb{Z}}$ for three series of triples and the exceptional case $(3, 3, 4)$. In Theorem 5.18 in Sect. 5.7 we will see that the $\mathrm{Br}_3 \ltimes \{\pm 1\}^3$ orbits of these three series and of the triple $(3, 3, 4)$ are the only triples $\underline{x}$ with $r(\underline{x}) \in \mathbb{Z}_{<0} \cup \mathbb{Z}_{>4}$, $(H_{\mathbb{Z}}, L, \underline{e})$ irreducible and *not* $G_{\mathbb{Z}}^M = G_{\mathbb{Z}} = \{\pm M^l \mid l \in \mathbb{Z}\}$.

Theorem 5.16 *For each $\underline{x} \in \mathbb{Z}^3$ below fix also the associated triple $(H_{\mathbb{Z}}, L, \underline{e})$.*

(a) Consider $\underline{x} = (x, x, x)$ with $x \in \mathbb{Z} - \{-1, 0, 1, 2, 3\}$ and $S = S(\underline{x})$. Then $\delta_3\sigma_2\sigma_1 \in (\mathrm{Br}_3 \ltimes \{\pm 1\}^3)_S$ and by Theorem 3.26 (c) (with $k = 0$)

$$M^{root,3} := Z(\delta_3\sigma_2\sigma_1) \in G_{\mathbb{Z}} \quad \text{with} \quad (M^{root,3})^3 = -M.$$

$M^{root,3}$ is cyclic with $M^{root,3}(f_3) = -f_3$. In the case $x = 4$ define also

$$M^{root,6} := -(M^{root,3})^2 - 2M^{root,3}.$$

Then

$$\begin{aligned}
G_{\mathbb{Z}} = G_{\mathbb{Z}}^M &= \{\pm (M^{root,3})^l \mid l \in \mathbb{Z}\} \quad \text{if } x \notin \{4, 5\}, \\
G_{\mathbb{Z}} = G_{\mathbb{Z}}^M &= \{\mathrm{id}, Q\} \times \{\pm (M^{root,3})^l \mid l \in \mathbb{Z}\} \quad \text{if } x = 5, \\
G_{\mathbb{Z}} = G_{\mathbb{Z}}^{(0)} &= G_{\mathbb{Z}}^{(1)} = \{\mathrm{id}, Q\} \times \{\pm (M^{root,3})^l \mid l \in \mathbb{Z}\} \\
&\overset{1:2}{\subset} G_{\mathbb{Z}}^M = \{\mathrm{id}, Q\} \times \{\pm (M^{root,6})^l \mid l \in \mathbb{Z}\} \quad \text{if } x = 4.
\end{aligned}$$

If $x = 4$ then $(M^{root,6})^2 = -M^{root,3}$.

(b) Consider $\underline{x} = (2y, 2y, 2y^2)$ with $y \in \mathbb{Z}_{\geq 2}$ and $S = S(\underline{x})$. Then $\sigma_2\sigma_1^2 \in (\mathrm{Br}_3 \ltimes \{\pm 1\}^3)_S$. By Lemma 3.25 (b)

$$M^{root,2} := Z(\sigma_2\sigma_1^2) \in G_{\mathbb{Z}}.$$

It satisfies $(M^{root,2})^2 = M$ *and* $M^{root,2}(f_3) = -f_3$. *In the case* $y = 2$ *define also*

$$M^{root,4} := -\frac{1}{4}M - 2M^{root,2} - \frac{3}{4}\,\mathrm{id}\,.$$

Then

$$\begin{aligned} G_{\mathbb{Z}} &= G_{\mathbb{Z}}^{M} = \{\pm(M^{root,2})^l \mid l \in \mathbb{Z}\} \quad \text{if } y \geq 3, \\ G_{\mathbb{Z}} &= G_{\mathbb{Z}}^{(0)} = G_{\mathbb{Z}}^{(1)} = \{\mathrm{id}, Q\} \times \{\pm(M^{root,2})^l \mid l \in \mathbb{Z}\} \\ &\overset{1:2}{\subset} G_{\mathbb{Z}}^{M} = \{\mathrm{id}, Q\} \times \{\pm(M^{root,4})^l \mid l \in \mathbb{Z}\} \quad \text{if } y = 2. \end{aligned}$$

If $y = 2$ *then* $(M^{root,4})^2 = -M^{root,2}$.

(c) Consider $\underline{x} = (x, x, 0)$ *with* $x \in \mathbb{Z}_{\geq 2}$ *and* $S = S(\underline{x})$. *In the case* $x = 2$ *define also*

$$M^{root,2} := \frac{1}{2}M + \frac{1}{2}\,\mathrm{id}\,.$$

Then

$$\begin{aligned} G_{\mathbb{Z}} &= G_{\mathbb{Z}}^{M} = \{\mathrm{id}, Q\} \times \{\pm M^l \mid l \in \mathbb{Z}\} \quad \text{if } x \geq 3, \\ G_{\mathbb{Z}} &= G_{\mathbb{Z}}^{(0)} = G_{\mathbb{Z}}^{(1)} = \{\mathrm{id}, Q\} \times \{\pm M^l \mid l \in \mathbb{Z}\} \\ &\overset{1:2}{\subset} G_{\mathbb{Z}}^{M} = \{\mathrm{id}, Q\} \times \{\pm(M^{root,2})^l \mid l \in \mathbb{Z}\} \quad \text{if } x = 2. \end{aligned}$$

If $x = 2$ *then* $(M^{root,2})^2 = M$.

(d) Consider $\underline{x} = (3, 3, 4)$. *Then*

$$G_{\mathbb{Z}} = G_{\mathbb{Z}}^{M} = \{\mathrm{id}, Q\} \times \{\pm M^l \mid l \in \mathbb{Z}\}.$$

(e) In all cases in (a)–(d) except for the four cases

$$\underline{x} \in \{(4, 4, 4), (5, 5, 5), (4, 4, 8), (3, 3, 4)\}$$

the map $Z : (\mathrm{Br}_3 \ltimes \{\pm 1\}^3)_S \to G_{\mathbb{Z}}$ *is surjective so* $G_{\mathbb{Z}} = G_{\mathbb{Z}}^{\mathcal{B}}$. *In the four exceptional cases* $Q \in G_{\mathbb{Z}} - G_{\mathbb{Z}}^{\mathcal{B}}$.

Proof In all cases in this theorem $r(\underline{x}) < 0$ or $r(\underline{x}) > 4$, so Lemma 5.15 applies, so $G_{\mathbb{Z}} = G_{\mathbb{Z}}^{(0)} = G_{\mathbb{Z}}^{(1)}$.

(a) S is as in Theorem 3.26 (c) with $k = 0$. Therefore $\delta_3\sigma_2\sigma_1$ is in the stabilizer of S and $M^{root,3}$ is in $G_{\mathbb{Z}}$, it is cyclic, and it satisfies $(M^{root,3})^3 = -M$. Explicitly (by Theorem 3.26 (a))

$$M^{root,3}(\underline{\epsilon}) = \underline{e} \cdot \begin{pmatrix} -x & -x & -1 \\ 1 & 0 & 0 \\ 0 & 1 & 0 \end{pmatrix}.$$

One sees $M^{root,3}(f_3) = -f_3$ where $f_3 = -e_1 + e_2 - e_3$, so its third eigenvalue is $\kappa_3 = -1$. The other two eigenvalues κ_1 and κ_2 are determined by the trace $-x = \kappa_1 + \kappa_2 - 1$ and the determinant $-1 = \kappa_1\kappa_2(-1)$ of $M^{root,3}$. The eigenvalues are

$$\kappa_{1/2} = \frac{1-x}{2} \pm \frac{1}{2}\sqrt{x^2 - 2x - 3}, \quad \kappa_3 = -1.$$

Because M and $M^{root,3}$ are regular, Lemma 5.1 applies. It gives an isomorphism of $\mathbb{Q}$-algebras

$$\begin{aligned} &\mathrm{End}(H_{\mathbb{Q}}, M) = \mathrm{End}(H_{\mathbb{Q}}, M^{root,3}) \\ &= \{p(M^{root,3}) \mid p(t) = \sum_{i=0}^{2} p_i t^i \in \mathbb{Q}[t]\} \to \mathbb{Q}[\kappa_1] \times \mathbb{Q} \\ &\qquad\qquad p(M^{root,3}) \mapsto (p(\kappa_1), p(-1)). \end{aligned}$$

The image of $-\mathrm{id}$ is $(-1, -1)$, the image of Q is $(-1, 1)$, the image of $M^{root,3}$ is $(\kappa_1, -1)$. The image of $G_{\mathbb{Z}}^M$ is a priori a subgroup of $\mathcal{O}^*_{\mathbb{Q}[\kappa_1]} \times \{\pm 1\}$. We have to find out which one. By Theorem 5.11 $Q \in G_{\mathbb{Z}}^M$ (and then also $Q \in G_{\mathbb{Z}}$) only for $x \in \{4, 5\}$.

Lemma 5.2 applies because $M^{root,3}$ is cyclic. It shows

$$\begin{aligned} G_{\mathbb{Z}}^M = \{&p(M^{root,3}) \mid p(t) = p_2 t^2 + p_1 t + p_0 \in \mathbb{Z}[t] \text{ with} \\ &(p(\kappa_1), p(-1)) \in \mathbb{Z}[\kappa_1]^* \times \{\pm 1\}\}, \\ G_{\mathbb{Z}} = \{&p(M^{root,3}) \mid p(t) = p_2 t^2 + p_1 t + p_0 \in \mathbb{Z}[t] \text{ with} \\ &(p(\kappa_1), p(-1)) \in \mathbb{Z}[\kappa_1]^* \times \{\pm 1\} \text{ and } p(\kappa_1)p(\kappa_2) = 1\}. \end{aligned}$$

Now Lemma C.1 (a) is useful. It says

$$\mathbb{Z}[\kappa_1]^* = \begin{cases} \{\pm \kappa_1^l \mid l \in \mathbb{Z}\} & \text{for } x \notin \{4, -2\}, \\ \{\pm(\kappa_1 - 1)^l \mid l \in \mathbb{Z}\} & \text{for } x = -2, \\ \{\pm(\kappa_1 + 1)^l \mid l \in \mathbb{Z}\} & \text{for } x = 4. \end{cases}$$

with

$$\begin{aligned} (\kappa_1 - 1)(\kappa_2 - 1) = -1 \quad &\text{and} \quad (\kappa_1 - 1)^2 = \kappa_1 \quad \text{for} \quad x = -2, \\ (\kappa_1 + 1)(\kappa_2 + 1) = -1 \quad &\text{and} \quad (\kappa_1 + 1)^2 = -\kappa_1 \quad \text{for} \quad x = 4. \end{aligned}$$

For $x \notin \{4, 5, -2\}$ the facts $-\mathrm{id}$, $M^{root,3} \in G_{\mathbb{Z}}$ and $Q \notin G_{\mathbb{Z}}$ show that the image of $G_{\mathbb{Z}}$ and $G_{\mathbb{Z}}^M$ in $\mathbb{Z}[\kappa_1]^* \times \{\pm 1\} = \{\pm\kappa_1^l \mid l \in \mathbb{Z}\} \times \{\pm 1\}$ has index 2. Therefore then $G_{\mathbb{Z}} = G_{\mathbb{Z}}^M$ is as claimed.

For $x = 5$ the facts $-\mathrm{id}$, $M^{root,3}$, $Q \in G_{\mathbb{Z}}$ show that the image of $G_{\mathbb{Z}}$ and $G_{\mathbb{Z}}^M$ is $\mathbb{Z}[\kappa_1]^* \times \{\pm 1\} = \{\pm\kappa_1^l \mid l \in \mathbb{Z}\} \times \{\pm 1\}$. Therefore then $G_{\mathbb{Z}} = G_{\mathbb{Z}}^M$ is as claimed.

Consider the case $x = -2$. If an automorphism $p(M^{root,3})$ is in $G_{\mathbb{Z}}^M$ which corresponds to a pair $(\kappa_1 - 1, \pm 1)$, then $p(\kappa_1) = \kappa_1 - 1$ means $p(t) = t - 1 + l_2(t^2 - 3t + 1)$ for some $l_2 \in \mathbb{Z}$. But then $p(-1) = -2 + l_2 \cdot 5 \notin \{\pm 1\}$. So $G_{\mathbb{Z}}^M$ does not contain such an automorphism. $G_{\mathbb{Z}}^M$ and $G_{\mathbb{Z}}$ are as claimed, because $Q \notin G_{\mathbb{Z}}^M$.

Consider the case $x = 4$. The polynomial $p(t) = -t^2 - 2t$ satisfies $p(-1) = 1$, $p(\kappa_1) = -\kappa_1^2 - 2\kappa_1 = -(-3\kappa_1 - 1) - 2\kappa_1 = \kappa_1 + 1$. Therefore $M^{root,6} = p(M^{root,3}) \in G_{\mathbb{Z}}^M$. Because $\kappa_1 + 1$ has norm -1, $M^{root,6}$ is not in $G_{\mathbb{Z}}$. The groups $G_{\mathbb{Z}}^M$ and $G_{\mathbb{Z}}$ are as claimed.

(b) Compare the sections 4.1 and 3.2 for the actions of $\mathrm{Br}_3 \ltimes \{\pm 1\}^3$ on $\mathbb{Z}^3$ and on $\mathcal{B}^{dist}$.

$$\begin{aligned}
&\sigma_2\sigma_1^2(2y, 2y, 2y^2)\\
&= \sigma_2\sigma_1(-2y, 2y^2 - 2y \cdot 2y, 2y) = \sigma_2\sigma_1(-2y, -2y^2, 2y)\\
&= \sigma_2(2y, 2y - (-2y)(-2y^2), -2y^2) = \sigma_2(2y, 2y - 4y^3, -2y^2)\\
&= ((2y - 4y^3) - 2y \cdot (-2y^2), 2y, 2y^2) = (2y, 2y, 2y^2),
\end{aligned}$$

so $\sigma_2\sigma_1^2$ is in the stabilizer of $(2y, 2y, 2y^2)$, so $M^{root,2} := Z(\sigma_2\sigma_1^2) \in G_{\mathbb{Z}}$.

$$\begin{aligned}
&M^{root,2}(\underline{e}) = \sigma_2\sigma_1^2(\underline{e})\\
&= \sigma_2\sigma_1(s_{e_1}^{(0)}(e_2), e_1, e_3)\\
&= \sigma_2\sigma_1(e_2 - 2ye_1, e_1, e_3)\\
&= \sigma_2(s_{e_2-2ye_1}^{(0)}(e_1), e_2 - 2ye_1, e_3)\\
&= \sigma_2(e_1 + 2y(e_2 - 2ye_1), e_2 - 2ye_1, e_3)\\
&= ((1 - 4y^2)e_1 + 2ye_2, s_{e_2-2ye_1}^{(0)}(e_3), e_2 - 2ye_1)\\
&= ((1 - 4y^2)e_1 + 2ye_2, e_3 + 2y^2(e_2 - 2ye_1), e_2 - 2ye_1)\\
&= \underline{e} \cdot \begin{pmatrix} 1 - 4y^2 & -4y^3 & -2y \\ 2y & 2y^2 & 1 \\ 0 & 1 & 0 \end{pmatrix} =: \underline{e} \cdot M^{root,2,mat}.
\end{aligned}$$

The map $Z : (\mathrm{Br}_3 \ltimes \{\pm 1\}^3)_S \to G_{\mathbb{Z}}$ in Lemma 3.25 is a group antihomomorphism. By Theorem 3.26 $Z(\sigma^{mon}) = M$. Therefore

$$\begin{aligned}
(M^{root,2})^2 &= Z(\sigma_2\sigma_1^2)Z(\sigma_2\sigma_1^2) = Z(\sigma_2\sigma_1(\sigma_1\sigma_2\sigma_1)\sigma_1)\\
&= Z(\sigma_2\sigma_1(\sigma_2\sigma_1\sigma_2)\sigma_1) = Z((\sigma_2\sigma_1)^3) = Z(\sigma^{mon}) = M.
\end{aligned}$$

One sees $M^{root,2}(f_3) = -f_3$, where $f_3 = -ye_1 + e_2 - e_3$, so its third eigenvalue is $\kappa_3 = -1$. The other two eigenvalues κ_1 and κ_2 are determined by the trace $1 - 2y^2 = \kappa_1 + \kappa_2 - 1$ and the product $\kappa_1\kappa_2 = 1$, which holds because of $M^{root,2} \in G_{\mathbb{Z}}$ (or because $\det M^{root,2} = -1$). The eigenvalues are

$$\kappa_{1,2} = (1 - y^2) \pm y\sqrt{y^2 - 2}, \quad \kappa_3 = -1.$$

Because M and $M^{root,2}$ are regular, Lemma 5.1 applies. It gives an isomorphism of $\mathbb{Q}$-algebras

$$\begin{aligned} \mathrm{End}(H_{\mathbb{Q}}, M) &= \mathrm{End}(H_{\mathbb{Q}}, M^{root,2}) \\ &= \{p(M^{root,2}) \mid p(t) = \sum_{i=0}^{2} p_i t^i \in \mathbb{Q}[t]\} \to \mathbb{Q}[\kappa_1] \times \mathbb{Q} \\ &\qquad p(M^{root,2}) \mapsto (p(\kappa_1), p(-1)). \end{aligned}$$

The image of $-\mathrm{id}$ is $(-1, -1)$, the image of Q is $(-1, 1)$, the image of $M^{root,2}$ is $(\kappa_1, -1)$. The image of $G_{\mathbb{Z}}^M$ is a priori a subgroup of $\mathcal{O}^*_{\mathbb{Q}[\kappa_1]} \times \{\pm 1\}$. We have to find out which one. By Theorem 5.11 $Q \in G_{\mathbb{Z}}^M$ (and then also $Q \in G_{\mathbb{Z}}$) only for $y = 2$.

Consider the decomposition $H_{\mathbb{Q}} = H_{\mathbb{Q},1} \oplus H_{\mathbb{Q},2}$ as in Definition 5.9 and the primitive sublattices $H_{\mathbb{Z},1} = H_{\mathbb{Q},1} \cap H_{\mathbb{Z}}$ and $H_{\mathbb{Z},2} = H_{\mathbb{Q},2} \cap H_{\mathbb{Z}} = \mathbb{Z}f_3$ in $H_{\mathbb{Z}}$. The sublattice $H_{\mathbb{Z},1}$ is the right L-orthogonal subspace $(\mathbb{Z}f_3)^{\perp} \subset H_{\mathbb{Z}}$ of $\mathbb{Z}f_3$, see (5.8):

$$\begin{aligned} H_{\mathbb{Z},1} &= \{\underline{e} \cdot \underline{z}^t \mid \underline{z} \in \mathbb{Z}^3, 0 = (-y, 1 - 2y^2, -1)\begin{pmatrix} z_1 \\ z_2 \\ z_3 \end{pmatrix}\} \\ &= \langle f_1, f_2 \rangle_{\mathbb{Z}} \quad \text{with} \quad f_1 = e_1 - ye_3, \quad f_2 = e_2 + (1 - 2y^2)e_3. \end{aligned}$$

Write $\underline{f} = (f_1, f_2, f_3) = \underline{e} \cdot M(\underline{e}, \underline{f})$,

$$M^{root,2}(\underline{e}) = \underline{e} \cdot M^{root,2,mat}, \quad M^{root,2}(\underline{f}) = \underline{f} M^{root,2,mat,\underline{f}}.$$

Then

$$\begin{aligned} M(\underline{e}, \underline{f}) &= \begin{pmatrix} 1 & 0 & -y \\ 0 & 1 & 1 \\ -y & 1 - 2y^2 & -1 \end{pmatrix}, \\ M(\underline{e}, \underline{f})^{-1} &= \frac{1}{y^2 - 2}\begin{pmatrix} 2y^2 - 2 & 2y^3 - y & y \\ -y & -y^2 - 1 & -1 \\ y & 2y^2 - 1 & 1 \end{pmatrix}, \end{aligned}$$

$$M^{root,2,mat,\underline{f}} = M(\underline{e},\underline{f})^{-1} M^{root,2,mat} M(\underline{e},\underline{f}) = \begin{pmatrix} -2y^2+1 & -2y & 0 \\ y & 1 & 0 \\ 0 & 0 & -1 \end{pmatrix}.$$

Any element $h = p(M) \in G_{\mathbb{Z}}^M$ with $p(t) \in \mathbb{Q}[t]$ restricts to an automorphism of $H_{\mathbb{Z},1}$ which commutes with $M^{root,2}|_{H_{\mathbb{Z},1}}$, so its restriction to $H_{\mathbb{Z},1}$ has the shape $h|_{H_{\mathbb{Z},1}} = a\,\mathrm{id} + bM^{root,2}|_{H_{\mathbb{Z},1}}$ with $a, b \in \mathbb{Q}$ with

$$a\begin{pmatrix} 1 & 0 \\ 0 & 1 \end{pmatrix} + b\begin{pmatrix} -2y^2+1 & -2y \\ y & 1 \end{pmatrix} \in GL_2(\mathbb{Z}).$$

This implies $by \in \mathbb{Z}$ and $a + b \in \mathbb{Z}$. The eigenvalue $p(\kappa_1)$ is

$$\begin{aligned} a + b\kappa_1 &= a + b(1 - y^2 + y\sqrt{y^2-2}) \\ &= (a+b) + by(-y + \sqrt{y^2-2}) \in \mathbb{Z}[\sqrt{y^2-2}]^*. \end{aligned}$$

Now Lemma C.1 (b) is useful. It says

$$\mathbb{Z}[\sqrt{y^2-2}]^* = \begin{cases} \{\pm\kappa_1^l \mid l \in \mathbb{Z}\} & \text{if } y \geq 3, \\ \{\pm(1+\sqrt{2})^l \mid l \in \mathbb{Z}\} & \text{if } y = 2. \end{cases}$$

Furthermore κ_1 has norm 1, $1+\sqrt{2}$ has norm -1, and $(1+\sqrt{2})^2 = -\kappa_2$ if $y = 2$.

Consider the cases $y \geq 3$. The map

$$G_{\mathbb{Z}}^M \to \mathcal{O}^*_{\mathbb{Q}[\kappa_1]} \times \{\pm 1\}, \quad p(M^{root,2}) \mapsto (p(\kappa_1), p(-1)),$$

has because of $Q \notin G_{\mathbb{Z}}^M$ as image the index 2 subgroup of $\{\pm\kappa_1^l \mid l \in \mathbb{Z}\} \times \{\pm 1\}$ which is generated by $(\kappa_1, -1)$ and $(-1,-1)$, because $Q \notin G_{\mathbb{Z}}^M$. Therefore then $G_{\mathbb{Z}}^M = G_{\mathbb{Z}} = \{\pm(M^{root,2})^l \mid l \in \mathbb{Z}\}$.

Consider the case $y = 2$. Then $Q \in G_{\mathbb{Z}} \subset G_{\mathbb{Z}}^M$. Therefore then $G_{\mathbb{Z}} = \{\mathrm{id}, Q\} \times \{\pm(M^{root,2})^l \mid l \in \mathbb{Z}\}$. The question remains whether $1+\sqrt{2}$ arises as eigenvalue $p(\kappa_1)$ for an element $p(M^{root,2}) \in G_{\mathbb{Z}}^M$. It does. $M^{root,4}$ has the first eigenvalue

$$-\frac{1}{4}(-3+2\sqrt{2})^2 - 2(-3+2\sqrt{2}) - \frac{3}{4} = 1 - \sqrt{2}$$

with $(1-\sqrt{2})^2 = 3 - 2\sqrt{2} = -\kappa_1$ and the third eigenvalue $-\frac{1}{4} - 2(-1) - \frac{3}{4} = 1$. Therefore $(M^{root,4})^2 = -M^{root,2}$. $M^{root,4}$ is in $G_{\mathbb{Z}}^M$ because

$$M^{root,4}(\underline{e}) = \underline{e}\left(-\frac{1}{4}\begin{pmatrix} 97 & 220 & 28 \\ -28 & -63 & -8 \\ 4 & 8 & 1 \end{pmatrix} - 2\begin{pmatrix} -15 & -32 & -4 \\ 4 & 8 & 1 \\ 0 & 1 & 0 \end{pmatrix} - \frac{3}{4}E_3\right)$$

$$= \underline{e}\begin{pmatrix} 5 & 9 & 1 \\ -1 & -1 & 0 \\ -1 & -4 & -1 \end{pmatrix}.$$

Therefore then $G_{\mathbb{Z}}^M = \{\mathrm{id}, Q\} \times \{\pm(M^{root,4})^l \mid l \in \mathbb{Z}\}$.

(c) Observe $r = 2x^2$. M has the eigenvalues $\lambda_1, \lambda_2, \lambda_3$ with

$$\lambda_{1/2} = (1 - x^2) \pm x\sqrt{x^2 - 2}, \quad \lambda_3 = 1.$$

Because M is regular, Lemma 5.1 applies. It gives an isomorphism of $\mathbb{Q}$-algebras

$$\begin{aligned} &\mathrm{End}(H_{\mathbb{Q}}, M) \\ &= \{p(M) \mid p(t) = p_2 t^2 + p_1 t + p_0 \in \mathbb{Q}[t]\} \to \mathbb{Q}[\lambda_1] \times \mathbb{Q} \\ &\qquad p(M) \mapsto (p(\lambda_1), p(1)). \end{aligned}$$

The image of $-\mathrm{id}$ is $(-1, -1)$, the image of Q is $(-1, 1)$, the image of M is $(\lambda_1, 1)$. The image of $G_{\mathbb{Z}}^M$ is a priori a subgroup of $\mathcal{O}_{\mathbb{Q}[\lambda_1]}^* \times \{\pm 1\}$. We have to find out which one. By Theorem 5.11 $Q \in G_{\mathbb{Z}} \subset G_{\mathbb{Z}}^M$.

Consider the decomposition $H_{\mathbb{Q}} = H_{\mathbb{Q},1} \oplus H_{\mathbb{Q},2}$ as in Definition 5.9 and the primitive sublattices $H_{\mathbb{Z},1} = H_{\mathbb{Q},1} \cap H_{\mathbb{Z}}$ and $H_{\mathbb{Z},2} = H_{\mathbb{Q},2} \cap H_{\mathbb{Z}} = \mathbb{Z} f_3$ in $H_{\mathbb{Z}}$. The sublattice $H_{\mathbb{Z},1}$ is the right L-orthogonal subspace $(\mathbb{Z} f_3)^{\perp} \subset H_{\mathbb{Z}}$ of $\mathbb{Z} f_3$, see (5.8):

$$\begin{aligned} H_{\mathbb{Z},1} &= \{\underline{e} \cdot \underline{z}^t \mid \underline{z} \in \mathbb{Z}^3, 0 = (0, 1, -1)\begin{pmatrix} z_1 \\ z_2 \\ z_3 \end{pmatrix}\} \\ &= \langle f_1, f_2 \rangle_{\mathbb{Z}} \quad \text{with} \quad f_1 = e_1, \quad f_2 = e_2 + e_3. \end{aligned}$$

Write $\underline{f} = (f_1, f_2, f_3) = \underline{e} \cdot M(\underline{e}, \underline{f})$,

$$M(\underline{e}) = \underline{e} \cdot M^{mat}, \quad M(\underline{f}) = \underline{f} M^{mat,\underline{f}}.$$

Then

$$M(\underline{e}, \underline{f}) = \begin{pmatrix} 1 & 0 & 0 \\ 0 & 1 & 1 \\ 0 & 1 & -1 \end{pmatrix}, \quad M(\underline{e}, \underline{f})^{-1} = \frac{1}{2}\begin{pmatrix} 2 & 0 & 0 \\ 0 & 1 & 1 \\ 0 & 1 & -1 \end{pmatrix},$$

$$M^{mat} = \begin{pmatrix} 1 - 2x^2 & -x & -x \\ x & 1 & 0 \\ x & 0 & 1 \end{pmatrix},$$

$$M^{mat,\underline{f}} = M(\underline{e}, \underline{f})^{-1} M^{mat} M(\underline{e}, \underline{f}) = \begin{pmatrix} 1 - 2x^2 & -2x & 0 \\ x & 1 & 0 \\ 0 & 0 & 1 \end{pmatrix}.$$

The upper left 2×2-matrix in $M^{mat,\underline{f}}$ coincides after identification of x and y with the upper left 2×2-matrix in $M^{root,2,mat,\underline{f}}$ in the proof of part (b). Therefore we can argue exactly as in the proof of part (b). Lemma C.1 (b) applies in the same way.

We obtain for $x \geq 3$ $G_{\mathbb{Z}}^M = G_{\mathbb{Z}} = \{\mathrm{id}, Q\} \times \{\pm M^l \mid l \in \mathbb{Z}\}$ and for $x = 2$ $G_{\mathbb{Z}} = \{\mathrm{id}, Q\} \times \{\pm M^l \mid l \in \mathbb{Z}\}$.

Consider the case $x = 2$. Then $M^{root,2}$ has the first eigenvalue

$$\frac{1}{2}\lambda_1 + \frac{1}{2} = -1 + \sqrt{2}$$

with $(-1+\sqrt{2})^2 = 3 - 2\sqrt{2} = -\lambda_1$ and the third eigenvalue $\frac{1}{2} + \frac{1}{2} = 1$. Therefore $(M^{root,2})^2 = QM$. $M^{root,2}$ is in $G_{\mathbb{Z}}^M$ because

$$M^{root,2}(\underline{e}) = \underline{e}\left(\frac{1}{2}\begin{pmatrix} -7 & -2 & -2 \\ 2 & 1 & 0 \\ 2 & 0 & 1 \end{pmatrix} + \frac{1}{2}E_3\right) = \underline{e}\begin{pmatrix} -3 & -1 & -1 \\ 1 & 1 & 0 \\ 1 & 0 & 1 \end{pmatrix}.$$

Therefore $G_{\mathbb{Z}}^M = \{\mathrm{id}, Q\} \times \{\pm (M^{root,2})^l \mid l \in \mathbb{Z}\}$ for $x = 2$.

(d) Here $r = -2$. M has the eigenvalues $\lambda_1, \lambda_2, \lambda_3$ with

$$\lambda_{1/2} = 2 \pm \sqrt{3}, \quad \lambda_3 = 1.$$

It is well known and can be seen easily either elementarily or with Theorem C.6 that

$$\mathcal{O}^*_{\mathbb{Q}[\lambda_1]} = \{\pm \lambda_1^l \mid l \in \mathbb{Z}\}.$$

$Q \in G_{\mathbb{Z}} \subset G_{\mathbb{Z}}^M$ by Theorem 5.11. Recall the proof of Lemma 5.15. The restriction of the map in (5.19) to the map

$$\{\mathrm{id}, Q\} \times \{M^l \mid l \in \mathbb{Z}\} \to \mathcal{O}^*_{\mathbb{Q}[\lambda_1]}$$

is an isomorphism. Therefore the map in (5.19) is an isomorphism and

$$G_{\mathbb{Z}}^M = G_{\mathbb{Z}} = \{\mathrm{id}, Q\} \times \{\pm M^l \mid l \in \mathbb{Z}\}.$$

(e) Observe in part (c)

$$\begin{aligned} Z(\sigma_2)(\underline{e}) &= \sigma_2(\underline{e}) = (e_1, e_3, e_2), \\ \text{so } Z(\sigma_2)(f_1, f_2, f_3) &= (f_1, f_2, -f_3), \\ \text{so } Z(\sigma_2) &= -Q. \end{aligned}$$

Now in all cases

$$-\,\mathrm{id} = Z(\delta_1\delta_2\delta_3), \quad M = Z(\sigma^{mon})$$

and

$$\text{in part (a):} \begin{cases} M^{root,3} &= Z(\delta_3\sigma_2\sigma_1), \\ G_{\mathbb{Z}} &= \{\pm(M^{root,3})^l \,|\, l \in \mathbb{Z}\} \text{ for } x \notin \{4,5\}, \end{cases}$$

$$\text{in part (b):} \begin{cases} M^{root,2} &= Z(\sigma_2\sigma_1^2), \\ G_{\mathbb{Z}} &= \{\pm(M^{root,2})^l \,|\, l \in \mathbb{Z}\} \text{ for } y \neq 2, \end{cases}$$

$$\text{in part (c):} \begin{cases} Q &= Z(\delta_1\delta_2\delta_3\sigma_2), \\ G_{\mathbb{Z}} &= \{\mathrm{id}, Q\} \times \{\pm M^l \,|\, l \in \mathbb{Z}\}. \end{cases}$$

This shows $G_{\mathbb{Z}} = G_{\mathbb{Z}}^{\mathcal{B}}$ in all but the four cases $\underline{x} \in \{(4,4,4), (5,5,5), (4,4,8), (3,3,4)\}$. In these four cases $Q \in G_{\mathbb{Z}}$. It remains to see $Q \notin G_{\mathbb{Z}}^{\mathcal{B}}$. We offer two proofs.

First proof: It uses that in these four cases the stabilizer of $\underline{e}$ in $\mathrm{Br}_3 \ltimes \{\pm 1\}^3$ is $\{\mathrm{id}\}$, which will be proved as part of Theorem 7.11. It also follows from $\Gamma^{(1)} = G^{free,3}$ in Theorem 6.18 (g) or $\Gamma^{(0)} = G^{fCox,3}$ in Theorem 6.11 (g) and from Example 3.4 (respectively Theorem 3.2 (a) or (b)). This implies that here $Z : (\mathrm{Br}_3 \ltimes \{\pm 1\}^3)_S \to G_{\mathbb{Z}}$ is injective. Observe furthermore $Q^2 = \mathrm{id}$. If $Q = Z(\beta)$ for some braid β, then $\beta^2 = \mathrm{id}$ as Z is a group antihomomorphism. But there is no braid of order two.

Second proof: By formula (5.5) in Theorem 5.11 in the four cases

$$\begin{aligned} Q(\underline{e}) &= -\underline{e} + 2f_3(1,3,1) && \text{in the case } \underline{x} = (4,4,4), \\ Q(\underline{e}) &= -\underline{e} + f_3(1,4,1) && \text{in the case } \underline{x} = (5,5,5), \\ Q(\underline{e}) &= -\underline{e} + f_3(2,7,1) && \text{in the case } \underline{x} = (4,4,8), \\ Q(\underline{e}) &= -\underline{e} + f_3(4,9,3) && \text{in the case } \underline{x} = (3,3,4). \end{aligned}$$

By Theorem 6.21 (g) the restriction to $\Delta^{(1)}$ of the projection $\mathrm{pr}^{H,(1)} : H_{\mathbb{Z}} \to \overline{H_{\mathbb{Z}}}^{(1)}$ is injective. Therefore in all four cases $Q(e_i) \notin \Delta^{(1)}$ for $i \in \{1,2,3\}$. But any automorphism in $G_{\mathbb{Z}}^{\mathcal{B}}$ maps each e_i to an odd vanishing cycle. Thus $Q \notin G_{\mathbb{Z}}^{\mathcal{B}}$. □

5.7 General Rank 3 Cases with Eigenvalues Not All in S^1

Theorem 5.18 below will show $G_{\mathbb{Z}} = G_{\mathbb{Z}}^M = \{\pm M^l \,|\, l \in \mathbb{Z}\}$ in all irreducible rank 3 with eigenvalues not all in S^1 which have not been treated in Theorem 5.16. This result is simple to write down, but the proof is long. It is a case discussion with many subcases. It builds on part (b) of the technical Lemma 5.17 which gives necessary and sufficient conditions when an endomorphism in $\mathrm{End}(H_{\mathbb{Q}})$ of a certain shape is in $\mathrm{End}(H_{\mathbb{Z}})$.

Any element of $\{h \in G_{\mathbb{Z}}^M \,|\, \mu_3 = 1\}$ can be written as $h = q(M)$ with $q(t) = 1 + q_0(t-1) + q_1(t-1)^2$ with unique coefficients $q_0, q_1 \in \mathbb{Q}$, but not all values $q_0, q_1 \in$

$\mathbb{Q}$ give such an element. Part (b) of the following lemma says which integrality conditions on q_0 and q_1 are necessary for $q(M) \in \mathrm{End}(H_{\mathbb{Z}})$. Part (a) is good to know in this context.

Lemma 5.17 *Fix $\underline{x} \in \mathbb{Z}^3 - \{(0,0,0)\}$ and the associated triple $(H_{\mathbb{Z}}, L, \underline{e})$. Recall $g = \gcd(x_1, x_2, x_3)$ and $\widetilde{x}_i = g^{-1}x_i$. Define*

$$\begin{aligned} g_1 &:= \gcd(2x_1 - x_2x_3, 2x_2 - x_1x_3, 2x_3 - x_1x_2) \in \mathbb{N} \cup \{0\}, \\ g_2 &:= \frac{g_1}{g} = \gcd(2\widetilde{x}_1 - g\widetilde{x}_2\widetilde{x}_3, 2\widetilde{x}_2 - g\widetilde{x}_1\widetilde{x}_3, 2\widetilde{x}_3 - g\widetilde{x}_1\widetilde{x}_2) \in \mathbb{N} \cup \{0\}. \end{aligned} \tag{5.20}$$

(a) We separate three cases.
(i) Case (three or two of x_1, x_2, x_3 are odd): Then g and g_2 are odd and

$$\gcd(g_2, \widetilde{x}_i) = \gcd(g_2, g) = 1, \quad g_2^2 \mid (r-4).$$

(ii) Case (exactly one of x_1, x_2, x_3 is odd): Then g is odd, $g_2 \equiv 2(4)$ and

$$\gcd(\frac{g_2}{2}, \widetilde{x}_i) = \gcd(\frac{g_2}{2}, g) = 1, \quad (\frac{g_2}{2})^2 \mid (r-4).$$

(iii) Case (none of x_1, x_2, x_3 is odd): Then g and g_2 are even. More precisely, $g_2 \equiv 0(4)$ only if $\frac{g}{2}$ and $\widetilde{x}_1, \widetilde{x}_2, \widetilde{x}_3$ are odd. Else $g_2 \equiv 2(4)$. Always

$$\gcd(\frac{g_2}{2}, \widetilde{x}_i) = \gcd(\frac{g_2}{2}, \frac{g}{2}) = 1, \quad g_2^2 \mid (r-4).$$

(b) Consider $q_0, q_1 \in \mathbb{Q}$, $q(t) := 1 + q_0(t-1) + q_1(t-1)^2 \in \mathbb{Q}[t]$ and $h = q(M) \in \mathbb{Q}[M]$. Define $q_2 := q_0 - 2q_1 \in \mathbb{Q}$.

Then $h \in \mathrm{End}(H_{\mathbb{Z}}, M)$ if and only if the following integrality conditions (5.21)–(5.24) *are satisfied.*

$$q_2 \cdot g^2 \in \mathbb{Z}, \tag{5.21}$$
$$q_1 \cdot gg_1 \in \mathbb{Z}, \tag{5.22}$$
$$q_0x_i - q_1x_jx_k \in \mathbb{Z} \text{for} \quad \{i,j,k\} = \{1,2,3\}, \tag{5.23}$$
$$q_1(x_i^2 - x_j^2) \in \mathbb{Z} \text{for} \quad \{i,j,k\} = \{1,2,3\}. \tag{5.24}$$

If these conditions hold, then also the following holds,

$$q_0 \cdot g_1 \in \mathbb{Z}. \tag{5.25}$$

(c) In part (b) the eigenvalue of $q(M)$ on $H_{\mathbb{C},\lambda_1}$ is

$$\mu_1 := q(\lambda_1) = (1 - rq_1) + (q_0 - rq_1)(\lambda_1 - 1)$$

$$= (1-q_0)+(q_0-rq_1)\lambda_1.$$

Proof (a) g is odd in the cases (i) and (ii) and even in case (iii). Therefore g_2 is odd in case (i) and even in the cases (ii) and (iii), and furthermore $\frac{g_2}{2}$ is odd in case (ii). Also $\frac{g_2}{2}$ odd in case (iii) almost always, namely except when $\frac{g}{2}$ and $\widetilde{x}_1,\widetilde{x}_2,\widetilde{x}_3$ are odd, as can be seen from the definition (5.20) of g_2. Here observe that at least one of $\widetilde{x}_1,\widetilde{x}_2,\widetilde{x}_3$ is odd because $\gcd(\widetilde{x}_1,\widetilde{x}_2,\widetilde{x}_3)=1$.

Now we consider first case (iii). A common divisor of $\frac{g_2}{2}$ and $\widetilde{x}_1$ would be odd. Because of the second term $2\widetilde{x}_2-g\widetilde{x}_1\widetilde{x}_3$ and the third term $2\widetilde{x}_3-g\widetilde{x}_1\widetilde{x}_2$ in (5.20) it would also divide $\widetilde{x}_2$ and $\widetilde{x}_3$. This is impossible because of $\gcd(\widetilde{x}_1,\widetilde{x}_2,\widetilde{x}_3)=1$. Therefore $\gcd(\frac{g_2}{2},\widetilde{x}_1)=1$. Analogously for $\widetilde{x}_2$ and $\widetilde{x}_3$.

A common divisor of $\frac{g_2}{2}$ and $\frac{g}{2}$ would be odd. Because of all three terms in (5.20) it would divide $\widetilde{x}_1,\widetilde{x}_2,\widetilde{x}_3$. Therefore $\gcd(\frac{g_2}{2},\frac{g}{2})=1$.

For i,j,k with $\{i,j,k\}=\{1,2,3\}$ observe

$$2(2\widetilde{x}_i-g\widetilde{x}_j\widetilde{x}_k)+g\widetilde{x}_k(2\widetilde{x}_j-g\widetilde{x}_i\widetilde{x}_k)=\widetilde{x}_i(4-x_k^2), \tag{5.26}$$

$$4(r-4)=g^2(2\widetilde{x}_i-g\widetilde{x}_j\widetilde{x}_k)^2-(4-x_j^2)(4-x_k^2). \tag{5.27}$$

(5.26) and $\gcd(\frac{g_2}{2},\widetilde{x}_i)=1$ imply in case (iii) that $\frac{g_2}{2}$ divides $4^{-1}(4-x_k^2)$. This and (5.27) imply that $(\frac{g_2}{2})^2$ divides $\frac{r-4}{4}$, so g_2^2 divides $r-4$.

The claims for the cases (i) and (ii) follow similarly.

(b) Recall the shape of $M^{mat}\in M_{3\times 3}(\mathbb{Z})$ with $M\underline{e}=\underline{e}M^{mat}$ from the beginning of Sect. 5.3. It gives

$$M^{mat}-E_3=\begin{pmatrix}-x_1^2-x_2^2+x_1x_2x_3 & -x_1-x_2x_3+x_1x_3^2 & x_1x_3-x_2\\ x_1-x_2x_3 & -x_3^2 & -x_3\\ x_2 & x_3 & 0\end{pmatrix},$$

$$\begin{pmatrix}1&x_1&x_2\\0&1&x_3\\0&0&1\end{pmatrix}(M^{mat}-E_3)=\begin{pmatrix}0&-x_1&-x_2\\x_1&0&-x_3\\x_2&x_3&0\end{pmatrix},$$

$$(M^{mat}-E_3)\begin{pmatrix}1&0&0\\0&1&0\\-x_2&-x_3&1\end{pmatrix}=\begin{pmatrix}-x_1^2&-x_1&x_1x_3-x_2\\x_1&0&-x_3\\x_2&x_3&0\end{pmatrix}.$$

Now

$$q(M)\in \mathrm{End}(H_{\mathbb{Z}}) \iff q(M^{mat})-E_3\in M_{3\times 3}(\mathbb{Z}),$$

and this is equivalent to the following matrix being in $M_{3\times 3}(\mathbb{Z})$,

$$\begin{pmatrix}1&x_1&x_2\\0&1&x_3\\0&0&1\end{pmatrix}(q(M^{mat})-E_3)\begin{pmatrix}1&0&0\\0&1&0\\-x_2&-x_3&1\end{pmatrix}$$

$$
\begin{aligned}
&= \begin{pmatrix} 0 & -x_1 & -x_2 \\ x_1 & 0 & -x_3 \\ x_2 & x_3 & 0 \end{pmatrix} \\
&\cdot \left[q_0 \begin{pmatrix} 1 & 0 & 0 \\ 0 & 1 & 0 \\ -x_2 & -x_3 & 1 \end{pmatrix} + q_1 \begin{pmatrix} -x_1^2 & -x_1 & x_1x_3 - x_2 \\ x_1 & 0 & -x_3 \\ x_2 & x_3 & 0 \end{pmatrix} \right] \\
&= q_0 \begin{pmatrix} x_2^2 & -x_1 + x_2x_3 & -x_2 \\ x_1 + x_2x_3 & x_3^2 & -x_3 \\ x_2 & x_3 & 0 \end{pmatrix} \\
&+ q_1 \begin{pmatrix} -x_1^2 - x_2^2 & -x_2x_3 & x_1x_3 \\ -x_1^3 - x_2x_3 & -x_1^2 - x_3^2 & x_1^2x_3 - x_1x_2 \\ -x_1^2x_2 + x_1x_3 & -x_1x_2 & x_1x_2x_3 - x_2^2 - x_3^2 \end{pmatrix}.
\end{aligned}
$$

This gives nine scalar conditions, which we denote by their place $[a, b]$ with $a, b \in \{1, 2, 3\}$ in the matrix, so for example $[2, 1]$ is the condition $q_0(x_1 + x_2x_3) + q_1(-x_1^3 - x_2x_3) \in \mathbb{Z}$. These nine conditions are sufficient and necessary for $q(M) \in \mathrm{End}(H_\mathbb{Z})$.

The following trick allows an easy derivation of implied conditions. Recall the cyclic action $\gamma : \mathbb{Z}^3 \to \mathbb{Z}^3, \underline{x} \mapsto (x_3, x_1, x_2)$. It lifts to an action of $\mathrm{Br}_3 \ltimes \{\pm 1\}^3$ on triangular bases of $(H_\mathbb{Z}, L)$. Therefore together with $M^{mat} = S(\underline{x})^{-1}S(\underline{x})^t$ also the matrix $\widetilde{M}^{mat} := S(\gamma(\underline{x}))^{-1}S(\gamma(\underline{x}))^t$ is a monodromy matrix. Integrality of $q(M^{mat})$ is equivalent to integrality of $q(\widetilde{M}^{mat})$. Therefore if the nine conditions hold, also the conditions hold which are obtained from the nine conditions by replacing (x_1, x_2, x_3) by (x_3, x_1, x_2) or by (x_2, x_3, x_1). In the following $[a, b]$ denotes all three so obtained conditions, so for example $[2, 1]$ denotes the conditions

$$
\begin{gathered}
q_0(x_i + x_jx_k) + q_1(-x_i^3 - x_jx_k) \in \mathbb{Z} \\
\text{for } (i, j, k) \in \{(1, 2, 3), (3, 1, 2), (2, 3, 1)\}.
\end{gathered}
$$

We have to show the following equivalence:

$$
\text{the conditions } [a, b] \text{ for } a, b \in \{1, 2, 3\} \iff \text{the conditions (5.21) – (5.24).}
$$

$\Longrightarrow$: $[1, 3]$ and $[3, 2]$ are equivalent to one another and to (5.23).

$[3, 3]$ says $q_1(x_i^2 - r) \in \mathbb{Z}$. One derives $q_1(x_i^2 - x_j^2) \in \mathbb{Z}$, which is (5.24).

$[1, 1]$ and (5.24) give $q_2x_i^2 \in \mathbb{Z}$, so $q_2 \gcd(x_1^2, x_2^2, x_3^2) = q_2g^2 \in \mathbb{Z}$ which is (5.21).

The derivation of $q_1gg_1 \in \mathbb{Z}$ is laborious and goes as follows:
$[3, 1]$&$[1, 3]$ imply $q_1x_i(2x_k - x_ix_j) \in \mathbb{Z}$.
$[3, 2]$&$[2, 3]$ imply $q_1x_i(2x_j - x_ix_k) \in \mathbb{Z}$.
$[3, 3]$&(5.24) imply $q_1x_i(2x_i - x_jx_k) \in \mathbb{Z}$.
One sees $q_1x_ig_1 \in \mathbb{Z}$ and then $q_1gg_1 \in \mathbb{Z}$.

$\Longleftarrow$: (5.23) gives $[1, 3]$ and $[3, 2]$.

(5.21) and (5.24) give $[1, 1]$ and $[2, 2]$.

(5.23) reduces [1, 2], [2, 1], [2, 3] and [3, 1] to $q_2 x_j x_k$, $q_0 x_j x_k - q_1 x_i^3$, $q_1 x_i(x_i x_k - 2x_j)$ and $q_1 x_i(-x_i x_j + 2x_k)$. The first follows from (5.21), the third and fourth follow from (5.22). The second reduces with (5.24) to $x_j(q_0 x_k - q_1 x_i x_j)$, which follows from (5.23).

[3, 3] reduces with (5.24) to $q_1 x_j(x_i x_k - 2x_j)$ which follows from (5.22).

The equivalence of the conditions $[a, b]$ with the conditions (5.21)–(5.24) is shown.

It remains to show how (5.21)-(5.24) imply (5.25). One combines two times (5.23), $2q_0 x_i - 2q_1 x_j x_k \in \mathbb{Z}$, with (5.21), $(q_0 - 2q_1)x_j x_k \in \mathbb{Z}$, and obtains $q_0(2x_i - x_j x_k) \in \mathbb{Z}$.

(c) Recall

$$\begin{aligned} (t-\lambda_1)(t-\lambda_2) &= t^2-(2-r)t+1, \\ \text{so}\ \ \lambda_1+\lambda_2 = 2-r, &\quad \lambda_1\lambda_2 = 1, \\ \lambda_1^2 = (2-r)\lambda_1 - 1, &\quad (\lambda_1-1)^2 = (-r)\lambda_1, \end{aligned}$$

so

$$\begin{aligned} \mu_1 &= 1 + q_0(\lambda_1 - 1) + q_1(\lambda_1 - 1)^2 \\ &= (1 - rq_1) + (q_0 - rq_1)(\lambda_1 - 1). \qquad \Box \end{aligned}$$

Theorem 5.18 *Consider a triple $\underline{x} \in \mathbb{Z}^3$ with $r(\underline{x}) \in \mathbb{Z}_{<0} \cup \mathbb{Z}_{>4}$ which is neither reducible nor in the* $\mathrm{Br}_3 \ltimes \{\pm 1\}^3$ *orbit of a triple in Theorem 5.16. More explicitly, the triple $\underline{x}$ is any triple in $\mathbb{Z}^3$ which is not in the* $\mathrm{Br}_3 \ltimes \{\pm 1\}^3$ *orbits of the triples in the following set,*

$$\begin{aligned} &\{(x,0,0)\,|\,x \in \mathbb{Z}\} \cup \{(x,x,x)\,|\,x \in \mathbb{Z}\} \cup \{(2y,2y,2y^2\,|\,y \in \mathbb{Z}_{\geq 2}\} \\ &\cup \{(x,x,0)\,|\,x \in \mathbb{Z}\} \cup \{(-l,2,-l)\,|\,l \in \mathbb{Z}_{\geq 3}\} \cup \{(3,3,4)\}. \end{aligned}$$

Consider the associated triple $(H_\mathbb{Z}, L, \underline{e})$ with $L(\underline{e}^t, \underline{e})^t = S(\underline{x})$. Then

$$G_\mathbb{Z} = G_\mathbb{Z}^M = \{\pm M^l\,|\,l \in \mathbb{Z}\}.$$

The map $Z : (\mathrm{Br}_3 \ltimes \{\pm 1\}^3)_S \to G_\mathbb{Z}$ is surjective, so $G_\mathbb{Z} = G_\mathbb{Z}^\mathcal{B}$.

Proof The surjectivity of Z follows from $G_\mathbb{Z} = \{\pm M^l\,|\,l \in \mathbb{Z}\}$, $-\mathrm{id} = Z(\delta_1\delta_2\delta_3)$ and $M = Z(\sigma^{mon})$. The main point is to prove $G_\mathbb{Z}^M = \{\pm M^l\,|\,l \in \mathbb{Z}\}$.

Theorem 5.11 says for which $\underline{x}$ the automorphism Q of $H_\mathbb{Q}$ in Definition 5.9 is in $G_\mathbb{Z}^M$. They are all excluded here. So here $Q \notin G_\mathbb{Z}^M$.

We use the notation $g = g(\underline{x})$ in Lemma 5.10 and the notations from the beginning of Sect. 5.3. Especially $r := r(\underline{x}) \in \mathbb{Z}_{<0} \cup \mathbb{Z}_{>4}$, and $\lambda_3 = 1$ and $\lambda_{1/2} = \frac{2-r}{2} \pm \sqrt{r(r-4)}$ are the eigenvalues of the monodromy.

The proof of Lemma 5.15 gives a certain control on $G_\mathbb{Z}^M$ and $G_\mathbb{Z}$. Recall the notations there. For $p(t) = \sum_{i=0}^2 p_i t^i \in \mathbb{Q}[t]$ write $\mu_j := p(\lambda_j)$ for the eigenvalues of $p(M) \in \mathrm{End}(H_\mathbb{Q})$. Recall (5.17), (5.18), the isomorphism of $\mathbb{Q}$-algebras

$$\mathrm{End}(H_{\mathbb{Q}}, M) = \{p(M) \mid p(t) = \sum_{i=0}^{2} p_i t^i \in \mathbb{Q}[t]\} \to \mathbb{Q}[\lambda_1] \times \mathbb{Q},$$
$$p(M) \mapsto (p(\lambda_1), p(1)),$$

and its restriction in (5.19), the injective group homomorphism

$$\{h \in G_{\mathbb{Z}}^M \mid \mu_3 = 1\} \to \mathcal{O}^*_{\mathbb{Q}[\lambda_1]}, \quad h = p(M) \mapsto \mu_1 = p(\lambda_1). \tag{5.28}$$

The image of $Q \in \mathrm{End}(H_{\mathbb{Q}}, M)$ in $\mathbb{Q}[\lambda_1] \times \mathbb{Q}$ is $(-1, 1)$. Because $Q \notin G_{\mathbb{Z}}^M$, the image in (5.28) does not contain -1, so it is a cyclic group. It contains λ_1 which is the image of M. Therefore the group $\{h \in G_{\mathbb{Z}}^M \mid \mu_3 = 1\}$ is cyclic. It has two generators which are inverse to one another. We denote by h_{gen} the generator such that a positive power of it is M, namely $(h_{gen})^{l_{gen}} = M$ for a unique number $l_{gen} \in \mathbb{N}$. We have to prove $l_{gen} = 1$.

We will argue indirectly. We will assume the existence of a root $h = p(M) \in G_{\mathbb{Z}}^M$ with $h^l = M$ for some $l \geq 2$, first eigenvalue $\mu_1 = p(\lambda_1)$ and third eigenvalue $\mu_3 = p(1) = 1$. Then $\mu_1^l = \lambda_1$ and $\mu_1 = (1 - q_0) + (q_0 - rq_1)\lambda_1$ for certain $q_0, q_1 \in \mathbb{Q}$ which must satisfy the properties in (5.21)–(5.24). We will come to a contradiction.

We can restrict to $\underline{x}$ in the following set, as the $\mathrm{Br}_3 \ltimes \{\pm 1\}^3$ orbits of the elements in this set are all $\underline{x}$ which we consider in this theorem:

Consider the two sets Y_I and $Y_{II} \subset \mathbb{Z}^3$,

$$\begin{aligned} Y_I := &\{\underline{x} \in \mathbb{Z}^3_{\leq 0} \mid x_1 \leq x_2 \leq x_3\} \\ &- \big[\{(x, 0, 0) \mid x \in \mathbb{Z}_{<0}\} \cup \{(x, x, x) \mid x \in \mathbb{Z}_{\leq 0}\} \\ &\quad \cup \{(x, x, 0) \mid x \in \mathbb{Z}_{<0}\}\big], \\ Y_{II} := &\{\underline{x} \in \mathbb{Z}^3_{\geq 3} \mid x_1 \leq x_2 \leq x_3, 2x_3 \leq x_1 x_2\} \\ &- \big[\{(x, x, x) \mid x \in \mathbb{Z}_{\geq 3}\} \\ &\quad \cup \{(2y, 2y, 2y^2) \mid y \in \mathbb{Z}_{\geq 2}\} \cup \{(3, 3, 4)\}\big]. \end{aligned}$$

All triples $\underline{x}$ in this theorem are in the $\mathrm{Br}_3 \ltimes \{\pm 1\}^3$ orbits of the triples in $Y_I \cup Y_{II} \cup \{\underline{x} \mid (x_2, x_1, x_3) \in Y_I \cup Y_{II}\}$. We will restrict to $\underline{x} \in Y_I \cup Y_{II}$. For $\underline{x}$ with $(x_2, x_1, x_3) \in Y_I \cup Y_{II}$, one can copy the following proof and exchange x_1 and x_2.

Because the triples (x, x, x) are excluded, $x_1 < x_3$ and $\widetilde{x}_1 < \widetilde{x}_3$.

We will assume the existence of a unit $\mu_1 \in \mathcal{O}^*_{\mathbb{Q}[\lambda_1]}$ with $\mu_1^l = \lambda_1$ for some $l \geq 2$ and some norm $\mathcal{N}(\mu_1) \in \{\pm 1\}$. The integrality conditions in Lemma 5.17 (b) for $q_0, q_1, q_2 \in \mathbb{Q}$ with $\mu_1 = (1 - q_0) + (q_0 - rq_1)\lambda_1$ and $q_2 = q_0 - 2q_1$ will lead to a contradiction. The proof is a case discussion. The cases split as follows.

Case I: $\underline{x} \in Y_I$.

Subcase I.1: $l \geq 3$ odd.

Subcase I.2: $l = 2$.

Case II: $\underline{x} \in Y_{II}$.

Subcase II.1: $l \geq 3$ odd.

Subcase II.2: $l = 2$.

Subcase II.2.1: $\mathcal{N}(\mu_1) = 1$:

Subcase II.2.1.1: $\mu_1 = \kappa_a$ for some $a \in \mathbb{Z}_{\geq 3}$.

Subcase II.2.1.2: $\mu_1 = -\kappa_a$ for some $a \in \mathbb{Z}_{\geq 3}$.

Subcase II.2.2: $\mathcal{N}(\mu_1) = -1$.

The treatment of the cases I, II.1 and II.2.2 will be fairly short. The treatment of the cases II.2.1 will be laborious.

Lemma D.2 prepares all cases with $\mathcal{N}(\mu_1) = 1$. Consider such a case. Suppose $\mu_1 = \kappa_a$ for some $a \in \mathbb{Z}_{\leq -3} \cup \mathbb{Z}_{\geq 3}$. Compare Lemma D.2 (c) and Lemma 5.17 (c):

$$q_0 = q_{0,l}(a), \quad q_1 = q_{1,l}(a), \quad q_2 = q_{2,l}(a), \quad r = r_l(a).$$

The integrality condition (5.21) $q_2 g^2 \in \mathbb{Z}$ together with (D.6) and (D.5) tells that $r/(2-a) = r_l(a)/(2-a)$ divides g^2.

For $l \geq 3$ odd $r/(2-a)$ is itself a square by (D.4), and g can be written as $g = \gamma_1\gamma_3$ with $\gamma_1, \gamma_3 \in \mathbb{N}$ and $\gamma_1^2 = r/(2-a)$. For $l = 2$ $r/(2-a) = a+2$, and g can be written as $g = \gamma_1\gamma_2\gamma_3$ with $\gamma_1, \gamma_2, \gamma_3 \in \mathbb{N}$, γ_2 squarefree, and $a + 2 = \gamma_1^2\gamma_2$, so $g^2 = (a+2)\gamma_2\gamma_3^2$.

On the other hand g^2 divides $r = r_l(a)$ by Lemma 1.3. (5.1) takes the shape

$$\widetilde{x}_1^2 + \widetilde{x}_2^2 + \widetilde{x}_3^2 - g\widetilde{x}_1\widetilde{x}_2\widetilde{x}_3 = \frac{r}{g^2} = \begin{cases} \frac{2-a}{\gamma_3^2} & \text{if } l \geq 3 \text{ is odd,} \\ \frac{2-a}{\gamma_2\gamma_3^2} & \text{if } l = 2. \end{cases} \tag{5.29}$$

This equation will be the key to contradictions in the cases discussed below. The absolute value of the left hand side will be large, the absolute value of the right hand side will be small. Now we start the case discussion.

Case I.1, $\underline{x} \in Y_l$, $l \geq 3$ odd: $\mathcal{N}(\mu_1)^l = \mathcal{N}(\lambda_1) = 1$ and l odd imply $\mathcal{N}(\mu_1) = 1$. Here $\lambda_1 \in (-1, 0)$, so $\mu_1 \in (-1, 0)$, so $\mu_1 = \kappa_a$ for some $a \in \mathbb{Z}_{\leq -3}$. By the discussion above $g = \gamma_1\gamma_3$, $\gamma_1 = |b_{(l+1)/2} + b_{(l-1)/2}|$, and (5.29) holds.

Case I.1.1, all $x_i < 0$: We excluded the triples (x, x, x). Therefore $\widetilde{x}_1 \leq -2$. For $l \geq 3$ $\gamma_1 = |b_{(l+1)/2} + b_{(l-1)/2}| \geq |b_2 + b_1| = |a| - 1$ by Lemma D.2 (b). Now by (5.29)

$$|a| + 2 \geq \frac{2-a}{\gamma_3^2} \geq 2g + 4 + 1 + 1 = 2\gamma_1\gamma_3 + 6 \geq 2(|a| - 1) + 6,$$

a contradiction.

Case I.1.2, $x_3 = 0$: The integrality conditions (5.23) and (5.24) say here

$$q_0x_1, q_0x_2, q_1x_1x_2 \in \mathbb{Z}, \quad q_1x_1^2, q_1x_2^2 \in \mathbb{Z}, \quad \text{so} \quad q_1g^2 \in \mathbb{Z}$$

(which is a bit stronger than (5.22)). $q_1g^2 \in \mathbb{Z}$ means

$$\frac{(b_l(a)-b_{l-1}(a)-1)g^2}{rb_l(a)} = \frac{b_l(a)-b_{l-1}(a)-1}{b_l(a)(2-a)/\gamma_3^2} \in \mathbb{Z}.$$

But $\frac{b_l(a)-b_{l-1}(a)-1}{b_l(a)} \in (1,2)$ because of Lemma D.2 (b), and $\frac{2-a}{\gamma_3^2} \in \mathbb{N}$, a contradiction.

Case I.2, $\underline{x} \in Y_I, l = 2$: Here $\lambda_1 < 0$. Therefore $\mu_1^2 = \lambda_1$ is impossible.

Case II.1, $\underline{x} \in Y_{II}, l \geq 3$ odd: $\mathcal{N}(\mu_1)^l = \mathcal{N}(\lambda_1) = 1$ and l odd imply $\mathcal{N}(\mu_1) = 1$. Here $\lambda_1 > 1$, so $\mu_1 > 1$, so $\mu_1 = \kappa_a$ for some $a \in \mathbb{Z}_{\geq 3}$. By the discussion above $g = \gamma_1\gamma_3$, $\gamma_1 = b_{(l+1)/2} + b_{(l-1)/2}$, and (5.29) holds. The proof of Lemma 5.10 (b) gives the first inequality below,

$$\begin{aligned} \frac{a-2}{\gamma_3^2} &= \frac{|r|}{g^2} \geq g\widetilde{x}_1\widetilde{x}_2^2 - \widetilde{x}_1^2 - 2\widetilde{x}_2^2 \geq \widetilde{x}_2^2(g\widetilde{x}_1 - 3), \\ \text{so}\quad a - 2 &\geq \gamma_3^2\widetilde{x}_2^2(\gamma_1\gamma_3\widetilde{x}_1 - 3), \\ \text{so}\quad (\gamma_3^2\widetilde{x}_2^2 - 1)3 &\geq (\gamma_3^2\widetilde{x}_2^2 - 1)\gamma_1\gamma_3\widetilde{x}_1 + \gamma_1\gamma_3\widetilde{x}_1 - (a+1). \end{aligned} \tag{5.30}$$

Observe with Lemma D.2 (b)

$$\gamma_1 = b_{(l+1)/2} + b_{(l-1)/2} \geq b_2 + b_1 = a + 1 \geq 4.$$

Therefore the inequality (5.30) can only hold if $\gamma_3 = \widetilde{x}_2 = \widetilde{x}_1 = 1$ and $\gamma_1 = a + 1$, so $l = 3$. Then also $g = \gamma_1\gamma_3 = a + 1$.

$$\begin{aligned} a - 2 = \frac{a-2}{\gamma_3^2} &= \frac{|r|}{g^2} = g\widetilde{x}_3 - \widetilde{x}_3^2 - 1 - 1 = (a + 1 - \widetilde{x}_3)\widetilde{x}_3 - 2, \\ 0 &= (\widetilde{x}_3 - 1)(\widetilde{x}_3 - a). \end{aligned}$$

We excluded the triples (x, x, x), so $\widetilde{x}_3 > 1$. But also $\widetilde{x}_3 \leq \frac{g}{2}\widetilde{x}_1\widetilde{x}_2 = \frac{a+1}{2} < a$. A contradiction.

Case II.2, $\underline{x} \in Y_{II}, l = 2$: Then $\mathcal{N}(\mu_1) = \varepsilon_1$ for some $\varepsilon_1 \in \{\pm 1\}$. Also $\lambda_1 > 1$ and $\varepsilon_2\mu_1 > 1$ for some $\varepsilon_2 \in \{\pm 1\}$. Then μ_1 is a zero of a polynomial $t^2 - \varepsilon_2 at + \varepsilon_1$, namely

$$\mu_1 = \varepsilon_2\left(\frac{a}{2} + \frac{1}{2}\sqrt{a^2 - 4\varepsilon_1}\right) \quad \text{with} \begin{cases} a \in \mathbb{Z}_{\geq 3} & \text{if } \varepsilon_1 = 1, \\ a \in \mathbb{N} & \text{if } \varepsilon_1 = -1, \end{cases}$$

$$\mu_1 + \mu_1^{conj} = \varepsilon_2 a, \quad \mu_1\mu_1^{conj} = \varepsilon_1, \quad \mu_1^2 = \varepsilon_2 a\mu_1 - \varepsilon_1.$$

Comparison with

$$\begin{aligned} \lambda_1 &= \frac{2-r}{2} + \frac{1}{2}\sqrt{r(r-4)} \\ &= \mu_1^2 = \frac{a^2 - 2\varepsilon_1}{2} + \frac{a}{2}\sqrt{a^2 - 4\varepsilon_1} \\ &= \varepsilon_2 a\mu_1 - \varepsilon_1 \end{aligned}$$

shows

$$\begin{aligned} r &= -a^2 + 2(\varepsilon_1 + 1), \\ \mu_1 &= \frac{\varepsilon_1\varepsilon_2}{a} + \frac{\varepsilon_2}{a}\lambda_1 \\ &= (1 - q_0) + (q_0 - rq_1)\lambda_1, \end{aligned}$$

with

$$\begin{aligned} q_0 &= \frac{a - \varepsilon_1\varepsilon_2}{a}, \\ q_1 &= \frac{\varepsilon_2(\varepsilon_1 + 1) - a}{a(a^2 - 2(\varepsilon_1 + 1))}. \end{aligned}$$

Case II.2.1, $\mathcal{N}(\mu_1) = 1$: Then $\varepsilon_1 = 1$ and

$$\begin{aligned} r &= -a^2 + 4 = (2 - a)(2 + a), \\ q_0 &= \frac{a - \varepsilon_2}{a}, \\ q_1 &= \frac{2\varepsilon_2 - a}{a(a^2 - 4)} = \frac{-1}{a(a + 2\varepsilon_2)}, \\ q_2 &= q_0 - 2q_1 = \frac{a + \varepsilon_2}{a + 2\varepsilon_2}. \end{aligned}$$

Write $a + 2\varepsilon_2 = \gamma_1^2\gamma_2$ with $\gamma_1, \gamma_2 \in \mathbb{N}$ and γ_2 squarefree. The integrality condition (5.21) $q_2g^2 \in \mathbb{Z}$ tells

$$g = \gamma_1\gamma_2\gamma_3 \quad \text{with} \quad a + 2\varepsilon_2 = \gamma_1^2\gamma_2, \quad g^2 = (a + 2\varepsilon_2)\gamma_2\gamma_3^2, \tag{5.31}$$

for some $\gamma_3 \in \mathbb{N}$. The conditions $r < 0$ and $g^2|r$ tell

$$\frac{a - 2\varepsilon_2}{\gamma_2\gamma_3^2} = \frac{-r}{g^2} = g\widetilde{x}_1\widetilde{x}_2\widetilde{x}_3 - \widetilde{x}_1^2 - \widetilde{x}_2^2 - \widetilde{x}_3^2 \in \mathbb{N}. \tag{5.32}$$

γ_2 divides $a + 2\varepsilon_2$ and $a - 2\varepsilon_2$, so it divides 4. As it is squarefree, $\gamma_2 \in \{1, 2\}$. Also $\gcd(g, a) = \gcd(\gamma_1\gamma_2\gamma_3, a) \in \{1, 2\}$ as $a + 2\varepsilon_2 = \gamma_1^2\gamma_2$ and $\gamma_2\gamma_3^3$ divides $a - 2\varepsilon_2$.

The integrality conditions (5.23), $q_0x_i - q_1x_jx_k \in \mathbb{Z}$ tell

$$\frac{a - \varepsilon_2}{a}x_i + \frac{1}{a(a + 2\varepsilon_2)}x_jx_k = x_i + \frac{1}{a}(-g\varepsilon_2\widetilde{x}_i + \gamma_2\gamma_3^2\widetilde{x}_j\widetilde{x}_k) \in \mathbb{Z},$$

$$\text{so after multiplying with } \gamma_1 \quad \frac{\gamma_3}{a}(-(a + 2\varepsilon_2)\varepsilon_2\widetilde{x}_i + g\widetilde{x}_j\widetilde{x}_k) \in \mathbb{Z},$$

$$\text{so} \quad \frac{\gamma_3}{a}(-2\widetilde{x}_i + g\widetilde{x}_j\widetilde{x}_k) \in \mathbb{Z},$$

$$\text{so} \quad \frac{\gamma_3}{a}g_2 \in \mathbb{Z}.$$

This is a bit stronger than the integrality condition (5.22) $q_1 g^2 g_2 \in \mathbb{Z}$ which says $\frac{\gamma_2\gamma_3^2}{a} g_2 \in \mathbb{Z}$. We can improve it even more, to

$$\frac{g_2}{a} \in \mathbb{Z}, \quad \frac{1}{a}(-2\widetilde{x}_i + g\widetilde{x}_j\widetilde{x}_k) \in \mathbb{Z}, \tag{5.33}$$

by the following case discussion: If $a \equiv 1(2)$, then $\gamma_3 \equiv 1(2)$, so $\gcd(a, \gamma_3) = 1$, so (5.33) holds. It $a \equiv 0(4)$, then $a - 2\varepsilon_2 \equiv 2(4)$, so $\gamma_3 \equiv 1(2)$, so $\gcd(a, \gamma_3) = 1$, so (5.33) holds. If $a \equiv 2(4)$, then a priori $\frac{2}{a} g_2 \in \mathbb{Z}$. But then $\frac{a}{2} \equiv 1(2)$ and $g \equiv 0(2)$, so $g_2 \equiv 0(2)$, so (5.33) holds.

Finally, the integrality conditions (5.24) $q_1(x_i^2 - x_j^2)$ say

$$\frac{\gamma_2\gamma_3^2}{a}(\widetilde{x}_i^2 - \widetilde{x}_j^2) \in \mathbb{Z}. \tag{5.34}$$

The following estimate which arises from (5.2) will also be useful:

$$\begin{aligned} \widetilde{x}_1^2 &\le \frac{(2+(4-r)^{1/3})^2}{g^2} = \frac{(2+a^{2/3})^2}{g^2} = \frac{4+4a^{2/3}+a^{4/3}}{g^2} \\ &\le \frac{4+2a^{1/3}+2a+a^{4/3}}{g^2} = \frac{(a+2)(2+a^{1/3})}{(a+2\varepsilon_2)\gamma_2\gamma_3^2}. \end{aligned} \tag{5.35}$$

Case II.2.1.1, $\varepsilon_2 = 1$, $\mu_1 = \kappa_a$ for some $a \in \mathbb{Z}_{\ge 3}$: The estimate (5.35) says

$$\widetilde{x}_1^2 \le \lfloor \frac{2+a^{1/3}}{\gamma_2\gamma_3^2} \rfloor \le \lfloor 2 + a^{1/3} \rfloor \le a \quad (\text{recall } a \in \mathbb{Z}_{\ge 3}). \tag{5.36}$$

Recall that $\underline{x}$ is a local minimum, so $2\widetilde{x}_3 \le g\widetilde{x}_1\widetilde{x}_2$ and that $\widetilde{x}_1 \le \widetilde{x}_2 \le \widetilde{x}_3$ and $1 \le \widetilde{x}_1 < \widetilde{x}_3$ (as (x, x, x) is excluded).

Case II.2.1.1.1, $2\widetilde{x}_3 < g\widetilde{x}_1\widetilde{x}_2$: Then (5.33) gives the existence of $\alpha \in \mathbb{N}$ with $g\widetilde{x}_1\widetilde{x}_2 = \alpha a + 2\widetilde{x}_3$. With this we go into (5.32),

$$\begin{aligned} \frac{a-2}{\gamma_2\gamma_3^2} &= \alpha a\widetilde{x}_3 - \widetilde{x}_1^2 + (\widetilde{x}_3^2 - \widetilde{x}_2^2) \\ &\overset{(5.36)}{\ge} \alpha a\widetilde{x}_3 - a + 0 \overset{\widetilde{x}_3 \ge 2}{\ge} a, \end{aligned}$$

a contradiction.

Case II.2.1.1.2, $2\widetilde{x}_3 = g\widetilde{x}_1\widetilde{x}_2$, $\widetilde{x}_1 \ge 2$: (5.32) takes the shape

$$\begin{aligned} \frac{a-2}{\gamma_2\gamma_3^2} &= \widetilde{x}_3^2 - \widetilde{x}_1^2 - \widetilde{x}_2^2 = g^2(\frac{\widetilde{x}_1}{2})^2\widetilde{x}_2^2 - \widetilde{x}_1^2 - \widetilde{x}_2^2 \\ &\ge (g^2-2)\widetilde{x}_2^2 \ge ((a+2)-2)\widetilde{x}_2^2 = a\widetilde{x}_2^2 \ge a, \end{aligned}$$

a contradiction.

Case II.2.1.1.3, $2\widetilde{x}_3 = g\widetilde{x}_1\widetilde{x}_2$, $\widetilde{x}_1 = 1$: Write $\gamma_4 := \gamma_2\gamma_3^2$. Then (5.32) takes the shape

$$\begin{aligned}\frac{a-2}{\gamma_4} &= \widetilde{x}_3^2 - 1 - \widetilde{x}_2^2 = (\frac{1}{4}(a+2)\gamma_4 - 1)\widetilde{x}_2^2 - 1,\\ &\overset{\widetilde{x}_2 \geq 1}{\geq} \frac{1}{4}(a+2)\gamma_4 - 2,\\ (\gamma_4^2 - 4)a &\leq -2(\gamma_4^2 - 4) + 8(\gamma_4 - 2).\end{aligned}$$

If $\gamma_4 > 2$ then $a \leq -2 + \frac{8}{\gamma_4+2} \leq -2 + \frac{8}{5}$, a contradiction. Therefore $\gamma_4 \in \{1, 2\}$. If $\gamma_4 = 1$ then (5.32) becomes

$$a - 2 = \frac{a-2}{4}\widetilde{x}_2^2 - 1, \quad \text{so} \quad 4 = (a-2)(\widetilde{x}_2^2 - 4),$$

which has no solution $(a, \widetilde{x}_2) \in \mathbb{Z}_{\geq 3} \times \mathbb{N}$, so a contradiction.

If $\gamma_4 = 2$ then (5.32) is solved with $\widetilde{x}_2 = 1$ and $a \in \mathbb{Z}_{\geq 3}$ arbitrary. Then $\widetilde{x}_1 = 1$, $\gamma_2 = 2$, $\gamma_3 = 1$, $g = 2\gamma_1$, $\widetilde{x}_3 = \frac{g}{2} = \gamma_1$, $\underline{x} = (2\gamma_1, 2\gamma_1, 2\gamma_1^2)$. These cases are excluded.

Case II.2.1.2, $\varepsilon_2 = -1$, $\mu_1 = -\kappa_a$ for some $a \in \mathbb{Z}_{\geq 3}$: The estimates (5.35) say here

$$\widetilde{x}_1^2 \leq \lfloor \frac{(2 + a^{2/3})^2}{(a-2)\gamma_2\gamma_3^2} \rfloor \leq \lfloor \frac{(a+2)(2 + a^{1/3})}{(a-2)\gamma_2\gamma_3^2} \rfloor. \tag{5.37}$$

This implies

$$\widetilde{x}_1^2 < a \quad \text{if} \quad a \geq 8. \tag{5.38}$$

We treat small a first. Recall $\gamma_2 \in \{1, 2\}$ and $a = \gamma_1^2\gamma_2 + 2$. So if $a \leq 9$ then $a \in \{3, 4, 6\}$. The following table lists constraints for $a \in \{3, 4, 6\}$. For $\widetilde{x}_1$ (5.37) gives an upper bound and $\frac{3}{g} \leq \frac{x_1}{g} = \widetilde{x}_1$ gives a lower bound. Recall the conditions $a - 2 = \gamma_1^2\gamma_2$ and $\frac{a+2}{\gamma_2\gamma_3^2} \in \mathbb{N}$.

a	3	4	6
$(\gamma_1, \gamma_2, \gamma_3)$	(1, 1, 1)	(1, 2, 1)	(2, 1, 1) or (2, 1, 2)
$\gamma_2\gamma_3^2$	1	2	1 or 4
g	1	2	2 or 4
$\frac{3}{g}$	3	$\frac{3}{2}$	$\frac{3}{2}$ or $\frac{3}{4}$
$\frac{(2+a^{2/3})^2}{a-2}$	16, 64..	10, 21..	7, 03..
$\widetilde{x}_1$	3 or 4	2	2 or 1

This gives five cases $(a, \widetilde{x}_1) \in \{(3, 3), (3, 4), (4, 2), (6, 2), (6, 1)\}$ with $a \le 9$. We treat these cases first and then all cases with $a \ge 10$. Because of (5.33) a number $\alpha \in \mathbb{Z}_{\ge 0}$ with $g\widetilde{x}_1\widetilde{x}_2 = \alpha a + 2\widetilde{x}_3$ exists. Then (5.32) becomes

$$\frac{a+2}{\gamma_2\gamma_3^2} = \alpha a\widetilde{x}_3 - \widetilde{x}_1^2 + (\widetilde{x}_3^2 - \widetilde{x}_2^2). \tag{5.39}$$

Also recall (5.34) $\frac{\gamma_2\gamma_3^2}{a}(\widetilde{x}_i^2 - \widetilde{x}_j^2) \in \mathbb{Z}$.

Case II.2.1.2.1, $(a, \gamma_1, \gamma_2, \gamma_3, g, \widetilde{x}_1) = (3, 1, 1, 1, 1, 3)$: (5.39) says

$$5 = 3\alpha\widetilde{x}_3 - 9 + (\widetilde{x}_3^2 - \widetilde{x}_2^2).$$

(5.34) says that 3 divides $\widetilde{x}_3^2 - \widetilde{x}_2^2$. A contradiction.

Case II.2.1.2.2, $(a, \gamma_1, \gamma_2, \gamma_3, g, \widetilde{x}_1) = (3, 1, 1, 1, 1, 4)$: (5.39) says

$$5 = 3\alpha\widetilde{x}_3 - 16 + (\widetilde{x}_3^2 - \widetilde{x}_2^2).$$

$g\widetilde{x}_1\widetilde{x}_2 = 4\widetilde{x}_2 = \alpha a + 2\widetilde{x}_3$ implies that α is even. This and $\widetilde{x}_3 > \widetilde{x}_1 = 4$ and (5.39) show $\alpha = 0$, so $2\widetilde{x}_2 = \widetilde{x}_3$, so $5 = 0 - 16 + 3\widetilde{x}_2^2$, a contradiction.

Case II.2.1.2.3, $(a, \gamma_1, \gamma_2, \gamma_3, g, \widetilde{x}_1) = (4, 1, 2, 1, 2, 2)$: (5.39) says

$$3 = 4\alpha\widetilde{x}_3 - 4 + (\widetilde{x}_3^2 - \widetilde{x}_2^2).$$

(5.34) says that 2 divides $\widetilde{x}_3^2 - \widetilde{x}_2^2$. A contradiction.

Case II.2.1.2.4, $(a, \gamma_1, \gamma_2, \gamma_3, g, \widetilde{x}_1) = (6, 2, 1, 1, 2, 2)$: (5.39) says

$$8 = 6\alpha\widetilde{x}_3 - 4 + (\widetilde{x}_3^2 - \widetilde{x}_2^2).$$

$\widetilde{x}_3 > \widetilde{x}_1 = 2$ and (5.39) imply $\alpha = 0$, so $12 = \widetilde{x}_3^2 - \widetilde{x}_2^2$. Only $\widetilde{x}_2 = 2$ and $\widetilde{x}_3 = 4$ satisfy this. But then $\gcd(\widetilde{x}_1, \widetilde{x}_2, \widetilde{x}_3) = 2 \neq 1$, a contradiction.

Case II.2.1.2.5, $(a, \gamma_1, \gamma_2, \gamma_3, g, \widetilde{x}_1) = (6, 2, 1, 2, 4, 1)$: (5.39) says

$$2 = 6\alpha\widetilde{x}_3 - 1 + (\widetilde{x}_3^2 - \widetilde{x}_2^2).$$

It implies $\alpha = 0$ and $\widetilde{x}_2 = 1$, $\widetilde{x}_3 = 2$, so $\underline{x} = (4, 4, 8)$. This case was excluded in Theorem 5.18.

Case II.2.1.2.6, $a \ge 10$:

Case II.2.1.2.6.1, $\alpha > 0$: (5.38) gives $\widetilde{x}_1^2 < a$. This and (5.39) and $\widetilde{x}_3 > \widetilde{x}_1 \ge 1$ show $\alpha = 1$, $\widetilde{x}_3 = 2$, $\widetilde{x}_1 = 1$, $\gamma_2 = \gamma_3 = 1$, so $a + 2 = 2a - 1 + (4 - \widetilde{x}_2^2)$. A contradiction to $a \ge 10$.

Case II.2.1.2.6.2, $\alpha = 0$, $\widetilde{x}_1 \ge 2$: (5.39) says

$$\frac{a+2}{\gamma_2\gamma_3^2} = \widetilde{x}_3^2 - \widetilde{x}_1^2 - \widetilde{x}_2^2 = ((a-2)\gamma_2\gamma_3^2\frac{1}{4}\widetilde{x}_1^2 - 1)\widetilde{x}_2^2 - \widetilde{x}_1^2,$$

$$\text{so}\quad a+2 \geq ((a-2)-1)\widetilde{x}_2^2 - \widetilde{x}_1^2 \geq ((a-2)-2)4,$$
$$\text{so}\quad 3a \leq 18, \quad a \leq 6,$$

a contradiction.

Case II.2.1.2.6.3, $\alpha = 0$, $\widetilde{x}_1 = 1$: Write $\gamma_4 := \gamma_2\gamma_3^2$. Then (5.39) says

$$\frac{a+2}{\gamma_4} = \widetilde{x}_3^2 - 1 - \widetilde{x}_2^2 = ((a-2)\gamma_4\frac{1}{4} - 1)\widetilde{x}_2^2 - 1,$$
$$\text{so}\quad \widetilde{x}_2^2 = \frac{4}{\gamma_4^2} \cdot \frac{a+2+\gamma_4}{a-2-\frac{4}{\gamma_4}}. \tag{5.40}$$

The right hand side must be ≥ 1. This means

$$\gamma_4^2(a-2-\frac{4}{\gamma_4}) \leq 4(a+2+\gamma_4),$$
$$(\gamma_4^2-4)a \leq 2(\gamma_4^2-4) + 8(\gamma_4+2).$$

If $\gamma_4 > 2$ then $a \leq 2 + \frac{8}{\gamma_4 - 2}$, which is in contradiction to $a \geq 10$, as $\gamma_4 = 3$ would mean $\gamma_2 = 3$, which is impossible. Therefore $\gamma_4 \in \{1, 2\}$.

If $\gamma_4 = 1$ then (5.40) says $\widetilde{x}_2^2 = 4\frac{a+3}{a-6} = 4 + \frac{36}{a-6}$. But the right hand side is not a square for any $a \geq 10$, a contradiction.

If $\gamma_4 = 2$ then (5.40) says $\widetilde{x}_2^2 = \frac{a+4}{a-4}$, which is also not a square for any $a \geq 10$, a contradiction.

Case II.2.2, $\mathcal{N}(\mu_1) = \varepsilon_1 = -1$: Recall the formulas for r, q_0 and q_1 at the beginning of case II.2. Now

$$r = -a^2,$$
$$q_0 = \frac{a+\varepsilon_2}{a},$$
$$q_1 = \frac{-1}{a^2},$$
$$q_2 = \frac{a^2+\varepsilon_2 a + 2}{a^2}.$$

The integrality condition (5.21) $q_2 g^2 \in \mathbb{Z}$ says $\frac{a}{2} \mid g$ if $a \equiv 2(4)$ and $a \mid g$ if $a \equiv 1(2)$ or $a \equiv 0(4)$. Also $g^2 \mid r = -a^2$. Therefore $g = a$ or $g = \frac{a}{2}$, and $g = \frac{a}{2}$ only if $a \equiv 2(4)$. The case $g = a$ means $\frac{r}{g^2} = -1$ which implies by Lemma 5.11 $Q \in G_{\mathbb{Z}}$. But all such cases are excluded in Theorem 5.18.

Therefore $g = \frac{a}{2}$ and $a \equiv 2(4)$. The integrality condition (5.22) $q_1 g^2 g_2 \in \mathbb{Z}$ says $\frac{g_2}{4} \in \mathbb{Z}$. But $g_2 \equiv 0(4)$ and $g \equiv 1(2)$ are together impossible in view of the definition $g_2 = \gcd(2\widetilde{x}_1 - g\widetilde{x}_2\widetilde{x}_3, 2\widetilde{x}_2 - g\widetilde{x}_1\widetilde{x}_3, 2\widetilde{x}_3 - g\widetilde{x}_1\widetilde{x}_2)$ and $\gcd(\widetilde{x}_1, \widetilde{x}_2, \widetilde{x}_3) = 1$. A contradiction.

Therefore in all cases the assumption that a nontrivial root μ_1 of λ_1 exists which satisfies the integrality conditions (5.21)–(5.24) leads to a contradiction. Theorem 5.18 is proved □

Remark 5.19 The results in this chapter give complete results on $G_{\mathbb{Z}}$ and $G_{\mathbb{Z}}^M \supset G_{\mathbb{Z}}$ for all unimodular bilinear lattices of rank 3.

The reducible cases: Lemma 5.4, Theorem 5.13.

The irreducible cases with $r \in \{0, 1, 2, 4\}$: Theorem 5.14

The irreducible cases with $r \in \mathbb{Z}_{<0} \cup \mathbb{Z}_{>4}$ and $G_{\mathbb{Z}}^M \supsetneqq \{\pm M^l \mid l \in \mathbb{Z}\}$: Theorem 5.16.

The irreducible cases with $r \in \mathbb{Z}_{<0} \cup \mathbb{Z}_{>4}$ and $G_{\mathbb{Z}}^M = \{\pm M^l \mid l \in \mathbb{Z}\}$: Theorem 5.18.

Chapter 6
Monodromy Groups and Vanishing Cycles

This chapter studies the monodromy groups $\Gamma^{(0)}$ and $\Gamma^{(1)}$ of the unimodular bilinear lattices $(H_{\mathbb{Z}}, L, \underline{e})$ with triangular basis $\underline{e}$ which have rank 2 or 3. In rank 3 the even as well as the odd cases split into many different case studies. They make the chapter long.

Section 6.1 considers for $k \in \{0; 1\}$ the quotient lattice $\overline{H_{\mathbb{Z}}}^{(1)} := H_{\mathbb{Z}}/\operatorname{Rad} I^{(k)}$ and the induced bilinear form $\overline{I}^{(k)}$ on it. Because $\Gamma^{(k)}$ acts trivially on $\operatorname{Rad} I^{(k)}$, it acts on this quotient lattice and respects $\overline{I}^{(k)}$. The homomorphism $\Gamma^{(k)} \to \operatorname{Aut}(\overline{H_{\mathbb{Z}}}^{(1)}, \overline{I}^{(k)})$ has an image $\Gamma_s^{(k)}$, the *simple part* of $\Gamma^{(k)}$ and a kernel $\Gamma_u^{(k)}$, the *unipotent part* of $\Gamma^{(k)}$. There is the exact sequence

$$\{\mathrm{id}\} \to \Gamma_u^{(k)} \to \Gamma^{(k)} \to \Gamma_s^{(k)} \to \{\mathrm{id}\}.$$

We will study $\Gamma^{(k)}$ together with $\Gamma_s^{(k)}$ and $\Gamma_u^{(k)}$. Also the natural homomorphism $j^{(k)} : H_{\mathbb{Z}} \to H_{\mathbb{Z}}^{\sharp} := \operatorname{Hom}_{\mathbb{Z}}(H_{\mathbb{Z}}, \mathbb{Z})$ and in the even case the spinor norm will be relevant. Section 6.1 fixes more or less well known general facts.

Section 6.2 treats the rank 2 cases. The even cases A_1^2 and A_2 are classical and easy. In the other even cases $\Gamma^{(0)} \cong G^{fCox,2}$. There we can characterize $\Delta^{(0)}$ arithmetically and geometrically. In the odd case A_2 $\Gamma^{(1)} \cong SL_2(\mathbb{Z})$. In the other irreducible odd cases $\Gamma^{(1)} \cong G^{free,2}$. The matrix group $\Gamma^{(1),mat} \subset SL_2(\mathbb{Z})$ is a Fuchsian group of the second kind, but has infinite index in $SL_2(\mathbb{Z})$ in most cases. We do not have a characterization of $\Delta^{(1)}$ which is as nice as in the even cases.

The long Theorem 6.11 in Sect. 6.3 states our results on the even monodromy group $\Gamma^{(0)}$ in the rank 3 cases. The results are detailed except for the local minima $\underline{x} \in \mathbb{Z}^3_{\geq 3}$ with $r(\underline{x}) \leq 0$ where we only state $\Gamma^{(0)} \cong \Gamma_s^{(0)} \cong G^{fCox,3}$ and $\Gamma_u^{(0)} = \{\mathrm{id}\}$. It is followed by Theorem 6.14 which gives the set $\Delta^{(0)}$ of even vanishing cycles in many, but not all cases. Especially in the cases of the local minima $\underline{x} \in \mathbb{Z}^3_{\geq 3}$ with $r(\underline{x}) \leq 0$ we know little and only state $\Delta^{(0)} = R^{(0)}$ in the case $(3, 3, 3)$, but

C. Hertling and K. Larabi, *Triangular Bases of Unimodular Bilinear Lattices and Induced Structures*, Lecture Notes in Mathematics 2398,
https://doi.org/10.1007/978-3-032-25997-4_6

$\Delta^{(0)} \subsetneq R^{(0)}$ in the four cases $(3, 3, 4)$, $(4, 4, 4)$, $(5, 5, 5)$ and $(4, 4, 8)$. The result $\Delta^{(0)} = R^{(0)}$ in the case $(3, 3, 3)$ seems to be new. Its proof is rather laborious.

Section 6.4 treats the odd monodromy group $\Gamma^{(1)}$ and the set of odd vanishing cycles $\Delta^{(1)}$ in the rank 3 cases. The long Theorem 6.18 fixes the results on $\Gamma^{(1)}$. The even longer Theorem 6.21 fixes the results on $\Delta^{(1)}$. Also their proofs are long. They are preceded by two technical lemmas, the second one helps to control $\Gamma_u^{(1)}$. Similar to the even rank 3 cases, in the case of a local minimum $\underline{x} \in \mathbb{Z}^3_{\geq 3}$ with $r(\underline{x}) \leq 0$ $\Gamma^{(1)} \cong \Gamma_s^{(1)} \cong G^{free,3}$ and $\Gamma_u^{(1)} = \{\text{id}\}$. In the same case, interestingly, the map $\Delta^{(1)} \to \overline{H_\mathbb{Z}}^{(1)}$ is injective. This leads in this case to the problem how to recover an odd vanishing cycle from its image in $\overline{H_\mathbb{Z}}^{(1)}$. One solution is offered in the most important case $\underline{x} = (3, 3, 3)$ in Lemma 6.26. One general application of the Theorems 6.18 and 6.21 is given in Corollary 6.23. It allows to separate many of the orbits of the bigger group $(G^{phi} \ltimes \widetilde{G}^{sign}) \rtimes \langle \gamma \rangle$ which acts on $T_3(\mathbb{Z})$ and $\mathbb{Z}^3$ in Lemma 4.18.

6.1 Basic Observations

Let $(H_\mathbb{Z}, L, \underline{e})$ be a unimodular bilinear lattice of rank $n \in \mathbb{N}$ with a triangular basis $\underline{e}$. Definition 2.8 gave two monodromy groups $\Gamma^{(0)}$ and $\Gamma^{(1)}$ and two sets $\Delta^{(0)}$ and $\Delta^{(1)}$ of vanishing cycles. Later in this chapter they shall be studied rather systematically in essentially all cases with $n = 2$ or $n = 3$. For that we need some notations and basic facts, which are collected here. Everything in this section is well known. Most of it is stated in the even case in [Eb84] and in the odd case in [Ja83].

Definition 6.1 Let $(H_\mathbb{Z}, L)$ be a unimodular bilinear lattice of rank $n \in \mathbb{N}$. In the following $k \in \{0; 1\}$. Denote

$$
\begin{aligned}
O^{(k)} &:= \text{Aut}(H_\mathbb{Z}, I^{(k)}) \quad \text{the group of automorphisms of } H_\mathbb{Z} \\
&\qquad \text{which respect } I^{(k)}. \\
H_\mathbb{Z}^\sharp &:= \text{Hom}(H_\mathbb{Z}, \mathbb{Z}) \quad \text{the dual lattice.} \\
j^{(k)} &: H_\mathbb{Z} \to H_\mathbb{Z}^\sharp, \quad a \mapsto (b \mapsto I^{(k)}(a, b)). \\
t^{(k)} &: O^{(k)} \to \text{Aut}(H_\mathbb{Z}^\sharp), \quad g \mapsto (l \mapsto l \circ g^{-1}).
\end{aligned}
$$

$$
\begin{aligned}
\overline{H_\mathbb{Z}}^{(k)} &:= H_\mathbb{Z}/\text{Rad}\, I^{(k)}, \quad \overline{H_\mathbb{R}}^{(k)} := H_\mathbb{R}/\text{Rad}_\mathbb{R}\, I^{(k)}. \\
\text{pr}^{H,(k)} &= \overline{(.)}^{(k)} : H_\mathbb{Z} \to \overline{H_\mathbb{Z}}^{(k)}, \quad a \mapsto \overline{a}^{(k)}, \quad \text{the projection.} \\
\overline{I}^{(k)} &: \overline{H_\mathbb{Z}}^{(k)} \times \overline{H_\mathbb{Z}}^{(k)} \to \mathbb{Z} \quad \text{the bilinear form on } \overline{H_\mathbb{Z}}^{(k)} \\
&\qquad \text{which is induced by } I^{(k)}, \\
\overline{H_\mathbb{Z}}^{(k),\sharp} &:= \text{Hom}(\overline{H_\mathbb{Z}}^{(k)}, \mathbb{Z}) \quad \text{the dual lattice.} \\
O^{(k),Rad} &:= \{g \in O^{(k)} \mid g|_{\text{Rad}\, I^{(k)}} = \text{id}\}.
\end{aligned}
$$

$$\mathrm{pr}^{A,(k)} = \overline{(.)} : O^{(k),Rad} \to \mathrm{Aut}(\overline{H_{\mathbb{Z}}}^{(k)}, \overline{I}^{(k)}), \quad g \mapsto \overline{g},$$
the natural map to the set of induced automorphisms.

For any subgroup $G^{(k)} \subset O^{(k),Rad}$ define

$$G_s^{(k)} := \mathrm{pr}^{A,(k)}(G^{(k)}) \subset \mathrm{Aut}(\overline{H_{\mathbb{Z}}}^{(k)}, \overline{I}^{(k)}),$$
$$G_u^{(k)} := \ker(\mathrm{pr}^{A,(k)} : G^{(k)} \to \mathrm{Aut}(\overline{H_{\mathbb{Z}}}^{(k)}, \overline{I}^{(k)})).$$

$G_s^{(k)}$ is called the *simple part* of $G^{(k)}$, and $G_u^{(k)}$ is called the *unipotent part* of $G^{(k)}$.

Lemma 6.2 *Let $(H_{\mathbb{Z}}, L, \underline{e})$ be a unimodular bilinear lattice of rank $n \in \mathbb{N}$ with a triangular basis $\underline{e}$.*

(a) The map $t^{(k)} : O^{(k)} \to \mathrm{Aut}(H_{\mathbb{Z}}^{\sharp})$ is a group homomorphism. For $g \in O^{(k)}$, $t^{(k)}(g)$ maps $j^{(k)}(H_{\mathbb{Z}}) \subset H_{\mathbb{Z}}^{\sharp}$ to itself, so it induces an automorphism $\tau^{(k)}(g) \in \mathrm{Aut}(H_{\mathbb{Z}}^{\sharp}/j^{(k)}(H_{\mathbb{Z}}))$. The map

$$\tau^{(k)} : O^{(k)} \to \mathrm{Aut}(H_{\mathbb{Z}}^{\sharp}/j^{(k)}(H_{\mathbb{Z}}))$$

is a group homomorphism.

(b) For $a \in R^{(0)}$ if $k = 0$ and for $a \in H_{\mathbb{Z}}$ if $k = 1$ the reflection or transvection $s_a^{(k)} \in O^{(k)}$ is in $\ker \tau^{(k)}$. Therefore $\Gamma^{(k)} \subset \ker \tau^{(k)}$.

(c) $\ker \tau^{(k)} \subset O^{(k),Rad}$.

(d) The horizontal lines of the following diagram are exact sequences.

$$\begin{array}{ccccccccc}
\{\mathrm{id}\} & \to & \Gamma_u^{(k)} & \to & \Gamma^{(k)} & \to & \Gamma_s^{(k)} & \to & \{\mathrm{id}\} \\
\| & & \cap & & \cap & & \cap & & \| \\
\{\mathrm{id}\} & \to & (\ker \tau^{(k)})_u & \to & \ker \tau^{(k)} & \to & (\ker \tau^{(k)})_s & \to & \{\mathrm{id}\} \\
\| & & \cap & & \cap & & \cap & & \| \\
\{\mathrm{id}\} & \to & O_u^{(k),Rad} & \to & O^{(k),Rad} & \to & O_s^{(k),Rad} & \to & \{\mathrm{id}\}
\end{array}$$

The second and third exact sequence split non-canonically.

$$O_s^{(k),Rad} = \mathrm{Aut}(\overline{H_{\mathbb{Z}}}^{(k)}, \overline{I}^{(k)}).$$

(e) The map

$$T : \overline{H_{\mathbb{Z}}}^{(k),\sharp} \otimes \mathrm{Rad}\, I^{(k)} \to O_u^{(k),Rad},$$
$$\sum_{i \in I} l_i \otimes r_i \mapsto \Big(a \mapsto a + \sum_{i \in I} l_i(\overline{a}^{(k)}) r_i\Big),$$
$$\text{shorter: } h \mapsto \big(a \mapsto a + h(\overline{a}^{(k)})\big),$$

is an isomorphism between abelian groups with

$$T(h_1+h_2)=T(h_1)\circ T(h_2),\quad T(h)^{-1}=T(-h).$$

It restricts to an isomorphism

$$T:\overline{j}^{(k)}(\overline{H_\mathbb{Z}}^{(k)})\otimes \mathrm{Rad}\, I^{(k)}\to (\ker\tau^{(k)})_u,$$

where $\overline{j}^{(k)}:\overline{H_\mathbb{Z}}^{(k)}\to\overline{H_\mathbb{Z}}^{(k),\sharp}$ *is the map*

$$a\mapsto \big(b\mapsto \overline{I}^{(k)}(a,b)\big)\quad\text{for an arbitrary } b\in\overline{H_\mathbb{Z}}^{(k)}.$$

(f) For $g\in O^{(k),Rad}$, $a\in\overline{H_\mathbb{Z}}^{(k)}$, $r\in\mathrm{Rad}\, I^{(k)}$

$$g\circ T(\overline{j}^{(k)}(a)\otimes r)\circ g^{-1}=T(\overline{j}^{(k)}(\overline{g}(a))\otimes r).$$

(g) Analogously to $t^{(k)}$ *and* $\tau^{(k)}$ *there are the group homomorphisms*

$$\overline{t}^{(k)}:O_s^{(k),Rad}=\mathrm{Aut}(\overline{H_\mathbb{Z}}^{(k)},\overline{I}^{(k)})\to\mathrm{Aut}(\overline{H_\mathbb{Z}}^{(k),\sharp}),\ g\mapsto(l\mapsto l\circ g^{-1}),$$
$$\text{and}\quad \overline{\tau}^{(k)}:O_s^{(k),Rad}\to\mathrm{Aut}(\overline{H_\mathbb{Z}}^{(k),\sharp}/\overline{j}^{(k)}(\overline{H_\mathbb{Z}}^{(k)})).$$

$\tau^{(k)}$ *and* $\overline{\tau}^{(k)}$ *satisfy*

$$(\ker\tau^{(k)})_s=\ker\overline{\tau}^{(k)}.$$

Proof (a) The map $t^{(k)}$ is a group homomorphism because

$$t^{(k)}(g_1g_2)(l)=l\circ(g_1g_2)^{-1}=l\circ g_2^{-1}\circ g_1^{-1}=t^{(k)}(g_1)t^{(k)}(g_2)(l).$$

$t^{(k)}(g)$ maps $j^{(k)}(H_\mathbb{Z})$ to itself because

$$\begin{aligned}t^{(k)}(g)(j^{(k)}(a))&=j^{(k)}(a)\circ g^{-1}=I^{(k)}(a,g^{-1}(.))\\&=I^{(k)}(g(a),(.))=j^{(k)}(g(a)).\end{aligned}$$

$\tau^{(k)}$ is a group homomorphism because $t^{(k)}$ is one.

(b) Choose $l\in H_\mathbb{Z}^\sharp$ and $b\in H_\mathbb{Z}$. Then

$$\begin{aligned}(t^{(k)}(s_a^{(k)})(l)-l)(b)&=l\circ(s_a^{(k)})^{-1}(b)-l(b)\\&=l(b-(-1)^kI^{(k)}(a,b)a)-l(b)\\&=(-1)^{k+1}I^{(k)}(a,b)l(a)\\&=j^{(k)}((-1)^{k+1}l(a)a)(b),\\ \text{so } t^{(k)}(s_a^{(k)})(l)-l&=j^{(k)}((-1)^{k+1}l(a)a)\in j^{(k)}(H_\mathbb{Z}),\\ \text{so } \tau^{(k)}(s_a^{(k)})&=\mathrm{id},\end{aligned}$$

so $s_a^{(k)} \in \ker \tau^{(k)}$.

(c) Let $g \in \ker \tau^{(k)}$ and let $r \in \text{Rad}\, I^{(k)}$. Also $g^{-1} \in \ker \tau^{(k)}$. Choose $l \in H_{\mathbb{Z}}^{\sharp}$. Now $\tau^{(k)}(g^{-1}) = \text{id}$ implies

$$t^{(k)}(g^{-1})(l) - l = j^{(k)}(a) \quad \text{for some } a \in H_{\mathbb{Z}},$$
$$0 = I^{(k)}(a, r) = j^{(k)}(a)(r) = \big(t^{(k)}(g^{-1})(l) - l\big)(r) = l((g - \text{id})(r)).$$

Because l is arbitrary, $(g - \text{id})(r) = 0$, so $g(r) = r$, so $g \in O^{(k),Rad}$.

(d) The exact sequences are obvious. Choose an arbitrary splitting of $H_{\mathbb{Z}}$ as $\mathbb{Z}$-module into $\text{Rad}\, I^{(k)}$ and a suitably chosen $\mathbb{Z}$-module $\widetilde{H_{\mathbb{Z}}}^{(k)}$,

$$H_{\mathbb{Z}} = \text{Rad}\, I^{(k)} \oplus \widetilde{H_{\mathbb{Z}}}^{(k)}.$$

The projection $\text{pr}^{H,(k)} : H_{\mathbb{Z}} \to \overline{H_{\mathbb{Z}}}^{(k)}$ restricts to an isomorphism

$$\text{pr}^{H,(k)} : (\widetilde{H_{\mathbb{Z}}}^{(k)}, I^{(k)}|_{\widetilde{H_{\mathbb{Z}}}^{(k)}}) \to (\overline{H_{\mathbb{Z}}}^{(k)}, \overline{I}^{(k)}).$$

Via this isomorphism, any element of $\text{Aut}(\overline{H_{\mathbb{Z}}}^{(k)}, \overline{I}^{(k)})$ lifts to an element of $O^{(k),Rad}$. This shows $O_s^{(k),Rad} = \text{Aut}(\overline{H_{\mathbb{Z}}}^{(k)}, \overline{I}^{(k)})$, and it gives a non-canonical splitting of the third exact sequence. The end of the proof of part (g) will show that this splitting restricts to a non-canonical splitting of the second exact sequence.

(e) The fact $\overline{r}^{(k)} = 0$ for $r \in \text{Rad}\, I^{(k)}$ easily implies that T is a group homomorphism with $T(h_1 + h_2) = T(h_1)T(h_2)$ and with image in $O_u^{(k),Rad}$.

Consider $g \in O_u^{(k),Rad}$. Then $g|_{\text{Rad}\, I^{(k)}} = \text{id}$ and $(g - \text{id})(a) \in \text{Rad}\, I^{(k)}$ for any $a \in H_{\mathbb{Z}}$. If $b \in a + \text{Rad}\, I^{(k)}$ then $(g - \text{id})(a - b) = 0$, so $(g - \text{id})(b) = (g - \text{id})(a)$. Thus there is an element $h \in \overline{H_{\mathbb{Z}}}^{(k),\sharp} \otimes \text{Rad}\, I^{(k)}$ with $h(\overline{a}^{(k)}) = (g - \text{id})(a)$ for any $a \in H_{\mathbb{Z}}$, so $T(h)(a) = a + h(\overline{a}^{(k)}) = g(a)$, so $T(h) = g$.

Choose a $\mathbb{Z}$-basis $r_1, ..., r_m$ of $\text{Rad}\, I^{(k)}$ and linear forms $l_1, ..., l_m \in H_{\mathbb{Z}}^{\sharp}$ with $l_i(r_j) = \delta_{ij}$. Then any $r \in \text{Rad}\, I^{(k)}$ satisfies $r = \sum_{i=1}^m l_i(r) r_i$.

Consider $h \in \overline{H_{\mathbb{Z}}}^{(k),\sharp} \otimes \text{Rad}\, I^{(k)}$ with $T(h) \in (\ker \tau^{(k)})_u$. Then

$$t^{(k)}(T(h))(l_i) - l_i = j^{(k)}(a_i) \quad \text{for some } a_i \in H_{\mathbb{Z}}, \text{ and also}$$
$$t^{(k)}(T(h))(l_i) - l_i = l_i \circ T(h)^{-1} - l_i = l_i \circ T(-h) - l_i$$
$$= l_i \circ (\text{id} - h(\overline{(.)}^{(k)})) - l_i = -l_i(h(\overline{(.)}^{(k)})).$$

For $b \in H_{\mathbb{Z}}$ $h(\overline{b}^{(k)}) \in \text{Rad}\, I^{(k)}$, so

$$h(\overline{b}^{(k)}) = \sum_{i=1}^m l_i(h(\overline{b}^{(k)})) r_i = -\sum_{i=1}^m j^{(k)}(a_i)(b) r_i,$$
$$\text{so } h \in \overline{j}^{(k)}(\overline{H_{\mathbb{Z}}}^{(k)}) \otimes \text{Rad}\, I^{(k)}.$$

Going backwards through these arguments, one sees that any $h \in \overline{j}^{(k)}(\overline{H_{\mathbb{Z}}}^{(k)}) \otimes \operatorname{Rad} I^{(k)}$ satisfies $T(h) \in (\ker \tau^{(k)})_u$.

(f) For $b \in H_{\mathbb{Z}}$

$$\begin{aligned}(g \circ T(\overline{j}^{(k)}(a) \otimes r) \circ g^{-1})(b) &= g(g^{-1}(b) + \overline{I}^{(k)}(a, \overline{g^{-1}(b)}^{(k)})r) \\ &= b + \overline{I}^{(k)}(\overline{g}(a), \overline{b}^{(k)})r \\ &= T(\overline{j}^{(k)}(\overline{g}(a)) \otimes r)(b).\end{aligned}$$

(g) The projection $\mathrm{pr}^{H,(k)} = \overline{()}^{(k)} : H_{\mathbb{Z}} \to \overline{H_{\mathbb{Z}}}^{(k)}$ induces the embedding

$$\begin{aligned} i^{(k)} : \overline{H_{\mathbb{Z}}}^{(k),\sharp} &\hookrightarrow H_{\mathbb{Z}}^{\sharp},\ l \mapsto l \circ \mathrm{pr}^{H,(k)}, \\ \text{with}\quad \operatorname{Im}(i^{(k)}) &= \{l \in H_{\mathbb{Z}}^{\sharp} \mid l|_{\operatorname{Rad} I^{(k)}} = 0\} \\ \text{and}\quad j^{(k)}(H_{\mathbb{Z}}) &= i^{(k)}(\overline{j}^{(k)}(\overline{H_{\mathbb{Z}}}^{(k)})) \subset \operatorname{Im}(i^{(k)}).\end{aligned}$$

The three lattices

$$H_{\mathbb{Z}}^{\sharp} \supset \operatorname{Im}(i^{(k)}) \supset j^{(k)}(H_{\mathbb{Z}})$$

have ranks $n, n - \operatorname{rk} \operatorname{Rad} I^{(k)}, n - \operatorname{rk} \operatorname{Rad} I^{(k)}$ and are for each $g \in O^{(k),\mathrm{Rad}}$ invariant under the map $t^{(k)}(g) = (l \mapsto l \circ g^{-1})$.

This map acts trivially on the quotient $H_{\mathbb{Z}}^{\sharp}/\operatorname{Im}(i^{(k)})$. It acts trivially on the quotient $\operatorname{Im}(i^{(k)})/j^{(k)}(H_{\mathbb{Z}})$ if and only if $\overline{g}^{(k)} \in \ker \overline{\tau}^{(k)}$. It acts trivially on the quotient $H_{\mathbb{Z}}^{\sharp}/j^{(k)}(H_{\mathbb{Z}})$ if and only if $g \in \ker \tau^{(k)}$. Therefore $(\ker \tau^{(k)})_s \subset \overline{\tau}^{(k)}$. It remains to find for each $\widetilde{g} \in \ker \overline{\tau}^{(k)}$ an element $g \in \ker \tau^{(k)}$ with $\overline{g}^{(k)} = \widetilde{g}$.

Choose a $\mathbb{Z}$-basis $r_1, ..., r_n$ of $H_{\mathbb{Z}}$ such that $r_1, ..., r_m$ (with $m = \operatorname{rk} \operatorname{Rad} I^{(k)}$) is a $\mathbb{Z}$-basis of $\operatorname{Rad} I^{(k)}$. Then $H_{\mathbb{Z}} = \operatorname{Rad} I^{(k)} \oplus \widetilde{H_{\mathbb{Z}}}^{(k)}$ with $\widetilde{H_{\mathbb{Z}}}^{(k)} = \bigoplus_{j=m+1}^n \mathbb{Z} \cdot r_j$ is a splitting of $H_{\mathbb{Z}}$ with $\widetilde{H_{\mathbb{Z}}}^{(k)} \cong \overline{H_{\mathbb{Z}}}^{(k)}$.

Consider the dual $\mathbb{Z}$-basis $l_1, ..., l_n$ of $H_{\mathbb{Z}}^{\sharp}$ with $l_i(r_j) = \delta_{ij}$. Then $\operatorname{Im}(i^{(k)}) = \bigoplus_{j=m+1}^n \mathbb{Z} \cdot l_j \supset j^{(k)}(H_{\mathbb{Z}})$.

An element $\widetilde{g} \in O_s^{(k),Rad}$ has a unique lift to an element $g \in O^{(k),Rad}$ with $g(\widetilde{H_{\mathbb{Z}}}^{(k)}) = \widetilde{H_{\mathbb{Z}}}^{(k)}$. This splitting of the third exact sequence in part (d) was used already in the proof of part (d). We claim that $g \in \ker \tau^{(k)}$ if $\widetilde{g} \in \ker \overline{\tau}^{(k)}$. We have

$$\begin{aligned} l_j - l_j \circ g^{-1} &\in j^{(k)}(H_{\mathbb{Z}}) \quad \text{for } j \in \{m+1, ..., n\} \\ &\qquad \text{because of } \widetilde{g} \in \ker \overline{\tau}^{(k)}, \\ l_i - l_i \circ g^{-1} &= 0 \quad \text{for } i \in \{1, ..., m\}.\end{aligned}$$

This shows the claim. Therefore $(\ker \tau^{(k)})_s = \ker \overline{\tau}^{(k)}$. The claim also shows that the non-canonical splitting of the third exact sequence in part (d) restricts to a non-canonical splitting of the second exact sequence in part (d). □

Remarks 6.3 (i) The exact sequence $\{\mathrm{id}\} \to \Gamma_u^{(k)} \to \Gamma^{(k)} \to \Gamma_s^{(k)} \to \{\mathrm{id}\}$ splits sometimes, sometimes not. When it splits and when $\Gamma_u^{(k)}$, $\Gamma_s^{(k)}$ and the splitting are known, then also $\Gamma^{(k)}$ is known.

(ii) Suppose that one has a presentation of $\Gamma_s^{(k)}$, namely an isomorphism

$$\Gamma_s^{(k)} \stackrel{\cong}{\longrightarrow} \langle g_1, ..., g_n \mid w_1(g_1, ..., g_m), ..., w_m(g_1, ..., g_n)\rangle,$$
$$\overline{s_{e_i}^{(k)}} \mapsto g_i,$$

where $w_1(g_1, ..., g_n), ..., w_m(g_1, ..., g_n)$ are certain words in $g_1^{\pm 1}, ..., g_n^{\pm 1}$. Then the group $\Gamma_u^{(k)}$ is the normal subgroup of $\Gamma^{(k)}$ generated by the elements $w_1(s_{e_1}^{(k)}, ..., s_{e_n}^{(k)}), ..., w_m(s_{e_1}^{(k)}, ..., s_{e_n}^{(k)})$. In many of the cases with $n = 2$ or $n = 3$ we have such a presentation.

(iii) The symmetric bilinear form $\overline{I}^{(0)}$ on $\overline{H_\mathbb{Z}}^{(0)}$ is nondegenerate. It is well known that for any $g \in \mathrm{Aut}(\overline{H_\mathbb{R}}^{(0)}, \overline{I}^{(0)})$ some $m \in \mathbb{N}$ and elements $a_1, ..., a_m \in \overline{H_\mathbb{R}}^{(0)}$ with $\overline{I}^{(0)}(a_i, a_i) \in \mathbb{R}^*$ and $g = \overline{s}_{a_1}^{(0)} ... \overline{s}_{a_m}^{(0)}$ exist and that the sign

$$\overline{\sigma}(g) := \prod_{i=1}^{m} \mathrm{sign}(\overline{I}^{(0)}(a_i, a_i)) \in \{\pm 1\}$$

is independent of m and $a_1, ..., a_m$. This sign $\overline{\sigma}(g) \in \{\pm 1\}$ is the *spinor norm* of g. The map $\overline{\sigma} : \mathrm{Aut}(\overline{H_\mathbb{R}}^{(0)}, \overline{I}^{(0)}) \to \{\pm 1\}$ is obviously a group homomorphism.

Definition 6.4 Keep the situation of Definition 6.1. Define the *spinor norm homomorphism*

$$\sigma : O^{(0),Rad} \to \{\pm 1\}, \quad \sigma(g) := \overline{\sigma}(\overline{g}).$$

Define the subgroup $O^{(k),*}$ of $O^{(k),Rad}$

$$O^{(k),*} := \begin{cases} \ker \tau^{(1)} & \text{if } k = 1, \\ \ker \tau^{(0)} \cap \ker \sigma & \text{if } k = 0. \end{cases}$$

Remarks 6.5 (i) For $a \in R^{(0)}$ of course $\sigma(s_a^{(0)}) = 1$. Therefore $\Gamma^{(0)} \subset \ker \sigma$. Thus

$$\Gamma^{(k)} \subset O^{(k),*} \quad \text{for } k \in \{0; 1\}.$$

(ii) For $g \in (\ker \tau^{(0)})_u$ $\sigma(g) = 1$ because $\overline{g} = \mathrm{id}$. Therefore

$$O_u^{(k),*} = (\ker \tau^{(k)})_u \quad \text{for } k \in \{0; 1\}.$$

Remarks 6.6 Finally we make some comments on the sets of vanishing cycles in a unimodular bilinear lattice $(H_\mathbb{Z}, L, \underline{e})$ with a triangular basis.

(i) Let $H_\mathbb{Z}^{prim}$ denote the set of primitive vectors in $H_\mathbb{Z}$, i.e. vectors $a \in H_\mathbb{Z} - \{0\}$ with $\mathbb{Z}a = \mathbb{Q}a \cap H_\mathbb{Z}$, and analogously $\overline{H_\mathbb{Z}}^{(k),prim}$. Then

$$\Delta^{(0)} \subset R^{(0)} \subset H_{\mathbb{Z}}^{prim}, \quad \Delta^{(1)} \subset H_{\mathbb{Z}}^{prim},$$
$$\overline{\Delta}^{(0)} \subset \overline{R}^{(0)} \subset \overline{H_{\mathbb{Z}}}^{(0),prim}, \quad \text{where } \overline{\Delta}^{(k)} := \mathrm{pr}^{H,(k)}(\Delta^{(k)}).$$

Here $R^{(0)} \subset H_{\mathbb{Z}}^{prim}$ and $\overline{R}^{(0)} \subset \overline{H_{\mathbb{Z}}}^{(0),prim}$ because of $2 = I^{(0)}(a,a) = \overline{I}^{(0)}(\overline{a}^{(0)}, \overline{a}^{(0)})$ for $a \in R^{(0)}$. Furthermore

$$\overline{e_i}^{(1)} \in \overline{H_{\mathbb{Z}}}^{(1),prim} \iff \Gamma_s^{(1)}\{\overline{e_i}^{(1)}\} \subset \overline{H_{\mathbb{Z}}}^{(1),prim}.$$

Whether or not $\overline{e_i}^{(1)} \in \overline{H_{\mathbb{Z}}}^{(1),prim}$, that depends on the situation. $\overline{\Delta}^{(1)} \subset \overline{H_{\mathbb{Z}}}^{(1),prim}$ may hold or not.

(ii) In general, an element $a \in H_{\mathbb{Z}}$ satisfies

$$\overline{a}^{(k)} \in \overline{H_{\mathbb{Z}}}^{(k),prim} \iff a + \mathrm{Rad}\, I^{(k)} \subset H_{\mathbb{Z}}^{prim}.$$

(iii) The set $\overline{\Delta}^{(k)} \subset \overline{H_{\mathbb{Z}}}^{(k)}$ is often simpler to describe than the set $\Delta^{(k)}$. The control of $\overline{\Delta}^{(k)}$ is a step towards the control of $\Delta^{(k)}$.

(iv) Given $a \in \Delta^{(k)}$, it is interesting to understand the three sets

$$(a + \mathrm{Rad}\, I^{(k)}) \cap \Delta^{(k)} \overset{(1)}{\supset} (a + \mathrm{Rad}\, I^{(k)}) \cap \Gamma^{(k)}\{a\} \overset{(2)}{\supset} \Gamma_u^{(k)}\{a\}.$$

The next lemma makes comments on the inclusions $\overset{(1)}{\supset}$ and $\overset{(2)}{\supset}$.

Lemma 6.7 *Keep the situation in Remark 6.6 (iv).*

(a) In $\overset{(1)}{\supset}$ equality holds if and only if the image in $\overline{H_{\mathbb{Z}}}^{(k)}$ of any $\Gamma^{(k)}$ orbit different from $\Gamma^{(k)}\{a\}$ is different from $\Gamma_s^{(k)}\{\overline{a}^{(k)}\}$.

(b) The following is an inclusion of groups,

$$\mathrm{Stab}_{\Gamma^{(k)}}(\overline{a}^{(k)}) \overset{(3)}{\supset} \Gamma_u^{(k)} \cdot \mathrm{Stab}_{\Gamma^{(k)}}(a).$$

In $\overset{(2)}{\supset}$ equality holds if and only if in $\overset{(3)}{\supset}$ equality holds.

Proof Trivial. □

6.2 The Rank 2 Cases

For $x \in \mathbb{Z} - \{0\}$ consider the matrix $S = S(x) = \begin{pmatrix} 1 & x \\ 0 & 1 \end{pmatrix} \in T_2(\mathbb{Z})$, and consider a unimodular bilinear lattice $(H_{\mathbb{Z}}, L)$ with a triangular basis $\underline{e} = (e_1, e_2)$ with $L(\underline{e}^t, \underline{e})^t = S$. Recall the formulas and the results in Sect. 5.2, especially $M^{root} : H_{\mathbb{Z}} \to H_{\mathbb{Z}}$ and its eigenvalues $\kappa_{1/2} = \frac{-x}{2} \pm \frac{1}{2}\sqrt{x^2 - 4}$.

We can restrict to $x<0$ because of $L((e_1,-e_2)^t,(e_1,-e_2))^t=\begin{pmatrix}1&-x\\0&1\end{pmatrix}$. So suppose $x<0$.

First we consider the even cases. Then $\Gamma^{(0)}$ is a Coxeter group, and $\Gamma^{(0)}$ and $\Delta^{(0)}$ are well known. Still we want to document and derive the facts in our way.

Theorem 6.8 *(a) We have*

$$\begin{aligned}
s^{(0)}_{e_i}\underline{e}&=\underline{e}\cdot s^{(0),mat}_{e_i}\quad\text{with}\\
s^{(0),mat}_{e_1}&=\begin{pmatrix}-1&-x\\0&1\end{pmatrix},\quad s^{(0),mat}_{e_2}=\begin{pmatrix}1&0\\-x&-1\end{pmatrix},\\
\Gamma^{(0)}&\cong\Gamma^{(0),mat}:=\langle s^{(0),mat}_{e_1},s^{(0),mat}_{e_2}\rangle\subset GL_2(\mathbb{Z}),\\
R^{(0)}&=\{y_1e_1+y_2e_2\in H_{\mathbb{Z}}\,|\,1=y_1^2+xy_1y_2+y_2^2\}.
\end{aligned}$$

(b) The case $x=-1$: $(H_{\mathbb{Z}},I^{(0)})$ is the A_2 root lattice. $\Gamma^{(0)}\cong D_6$ is a dihedral group with six elements, the identity, three reflections and two rotations,

$$\begin{aligned}
\Gamma^{(0)}&=\langle \mathrm{id},\ s^{(0)}_{e_1},\ s^{(0)}_{e_2},\ s^{(0)}_{e_1}s^{(0)}_{e_2}s^{(0)}_{e_1},\ s^{(0)}_{e_1}s^{(0)}_{e_2},\ s^{(0)}_{e_2}s^{(0)}_{e_1}\rangle\cong\Gamma^{(0),mat}\\
&=\langle E_2,\begin{pmatrix}-1&1\\0&1\end{pmatrix},\begin{pmatrix}1&0\\1&-1\end{pmatrix},\begin{pmatrix}0&-1\\-1&0\end{pmatrix},\begin{pmatrix}0&-1\\1&-1\end{pmatrix},\begin{pmatrix}-1&1\\-1&0\end{pmatrix}\rangle.
\end{aligned}$$

The set $\Delta^{(0)}$ of vanishing cycles coincides with the set $R^{(0)}$ of L1-cycles and is

$$\Delta^{(0)}=R^{(0)}=\{\pm e_1,\pm e_2,\pm(e_1+e_2)\}.$$

Figure 6.1 shows the action of $\Gamma^{(0)}$ on $\Delta^{(0)}$. One sees the action of D_6 on the vertices of a regular 6-gon.

One sees also the action of D_{12} on the regular 6-gon. This fits to the following.

$$D_6\cong\Gamma^{(0)}=\ker\tau^{(0)}=O^{(0),*}\overset{1:2}{\subset}O^{(0)}\cong D_{12}.$$

(c) The case $x=-2$: $\Gamma^{(0)}\cong G^{fCox,2}$ is a free Coxeter group with the two generators $s^{(0)}_{e_1}$ and $s^{(0)}_{e_2}$. Here

$$\operatorname{Rad}I^{(0)}=\mathbb{Z}f_1\quad\text{with}\quad f_1=e_1+e_2.$$

The set $\Delta^{(0)}$ of vanishing cycles coincides with the set $R^{(0)}$ of L1-cycles and is

$$\begin{aligned}
\Delta^{(0)}=R^{(0)}&=\{y_1e_1+y_2e_2\in H_{\mathbb{Z}}\,|\,1=(y_1-y_2)^2\}\\
&=(e_1+\mathbb{Z}f_1)\,\dot\cup\,(e_2+\mathbb{Z}f_1).
\end{aligned}$$

It splits into the two disjoint orbits

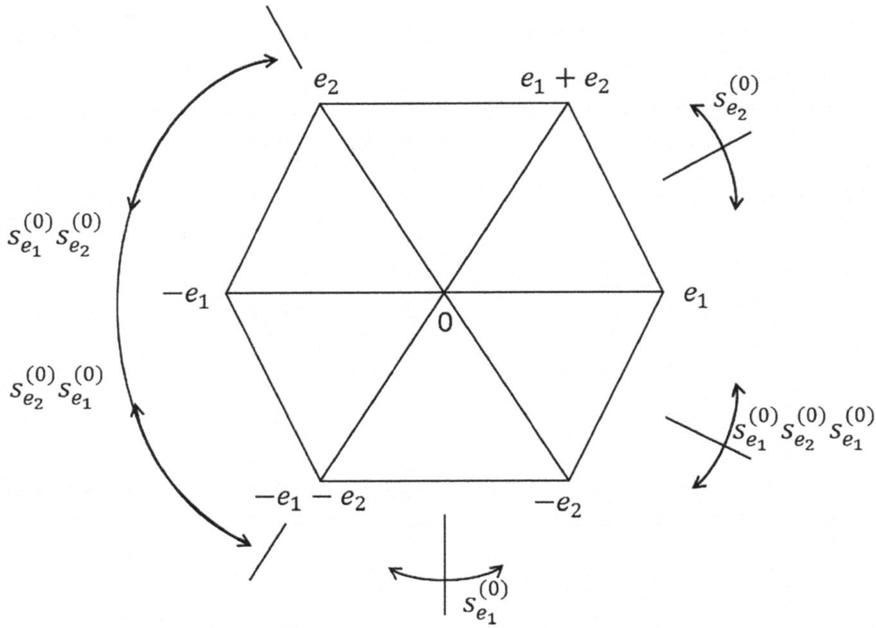

Fig. 6.1 A regular 6-gon, actions of D_6 and D_{12}

$$\Gamma^{(0)}\{e_1\} = \Gamma^{(0)}\{-e_1\} = (e_1 + \mathbb{Z}2f_1) \,\dot{\cup}\, (-e_1 + \mathbb{Z}2f_1),$$
$$\Gamma^{(0)}\{e_2\} = \Gamma^{(0)}\{-e_2\} = (e_2 + \mathbb{Z}2f_1) \,\dot{\cup}\, (-e_2 + \mathbb{Z}2f_1).$$

$s^{(0)}_{e_1}$ acts on $\Delta^{(0)}$ by permuting vanishing cycles horizontally, so by adding $\pm 2e_1$. $s^{(0)}_{e_2}$ acts on $\Delta^{(0)}$ by permuting vanishing cycles vertically, so by adding $\pm 2e_2$, see the following formulas and Fig. 6.2. For $\varepsilon \in \{\pm 1\}$ and $m \in \mathbb{Z}$

$$s^{(0)}_{e_1}(\varepsilon e_1 + mf_1) = -\varepsilon e_1 + mf_1, \quad s^{(0)}_{e_1}(\varepsilon e_2 + mf_1) = -\varepsilon e_2 + (m + 2\varepsilon) f_1,$$
$$s^{(0)}_{e_2}(\varepsilon e_1 + mf_1) = -\varepsilon e_1 + (m + 2\varepsilon) f_1, \quad s^{(0)}_{e_2}(\varepsilon e_2 + mf_1) = -\varepsilon e_2 + mf_1.$$

The matrix group $\Gamma^{(0),mat}$ is given by the following formulas for $m \in \mathbb{Z}$,

$$(s^{(0)}_{e_1} s^{(0)}_{e_2})^m(\underline{e}) = \underline{e} \begin{pmatrix} 2m+1 & -2m \\ 2m & -2m+1 \end{pmatrix},$$
$$(s^{(0)}_{e_1} s^{(0)}_{e_2})^m s^{(0)}_{e_1}(\underline{e}) = \underline{e} \begin{pmatrix} -2m-1 & 2m+2 \\ -2m & 2m+1 \end{pmatrix}.$$

(d) The cases $x \leq -3$: $\Gamma^{(0)} \cong G^{fCox,2}$ is a free Coxeter group with the two generators $s^{(0)}_{e_1}$ and $s^{(0)}_{e_2}$. The set $\Delta^{(0)}$ of vanishing cycles coincides with the set $R^{(0)}$ of L1-cycles. More information on $\Delta^{(0)}$:

Fig. 6.2 Even vanishing cycles in the case $S = \begin{pmatrix} 1 & -2 \\ 0 & 1 \end{pmatrix}$

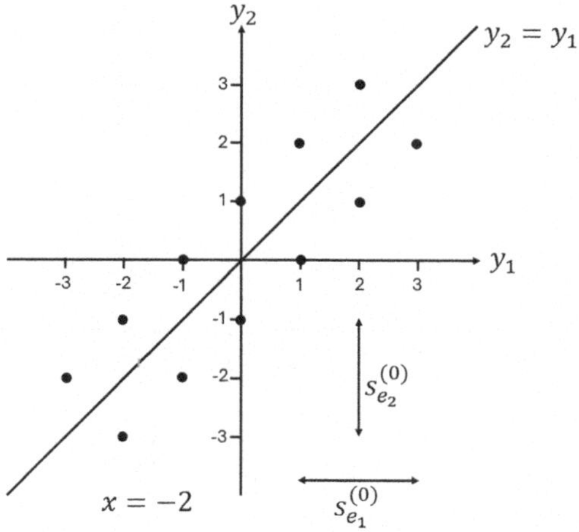

(i) Recall Lemma C.1 (a). The map

$$u : \Delta^{(0)} \to \{\text{units in } \mathbb{Z}[\kappa_1] \text{ with norm } 1\} = \{\pm\kappa_1^l \mid l \in \mathbb{Z}\}$$
$$y_1e_1 + y_2e_2 \mapsto y_1 - \kappa_1 y_2,$$

is well defined and a bijection with

$$s^{(0)}_{e_1}(u^{-1}(\varepsilon\kappa_1^l)) = u^{-1}(-\varepsilon\kappa_1^{-l}), \quad s^{(0)}_{e_2}(u^{-1}(\varepsilon\kappa_1^l)) = u^{-1}(-\varepsilon\kappa_1^{2-l}),$$

for $\varepsilon \in \{\pm 1\}$, $l \in \mathbb{Z}$. *Especially,* $\Delta^{(0)}$ *splits into the two disjoint orbits* $\Gamma^{(0)}\{e_1\}$ *and* $\Gamma^{(0)}\{e_2\}$.

(ii) The matrix group $\Gamma^{(0),mat}$ *is given by the following formulas for* $m \in \mathbb{Z}$,

$$(s^{(0)}_{e_1}s^{(0)}_{e_2})^m(\underline{e}) = \underline{e}\begin{pmatrix} y_1 & -y_2 \\ y_2 & y_1 + xy_2 \end{pmatrix} = (u^{-1}(\kappa_1^{-2m}), u^{-1}(-\kappa_1^{-2m+1})),$$
$$\text{where } u^{-1}(\kappa_1^{-2m}) = y_1 - \kappa_1 y_2,$$
$$(s^{(0)}_{e_1}s^{(0)}_{e_2})^m s^{(0)}_{e_1}(\underline{e}) = \underline{e}\begin{pmatrix} y_1 & xy_1 + y_2 \\ y_2 & -y_1 \end{pmatrix} = (u^{-1}(-\kappa_1^{-2m}), u^{-1}(\kappa_1^{-2m-1}))$$
$$\text{where } u^{-1}(-\kappa_1^{-2m}) = y_1 - \kappa_1 y_2.$$

(iii) $\Delta^{(0)} \subset H_{\mathbb{R}} \cong \mathbb{R}^2$ *is part of the hyperbola* $\{y_1e_1 + y_2e_2 \in H_{\mathbb{R}} \mid (y_1 - \kappa_1 y_2)(y_1 - \kappa_2 y_2) = 1\}$ *with asymptotic lines* $y_2 = \kappa_2 y_1$ *and* $y_2 = \kappa_1 y_1$. *Both branches of this hyperbola are strictly monotonously increasing. The lower right branch is concave and contains the points* $u^{-1}(\kappa_1^l)$ *for* $l \in \mathbb{Z}$, *the upper left branch is convex and contains the points* $u^{-1}(-\kappa^l)$ *for* $l \in \mathbb{Z}$. *The horizontal respectively vertical line through a vanishing cycle* $a \in \Delta^{(0)}$ *intersects the other branch of the hyperbola in* $s^{(0)}_{e_1}(a)$ *respectively* $s^{(0)}_{e_2}(a)$. *See Fig. 6.3.*

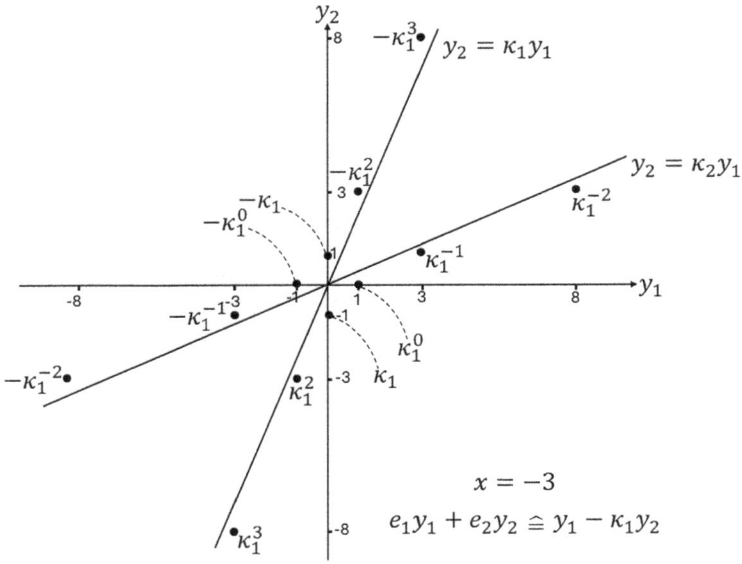

Fig. 6.3 Even vanishing cycles in the case $S = \begin{pmatrix} 1 & -3 \\ 0 & 1 \end{pmatrix}$

(iv) Denote by $\widetilde{s}_{e_i}^{(0)}$, $\widetilde{\Gamma}^{(0)}$ *and* $\widetilde{\Delta}^{(0)}$ *the objects for* $x = -2$ *and as usual by* $s_{e_i}^{(0)}$, $\Gamma^{(0)}$ *and* $\Delta^{(0)}$ *the objects for an* $x \leq -3$.

The map $\widetilde{s}_{e_i}^{(0)} \mapsto s_{e_i}^{(0)}$ *extends to a group isomorphism* $\widetilde{\Gamma}^{(0)} \to \Gamma^{(0)}$. *The map*

$$\begin{aligned} &\widetilde{\Delta}^{(0)} \to \Delta^{(0)}, \\ &\underline{e}\begin{pmatrix} 1-l \\ -l \end{pmatrix} \mapsto u^{-1}(\kappa_1^l), \quad \underline{e}\begin{pmatrix} l-1 \\ l \end{pmatrix} \mapsto u^{-1}(-\kappa_1^l) \quad \text{for } l \in \mathbb{Z}, \end{aligned}$$

is a bijection. The bijections $\widetilde{\Gamma}^{(0)} \to \Gamma^{(0)}$ *and* $\widetilde{\Delta}^{(0)} \to \Delta^{(0)}$ *are compatible with the action of* $\widetilde{\Gamma}^{(0)}$ *on* $\widetilde{\Delta}^{(0)}$ *and of* $\Gamma^{(0)}$ *on* $\Delta^{(0)}$.

(e) More on the cases $x \leq -2$*: The automorphism* $g_{1,2} : H_{\mathbb{Z}} \to H_{\mathbb{Z}}$ *with* $g_{1,2} : e_1 \leftrightarrow e_2$ *is in* $O^{(0)}$. *The set* $\{\pm \mathrm{id}, \pm g_{1,2}\}$ *is a subgroup of* $O^{(0)}$ *with*

$$\Gamma^{(0)} = \ker \tau^{(0)} = O^{(0),*} \overset{1:4}{\subset} O^{(0)} = \Gamma^{(0)} \rtimes \{\pm \mathrm{id}, \pm g_{1,2}\}.$$

Proof (a) Everything except possibly the shape of $R^{(0)}$ is obvious.

$$\begin{aligned} R^{(0)} &= \{y_1 e_1 + y_2 e_2 \in H_{\mathbb{Z}} \,|\, 2 = I^{(0)}(y_1 e_1 + y_2 e_2, y_1 e_1 + y_2 e_2)\} \\ &= \{y_1 e_1 + y_2 e_2 \in H_{\mathbb{Z}} \,|\, 2 = \begin{pmatrix} y_1 & y_2 \end{pmatrix} \begin{pmatrix} 2 & x \\ x & 2 \end{pmatrix} \begin{pmatrix} y_1 \\ y_2 \end{pmatrix}\} \\ &= \{y_1 e_1 + y_2 e_2 \in H_{\mathbb{Z}} \,|\, 1 = y_1^2 + x y_1 y_2 + y_2^2\}. \end{aligned}$$

(b) This is classical and elementary. $R^{(0)} = \{\pm e_1, \pm e_2, \pm(e_1 + e_2)\}$. The actions of $s_{e_1}^{(0)}$ and $s_{e_2}^{(0)}$ on this set extend to the action of the dihedral group D_6 on the vertices of a regular 6-gon. Therefore $\Delta^{(0)} = R^{(0)}$ and $\Gamma^{(0)} \cong D_6$.

$O^{(0)} \cong D_{12}$ is obvious as the vanishing cycles form the vertices of a regular 6-gon in $(H_{\mathbb{Z}}, I^{(0)})$. It remains to show for some element $g \in O^{(0)} - \Gamma^{(0)}$ $g \notin \ker \tau^{(0)}$.

Consider the reflection $g \in O^{(0)}$ with $g(\underline{e}) = (e_1, -e_1 - e_2)$ and the linear form $l : H_{\mathbb{Z}} \to \mathbb{Z}$ with $l(\underline{e}) = (1, 0)$. Then

$$t^{(0)}(g)(l) = l \circ g^{-1}, \quad (l \circ g^{-1})(\underline{e}) = (l(e_1), l(-e_1 - e_2)) = (1, -1),$$
$$(l - l \circ g^{-1})(\underline{e}) = (0, 1),$$
$$l - l \circ g^{-1} \notin j^{(0)}(H_{\mathbb{Z}}) = \langle (\underline{e} \mapsto (2, -1)), (\underline{e} \mapsto (-1, 2)) \rangle,$$

so $g \notin \ker \tau^{(0)}$.

(c) and (d) The group $\Gamma^{(0)}$ for $x \leq -2$: Recall the Remarks and Notations A.1. The matrices $s_{e_1}^{(0),mat} = \begin{pmatrix} -1 & -x \\ 0 & 1 \end{pmatrix}$ and $s_{e_2}^{(0),mat} = \begin{pmatrix} 1 & 0 \\ -x & -1 \end{pmatrix}$ have the eigenvectors $\begin{pmatrix} 1 \\ 0 \end{pmatrix}$ respectively $\begin{pmatrix} 0 \\ 1 \end{pmatrix}$ with eigenvalue -1 and the eigenvectors $\begin{pmatrix} -x/2 \\ 1 \end{pmatrix}$ respectively $\begin{pmatrix} -2/x \\ 1 \end{pmatrix}$ with eigenvalue 1

Therefore $\mu(s_{e_1}^{(0),mat})$ and $\mu(s_{e_2}^{(0),mat}) \in \mathrm{Isom}(\mathbb{H})$ are reflections along the hyperbolic line $A(\infty, -\frac{x}{2})$ respectively $A(0, -\frac{2}{x})$. As $x \leq -2$, we have $-\frac{2}{x} \leq -\frac{x}{2}$, so $A(0, -\frac{2}{x}) \cap A(-\frac{x}{2}, \infty) = \emptyset$, see the pictures in Fig. 6.4.

Theorem A.2 (a) applies and shows that $\langle \mu(s_{e_1}^{(0),mat}), \mu(s_{e_2}^{(0),mat}) \rangle \subset \mathrm{Isom}(\mathbb{H})$ is a free Coxeter group with the two given generators. Therefore also $\Gamma^{(0)}$ is a free Coxeter group with the two generators $s_{e_1}^{(0)}$ and $s_{e_2}^{(0)}$.

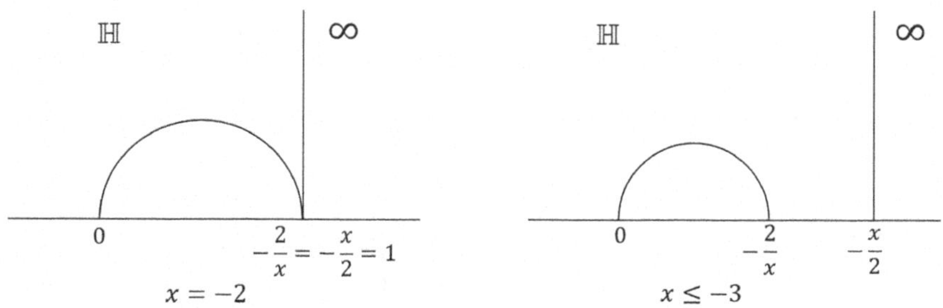

Fig. 6.4 Fundamental domain in $\mathbb{H}$ of $\Gamma^{(0),mat}$ for $x = -2$ and $x \leq -3$

(c) The set $\Delta^{(0)}$ for $x = -2$:

$$\begin{aligned} R^{(0)} &= \{y_1e_1 + y_2e_2 \in H_{\mathbb{Z}} \mid 1 = (y_1 - y_2)^2\} \\ &= (e_1 + \mathbb{Z}f_1) \,\dot\cup\, (e_2 + \mathbb{Z}f_1). \end{aligned}$$

For $m \in \mathbb{Z}$, $\varepsilon \in \{\pm 1\}$,

$$\begin{aligned} s_{e_1}^{(0)}(\varepsilon e_1 + mf_1) &= -\varepsilon e_1 + mf_1, \\ s_{e_2}^{(0)}(\varepsilon e_2 + mf_1) &= -\varepsilon e_2 + mf_1, \\ s_{e_1}^{(0)}s_{e_2}^{(0)}(e_1 + mf_1, e_2 + mf_1) &= (e_1 + (m+2)f_1, e_2 + (m-2)f_1). \end{aligned}$$

This shows all claims on $\Delta^{(0)}$ in part (c).

(d) (i) Recall $\kappa_1 + \kappa_2 = -x$, $\kappa_1\kappa_2 = 1$, $0 = \kappa_i^2 + x\kappa_i + 1$, $\kappa_2 = \kappa_1^{-1} = -\kappa_1 - x$, $\kappa_1 = -\kappa_2 - x$.

Because of $R^{(0)} = \{y_1e_1 + y_2e_2 \in H_{\mathbb{Z}} \mid 1 = y_1^2 + xy_1y_2 + y_2^2\}$ the map

$$\begin{aligned} u : R^{(0)} &\to \{\text{the units with norm 1 in } \mathbb{Z}[\kappa_1]\} \\ y_1e_1 + y_2e_2 &\mapsto y_1 - \kappa_1 y_2 \end{aligned}$$

is well defined and a bijection. Because of Lemma C.1 (a)

$$\{\text{the units with norm 1 in } \mathbb{Z}[\kappa_1]\} = \{\pm\kappa_1^l \mid l \in \mathbb{Z}\}.$$

Now

$$s_{e_1}^{(0)}(u^{-1}(\varepsilon\kappa_1^l)) = u^{-1}(-\varepsilon\kappa_1^{-l}) \quad \text{and} \quad s_{e_2}^{(0)}(u^{-1}(\varepsilon\kappa_1^l)) = u^{-1}(-\varepsilon\kappa_1^{2-l})$$

follow from

$$\begin{aligned} s_{e_1}^{(0)}\underline{e}\begin{pmatrix} y_1 \\ y_2 \end{pmatrix} &= \underline{e}\begin{pmatrix} -1 & -x \\ 0 & 1 \end{pmatrix}\begin{pmatrix} y_1 \\ y_2 \end{pmatrix} = \underline{e}\begin{pmatrix} -y_1 - xy_2 \\ y_2 \end{pmatrix}, \\ (-y_1 - xy_2) - \kappa_1 y_2 &= -(y_1 - \kappa_2 y_2) = -(y_1 - \kappa_1 y_2)^{-1}, \\ s_{e_2}^{(0)}\underline{e}\begin{pmatrix} y_1 \\ y_2 \end{pmatrix} &= \underline{e}\begin{pmatrix} 1 & 0 \\ -x & -1 \end{pmatrix}\begin{pmatrix} y_1 \\ y_2 \end{pmatrix} = \underline{e}\begin{pmatrix} y_1 \\ -xy_1 - y_2 \end{pmatrix}, \\ y_1 - \kappa_1(-xy_1 - y_2) &= \kappa_1((\kappa_2 + x)y_1 + y_2) = \kappa_1(-\kappa_1 y_1 + y_2) \\ &= -\kappa_1^2(y_1 - \kappa_2 y_2) = -\kappa_1^2(y_1 - \kappa_1 y_2)^{-1}. \end{aligned}$$

This shows $\Delta^{(0)} = R^{(0)}$ and that $\Delta^{(0)}$ splits into the two disjoint orbits $\Gamma^{(0)}\{e_1\}$ and $\Gamma^{(0)}\{e_2\}$.

(ii) The formulas in part (i) show immediately $(s_{e_1}^{(0)}s_{e_2}^{(0)})^m(u^{-1}(\varepsilon\kappa_1^l)) = u^{-1}(\varepsilon\kappa_1^{l-2m})$ for $l, m \in \mathbb{Z}$, $\varepsilon \in \{\pm 1\}$. Together with $u(e_1) = 1$ and $u(e_2) = -\kappa_1$ this implies the formulas in part (ii).

(iv) This follows from the formulas for the action of $\widetilde{s}^{(0)}_{e_i}$ on $\widetilde{R}^{(0)}$ and of $s^{(0)}_{e_i}$ on $R^{(0)}$.

(iii) First we consider the lower right branch of the hyperbola. There $y_1-\kappa_1 y_2>0$ and $y_1-\kappa_2 y_2>0$. We consider y_2 as an implicit function in y_1. The equation

$$1=y_1^2+xy_1y_2+y_2^2=(y_1-\kappa_1y_2)(y_1-\kappa_2y_2)$$

implies

$$\begin{aligned}
0&=(y_1-\kappa_1y_2)(1-\kappa_2y_2')+(y_1-\kappa_2y_2)(1-\kappa_1y_2')\\
&=[(y_1-\kappa_1y_2)+(y_1-\kappa_2y_2)]-[(y_1-\kappa_1y_2)\kappa_2+(y_1-\kappa_2y_2)\kappa_1]y_2',\\
\text{so}\quad & y_2'>0\quad\text{and}\quad(1-\kappa_1y_2')(1-\kappa_2y_2')<0,\\
0&=(y_1-\kappa_1y_2)(-\kappa_2y_2'')+2(1-\kappa_1y_2')(1-\kappa_2y_2')\\
&\quad+(y_1-\kappa_2y_2)(-\kappa_1y_2'')\\
&=-[(y_1-\kappa_1y_2)\kappa_2+(y_1-\kappa_2y_2)\kappa_1]y_2''+2(1-\kappa_1y_2')(1-\kappa_2y_2'),\\
\text{so}\quad & y_2''<0.
\end{aligned}$$

Therefore the lower right branch of the hyperbola is strictly monotonously increasing and concave. The upper left branch is obtained from the lower right branch by the reflection $H_\mathbb{R}\to H_\mathbb{R}$, $(y_1,y_2)\mapsto(y_2,y_1)$, along the diagonal. Therefore it is strictly monotonously increasing and convex.

By definition $s^{(0)}_{e_1}$ maps each horizontal line in $H_\mathbb{R}$ to itself, and $s^{(0)}_{e_2}$ maps each vertical line in $H_\mathbb{R}$ to itself. As they map $\Delta^{(0)}=R^{(0)}$ to itself, this shows all statements in (iii).

(e) Obviously $g_{1,2}\in O^{(0)}$ and $\{\pm\,\mathrm{id},\pm g_{1,2}\}$ is a subgroup of $O^{(0)}$. Recall

$$\begin{aligned}
&H_\mathbb{Z}^\sharp\supset j^{(0)}(H_\mathbb{Z})=\langle j^{(0)}(e_1),j^{(0)}(e_2)\rangle\\
\text{with}\quad & j^{(0)}(e_1)(\underline{e})=(2,x),\ j^{(0)}(e_2)(\underline{e})=(x,2).
\end{aligned}$$

Define $l\in H_\mathbb{Z}^\sharp$ with $l(\underline{e})=(1,0)$. Then

$$\begin{aligned}
(l-l\circ(-\mathrm{id})^{-1})(\underline{e})&=2l(\underline{e})=(2,0),\\
\text{so}\quad l-l\circ(-\mathrm{id})^{-1}\notin j^{(0)}(H_\mathbb{Z}),\ \text{so}&\quad -\mathrm{id}\notin\ker\tau^{(0)}.\\
(l-l\circ g_{1,2}^{-1})(\underline{e})&=(1,-1),\\
\text{so}\quad l-l\circ g_{1,2}^{-1}\notin j^{(0)}(H_\mathbb{Z}).\ \text{so}&\quad g_{1,2}\notin\ker\tau^{(0)}.
\end{aligned}$$

Denote for a moment by $\widetilde{O}^{(0)}$ the subgroup of $O^{(0)}$ which is generated by $\{\pm\,\mathrm{id},\pm g_{1,2}\}$ and $\Gamma^{(0)}$. We just saw

$$\begin{aligned}
&(\ker\tau^{(0)})\cap\widetilde{O}^{(0)}=\Gamma^{(0)},\\
\text{so}\quad &\widetilde{O}^{(0)}=\Gamma^{(0)}\rtimes\{\pm\,\mathrm{id},\pm g_{1,2}\}.
\end{aligned}$$

It remains to show that this subgroup is $O^{(0)}$.

By the parts (c) and (d) (iii) the set $\Delta^{(0)} = R^{(0)}$ splits into two $\Gamma^{(0)}$ orbits, $\Gamma^{(0)}\{e_1\}$ and $\Gamma^{(0)}\{e_2\}$. The element $g_{1,2}$ interchanges them, so $\Delta^{(0)} = R^{(0)}$ is a single $\widetilde{O}^{(0)}$ orbit.

Therefore each element of $O^{(0)}$ can be written as a product of an element in $\widetilde{O}^{(0)}$ and an element $g \in O^{(0)}$ with $g(e_1) = e_1$. It is sufficient to show $g \in \widetilde{O}^{(0)}$. Observe that the set $\{v \in H_{\mathbb{Z}} \mid I^{(0)}(v, v) > 0\}$ consists of two components, with e_1 in one component and e_2 in the other component. Because of $g(e_1) = e_1$, $g(e_2)$ is in the same component as e_2. Now $H_{\mathbb{Z}} = \mathbb{Z}e_1 + \mathbb{Z}g(e_2)$ shows $g(e_2) \in \{e_2, s^{(0)}_{e_1}(-e_2)\}$ and $g \in \{\mathrm{id}, -s^{(0)}_{e_1}\}$. Therefore $\widetilde{O}^{(0)} = O^{(0)}$. □

Remarks 6.9 (i) By part (iv) of Theorem 6.8 (d) the pairs $(\Gamma^{(0)}, \Delta^{(0)})$ with the action of $\Gamma^{(0)}$ on $\Delta^{(0)}$ are isomorphic for all $x \leq -2$. This is interesting as in the case $x = -2$ the set $\Delta^{(0)}$ and this action could be written down in a very simple way.

The parts (i) and (iii) of Theorem 6.8 (d) offered two ways to control the set $\Delta^{(0)}$ and this action also for $x \leq -3$, a number theoretic way and a geometric way. But both ways are less simple than $\Delta^{(0)}$ in the case $x = -2$.

(ii) In the odd cases the situation will be partly similar, partly different. The pairs $(\Gamma^{(1)}, \Delta^{(1)})$ with the action of $\Gamma^{(1)}$ on $\Delta^{(1)}$ are isomorphic for all $x \leq -2$. In the case $x = -2$ the set $\Delta^{(1)}$ and this action can be written down in a fairly simple way. But we lack analogs of the parts (i) and (iii) in Theorem 6.8 (d). We do not have a good control on the sets $\Delta^{(1)}$ for $x \leq -3$.

(iii) For each $x \leq -2$ and $i \in \{1, 2\}$ $\mathrm{Stab}_{\Gamma^{(0)}}(e_i) = \{\mathrm{id}\}$, so the map $\Gamma^{(0)} \to \Gamma^{(0)}\{e_i\}$, $\gamma \mapsto \gamma(e_i)$, is a bijection. The action of $\Gamma^{(0)}$ on $\Gamma^{(0)}\{e_i\}$ shows again immediately that $\Gamma^{(0)}$ is a free Coxeter group with generators $s^{(0)}_{e_1}$ and $s^{(0)}_{e_2}$.

Now we come to the odd cases. As before we restrict to $x \in \mathbb{Z}_{<0}$.

Theorem 6.10 *(a) We have*

$$O^{(1)} \cong SL_2(\mathbb{Z}),$$
$$\ker \tau^{(1)} \cong \Gamma(x) := \{A \in SL_2(\mathbb{Z}) \mid A \equiv E_2 \bmod x\},$$

$$s^{(1)}_{e_i} \underline{e} = \underline{e} \cdot s^{(1),mat}_{e_i} \quad \text{with}$$
$$s^{(1),mat}_{e_1} = \begin{pmatrix} 1 & -x \\ 0 & 1 \end{pmatrix}, \quad s^{(1),mat}_{e_2} = \begin{pmatrix} 1 & 0 \\ x & 1 \end{pmatrix},$$
$$\Gamma^{(1)} \cong \Gamma^{(1),mat} := \langle s^{(1),mat}_{e_1}, s^{(1),mat}_{e_2} \rangle \subset SL_2(\mathbb{Z}),$$

The map from $\Delta^{(1)}$ to its image in $\widehat{\mathbb{R}} = \mathbb{R} \cup \{\infty\}$ under the composition $C : \Delta^{(1)} \to \widehat{\mathbb{R}}$ of maps

$$\Delta^{(1)} \to (\mathbb{R}^*\Delta^{(1)})/\mathbb{R}^* = \{\text{lines through vanishing cycles}\}$$

$$\hookrightarrow (H_{\mathbb{R}} - \{0\})/\mathbb{R}^* = \{\text{lines in } H_{\mathbb{R}}\} \xrightarrow{\cong} \widehat{\mathbb{R}},$$
$$\mathbb{R}^*\underline{e}\begin{pmatrix} y_1 \\ 1 \end{pmatrix} \mapsto y_1, \quad \mathbb{R}^* e_1 \mapsto \infty,$$

is two-to-one.

(b) The case $x = -1$: $\Gamma^{(1),mat} = SL_2(\mathbb{Z})$.

$$\Delta^{(1)} = \Gamma^{(1)}\{e_1\} = \{y_1e_1 + y_2e_2 \in H_{\mathbb{Z}} \mid \gcd(y_1, y_2) = 1\} = H_{\mathbb{Z}}^{prim},$$

where $H_{\mathbb{Z}}^{prim}$ denotes the set of primitive vectors in $H_{\mathbb{Z}}$. The image $C(\Delta^{(1)}) \subset \widehat{\mathbb{R}}$ is $\widehat{\mathbb{Q}} = \mathbb{Q} \cup \{\infty\}$.

(c) The case $x = -2$: $\Gamma^{(1)} \cong G^{free,2}$ is a free group with the two generators $s_{e_1}^{(1)}$ and $s_{e_2}^{(1)}$. The (isomorphic) matrix group $\Gamma^{(1),mat}$ is

$$\Gamma^{(1),mat} = \{\begin{pmatrix} a & b \\ c & d \end{pmatrix} \in SL_2(\mathbb{Z}) \mid a \equiv d \equiv 1(4),\ b \equiv c \equiv 0(2)\}.$$

It is a subgroup of index 2 in the principal congruence subgroup

$$\Gamma(2) = \{\begin{pmatrix} a & b \\ c & d \end{pmatrix} \mid a \equiv d \equiv 1(2),\ b \equiv c \equiv 0(2)\}$$

of $SL_2(\mathbb{Z})$. The set $\Delta^{(1)} = \Gamma^{(1)}\{\pm e_1, \pm e_2\}$ is

$$\Delta^{(1)} = \{y_1e_1 + y_2e_2 \in H_{\mathbb{Z}}^{prim} \mid y_1 + y_2 \equiv 1(2)\}.$$

It splits into the four disjoint orbits

$$\begin{aligned}
\Gamma^{(1)}\{e_1\} &= \{y_1e_1 + y_2e_2 \in H_{\mathbb{Z}}^{prim} \mid y_1 \equiv 1(4), y_2 \equiv 0(2)\} \\
\Gamma^{(1)}\{-e_1\} &= \{y_1e_1 + y_2e_2 \in H_{\mathbb{Z}}^{prim} \mid y_1 \equiv 3(4), y_2 \equiv 0(2)\} \\
\Gamma^{(1)}\{e_2\} &= \{y_1e_1 + y_2e_2 \in H_{\mathbb{Z}}^{prim} \mid y_1 \equiv 0(2), y_2 \equiv 1(4)\} \\
\Gamma^{(1)}\{-e_2\} &= \{y_1e_1 + y_2e_2 \in H_{\mathbb{Z}}^{prim} \mid y_1 \equiv 0(2), y_2 \equiv 3(4)\}.
\end{aligned}$$

The set $H_{\mathbb{Z}}^{prim}$ of primitive vectors is the disjoint union of $\Delta^{(1)}$ and the set

$$\{y_1e_1 + y_2e_2 \in H_{\mathbb{Z}}^{prim} \mid y_1 \equiv y_2 \equiv 1(2)\}.$$

The image $C(\Delta^{(1)}) \subset \widehat{\mathbb{R}}$ is

$$\{\infty\} \cup \{\frac{a}{b} \mid a \in \mathbb{Z}, b \in \mathbb{N}, \gcd(a, b) = 1, a \equiv 0(2) \text{ or } b \equiv 0(2)\} \subset \widehat{\mathbb{Q}},$$

and is dense in $\widehat{\mathbb{R}}$.

(d) The cases $x \leq -3$.

(i) $\Gamma^{(1)} \cong G^{free,2}$ is a free group with the two generators $s^{(1)}_{e_1}$ and $s^{(1)}_{e_2}$.

(ii) The matrix group $\Gamma^{(1),mat}$ is a Fuchsian group of the second kind. It has infinite index in the group

$$\{\begin{pmatrix} a & b \\ c & d \end{pmatrix} \in SL_2(\mathbb{Z}) \mid a \equiv d \equiv 1(x^2), b \equiv c \equiv 0(x)\},$$

which has finite index in $SL_2(\mathbb{Z})$.

(iii) The image $C(\Delta^{(1)}) \subset \widehat{\mathbb{R}}$ is a subset of $\widehat{\mathbb{Q}}$ which is nowhere dense in $\widehat{\mathbb{R}}$.

(iv) Denote by $\widetilde{s}^{(1)}_{e_i}$, $\widetilde{\Gamma}^{(1)}$ and $\widetilde{\Delta}^{(1)}$ the objects for $x = -2$ and as before by $s^{(1)}_{e_i}$, $\Gamma^{(1)}$ and $\Delta^{(1)}$ the objects for an $x \leq -3$.

The map $\widetilde{s}^{(1)}_{e_i} \mapsto s^{(1)}_{e_i}$ extends to a group isomorphism $\widetilde{\Gamma}^{(1)} \to \Gamma^{(1)}$, which maps the stabilizer $\langle \widetilde{s}^{(1)}_{e_i} \rangle$ of e_i in $\widetilde{\Gamma}^{(1)}$ to the stabilizer $\langle s^{(1)}_{e_i} \rangle$ of e_i in $\Gamma^{(1)}$. The induced map

$$\widetilde{\Delta}^{(1)} \to \Delta^{(1)}, \ \widetilde{\gamma}(\varepsilon e_i) \mapsto \gamma(\varepsilon e_i) \quad \text{for } \varepsilon \in \{\pm 1\}, \widetilde{\gamma} \mapsto \gamma,$$

is a bijection. The bijections $\widetilde{\Gamma}^{(1)} \to \Gamma^{(1)}$ and $\widetilde{\Delta}^{(1)} \to \Delta^{(1)}$ are compatible with the actions of $\widetilde{\Gamma}^{(1)}$ on $\widetilde{\Delta}^{(1)}$ and of $\Gamma^{(1)}$ on $\Delta^{(1)}$. Especially, $\Delta^{(1)}$ splits into the four disjoint orbits $\Gamma^{(1)}\{e_1\}$, $\Gamma^{(1)}\{-e_1\}$, $\Gamma^{(1)}\{e_2\}$, $\Gamma^{(1)}\{-e_2\}$.

(e) In the case A_1^2 $\Delta^{(0)} = \Delta^{(1)} = \{\pm e_1, \pm e_2\}$. In all other rank 2 cases $\Delta^{(0)} \subsetneqq \Delta^{(1)}$.

Proof (a) Because of $I^{(1)}(\underline{e}^t, \underline{e}) = \begin{pmatrix} 0 & x \\ -x & 0 \end{pmatrix}$ we have $O^{(1)} \cong SL_2(\mathbb{Z})$. In order to see $\ker \tau^{(1)} \cong \Gamma(x)$, consider the generators $l_1, l_2 \in H^\sharp_{\mathbb{Z}}$ of $H^\sharp_{\mathbb{Z}}$ with $l_1(\underline{e}) = (1, 0)$ and $l_2(\underline{e}) = (0, 1)$. Observe first

$$j^{(1)}(e_1)(\underline{e}) = (0, x), \ j^{(1)}(e_2)(\underline{e}) = (-x, 0), \quad \text{so } j^{(1)}(H_{\mathbb{Z}}) = xH^\sharp_{\mathbb{Z}},$$

and second that $g \in O^{(1)}$ with $g^{-1}(\underline{e}) = \underline{e} \cdot \begin{pmatrix} a & b \\ c & d \end{pmatrix}$ satisfies

$$(l_1 - l_1 \circ g^{-1})(\underline{e}) = (1 - a, b),$$
$$(l_2 - l_2 \circ g^{-1})(\underline{e}) = (-c, 1 - d),$$

so $g \in \ker \tau^{(1)}$ if and only if $\begin{pmatrix} a & b \\ c & e \end{pmatrix} \equiv E_2 \bmod x$.

It remains to prove that the line $\mathbb{R} \cdot \delta \subset H_{\mathbb{R}}$ through a vanishing cycle $\delta \in \Delta^{(1)}$ intersects $\Delta^{(1)}$ only in $\pm\delta$. To prove this we can restrict to $\delta = e_i$. There it follows from the fact that any matrix $A \in SL_2(\mathbb{Z})$ with a zero in an entry A_{ij} has entries $A_{i,j+1(2)}, A_{i+1(2),j} \in \{\pm 1\}$.

(b) $\Gamma^{(1),mat} = SL_2(\mathbb{Z})$ is well known. The standard arguments for this are as follows. The group $\mu(\Gamma^{(1),mat}) \subset \text{Isom}(\mathbb{H})$ is generated by

$$\mu(s_{e_1}^{(1),mat}) = \mu(\begin{pmatrix} 1 & 1 \\ 0 & 1 \end{pmatrix}) = (z \mapsto z+1),$$

$$\mu(s_{e_1}^{(1),mat} s_{e_2}^{(1),mat} s_{e_1}^{(1),mat}) = \mu(\begin{pmatrix} 0 & 1 \\ -1 & 0 \end{pmatrix}) = (z \mapsto -z^{-1}).$$

One sees almost immediately that it acts transitively on $\widehat{\mathbb{Q}}$ and that the stabilizer $\langle\mu(s_{e_1}^{(1),mat})\rangle$ of ∞ in $\mu(\Gamma^{(1),mat})$ coincides with the stabilizer of ∞ in $\mu(SL_2(\mathbb{Z}))$. Therefore $\mu(\Gamma^{(1),mat}) = \mu(SL_2(\mathbb{Z}))$. But $-E_2 \in \Gamma^{(1),mat}$ because of $\begin{pmatrix} 0 & 1 \\ -1 & 0 \end{pmatrix}^2 = -E_2$. Therefore $\Gamma^{(1),mat} = SL_2(\mathbb{Z})$.

The fact that $\mu(SL_2(\mathbb{Z}))$ acts transitively on $\widehat{\mathbb{Q}}$ shows $C(\Delta^{(1)}) = \widehat{\mathbb{Q}}$. Together with $-E_2 \in SL_2(\mathbb{Z})$ this shows

$$\Delta^{(1)} = \Gamma^{(1)}\{e_1\} = H_{\mathbb{Z}}^{prim}.$$

(c) and (d) The group $\Gamma^{(1)}$ for $x \le -2$: Recall the Remarks and Notations A.1. The elements

$$\mu(s_{e_1}^{(1),mat}) = \mu(\begin{pmatrix} 1 & -x \\ 0 & 1 \end{pmatrix}) = (z \mapsto z - x) \quad \text{and}$$

$$\mu(s_{e_2}^{(1),mat}) = \mu(\begin{pmatrix} 1 & 0 \\ x & 1 \end{pmatrix}) = (z \mapsto \frac{z}{xz+1})$$

of $\mathrm{Isom}(\mathbb{H})$ are parabolic with fixed points ∞ respectively 0 on $\widehat{\mathbb{R}}$. Observe

$$\mu(s_{e_1}^{(1),mat})^{-1}(1) = 1+x, \quad \mu(s_{e_2}^{(1),mat})(1) = (1+x)^{-1},$$
$$(1+x)^{-1} \ge 1+x \quad \text{for } x \le -2.$$

Therefore $\mu(s_{e_1}^{(1),mat})^{-1}(A(\infty,1)) = A(\infty, 1+x)$ and $\mu(s_{e_2}^{(1),mat})(A(0,1)) = A(0,(1+x)^{-1})$ do not intersect. See Fig. 6.5

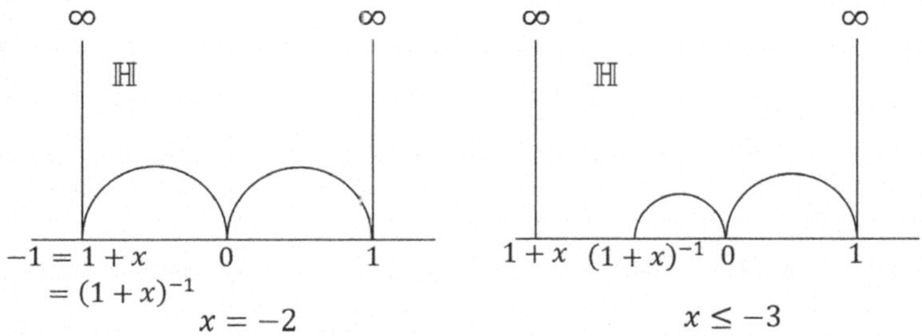

Fig. 6.5 Fundamental domain in $\mathbb{H}$ of $\Gamma^{(1),mat}$ for $x = -2$ and $x \le -3$

Theorem A.2 (c) applies and shows that $\mu(\Gamma^{(1),mat})$ is a free group with the two generators $\mu(s_{e_1}^{(1),mat})$ and $\mu(s_{e_2}^{(1),mat})$. Therefore also $\Gamma^{(1)}$ is a free group with the two generators $s_{e_1}^{(1)}$ and $s_{e_2}^{(1)}$. As therefore the map $\Gamma^{(1),mat} \to \mu(\Gamma^{(1),mat})$ is an isomorphism, $-E_2 \notin \Gamma^{(1),mat}$ and $-\text{id} \notin \Gamma^{(1)}$.

Theorem A.2 (c) says also that the contractible open set $\mathcal{F}$ whose hyperbolic boundary consists of the four hyperbolic lines which were used above, $A(1+x,\infty)$, $A(\infty,1)$, $A(1,0)$, $A(0,(1+x)^{-1})$, is a fundamental domain for $\mu(\Gamma^{(1),mat})$. If $x = -2$ then its euclidean boundary in $\widehat{\mathbb{C}}$ consists of these four hyperbolic lines and the four points ∞, 1, 0, $-1 = 1+x = (1+x)^{-1}$. If $x \leq -3$ then its euclidean boundary consists of these four hyperbolic lines, the three points ∞, 1, 0, and the interval $[1+x, (1+x)^{-1}]$.

(c) $\Gamma^{(1),mat}$ and $\Delta^{(1)}$ for $x = -2$: The following facts together imply $\mu(\Gamma^{(1),mat}) = \mu(\Gamma(2))$:

$$\Gamma^{(1),mat} \subset \Gamma(2), \quad [SL_2(\mathbb{Z}) : \Gamma(2)] = 6, \quad -E_2 \in \Gamma(2),$$
$$(\text{hyperbolic area of the fundamental domain } \mathcal{F} \text{ of } \mu(\Gamma^{(1),mat})) = 2\pi,$$
$$(\text{hyperbolic area of a fundamental domain of } \mu(SL_2(\mathbb{Z}))) = \frac{\pi}{3}.$$

Therefore either $\Gamma^{(1),mat} = \Gamma(2)$ or $\Gamma^{(1),mat}$ is a subgroup of index 2 in $\Gamma(2)$. But $\Gamma^{(1),mat}$ is certainly a subgroup of the subgroup

$$\begin{pmatrix} a & b \\ c & d \end{pmatrix} \in SL_2(\mathbb{Z}) \,|\, a \equiv d \equiv 1(4), b \equiv c \equiv 0(2)\}$$

of $\Gamma(2)$ and does not contain $-E_2$. Therefore $\Gamma^{(1),mat}$ coincides with this subgroup of $\Gamma(2)$ and has index 2 in $\Gamma(2)$.

Therefore the orbits $\Gamma^{(1)}\{e_1\}$, $\Gamma^{(1)}\{-e_1\}$, $\Gamma^{(1)}\{e_2\}$, $\Gamma^{(1)}\{-e_2\}$ are contained in the right hand sides of the equations in part (c) which describe them and are disjoint. It remains to show equality.

We restrict to $\Gamma^{(1)}\{e_1\}$. The argument for $\Gamma^{(1)}\{e_2\}$ is analogous, and the equations for $\Gamma^{(1)}\{-e_1\}$ and $\Gamma^{(1)}\{-e_2\}$ follow immediately.

Suppose $y_1, y_3 \in \mathbb{Z}$ with $y_1 \equiv 1(4)$, $y_3 \equiv 0(2)$, $\gcd(y_1, y_3) = 1$. We have to show $\underline{e}\begin{pmatrix} y_1 \\ y_3 \end{pmatrix} \in \Gamma^{(1)}\{e_1\}$. For that we have to find $y_2, y_4 \in \mathbb{Z}$ with $\begin{pmatrix} y_1 & y_2 \\ y_3 & y_4 \end{pmatrix} \in \Gamma^{(1)}$, so with $1 = y_1y_4 - y_2y_3$, $y_4 \equiv 1(4)$, $y_2 \equiv 0(2)$.

The condition $\gcd(y_1, y_3)$ implies existence of $\widetilde{y}_2, \widetilde{y}_4 \in \mathbb{Z}$ with $1 = y_1\widetilde{y}_4 - \widetilde{y}_2y_3$.

1st case, $\widetilde{y}_2 \equiv 0(2)$: Then $1 = y_1\widetilde{y}_4 - \widetilde{y}_2y_3$ shows $\widetilde{y}_4 \equiv 1(4)$, and $(y_2, y_4) = (\widetilde{y}_2, \widetilde{y}_4)$ works.

2nd case, $\widetilde{y}_2 \equiv 1(2)$: Then $(y_2, y_4) = (\widetilde{y}_2 + y_1, \widetilde{y}_4 + y_3)$ satisfies $1 = y_1y_4 - y_2y_3$ and $y_2 \equiv 0(2)$, so we are in the 1st case, so (y_2, y_4) works.

Therefore the orbit $\Gamma^{(1)}\{e_1\}$ is as claimed the set $\{y_1e_1 + y_2e_2 \in H_{\mathbb{Z}}^{prim} \,|\, y_1 \equiv 1(4), y_2 \equiv 0(2)\}$.

The statements on $H_{\mathbb{Z}}^{prim}$ and $C(\Delta^{(1)})$ are clear now, too.

(d) $\Gamma^{(1),mat}$ and $\Delta^{(1)}$ for $x \leq -3$: Part (i) was shown above.

(ii) The euclidean boundary of the fundamental domain $\mathcal{F}$ above of $\mu(\Gamma^{(1),mat})$ contains the real interval $[(1+x)^{-1}, 1+x]$. Therefore its hyperbolic area is ∞, $\Gamma^{(1),mat}$ is a Fuchsian group of the second kind, and the index of $\Gamma^{(1),mat}$ in $SL_2(\mathbb{Z})$ is ∞ (see e.g. [Fo51, 34.] [Be83, §5.3, §8.1]).

(iii) The set $C(\Delta^{(1)}) \subset \widehat{\mathbb{R}}$ is the union of the $\mu(\Gamma^{(1),mat})$ orbits of ∞ and 0 in $\widehat{\mathbb{R}}$. Because $\Gamma^{(1),mat}$ is a Fuchsian group of the second kind, these two orbits are nowhere dense in $\widehat{\mathbb{R}}$ (see e.g. [Fo51] [Be83]). For example, they contain no point of the open interval $((1+x)^{-1}, (1+x))$ and of its $\mu(\Gamma^{(1),mat})$ orbit.

(iv) The groups $\widetilde{\Gamma}^{(1)}$ and $\Gamma^{(1)}$ are free groups with generators $\widetilde{s}^{(1)}_{e_1}, \widetilde{s}^{(1)}_{e_2}$ and $s^{(1)}_{e_1}, s^{(1)}_{e_2}$. Therefore there is a unique isomorphism $\widetilde{\Gamma}^{(1)} \to \Gamma^{(1)}$ with $\widetilde{s}^{(1)}_{e_i} \mapsto s^{(1)}_{e_i}$. The other statements follow immediately.

(e) The case $x = 0$, so A_1^2:

$$\Delta^{(0)} = \Delta^{(1)} = \{\pm e_1, \pm e_2\}.$$

The case $x = -1$, so A_2:

$$\Delta^{(0)} = \{\pm e_1, \pm e_2, \pm(e_1 + e_2)\} \subsetneqq H_{\mathbb{Z}}^{prim} = \Delta^{(1)}.$$

The case $x = -2$, so $\mathbb{P}^1 A_1$:

$$\begin{aligned}\Delta^{(0)} &= (e_1 + \mathbb{Z} f_1) \,\dot{\cup}\, (e_2 + \mathbb{Z} f_1) \\ &\subsetneqq \{y_1 e_1 + y_2 e_2 \in H_{\mathbb{Z}}^{prim} \mid y_1 + y_2 \equiv 1(2)\} = \Delta^{(1)}.\end{aligned}$$

The cases $x \leq -3$: Recall $s^{(0)}_{e_1} s^{(0)}_{e_2} = -M = -s^{(1)}_{e_1} s^{(1)}_{e_2}$. Therefore for $m \in \mathbb{Z}, \varepsilon \in \{\pm 1\}$

$$\begin{aligned}(s^{(0)}_{e_1} s^{(0)}_{e_2})^m(\underline{e}) &= (-s^{(1)}_{e_1} s^{(1)}_{e_2})^m(\underline{e}) \in (\Delta^{(1)})^2, \\ (s^{(0)}_{e_1} s^{(0)}_{e_2})^m s^{(0)}_{e_1}(e_1) &= -(-s^{(1)}_{e_1} s^{(1)}_{e_2})^m(e_1) \in \Delta^{(1)}, \\ (s^{(0)}_{e_1} s^{(0)}_{e_2})^m s^{(0)}_{e_1}(e_2) &= (s^{(0)}_{e_1} s^{(0)}_{e_2})^{m+1} s^{(0)}_{e_2}(e_2) \in \Delta^{(1)}, \\ &= -(-s^{(1)}_{e_1} s^{(1)}_{e_2})^{m+1}(e_2) \in \Delta^{(1)}.\end{aligned}$$

This shows $\Delta^{(0)} \subset \Delta^{(1)}$. For example $(s^{(1)}_{e_1})^{-1}(e_2) = x e_1 + e_2$ satisfies

$$L(xe_1 + e_2, xe_1 + e_2) = (x\ 1)\begin{pmatrix}1 & 0 \\ x & 1\end{pmatrix}\begin{pmatrix}x \\ 1\end{pmatrix} = 2x^2 + 1 \neq 1,$$

so $\Delta^{(0)} \subsetneqq \Delta^{(1)}$. □

6.3 The Even Rank 3 Cases

For $\underline{x} = (x_1, x_2, x_3) \in \mathbb{Z}^3$ consider the matrix $S = S(\underline{x}) = \begin{pmatrix} 1 & x_1 & x_2 \\ 0 & 1 & x_3 \\ 0 & 0 & 1 \end{pmatrix} \in T_3(\mathbb{Z})$, and consider a unimodular bilinear lattice $(H_\mathbb{Z}, L)$ with a triangular basis $\underline{e} = (e_1, e_2, e_3)$ with $L(\underline{e}^t, \underline{e})^t = S$. In this section we will determine in all cases the even monodromy group $\Gamma^{(0)} = \langle s^{(0)}_{e_1}, s^{(0)}_{e_2}, s^{(0)}_{e_3} \rangle$ and in many, but not all, cases the set $\Delta^{(0)} = \Gamma^{(0)}\{\pm e_1, \pm e_2, \pm e_3\}$ of even vanishing cycles. The cases where we control $\Delta^{(0)}$ well contain all cases with $r(\underline{x}) \in \{0, 1, 2, 4\}$ (3 does not turn up).

The group $\mathrm{Br}_3 \ltimes \{\pm 1\}^3$ acts on the set $\mathcal{B}^{tri}$ of triangular bases of $(H_\mathbb{Z}, L)$, but this action does not change $\Gamma^{(0)}$ and $\Delta^{(0)}$. Therefore the analysis of the action of $\mathrm{Br}_3 \ltimes \{\pm 1\}^3$ on $T_3(\mathbb{Z})$ in Theorem 4.6 allows to restrict to the matrices $S(\underline{x})$ with $\underline{x}$ in the following list:

$$\begin{aligned} &S(\underline{x}) \text{ with } \underline{x} \in \mathbb{Z}^3_{\leq 0} \text{ and } r(\underline{x}) > 4, \\ &S(A_1^3),\ S(\mathbb{P}^2),\ S(A_2A_1),\ S(A_3),\ S(\mathbb{P}^1A_1),\ S(\widehat{A}_2),\ S(\mathcal{H}_{1,2}), \\ &S(-l, 2, -l) \text{ for } l \geq 3, \\ &\begin{cases} S(\underline{x}) \text{ with } \underline{x} \in \mathbb{Z}^3_{\geq 3} \text{ and } r(\underline{x}) < 0 \text{ and} \\ x_i \leq \frac{1}{2}x_jx_k \text{ for } \{i, j, k\} = \{1, 2, 3\}. \end{cases} \end{aligned}$$

The following of these matrices satisfy $\underline{x} \in \mathbb{Z}^3_{\leq 0}$:

$$\begin{aligned} &S(\underline{x}) \text{ with } \underline{x} \in \mathbb{Z}^3_{\leq 0} \text{ and } r(\underline{x}) > 4, \\ &S(A_1^3),\ S(A_2A_1),\ S(A_3),\ S(\mathbb{P}^1A_1),\ S(\widehat{A}_2). \end{aligned}$$

These are all Coxeter matrices. Their even monodromy groups $\Gamma^{(0)}$ are Coxeter groups and are well known. The cases with $\underline{x} \in \{0, 1, 2\}^3$ are classical, the extension to all $\underline{x} \in \mathbb{Z}^3_{\leq 0}$ has been done by Vinberg [Vi71, Prop. 6, Thm. 1, Thm. 2, Prop. 17]. If we write $(x_1, x_2, x_3) = (S_{12}, S_{13}, S_{23})$ then the following holds [Hu90, 5.3+5.4] [Vi71] [BB05, 4.1+4.2]: All relations in $\Gamma^{(0)} = \langle s^{(0)}_{e_1}, s^{(0)}_{e_2}, s^{(0)}_{e_3} \rangle$ are generated by the relations

$$(s^{(0)}_{e_i})^2 = \mathrm{id} \quad \text{for } i \in \{1, 2, 3\}, \tag{6.1}$$

$$(s^{(0)}_{e_i} s^{(0)}_{e_j})^2 = \mathrm{id} \quad \text{for } \{i, j, k\} = \{1, 2, 3\} \text{ with } s_{ij} = 0 \tag{6.2}$$

$$[\text{equivalent: } s^{(0)}_{e_i} \text{ and } s^{(0)}_{e_j} \text{ commute}],$$

$$(s^{(0)}_{e_i} s^{(0)}_{e_j})^3 = \mathrm{id} \quad \text{for } \{i, j, k\} = \{1, 2, 3\} \text{ with } s_{ij} = -1, \tag{6.3}$$

$$\text{no relation} \quad \text{for } \{i, j, k\} = \{1, 2, 3\} \text{ with } s_{ij} \leq -2. \tag{6.4}$$

Especially, $\Gamma^{(0)} \cong G^{fCox,3}$ is a free Coxeter group with three generators if $\underline{x} \in \mathbb{Z}^3_{\leq -2}$.

In Theorem 6.11 we recover this result, we say more about the Coxeter groups with $r \in \{0, 1, 2, 4\}$, so the cases $S(A_1^3)$, $S(A_2A_1)$, $S(A_3)$, $S(\mathbb{P}^1 A_1)$, $S(\widehat{A}_2)$, and we treat also the other cases where $\Gamma^{(0)}$ is not a Coxeter group.

The only cases where $\mathrm{Rad}\, I^{(0)} \supsetneqq \{0\}$ are the cases with $r(\underline{x}) = 4$, so the cases $S(\mathbb{P}^1 A_1)$, $S(\widehat{A}_2)$, $S(\mathcal{H}_{1,2})$ and $S(-l, 2, -l)$ with $l \geq 3$. In these cases, we have the exact sequence

$$\{1\} \to \Gamma_u^{(0)} \to \Gamma^{(0)} \to \Gamma_s^{(0)} \to \{1\} \tag{6.5}$$

in Lemma 6.2 (d).

Theorem 6.11 *(a) We have*

$$\begin{aligned}
s_{e_i}^{(0)}\underline{e} &= \underline{e} \cdot s_{e_i}^{(0),mat} \quad \text{with} \quad s_{e_1}^{(0),mat} = \begin{pmatrix} -1 & -x_1 & -x_2 \\ 0 & 1 & 0 \\ 0 & 0 & 1 \end{pmatrix}, \\
s_{e_2}^{(0),mat} &= \begin{pmatrix} 1 & 0 & 0 \\ -x_1 & -1 & -x_3 \\ 0 & 0 & 1 \end{pmatrix}, \ s_{e_3}^{(0),mat} = \begin{pmatrix} 1 & 0 & 0 \\ 0 & 1 & 0 \\ -x_2 & -x_3 & -1 \end{pmatrix}, \\
\Gamma^{(0)} &\cong \Gamma^{(0),mat} := \langle s_{e_1}^{(0),mat}, s_{e_2}^{(0),mat}, s_{e_3}^{(0),mat} \rangle \subset GL_3(\mathbb{Z}), \\
R^{(0)} &= \{y_1e_1 + y_2e_2 + y_3e_3 \in H_{\mathbb{Z}} \mid \\
&\qquad 1 = y_1^2 + y_2^2 + y_3^2 + x_1y_1y_2 + x_2y_1y_3 + x_3y_2y_3\}.
\end{aligned}$$

(b) In the cases $S(\underline{x})$ with $\underline{x} \in \mathbb{Z}^3_{\leq 0}$ and $r(\underline{x}) > 4$ and in the reducible cases $S(A_1^3)$, $S(A_2A_1)$, $S(\mathbb{P}^1 A_1)$, all relations in $\Gamma^{(0)}$ are generated by the relations in (6.1)–(6.4). *Especially*

$$\begin{aligned}
\Gamma^{(0)}(A_1^3) &\cong \Gamma^{(0)}(A_1) \times \Gamma^{(0)}(A_1) \times \Gamma^{(0)}(A_1) \cong (G^{fCox,1})^3 \cong \{\pm 1\}^3, \\
\Gamma^{(0)}(A_2A_1) &\cong \Gamma^{(0)}(A_2) \times \Gamma^{(0)}(A_1) \cong D_6 \times \{\pm 1\} \cong S_3 \times \{\pm 1\}, \\
\Gamma^{(0)}(\mathbb{P}^1 A_1) &\cong \Gamma^{(0)}(\mathbb{P}^1) \times \Gamma^{(0)}(A_1) \cong G^{fCox,2} \times \{\pm 1\}.
\end{aligned}$$

(c) In the case $S(A_3)$ the group $\Gamma^{(0)}$ is the Weyl group of the root system A_3, so $\Gamma^{(0)} = \ker \tau^{(0)} = O^{(0),*} \cong S_4$.

(d) In the case $S(\widehat{A}_2)$ the group $\Gamma^{(0)}$ is the Weyl group of the affine root system $\widehat{A}_2$. More concretely, the following holds.

$$\begin{aligned}
\mathrm{Rad}\, I^{(0)} &= \mathbb{Z} f_1 \text{ with } f_1 = e_1 + e_2 + e_3, \\
\overline{H_{\mathbb{Z}}}^{(0)} &= \mathbb{Z}\overline{e_1}^{(0)} \oplus \mathbb{Z}\overline{e_2}^{(0)}, \\
\Gamma_u^{(0)} &= \left(\ker \tau^{(0)}\right)_u = T(\overline{j}^{(0)}(\overline{H_{\mathbb{Z}}}^{(0)}) \otimes \mathbb{Z} f_1) \\
&= \langle T(\overline{j}^{(0)}(\overline{e_1}^{(0)}) \otimes f_1), T(\overline{j}^{(0)}(\overline{e_2}^{(0)}) \otimes f_1)\rangle \cong \mathbb{Z}^2 \quad \text{with} \\
&\quad T(\overline{j}^{(0)}(\overline{e_1}^{(0)}) \otimes f_1)(\underline{e}) = \underline{e} + f_1(2, -1, -1),
\end{aligned}$$

$$
\begin{aligned}
&\qquad T(\overline{j}^{(0)}(\overline{e_2}^{(0)})\otimes f_1)(\underline{e}) = \underline{e} + f_1(-1,2,-1),\\
\Gamma_u^{(0)} &= \big(\ker\tau^{(0)}\big)_u \overset{1:3}{\subset} O_u^{(0),Rad} = T(\overline{H_{\mathbb{Z}}}^{(0),\sharp}\otimes\mathbb{Z}f_1),\\
\Gamma_s^{(0)} &= (\ker\tau^{(0)})_s \cong \Gamma^{(0)}(A_2) \cong D_6 \cong S_3,\\
\Gamma_s^{(0)} &\overset{1:2}{\subset} O_s^{(0),Rad} = \mathrm{Aut}(\overline{H_{\mathbb{Z}}}^{(0)}, \overline{I}^{(0)}) \cong D_{12},\\
\Gamma^{(0)} &= \ker\tau^{(0)} = O^{(0),*} \overset{1:6}{\subset} O^{(0),Rad}.
\end{aligned}
$$

The exact sequence (6.5) *splits non-canonically with* $\Gamma_s^{(0)} \cong \langle s_{e_1}^{(0)}, s_{e_2}^{(0)}\rangle \subset \Gamma^{(0)}$ *(for example).*

(e) The case $S(\mathcal{H}_{1,2})$*: The following holds.*

$$
\begin{aligned}
H_{\mathbb{Z}} &= \mathbb{Z}f_3 \oplus \mathrm{Rad}\, I^{(0)} \text{ with } f_3 = e_1+e_2+e_3,\\
\overline{H_{\mathbb{Z}}}^{(0)} &= \mathbb{Z}\overline{f_3}^{(0)},\\
\mathrm{Rad}\, I^{(0)} &= \mathbb{Z}f_1 \oplus \mathbb{Z}f_2 \text{ with } f_1 = e_1+e_2,\ f_2 = e_2+e_3,\\
\Gamma_u^{(0)} &= (\ker\tau^{(0)})_u = T(\overline{j}^{(0)}(\overline{H_{\mathbb{Z}}}^{(0)})\otimes \mathrm{Rad}\, I^{(0)})\\
&= \langle T(\overline{j}^{(0)}(\overline{f_3}^{(0)})\otimes f_1), T(\overline{j}^{(0)}(\overline{f_3}^{(0)})\otimes f_2)\rangle \cong \mathbb{Z}^2 \quad\text{with}\\
&\qquad T(\overline{j}^{(0)}(\overline{f_3}^{(0)})\otimes f_1)(f_1,f_2,f_3) = (f_1,f_2,f_3+2f_1),\\
&\qquad T(\overline{j}^{(0)}(\overline{f_3}^{(0)})\otimes f_2)(f_1,f_2,f_3) = (f_1,f_2,f_3+2f_2),\\
\Gamma_u^{(0)} &\overset{1:4}{\subset} O_u^{(0),\mathrm{Rad}} = T(\overline{H_{\mathbb{Z}}}^{(0),\sharp}\otimes \mathrm{Rad}\, I^{(0)}),\\
\Gamma_s^{(0)} &= (\ker\tau^{(0)})_s = O_s^{(0),Rad} \cong \Gamma^{(0)}(A_1) \cong \{\pm 1\},\\
\Gamma^{(0)} &= \ker\tau^{(0)} = O^{(0),*} \overset{1:4}{\subset} O^{(0),Rad}.
\end{aligned}
$$

The exact sequence (6.5) *splits non-canonically with* $\Gamma_s^{(0)} \cong \langle -M\rangle \subset \Gamma^{(0)}$ *and* $-M(f_1,f_2,f_3) = (f_1,f_2,-f_3)$*. Therefore*

$$
\Gamma^{(0)} = \{(f_1,f_2,f_3) \mapsto (f_1,f_2,\varepsilon f_3 + 2\beta_1 f_1 + 2\beta_2 f_2) \,|\, \varepsilon\in\{\pm 1\}, \beta_1,\beta_2\in\mathbb{Z}\}.
$$

(f) The cases $S(-l,2,-l)$ *with* $l \geq 3$*: The following holds.*

$$
\begin{aligned}
\mathrm{Rad}\, I^{(0)} &= \mathbb{Z}f_1 \text{ with } f_1 = e_1 - e_3,\\
\overline{H_{\mathbb{Z}}}^{(0)} &= \mathbb{Z}\overline{e_1}^{(0)} \oplus \mathbb{Z}\overline{e_2}^{(0)},\\
&\qquad \overline{I}^{(0)}((\overline{e_1}^{(0)},\overline{e_2}^{(0)})^t, (\overline{e_1}^{(0)},\overline{e_2}^{(0)})) = \begin{pmatrix} 2 & -l\\ -l & 2\end{pmatrix},\\
\Gamma_u^{(0)} &= \langle T(\overline{j}^{(0)}(\overline{e_1}^{(0)})\otimes f_1), T(\overline{j}^{(0)}(l\overline{e_2}^{(0)})\otimes f_1)\rangle \cong \mathbb{Z}^2 \quad\text{with}\\
&\qquad T(\overline{j}^{(0)}(\overline{e_1}^{(0)})\otimes f_1)(\underline{e}) = \underline{e} + f_1(2,-l,2),\\
&\qquad T(\overline{j}^{(0)}(l\overline{e_2}^{(0)})\otimes f_1)(\underline{e}) = \underline{e} + f_1(-l^2,2l,-l^2),
\end{aligned}
$$

$$\Gamma_u^{(0)} \overset{1:l}{\subset} (\ker \tau^{(0)})_u \overset{1:(l^2-4)}{\subset} O_u^{(0),Rad} \cong \mathbb{Z}^2,$$
$$\Gamma_s^{(0)} \cong \Gamma^{(0)}(S(-l)) \cong G^{fCox,2},$$
$$\Gamma_s^{(0)} = (\ker \tau^{(0)})_s \cap \ker \overline{\sigma} \overset{1:4}{\subset} O_s^{(0),Rad},$$
$$\Gamma^{(0)} \overset{1:l}{\subset} O^{(0),*} \overset{1:4(l^2-4)}{\subset} O^{(0),Rad}.$$

The exact sequence (6.5) *splits non-canonically with* $\Gamma_s^{(0)} \cong \langle s_{e_1}^{(0)}, s_{e_2}^{(0)}\rangle \subset \Gamma^{(0)}$ *(for example).*

(g) The case $S(\mathbb{P}^2)$ *and the cases* $S(\underline{x})$ *with* $\underline{x} \in \mathbb{Z}^3_{\geq 3}$, $r(\underline{x}) < 0$ *and* $x_i \leq \frac{1}{2}x_jx_k$ *for* $\{i,j,k\} = \{1,2,3\}$*:* $\Gamma^{(0)} \cong G^{fCox,3}$ *is a free Coxeter group with the three generators* $s_{e_1}^{(0)}, s_{e_2}^{(0)}, s_{e_3}^{(0)}$.

Proof (a) This follows from the definitions in Lemma 2.6 (a) and in Definition 2.8. Especially

$$R^{(0)} = \{\underline{e}\begin{pmatrix} y_1\\ y_2\\ y_3\end{pmatrix} \in H_{\mathbb{Z}} \,|\, 2 = I^{(0)}((\underline{e}\begin{pmatrix} y_1\\ y_2\\ y_3\end{pmatrix})^t, \underline{e}\begin{pmatrix} y_1\\ y_2\\ y_3\end{pmatrix})$$
$$= \{\underline{e}\begin{pmatrix} y_1\\ y_2\\ y_3\end{pmatrix} \in H_{\mathbb{Z}} \,|\, 2 = (y_1\ y_2\ y_3)\begin{pmatrix} 2 & x_1 & x_2\\ x_1 & 2 & x_3\\ x_2 & x_3 & 2\end{pmatrix}\begin{pmatrix} y_1\\ y_2\\ y_3\end{pmatrix}\}.$$

(b) First we consider the cases $S(\underline{x})$ with $\underline{x} \in \mathbb{Z}^3_{\leq 0}$ and $r(\underline{x}) > 4$. By Lemma 5.7 (b) sign $I^{(0)} = (++-)$. We will apply Theorem A.4 with $I^{[0]} := -I^{(0)}$, which has signature sign $I^{[0]} = (+--)$.

The vectors e_1, e_2, e_3 are negative with respect to $I^{[0]}$. By Theorem A.4 (c) (vi) the reflection $s_{e_i}^{(0)}$ acts on the model $\mathcal{K}/\mathbb{R}^*$ of the hyperbolic plane as reflection along the hyperbolic line $((\mathbb{R}e_i)^\perp \cap \mathcal{K})/\mathbb{R}^* \subset \mathcal{K}/\mathbb{R}^*$. The corresponding three planes $(\mathbb{R}e_1)^\perp$, $(\mathbb{R}e_2)^\perp$ and $(\mathbb{R}e_3)^\perp$ in $H_{\mathbb{R}}$ intersect pairwise in the following three lines

$$(\mathbb{R}e_1)^\perp \cap (\mathbb{R}e_2)^\perp = \mathbb{R}y^{[1]},\ \ y^{[1]} = (-2x_2 + x_1x_3, -2x_3 + x_1x_2, 4 - x_1^2),$$
$$(\mathbb{R}e_1)^\perp \cap (\mathbb{R}e_3)^\perp = \mathbb{R}y^{[2]},\ \ y^{[2]} = (-2x_1 + x_2x_3, 4 - x_2^2, -2x_3 + x_1x_2),$$
$$(\mathbb{R}e_2)^\perp \cap (\mathbb{R}e_3)^\perp = \mathbb{R}y^{[3]},\ \ y^{[3]} = (4 - x_3^2, -2x_1 + x_2x_3, -2x_2 + x_1x_3).$$

$\underline{x} \in \mathbb{Z}^3_{\leq 0}$ and $y^{[1]} = 0$ would imply $x_2 = x_3 = 0$, $x_1 = -2$, $r(\underline{x}) = 4$. But $r(\underline{x}) > 4$ by assumption. Therefore $y^{[1]} \neq 0$. Analogously $y^{[2]} \neq 0$ and $y^{[3]} \neq 0$.

One calculates

$$I^{[0]}(y^{[i]}, y^{[i]}) = 2(4 - x_i^2)(r(\underline{x}) - 4) \quad \text{for } i \in \{1,2,3\}$$
$$\begin{cases} \leq 0 & \text{for } x_i \leq -2,\\ > 0 & \text{for } x_i \in \{0,-1\}.\end{cases}$$

Therefore two of the three hyperbolic lines $((\mathbb{R}e_j)^\perp \cap \mathcal{K})/\mathbb{R}^*$ $(j \in \{1,2,3\})$ intersect in $\mathcal{K}/\mathbb{R}^*$ if and only if the corresponding x_i is 0 or -1.

Claim: *If* $x_i \in \{0,-1\}$ *then the angle between the two hyperbolic lines at the intersection point* $\mathbb{R}^* y^{[i]} \in \mathcal{K}/\mathbb{R}^*$ *is* $\frac{\pi}{2}$ if $x_i = 0$ *and* $\frac{\pi}{3}$ if $x_i = -1$.

We prove the claim in an indirect way. Observe in general

$$x_1 = 0 \Rightarrow \big(s_{e_1}^{(0),mat} s_{e_2}^{(0),mat}\big)^2 = \left(\begin{pmatrix} -1 & 0 & -x_2 \\ 0 & 1 & 0 \\ 0 & 0 & 1 \end{pmatrix} \begin{pmatrix} 1 & 0 & 0 \\ 0 & -1 & -x_3 \\ 0 & 0 & 1 \end{pmatrix}\right)^2$$
$$= \begin{pmatrix} -1 & 0 & -x_2 \\ 0 & -1 & -x_3 \\ 0 & 0 & 1 \end{pmatrix}^2 = E_3, \tag{6.6}$$

$$x_1 = -1 \Rightarrow \big(s_{e_1}^{(0),mat} s_{e_2}^{(0),mat}\big)^3 = \left(\begin{pmatrix} -1 & 1 & -x_2 \\ 0 & 1 & 0 \\ 0 & 0 & 1 \end{pmatrix} \begin{pmatrix} 1 & 0 & 0 \\ 0 & -1 & -x_3 \\ 0 & 0 & 1 \end{pmatrix}\right)^3$$
$$= \begin{pmatrix} -1 & -1 & -x_2-x_3 \\ 0 & -1 & -x_3 \\ 0 & 0 & 1 \end{pmatrix}^3 = E_3, \tag{6.7}$$

and analogously for x_2 and x_3. Therefore the angle between the hyperbolic lines must be $\frac{\pi}{2}$ if $x_i = 0$ and $\frac{\pi}{3}$ or $\frac{2\pi}{3}$ if $x_i = -1$. But in the case $x_1 = x_2 = x_3 = -1$ the three intersection points are the vertices of a hyperbolic triangle, so then the angles are all $\frac{\pi}{3}$. Deforming x_2 and x_3 does not change the angle at $\mathbb{R}^* y^{[1]}$, so it is $\frac{\pi}{3}$ if $x_1 = -1$. This proves the Claim. (□)

Now Theorem A.2 (a) shows that in the group of automorphisms of $\mathcal{K}/\mathbb{R}^*$ which is induced by $\Gamma^{(0)}$ all relations are generated by the relations in (6.1)–(6.4). Therefore this holds also for $\Gamma^{(0)}$ itself.

Now we consider the three reducible cases $S(A_1^3)$, $S(A_2A_1)$ and $S(\mathbb{P}^1A_1)$. Lemma 2.11 gives the first isomorphisms in part (b) for $\Gamma^{(0)}(A_1^3)$, $\Gamma^{(0)}(A_2A_1)$ and $\Gamma^{(0)}(\mathbb{P}^1A_1)$. Lemma 2.12 (for A_1) and Theorem 6.8 (b) and (c) give the second isomorphisms in part (b). The isomorphisms show in these three cases that all relations in $\Gamma^{(0)}$ are generated by the relations in (6.1)–(6.4).

(c) It is classical that in the case of the A_3 root lattice the monodromy group $\Gamma^{(0)}$ is the Weyl group and is $\ker \tau^{(0)} \cong S_4$.

(d) The proof of Theorem 5.14 (b) (iii) shows

$$\begin{aligned} \operatorname{Rad} I^{(0)} &= \ker \Phi_2(M) = \ker \Phi_1(M^{root}) = \mathbb{Z}f_1, \\ \overline{H_{\mathbb{Z}}}^{(0)} &= \mathbb{Z}\overline{e_1}^{(0)} \oplus \mathbb{Z}\overline{e_2}^{(0)}, \\ \overline{I}^{(0)}&((\overline{e_1}^{(0)}, \overline{e_2}^{(0)})^t, (\overline{e_1}^{(0)}, \overline{e_2}^{(0)})) = \begin{pmatrix} 2 & -1 \\ -1 & 2 \end{pmatrix}, \end{aligned}$$

so $(\overline{H_{\mathbb{Z}}}^{(0)}, \overline{I}^{(0)})$ is an A_2 root lattice. This was treated in Theorem 6.8 (b). We have

$$\begin{aligned} O_s^{(0),Rad} &= \operatorname{Aut}(\overline{H_{\mathbb{Z}}}^{(0)}, \overline{I}^{(0)}) \cong D_{12}, \\ \overline{e_3}^{(0)} &= -\overline{e_1}^{(0)} - \overline{e_2}^{(0)}, \\ R^{(0)}(\overline{H_{\mathbb{Z}}}^{(0)}, \overline{I}^{(0)}) &= \{\pm\overline{e_1}^{(0)}, \pm\overline{e_2}^{(0)}, \pm\overline{e_3}^{(0)}\}, \\ \overline{\Gamma^{(0)}} = \Gamma_s^{(0)} &= \langle \overline{s_{e_1}^{(0)}}, \overline{s_{e_2}^{(0)}} \rangle \cong \Gamma^{(0)}(A_2) \cong D_6 \cong S_3, \\ \Gamma_s^{(0)} &= (\ker \tau^{(0)})_s \overset{1:2}{\subset} O_s^{(0),Rad}. \end{aligned}$$

Observe also

$$\begin{aligned} & s_{e_2}^{(0)} s_{e_1}^{(0)} s_{e_2}^{(0)} s_{e_3}^{(0)}(\underline{e}) \\ &= \underline{e} \begin{pmatrix} 1 & 0 & 0 \\ 1 & -1 & 1 \\ 0 & 0 & 1 \end{pmatrix} \begin{pmatrix} -1 & 1 & 1 \\ 0 & 1 & 0 \\ 0 & 0 & 1 \end{pmatrix} \begin{pmatrix} 1 & 0 & 0 \\ 1 & -1 & 1 \\ 0 & 0 & 1 \end{pmatrix} \begin{pmatrix} 1 & 0 & 0 \\ 0 & 1 & 0 \\ 1 & 1 & -1 \end{pmatrix} \\ &= \underline{e} + f_1(1, 1, -2) = T(\overline{j}^{(0)}(-\overline{e_3}^{(0)}) \otimes f_1)(\underline{e}), \end{aligned}$$

so $T(j^{(0)}(-\overline{e_3}^{(0)}) \otimes f_1) \in \Gamma_u^{(0)}$. Compare Lemma 6.2 (f) and recall that $\Gamma_s^{(0)}$ acts transitively on $\{\pm\overline{e_1}^{(0)}, \pm\overline{e_2}^{(0)}, \pm\overline{e_3}^{(0)}\}$. Therefore $T(\overline{j}^{(0)}(\overline{e_j}^{(0)}) \otimes f_1) \in \Gamma_u^{(0)}$ for $j \in \{1, 2, 3\}$ with

$$\begin{aligned} T(\overline{j}^{(0)}(\overline{e_1}^{(0)}) \otimes f_1)(\underline{e}) &= \underline{e} + f_1(2, -1, -1), \\ T(\overline{j}^{(0)}(\overline{e_2}^{(0)}) \otimes f_1)(\underline{e}) &= \underline{e} + f_1(-1, 2, -1), \end{aligned}$$

so

$$\begin{aligned} \Gamma_u^{(0)} &= T(\overline{j}^{(0)}(\overline{H_{\mathbb{Z}}}^{(0)}) \otimes f_1) = (\ker \tau^{(0)})_u \\ &\subset T(\overline{H_{\mathbb{Z}}}^{(0),\sharp} \otimes f_1) = O_u^{(0),Rad} \cong \mathbb{Z}^2. \end{aligned}$$

$T(\overline{H_{\mathbb{Z}}}^{(0),\sharp} \otimes f_1)$ is generated by

$$(\underline{e} \mapsto \underline{e} + f_1(1, -1, 0)) \quad \text{and} \quad (\underline{e} \mapsto \underline{e} + f_1(0, 1, -1)).$$

Therefore $\Gamma_u^{(0)} \overset{1:3}{\subset} O_u^{(0),Rad}$.

Together $\Gamma_s^{(0)} = (\ker \tau^{(0)})_s \overset{1:2}{\subset} O_s^{(0),Rad}$ and $\Gamma_u^{(0)} = (\ker \tau^{(0)})_u \overset{1:3}{\subset} O_u^{(0),Rad}$ show

$$\Gamma^{(0)} = \ker \tau^{(0)} \overset{1:6}{\subset} O^{(0),Rad}$$

(6.7) and $x_1 = -1$ show $\langle s_{e_1}^{(0)}, s_{e_2}^{(0)} \rangle \cong D_6$, so $\Gamma_s^{(0)} \cong \langle s_{e_1}^{(0)}, s_{e_2}^{(0)} \rangle \subset \Gamma^{(0)}$, so the exact sequence (6.5) splits non-canonically.

(e) Recall from the proof of Theorem 5.14 (a) (i) that

$$\underline{f} = (f_1, f_2, f_3) := \underline{e}\begin{pmatrix}1&0&1\\1&1&1\\0&1&1\end{pmatrix}$$

is a $\mathbb{Z}$-basis of $H_\mathbb{Z}$ and

$$\begin{aligned}\mathrm{Rad}\, I^{(0)} &= \mathbb{Z}f_1 \oplus \mathbb{Z}f_2,\\ H_\mathbb{Z} &= \mathbb{Z}f_3 \oplus \mathrm{Rad}\, I^{(0)},\\ \overline{H_\mathbb{Z}}^{(0)} &= \mathbb{Z}\overline{f_3}^{(0)}.\end{aligned}$$

Also observe

$$\begin{aligned}s_{e_j}^{(0)}|_{\mathrm{Rad}\, I^{(0)}} &= \mathrm{id} \quad \text{for } i \in \{1,2,3\},\\ s_{e_1}^{(0)}(f_3) &= -f_3 + 2f_2,\\ s_{e_2}^{(0)}(f_3) &= -f_3 + 2f_1 + 2f_2,\\ s_{e_3}^{(0)}(f_3) &= -f_3 + 2f_1,\\ \overline{\Gamma^{(0)}} = \Gamma_s^{(0)} &= \{\pm\,\mathrm{id}\} = (\ker \tau^{(0)})_s = O_s^{(0),Rad} \cong \Gamma^{(0)}(A_1) \cong \{\pm 1\}.\end{aligned}$$

Therefore

$$\begin{aligned}\Gamma_u^{(0)} \ni s_{e_1}^{(0)} s_{e_2}^{(0)} &= (\underline{f} \mapsto \underline{f} + 2f_1(0,0,1)) = T(\overline{j}^{(0)}(\overline{f_3}^{(0)}) \otimes f_1),\\ \Gamma_u^{(0)} \ni s_{e_3}^{(0)} s_{e_2}^{(0)} &= (\underline{f} \mapsto \underline{f} + 2f_2(0,0,1)) = T(\overline{j}^{(0)}(\overline{f_3}^{(0)}) \otimes f_2),\end{aligned}$$

so

$$\begin{aligned}\Gamma_u^{(0)} &= (\ker \tau^{(0)})_u = T(\overline{j}^{(0)}(\overline{H_\mathbb{Z}}^{(0)}) \otimes \mathrm{Rad}\, I^{(0)})\\ &\overset{1:4}{\subset} O^{(0),Rad} = T(\overline{H_\mathbb{Z}}^{(0),\sharp} \otimes \mathrm{Rad}\, I^{(0)})\\ &= \langle (\underline{f} \mapsto \underline{f} + f_1(0,0,1)), (\underline{f} \mapsto \underline{f} + f_2(0,0,1))\rangle \cong \mathbb{Z}^2.\end{aligned}$$

Together the statements on $\Gamma_s^{(0)}$ and $\Gamma_u^{(0)}$ imply

$$\Gamma^{(0)} = \ker \tau^{(0)} \overset{1:4}{\subset} O^{(0),Rad}.$$

The exact sequence (6.5) splits non-canonically with $\Gamma_s^{(0)} = \{\pm\mathrm{id}\} \cong \langle -\mathrm{M}\rangle \subset \Gamma^{(0)}$ (for example).

(f) Recall from the proof of Theorem 5.14 (a) (ii) and (b) (iv)

$$\mathrm{Rad}\, I^{(0)} = \mathbb{Z}f_1 \quad \text{with} \quad f_1 = e_1 - e_3, \quad \text{so}$$

$$\overline{H_{\mathbb{Z}}}^{(0)} = \mathbb{Z}\overline{e_1}^{(0)} \oplus \mathbb{Z}\overline{e_2}^{(0)},$$

$$\overline{I}^{(0)}((\overline{e_1}^{(0)}, \overline{e_2}^{(0)})^t, (\overline{e_1}^{(0)}, \overline{e_2}^{(0)})) = \begin{pmatrix} 2 & -l \\ -l & 2 \end{pmatrix}.$$

Observe

$$s_{e_1}^{(0),mat} = \begin{pmatrix} -1 & l & -2 \\ 0 & 1 & 0 \\ 0 & 0 & 1 \end{pmatrix}, \quad s_{e_2}^{(0),mat} = \begin{pmatrix} 1 & 0 & 0 \\ l & -1 & l \\ 0 & 0 & 1 \end{pmatrix},$$

$$s_{e_3}^{(0),mat} = \begin{pmatrix} 1 & 0 & 0 \\ 0 & 1 & 0 \\ -2 & l & -1 \end{pmatrix},$$

$$\overline{s_{e_1}^{(0)}}(\overline{e_1}^{(0)}, \overline{e_2}^{(0)}) = \overline{s_{e_3}^{(0)}}(\overline{e_1}^{(0)}, \overline{e_2}^{(0)}) = (\overline{e_1}^{(0)}, \overline{e_2}^{(0)}) \begin{pmatrix} -1 & l \\ 0 & 1 \end{pmatrix},$$

$$\overline{s_{e_2}^{(0)}}(\overline{e_1}^{(0)}, \overline{e_2}^{(0)}) = \begin{pmatrix} 1 & 0 \\ l & -1 \end{pmatrix}.$$

Theorem 6.8 (d) shows

$$\overline{\Gamma^{(0)}} = \Gamma_s^{(0)} \cong \Gamma^{(0)}(S(-l)) \cong G^{fCox,2}.$$

Therefore with respect to the generators $\overline{s_{e_1}^{(0)}}, \overline{s_{e_2}^{(0)}}$ and $\overline{s_{e_3}^{(0)}}$, all relations in $\Gamma_s^{(0)}$ are generated by the relations

$$(\overline{s_{e_1}^{(0)}})^2 = (\overline{s_{e_2}^{(0)}})^2 = \overline{s_{e_1}^{(0)}}\,\overline{s_{e_3}^{(0)}} = \mathrm{id}.$$

Therefore $\Gamma_u^{(0)}$ is generated by the set $\{g s_{e_1}^{(0)} s_{e_3}^{(0)} g^{-1} \,|\, g \in \Gamma^{(0)}\}$ of conjugates of $s_{e_1}^{(0)} s_{e_3}^{(0)}$. Observe

$$s_{e_1}^{(0)} s_{e_3}^{(0)}(\underline{e}) = \underline{e} \begin{pmatrix} -1 & l & -2 \\ 0 & 1 & 0 \\ 0 & 0 & 1 \end{pmatrix} \begin{pmatrix} 1 & 0 & 0 \\ 0 & 1 & 0 \\ -2 & l & -1 \end{pmatrix}$$

$$= \underline{e} \begin{pmatrix} 3 & -l & 2 \\ 0 & 1 & 0 \\ -2 & l & -1 \end{pmatrix} = \underline{e} + f_1(2, -l, 2),$$

$$\text{so } s_{e_1}^{(0)} s_{e_3}^{(0)} = T(\overline{j}^{(0)}(\overline{e_1}^{(0)}) \otimes f_1),$$

and recall Lemma 6.2 (f). The $\mathbb{Z}$-lattice generated by the $\Gamma^{(0)}$ orbit of e_1 is $\mathbb{Z}e_1 \oplus \mathbb{Z}le_2 \oplus \mathbb{Z}\gcd(2, l)e_3$. Therefore

$$\Gamma_u^{(0)} = \langle T(\overline{j}^{(0)}(\overline{e_1}^{(0)}) \otimes f_1), T(\overline{j}^{(0)}(l\overline{e_2}^{(0)}) \otimes f_1)\rangle \cong \mathbb{Z}^2 \text{ with}$$

$$T(\overline{j}^{(0)}(l\overline{e_2}^{(0)})) \otimes f_1)(\underline{e}) = \underline{e} + f_1(-l^2, 2l, -l^2).$$

Compare

$$\begin{aligned}
(\ker \tau^{(0)})_u &= T(\overline{j}^{(0)}(\overline{H_\mathbb{Z}}^{(0)}) \otimes f_1) \\
&= \langle T(\overline{j}^{(0)}(\overline{e_1}^{(0)}) \otimes f_1), T(\overline{j}^{(0)}(\overline{e_2}^{(0)}) \otimes f_1)\rangle, \\
O^{(0),Rad} &= T(\overline{H_\mathbb{Z}}^{(0),\sharp} \otimes f_1) \\
&= \langle (\underline{e} \mapsto \underline{e} + f_1(1,0,1)), (\underline{e} \mapsto \underline{e} + f_1(0,1,0))\rangle.
\end{aligned}$$

Therefore

$$\Gamma_u^{(0)} \overset{1:l}{\subset} (\ker \tau^{(0)})_u \overset{1:(l^2-4)}{\subset} O_u^{(0),Rad} \cong \mathbb{Z}^2.$$

Theorem 6.8 (e) shows also

$$\Gamma_s^{(0)} = (\ker \tau^{(0)})_s \cap \ker \overline{\sigma} \overset{1:4}{\subset} O_s^{(0),Rad}.$$

Therefore

$$\Gamma^{(0)} \overset{1:l}{\subset} O^{(0),*} \overset{1:4(l^2-4)}{\subset} O^{(0),Rad}.$$

(g) By Lemma 5.7 (b) sign $I^{(0)} = (+--)$. We will apply Theorem A.4 with $I^{[0]} = I^{(0)}$ and Theorem A.2 (b). The vectors e_1, e_2 and e_3 are positive. By Theorem A.4 (c) (vii) the reflection $s_{e_i}^{(0)}$ acts on the model $\mathcal{K}/\mathbb{R}^*$ of the hyperbolic plane as an elliptic element of order 2 with fixed point $\mathbb{R}^* e_i \in \mathcal{K}/\mathbb{R}^*$.

Consider the three vectors $v_1, v_2, v_3 \in H_\mathbb{Z} \subset H_\mathbb{R}$

$$\begin{aligned}
v_1 &:= -x_3 e_1 + x_2 e_2 + x_1 e_3, \\
v_2 &:= x_3 e_1 - x_2 e_2 + x_1 e_3, \\
v_3 &:= x_3 e_1 + x_2 e_2 - x_1 e_3,
\end{aligned}$$

and observe

$$\begin{aligned}
v_1 + v_2 = 2x_1 e_3,\ v_1 + v_3 = 2x_2 e_2,\ v_2 + v_3 = 2x_3 e_1, \\
I^{(0)}(v_i, v_i) = r(\underline{x}) \le 0.
\end{aligned}$$

The three planes $\mathbb{R}v_1 \oplus \mathbb{R}v_2$, $\mathbb{R}v_1 \oplus \mathbb{R}v_3$ and $\mathbb{R}v_2 \oplus \mathbb{R}v_3$ contain the lines $\mathbb{R}e_3$, $\mathbb{R}e_2$ respectively $\mathbb{R}e_1$. Two of the three planes intersect in one of the lines $\mathbb{R}v_1$, $\mathbb{R}v_2$ and $\mathbb{R}v_3$, and these three lines do not meet $\mathcal{K}$. Therefore the three hyperbolic lines $((\mathbb{R}v_1 \oplus \mathbb{R}v_2) \cap \mathcal{K})/\mathbb{R}^*$, $((\mathbb{R}v_1 \oplus \mathbb{R}v_3) \cap \mathcal{K})/\mathbb{R}^*$ and $((\mathbb{R}v_2 \oplus \mathbb{R}v_3) \cap \mathcal{K})/\mathbb{R}^*$ in $\mathcal{K}/\mathbb{R}^*$ contain the points $\mathbb{R}^* e_3$, $\mathbb{R}^* e_2$ respectively $\mathbb{R}^* e_1$ and do not meet.

Now Theorem A.2 (b) shows that the group of automorphisms of $\mathcal{K}/\mathbb{R}^*$ which is induced by $\Gamma^{(0)}$ is isomorphic to $G^{fCox,3}$. Therefore also $\Gamma^{(0)}$ itself is isomorphic to $G^{fCox,3}$. □

Remarks 6.12 (i) In part (g) of Theorem 6.12 we have less informations than in the other cases. We do not even know in which cases in part (g) $\Gamma^{(0)} = O^{(0),*}$ respectively $\Gamma^{(0)} \subsetneqq O^{(0),*}$ holds.

(ii) In the case of $S(\mathbb{P}^2)$, the proof of Theorem 6.11(g) gave three hyperbolic lines in the model $\mathcal{K}/\mathbb{R}^*$ which form a degenerate hyperbolic triangle, so with vertices on the euclidean boundary of the hyperbolic plane. These vertices are the lines $\mathbb{R}^* v_1$, $\mathbb{R}^* v_2$, $\mathbb{R}^* v_3$, which are isotropic in the case of $S(\mathbb{P}^2)$ because there $I^{(0)}(v_i, v_i) = r(\underline{x}) = 0$. The reflections $s_{e_1}^{(0)}, s_{e_2}^{(0)}, s_{e_3}^{(0)}$ act as elliptic elements of order 2 with fixed points on these hyperbolic lines. Therefore the degenerate hyperbolic triangle is a fundamental domain of this action of $\Gamma^{(0)}$.

(iii) Milanov [Mi19, 4.1] had a different point of view on $\Gamma^{(0)}$ in the case of $S(\mathbb{P}^2)$. He gave an isomorphism $\Gamma^{(0)} \cong U$ to a certain subgroup U of index 3 in $PSL_2(\mathbb{Z})$. First we describe U in (iv), then we present our way to see this isomorphism in (v).

(iv) The class in $PSL_2(\mathbb{Z})$ of a matrix $A \in SL_2(\mathbb{Z})$ is denoted by $[A]$. It is well known that there is an isomorphism of the free product of $\mathbb{Z}/2\mathbb{Z}$ and $\mathbb{Z}/3\mathbb{Z}$ with $PSL_2(\mathbb{Z})$,

$$\langle \alpha \,|\, \alpha^2 = e \rangle * \langle \beta \,|\, \beta^3 = e \rangle \to PSL_2(\mathbb{Z}),$$
$$\alpha \mapsto [\begin{pmatrix} 0 & -1 \\ 1 & 0 \end{pmatrix}],\ \beta \mapsto [\begin{pmatrix} -1 & -1 \\ 1 & 0 \end{pmatrix}].$$

Consider the character

$$\chi : \langle \alpha \,|\, \alpha^2 = e \rangle * \langle \beta \,|\, \beta^3 = e \rangle \to \{1, e^{2\pi i/3}, e^{2\pi i 2/3}\},\ \alpha \mapsto 1,\ \beta \mapsto e^{2\pi i/3},$$

and the corresponding character $\widetilde{\chi}$ on $PSL_2(\mathbb{Z})$. Then

$$\begin{aligned} U &= \ker \widetilde{\chi} \overset{1:3}{\subset} PSL_2(\mathbb{Z}), \\ \ker \chi &= \langle \alpha, \beta\alpha\beta^2, \beta^2\alpha\beta \rangle \\ \text{with} &\quad \alpha \simeq [F_1], \beta\alpha\beta^2 \simeq [F_2], \beta^2\alpha\beta \simeq [F_3] \text{ and} \\ F_1 &= \begin{pmatrix} 0 & -1 \\ 1 & 0 \end{pmatrix},\quad F_2 = \begin{pmatrix} -1 & -2 \\ 1 & 1 \end{pmatrix},\quad F_3 = \begin{pmatrix} -1 & -1 \\ 2 & 1 \end{pmatrix}. \end{aligned}$$

It is easy to see $\ker \chi \cong G^{fCox,3}$, with the three generators $\alpha, \beta\alpha\beta^2, \beta^2\alpha\beta$. It is well known and easy to see that

$$\langle F_1, F_2, F_3 \rangle = \{\begin{pmatrix} a & b \\ c & d \end{pmatrix} \in SL_2(\mathbb{Z}) \,|\, a \equiv d \equiv 0 \bmod 3 \text{ or } b \equiv c \bmod 3\}.$$

The Möbius transformations $\mu(F_1), \mu(F_2), \mu(F_3)$ are elliptic of order 2 with fixed points $z_1 = i, z_2 = -1 + i, z_3 = -\frac{1}{2} + \frac{1}{2}i$.

The hyperbolic lines $\widetilde{l_1} := A(\infty, 0)$ $\widetilde{l_2} := A(-1, \infty)$, $\widetilde{l_3} := A(0, -1)$ (notations from the Remarks and Notations A.1 (iii)) form a degenerate hyperbolic triangle, and $\widetilde{l_i}$ contains z_i.

(v) Consider the matrices

$$B := \begin{pmatrix} z_1\overline{z_1} & -z_2\overline{z_2} & 2z_3\overline{z_3} \\ \mathrm{Re}(z_1) & -\mathrm{Re}(z_2) & 2\,\mathrm{Re}(z_3) \\ 1 & -1 & 2 \end{pmatrix} = \begin{pmatrix} 1 & -2 & 1 \\ 0 & 1 & -1 \\ 1 & -1 & 2 \end{pmatrix} \quad \text{and}$$

$$B^{-1} = \frac{1}{2}\begin{pmatrix} 1 & 3 & 1 \\ -1 & 1 & 1 \\ -1 & -1 & 1 \end{pmatrix}.$$

One checks

$$B^t \begin{pmatrix} 0 & 0 & 1 \\ 0 & -2 & 0 \\ 1 & 0 & 0 \end{pmatrix} B = \begin{pmatrix} 2 & -3 & 3 \\ -3 & 2 & -3 \\ 3 & -3 & 2 \end{pmatrix} = S(\mathbb{P}^2) + S(\mathbb{P}^2)^t,$$

$$B^{-1}\Theta(F_i)B = -s_{e_i}^{(0),mat} \quad \text{for } i \in \{1, 2, 3\}$$

(see Theorem A.4 (i) for Θ). The $\mathbb{Z}$-basis $\underline{e}$ of $H_\mathbb{Z}$ and the $\mathbb{R}$-basis $\underline{f}$ in Theorem A.4 of $H_\mathbb{R}$ are related by $\underline{e} = \underline{f} \cdot B$. The tuple (v_1, v_2, v_3) in the proof of Theorem 6.11 (g) is

$$\begin{aligned}(v_1, v_2, v_3) &= \underline{e}\begin{pmatrix} 3 & -3 & -3 \\ 3 & -3 & 3 \\ -3 & -3 & 3 \end{pmatrix} \\ &= \underline{f} \cdot B \cdot \begin{pmatrix} 3 & -3 & -3 \\ 3 & -3 & 3 \\ -3 & -3 & 3 \end{pmatrix} = \underline{f} \cdot (-6)\begin{pmatrix} 1 & 0 & 1 \\ -1 & 0 & 0 \\ 1 & 1 & 0 \end{pmatrix}.\end{aligned}$$

Finally observe that $\vartheta : \mathbb{H} \to \mathcal{K}/\mathbb{R}^*$ in Theorem A.4 extends to the euclidean boundary with

$$(\vartheta(-1), \vartheta(0), \vartheta(\infty)) = \mathbb{R}^* \cdot \underline{f} \cdot \begin{pmatrix} 1 & 0 & 1 \\ -1 & 0 & 0 \\ 1 & 1 & 0 \end{pmatrix}.$$

So the points $-1, 0, \infty$ are mapped to the points $\mathbb{R}^* v_1$, $\mathbb{R}^* v_2$, $\mathbb{R}^* v_3$. The groups $\Gamma^{(0),mat} = \langle s_{e_i}^{(0),mat} \mid i \in \{1, 2, 3\}\rangle$ and $\langle -s_{e_i}^{(0),mat} \mid i \in \{1, 2, 3\}\rangle$ are isomorphic because $\Gamma^{(0),mat}$ does not contain $-E_3$ because else it would not be a free Coxeter group with three generators.

Therefore the group $U = \langle [F_1], [F_2], [F_3] \rangle \subset PSL_2(\mathbb{Z})$ is isomorphic to the group $\langle B^{-1}\Theta(F_i)B \mid i \in \{1,2,3\}\rangle = \langle -s_{e_i}^{(0),mat} \mid i \in \{1,2,3\}\rangle$ and to the groups $\Gamma^{(0),mat}$ and $\Gamma^{(0)}$.

Now we turn to the study of the set $R^{(0)}$ of L1-cycles and the subset $\Delta^{(0)} \subset R^{(0)}$ of vanishing cycles. For the set $R^{(0)}$ Theorem 6.11 (a) gave the general formula $R^{(0)} = \{y_1e_1 + y_2e_2 + y_3e_3 \in H_{\mathbb{Z}} \mid 1 = Q_3(y_1, y_2, y_3)\}$ with the quadratic form

$$Q_3 : \mathbb{Z}^3 \to \mathbb{Z}, \quad (y_1, y_2, y_3) \mapsto y_1^2 + y_2^2 + y_3^2 + x_1y_1y_2 + x_2y_1y_3 + x_3y_2y_3.$$

It gave also a good control on $\Gamma^{(0)}$ for all cases $S(\underline{x})$ with $\underline{x} \in \mathbb{Z}^3$.

With respect to $\Delta^{(0)}$ and $R^{(0)}$ we know less. We have a good control on them for the cases with $r(\underline{x}) \in \{0, 1, 2, 4\}$ and the reducible cases, but not for all other cases. Theorem 6.14 treats all cases except those in the Remarks 6.13 (ii).

Remarks 6.13 (i) The cases $S(\mathcal{H}_{1,2})$, $S(-l, 2, -l)$ for $l \geq 3$ and the four cases $S(3, 3, 4)$, $S(4, 4, 4)$, $S(5, 5, 5)$, $S(4, 4, 8)$ (more precisely their $\mathrm{Br}_3 \ltimes \{\pm 1\}^3$ orbits) are the only cases in rank 3 where we know $\Delta^{(0)} \subsetneq R^{(0)}$.

(ii) We do not know whether $\Delta^{(0)} = R^{(0)}$ or $\Delta^{(0)} \subsetneq R^{(0)}$ in the following cases:

(a) All cases $S(\underline{x})$ with $r(\underline{x}) < 0$ except the four cases $S(3, 3, 4)$, $S(4, 4, 4)$, $S(5, 5, 5)$, $S(4, 4, 8)$. With the action of $\mathrm{Br}_3 \ltimes \{\pm 1\}^3$ and Theorem 4.6 (c) they can be reduced to the cases $S(\underline{x})$ with $\underline{x} \in \mathbb{Z}^3_{\geq 3}$, $r(\underline{x}) < 0$, $x_i \leq \frac{1}{2}x_jx_k$ for $\{i, j, k\} = \{1, 2, 3\}$.

(b) The irreducible cases $S(\underline{x})$ with $\underline{x} \in \mathbb{Z}^3_{\leq 0}$, $r(\underline{x}) > 4$ and $\underline{x} \notin \{0, -1, -2\}^3$.

Theorem 6.14 *(a) Consider the reducible cases (these include $S(A_1^3)$, $S(A_2A_1)$, $S(\mathbb{P}^1A_1)$). More precisely, suppose that $\underline{x} = (x_1, 0, 0)$. Then the tuple $(H_{\mathbb{Z}}, L, \underline{e})$ splits into the two tuples $(\mathbb{Z}e_1 \oplus \mathbb{Z}e_2, L|_{\mathbb{Z}e_1\oplus\mathbb{Z}e_2}, (e_1, e_2))$ and $(\mathbb{Z}e_3, L|_{\mathbb{Z}e_3}, e_3)$ with sets $\Delta_1^{(0)} = R_1^{(0)} \subset \mathbb{Z}e_1 \oplus \mathbb{Z}e_2$ and $\Delta_2^{(0)} = R_2^{(0)} = \{\pm e_3\} \subset \mathbb{Z}e_3$ of vanishing cycles and L1-cycles, and*

$$\Delta^{(0)} = \Delta_1^{(0)} \,\dot\cup\, \Delta_2^{(0)} = R^{(0)} = R_1^{(0)} \,\dot\cup\, R_2^{(0)}.$$

$\Delta_1^{(0)} = R_1^{(0)}$ is given in Theorem 5.8.

(b) Consider $S(\underline{x})$ with $\underline{x} \in \{0, -1, -2\}^3$ and $r(\underline{x}) > 4$. Then $\Delta^{(0)} = R^{(0)}$.

(c) The case $S(A_3)$ is classical. There

$$\Delta^{(0)} = R^{(0)} = \{\pm e_1, \pm e_2, \pm e_3, \pm(e_1 + e_2), \pm(e_2 + e_3), \pm(e_1 + e_2 + e_3)\}.$$

(d) The case $S(\widehat{A}_2)$: Recall $\mathrm{Rad}\, I^{(0)} = \mathbb{Z}f_1$ *with $f_1 = e_1 + e_2 + e_3$. There*

$$\begin{aligned}\Delta^{(0)} = R^{(0)} &= \Gamma^{(0)}\{e_1\} \\ &= (\pm e_1 + \mathbb{Z}f_1) \,\dot\cup\, (\pm e_2 + \mathbb{Z}f_1) \,\dot\cup\, (\pm e_3 + \mathbb{Z}f_1).\end{aligned}$$

(e) The case $S(\mathcal{H}_{1,2})$*: Recall* $(f_1, f_2, f_3) = \underline{e}\begin{pmatrix}1&0&1\\1&1&1\\0&1&1\end{pmatrix}$ *and* $\mathrm{Rad}\, I^{(0)} = \mathbb{Z}f_1 \oplus \mathbb{Z}f_2$*. The set of L1-cycles is*

$$R^{(0)} = \pm e_1 + \mathrm{Rad}\, I^{(0)} = (e_1 + \mathrm{Rad}\, I^{(0)}) \,\dot\cup\, (-e_1 + \mathrm{Rad}\, I^{(0)}),$$

with

$$e_1 + \mathrm{Rad}\, I^{(0)} = -e_2 + \mathrm{Rad}\, I^{(0)} = e_3 + \mathrm{Rad}\, I^{(0)} = f_3 + \mathrm{Rad}\, I^{(0)}.$$

It splits into the four $\Gamma^{(0)}$ *orbits*

$$\Gamma^{(0)}\{e_1\} = \pm e_1 + 2\,\mathrm{Rad}\, I^{(0)},\quad \Gamma^{(0)}\{e_2\} = \pm e_2 + 2\,\mathrm{Rad}\, I^{(0)},$$
$$\Gamma^{(0)}\{e_3\} = \pm e_3 + 2\,\mathrm{Rad}\, I^{(0)},\quad \Gamma^{(0)}\{f_3\} = \pm f_3 + 2\,\mathrm{Rad}\, I^{(0)}.$$

The set $\Delta^{(0)}$ *of vanishing cycles consists of the first three of these sets,*

$$\Delta^{(0)} = \Gamma^{(0)}\{e_1\} \,\dot\cup\, \Gamma^{(0)}\{e_2\} \,\dot\cup\, \Gamma^{(0)}\{e_3\},$$

so $\Delta^{(0)} \subsetneqq R^{(0)}$.

(f) The cases $S(-l, 2, -l)$ *with* $l \geq 3$*: Recall* $\mathrm{Rad}\, I^{(0)} = \mathbb{Z}f_1$ *with* $f_1 = e_1 - e_3$*. As the tuple* $(\overline{H_{\mathbb{Z}}}^{(0)}, \overline{I}^{(0)}, (\overline{e_1}^{(0)}, \overline{e_2}^{(0)}))$ *is isomorphic to the corresponding tuple from the* 2×2 *matrix* $S(-l) = \begin{pmatrix}1&-l\\0&1\end{pmatrix}$*, its sets of L1-cycles and its set of even vanishing cycles coincide because of Theorem 6.8. These sets are called* $R^{(0)}(S(-l))$*. Then*

$$R^{(0)} = \{\widetilde{y}_1 e_1 + \widetilde{y}_2 e_2 \in H_{\mathbb{Z}}^{(0)} \mid \widetilde{y}_1 \overline{e_1}^{(0)} + \widetilde{y}_2 \overline{e_2}^{(0)} \in R^{(0)}(S(-l))\} + \mathrm{Rad}\, I^{(0)}.$$

The cases with l *even:* $R^{(0)}$ *splits into the following* $l + 2$ $\Gamma^{(0)}$ *orbits,*

$$\Gamma^{(0)}\{e_1\},\quad \Gamma^{(0)}\{e_3\},\quad \Gamma^{(0)}\{e_2 + mf_1\} \text{ for } m \in \{0, 1, ..., l-1\}.$$

The set $\Delta^{(0)}$ *of vanishing cycles consists of the first three* $\Gamma^{(0)}$ *orbits,*

$$\Delta^{(0)} = \Gamma^{(0)}\{e_1\} \,\dot\cup\, \Gamma^{(0)}\{e_3\} \,\dot\cup\, \Gamma^{(0)}\{e_2\}.$$

The cases with l *odd:* $R^{(0)}$ *splits into the following* $l + 1$ $\Gamma^{(0)}$ *orbits,*

$$\Gamma^{(0)}\{e_1\} = \Gamma^{(0)}\{e_3\},\quad \Gamma^{(0)}\{e_2 + mf_1\} \text{ for } m \in \{0, 1, ..., l-1\}.$$

The set $\Delta^{(0)}$ *of vanishing cycles consists of the first two* $\Gamma^{(0)}$ *orbits,*

$$\Delta^{(0)} = \Gamma^{(0)}\{e_1\} \,\dot\cup\, \Gamma^{(0)}\{e_2\}.$$

In both cases $\Delta^{(0)} \subsetneqq R^{(0)}$.

(g) The case $S(\mathbb{P}^2)$. *Then* $\Delta^{(0)} = R^{(0)}$, *and* $R^{(0)}$ *splits into three* $\Gamma^{(0)}$ *orbits,*

$$R^{(0)} = \Gamma^{(0)}\{e_1\} \dot\cup \Gamma^{(0)}\{e_2\} \dot\cup \Gamma^{(0)}\{e_3\} \quad \text{with}$$
$$\Gamma^{(0)}\{e_i\} \subset \pm e_i + 3H_{\mathbb{Z}} \quad \text{for } i \in \{1, 2, 3\}$$

(but we would like to have a better control on $R^{(0)}$*).*

(h) The cases $S(3, 3, 4)$, $S(4, 4, 4)$, $S(5, 5, 5)$ *and* $S(4, 4, 8)$. *Then*

$$\Delta^{(0)} \subsetneqq R^{(0)}.$$

Proof (a) The splittings $\Delta^{(0)} = \Delta_1^{(0)} \dot\cup \Delta_2^{(0)}$ and $R^{(0)} = R_1^{(0)} \dot\cup R_2^{(0)}$ are part of Lemma 2.11. Lemma 2.12 gives for A_1 $\Delta_2^{(0)} = R_2^{(0)} = \{\pm e_3\}$. Theorem 6.8 gives for any rank 2 case $\Delta_1^{(0)} = R_1^{(0)}$.

(b) The cases where $(H_{\mathbb{Z}}, L, \underline{e})$ is reducible are covered by part (a). In any irreducible case, the bilinear lattice $(H_{\mathbb{Z}}, L, \underline{e})$ with triangular basis is *hyperbolic* in the sense of the definition before Theorem 3.12 in [HK16], because $I^{(0)}$ is indefinite, but the submatrices (2), $\begin{pmatrix} 2 & x_1 \\ x_1 & 2 \end{pmatrix}$, $\begin{pmatrix} 2 & x_2 \\ x_2 & 2 \end{pmatrix}$, $\begin{pmatrix} 2 & x_3 \\ x_3 & 2 \end{pmatrix}$ of the matrix $I^{(0)}(\underline{e}^t, \underline{e})$ are positive definite or positive semidefinite. Theorem 3.12 in [HK16] applies and gives $\Delta^{(0)} = R^{(0)}$.

(c) This is classical. It follows also with

$$\begin{aligned} Q_3(y_1, y_2, y_3) &= 2(y_1^2 + y_2^2 + y_3^2 - y_1 y_2 - y_2 y_3) \\ &= y_1^2 + (y_1 - y_2)^2 + (y_2 - y_3)^2 + y_3^2 \end{aligned}$$

and the transitivity of the action of $\Gamma^{(0)}$ on $R^{(0)}$.

(d) The quotient lattice $(\overline{H_{\mathbb{Z}}}^{(0)}, \overline{I}^{(0)})$ is an A_2 lattice with set of roots $\{\pm\overline{e_1}^{(0)}, \pm\overline{e_2}^{(0)}, \pm\overline{e_3}^{(0)}\}$. Therefore

$$R^{(0)} = (\pm e_1 + \mathbb{Z} f_1) \dot\cup (\pm e_2 + \mathbb{Z} f_1) \dot\cup (\pm e_3 + \mathbb{Z} f_1).$$

(One can prove this also using $2Q_3 = (y_1 - y_2)^2 + (y_1 - y_3)^2 + (y_2 - y_3)^2$.) $\Gamma_s^{(0)} \cong D_6$ acts transitively on the set $\{\pm\overline{e_1}^{(0)}, \pm\overline{e_2}^{(0)}, \pm\overline{e_3}^{(0)}\}$. The group $\Gamma_u^{(0)} \cong \mathbb{Z}^2$ contains the elements

$$(\underline{e} \mapsto \underline{e} + f_1(2, -1, -1)) \quad \text{and} \quad (\underline{e} \mapsto \underline{e} + f_1(-1, 2, -1)).$$

Therefore it acts transitively on each of the six sets $\varepsilon e_i + \mathbb{Z} f_1$ with $\varepsilon \in \{\pm 1\}$, $i \in \{1, 2, 3\}$. Thus $\Gamma^{(0)}$ acts transitively on $R^{(0)}$, so $\Delta^{(0)} = R^{(0)} = \Gamma^{(0)}\{e_1\}$.

(e) The quotient lattice $(\overline{H_{\mathbb{Z}}}^{(0)}, \overline{I}^{(0)})$ is an A_1 lattice with set of roots $\{\pm\overline{e_1}^{(0)}\}$. Therefore

$$R^{(0)} = \pm e_1 + \operatorname{Rad} I^{(0)} = (e_1 + \operatorname{Rad} I^{(0)}) \dot\cup (-e_1 + \operatorname{Rad} I^{(0)}).$$

(One can prove this also using $Q_3 = (y_1 - y_2 + y_3)^2$.) $s_{e_i}^{(0)}$ exchanges e_i and $-e_i$, and $s_{e_1}^{(0)}$ maps f_3 to $-f_3 + 2f_2$. $\Gamma_u^{(0)} \cong \mathbb{Z}^2$ is generated by the elements

$$((f_1, f_2, f_3) \mapsto (f_1, f_2, f_3 + 2f_1)) = (\underline{e} \mapsto \underline{e} + f_1(2, -2, 2)),$$
$$((f_1, f_2, f_3) \mapsto (f_1, f_2, f_3 + 2f_2)) = (\underline{e} \mapsto \underline{e} + f_2(2, -2, 2)).$$

Therefore $R^{(0)}$ splits into the four $\Gamma^{(0)}$ orbits

$$\Gamma^{(0)}\{e_1\} = \pm e_1 + 2\,\mathrm{Rad}\, I^{(0)}, \quad \Gamma^{(0)}\{e_2\} = \pm e_2 + 2\,\mathrm{Rad}\, I^{(0)},$$
$$\Gamma^{(0)}\{e_3\} = \pm e_3 + 2\,\mathrm{Rad}\, I^{(0)}, \quad \Gamma^{(0)}\{f_3\} = \pm f_3 + 2\,\mathrm{Rad}\, I^{(0)}.$$

$\Delta^{(0)}$ consists of the first three of them.

(f) The set of L1-cycles of the quotient lattice $(\overline{H_{\mathbb{Z}}}^{(0)}, \overline{I}^{(0)})$ is called $R^{(0)}(S(-l))$. Theorem 6.8 describes it. Therefore

$$R^{(0)} = \{\widetilde{y}_1 e_1 + \widetilde{y}_2 e_2 \in H_{\mathbb{Z}} \mid \widetilde{y}_1 \overline{e_1}^{(0)} + \widetilde{y}_2 \overline{e_2}^{(0)} \in R^{(0)}(S(-l))\} + \mathrm{Rad}\, I^{(0)}.$$

By Theorem 6.8 (d) (iv) and (c), $R^{(0)}(S(-l))$ splits into the two $\Gamma_s^{(0)}$ orbits $\Gamma_s^{(0)}\{\overline{e_1}^{(0)}\}$ and $\Gamma_s^{(0)}\{\overline{e_2}^{(0)}\}$, and the action of $\Gamma_s^{(0)}$ on each of these two orbits is simply transitive. $\Gamma_u^{(0)} \cong \mathbb{Z}^2$ is generated by the elements

$$(\underline{e} \mapsto \underline{e} + f_1(2, -l, 2)) \text{ and } (\underline{e} \mapsto \underline{e} + f_1(-l^2, 2l, -l^2)).$$

Therefore for $m \in \{0, 1, ..., l-1\}$

$$\Gamma^{(0)}\{e_2 + mf_1\} \cap (e_2 + \mathbb{Z} f_1) = e_2 + mf_1 + \mathbb{Z} l f_1.$$

If l is odd then $1 = \gcd(2, l^2)$ and

$$\Gamma^{(0)}\{e_1\} \supset e_1 + \mathbb{Z} f_1 \ni e_3 = e_1 - f_1, \quad \text{so } \Gamma^{(0)}\{e_1\} = \Gamma^{(0)}\{e_3\}.$$

If l is even then $2 = \gcd(2, l^2)$ and

$$\Gamma^{(0)}\{e_1\} \cap (e_1 + 2\mathbb{Z} f_1) = e_1 + \mathbb{Z} 2 f_1 \not\ni e_3, \quad \text{so } \Gamma^{(0)}\{e_1\} \cap \Gamma^{(0)}\{e_3\} = \emptyset.$$

This shows all claims.

(g) The matrices $s_{e_i}^{(0),mat} \in M_{3\times 3}(\mathbb{Z})$ with $s_{e_i}^{(0)}(\underline{e}) = \underline{e} \cdot s_{e_i}^{(0),mat}$ are

$$s_{e_1}^{(0),mat} = \begin{pmatrix} -1 & 3 & -3 \\ 0 & 1 & 0 \\ 0 & 0 & 1 \end{pmatrix},$$
$$s_{e_2}^{(0),mat} = \begin{pmatrix} 1 & 0 & 0 \\ 3 & -1 & 3 \\ 0 & 0 & 1 \end{pmatrix}, \quad s_{e_3}^{(0),mat} = \begin{pmatrix} 1 & 0 & 0 \\ 0 & 1 & 0 \\ -3 & 3 & -1 \end{pmatrix}.$$

One sees $\Gamma^{(0)}\{e_i\} \subset \pm e_i + 3H_{\mathbb{Z}}$ for $i \in \{1, 2, 3\}$. Therefore $\Delta^{(0)}$ splits into three $\Gamma^{(0)}$ orbits,

$$\Delta^{(0)} = \Gamma^{(0)}\{e_1\} \,\dot\cup\, \Gamma^{(0)}\{e_2\} \,\dot\cup\, \Gamma^{(0)}\{e_3\}.$$

It remains to show $\Delta^{(0)} = R^{(0)}$. Write $\underline{\widetilde{e}} = (e_1, -e_2, e_3)$, so that

$$L(\underline{\widetilde{e}}^t, \underline{\widetilde{e}})^t = \widetilde{S} = \begin{pmatrix} 1 & 3 & 3 \\ 0 & 1 & 3 \\ 0 & 0 & 1 \end{pmatrix} \quad \text{and} \quad I^{(0)}(\underline{\widetilde{e}}^t, \underline{\widetilde{e}})^t = \widetilde{S} + \widetilde{S}^t = \begin{pmatrix} 2 & 3 & 3 \\ 3 & 2 & 3 \\ 3 & 3 & 2 \end{pmatrix}.$$

The quadratic form $\widetilde{Q}_3 : \mathbb{Z}^3 \to \mathbb{Z}$ with

$$\begin{aligned} \widetilde{Q}_3(\underline{y}) &= \frac{1}{2} I^{(0)}\Big((\underline{\widetilde{e}} \begin{pmatrix} y_1 \\ y_2 \\ y_3 \end{pmatrix})^t, \underline{\widetilde{e}} \begin{pmatrix} y_1 \\ y_2 \\ y_3 \end{pmatrix}\Big) = \frac{1}{2}(y_1\ y_2\ y_3) \begin{pmatrix} 2 & 3 & 3 \\ 3 & 2 & 3 \\ 3 & 3 & 2 \end{pmatrix} \begin{pmatrix} y_1 \\ y_2 \\ y_3 \end{pmatrix} \\ &= y_1^2 + y_2^2 + y_3^2 + 3(y_1y_2 + y_1y_3 + y_2y_3) \end{aligned}$$

can also be written in the following two ways which will be useful below,

$$\begin{aligned} \widetilde{Q}_3(\underline{y}) &= (y_1 + y_2)(y_1 + y_3) + (y_1 + y_2)(y_2 + y_3) \\ &\quad + (y_1 + y_3)(y_2 + y_3), & (6.8) \\ \widetilde{Q}_3(\underline{y}) &= \frac{3}{2}(y_1 + y_2 + y_3)^2 - \frac{1}{2}(y_1^2 + y_2^2 + y_3^2). & (6.9) \end{aligned}$$

We have $R^{(0)} = \{y_1\widetilde{e_1} + y_2\widetilde{e_2} + y_3\widetilde{e_3} \in H_{\mathbb{Z}} \mid \widetilde{Q}_3(\underline{y}) = 1\}$. Define

$$\|a\| := \sqrt{y_1^2 + y_2^2 + y_3^2} \quad \text{for} \quad a = \sum_{i=1}^{3} y_i e_i \in H_{\mathbb{Z}}.$$

Claim: *For any* $a \in R^{(0)} - \{\pm e_1, \pm e_2, \pm e_3\}$ *an index* $i \in \{1, 2, 3\}$ *with*

$$\|s_{e_i}^{(0)}(a)\| < \|a\|$$

exists.

The Claim implies $\Delta^{(0)} = R^{(0)}$ because it says that any $a \in R^{(0)}$ can be mapped by a suitable sequence of reflections in $\{s_{e_1}^{(0)}, s_{e_2}^{(0)}, s_{e_3}^{(0)}\}$ to e_1 or e_2 or e_3. It remains to prove the Claim.

Proof of the Claim: Suppose $a \in R^{(0)} - \{\pm e_1, \pm e_2, \pm e_3\}$ satisfies $\|s_{e_i}^{(0)}(a)\| \geq a$ for any $i \in \{1, 2, 3\}$. Write $a = y_1\widetilde{e_1} + y_2\widetilde{e_2} + y_3\widetilde{e_3}$. For j and k with $\{i, j, k\} = \{1, 2, 3\}$

$$
\begin{aligned}
&\|s_{e_i}^{(0)}(a)\| \geq \|a\| \iff \|s_{e_i}^{(0)}(a)\|^2 \geq \|a\|^2 \\
\iff &(-y_i - 3y_j - 3y_k)^2 + y_j^2 + y_k^2 \geq y_i^2 + y_j^2 + y_k^2 \\
\iff &6y_i(y_j + y_k) + 9(y_j + y_k)^2 \geq 0 \\
\iff &\begin{cases} 3(y_j + y_k) \geq -2y_i & \text{if } y_j + y_k > 0, \\ 3(y_j + y_k) \leq -2y_i & \text{if } y_j + y_k < 0, \\ \text{no condition} & \text{if } y_j + y_k = 0. \end{cases}
\end{aligned}
$$

$y_j + y_k = 0$ is impossible because else by formula (6.8)

$$
\begin{aligned}
1 = \widetilde{Q}_3(\underline{y}) = (y_i + y_j)(y_i + y_k) = (y_i + y_j)(y_i - y_j) = y_i^2 - y_j^2, \\
\text{so } y_i = \pm 1,\ y_j = y_k = 0,
\end{aligned}
$$

which is excluded by $a \in R^{(0)} - \{\pm e_1, \pm e_2, \pm e_3\}$. Also $(y_1 + y_2 > 0, y_1 + y_3 > 0, y_2 + y_3 > 0)$ and $(y_1 + y_2 < 0, y_1 + y_3 < 0, y_2 + y_3 < 0)$ are impossible because of $1 = \widetilde{Q}_3(\underline{y})$ and (6.8).

We can suppose

$$
y_1 + y_2 > 0, \quad y_1 + y_3 > 0, \quad y_2 + y_3 < 0, \quad y_1 \geq y_2 \geq y_3.
$$

Then

$$
\begin{aligned}
y_1 > 0, \quad y_3 \in \mathbb{Z} \cap [-y_1 + 1, -1], \quad y_2 \in [-y_3 - 1, y_3], \\
3(y_1 + y_2) \geq -2y_3 \quad \text{because of } y_1 + y_2 > 0, \\
3(y_2 + y_3) \leq -2y_1 \quad \text{because of } y_2 + y_3 < 0,
\end{aligned}
$$

so

$$
\begin{aligned}
y_1 &\geq 3(y_1 + y_2 + y_3) \geq y_3 \geq -y_1 + 1 > -y_1, \\
|y_1| &\geq 3|y_1 + y_2 + y_3|, \\
y_1^2 &\geq 9(y_1 + y_2 + y_3)^2, \\
\widetilde{Q}_3(\underline{y}) &\overset{(6.9)}{=} \frac{3}{2}(y_1 + y_2 + y_3)^2 - \frac{1}{2}(y_1^2 + y_2^2 + y_3^2) \\
&\leq \frac{1}{6}y_1^2 - \frac{1}{2}(y_1^2 + y_2^2 + y_3^2) \leq 0,
\end{aligned}
$$

a contradiction. Therefore an $a \in R^{(0)} - \{\pm e_1, \pm e_2, \pm e_3\}$ with $\|s_{e_i}^{(0)}(a)\| \geq \|a\|$ for each $i \in \{1, 2, 3\}$ does not exist. The Claim is proved.

(h) By Theorem 6.11 (g) $\Gamma^{(0)}$ is a free Coxeter group with generators $s_{e_1}^{(0)}, s_{e_2}^{(0)}$ and $s_{e_3}^{(0)}$. By Example 3.23 (iv) equality holds in (3.3), so

$$
\mathcal{B}^{dist} = \{\underline{v} \in (\Delta^{(0)})^3 \mid s_{v_1}^{(0)} s_{v_2}^{(0)} s_{v_3}^{(0)} = -M\}
$$

(see also Theorem 7.2 (a)). By Theorem 5.16 (a)+(b)+(d)+(e) $Q \in G_{\mathbb{Z}} - G_{\mathbb{Z}}^{\mathcal{B}}$. Lemma 3.22 (a) and $QMQ^{-1} = M$ give

$$s^{(0)}_{Q(e_1)} s^{(0)}_{Q(e_2)} s^{(0)}_{Q(e_3)} = Q s^{(0)}_{e_1} s^{(0)}_{e_2} s^{(0)}_{e_3} Q^{-1} = Q(-M)Q^{-1} = -M.$$

If $Q(e_1), Q(e_2), Q(e_3)$ were all in $\Delta^{(0)}$ then equality in (3.3) would imply $(Q(e_1), Q(e_2), Q(e_3)) \in \mathcal{B}^{dist}$ and $Q \in G_{\mathbb{Z}}^{\mathcal{B}}$, a contradiction. So $Q(e_1), Q(e_2), Q(e_3)$ are not all in $\Delta^{(0)}$. But of course they are in $R^{(0)}$. □

Remarks 6.15 (i) In view of Remark 6.13 (ii) it would be desirable to extend the proof of $\Delta^{(0)} = R^{(0)}$ in part (g) of Theorem 6.11 to other cases. The useful formulas (6.8) and (6.9) generalize as follows:

$$\begin{aligned}(x_1 + x_2 + x_3)Q_3(\underline{y}) &= (x_1 + x_2 + x_3)(y_1^2 + y_2^2 + y_3^2)\\ -x_1x_2y_1^2 - x_1x_3y_2^2 - x_2x_3y_3^2 &+ (x_1y_2 + x_2y_3)(x_1y_1 + x_3y_3)\\ +(x_1y_2 + x_2y_3)(x_2y_1 + x_3y_2) &+ (x_1y_1 + x_3y_3)(x_2y_1 + x_3y_2).\end{aligned}$$

If $\underline{x} \in (\mathbb{Z} - \{0\})^3$ then

$$\begin{aligned}2Q_3(\underline{y}) = x_1x_2x_3\Big(\frac{y_1}{x_3} + \frac{y_2}{x_2} + \frac{y_3}{x_1}\Big)^2 - (x_1x_2x_3 - 2x_3^2)\Big(\frac{y_1}{x_3}\Big)^2\\ - (x_1x_2x_3 - 2x_2^2)\Big(\frac{y_2}{x_2}\Big)^2 - (x_1x_2x_3 - 2x_1^2)\Big(\frac{y_3}{x_1}\Big)^2.\end{aligned}$$

Also the rephrasing in the proof of part (g) of the inequality $\|s^{(0)}_{e_i}(a)\| \geq \|a\|$ generalizes naturally. But the further arguments do not seem to generalize easily.

(ii) In view of Theorem 6.14, we know the following for $n = 3$:

$\Delta^{(0)} = R^{(0)}$ in the cases $S(\underline{x})$ with $\underline{x} \in \{0, -1, -2\}^3$ with $r(\underline{x}) > 4$,
in all reducible cases
and in the cases A_3, $\widehat{A}_2$, $\mathbb{P}^2$.

$\Delta^{(0)} \subsetneqq R^{(0)}$ in the cases $\mathcal{H}_{1,2}$, $S(-l, 2, -l)$ with $l \geq 3$,
$S(3, 3, 3)$, $S(4, 4, 4)$, $S(5, 5, 5)$ and $S(4, 4, 8)$.

In the following cases with $n = 3$ we do not know whether $\Delta^{(0)} = R^{(0)}$ or $\Delta^{(0)} \subsetneqq R^{(0)}$ holds: All cases $\underline{x} \in \mathbb{Z}^3$ with $r(\underline{x}) < 0$ except four cases and many cases $\underline{x} \in \mathbb{Z}^3$ with $r(\underline{x}) > 4$.

6.4 The Odd Rank 3 Cases

For $\underline{x} \in \mathbb{Z}^3$ consider the matrix $S = S(\underline{x}) = \begin{pmatrix} 1 & x_1 & x_2 \\ 0 & 1 & x_3 \\ 0 & 0 & 1 \end{pmatrix} \in T_3(\mathbb{Z})$, and consider a unimodular bilinear lattice $(H_\mathbb{Z}, L)$ with a triangular basis $\underline{e} = (e_1, e_2, e_3)$ with $L(\underline{e}^t, \underline{e})^t = S$.

In this section we will determine in all cases the odd monodromy group $\Gamma^{(1)} = \langle s^{(1)}_{e_1}, s^{(1)}_{e_2}, s^{(1)}_{e_3}\rangle \subset O^{(1)}$ and in many, but not all cases the set $\Delta^{(1)} = \Gamma^{(1)}\{\pm e_1, \pm e_2, \pm e_3\}$ of odd vanishing cycles.

Recall Remark 4.17. The group $\Gamma^{(1)}$ and the set $\Delta^{(1)}$ are determined by the triple $(H_\mathbb{Z}, I^{(1)}, \underline{e})$, and here $I^{(1)}$ is needed only up to the sign. By Remark 4.17 and Lemma 4.18 we can restrict to $\underline{x}$ in the union of the following three families. It will be useful to split each of the first two families into the three subfamilies on the right hand side.

$$(x_1, x_2, 0) \text{ with } x_1 \geq x_2 \geq 0, \quad \begin{cases} (x_1, 0, 0): & \text{reducible cases,} \\ (1, 1, 0): & A_3 \text{ and } \widehat{A}_2, \\ (x_1, x_2, 0) & \text{with } 2 \leq x_1 \geq x_2 > 0, \end{cases}$$

$$(-l, 2, -l) \text{ with } l \geq 2, \quad \begin{cases} (-l, 2, -l) \text{ with } l \equiv 0(4), \\ (-l, 2, -l) \text{ with } l \equiv 2(4) (\text{ this includes } \mathcal{H}_{1,2}), \\ (-l, 2, -l) \text{ with } l \equiv 1(2), \end{cases}$$

$$(x_1, x_2, x_3) \in \mathbb{Z}^3_{\geq 3} \text{ with } 2x_i \leq x_j x_k \text{ for } \{i, j, k\} = \{1, 2, 3\}$$
$$(\text{this includes } \mathbb{P}^2).$$

Recall

$$\underline{\widetilde{x}} = (\widetilde{x}_1, \widetilde{x}_2, \widetilde{x}_3) := \gcd(x_1, x_2, x_3)^{-1}(x_1, x_2, x_3) \quad \text{for } \underline{x} \neq (0, 0, 0).$$

Recall from section 5.3 the definition

$$f_3 := -\widetilde{x}_3 e_1 + \widetilde{x}_2 e_2 - \widetilde{x}_1 e_3 \in H_\mathbb{Z}^{prim} \quad \text{for } \underline{x} \neq (0, 0, 0)$$

and the fact

$$\text{Rad}\, I^{(1)} = \begin{cases} \mathbb{Z} f_3 & \text{if } \underline{x} \neq (0, 0, 0), \\ H_\mathbb{Z} & \text{if } \underline{x} = (0, 0, 0). \end{cases}$$

Therefore in all cases except $\underline{x} = (0, 0, 0)$ the exact sequence

$$\{1\} \to \Gamma^{(1)}_u \to \Gamma^{(1)} \to \Gamma^{(1)}_s \to \{1\} \tag{6.10}$$

in Lemma 6.2 (d) is interesting.

Lemma 6.16 *Suppose $x_1 \neq 0$ (this holds in the three families above except for the case $\underline{x} = (0, 0, 0)$).*

(a) The sublattice $\mathbb{Z}\overline{e_1}^{(1)} + \mathbb{Z}\overline{e_2}^{(1)} \subset \overline{H_\mathbb{Z}}^{(1)}$ has index $\widetilde{x}_1$ in $\overline{H_\mathbb{Z}}^{(1)}$.

(b) $\overline{I}^{(1)}$ is nondegenerate. For each $\mathbb{Z}$-basis $\underline{b} = (b_1, b_2)$ of $\overline{H_\mathbb{Z}}^{(1)}$

$$\overline{I}^{(1)}(\underline{b}^t, \underline{b}) = \varepsilon \gcd(x_1, x_2, x_3) \begin{pmatrix} 0 & -1 \\ 1 & 0 \end{pmatrix}$$

for some $\varepsilon \in \{\pm 1\}$. Also $O_s^{(1),Rad} \cong SL_2(\mathbb{Z})$.

(c)

$$\overline{e_1}^{(1)} \in \gcd(\widetilde{x}_1, \widetilde{x}_2)\overline{H_\mathbb{Z}}^{(1),prim}, \quad \overline{e_2}^{(1)} \in \gcd(\widetilde{x}_1, \widetilde{x}_3)\overline{H_\mathbb{Z}}^{(1),prim},$$
$$\overline{e_3}^{(1)} \in \gcd(\widetilde{x}_2, \widetilde{x}_3)\overline{H_\mathbb{Z}}^{(1),prim} \text{ if } (\widetilde{x}_2, \widetilde{x}_3) \neq (0,0), \quad \text{else } \overline{e_3}^{(1)} = 0.$$

Proof (a)

$$\begin{aligned}
\overline{H_\mathbb{Z}}^{(1)} &= \mathbb{Z}\overline{e_1}^{(1)} + \mathbb{Z}\overline{e_2}^{(1)} + \mathbb{Z}\overline{e_3}^{(1)} \\
&= \mathbb{Z}\overline{e_1}^{(1)} + \mathbb{Z}\overline{e_2}^{(1)} + \mathbb{Z}\frac{1}{\widetilde{x}_1}(-\widetilde{x}_3\overline{e_1}^{(1)} + \widetilde{x}_2\overline{e_2}^{(1)}) \\
&= \mathbb{Z}\overline{e_1}^{(1)} + \mathbb{Z}\overline{e_2}^{(1)} + \mathbb{Z}\frac{\xi}{\widetilde{x}_1}h_2 \quad \text{with} \\
\xi &:= \gcd(\widetilde{x}_2, \widetilde{x}_3), \quad h_2 := -\frac{\widetilde{x}_3}{\xi}\overline{e_1}^{(1)} + \frac{\widetilde{x}_2}{\xi}\overline{e_2}^{(1)}.
\end{aligned}$$

The element h_2 is in $(\mathbb{Z}\overline{e_1}^{(1)} + \mathbb{Z}\overline{e_2}^{(1)})^{prim}$. One can choose a second element $h_1 \in \mathbb{Z}\overline{e_1}^{(1)} + \mathbb{Z}\overline{e_2}^{(1)}$ with $\mathbb{Z}\overline{e_1}^{(1)} \oplus \mathbb{Z}\overline{e_2}^{(1)} = \mathbb{Z}h_1 \oplus \mathbb{Z}h_2$. Then

$$\mathbb{Z}h_2 + \mathbb{Z}\frac{\xi}{\widetilde{x}_1}h_2 = \mathbb{Z}\frac{1}{\widetilde{x}_1}h_2$$

because $\gcd(\widetilde{x}_1, \xi) = 1$. Therefore

$$\overline{H_\mathbb{Z}}^{(1)} = (\mathbb{Z}h_1 + \mathbb{Z}h_2) + \mathbb{Z}\frac{\xi}{\widetilde{x}_1}h_2 = \mathbb{Z}h_1 \oplus \mathbb{Z}\frac{1}{\widetilde{x}_1}h_2 \overset{\widetilde{x}_1:1}{\supset} \mathbb{Z}h_1 \oplus \mathbb{Z}h_2.$$

(b) $\overline{I}^{(1)}(\overline{e_1}^{(1)}, \overline{e_2}^{(1)}) = x_1 \neq 0$. With part (a) one sees

$$I^{(1)}(b_1, b_2) = \pm\frac{x_1}{\widetilde{x}_1} = \pm\gcd(x_1, x_2, x_3).$$

A rank two lattice with a nondegenerate skew-symmetric bilinear form has an automorphism group isomorphic to $SL_2(\mathbb{Z})$.

(c) The proof of part (a) and $1 = \gcd(\widetilde{x}_3, \gcd(\widetilde{x}_1, \widetilde{x}_2))$ show

$$\mathbb{Q}\overline{e_1}^{(1)} \cap \overline{H_\mathbb{Z}}^{(1)} = \mathbb{Z}\overline{e_1}^{(1)} + \mathbb{Z}\frac{-\widetilde{x}_3}{\gcd(\widetilde{x}_1, \widetilde{x}_2)}\overline{e_1}^{(1)} = \mathbb{Z}\frac{1}{\gcd(\widetilde{x}_1, \widetilde{x}_2)}\overline{e_1}^{(1)}.$$

This shows $\overline{e_1}^{(1)} \in \gcd(\widetilde{x}_1, \widetilde{x}_2)\overline{H_\mathbb{Z}}^{(1),prim}$. Analogously $\overline{e_2}^{(1)} \in \gcd(\widetilde{x}_1, \widetilde{x}_3)\overline{H_\mathbb{Z}}^{(1),prim}$.

If $(\widetilde{x}_2, \widetilde{x}_3) = (0, 0)$ then $-\widetilde{x}_1\overline{e_3}^{(1)} = \overline{f_3}^{(1)} = 0$, so $\overline{e_3}^{(1)} = 0$. If $\widetilde{x}_2 \neq 0$ or $\widetilde{x}_3 \neq 0$, formulas as in the proof of part (a) hold also for $\mathbb{Z}\overline{e_1}^{(1)} + \mathbb{Z}\overline{e_3}^{(1)}$ respectively $\mathbb{Z}\overline{e_2}^{(1)} + \mathbb{Z}\overline{e_3}^{(1)}$. In both cases one shows $\overline{e_3}^{(1)} \in \gcd(\widetilde{x}_2, \widetilde{x}_3)\overline{H_\mathbb{Z}}^{(1),prim}$ as above. □

In Theorem 6.18 we will consider in many cases the three groups $\Gamma_u^{(1)} \subset (\ker \tau^{(1)})_u \subset O_u^{(1),Rad}$. Their descriptions in Lemma 6.2 (e) simplify because now $\mathrm{Rad}\, I^{(1)} = \mathbb{Z}f_3$ if $\underline{x} \neq (0, 0, 0)$. In the cases $\underline{x} = (-l, 2, -l)$ with $l \equiv 2(4)$ also the larger group

$$O_\pm^{(1),Rad} := \{g \in O^{(1),Rad} \mid \overline{g} = \pm \mathrm{id}\} \overset{2:1}{\supset} O_u^{(1),Rad}$$

will be considered. The following Lemma fixes notations and gives a description of $O_\pm^{(1),Rad}$ similar to the one for $O_u^{(1),Rad}$ in Lemma 6.2 (e). It makes also $O_u^{(1),Rad}$ and $(\ker \tau^{(1)})_u$ more explicit, and, under some condition, $\Gamma_u^{(1)}$ and $\Gamma^{(1)} \cap O_\pm^{(1),Rad}$.

Lemma 6.17 *Suppose $x_1 \neq 0$. Denote*

$$\begin{aligned}
&\mathrm{Hom}_0(H_\mathbb{Z}, \mathbb{Z}) := \{\lambda : H_\mathbb{Z} \to \mathbb{Z} \mid \lambda \text{ is } \mathbb{Z}\text{-linear}, \lambda(f_3) = 0\},\\
&\mathrm{Hom}_2(H_\mathbb{Z}, \mathbb{Z}) := \{\lambda : H_\mathbb{Z} \to \mathbb{Z} \mid \lambda \text{ is } \mathbb{Z}\text{-linear}, \lambda(f_3) = 2\},\\
&t_\lambda^+ : H_\mathbb{Z} \to H_\mathbb{Z} \quad \text{with } t_\lambda^+(a) = a + \lambda(a)f_3 \quad \text{for } \lambda \in \mathrm{Hom}_0(H_\mathbb{Z}, \mathbb{Z}),\\
&t_\lambda^- : H_\mathbb{Z} \to H_\mathbb{Z} \quad \text{with } t_\lambda^-(a) = -a + \lambda(a)f_3 \quad \text{for } \lambda \in \mathrm{Hom}_2(H_\mathbb{Z}, \mathbb{Z}).
\end{aligned}$$

(a) Then $t_\lambda^+ \in O_u^{(1),Rad}$ for $\lambda \in \mathrm{Hom}_0(H_\mathbb{Z}, \mathbb{Z})$, $t_\lambda^- \in O_\pm^{(1),Rad} - O_u^{(1),\mathrm{Rad}}$ for $\lambda \in \mathrm{Hom}_2(H_\mathbb{Z}, \mathbb{Z})$. The maps

$$\begin{aligned}
&\mathrm{Hom}_0(H_\mathbb{Z}, \mathbb{Z}) \to O_u^{(1),Rad}, \quad \lambda \mapsto t_\lambda^+,\\
&\mathrm{Hom}_2(H_\mathbb{Z}, \mathbb{Z}) \to O_\pm^{(1),Rad} - O_u^{(1),Rad}, \quad \lambda \mapsto t_\lambda^-,
\end{aligned}$$

are bijections, and the first one is a group isomorphism. For $\lambda_1, \lambda_2 \in \mathrm{Hom}_0(H_\mathbb{Z}, \mathbb{Z})$ and $\lambda_3, \lambda_4 \in \mathrm{Hom}_2(H_\mathbb{Z}, \mathbb{Z})$

$$\begin{aligned}
&t_{\lambda_2}^+ \circ t_{\lambda_1}^+ = t_{\lambda_2+\lambda_1}^+, \quad t_{\lambda_3}^- \circ t_{\lambda_1}^+ = t_{\lambda_3+\lambda_1}^-,\\
&t_{\lambda_1}^+ \circ t_{\lambda_3}^- = t_{-\lambda_1+\lambda_3}^-, \quad t_{\lambda_4}^- \circ t_{\lambda_3}^- = t_{-\lambda_4+\lambda_3}^+,\\
&\text{and especially } (t_{\lambda_3}^-)^2 = \mathrm{id}\,.
\end{aligned}$$

(b)

$$(\ker \tau^{(1)})_u = \{t_\lambda^+ \mid \lambda(\underline{e}) \in \langle (0, x_1, x_2), (-x_1, 0, x_3), (x_2, x_3, 0)\rangle_\mathbb{Z}\}.$$

(c) If $\Gamma_u^{(1)}$ is the normal subgroup generated by $t_{\lambda_1}^+$ for some $\lambda_1 \in \mathrm{Hom}_0(H_\mathbb{Z}, \mathbb{Z})$, then $\Gamma_u^{(1)} = \{t_\lambda^+ \mid \lambda(\underline{e}) \in L\}$ where $L \subset \mathbb{Z}^3$ is the smallest sublattice with $\lambda_1(\underline{e}) \in L$ and $L \cdot (s_{e_i}^{(1),mat})^{\pm 1} \subset L$ for $i \in \{1, 2, 3\}$.

(d) If $\Gamma^{(1)} \cap O_\pm^{(1),Rad}$ is the normal subgroup generated by $t_{\lambda_1}^-$ for some $\lambda_1 \in \mathrm{Hom}_2(H_\mathbb{Z}, \mathbb{Z})$, then

$$\Gamma_u^{(1)} = \{t_\lambda^+ \mid \lambda(\underline{e}) \in L\} \quad \text{and}$$
$$\Gamma^{(1)} \cap O_\pm^{(1),Rad} - \Gamma_u^{(1)} = \{t_\lambda^- \mid \lambda(\underline{e}) \in \lambda_1(\underline{e}) + L\}$$

where $L \subset \mathbb{Z}^3$ is the smallest sublattice with $\lambda_1(\underline{e}) - \lambda_1(\underline{e}) \cdot (s_{e_i}^{(1),mat})^{\pm 1} \in L$ and $L \cdot (s_{e_i}^{(1),mat})^{\pm 1} \subset L$ for $i \in \{1, 2, 3\}$.

Proof (a) By definition $t_\lambda^+ = T(\overline{\lambda} \otimes f_3)$ where $\overline{\lambda} \in \overline{H_\mathbb{Z}}^{(1),\sharp}$ denotes the element which is induced by λ. The map $\mathrm{Hom}_0(H_\mathbb{Z}, \mathbb{Z}) \to O_u^{(1),Rad}$ is an isomorphism by Lemma 6.2 (e). The proofs of the other statements are similar or easy.

(b) The row vectors $(0, x_1, x_2)$, $(-x_1, 0, x_3)$, $(-x_2, -x_3, 0)$ are the rows of the matrix $I^{(1)}(\underline{e}^t, \underline{e})$. Because of this and Lemma 6.2 (e) $(\ker \tau^{(1)})_u$ is as claimed.

(c) This follows from

$$\Gamma_u^{(1)} = \langle g^{-1} \circ t_{\lambda_1}^+ \circ g \mid g \in \Gamma^{(1)} \rangle,$$
$$g^{-1} \circ t_\lambda^+ \circ g = t_{\lambda \circ g}^+,$$
$$\text{and } \lambda \circ (s_{e_i}^{(1)})^{\pm 1}(\underline{e}) = \lambda(\underline{e}) \cdot (s_{e_i}^{(1),mat})^{\pm 1},$$

for $\lambda \in \mathrm{Hom}_0(H_\mathbb{Z}, \mathbb{Z})$.

(d) Similar to the proof of part (c). □

The following theorem gives $\Gamma^{(1)}$ for $\underline{x}$ in one of the three families above and thus via Remark 4.17 and Lemma 4.18 in principle for all $\underline{x} \in \mathbb{Z}^3$. Though recall that it is nontrivial to find for a given $\underline{x} \in \mathbb{Z}^3$ an element in one of the three families above which is in the $(G^{phi} \ltimes \tilde{G}^{sign}) \rtimes \langle \gamma \rangle$ orbit of $\underline{x}$.

Theorem 6.18 *(a) We have*

$$s_{e_i}^{(1)} \underline{e} = \underline{e} \cdot s_{e_i}^{(1),mat} \quad \text{with} \quad s_{e_1}^{(1),mat} = \begin{pmatrix} 1 & -x_1 & -x_2 \\ 0 & 1 & 0 \\ 0 & 0 & 1 \end{pmatrix},$$

$$s_{e_2}^{(1),mat} = \begin{pmatrix} 1 & 0 & 0 \\ x_1 & 1 & -x_3 \\ 0 & 0 & 1 \end{pmatrix}, \quad s_{e_3}^{(1),mat} = \begin{pmatrix} 1 & 0 & 0 \\ 0 & 1 & 0 \\ x_2 & x_3 & 1 \end{pmatrix},$$

$$\Gamma^{(1)} \cong \Gamma^{(1),mat} = \langle s_{e_1}^{(1),mat}, s_{e_2}^{(1),mat}, s_{e_3}^{(1),mat} \rangle \subset SL_3(\mathbb{Z}).$$

(b) In each reducible case $\underline{x} = (x_1, 0, 0)$

$$\Gamma^{(1)} \cong \Gamma^{(1)}(S(-x_1)) \times \Gamma^{(1)}(A_1) \cong \Gamma^{(1)}(S(-x_1)) \times \{1\},$$

and $\Gamma^{(1)}(S(-x_1))$ *is given in Theorem 6.10, with*

$$\begin{aligned}\Gamma^{(1)}(S(0)) &\cong \Gamma^{(1)}(A_1^2) \cong \{1\},\\ \Gamma^{(1)}(S(-1)) &\cong \Gamma^{(1)}(A_2) \cong SL_2(\mathbb{Z}),\\ \Gamma^{(1)}(S(-x_1)) &\cong G^{free,2} \quad \text{for } x_1 \geq 2.\end{aligned}$$

Also $\Gamma^{(1)} \cong \Gamma_s^{(1)}$ *and* $\Gamma_u^{(1)} = \{\mathrm{id}\}$.

(c) The case $\underline{x} = (1, 1, 0)$*: (This is the case of* A_3 *and* $\widehat{A}_2$*.)*

$$\begin{aligned}\mathrm{Rad}\, I^{(1)} &= \mathbb{Z} f_3 \quad \text{with} \quad f_3 = e_2 - e_3,\\ \overline{H_{\mathbb{Z}}}^{(1)} &= \mathbb{Z}\overline{e_1}^{(1)} \oplus \mathbb{Z}\overline{e_2}^{(1)},\\ \Gamma_u^{(1)} &= (\ker \tau^{(1)})_u = O_u^{(1),Rad} = \{t_\lambda^+ \mid \lambda \in \langle \lambda_1, \lambda_2\rangle_{\mathbb{Z}}\} \cong \mathbb{Z}^2,\\ &\quad \text{with} \quad \lambda_1(\underline{e}) = (1, 0, 0), \quad \lambda_2(\underline{e}) = (0, 1, 1),\\ \Gamma_s^{(1)} &= (\ker \tau^{(1)})_s = O_s^{(1),Rad} \cong SL_2(\mathbb{Z}),\\ \Gamma^{(1)} &= \ker \tau^{(1)} = O^{(1),Rad}.\end{aligned}$$

The exact sequence (6.10) *splits non-canonically with* $\Gamma_s^{(1)} \cong \langle s_{e_1}^{(1)}, s_{e_2}^{(1)}\rangle \subset \Gamma^{(1)}$ *(for example).*

(d) The cases $\underline{x} = (x_1, x_2, 0)$ *with* $2 \leq x_1 \geq x_2 > 0$*: Write*

$$x_{12} := \gcd(x_1, x_2) = \frac{x_1}{\widetilde{x}_1} = \frac{x_2}{\widetilde{x}_2}.$$

Then

$$\begin{aligned}\mathrm{Rad}\, I^{(1)} &= \mathbb{Z} f_3 \quad \text{with} \quad f_3 = \widetilde{x}_2 e_2 - \widetilde{x}_1 e_3,\\ \overline{H_{\mathbb{Z}}}^{(1)} &= \mathbb{Z}\overline{e_1}^{(1)} \oplus \mathbb{Z} g_2 \quad \text{with } g_2 := \frac{1}{\widetilde{x}_1}\overline{e_2}^{(1)} = \frac{1}{\widetilde{x}_2}\overline{e_3}^{(1)} \in \overline{H_{\mathbb{Z}}}^{(1)},\end{aligned}$$

$$\begin{aligned}\Gamma_u^{(1)} &= \{t_\lambda^+ \mid \lambda \in \langle \lambda_1, \lambda_2\rangle_{\mathbb{Z}}\} \cong \mathbb{Z}^2 \quad \text{with}\\ &\quad \lambda_1(\underline{e}) = x_{12}\widetilde{x}_1\widetilde{x}_2(1, 0, 0), \quad \lambda_2(\underline{e}) = x_1 x_2(0, \widetilde{x}_1, \widetilde{x}_2),\\ (\ker \tau^{(1)})_u &= \{t_\lambda^+ \mid \lambda(\underline{e}) \in \langle x_{12}(1, 0, 0), (0, x_1, x_2)\rangle_{\mathbb{Z}}\},\\ O_u^{(1),Rad} &= \{t_\lambda^+ \mid \lambda(\underline{e}) \in \langle (1, 0, 0), (0, \widetilde{x}_1, \widetilde{x}_2)\rangle_{\mathbb{Z}}\},\end{aligned}$$

$$\Gamma_s^{(1)} \cong \Gamma^{(1)}(S(-x_{12})) \cong \langle \begin{pmatrix} 1 & -x_{12} \\ 0 & 1 \end{pmatrix}, \begin{pmatrix} 1 & 0 \\ x_{12} & 1 \end{pmatrix} \rangle$$

$$\cong \begin{cases} G^{free,2} & \text{if } x_{12} > 1 \\ SL_2(\mathbb{Z}) & \text{if } x_{12} = 1. \end{cases}$$

This matrix group has finite index in $SL_2(\mathbb{Z})$ if and only if $x_{12} \in \{1, 2\}$. The exact sequence (6.10) *splits non-canonically.*

(e) The cases $\underline{x} = (-l, 2, -l)$ with $l \geq 2$ even:

(This includes the case $\underline{x} = (-2, 2, -2)$ which is the case of $\mathcal{H}_{1,2}$.)

$$\begin{aligned} \operatorname{Rad} I^{(1)} &= \mathbb{Z} f_3 \quad \text{with} \quad f_3 = \frac{l}{2}(e_1 + e_3) + e_2, \\ \overline{H_{\mathbb{Z}}}^{(1)} &= \mathbb{Z}\overline{e_1}^{(1)} \oplus \mathbb{Z}\overline{e_3}^{(1)}, \quad \overline{e_2}^{(1)} = -\frac{l}{2}(\overline{e_1}^{(1)} + \overline{e_3}^{(1)}), \end{aligned}$$

$$\begin{aligned} &\langle s_{e_1}^{(1)}, s_{e_3}^{(1)} \rangle \cong \langle \overline{s_{e_1}^{(1)}}, \overline{s_{e_3}^{(1)}} \rangle \cong \langle \begin{pmatrix} 1 & -2 \\ 0 & 1 \end{pmatrix}, \begin{pmatrix} 1 & 0 \\ 2 & 1 \end{pmatrix} \rangle \overset{1:2}{\subset} \Gamma(2), \\ &\langle s_{e_1}^{(1)}, s_{e_3}^{(1)} \rangle \cong G^{free,2}, \end{aligned}$$

$$\begin{aligned} (\ker \tau^{(1)})_u &= \{ t_\lambda^+ \mid \lambda(\underline{e}) \in \langle (-2, 0, 2), (-2, l, 0) \rangle_{\mathbb{Z}} \}, \\ O_u^{(1),Rad} &= \{ t_\lambda^+ \mid \lambda(\underline{e}) \in \langle (-1, 0, 1), (-1, \frac{l}{2}, 0) \rangle_{\mathbb{Z}} \}. \end{aligned}$$

(i) The subcases with $l \equiv 0(4)$: $\Gamma_s^{(1)} \cong \langle \overline{s_{e_1}^{(1)}}, \overline{s_{e_3}^{(1)}} \rangle \cong G^{free,2}$. The isomorphism $\Gamma_s^{(1)} \cong \langle s_{e_1}^{(1)}, s_{e_3}^{(1)} \rangle \subset \Gamma^{(1)}$ gives a splitting of the exact sequence (6.10)*. Here* $-\operatorname{id} \notin \Gamma_s^{(1)}$.

$$\begin{aligned} \Gamma_u^{(1)} = \{ t_\lambda^+ \mid \lambda \in \langle \lambda_1, \lambda_2 \rangle_{\mathbb{Z}} \} \cong \mathbb{Z}^2 \quad \text{with} \\ \lambda_1(\underline{e}) = (-l, 0, l), \quad \lambda_2(\underline{e}) = (2l, -l^2, 0). \end{aligned}$$

(ii) The subcases with $l \equiv 2(4)$: Here $-\operatorname{id} \in \Gamma_s^{(1)}$.

$$\begin{aligned} \Gamma_s^{(1)} &\cong \langle \overline{s_{e_1}^{(1)}}, \overline{s_{e_3}^{(1)}}, -\operatorname{id} \rangle \cong \langle \begin{pmatrix} 1 & -2 \\ 0 & 1 \end{pmatrix}, \begin{pmatrix} 1 & 0 \\ 2 & 1 \end{pmatrix}, \begin{pmatrix} -1 & 0 \\ 0 & -1 \end{pmatrix} \rangle = \Gamma(2) \\ &\cong G^{free,2} \times \{\pm 1\}. \end{aligned}$$

The isomorphism $\Gamma_s^{(1)}/\{\pm \operatorname{id}\} \cong \langle s_{e_1}^{(1)}, s_{e_2}^{(1)} \rangle \subset \Gamma^{(1)}$ gives a splitting of the exact sequence

$$\{1\} \to \Gamma^{(1)} \cap O_{\pm}^{(1),Rad} \to \Gamma^{(1)} \to \Gamma_s^{(1)}/\{\pm \operatorname{id}\} \to \{1\}.$$

$$\begin{aligned}
\Gamma_u^{(1)} &= \{t_\lambda^+ \mid \lambda \in \langle 2\lambda_1, \lambda_2\rangle_{\mathbb{Z}}\} \cong \mathbb{Z}^2 \quad \text{with} \\
&\qquad \lambda_1(\underline{e}) = (-l, 0, l), \quad \lambda_2(\underline{e}) = (2l, -l^2, 0), \\
\Gamma^{(1)} \cap O_\pm^{(1),Rad} - \Gamma_u^{(1)} &= \{t_\lambda^- \mid \lambda \in \lambda_3 + \langle 2\lambda_1, \lambda_2\rangle_{\mathbb{Z}}\} \quad \text{with} \\
&\qquad \lambda_3(\underline{e}) = (-l, 2, l), \quad \lambda_3(f_3) = 2.
\end{aligned}$$

(f) The cases $\underline{x} = (-l, 2, -l)$ with $l \geq 3$ odd:

$$\begin{aligned}
\operatorname{Rad} I^{(1)} &= \mathbb{Z} f_3 \quad \text{with} \quad f_3 = l(e_1 + e_3) + 2e_2, \\
\overline{H_{\mathbb{Z}}}^{(1)} &= \mathbb{Z}\overline{e_1}^{(1)} \oplus \mathbb{Z}\overline{g_2}^{(1)} \quad \text{with} \quad g_2 := \frac{1}{2}(e_1 + e_3) - \frac{l}{2} f_3 \in H_{\mathbb{Z}}, \\
\underline{\widetilde{e}} &:= (e_1, g_2, f_3) \quad \text{is a } \mathbb{Z}\text{-basis of } H_{\mathbb{Z}}.
\end{aligned}$$

Consider

$$s_4 := \big(s_{e_3}^{(1)} s_{e_1}^{(1)}\big)^{\frac{l^2-1}{4}} s_{e_2}^{(1)} \in \Gamma^{(1)}.$$

Then

$$\Gamma_s^{(1)} \cong \langle \overline{s_{e_1}^{(1)}}, \overline{s_4}\rangle \cong \langle s_{e_1}^{(1)}, s_4\rangle \cong SL_2(\mathbb{Z}),$$

and the isomorphism $\Gamma_s^{(1)} \cong \langle s_{e_1}^{(1)}, s_4\rangle \subset \Gamma^{(1)}$ gives a splitting of the exact sequence (6.10).

$$\begin{aligned}
\Gamma_u^{(1)} &= \{t_\lambda^+ \mid \lambda \in \langle \lambda_1, \lambda_2\rangle_{\mathbb{Z}}\} \cong \mathbb{Z}^2 \quad \text{with} \\
&\qquad \lambda_1(\underline{e}) = (-l, 0, l), \quad \lambda_2(\underline{e}) = (2l, -l^2, 0), \\
(\ker \tau^{(1)})_u &= O_u^{(1),Rad} = \{t_\lambda^+ \mid \lambda(\underline{e}) \in \langle (-1, 0, 1), (2, -l, 0)\rangle_{\mathbb{Z}}\}.
\end{aligned}$$

(g) The cases $\underline{x} \in \mathbb{Z}^3_{\geq 3}$ with $2x_i \leq x_j x_k$ for $\{i, j, k\} = \{1, 2, 3\}$:
(This includes the case $\underline{x} = (3, 3, 3)$ which is the case of $\mathbb{P}^2$.)

$$\begin{aligned}
\operatorname{Rad} I^{(1)} &= \mathbb{Z} f_3 \quad \text{with} \quad f_3 = -\widetilde{x}_3 e_1 + \widetilde{x}_2 e_2 - \widetilde{x}_1 e_3, \\
\Gamma_u^{(1)} &= \{\mathrm{id}\} \subsetneqq (\ker \tau^{(1)})_u \cong \mathbb{Z}^2, \\
\Gamma^{(1)} &\cong \Gamma_s^{(1)} \cong G^{free,3},
\end{aligned}$$

$\Gamma^{(1)}$ and $\Gamma_s^{(1)}$ are free groups with the three generators $s_{e_1}^{(1)}, s_{e_2}^{(1)}, s_{e_3}^{(1)}$ respectively $\overline{s_{e_1}^{(1)}}$, $\overline{s_{e_2}^{(1)}}$, $\overline{s_{e_3}^{(1)}}$.

Proof (a) This follows from the definitions in Lemma 2.6 (a) and in Definition 2.8.

(b) This follows from Lemma 2.11 and Lemma 2.12.

(c)–(f) In (c)–(f) $(\ker \tau^{(1)})_u$ and $O_u^{(1),Rad}$ are calculated with Lemma 6.17 (a) and (b).

(c) The first statements $\mathrm{Rad}\, I^{(1)} = \mathbb{Z} f_3$ and $\overline{H_{\mathbb{Z}}}^{(1)} = \mathbb{Z}\overline{e_1}^{(1)} \oplus \mathbb{Z}\overline{e_2}^{(1)}$ are known respectively obvious.

Also (e_1, e_2, f_3) is a $\mathbb{Z}$-basis of $H_{\mathbb{Z}}$. With respect to this basis

$$s_{e_1}^{(1)}(e_1, e_2, f_3) = (e_1, e_2, f_3) \begin{pmatrix} 1 & -1 & 0 \\ 0 & 1 & 0 \\ 0 & 0 & 1 \end{pmatrix},$$

$$s_{e_2}^{(1)}(e_1, e_2, f_3) = (e_1, e_2, f_3) \begin{pmatrix} 1 & 0 & 0 \\ 1 & 1 & 0 \\ 0 & 0 & 1 \end{pmatrix}.$$

This shows $\langle s_{e_1}^{(1)}, s_{e_2}^{(1)} \rangle \cong \langle \overline{s_{e_1}^{(1)}}, \overline{s_{e_2}^{(1)}} \rangle \cong SL_2(\mathbb{Z})$. Together with Lemma 6.16 (b) and Lemma 6.2 (d) we obtain

$$\Gamma_s^{(1)} = \langle \overline{s_{e_1}^{(1)}}, \overline{s_{e_2}^{(1)}} \rangle = (\ker \tau^{(1)})_s = O_s^{(1),Rad} \cong SL_2(\mathbb{Z}),$$

and that the exact sequence (6.10) splits non-canonically with $\Gamma_s^{(1)} \cong \langle s_{e_1}^{(1)}, s_{e_2}^{(1)} \rangle \subset \Gamma^{(1)}$.

From the actions of $s_{e_1}^{(1)}$ and $s_{e_2}^{(1)}$ on (e_1, e_2, f_3) and from

$$s_{e_3}^{(1)}((e_1, e_2, f_3)) = (e_1, e_2, f_3) \begin{pmatrix} 1 & 0 & 0 \\ 1 & 1 & 0 \\ -1 & 0 & 1 \end{pmatrix},$$

one sees that the map

$$\left((e_1, e_2, f_3) \mapsto (e_1, e_2, f_3) \begin{pmatrix} 1 & 0 & 0 \\ 0 & 1 & 0 \\ -1 & 0 & 1 \end{pmatrix}\right) = t_{-\lambda_1}^{+}$$

is in $\Gamma_u^{(1)}$ and that

$$\Gamma^{(1)} = \langle s_{e_1}^{(1)}, s_{e_2}^{(1)}, t_{\lambda_1}^{+} \rangle.$$

Also

$$(s_{e_1}^{(1)})^{-1} \circ t_{\lambda_1}^{+} \circ s_{e_1}^{(1)} = t_{\lambda_1 \circ s_{e_1}^{(1)}}^{+}, \quad \text{with} \quad t_{\lambda_1 \circ s_{e_1}^{(1)}}^{+}(\underline{e}) = (1, -1, -1),$$

and therefore $t_{\lambda_2}^{+} = t_{\lambda_1}^{+} - t_{\lambda_1 \circ s_{e_1}^{(1)}}^{+} \in \Gamma_u^{(1)}$. But $O_u^{(1),Rad} = \langle t_{\lambda_1}^{+}, t_{\lambda_2}^{+} \rangle_{\mathbb{Z}}$, so

$$\Gamma_u^{(1)} = (\ker \tau^{(1)})_u = O_u^{(1),Rad} = \langle t_{\lambda_1}^{+}, t_{\lambda_2}^{+} \rangle_{\mathbb{Z}}.$$

In the diagram of exact sequences in Lemma 6.2 (d) the inclusions in the second and fourth columns are bijections, so also the inclusions in the middle column,

$$\Gamma^{(1)} = \ker \tau^{(1)} = O^{(1),Rad}.$$

(d) $f_3 = \widetilde{x}_2 e_2 - \widetilde{x}_1 e_3$ implies $\widetilde{x}_2 \overline{e_2}^{(1)} = \widetilde{x}_1 \overline{e_3}^{(1)}$. Lemma 6.16 (c) gives $\overline{e_2}^{(1)} \in \widetilde{x}_1 \overline{H_\mathbb{Z}}^{(1),prim}$, so $g_2 := \frac{1}{\widetilde{x}_1}\overline{e_2}^{(1)} = \frac{1}{\widetilde{x}_2}\overline{e_3}^{(1)}$ is in $\overline{H_\mathbb{Z}}^{(1),prim}$. Thus

$$\overline{H_\mathbb{Z}}^{(1)} = \mathbb{Z}\overline{e_1}^{(1)} + \mathbb{Z}\overline{e_2}^{(1)} + \mathbb{Z}\overline{e_3}^{(1)} = \mathbb{Z}\overline{e_1}^{(1)} \oplus \mathbb{Z} g_2.$$

First we consider $\Gamma_s^{(1)}$. Define $\underline{g} := (\overline{e_1}^{(1)}, g_2)$. One sees

$$\overline{s_{e_1}^{(1)}}\,\underline{g} = \underline{g}\begin{pmatrix} 1 & -x_{12} \\ 0 & 1 \end{pmatrix}, \quad \overline{s_{e_2}^{(1)}}\,\underline{g} = \underline{g}\begin{pmatrix} 1 & 0 \\ x_1\widetilde{x}_1 & 1 \end{pmatrix}, \quad \overline{s_{e_3}^{(1)}}\,\underline{g} = \underline{g}\begin{pmatrix} 1 & 0 \\ x_2\widetilde{x}_2 & 1 \end{pmatrix}.$$

Choose $y_1, y_2 \in \mathbb{Z}$ with $1 = y_1\widetilde{x}_1^2 + y_2\widetilde{x}_2^2$ and define

$$s_4 := (s_{e_2}^{(1)})^{y_1}(s_{e_3}^{(1)})^{y_2} \in \Gamma^{(1)}.$$

Then

$$\overline{s_4}\,\underline{g} = \underline{g}\begin{pmatrix} 1 & 0 \\ x_{12} & 1 \end{pmatrix}, \quad s_4\,\underline{e} = \underline{e}\cdot s_4^{mat} \text{ with } s_4^{mat} = \begin{pmatrix} 1 & 0 & 0 \\ y_1x_1 & 1 & 0 \\ y_2x_2 & 0 & 1 \end{pmatrix}.$$

$\overline{s_{e_3}^{(1)}}$ and $\overline{s_{e_2}^{(1)}}$ are powers of $\overline{s_4}$, so $\Gamma_s^{(1)} = \langle \overline{s_{e_1}^{(1)}}, \overline{s_4}\rangle$, so

$$\Gamma_s^{(1)} \cong \langle \begin{pmatrix} 1 & -x_{12} \\ 0 & 1 \end{pmatrix}, \begin{pmatrix} 1 & 0 \\ x_{12} & 1 \end{pmatrix}\rangle = \Gamma^{(1),mat}(S(-x_{12})) \subset SL_2(\mathbb{Z}).$$

The matrix group $\Gamma^{(1),mat}(S(-x_{12}))$ was treated in Theorem 6.10 (b)–(d):

$$\Gamma^{(1),mat}(S(-x_{12})) \cong \langle \begin{pmatrix} 1 & -x_{12} \\ 0 & 1 \end{pmatrix}, \begin{pmatrix} 1 & 0 \\ x_{12} & 1 \end{pmatrix}\rangle \begin{cases} \cong G^{free,2} & \text{if } x_{12} > 1, \\ = SL_2(\mathbb{Z}) & \text{if } x_{12} = 1. \end{cases}$$

The case $x_{12} > 1$: Then $\Gamma_s^{(1)} = \langle \overline{s_{e_1}^{(1)}}, \overline{s_4}\rangle$ and $\langle s_{e_1}^{(1)}, s_4\rangle \subset \Gamma^{(1)}$ are free groups with the two given generators. Then $\langle s_{e_1}^{(1)}, s_4\rangle \subset \Gamma^{(1)}$ gives a splitting of the exact sequence (6.10). The generating relations in $\Gamma_s^{(1)}$ with respect to the four elements $\overline{s_{e_1}^{(1)}}$, $\overline{s_{e_2}^{(1)}}$, $\overline{s_{e_3}^{(1)}}$ and $\overline{s_4}$ are

$$\overline{s_{e_2}^{(1)}} = (\overline{s_4})^{\widetilde{x}_1^2}, \quad \overline{s_{e_3}^{(1)}} = (\overline{s_4})^{\widetilde{x}_2^2}. \tag{6.11}$$

Therefore $\Gamma_u^{(1)}$ is the normal subgroup of $\Gamma^{(1)}$ generated by the elements $s_4^{\widetilde{x}_1^2}(s_{e_2}^{(1)})^{-1}$ and $s_4^{\widetilde{x}_2^2}(s_{e_3}^{(1)})^{-1}$.

The case $x_{12}=1$: Then $\Gamma_s^{(1)}=\langle\overline{s_{e_1}^{(1)}},\overline{s_4}\rangle\cong SL_2(\mathbb{Z})$

Claim: Also $\langle s_{e_1}^{(1)},s_4\rangle\cong SL_2(\mathbb{Z})$.

Proof of the Claim: The generating relations in $SL_2(\mathbb{Z})$ with respect to the generators $\overline{s_{e_1}^{(1)}}^{mat}=\begin{pmatrix}1&-1\\0&1\end{pmatrix}$ and $\overline{s_4}^{mat}=\begin{pmatrix}1&0\\1&1\end{pmatrix}$ are

$$\overline{s_{e_1}^{(1)}}^{mat}\,\overline{s_4}^{mat}\,\overline{s_{e_1}^{(1)}}^{mat}=\overline{s_4}^{mat}\,\overline{s_{e_1}^{(1)}}^{mat}\,\overline{s_4}^{mat}\quad\text{and}\quad E_2=\left(\overline{s_{e_1}^{(1)}}^{mat}\,\overline{s_4}^{mat}\right)^6.$$

One checks with calculations that they lift to $\Gamma^{(1),mat}$,

$$s_{e_1}^{(1),mat}s_4^{mat}s_{e_1}^{(1),mat}=s_4^{mat}s_{e_1}^{(1),mat}s_4^{mat}\quad\text{and}\quad E_3=\left(s_{e_1}^{(1),mat}s_4^{mat}\right)^6.\qquad(\Box)$$

Because of the Claim, $\langle s_{e_1}^{(1)},s_4\rangle\subset\Gamma^{(1)}$ gives a splitting of the exact sequence (6.10). The generating relations in $\Gamma_s^{(1)}$ with respect to the four elements $\overline{s_{e_1}^{(1)}}$, $\overline{s_{e_2}^{(1)}}$, $\overline{s_{e_3}^{(1)}}$ and $\overline{s_4}$ are the relations between $\overline{s_{e_1}^{(1)}}$ and $\overline{s_4}$ in the proof of the Claim and the relations in (6.11). The relations in the proof of the Claim lift to $\Gamma^{(1)}$. Therefore again $\Gamma_u^{(1)}$ is the normal subgroup of $\Gamma^{(1)}$ generated by the elements $s_4^{\widetilde{x}_1^2}(s_{e_2}^{(1)})^{-1}$ and $s_4^{\widetilde{x}_2^2}(s_{e_3}^{(1)})^{-1}$.

Back to both cases $x_{12}\geq 1$ **together**: We have to determine these two elements of $\Gamma_u^{(1)}$. The first one is given by the following calculation,

$$\begin{aligned}&s_4^{\widetilde{x}_1^2}(s_{e_2}^{(1)})^{-1}(\underline{e})\\
&=\underline{e}\begin{pmatrix}1&0&0\\y_1x_1\widetilde{x}_1^2&1&0\\y_2x_2\widetilde{x}_1^2&0&1\end{pmatrix}\begin{pmatrix}1&0&0\\-x_1&1&0\\0&0&1\end{pmatrix}=\underline{e}\begin{pmatrix}1&0&0\\-x_1+y_1x_1\widetilde{x}_1^2&1&0\\y_2x_2\widetilde{x}_1^2&0&1\end{pmatrix}\\
&=\underline{e}\begin{pmatrix}1&0&0\\-y_2x_1\widetilde{x}_2^2&1&0\\y_2x_2\widetilde{x}_1^2&0&1\end{pmatrix}=\underline{e}+f_3(-y_2x_{12}\widetilde{x}_1\widetilde{x}_2,0,0),\\
&\text{so } s_4^{\widetilde{x}_1^2}(s_{e_2}^{(1)})^{-1}(\underline{e})=t^+_{-y_2\lambda_1}.\end{aligned}$$

A similar calculation gives

$$s_4^{\widetilde{x}_2^2}(s_{e_3}^{(1)})^{-1}=t^+_{y_1\lambda_1}.$$

Observe $\gcd(y_1,y_2)=1$. Therefore $\Gamma_u^{(1)}$ is the normal subgroup of $\Gamma^{(1)}$ generated by $t^+_{\lambda_1}$. Lemma 6.17 (c) and the calculations

$$\lambda_1 \circ s_{e_1}^{(1)}(\underline{e}) = x_{12}\widetilde{x}_1\widetilde{x}_2(1, -x_1, -x_2),$$
$$\text{so } \langle \lambda_1, \lambda_1 \circ s_{e_1}^{(1)}\rangle_{\mathbb{Z}} = \langle \lambda_1, \lambda_2\rangle_{\mathbb{Z}},$$
$$\lambda_1 \circ (s_{e_i}^{(1)})^{\pm 1}, \lambda_2 \circ (s_{e_i}^{(1)})^{\pm 1} \in \langle \lambda_1, \lambda_2\rangle_{\mathbb{Z}},$$

give

$$\Gamma_u^{(1)} = \{t_\lambda^+ \mid \lambda \in \langle \lambda_1, \lambda_2\rangle_{\mathbb{Z}}\}.$$

(e) The first statements $\operatorname{Rad} I^{(1)} = \mathbb{Z}f_3,\ f_3 = \frac{l}{2}(e_1 + e_3) + e_2$ and $\overline{H_{\mathbb{Z}}}^{(1)} = \mathbb{Z}\overline{e_1}^{(1)} \oplus \mathbb{Z}\overline{e_3}^{(1)}$ are obvious, also

$$\overline{s_{e_i}^{(1)}}(\overline{e_1}^{(1)}, \overline{e_3}^{(1)}) = (\overline{e_1}^{(1)}, \overline{e_3}^{(1)})\overline{s_{e_i}^{(1)}}^{mat} \quad \text{with } \overline{s_{e_1}^{(1)}}^{mat} = \begin{pmatrix} 1 & -2 \\ 0 & 1 \end{pmatrix},$$
$$\overline{s_{e_2}^{(1)}}^{mat} = \begin{pmatrix} 1 + \frac{l^2}{2} & -\frac{l^2}{2} \\ \frac{l^2}{2} & 1 - \frac{l^2}{2} \end{pmatrix}, \quad \overline{s_{e_3}^{(1)}}^{mat} = \begin{pmatrix} 1 & 0 \\ 2 & 1 \end{pmatrix}.$$

By Theorem 6.10 (c) $\langle \overline{s_{e_1}^{(1)}}^{mat}, \overline{s_{e_3}^{(1)}}^{mat}\rangle \cong G^{free,2}$ and therefore

$$\langle s_{e_1}^{(1)}, s_{e_3}^{(1)}\rangle \cong G^{free,2}.$$

We have

$$\overline{s_{e_2}^{(1)}}^{mat} \equiv \begin{pmatrix} 1 & 0 \\ 0 & 1 \end{pmatrix} \bmod 4 \quad \text{if } l \equiv 0(4),$$
$$\overline{s_{e_2}^{(1)}}^{mat} \equiv \begin{pmatrix} 3 & 2 \\ 2 & 3 \end{pmatrix} \bmod 4 \quad \text{if } l \equiv 2(4).$$

If $l \equiv 0(4)$ then by Theorem 6.10 (c) $\overline{s_{e_2}^{(1)}}^{mat} \in \langle \overline{s_{e_1}^{(1)}}^{mat}, \overline{s_{e_3}^{(1)}}^{mat}\rangle$, so then $\Gamma_s^{(1)} = \langle \overline{s_{e_1}^{(1)}}, \overline{s_{e_3}^{(1)}}\rangle$, and the isomorphism $\Gamma_s^{(1)} \cong \langle s_{e_1}^{(1)}, s_{e_3}^{(1)}\rangle \subset \Gamma^{(1)}$ gives a splitting of the exact sequence (6.10).

If $l \equiv 2(4)$ then by Theorem 6.10 (c) $\overline{s_{e_2}^{(1)}}^{mat} \notin \langle \overline{s_{e_1}^{(1)}}^{mat}, \overline{s_{e_3}^{(1)}}^{mat}\rangle$, so then $\langle \overline{s_{e_i}^{(1)}}^{mat} \mid i \in \{1, 2, 3\}\rangle = \Gamma^{(2)} \cong G^{free,2} \times \{\pm 1\}$, $-\operatorname{id} \in \Gamma_s^{(1)}$, and the isomorphism $\Gamma_s^{(1)}/\{\pm \operatorname{id}\} \cong \langle s_{e_1}^{(1)}, s_{e_3}^{(1)}\rangle \subset \Gamma^{(1)}$ gives a splitting of the exact sequence in part (ii).

Claim: $\Gamma_u^{(1)}$ if $l \equiv 0(4)$ and $\Gamma^{(1)} \cap O_{\pm}^{(1),Rad}$ if $l \equiv 2(4)$ is the normal subgroup generated by $s_4 := \left(s_{e_3}^{(1)} s_{e_1}^{(1)}\right)^{l^2/4} s_{e_2}^{(1)}$.

Proof of the Claim: Consider the $\mathbb{Z}$-basis $\underline{\widetilde{e}} := (e_1, e_1 + e_3, f_3)$ of $H_{\mathbb{Z}}$.

$$s_{e_1}^{(1)}\underline{\widetilde{e}} = \underline{\widetilde{e}}\begin{pmatrix}1 & -2 & 0\\ 0 & 1 & 0\\ 0 & 0 & 1\end{pmatrix},\ s_{e_2}^{(1)}\underline{\widetilde{e}} = \underline{\widetilde{e}}\begin{pmatrix}1 & 0 & 0\\ \frac{l^2}{2} & 1 & 0\\ -l & 0 & 1\end{pmatrix},\ s_{e_3}^{(1)}\underline{\widetilde{e}} = \underline{\widetilde{e}}\begin{pmatrix}-1 & -2 & 0\\ 2 & 3 & 0\\ 0 & 0 & 1\end{pmatrix},$$

$$\begin{aligned}
s_4\underline{\widetilde{e}} &= \underline{\widetilde{e}}\begin{pmatrix}-1 & 0 & 0\\ 2 & -1 & 0\\ 0 & 0 & 1\end{pmatrix}^{l^2/4}\begin{pmatrix}1 & 0 & 0\\ \frac{l^2}{2} & 1 & 0\\ -l & 0 & 1\end{pmatrix}\\
&= \underline{\widetilde{e}}\begin{pmatrix}-1 & 0 & 0\\ 0 & -1 & 0\\ 0 & 0 & 1\end{pmatrix}^{l/2}\begin{pmatrix}1 & 0 & 0\\ -\frac{l^2}{2} & 1 & 0\\ 0 & 0 & 1\end{pmatrix}\begin{pmatrix}1 & 0 & 0\\ \frac{l^2}{2} & 1 & 0\\ -l & 0 & 1\end{pmatrix}\\
&= \underline{\widetilde{e}}\begin{pmatrix}-1 & 0 & 0\\ 0 & -1 & 0\\ 0 & 0 & 1\end{pmatrix}^{l/2}\begin{pmatrix}1 & 0 & 0\\ 0 & 1 & 0\\ -l & 0 & 1\end{pmatrix} = \underline{\widetilde{e}}\begin{pmatrix}(-1)^{l/2} & 0 & 0\\ 0 & (-1)^{l/2} & 0\\ -l & 0 & 1\end{pmatrix},
\end{aligned}$$

$$\begin{aligned}
s_4\underline{e} &= (-1)^{l/2}\underline{e} + f_3(-l, 1-(-1)^{l/2}, l),\\
s_4 &= \begin{cases} t_{\lambda_1}^+ \in \Gamma_u^{(1)} & \text{if } l \equiv 0(4),\\ t_{\lambda_3}^- \in \Gamma^{(1)} \cap O_\pm^{(1),Rad} & \text{if } l \equiv 2(4).\end{cases}
\end{aligned}\tag{6.12}$$

The splittings above of the exact sequences give semidirect products

$$\Gamma^{(1)} = \begin{cases} \Gamma_u^{(1)} \rtimes \langle s_{e_1}^{(1)}, s_{e_3}^{(1)}\rangle & \text{if } l \equiv 0(4),\\ \Gamma^{(1)} \cap O_\pm^{(1),Rad} \rtimes \langle s_{e_1}^{(1)}, s_{e_3}^{(1)}\rangle & \text{if } l \equiv 2(4).\end{cases}$$

$s_{e_2}^{(1)}$ turns up linearly in s_4. Therefore $\Gamma^{(1)} = \langle s_{e_1}^{(1)}, s_{e_3}^{(1)}, s_4\rangle$. Together these facts show that $\Gamma_u^{(1)}$ respectively $\Gamma^{(1)} \cap O_\pm^{(1),Rad}$ is the normal subgroup generated by s_4. This finishes the proof of the Claim. (□)

Now we can apply Lemma 6.17 (c) if $l \equiv 0(4)$ and Lemma 6.17 (d) if $l \equiv 2(4)$. The following calculations show the claims on $\Gamma_u^{(1)}$ and (in the case $l \equiv 2(4)$) $\Gamma^{(1)} \cap O_\pm^{(1),Rad}$.

The case $l \equiv 0(4)$:

$$\begin{aligned}
\lambda_1 \circ s_{e_1}^{(1)}(\underline{e}) &= (-l, -l^2, 3l),\\
\text{so } \lambda_1 \circ s_{e_1}^{(1)} &= 3\lambda_1 + \lambda_2,\\
\lambda_1 \circ (s_{e_i}^{(1)})^{\pm 1}, \lambda_2 \circ (s_{e_i}^{(1)})^{\pm 1} &\in \langle \lambda_1, \lambda_2\rangle_{\mathbb{Z}}.
\end{aligned}$$

The case $l \equiv 2(4)$:

$$\begin{aligned}
\lambda_3 \circ s_{e_1}^{(1)}(\underline{e}) &= (-l, 2-l^2, 3l),\\
\text{so } \lambda_3 \circ s_{e_1}^{(1)} &= \lambda_3 + 2\lambda_1 + \lambda_2,\\
\lambda_3 \circ (s_{e_2}^{(1)})^{-1}(\underline{e}) &= (l, 2, -l),
\end{aligned}$$

$$\text{so } \lambda_3 \circ (s^{(1)}_{e_2})^{-1} = \lambda_3 - 2\lambda_1,$$
$$\lambda_3 \circ (s^{(1)}_{e_i})^{\pm 1} \in \lambda_3 + \langle 2\lambda_1, \lambda_2\rangle_{\mathbb{Z}},$$
$$2\lambda_1 \circ (s^{(1)}_{e_i})^{\pm 1}, \lambda_2 \circ (s^{(1)}_{e_i})^{\pm 1} \in \langle 2\lambda_1, \lambda_2\rangle_{\mathbb{Z}}.$$

(f) Rad $I^{(1)} = \mathbb{Z} f_3$ is known. $e_2 = -\frac{l}{2}(e_1 + e_3) + \frac{1}{2} f_3$ shows $\overline{H_{\mathbb{Z}}}^{(1)} = \mathbb{Z}\overline{e_1}^{(1)} \oplus \mathbb{Z}(\frac{1}{2}(\overline{e_1}^{(1)} + \overline{e_3}^{(1)}))$. We have

$$\underline{\widetilde{e}} = \underline{e} \begin{pmatrix} 1 & \frac{1-l^2}{2} & l \\ 0 & -l & 2 \\ 0 & \frac{1-l^2}{2} & l \end{pmatrix}, \quad \underline{e} = \underline{\widetilde{e}} \begin{pmatrix} 1 & 0 & -1 \\ 0 & -l & 2 \\ 0 & \frac{1-l^2}{2} & l \end{pmatrix},$$

so $\underline{\widetilde{e}}$ is a $\mathbb{Z}$-basis of $H_{\mathbb{Z}}$.

First we calculate s_4 with respect to the $\mathbb{Q}$-basis $\underline{\widetilde{f}} := (e_1, \frac{1}{2}(e_1 + e_3), f_3)$ of $H_{\mathbb{Q}}$,

$$\begin{aligned} s_4(\underline{\widetilde{f}}) &= \underline{\widetilde{f}}\left(\begin{pmatrix} -1 & -1 & 0 \\ 4 & 3 & 0 \\ 0 & 0 & 1 \end{pmatrix} \begin{pmatrix} 1 & -1 & 0 \\ 0 & 1 & 0 \\ 0 & 0 & 1 \end{pmatrix}\right)^{\frac{l^2-1}{4}} \begin{pmatrix} 1 & 0 & 0 \\ l^2 & 1 & 0 \\ -\frac{l}{2} & 0 & 1 \end{pmatrix} \\ &= \underline{\widetilde{f}} \begin{pmatrix} -1 & 0 & 0 \\ 4 & -1 & 0 \\ 0 & 0 & 1 \end{pmatrix}^{\frac{l^2-1}{4}} \begin{pmatrix} 1 & 0 & 0 \\ l^2 & 1 & 0 \\ -\frac{l}{2} & 0 & 1 \end{pmatrix} \\ &= \underline{\widetilde{f}} \begin{pmatrix} 1 & 0 & 0 \\ 1-l^2 & 1 & 0 \\ 0 & 0 & 1 \end{pmatrix} \begin{pmatrix} 1 & 0 & 0 \\ l^2 & 1 & 0 \\ -\frac{l}{2} & 0 & 1 \end{pmatrix} = \underline{\widetilde{f}} \begin{pmatrix} 1 & 0 & 0 \\ 1 & 1 & 0 \\ -\frac{l}{2} & 0 & 1 \end{pmatrix}. \end{aligned}$$

Therefore

$$s_4(\underline{\widetilde{e}}) = \underline{\widetilde{e}} \begin{pmatrix} 1 & 0 & 0 \\ 1 & 1 & 0 \\ 0 & 0 & 1 \end{pmatrix}, \quad s^{(1)}_{e_1}(\underline{\widetilde{e}}) = \underline{\widetilde{e}} \begin{pmatrix} 1 & -1 & 0 \\ 0 & 1 & 0 \\ 0 & 0 & 1 \end{pmatrix}. \tag{6.13}$$

Thus

$$\Gamma^{(1)}_s = \langle \overline{s^{(1)}_{e_1}}, \overline{s_4}\rangle \cong \langle s^{(1)}_{e_1}, s_4\rangle \subset \Gamma^{(1)},$$

and this isomorphism gives a splitting of the exact sequence (6.10).

Equation (6.13) shows that the group $\langle s^{(1)}_{e_1}, s_4\rangle$ contains an element s_5 with

$$s_5(\underline{\widetilde{e}}) = \underline{\widetilde{e}} \begin{pmatrix} 3 & 1 & 0 \\ -4 & -1 & 0 \\ 0 & 0 & 1 \end{pmatrix},$$

thus

$$s_5 s_{e_3}^{(1)}(\underline{\widetilde{e}}) = \underline{\widetilde{e}} \begin{pmatrix} 3 & 1 & 0 \\ -4 & -1 & 0 \\ 0 & 0 & 1 \end{pmatrix} \begin{pmatrix} -1 & -1 & 0 \\ 4 & 3 & 0 \\ 2l & l & 1 \end{pmatrix} = \underline{\widetilde{e}} \begin{pmatrix} 1 & 0 & 0 \\ 0 & 1 & 0 \\ 2l & l & 1 \end{pmatrix},$$

$$s_5 s_{e_3}^{(1)}(\underline{e}) = \underline{e} + f_3(2l, -l^2, 0),$$

$$s_5 s_{e_3}^{(1)} = t_{\lambda_2}^+ \quad \text{with} \quad \lambda_2(\underline{e}) = (2l, -l^2, 0).$$

Also

$$\Gamma^{(1)} = \langle s_{e_1}^{(1)}, s_{e_2}^{(1)}, s_{e_3}^{(1)} \rangle = \langle s_4, s_{e_1}^{(1)}, s_{e_3}^{(1)} \rangle = \langle s_4, s_{e_1}^{(1)}, t_{\lambda_2}^+ \rangle,$$

so $\Gamma_u^{(1)}$ is the normal subgroup generated by $t_{\lambda_2}^+$.

Now we can apply Lemma 6.17 (c). The following formulas show $\Gamma_u^{(1)} = \{t_\lambda^+ \mid \lambda \in \langle \lambda_1, \lambda_2 \rangle_{\mathbb{Z}}\}$.

$$\begin{aligned} (2l, -l^2, 0) s_{e_1}^{(1),mat} &= (2l, l^2, -4l) = (2l, -l^2, 0) + (0, 2l^2, -4l), \\ 2(2l, -l^2, 0) + (0, 2l^2, -4l) &= (4l, 0, -4l), \\ (2l, -l^2, 0) s_{e_2}^{(1),mat} &= (2l, -l^2, 0) + (l^3, 0, -l^3), \\ \gcd(4l, l^3) = l, &\quad \text{so } \lambda_1 \in \Gamma_u^{(1)}, \\ \lambda_1 \circ (s_{e_i}^{(1)})^{\pm 1}, \lambda_2 \circ (s_{e_i}^{(1)})^{\pm 1} &\in \langle \lambda_1, \lambda_2 \rangle_{\mathbb{Z}}. \end{aligned}$$

(g) Because of the cyclic action of $\gamma \in (G^{phi} \ltimes \widetilde{G}^{sign}) \rtimes \langle \gamma \rangle$ on $\mathbb{Z}^3$, we can suppose $x_1 = \max(x_1, x_2, x_3)$. With respect to the $\mathbb{Q}$-basis $(\overline{e_1}^{(1)}, \overline{e_2}^{(1)})$ of $\overline{H_{\mathbb{Q}}}^{(1)}$, $\overline{s_{e_1}^{(1)}}, \overline{s_{e_2}^{(1)}}$ and $\overline{s_{e_3}^{(1)}}$ take the shape $\overline{s_{e_i}^{(1)}}(\overline{e_1}^{(1)}, \overline{e_2}^{(1)}) = (\overline{e_1}^{(1)}, \overline{e_2}^{(1)}) B_i$ with

$$B_1 = \begin{pmatrix} 1 & -x_1 \\ 0 & 1 \end{pmatrix}, \quad B_2 = \begin{pmatrix} 1 & 0 \\ x_1 & 1 \end{pmatrix}, \quad B_3 = \begin{pmatrix} 1 - \frac{x_3 x_2}{x_1} & -\frac{x_3^2}{x_1} \\ \frac{x_2^2}{x_1} & 1 + \frac{x_3 x_2}{x_1} \end{pmatrix}.$$

We will show that the group $\langle \mu_1, \mu_2, \mu_3 \rangle$ of Möbius transformations $\mu_i := \mu(B_i)$ is a free group with the three generators μ_1, μ_2, μ_3. This implies first $\Gamma_s^{(1)} \cong G^{free,3}$ and then $\Gamma^{(1)} \cong \Gamma_s^{(1)} \cong G^{free,3}$ and $\Gamma_u^{(1)} = \{\mathrm{id}\}$.

We will apply Theorem A.2 (c). μ_1, μ_2 and μ_3 are parabolic,

$$\begin{aligned} \mu_1 &= (z \mapsto z - x_1) \quad \text{with fixed point } \infty, \\ \mu_2 &= \left(z \mapsto \frac{z}{x_1 z + 1}\right) \quad \text{with fixed point } 0, \\ \mu_3 &= \left(z \mapsto \frac{(1 - \frac{x_3 x_2}{x_1}) z - \frac{x_3^2}{x_1}}{\frac{x_2^2}{x_1} z + (1 + \frac{x_3 x_2}{x_1})} = \frac{(x_1 - x_2 x_3) z - x_3^2}{x_2^2 z + (x_1 + x_2 x_3)}\right) \\ &\qquad \text{with fixed point } -\frac{x_3}{x_2}. \end{aligned}$$

Consider

$$r_1 := \mu_1(1) = 1 - x_1, \quad r_2 := \mu_2^{-1}(1) = \frac{1}{1-x_1},$$
$$r_3 := \mu_3(r_1) = \frac{(x_1 - x_2x_3)(1-x_1) - x_3^2}{x_2^2(1-x_1) + (x_1 + x_2x_3)}.$$

It is sufficient to show the inequalities

$$-\infty < r_1 < -\frac{x_3}{x_2} < r_3 \le r_2 < 0 < 1 < \infty. \tag{6.14}$$

Then Theorem A.2 (c) applies. Compare Fig. 6.6.

We treat the case $\underline{x} = (3, 3, 3)$ separately and first. Then

$$(r_1, -\frac{x_3}{x_2}, r_3, r_2) = (-2, -1, -\frac{1}{2}, -\frac{1}{2}),$$

so then (6.14) holds.

From now on we suppose $x_1 \ge 4$. The inequality $r_2 < 0$ is trivial. Consider the number

$$r_4 := -\frac{x_1 + x_2x_3}{x_2^2} = -\frac{x_1}{x_2^2} - \frac{x_3}{x_2} < -\frac{x_3}{x_2}, \quad \text{with } \mu_3(r_4) = \infty.$$

We will show in this order the following claims:

(i) $r_1 < r_4$, which implies $r_1 < -\frac{x_3}{x_2}$.
(ii) $r_3 \in (-\frac{x_3}{x_2}, \infty)$.
(iii) The numerator $(x_2x_3 - x_1)(x_1 - 1) - x_3^2$ of r_3 is positive.

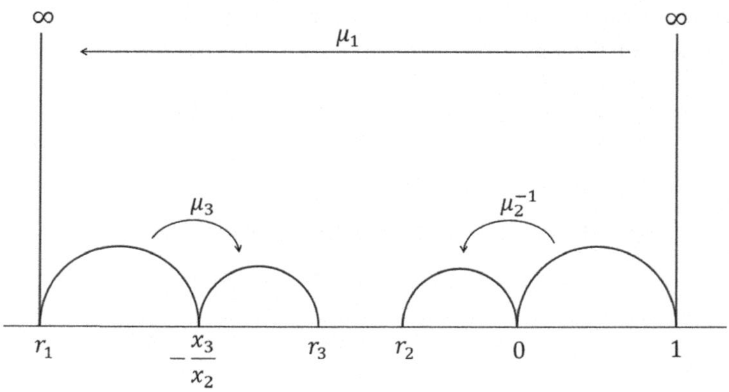

Fig. 6.6 $G^{free,3}$ generated by three parabolic Möbius transformations, an application of Theorem A.2 (c)

(iv) The denominator $x_2^2(1-x_1)+(x_1+x_2x_3)$ of r_3 is negative.
(v) $r_3 \leq r_2$.

Together (i)–(v) and $r_2 < 0$ show (6.14).

Two of the three inequalities $2x_i \leq x_jx_k$ for $\{i, j, k\} = \{1, 2, 3\}$ in the assumption on $\underline{x}$ can be improved,

$$x_1x_2 \geq 3x_1 \geq 3x_3, \quad x_1x_3 \geq 3x_1 \geq 3x_2. \quad \text{we keep} \quad x_2x_3 \geq 2x_1.$$

(i) holds because

$$r_1 < r_4 \iff 1 < x_1 - \frac{x_1}{x_2^2} - \frac{x_3}{x_2} \quad \text{and}$$
$$x_1 - \frac{x_1}{x_2^2} - \frac{x_3}{x_2} \geq x_1 - \frac{x_1}{9} - \frac{x_1}{3} = \frac{5x_1}{9} \geq \frac{20}{9}.$$

(ii) $r_1 < r_4$ and $\mu_3(r_4) = \infty$ imply $r_3 = \mu_3(r_1) \in (-\frac{x_3}{x_2}, \infty)$.
(iii) holds because

$$(x_2x_3 - x_1)(x_1 - 1) - x_3^2 > 0 \iff x_1 - 1 > \frac{x_3^2}{x_2x_3 - x_1} = \frac{x_3}{x_2}\frac{1}{1 - \frac{x_1}{x_2x_3}}$$
$$\text{and } \frac{x_3}{x_2}\frac{1}{1 - \frac{x_1}{x_2x_3}} \leq \frac{x_1}{3}\frac{1}{1 - \frac{1}{2}} = \frac{2x_1}{3}, \quad x_1 - 1 > \frac{2x_1}{3} \Leftarrow x_1 \geq 4.$$

(iv) holds because

$$x_2^2(x_1 - 1) - (x_1 + x_2x_3) > 0 \iff x_1 - 1 > \frac{x_1 + x_2x_3}{x_2^2} = \frac{x_3}{x_2}(1 + \frac{x_1}{x_2x_3})$$
$$\text{and } \frac{x_3}{x_2}(1 + \frac{x_1}{x_2x_3}) \leq \frac{x_1}{3}(1 + \frac{1}{2}) = \frac{x_1}{2}, \quad x_1 - 1 > \frac{x_1}{2} \Leftarrow x_1 \geq 4.$$

(ii)–(iv) show $r_3 \in (-\frac{x_3}{x_2}, 0)$. Now

$$r_3 \leq r_2 \iff \big[(x_2x_3 - x_1)(x_1 - 1) - x_3^2\big](x_1 - 1)$$
$$-\big[x_2^2(x_1 - 1) - (x_1 + x_2x_3)\big] \geq 0.$$

The right hand side is

$$g(x_1, x_2, x_3) := (x_2x_3 - x_1)(x_1 - 1)^2 - (x_2^2 + x_3^2)(x_1 - 1) + (x_1 + x_2x_3).$$

This is symmetric and homogeneous of degree 2 in x_2 and x_3.

$$\frac{\partial g}{\partial x_2}(\underline{x}) = (x_1 - 1)^2x_3 - 2(x_1 - 1)x_2 + x_3,$$

so for fixed x_1 and x_3 $g(\underline{x})$ takes its maximum in

$$x_2^0 = \frac{(x_1-1)^2x_3 + x_3}{2(x_1-1)} = \frac{x_3(x_1-1)}{2} + \frac{x_3}{2(x_1-1)} > \frac{3(x_1-1)}{2} > x_1,$$

and is monotonously increasing left of x_2^0.

Because of the symmetry we can restrict to the cases $x_1 \geq x_2 \geq x_3$. Then $g(\underline{x})$ takes for fixed x_1 and x_3 its minimum in $x_2 = x_3$.

$$\widetilde{g}(x_1, x_3) := g(x_1, x_3, x_3) = x_3^2(x_1-2)^2 - (x_1-1)^2x_1 + x_1.$$

$\widetilde{g}$ takes for fixed x_1 its minimum at the minimal possible x_3, which is $x_3^0 := \max(3, \sqrt{2x_1})$ because of $2x_1 \leq x_2x_3 = x_3^2$.

$$\widetilde{g}(x_1, x_3^0) = \begin{cases} \widetilde{g}(4,3) = 4 > 0 & \text{if } x_1 = 4, \\ \widetilde{g}(x_1, \sqrt{2x_1}) = x_1^3 - 6x_1^2 + 8x_1 & \\ \qquad = x_1(x_1-2)(x_1-4) > 0 & \text{if } x_1 \geq 5. \end{cases}$$

Therefore $r_3 < r_2$ and (6.14) is proved. □

Remarks 6.19 (i) In the proof of part (g) of Theorem 6.18 the hyperbolic polygon P whose relative boundary is the union of the six arcs

$$A(\infty, r_1),\ A(r_1, -\frac{x_3}{x_2}),\ A(-\frac{x_3}{x_2}, r_3),\ A(r_2, 0),\ A(0,1),\ A(1,\infty),$$

is a fundamental domain for the action of the group $\langle \mu_1, \mu_2, \mu_3\rangle$ on the upper half plane $\mathbb{H}$.

(ii) In part (g) of Theorem 6.18 the case $\underline{x} = (3,3,3)$ is especially interesting. It is the only case within part (g) where $r_3 = r_2$, so the only case in part (g) where the hyperbolic polygon P has finite hyperbolic area. In this case

$$\langle B_1, B_2, B_3\rangle = \langle \begin{pmatrix} 1 & -3 \\ 0 & 1 \end{pmatrix}, \begin{pmatrix} 1 & 0 \\ 3 & 1 \end{pmatrix}, \begin{pmatrix} -2 & -3 \\ 3 & 4 \end{pmatrix}\rangle = \Gamma(3).$$

Now we turn to the study of the set $\Delta^{(1)}$ of odd vanishing cycles. The shape of $\Delta^{(1)}$ and our knowledge on it are very different for different $\underline{x} \in \mathbb{Z}^3$.

Remarks 6.20 (i) In the cases in the parts (c)–(f) of Theorem 6.21 we will give $\Delta^{(1)}$ rather explicitly. There $\overline{\Delta^{(1)}}$ is known, and we can give a subset of $\Delta^{(1)}$ explicitly which maps bijectively to $\overline{\Delta^{(1)}}$.

Furthermore $\Gamma_u^{(1)} \cong \mathbb{Z}^2$, and for $\varepsilon \in \{\pm 1\}$ and $i \in \{1,2,3\}$ there is a subset $F_{\varepsilon,i} \subset \mathbb{Z}f_3$ with

$$\Delta^{(1)} \cap (\varepsilon e_i + \mathbb{Z}f_3) = \varepsilon e_i + F_{\varepsilon,i} \quad \text{and}$$

$$\Delta^{(1)} \cap (a + \mathbb{Z}f_3) = a + F_{\varepsilon,i} \quad \text{for any } a \in \Gamma^{(1)}\{\varepsilon e_i\}.$$

In many, but not all cases $F_{\varepsilon,i} \subset \mathbb{Z}f_3$ is a sublattice.

(ii) On the contrary, in the cases in part (g) of Theorem 6.21 we know rather little. There $\Gamma_u^{(1)} = \{\mathrm{id}\}$, and remarkably the projection $\Delta^{(1)} \to \overline{\Delta^{(1)}} \subset \overline{H_{\mathbb{Z}}}^{(1)}$ is a bijection. In the case $\underline{x} = (3,3,3)$ we know $\overline{\Delta^{(1)}}$. But the lift $a \in \Delta^{(1)}$ of an element $\overline{a}^{(1)} \in \overline{\Delta^{(1)}}$ is difficult to determine. See Lemma 6.26.

Theorem 6.21 *(a) (Empty: (b)–(g) shall correspond to (b)–(g) in Theorem 6.18.)*

(b) In each reducible case $\underline{x} = (x_1, 0, 0)$

$$\begin{aligned} \Delta^{(1)} &= \Delta^{(1)} \cap (\mathbb{Z}e_1 \oplus \mathbb{Z}e_2)\ \dot\cup\ \{\pm e_3\} \\ \text{with} \quad \Delta^{(1)} &\cap (\mathbb{Z}e_1 \oplus \mathbb{Z}e_2) \cong \Delta^{(1)}(S(-x_1)), \end{aligned}$$

and $\Delta^{(1)}(S(-x_1))$ *is given in Theorem 6.10.*

(c) The case $\underline{x} = (1,1,0)$*:*

$$\begin{aligned} \overline{\Delta^{(1)}} &= \overline{H_{\mathbb{Z}}}^{(1),prim}, \\ \Delta^{(1)} &= (pr^{H,(1)})^{-1}(\overline{\Delta^{(1)}}) = (\mathbb{Z}e_1 \oplus \mathbb{Z}e_2)^{prim} + \mathbb{Z}f_3 \subset H_{\mathbb{Z}}^{prim}, \\ \Delta^{(1)} &= \Gamma^{(1)}\{e_1\}. \end{aligned}$$

$\Delta^{(1)}$ *and* $\overline{\Delta^{(1)}}$ *consist each of one orbit.*

(d) The cases $\underline{x} = (x_1, x_2, 0)$ *with* $2 \le x_1 \ge x_2 > 0$*: Recall* $x_{12} := \gcd(x_1, x_2)$.

(i) The cases with $x_{12} = 1$ *and* $x_2 > 1$*: Then* $x_1 = \widetilde{x}_1$, $x_2 = \widetilde{x}_2$ *and* $x_1 > x_2 > 1$. *Choose* $y_1, y_2 \in \mathbb{Z}$ *with* $1 = y_1 x_1^2 + y_2 x_2^2$. $\overline{\Delta^{(1)}}$ *consists of the three orbits*

$$\begin{aligned} \Gamma_s^{(1)}\{\overline{e_1}^{(1)}\} &= \overline{H_{\mathbb{Z}}}^{(1),prim}, \\ \Gamma_s^{(1)}\{\overline{e_2}^{(1)}\} &= x_1 \overline{H_{\mathbb{Z}}}^{(1),prim}, \\ \Gamma_s^{(1)}\{\overline{e_3}^{(1)}\} &= x_2 \overline{H_{\mathbb{Z}}}^{(1),prim}. \end{aligned}$$

$\Delta^{(1)}$ *consists of five orbits:*

$$\Gamma^{(1)}\{e_1\} = \Gamma^{(1)}\{\pm e_1\}, \quad \Gamma^{(1)}\{\varepsilon e_2\}, \quad \Gamma^{(1)}\{\varepsilon e_3\} \quad \text{with } \varepsilon \in \{\pm 1\}.$$

Here

$$\begin{aligned} &\Delta^{(1)} \cap (e_1 + \mathbb{Z}f_3) = \Gamma^{(1)}\{e_1\} \cap (e_1 + \mathbb{Z}f_3) \\ &\quad = \Gamma_u^{(1)}\{e_1\} = e_1 + \mathbb{Z}x_1x_2 f_3, \\ &\Delta^{(1)} \cap (\varepsilon e_2 + \mathbb{Z}f_3) \\ &\quad = \Big(\Gamma^{(1)}\{\varepsilon e_2\} \cap (\varepsilon e_2 + \mathbb{Z}f_3)\Big)\ \dot\cup\ \Big(\Gamma^{(1)}\{-\varepsilon e_2\} \cap (\varepsilon e_2 + \mathbb{Z}f_3)\Big) \\ &\quad = \Big(\varepsilon e_2 + \mathbb{Z}x_1^2 x_2 f_3\Big)\ \dot\cup\ \Big(\varepsilon e_2 - \varepsilon 2 y_2 x_2 f_3 + \mathbb{Z}x_1^2 x_2 f_3\Big), \end{aligned}$$

$$\begin{aligned}
&\Delta^{(1)} \cap (\varepsilon e_3 + \mathbb{Z} f_3) \\
&\quad = \Big(\Gamma^{(1)}\{\varepsilon e_3\} \cap (\varepsilon e_3 + \mathbb{Z} f_3)\Big) \,\dot\cup\, \Big(\Gamma^{(1)}\{-\varepsilon e_3\} \cap (\varepsilon e_3 + \mathbb{Z} f_3)\Big) \\
&\quad = \Big(\varepsilon e_3 + \mathbb{Z} x_1 x_2^2 f_3)\Big) \,\dot\cup\, \Big(\varepsilon e_3 + \varepsilon 2 y_1 x_1 f_3 + \mathbb{Z} x_1 x_2^2 f_3\Big).
\end{aligned}$$

(ii) The cases with $x_2 = 1$: Then $x_1 = \widetilde{x}_1 > x_2 = \widetilde{x}_2 = x_{12} = 1$. $\overline{\Delta^{(1)}}$ consists of the two orbits

$$\begin{aligned}
\Gamma_s^{(1)}\{\overline{e_1}^{(1)}\} &= \Gamma_s^{(1)}\{\pm\overline{e_1}^{(1)}, \pm\overline{e_3}^{(1)}\} = \overline{H_{\mathbb{Z}}}^{(1),prim}, \\
\Gamma_s^{(1)}\{\overline{e_2}^{(1)}\} &= \Gamma_s^{(1)}\{\pm\overline{e_2}^{(1)}\} = x_1 \overline{H_{\mathbb{Z}}}^{(1),prim}.
\end{aligned}$$

$\Delta^{(1)}$ consists of three orbits,

$$\Gamma^{(1)}\{e_1\} = \Gamma^{(1)}\{\pm e_1, \pm e_3\}, \quad \Gamma^{(1)}\{\varepsilon e_2\}, \quad \text{with } \varepsilon \in \{\pm 1\}.$$

Here

$$\begin{aligned}
&\Delta^{(1)} \cap (e_1 + \mathbb{Z} f_3) = \Gamma^{(1)}\{e_1\} \cap (e_1 + \mathbb{Z} f_3) = \Gamma_u^{(1)}\{e_1\} = e_1 + \mathbb{Z} x_1 f_3, \\
&\Delta^{(1)} \cap (\varepsilon e_2 + \mathbb{Z} f_3) \\
&\quad = \Big(\Gamma^{(1)}\{\varepsilon e_2\} \cap (\varepsilon e_2 + \mathbb{Z} f_3)\Big) \,\dot\cup\, \Big(\Gamma^{(1)}\{-\varepsilon e_2\} \cap (\varepsilon e_2 + \mathbb{Z} f_3)\Big) \\
&\quad = \Big(\varepsilon e_2 + \mathbb{Z} x_1^2 f_3\Big) \,\dot\cup\, \Big(\varepsilon e_2 - \varepsilon 2 f_3 + \mathbb{Z} x_1^2 f_3\Big).
\end{aligned}$$

(iii) The cases with $x_{12} > 1$ and $x_1 > x_2 > 1$: Then $\widetilde{x}_1 > \widetilde{x}_2 \geq 1$. Recall from Theorem 6.18 (s) $g_2 = \frac{1}{\widetilde{x}_1}\overline{e_2}^{(1)} = \frac{1}{\widetilde{x}_2}\overline{e_3}^{(1)} \in \overline{H_{\mathbb{Z}}}^{(1)}$. Here $\overline{\Delta^{(1)}} \subset \overline{H_{\mathbb{Z}}}^{(1)}$ consists of the six orbits (with $\varepsilon \in \{\pm 1\}$)

$$\begin{aligned}
&\Gamma_s^{(1)}\{\varepsilon\overline{e_1}^{(1)}\} \subset \overline{H_{\mathbb{Z}}}^{(1),prim}, \\
&\Gamma_s^{(1)}\{\varepsilon\overline{e_2}^{(1)}\} = \widetilde{x}_1 \Gamma_s^{(1)}\{\varepsilon g_2\} \subset \widetilde{x}_1 \overline{H_{\mathbb{Z}}}^{(1),prim}, \\
&\Gamma_s^{(1)}\{\varepsilon\overline{e_3}^{(1)}\} = \widetilde{x}_2 \Gamma_s^{(1)}\{\varepsilon g_2\} \subset \widetilde{x}_2 \overline{H_{\mathbb{Z}}}^{(1),prim}.
\end{aligned}$$

Theorem 6.10 (c) and (d) describe the orbits $\Gamma_s^{(1)}\{\overline{e_1}^{(1)}\}$ and $\Gamma_s^{(1)}\{g_2\}$. Also $\Delta^{(1)}$ consists of six orbits.

$$\begin{aligned}
\Delta^{(1)} \cap (\varepsilon e_1 + \mathbb{Z} f_3) &= \varepsilon e_1 + \mathbb{Z} x_{12} \widetilde{x}_1 \widetilde{x}_2 f_3, \\
\Delta^{(1)} \cap (\varepsilon e_2 + \mathbb{Z} f_3) &= \varepsilon e_2 + \mathbb{Z} x_1 x_2 \widetilde{x}_1 f_3, \\
\Delta^{(1)} \cap (\varepsilon e_3 + \mathbb{Z} f_3) &= \varepsilon e_3 + \mathbb{Z} x_1 x_2 \widetilde{x}_2 f_3.
\end{aligned}$$

(iv) The cases with $x_1 = x_2 \geq 2$: Then $x_{12} = x_1 = x_2$, $\widetilde{x}_1 = \widetilde{x}_2 = 1$, $\overline{e_2}^{(1)} = \overline{e_3}^{(1)} = g_2$. Then $\overline{\Delta^{(1)}} \subset \overline{H_{\mathbb{Z}}}^{(1)}$ consists of the four orbits (with $\varepsilon \in \{\pm 1\}$)

$$\Gamma_s^{(1)}\{\varepsilon\overline{e_1}^{(1)}\} \subset \overline{H_{\mathbb{Z}}}^{(1),prim},$$
$$\Gamma_s^{(1)}\{\varepsilon\overline{e_2}^{(1)}\} = \Gamma_s^{(1)}\{\varepsilon g_2\} \subset \overline{H_{\mathbb{Z}}}^{(1),prim}.$$

Theorem 6.10 (c) and (d) describe these orbits. $\Delta^{(1)}$ consists of six orbits.

$$\begin{aligned}
\Delta^{(1)} \cap (\varepsilon e_1 + \mathbb{Z}f_3) &= \Gamma^{(1)}\{\varepsilon e_1\} \cap (\varepsilon e_1 + \mathbb{Z}f_3) = \Gamma_u^{(1)}\{\varepsilon e_1\} \\
&= \varepsilon e_1 + \mathbb{Z}x_1 f_3, \\
\Delta^{(1)} \cap (\varepsilon e_2 + \mathbb{Z}f_3) &= \Delta^{(1)} \cap (\varepsilon e_3 + \mathbb{Z}f_3) \\
&= \Big(\Gamma^{(1)}\{\varepsilon e_2\} \cap (\varepsilon e_2 + \mathbb{Z}f_3)\Big) \dot\cup \Big(\Gamma^{(1)}\{\varepsilon e_3\} \cap (\varepsilon e_2 + \mathbb{Z}f_3)\Big) \\
&= \Big(\varepsilon e_2 + \mathbb{Z}x_1^2 f_3\Big) \dot\cup \Big(\varepsilon e_2 - \varepsilon f_3 + \mathbb{Z}x_1^2 f_3\Big).
\end{aligned}$$

(e) The cases $\underline{x} = (-l, 2, -l)$ with $l \geq 2$ even: Consider $\varepsilon \in \{\pm 1\}$.
(i) The cases with $l \equiv 0(4)$: $\Delta^{(1)}$ consists of six orbits,

$$\begin{aligned}
\Gamma^{(1)}\{\varepsilon e_1\} &= \{y_1 e_1 + y_2 e_3 \in H_{\mathbb{Z}}^{prim} \mid y_1 \equiv \varepsilon(4),\, y_2 \equiv 0(2)\} + \mathbb{Z}l f_3, \\
\Gamma^{(1)}\{\varepsilon e_3\} &= \{y_1 e_1 + y_2 e_3 \in H_{\mathbb{Z}}^{prim} \mid y_1 \equiv 0(2),\, y_2 \equiv \varepsilon(4)\} + \mathbb{Z}l f_3, \\
\Gamma^{(1)}\{\varepsilon e_2\} &= \frac{l}{2}\{y_1 e_1 + y_2 e_3 \in H_{\mathbb{Z}}^{prim} \mid y_1 \equiv 1(2),\, y_2 \equiv 1(2)\} \\
&\quad + \varepsilon f_3 + \mathbb{Z}l^2 f_3.
\end{aligned}$$

$\overline{\Delta^{(1)}}$ consists of five orbits, $\Gamma_s^{(1)}\{\overline{e_2}^{(1)}\} = \Gamma_s^{(1)}\{-\overline{e_2}^{(1)}\}$.
(ii) The cases with $l \equiv 2(4)$: $\Delta^{(1)}$ consists of six orbits,

$$\begin{aligned}
\Gamma^{(1)}\{\varepsilon e_1\} &= \Big(\{y_1 e_1 + y_2 e_3 \in H_{\mathbb{Z}}^{prim} \mid y_1 \equiv \varepsilon(4),\, y_2 \equiv 0(2)\} + \mathbb{Z}2l f_3\Big) \\
&\quad \dot\cup \Big(\{y_1 e_1 + y_2 e_3 \in H_{\mathbb{Z}}^{prim} \mid y_1 \equiv -\varepsilon(4),\, y_2 \equiv 0(2)\} + l f_3 + \mathbb{Z}2l f_3\Big), \\
\Gamma^{(1)}\{\varepsilon e_3\} &= \Big(\{y_1 e_1 + y_2 e_3 \in H_{\mathbb{Z}}^{prim} \mid y_1 \equiv 0(2),\, y_2 \equiv \varepsilon(4)\} + \mathbb{Z}2l f_3\Big) \\
&\quad \dot\cup \Big(\{y_1 e_1 + y_2 e_3 \in H_{\mathbb{Z}}^{prim} \mid y_1 \equiv 0(2),\, y_2 \equiv -\varepsilon(4)\} + l f_3 + \mathbb{Z}2l f_3\Big), \\
\Gamma^{(1)}\{\varepsilon e_2\} &= \frac{l}{2}\{y_1 e_1 + y_2 e_3 \in H_{\mathbb{Z}}^{prim} \mid y_1 \equiv 1(2),\, y_2 \equiv 1(2)\} \\
&\quad + \varepsilon f_3 + \mathbb{Z}l^2 f_3.
\end{aligned}$$

Especially

$$\begin{aligned}
&\Delta^{(1)} \cap (\varepsilon e_1 + \mathbb{Z}f_3) = \varepsilon e_1 + \mathbb{Z}l f_3 \\
&\supsetneqq \Gamma^{(1)}\{\varepsilon e_1\} \cap (\varepsilon e_1 + \mathbb{Z}f_3) = \varepsilon e_1 + \mathbb{Z}2l f_3,
\end{aligned}$$

and similarly for εe_3. $\overline{\Delta^{(1)}}$ *consists of three orbits,* $\Gamma_s^{(1)}\{\overline{e_i}^{(1)}\} = \Gamma_s^{(1)}\{-\overline{e_i}^{(1)}\}$ *for* $i \in \{1, 2, 3\}$.

(f) The cases $\underline{x} = (-l, 2, -l)$ *with* $l \geq 3$ *odd:* $\Delta^{(1)}$ *consists of three orbits (with* $\varepsilon \in \{\pm 1\}$*)*

$$\Gamma^{(1)}\{e_1\} = \Gamma^{(1)}\{\pm e_1, \pm e_3\} = (\mathbb{Z}e_1 \oplus \mathbb{Z}g_2)^{prim} + \mathbb{Z}lf_3,$$
$$\Gamma^{(1)}\{\varepsilon e_2\} = l(\mathbb{Z}e_1 \oplus \mathbb{Z}g_2)^{prim} + \varepsilon\frac{1-l^2}{2}f_3 + \mathbb{Z}l^2 f_3.$$

$\overline{\Delta^{(1)}}$ *consists of two orbits,* $\Gamma_s^{(1)}\{\overline{e_2}^{(1)}\} = \Gamma_s^{(1)}\{-\overline{e_2}^{(1)}\}$.

(g) The cases $\underline{x} \in \mathbb{Z}^3_{\geq 3}$ *with* $2x_i \leq x_j x_k$ *for* $\{i, j, k\} = \{1, 2, 3\}$*:* $\Delta^{(1)}$ *and* $\overline{\Delta^{(1)}}$ *consist each of six orbits,*

$$\Gamma^{(1)}\{\varepsilon e_i\} \quad \text{respectively} \quad \Gamma_s^{(1)}\{\varepsilon\overline{e_i}^{(1)}\} \quad \text{for } (\varepsilon, i) \in \{\pm 1\} \times \{1, 2, 3\}.$$

The projection $\Delta^{(1)} \to \overline{\Delta^{(1)}}$ *is a bijection.*

(h) The case $\underline{x} = (3, 3, 3)$*: In this subcase of (g) the statements in (g) hold, and*

$$\Gamma^{(1)}\{\varepsilon e_i\} \subset \varepsilon e_i + 3H_{\mathbb{Z}}, \quad \Gamma_s^{(1)}\{\varepsilon\overline{e_i}^{(1)}\} = (\varepsilon\overline{e_i}^{(1)} + 3\overline{H_{\mathbb{Z}}}^{(1)})^{prim}$$

for $\varepsilon \in \{\pm 1\}$*,* $i \in \{1, 2, 3\}$.

Proof (b) The splitting of $\Delta^{(1)}$ follows with Lemma 2.11. The first and second subset are treated by Theorem 6.10 and Lemma 2.12.

(c) $\overline{\Delta^{(1)}} = \overline{H_{\mathbb{Z}}}^{prim}$ follows with $\Gamma_s^{(1)} \cong SL_2(\mathbb{Z})$ and Theorem 6.10 (b). $\Delta^{(1)}$ is the full preimage in $H_{\mathbb{Z}}$, because $\Gamma_u^{(1)} = O_u^{(1),Rad} = \{t_\lambda^+ \mid \lambda(\underline{e}) \in \langle(1, 0, 0), (0, 1, 1)\rangle_{\mathbb{Z}}\}$.

(d) (i) All statements on $\overline{\Delta^{(1)}}$ follow from $\Gamma_s^{(1)} \cong SL_2(\mathbb{Z})$ and $\overline{e_2}^{(1)} = x_1 g_2, \overline{e_3}^{(1)} = x_2 g_2$.

Recall the definition and the properties of $s_4 \in \Gamma^{(1)}$ in the proof of Theorem 6.18. For $\Delta^{(1)}$ we have to use Lemma 6.7.

Claim 1: For $a \in \{\pm e_1, \pm e_2, \pm e_3\}$ equality in $\overset{(3)}{\supset}$ in Lemma 6.7 holds.

Proof of Claim 1: Suppose $h \in \mathrm{Stab}_{\Gamma^{(1)}}(\overline{a}^{(1)})$. Then

$$\overline{h} \in \mathrm{Stab}_{\Gamma_s^{(1)}}(\overline{a}^{(1)}) = \begin{cases} \langle\overline{s_{e_1}^{(1)}}\rangle & \text{if } a \in \{\pm e_1\}, \\ \langle\overline{s_4}\rangle & \text{if } a \in \{\pm e_2, \pm e_3\}, \end{cases}$$

so for a suitable $l \in \mathbb{Z}$

$$\overline{h}(\overline{s_{e_1}^{(1)}})^l = \mathrm{id} \text{ and } h(s_{e_1}^{(1)})^l \in \Gamma_u^{(1)} \quad \text{for } a \in \{\pm e_1\},$$
$$\overline{h}(\overline{s_4})^l = \mathrm{id} \text{ and } h(s_4)^l \in \Gamma_u^{(1)} \quad \text{for } a \in \{\pm e_2, \pm e_3\}.$$

As $s_{e_1}^{(1)} \in \mathrm{Stab}_{\Gamma^{(1)}}(\varepsilon e_1)$ and $s_4 \in \mathrm{Stab}_{\Gamma^{(1)}}(\varepsilon e_2) = \mathrm{Stab}_{\Gamma^{(1)}}(\varepsilon e_3)$

$$h \in \Gamma_u^{(1)} \mathrm{Stab}_{\Gamma^{(1)}}(a).$$

(□)

By Lemma 6.7 (b)

$$\Gamma^{(1)}\{\varepsilon e_i\} \cap (\varepsilon e_i + \mathbb{Z} f_3) = \Gamma_u^{(1)}\{\varepsilon e_i\}. \tag{6.15}$$

Claim 2: $(s_{e_1}^{(1)} s_4 s_{e_1}^{(1)})^2 = t_{\lambda_3}^-$ with $\lambda_3 \in \mathrm{Hom}_2(H_\mathbb{Z}, \mathbb{Z}), \lambda_3(\underline{e}) = (0, 2y_2x_2, -2y_1x_1)$.

Proof of Claim 2: This is a straightforward calculation with $s_{e_1}^{(1),mat}$ and s_4^{mat}.(□)

Therefore

$$\begin{aligned}
\Gamma^{(1)}\{e_1\} &= \Gamma^{(1)}\{-e_1\}, \\
\Delta^{(1)} \cap (e_1 + \mathbb{Z} f_3) &= \Gamma^{(1)}\{e_1\} \cap (e_1 + \mathbb{Z} f_3) = \Gamma_u^{(1)}\{e_1\} \\
&= e_1 + \mathbb{Z} x_1 x_2 f_3.
\end{aligned}$$

Also

$$\Gamma^{(1)}\{e_2\} \ni -e_2 + 2y_2x_2 f_3, \quad \Gamma^{(1)}\{e_3\} \ni -e_3 - 2y_1x_1 f_3.$$

This together with (6.15) and the shape of $\Gamma_u^{(1)}$ shows the statements on $\Delta^{(1)} \cap (\varepsilon e_2 + \mathbb{Z} f_3)$ and $\Delta^{(1)} \cap (\varepsilon e_3 + \mathbb{Z} f_3)$.

(ii) All statements on $\overline{\Delta^{(1)}}$ follow from $\Gamma_s^{(1)} \cong SL_2(\mathbb{Z})$ and $\overline{e_2}^{(1)} = x_1 g_2$ and $\overline{e_3}^{(1)} = g_2$.

For $\Delta^{(1)}$ we have to use Lemma 6.7. Claim 1 in the proof of part (i), its proof and the implication (6.15) still hold. Now we can choose $(y_1, y_2) = (0, 1)$, so $s_4 = s_{e_3}^{(1)}$. One calculates

$$s_{e_1}^{(1)} s_{e_3}^{(1)} s_{e_1}^{(1)} \underline{e} = \underline{e} \begin{pmatrix} 0 & -x_1 & -1 \\ 0 & 1 & 0 \\ 1 & -x_1 & 0 \end{pmatrix}.$$

Therefore $\Gamma^{(1)}\{e_1\} = \Gamma^{(1)}\{\pm e_1, \pm e_3\}$.

Claim 2 in the proof of part (i) still holds. It gives $(s_{e_1}^{(1)} s_{e_3}^{(1)} s_{e_1}^{(1)})^2 = t_{\lambda_3}^-$ with $\lambda_3 \in \mathrm{Hom}_2(H_\mathbb{Z}, \mathbb{Z}), \lambda_3(\underline{e}) = (0, 2, 0)$. Especially $t_{\lambda_3}^-(-e_2) = e_2 - 2f_3$.

This fact, (6.15) and the shape of $\Gamma_u^{(1)}$ imply the statements on $\Delta^{(1)} \cap (e_1 + \mathbb{Z} f_3)$ and $\Delta^{(1)} \cap (\varepsilon e_2 + \mathbb{Z} f_3)$.

(iii) and (iv) By Theorem 6.18 (d) $\Gamma_s^{(1)} \cong \Gamma^{(1)}(S(-x_{12}))$. By Theorem 6.10 (c) and (d) $\Gamma_s^{(1)}\{\pm \overline{e_1}^{(1)}, \pm g_2\}$ consists of the four orbits $\Gamma_s^{(1)}\{\varepsilon \overline{e_1}^{(1)}\}$ and $\Gamma_s^{(1)}\{\varepsilon g_2\}$ with $\varepsilon \in \{\pm 1\}$. Therefore if $x_1 > x_2$ then $\overline{\Delta^{(1)}}$ consists of the six orbits in (iii), and if $x_1 = x_2$ then $\overline{\Delta^{(1)}}$ consists of the four orbits $\Gamma_s^{(1)}\{\varepsilon \overline{e_1}^{(1)}\}$ and $\Gamma_s^{(1)}\{\varepsilon \overline{e_2}^{(1)}\} = \Gamma_s^{(1)}\{\varepsilon \overline{e_3}^{(1)}\} = \Gamma_s^{(1)}\{\varepsilon g_2\}$.

For $\Delta^{(1)}$ we have to use Lemma 6.7. Claim 1 in the proof of part (i), its proof and the implication (6.15) still hold.

In the case $x_1 > x_2$, $\overline{\Delta^{(1)}}$ consists of six orbits. Part (a) of Lemma 6.7, (6.15) and the shape of $\Gamma_u^{(1)}$ imply the statements on $\Delta^{(1)} \cap (\varepsilon e_i + \mathbb{Z} f_3)$ in part (iii).

In the case $x_1 = x_2$, $\overline{\Delta^{(1)}}$ consists of four orbits. Then $e_3 = e_2 - f_3$, (6.15) and the shape of $\Gamma_u^{(1)}$ imply the statements on $\Delta^{(1)} \cap (\varepsilon e_i + \mathbb{Z} f_3)$ in part (iv).

(e) Theorem 6.10 (c) and

$$s_{e_1}^{(1)}(e_1, e_3) = (e_1, e_3)\begin{pmatrix} 1 & -2 \\ 0 & 1 \end{pmatrix}, \quad s_{e_3}^{(1)}(e_1, e_3) = (e_1, e_3)\begin{pmatrix} 1 & 0 \\ 2 & 1 \end{pmatrix},$$

imply

$$\langle s_{e_1}^{(1)}, s_{e_3}^{(1)}\rangle\{\varepsilon e_1\} = \{y_1 e_1 + y_2 e_3 \in H_{\mathbb{Z}}^{prim} \mid y_1 \equiv \varepsilon(4), y_2 \equiv 0(2)\},$$
$$\langle s_{e_1}^{(1)}, s_{e_3}^{(1)}\rangle\{\varepsilon e_3\} = \{y_1 e_1 + y_2 e_3 \in H_{\mathbb{Z}}^{prim} \mid y_1 \equiv 0(2)), y_2 \equiv \varepsilon(4)\}.$$

In the case (i), the semidirect product $\Gamma^{(1)} = \Gamma_u^{(1)} \rtimes \langle s_{e_1}^{(1)}, s_{e_3}^{(1)}\rangle$ and the shape of $\Gamma_u^{(1)}$ show that $\Gamma^{(1)}\{\varepsilon e_1\}$ and $\Gamma^{(1)}\{\varepsilon e_3\}$ are as claimed.

In the case (ii), the semidirect product $\Gamma^{(1)} = \Gamma^{(1)} \cap O_{\pm}^{(1),Rad} \rtimes \langle s_{e_1}^{(1)}, s_{e_3}^{(1)}\rangle$ and the shape of $\Gamma^{(1)} \cap O_{\pm}^{(1),Rad}$ show that $\Gamma^{(1)}\{\varepsilon e_1\}$ and $\Gamma^{(1)}\{\varepsilon e_3\}$ are as claimed.

The following fact was not mentioned in Theorem 6.10 (c):

$$\{y_1 e_1 + y_2 e_3 \in H_{\mathbb{Z}}^{prim} \mid y_1 \equiv y_2 \equiv 1(2)\} = \langle s_{e_1}^{(1)}, s_{e_3}^{(1)}\rangle\{e_1 + e_3\},$$

so this set is a single $\langle s_{e_1}^{(1)}, s_{e_3}^{(1)}\rangle$ orbit. We skip its proof (it contains the observation $s_{e_3}^{(1)} s_{e_1}^{(1)}(e_1 + e_3) = -e_1 - e_3$).

This fact, the semidirect products above of $\Gamma^{(1)}$, the shape of $\Gamma_u^{(1)}$ in case (i) and of $\Gamma^{(1)} \cap O_{\pm}^{(1),Rad}$ in case (ii), and $e_2 = -\frac{l}{2}(e_1 + e_3) + f_3$ show in case (i) and case (ii) that $\Gamma^{(1)}\{\varepsilon e_2\}$ is as claimed.

(f) Theorem 6.10 (b), $\underline{\widetilde{e}} = (e_1, g_2, f_3)$,

$$s_{e_1}^{(1)}(\underline{\widetilde{e}}) = \underline{\widetilde{e}}\begin{pmatrix} 1 & -1 & 0 \\ 0 & 1 & 0 \\ 0 & 0 & 1 \end{pmatrix}, \quad s_4(\underline{\widetilde{e}}) = \underline{\widetilde{e}}\begin{pmatrix} 1 & 0 & 0 \\ 1 & 1 & 0 \\ 0 & 0 & 1 \end{pmatrix},$$
$$e_3 = -e_1 + 2g_2 + l f_3 \quad \text{and} \quad e_2 = -l g_2 + \frac{1 - l^2}{2} f_3$$

imply

$$\langle s_{e_1}^{(1)}, s_4\rangle(e_1) = (\mathbb{Z}e_1 \oplus \mathbb{Z}g_2)^{prim},$$
$$\langle s_{e_1}^{(1)}, s_4\rangle(e_3) = (\mathbb{Z}e_1 \oplus \mathbb{Z}g_2)^{prim} + l f_3,$$

$$\langle s_{e_1}^{(1)}, s_4\rangle(e_2) = l(\mathbb{Z}e_1 \oplus \mathbb{Z}g_2)^{prim} + \frac{1-l^2}{2} f_3.$$

The semidirect product $\Gamma^{(1)} = \Gamma_u^{(1)} \rtimes \langle s_{e_1}^{(1)}, s_4\rangle$ and the shape of $\Gamma_u^{(1)}$ show (with $\varepsilon \in \{\pm 1\}$)

$$\Gamma^{(1)}\{e_1\} = \Gamma^{(1)}\{\pm e_1, \pm e_3\} = (\mathbb{Z}e_1 \oplus \mathbb{Z}g_2)^{prim} + \mathbb{Z}lf_3,$$
$$\Gamma^{(1)}\{\varepsilon e_2\} = l(\mathbb{Z}e_1 \oplus \mathbb{Z}g_2)^{prim} + \varepsilon\frac{1-l^2}{2} f_3 + \mathbb{Z}l^2 f_3.$$

(g) In the proof of part (g) of Theorem 6.18 the hyperbolic polygon P with the six arcs in Remark 6.19 was used. It is a fundamental polygon of the action of the group $\langle \mu_1, \mu_2, \mu_3\rangle$ on $\mathbb{H}$. Here μ_1, μ_2, μ_3 are parabolic Möbius transformations with fixed points ∞, 0 and $-\frac{x_3}{x_2}$. These fixed points are cusps of P. This geometry implies

$$\mathrm{Stab}_{\langle\mu_1,\mu_2,\mu_3\rangle}(\infty) = \langle\mu_1\rangle,$$
$$\mathrm{Stab}_{\langle\mu_1,\mu_2,\mu_3\rangle}(0) = \langle\mu_2\rangle,$$
$$\mathrm{Stab}_{\langle\mu_1,\mu_2,\mu_3\rangle}(-\frac{x_3}{x_2}) = \langle\mu_3\rangle.$$

As $\langle \mu_1, \mu_2, \mu_3\rangle \cong \Gamma_s^{(1)}(\cong G^{free,3})$ with $\mu_i \sim \overline{s_{e_i}^{(1)}}$, this implies

$$\mathrm{Stab}_{\Gamma_s^{(1)}}(\{\pm\overline{e_i}^{(1)}\}) = \langle\overline{s_{e_i}^{(1)}}\rangle = \mathrm{Stab}_{\Gamma_s^{(1)}}(\overline{e_i}^{(1)}) \quad \text{for} \quad i \in \{1,2,3\}.$$

Especially $-\overline{e_i}^{(1)} \notin \Gamma_s^{(1)}(\overline{e_i}^{(1)})$, so the orbits $\Gamma_s^{(1)}\{\overline{e_i}^{(1)}\}$ and $\Gamma_s^{(1)}\{-\overline{e_i}^{(1)}\}$ are disjoint.

The cusps ∞, 0 and $-\frac{x_3}{x_2}$ of the fundamental domain P of the group $\langle\mu_1, \mu_2, \mu_3\rangle$ are in disjoint orbits of $\langle\mu_1, \mu_2, \mu_3\rangle$. Therefore the sets $\Gamma_s^{(1)}\{\pm\overline{e_1}^{(1)}\}$, $\Gamma_s^{(1)}\{\pm\overline{e_2}^{(1)}\}$ and $\Gamma_s^{(1)}\{\pm\overline{e_3}^{(1)}\}$ are disjoint. Therefore $\overline{\Delta^{(1)}}$ consists of the six disjoint orbits $\Gamma_s^{(1)}\{\varepsilon\overline{e_i}^{(1)}\}$ with $(\varepsilon, i) \in \{\pm 1\} \times \{1,2,3\}$. Therefore also $\Delta^{(1)}$ consists of the six orbits $\Gamma^{(1)}\{\varepsilon e_i\}$ with $(\varepsilon, i) \in \{\pm 1\} \times \{1,2,3\}$.

Claim: For $a \in \{\pm e_1, \pm e_2, \pm e_3\}$ equality in $\overset{(1)}{\supset}$ and $\overset{(2)}{\supset}$ before Lemma 6.7 holds.

Proof of the Claim: Equality in $\overset{(1)}{\supset}$ holds because of Lemma 6.7 (a) and because $\Delta^{(1)}$ and $\overline{\Delta^{(1)}}$ each consist of six orbits.

Equality in $\overset{(2)}{\supset}$ is by Lemma 6.7 (b) equivalent to equality in $\overset{(3)}{\supset}$.

Here $\Gamma^{(1)} \cong \Gamma_s^{(1)}(\cong G^{free,3})$, $\Gamma_u^{(1)} = \{\mathrm{id}\}$, the lift $s_{e_i}^{(1)} \in \Gamma^{(1)}$ of $\overline{s_{e_i}^{(1)}} \in \Gamma_s^{(1)}$ is in $\mathrm{Stab}_{\Gamma^{(1)}}(\varepsilon e_i)$, and therefore

$$\mathrm{Stab}_{\Gamma^{(1)}}(\varepsilon\overline{e_i}^{(1)}) = \langle s_{e_i}^{(1)}\rangle = \mathrm{Stab}_{\Gamma^{(1)}}(\varepsilon e_i) = \Gamma_u^{(1)} \cdot \mathrm{Stab}_{\Gamma^{(1)}}(\varepsilon e_i),$$

so equality in $\overset{(3)}{\supset}$ holds. The Claim is proved. ($\square$)

The Claim and $\Gamma_u^{(1)} = \{\mathrm{id}\}$ show

$$(\varepsilon e_i + \mathbb{Z} f_3) \cap \Delta^{(1)} = \varepsilon e_i.$$

Therefore the projection $\Delta^{(1)} \to \overline{\Delta^{(1)}}$ is a bijection.

(h) In the case $\underline{x} = (3, 3, 3)$

$$\underline{\widetilde{x}} = (1, 1, 1),\ f_3 = -e_1 + e_2 - e_3,\ \overline{e_3}^{(1)} = -\overline{e_1}^{(1)} + \overline{e_2}^{(1)},$$
$$\overline{H_{\mathbb{Z}}}^{(1)} = \mathbb{Z}\overline{e_1}^{(1)} \oplus \mathbb{Z}\overline{e_2}^{(1)}.$$

Recall that the matrices B_1, B_2, B_3 in the proof of Theorem 6.18 (g) with $\overline{s_{e_i}^{(1)}}(\overline{e_1}^{(1)}, \overline{e_2}^{(1)}) = (\overline{e_1}^{(1)}, \overline{e_2}^{(1)}) B_i$ are here

$$B_1 = \begin{pmatrix} 1 & -3 \\ 0 & 1 \end{pmatrix},\ B_1 = \begin{pmatrix} 1 & 0 \\ 3 & 1 \end{pmatrix},\ B_3 = \begin{pmatrix} -2 & -3 \\ 3 & 4 \end{pmatrix},$$

and generate $\Gamma(3)$. Because $s_{e_i}^{(1),mat} \equiv E_3 \bmod 3$ and $B_i \equiv E_2 \bmod 3$,

$$\Gamma^{(1)}\{\varepsilon e_i\} \subset \varepsilon e_i + 3H_{\mathbb{Z}} \quad \text{and} \quad \Gamma_s^{(1)}\{\varepsilon \overline{e_i}^{(1)}\} \subset \varepsilon \overline{e_i}^{(1)} + 3\overline{H_{\mathbb{Z}}}^{(1)}.$$

It remains to show $\Gamma_s^{(1)}\{\overline{e_i}^{(1)}\} = (\overline{e_i}^{(1)} + 3\overline{H_{\mathbb{Z}}}^{(1)})^{prim}$. This is equivalent to the following three statements:

(i) For $(a_1, a_3) \in \mathbb{Z}^2$ with $(a_1, a_3) \equiv (1, 0) \bmod 3$ and $\gcd(a_1, a_3) = 1$ a pair $(a_2, a_4) \in \mathbb{Z}^2$ with $\begin{pmatrix} a_1 & a_2 \\ a_3 & a_4 \end{pmatrix} \in \Gamma(3)$ exists.

(ii) For $(a_2, a_4) \in \mathbb{Z}^2$ with $(a_2, a_4) \equiv (0, 1) \bmod 3$ and $\gcd(a_2, a_4) = 1$ a pair $(a_1, a_3) \in \mathbb{Z}^2$ with $\begin{pmatrix} a_1 & a_2 \\ a_3 & a_4 \end{pmatrix} \in \Gamma(3)$ exists.

(iii) For $(b_1, b_2) \in \mathbb{Z}^2$ with $(b_1, b_2) \equiv (-1, 1) \bmod 3$ and $\gcd(b_1, b_2) = 1$ a matrix $\begin{pmatrix} a_1 & a_2 \\ a_3 & a_4 \end{pmatrix} \in \Gamma(3)$ with $\begin{pmatrix} -a_1 + a_2 \\ -a_3 + a_4 \end{pmatrix} = \begin{pmatrix} b_1 \\ b_2 \end{pmatrix}$ exists.

(i) is proved as follows. There exist $(\widetilde{a_2}, \widetilde{a_4}) \in \mathbb{Z}^2$ with $1 = a_1\widetilde{a_4} - a_3\widetilde{a_2}$. Thus $\widetilde{a_4} \equiv 1(3)$. Let $\widetilde{a_2} \equiv r(3)$ with $r \in \{0, 1, 2\}$. Choose $(a_2, a_4) := (\widetilde{a_2} - ra_1, \widetilde{a_4} - ra_3)$. The proofs of (ii) and (iii) are similar.

The proof of Theorem 6.21 (h) is finished. □

In quite some cases $\Delta^{(0)} \subset \Delta^{(1)}$, but nevertheless in general $\Delta^{(0)} \not\subset \Delta^{(1)}$. Corollary 6.22 gives some details.

Corollary 6.22 *(a) $\Delta^{(0)} = \Delta^{(1)}$ holds only in the cases A_1^n, so the cases with $S = E_n$ for some $n \in \mathbb{N}$. In all other cases $\Delta^{(1)} \not\subset R^{(0)}$.*

(b) $\Delta^{(0)} \subsetneqq \Delta^{(1)}$ holds in the cases $n = 2$ except A_1^2, in the reducible cases with $n = 3$ except A_1^3 and in the case A_3.

(c) $\Delta^{(0)} \not\subset \Delta^{(1)}$ *holds in the following cases with* $n = 3$*:* $\widehat{A}_2$, $\mathcal{H}_{1,2}$, $S(-l, 2, -l)$ *with* $l \geq 3$ *and* $\mathbb{P}^2$.

Proof (a) In the cases A_1^n $\Delta^{(0)} = \Delta^{(1)} = \{\pm e_1, ..., \pm e_n\}$ by Lemma 2.12. In a case with $S \in T_n(\mathbb{Z}) - \{E_n\}$ there is an entry $S_{ij} \neq 0$ for some $i < j$, so $L(e_j, e_i) \neq 0$. We can restrict to the rank 2 unimodular bilinear lattice $(\mathbb{Z}e_i + \mathbb{Z}e_j, L|_{\mathbb{Z}e_i+\mathbb{Z}e_j}, (e_i, e_j))$ with triangular basis (e_i, e_j). Part (e) of Theorem 6.10 shows that it has odd vanishing cycles which are not L1-cycles. They are also odd vanishing cycles of $(H_\mathbb{Z}, L, \underline{e})$, so then $\Delta^{(1)} \not\subset R^{(0)} \supset \Delta^{(0)}$.

(b) For the cases $n = 2$ see Theorem 6.10 (e). The reducible cases with $n = 3$ follow from the case A_1 and the cases with $n = 2$. In the case A_3 the twelve elements of $\Delta^{(0)}$ are given in Theorem 6.14 (c). The set $\Delta^{(1)}$ is by Theorem 6.21 (c)

$$(\mathrm{pr}^{H,(1)})^{-1}(\overline{H_\mathbb{Z}}^{(1),prim}) + \mathbb{Z}f_3.$$

which contains $\Delta^{(0)}$ as a strict subset.

(c) The case $\widehat{A}_2$: With $\underline{x} = (-1, -1, -1)$ we have $f_1 = e_1 + e_2 + e_3$ and $f_3 = e_1 - e_2 + e_3$. By Theorem 6.14 (d)

$$\Delta^{(0)} = (\pm e_1 + \mathbb{Z}f_1) \,\dot\cup\, (\pm e_2 + \mathbb{Z}f_1) \,\dot\cup\, (\pm e_3 + \mathbb{Z}f_1).$$

By Theorem 6.21 (c)

$$\Delta^{(1)} = (\mathrm{pr}^{H,(1)})^{-1}(\overline{H_\mathbb{Z}}^{(1),prim}) + \mathbb{Z}f_3.$$

Here for example for $m \in \mathbb{Z} - \{0, -1\}$

$$\Delta^{(0)} \ni e_2 + mf_1 = (2m + 1)e_2 + mf_3 \notin \Delta^{(1)}.$$

The case $\mathcal{H}_{1,2}$: With $\underline{x} = (-2, 2, -2)$ we have $\mathrm{Rad}\, I^{(0)} = \mathbb{Z}(e_1 + e_2) \oplus \mathbb{Z}(e_2 + e_3)$ and $f_3 = e_1 + e_2 + e_3$. By Theorem 6.14 (e)

$$\Delta^{(0)} = (\pm e_1 + 2\,\mathrm{Rad}\, I^{(0)}) \,\dot\cup\, (\pm e_2 + 2\,\mathrm{Rad}\, I^{(0)}) \,\dot\cup\, (\pm e_3 + 2\,\mathrm{Rad}\, I^{(0)}).$$

By Theorem 6.21 (e) (ii)

$$\Delta^{(1)} \subset (\mathbb{Z}e_1 + \mathbb{Z}e_3)^{prim} + \mathbb{Z}f_3.$$

Here for example

$$\begin{aligned}\Delta^{(0)} &\ni e_1 + 6(e_1 + e_2) + 4(e_2 + e_3) \\ &= (-3)e_1 + (-6)e_3 + 10f_3 \notin (\mathbb{Z}e_1 + \mathbb{Z}e_3)^{prim} + \mathbb{Z}f_3.\end{aligned}$$

This element is not contained in $\Delta^{(1)}$ because by Theorem 6.21 (e)

$$\Delta^{(1)} \subset \Big((\mathbb{Z}e_1 + \mathbb{Z}e_3)^{prim} + \mathbb{Z}f_3 \Big) \dot\cup \Big(\frac{l}{2}(\mathbb{Z}e_1 + \mathbb{Z}e_3)^{prim} + \mathbb{Z}f_3 \Big).$$

The cases $S(-l,2,-l)$: Recall $\mathrm{Rad}\, I^{(0)} = \mathbb{Z}f_1$, $f_1 = e_1 - e_3$,

$$f_3 = \frac{1}{2}(le_1 + 2e_2 + le_3) \quad \text{if } l \geq 3 \text{ is even},$$

$$\left.\begin{array}{l} f_3 = le_1 + 2e_2 + le_3 \\ g_2 = \frac{1}{2}(e_1 + e_3) - \frac{l}{2}e_2 \end{array}\right\} \text{ if } l \geq 3 \text{ is odd.}$$

Consider the element

$$\begin{aligned} h_1(e_1) - la_1 f_1 &= h_1(e_1 - la_1 f_1) \in H_{\mathbb{Z}} \quad \text{with} \\ h_1 &:= (s_{e_1}^{(0)} s_{e_2}^{(0)})^3 \in \Gamma^{(0)}, \\ a_1 &:= \frac{1}{2}(l^5 - 4l^3 + 3l) \in \mathbb{Z} \end{aligned}$$

By Theorem 6.11 (f) $T(\overline{j}^{(0)}(\overline{e_1}^{(0)}) \otimes f_1)$ and $T(\overline{j}^{(0)}(l\overline{e_2}^{(0)}) \otimes f_1) \in \Gamma_u^{(0)}$ with

$$\begin{aligned} T(\overline{j}^{(0)}(\overline{e_1}^{(0)}) \otimes f_1)(e_1) &= e_1 + 2f_1, \\ T(\overline{j}^{(0)}(l\overline{e_2}^{(0)}) \otimes f_1)(e_1) &= e_1 - l^2 f_1, \end{aligned}$$

so

$$\Delta^{(0)} \supset \begin{cases} e_1 + 2\mathbb{Z}f_1 & \text{if } l \text{ is even,} \\ e_1 + \mathbb{Z}f_1 & \text{if is odd.} \end{cases}$$

Therefore $h_1(e_1) - la_1 f_1 = h_1(e_1 - la_1 f_1) \in \Delta^{(0)}$. One calculates

$$\begin{aligned} h_1(e_1) - la_1 f_1 &= (s_{e_1}^{(0)} s_{e_2}^{(0)})^3(e_1) - la_1 f_1 \\ &= \underline{e}(\begin{pmatrix} -1 & l & 2 \\ 0 & 1 & 0 \\ 0 & 0 & 1 \end{pmatrix} \begin{pmatrix} 1 & 0 & 0 \\ l & -1 & l \\ 0 & 0 & 1 \end{pmatrix})^2 \begin{pmatrix} 1 \\ 0 \\ 0 \end{pmatrix} - la_1 f_1 \\ &= \underline{e} \begin{pmatrix} l^2-1 & -l & l^2-2 \\ l & -1 & l \\ 0 & 0 & 1 \end{pmatrix}^3 \begin{pmatrix} 1 \\ 0 \\ 0 \end{pmatrix} - la_1 f_1 \\ &= \underline{e} \begin{pmatrix} l^6 - 5l^4 + 6l^2 - 1 \\ l^5 - 4l^3 + 3l \\ 0 \end{pmatrix} - la_1 f_1 \\ &= (-l^4 + 3l^2 - 1)e_1 + \begin{cases} 2a_1 f_3 & \text{if } l \text{ is even,} \\ a_1 f_3 & \text{if } l \text{ is odd.} \end{cases} \end{aligned}$$

This element is not contained in $\Delta^{(1)}$ because $(-l^4+3l^2-1) \notin \{\pm 1, \pm \frac{l}{2}, \pm l\}$ and because by Theorem 6.21 (e) and (f)

$$\begin{aligned}
\Delta^{(1)} &\subset \Big((\mathbb{Z}e_1+\mathbb{Z}e_3)^{prim}+\mathbb{Z}f_3\Big)\,\dot\cup\,\Big(\frac{l}{2}(\mathbb{Z}e_1+\mathbb{Z}e_3)^{prim}+\mathbb{Z}f_3\Big) \\
&\quad \text{if } l \text{ is even}, \\
\Delta^{(1)} &\subset \Big((\mathbb{Z}e_1+\mathbb{Z}g_2)^{prim}+\mathbb{Z}f_3\Big)\,\dot\cup\,\Big(l(\mathbb{Z}e_1+\mathbb{Z}g_2)^{prim}+\mathbb{Z}f_3\Big) \\
&\quad \text{if } l \text{ is odd}.
\end{aligned}$$

The case $\mathbb{P}^2$: With $\underline{x}=(-3,3,-3)$ we have $f_3=e_1+e_2+e_3$.

$$\begin{aligned}
\Delta^{(1)} &\ni s^{(1)}_{e_3}(s^{(1)}_{e_2})^{-2}(e_1) \\
&= \underline{e}\begin{pmatrix}1&0&0\\0&1&0\\3&-3&1\end{pmatrix}\begin{pmatrix}1&0&0\\3&1&-3\\0&0&1\end{pmatrix}\begin{pmatrix}1&0&0\\3&1&-3\\0&0&1\end{pmatrix}\begin{pmatrix}1\\0\\0\end{pmatrix}=\underline{e}\begin{pmatrix}1\\6\\-15\end{pmatrix}.
\end{aligned}$$

By Theorem 6.21 (g) the projection $\Delta^{(1)}\to\overline{\Delta^{(1)}}$ is a bijection. Therefore

$$\Delta^{(1)} \not\ni (e_1+6e_2-15e_3)+9f_3 = 10e_1+15e_2-6e_3.$$

On the other hand

$$\begin{aligned}
&L(10e_1+15e_2-6e_3,\,10e_1+15e_2-6e_3) \\
&= \begin{pmatrix}10 & 15 & -6\end{pmatrix}\begin{pmatrix}1&0&0\\-3&1&0\\3&-3&1\end{pmatrix}\begin{pmatrix}10\\15\\-6\end{pmatrix}=1,
\end{aligned}$$

so $10e_1+15e_2-6e_3\in R^{(0)}=\Delta^{(0)}$. □

Consider a triple $\underline{x}\in\mathbb{Z}^3$ and a corresponding unimodular bilinear lattice $(H_\mathbb{Z}, L, \underline{e})$ with a triangular basis $\underline{e}$ with $L(\underline{e}^t,\underline{e})^t=S(\underline{x})$. The Remarks 4.17 explained that the tuple $(H_\mathbb{Z}, \pm I^{(1)}, \Gamma^{(1)}, \Delta^{(1)})$ depends only on the $(G^{phi}\ltimes\widetilde{G}^{sign})\rtimes\langle\gamma\rangle$ orbit of $\underline{x}\in\mathbb{Z}^3$. Lemma 4.18 gave at least one element of each orbit of this group in $\mathbb{Z}^3$. Theorem 6.18 and Theorem 6.21 gave detailed information on the tuple $(H_\mathbb{Z}, \pm I^{(1)}, \Gamma^{(1)}, \Delta^{(1)})$ for the elements in Lemma 4.18 (b)+(c) and rather coarse information for the elements in Lemma 4.18 (a).

The next corollary uses this information to conclude that the $(G^{phi}\ltimes\widetilde{G}^{sign})\rtimes\langle\gamma\rangle$ orbits of the elements in Lemma 4.18 (b)+(c) are pairwise different and also different from the orbits of the elements in Lemma 4.18 (a), because the corresponding tuples $(H_\mathbb{Z}, \pm I^{(1)}, \Gamma^{(1)}, \Delta^{(1)})$ are not isomorphic. As Theorem 6.18 and Theorem 6.21 give only coarse information on the cases in Lemma 4.18 (a), also Corollary 6.23 is vague about them.

The set of local minima in Lemma 4.18 (b)+(c) and (3, 3, 3) is called Λ_1, the set of local minima in Lemma 4.18 (a) without (3, 3, 3) is called Λ_2,

$$\begin{aligned}\Lambda_1 &:= \{(3,3,3)\} \cup \{(-l, 2, -l) \mid l \geq 2\} \\ &\cup \{(x_1, x_2, 0) \mid x_1, x_2 \in \mathbb{Z}_{\geq 0}, x_1 \geq x_2\}, \\ \Lambda_2 &:= \{\underline{x} \in \mathbb{Z}^3_{\geq 3} \mid 2x_i \leq x_j x_k \text{ for } \{i,j,k\} = \{1,2,3\}\} - \{(3,3,3)\}.\end{aligned}$$

Corollary 6.23 *Consider $\underline{x}$ and $\underline{\widetilde{x}} \in \Lambda_1$ or $\underline{x} \in \Lambda_1$ and $\underline{\widetilde{x}} \in \Lambda_2$. Suppose $\underline{x} \neq \underline{\widetilde{x}}$. Then the tuples $(H_{\mathbb{Z}}, \pm I^{(1)}, \Gamma^{(1)}, \Delta^{(1)})$ of $\underline{x}$ and $\underline{\widetilde{x}}$ are not isomorphic. Consequently, the $(G^{phi} \ltimes \widetilde{G}^{sign}) \rtimes \langle\gamma\rangle$ orbits of $\underline{x}$ and $\underline{\widetilde{x}}$ are disjoint.*

Proof In the following, (b), (c), (d), (d)(i), (d)(ii), (d)(iii), (d)(iv), (e), (e)(i), (e)(ii), (f), (g), (h)($\subset$ (g)) mean the corresponding families of cases in Theorem 6.21. Of course, (c) and (h) are single cases. We will first discuss how to separate the families by properties of the isomorphism classes of the tuples $(H_{\mathbb{Z}}, \pm I^{(1)}, \Gamma^{(1)}, \Delta^{(1)})$, and then how to separate the cases within one family.

The pair $(\Gamma^{(1)}_u, \Gamma^{(1)}_s)$ gives the following incomplete separation of families,

$\Gamma^{(1)}_u \cong ?$	$\Gamma^{(1)}_s \cong ?$	families
$\{\mathrm{id}\}$	$G^{free,3}$	(g)
$\{\mathrm{id}\}$	$\not\cong G^{free,3}$	(b)
$\mathbb{Z}^2$	$SL_2(\mathbb{Z})$	$(c), (d)(i) + (ii), (f)$
$\mathbb{Z}^2$	$SL_2(\mathbb{Z}) \times \{\pm 1\}$	$(e)(ii)$
$\mathbb{Z}^2$	$G^{free,2}$	$(d)(iii) + (iv), (e)(i)$

The fundamental polygon P in Remark 6.19 (ii) has finite area in the case (h) (i.e. $\underline{x} = (3,3,3)$) and infinite area in the other cases in (g). So it separates the case (h) from the other cases in (g).

The number of $\mathrm{Br}_3 \ltimes \{\pm 1\}^3$ orbits in $\overline{\Delta^{(1)}}$ separates the families (c),(d),(e)(i),(f) almost completely:

$\lvert\{\text{orbits in } \overline{\Delta^{(1)}}\}\rvert$	1	2	3	4	5	6
families	(c)	$(d)(ii), (f)$	$(d)(i)$	$(d)(iv)$	$(e)(i)$	$(d)(iii)$

The separation of the family (d)(ii) from the family (f) is more difficult and can be done as follows. In both families of cases $\Delta^{(1)}$ consists of three orbits, and $\overline{\Delta^{(1)}}$ consists of two orbits. The two orbits $\Gamma^{(1)}\{e_2\}$ and $\Gamma^{(1)}\{-e_2\}$ unite to a single orbit $\Gamma^{(1)}_s\{\overline{e_2}^{(1)}\} = \Gamma^{(1)}_s\{-\overline{e_2}^{(1)}\}$. The set

$$\{n \in \mathbb{N} \mid \text{there exists } a_1 \in \Gamma^{(1)}\{e_2\} \text{ and } \varepsilon \in \{\pm 1\} \\ \text{with } a_1 + \varepsilon n f_3 \in \Gamma^{(1)}\{-e_2\}\}$$

is well defined. Its minimum is 2 in each case in (d)(ii) because there $x_1 > x_2 = 1$ and

$$\Gamma^{(1)}\{e_2\} \cap (e_2 + \mathbb{Z} f_3) = e_2 + \mathbb{Z} x_1^2 f_3,$$
$$\Gamma^{(1)}\{-e_2\} \cap (e_2 + \mathbb{Z} f_3) = e_2 - 2f_3 + \mathbb{Z} x_1^2 f_3.$$

Its minimum is 1 in each case in (f) because there

$$\Gamma^{(1)}\{e_2\} \cap (le_1 + \mathbb{Z} f_3) = le_1 + \frac{1+l^2}{2} f_3 + \mathbb{Z} l^2 f_3,$$
$$\Gamma^{(1)}\{-e_2\} \cap (le_1 + \mathbb{Z} f_3) = le_1 + \frac{-1+l^2}{2} + \mathbb{Z} l^2 f_3.$$

It remains to separate within each family (b), (d), (e) and (f) the cases ((c) and (h) are single cases). The pair $(\overline{H_{\mathbb{Z}}}^{(1)}, \pm\overline{I}^{(1)})$ and Lemma 6.16 (b) allow to recover $\gcd(x_1, x_2, x_3)$ which is as follows in these families,

family of cases	(b)	(d)	(e)	(f)
$\gcd(x_1, x_2, x_3)$	x_1	x_{12}	2	1

Within the family (b) this separates the cases. For the family (d) we need additionally the pair $(\widetilde{x}_1, \widetilde{x}_2)$ because $(x_1, x_2) = (x_{12}\widetilde{x}_1, x_{12}\widetilde{x}_2)$. The pair $(\widetilde{x}_1, \widetilde{x}_2)$ can be read off from $\overline{\Delta}^{(1)} \subset \overline{H_{\mathbb{Z}}}^{(1)}$, more precisely, from the relation of the $\Gamma_s^{(1)}$ orbits in $\overline{\Delta^{(1)}}$ to the set $\overline{H_{\mathbb{Z}}}^{(1),prim} \subset \overline{H_{\mathbb{Z}}}^{(1)}$. In the family (e) one can read off $\frac{l}{2}$, and in the family (f) one can read off l from the relation of the $\Gamma_s^{(1)}$ orbits in $\overline{\Delta}^{(1)}$ to the set $\overline{H_{\mathbb{Z}}}^{(1),prim} \subset \overline{H_{\mathbb{Z}}}^{(1)}$. □

Remarks 6.24 Let $(H_{\mathbb{Z}}, L)$ be a unimodular bilinear lattice of rank $n \geq 2$, and let $\underline{e}$ be a triangular basis with matrix $S = L(\underline{e}^t, \underline{e})^t \in T_n(\mathbb{Z})$. Recall Theorem 3.7 (a). If $S_{ij} \leq 0$ for $i < j$ then $(\Gamma^{(0)}, \{s_{e_1}^{(0)}, ..., s_{e_n}^{(0)}\})$ is a Coxeter system, and the presentation in Definition 3.15 of the Coxeter group $\Gamma^{(0)}$ is determined by S. Especially $\Gamma^{(0)} \cong G^{fCox,n}$ if $S_{ij} \leq -2$ for $i < j$.

One might hope for a similar easy control of $\Gamma^{(1)}$ if $S_{ij} \leq 0$ for $i < j$. In the cases with $n = 2$ this works by Lemma 2.12 and Theorem 6.10:

$$\Gamma^{(1)} \cong \begin{cases} \{\text{id}\} & \text{if } x = 0, \\ SL_2(\mathbb{Z}) & \text{if } x = -1, \\ G^{free,2} & \text{if } x \leq -2. \end{cases}$$

But in the cases with $n = 3$ this fails. The Remarks 4.17 show $\Gamma^{(1)}(S(\underline{x})) \cong \Gamma^{(1)}(S(-\underline{x}))$ for any $\underline{x} \in \mathbb{Z}^3$. The cases $S(\underline{\widetilde{x}})$ with $\underline{\widetilde{x}} \in \mathbb{Z}^3_{\geq 0}$ lead by the action of $(G^{phi} \ltimes \widetilde{G}^{sign}) \rtimes \langle \gamma \rangle$ to all cases in Theorem 6.18.

Especially, the cases $S(\underline{\widetilde{x}})$ with $\underline{\widetilde{x}} \in \mathbb{Z}^3_{\geq 2}$ contain the nice cases in Theorem 6.18 (g) with $\Gamma^{(1)} \cong G^{free,3}$, but also many other cases. Compare the family $\{(3, 3, l) \mid l \geq 2\}$ in the Examples 4.20 (iv) or the case $S = S(2, 2, 2) \sim S(-2, -2, -2) \sim S(\mathcal{H}_{1,2})$ with $\Gamma^{(1)}$ far from $G^{free,3}$.

Remarks 6.25 In the cases $\underline{x} \in \mathbb{Z}^3$ in Lemma 4.18 (a), so $\underline{x} \in \mathbb{Z}^3_{\geq 2}$ with $2x_i \leq x_j x_k$ for $\{i, j, k\} = \{1, 2, 3\}$, Theorem 6.18 and Theorem 6.21 give rather coarse information,

$$\Gamma_u^{(1)} = \{\text{id}\} \quad \text{and} \quad \Gamma^{(1)} \cong \Gamma_s^{(1)} \cong G^{free,3} \quad \text{by Theorem 6.18},$$
$$\Delta^{(1)} \to \overline{\Delta^{(1)}} \quad \text{is a bijection} \quad \text{by Theorem 6.21}.$$

But it is nontrivial to determine the unique preimage in $\Gamma^{(1),mat}$ of an element of $\Gamma_s^{(1)}$ and the unique preimage in $\Delta^{(1)}$ of an element of $\overline{\Delta^{(1)}}$. This holds especially for the case $\underline{x} = (3, 3, 3)$ where $\Gamma_s^{(1)} \cong \Gamma(3)$ and $\overline{\Delta^{(1)}}$ are known. Part (c) of the following lemma gives for $\underline{x} = (3, 3, 3)$ at least an inductive way to determine the preimage in $\Gamma^{(1),mat}$ of a matrix in $\Gamma(3) \cong \Gamma_s^{(1)}$.

Lemma 6.26 *Consider the case $\underline{x} = (3, 3, 3)$. Denote by $L_{\mathbb{P}^2} : \Gamma(3) \to \Gamma^{(1),mat}$ the inverse of the natural group isomorphism*

$$\begin{array}{ccccccc}
\Gamma^{(1),mat} & \longrightarrow & \Gamma^{(1)} & \longrightarrow & \Gamma_s^{(1)} & \longrightarrow & \Gamma(3), \\
s_{e_i}^{(1),mat} & \longrightarrow & s_{e_i}^{(1)} & \longrightarrow & \overline{s_{e_i}^{(1)}} & \longrightarrow & B_i \\
g^{mat} & \longleftarrow & g & \longrightarrow & \overline{g} & \longrightarrow & \overline{g}^{mat}
\end{array}$$

with

$$g(\underline{e}) = \underline{e} \cdot g^{mat} \quad \text{and}$$
$$\overline{g}(\overline{e_1}^{(1)}, \overline{e_2}^{(1)}) = (\overline{e_1}^{(1)}, \overline{e_2}^{(1)}) \cdot \overline{g}^{mat}$$

for $g \in \Gamma^{(1)}$. Define the subgroup of $SL_3(\mathbb{Z})$

$$G^{(3,3,3)} := \{F \in SL_3(\mathbb{Z}) \mid F \equiv E_3 \bmod 3,\ F \begin{pmatrix} -1 \\ 1 \\ -1 \end{pmatrix} = \begin{pmatrix} -1 \\ 1 \\ -1 \end{pmatrix}\}.$$

Define the map (st for standard)

$$L_{st} : \Gamma(3) \to M_{3\times 3}(\mathbb{Z}), \quad \begin{pmatrix} a & b \\ c & d \end{pmatrix} \mapsto \begin{pmatrix} a & b & 1-a+b \\ c & d & -1-c+d \\ 0 & 0 & 1 \end{pmatrix},$$

and the three matrices

$$K_1 := \begin{pmatrix} 3 & 0 & -3 \\ -3 & 0 & 3 \\ 3 & 0 & -3 \end{pmatrix}, \quad K_2 := \begin{pmatrix} 0 & 3 & 3 \\ 0 & -3 & -3 \\ 0 & 3 & 3 \end{pmatrix},$$

$$K_3 := K_1 + K_2 = \begin{pmatrix} 3 & 3 & 0 \\ -3 & -3 & 0 \\ 3 & 3 & 0 \end{pmatrix}.$$

(a) L_{st} is an injective group homomorphism $L_{st} : \Gamma(3) \to G^{(3,3,3)}$,

$$\begin{aligned} K_i K_j &= 0 \quad \text{for } i, j \in \{1, 2, 3\}, \\ L_{st}(C)K_i &= K_i \quad \text{for } C \in \Gamma(3),\ i \in \{1, 2, 3\}, \\ G^{(3,3,3)} &= \{L_{st}(C) + \alpha K_1 + \beta K_2 \mid C \in \Gamma(3), \alpha, \beta \in \mathbb{Z}\}. \end{aligned}$$

The following sequence is an exact sequence of group homomorphisms,

$$\begin{array}{ccccccccc} \{1\} & \longrightarrow & \mathbb{Z}^2 & \longrightarrow & G^{(3,3,3)} & \longrightarrow & \Gamma(3) & \longrightarrow & \{1\} \\ & & (\alpha, \beta) & \longrightarrow & E_3 + \alpha K_1 + \beta K_2 & & & & \\ & & & & L_{st}(C) + \alpha K_1 + \beta K_2 & \longrightarrow & C & & \end{array}$$

L_{st} is a splitting of this exact sequence.

(b) $\Gamma^{(1),mat} \subset G^{(3,3,3)}$, and $L_{\mathbb{P}^2} : \Gamma(3) \to \Gamma^{(1),mat}$ is another splitting of the exact sequence in part (b). It satisfies

$$L_{\mathbb{P}^2}(B_1) = L_{st}(B_1) = s^{(1),mat}_{e_1} = \begin{pmatrix} 1 & -3 & -3 \\ 0 & 1 & 0 \\ 0 & 0 & 1 \end{pmatrix} \quad \text{for } B_1 = \begin{pmatrix} 1 & -3 \\ 0 & 1 \end{pmatrix},$$

$$L_{\mathbb{P}^2}(B_2) = L_{st}(B_2) = s^{(1),mat}_{e_2} = \begin{pmatrix} 1 & 0 & 0 \\ 3 & 1 & -3 \\ 0 & 0 & 1 \end{pmatrix} \quad \text{for } B_2 = \begin{pmatrix} 1 & 0 \\ 3 & 1 \end{pmatrix},$$

$$L_{\mathbb{P}^2}(B_3) = L_{st}(B_3) + K_3 = s^{(1),mat}_{e_3} = \begin{pmatrix} 1 & 0 & 0 \\ 0 & 1 & 0 \\ 3 & 3 & 1 \end{pmatrix} \quad \text{for } B_3 = \begin{pmatrix} -2 & -3 \\ 3 & 4 \end{pmatrix},$$

$$L_{\mathbb{P}^2}(B_3^{-1}) = L_{st}(B_3^{-1}) - K_3.$$

(c) An arbitrary element $C \in \Gamma(3)$ can be written in a unique way as a product

$$C = C_1 B_3^{\varepsilon_1} C_2 B_3^{\varepsilon_2} C_3 ... C_m B_3^{\varepsilon_m} C_{m+1}$$
$$\text{with} \quad m \in \mathbb{Z}_{\geq 0},\ C_1, ..., C_{m+1} \in \langle B_1^{\pm 1}, B_2^{\pm 1} \rangle,\ \varepsilon_1, ..., \varepsilon_m \in \{\pm 1\}.$$

Then

$$\begin{aligned} L_{\mathbb{P}^2}(C) = L_{st}(C) + K_3 \Big(& \varepsilon_1 L_{st}(C_2 B_3^{\varepsilon_2} C_3 ... C_m B_3^{\varepsilon_m} C_{m+1}) \\ & + \varepsilon_2 L_{st}(C_3 B_3^{\varepsilon_3} C_4 ... C_m B_3^{\varepsilon_m} C_{m+1}) \\ & + ... + \varepsilon_m L_{st}(C_{m+1}) \Big). \end{aligned}$$

Proof The parts (a) and (b) are easy.

(c) By part (b) $L_{\mathbb{P}^2}(C_j) = L_{st}(C_j)$ and $L_{\mathbb{P}^2}(B_3^{\varepsilon_j}) = L_{st}(B_3^{\varepsilon_j}) + \varepsilon_j K_3$, so

$$\begin{aligned} L_{\mathbb{P}^2}(C) &= L_{\mathbb{P}^2}(C_1)L_{\mathbb{P}^2}(B_3^{\varepsilon_1})L_{\mathbb{P}^2}(C_2)...L_{\mathbb{P}^2}(C_m)L_{\mathbb{P}^2}(B_3^{\varepsilon_m})L_{\mathbb{P}^2}(C_{m+1}) \\ &= L_{st}(C_1)(L_{st}(B_3^{\varepsilon_1}) + \varepsilon_1 K_3)L_{st}(C_2)... \\ &\quad L_{st}(C_m)(L_{st}(B_3^{\varepsilon_m}) + \varepsilon_m K_3)L_{st}(C_{m+1}). \end{aligned}$$

Observe $K_3 L_{st}(\widetilde{C})K_3 = K_3K_3 = 0$ for $\widetilde{C} \in \Gamma(3)$. Therefore if one writes the product above as a sum of 2^m terms, only the $1+m$ terms do not vanish in which K_3 turns up at most once. This leads to the claimed formula for $L_{\mathbb{P}^2}(C)$. □

Chapter 7
Distinguished Bases in the Rank 2 and Rank 3 Cases

In Sect. 3.3 we introduced the set of distinguished bases of a unimodular bilinear lattice $(H_{\mathbb{Z}}, L, \underline{e})$ with a triangular basis. It is the orbit $\mathcal{B}^{dist} = \mathrm{Br}_n \ltimes \{\pm 1\}^n(\underline{e})$ of $\underline{e}$ under the group $\mathrm{Br}_n \ltimes \{\pm 1\}^n$. We also posed the question when this set can be characterized in an easy way, more precisely, when the inclusions in (3.3) or (3.4) are equalities,

$$\mathcal{B}^{dist} \subset \{\underline{v} \in (\Delta^{(0)})^n \mid s^{(0)}_{v_1}\ldots s^{(0)}_{v_n} = -M\}, \tag{3.3}$$
$$\mathcal{B}^{dist} \subset \{\underline{v} \in (\Delta^{(1)})^n \mid s^{(1)}_{v_1}\ldots s^{(1)}_{v_n} = M\}. \tag{3.4}$$

Theorem 3.2 (a) and (b) imply that (3.4) is an equality if $\Gamma^{(1)}$ is a free group with generators $s^{(1)}_{e_1}, \ldots, s^{(1)}_{e_n}$ and that (3.3) is an equality if $\Gamma^{(0)}$ is a free Coxeter group with generators $s^{(0)}_{e_1}, \ldots, s^{(0)}_{e_n}$, see the Examples (3.23) (iv). More generally, if $(\Gamma^{(0)}, s^{(0)}_{e_1}, \ldots, s^{(0)}_{e_n})$ is a Coxeter system (Definition 3.5) then by Theorem 3.6 (3.3) is an equality, see the Examples 3.23 (v). It is remarkable that the property $\sum_{i=1}^n \mathbb{Z}v_i = H_{\mathbb{Z}}$, which each distinguished basis $\underline{v} \in \mathcal{B}^{dist}$ satisfies is not needed in the characterization in these cases.

These are positive results. In the Sects. 7.1–7.3 we study systematically all cases of rank 2 and 3 and find also negative results.

In rank 2 in Sect. 7.2 (3.3) is always an equality, and (3.4) is an equality in all cases except the case A_1^2.

In the even rank 3 cases in Sect. 7.2 (3.3) is in all cases except the case $\mathcal{H}_{1,2}$ an equality. In the case $\mathcal{H}_{1,2}$ the set on the right hand side contains $\mathrm{Br}_3 \ltimes \{\pm 1\}^3$ orbits of tuples $\underline{v}$ with arbitrarily large finite index $[H_{\mathbb{Z}} : \sum_{i=1}^3 \mathbb{Z}v_i]$ and two orbits with index 1, $\mathcal{B}^{dist}$ and one other orbit.

In the odd rank 3 cases in Sect. 7.3 we understand the set $B_1 \cup B_2$ of triples $\underline{x} \in \mathbb{Z}^3$ such that (3.4) is an equality, and we also know a set $B_3 \cup B_4$ of triples $\underline{x} \in \mathbb{Z}^3$ such that (3.4) becomes an equality if one adds the condition $H_{\mathbb{Z}} = \sum_{i=1}^3 \mathbb{Z}v_i$. But for $\underline{x} \in \mathbb{Z}^3 - \bigcup_{j=1}^4 B_j$, we know little.

C. Hertling and K. Larabi, *Triangular Bases of Unimodular Bilinear Lattices and Induced Structures*, Lecture Notes in Mathematics 2398,
https://doi.org/10.1007/978-3-032-25997-4_7

Section 7.4 builds on Sect. 4.4 where for a unimodular bilinear lattice $(H_{\mathbb{Z}}, L, \underline{e})$ the stabilizer $(\mathrm{Br}_3)_{\underline{x}/\{\pm 1\}^3}$ had been determined. It determines the stabilizer $(\mathrm{Br}_3)_{\underline{e}/\{\pm 1\}^3}$. It uses the systematic results in Chap. 5 on the group $G_{\mathbb{Z}}$ and on the map $Z : (\mathrm{Br}_n \ltimes \{\pm 1\}^n)_S \to G_{\mathbb{Z}}$ in the rank 3 cases.

In the Sects. 4.3 and 4.4 the pseudo-graph $\mathcal{G}(\underline{x})$ with vertex set an orbit $\mathrm{Br}_3(\underline{x}/\{\pm 1\})$ and edge set from generators of the group $G^{phi} \rtimes \langle\gamma\rangle$ had been crucial. In Sect. 7.4 we introduce a variant with the same vertex set, but different edge set, namely (now) oriented edges coming from the elementary braids $\sigma_i^{\pm 1}$. We also define the much larger σ-pseudo-graph with vertex set a set $\mathcal{B}^{dist}/\{\pm 1\}^3$ of distinguished bases up to signs and oriented edges coming from the elementary braids $\sigma_i^{\pm 1}$. We consider especially the examples where the set $\mathrm{Br}_3(\underline{x}/\{\pm 1\}^3)$ is finite.

7.1 Distinguished Bases in the Rank 2 Cases

In the rank 2 cases the inclusion (3.3) is always an equality, and the inclusion (3.4) is almost always an equality, namely in all cases except the case A_1^2.

Theorem 7.1 *Let $(H_{\mathbb{Z}}, L, \underline{e})$ be a unimodular bilinear lattice of rank 2 with a triangular basis $\underline{e} = (e_1, e_2)$ with matrix $S = S(x) = \begin{pmatrix} 1 & x \\ 0 & 1 \end{pmatrix} = L(\underline{e}^t, \underline{e})^t$ with $x \in \mathbb{Z}$. Fix $k \in \{0, 1\}$.*

(a) The inclusion (3.3) respectively (3.4) in Remark 3.19 is an equality in all cases except the odd case A_1^2*, so the case $(k, x) = (1, 0)$. In that case the right hand side in (3.4) splits into the orbits of the three pairs (e_1, e_1), (e_1, e_2), (e_2, e_2).*

(b) The stabilizers in Br_2 *of $S/\{\pm 1\}^2$ and of $\underline{e}/\{\pm 1\}^2$ are*

$$(\mathrm{Br}_2)_{S/\{\pm 1\}^2} = \mathrm{Br}_2 \quad \text{and}$$

$$(\mathrm{Br}_2)_{\underline{e}/\{\pm 1\}^2} = \begin{cases} \langle \sigma_1^2 \rangle & \text{if } x = 0, \\ \langle \sigma_1^3 \rangle & \text{if } x \in \{\pm 1\}, \\ \{\mathrm{id}\} & \text{if } |x| \geq 2. \end{cases}$$

Proof (a) The even and odd cases A_1^2: See the Examples 3.23 (iii).

The cases with $|x| \geq 2$: Theorem 6.8 (c)+(d) and Theorem 6.10 (c)+(d) show

$$\Gamma^{(k)} \cong \begin{cases} G^{fCox,2} \text{ with generators } s_{e_1}^{(0)}, s_{e_2}^{(0)} \text{ if } k = 0, \\ G^{free,2} \text{ with generators } s_{e_1}^{(1)}, s_{e_2}^{(1)} \text{ if } k = 1. \end{cases}$$

The Examples 3.23 (iv) apply and give equality in (3.3) and (3.4).

The cases with $x = \pm 1$: We can restrict to the case $x = -1$. The even case is a simple case of Example 3.23 (v) (in the Remarks 7.2 we will offer an elementary proof for the even case).

It remains to show equality in (3.4) in the odd case $(k, x) = (1, -1)$. Consider $\underline{v} \in (\Delta^{(1)})^2$ with $s_{v_1}^{(1)} s_{v_2}^{(1)} = M$. Let $b := I^{(1)}(v_1, v_2) \in \mathbb{Z}$. If $b = 0$ then $v_2 = \pm v_1$

and $M = (s^{(1)}_{v_1})^2$ would have an eigenvalue 1, a contradiction. Therefore $b \neq 0$ and $\mathbb{Z}v_1 + \mathbb{Z}v_2$ has rank 2. Then

$$\begin{aligned} M(\underline{v}) &= s^{(1)}_{v_1} s^{(1)}_{v_2}(\underline{v}) = s^{(1)}_{v_1}(v_1 + bv_2, v_2) \\ &= (v_1 + bv_2 - b^2 v_1, v_2 - bv_1) = \underline{v}\begin{pmatrix} 1-b^2 & -b \\ b & 1 \end{pmatrix}, \end{aligned}$$

$1 = \operatorname{tr} M = (1-b^2) + 1$, so $b = \pm 1$. By possibly changing the sign of v_2, we can suppose $b = -1 = x$. Then

$$I^{(1)}(\underline{v}^t, \underline{v}) = \begin{pmatrix} 0 & -1 \\ 1 & 0 \end{pmatrix}.$$

Therefore $\underline{v}$ is a $\mathbb{Z}$-basis of $H_{\mathbb{Z}}$, and the automorphism $g \in \operatorname{Aut}(H_{\mathbb{Z}})$ with $(g(e_1), g(e_2)) = (v_1, v_2)$ is in $O^{(1)}$. By Lemma 3.22 (a)

$$gMg^{-1} = g \circ ((\pi_2 \circ \pi_2^{(1)})(\underline{e})) \circ g^{-1} = (\pi_2 \circ \pi_2^{(1)})(\underline{v}) = s^{(1)}_{v_1} s^{(1)}_{v_2} = M,$$

so $gMg^{-1} = M$, so $g \in G^{(1)}_{\mathbb{Z}} = G^M_{\mathbb{Z}} \cap O^{(1)}$. Theorem 5.5 can be applied and gives $\overset{(*)}{=}$,

$$G^{(1)}_{\mathbb{Z}} \overset{(*)}{=} G_{\mathbb{Z}} \overset{(*)}{=} \{\pm (M^{root})^l \mid l \in \mathbb{Z}\} \overset{(*)}{=} Z(\mathrm{Br}_2 \ltimes \{\pm 1\}^2).$$

Therefore $\underline{v} \in \mathcal{B}^{dist}$. This shows equality in (3.4).

(b) Because of $\sigma_1 \begin{pmatrix} 1 & x \\ 0 & 1 \end{pmatrix} = \begin{pmatrix} 1 & -x \\ 0 & 1 \end{pmatrix}$, the stabilizer $(\mathrm{Br}_2)_{S/\{\pm 1\}^2}$ is the whole group $\mathrm{Br}_2 = \langle \sigma_1 \rangle$. If $x = 0$,

$$(e_1, e_2) \overset{\sigma_1}{\mapsto} (e_2, e_1) \overset{\sigma_1}{\mapsto} (e_1, e_2), \quad \text{so } (\mathrm{Br}_2)_{\underline{e}/\{\pm 1\}^2} = \langle \sigma_1^2 \rangle.$$

If $x = -1$,

$$\begin{aligned} (e_1, e_2) \overset{\sigma_1}{\mapsto} (e_1 + e_2, e_1) \overset{\sigma_1}{\mapsto} (-e_2, e_1 + e_2) \overset{\sigma_1}{\mapsto} (e_1, -e_2), \\ \text{so } (\mathrm{Br}_2)_{\underline{e}/\{\pm 1\}^2} = \langle \sigma_1^3 \rangle. \end{aligned}$$

If $|x| \geq 2$ Theorem 3.2 (a) or (b) and $\Gamma^{(1)} \cong G^{free,2}$ or $\Gamma^{(0)} \cong G^{fCox,2}$ show $(\mathrm{Br}_2)_{\underline{e}/\{\pm 1\}^2} = \{\mathrm{id}\}$. □

Remarks 7.2 (i) A direct elementary proof of equality in (3.3) for the even case A_2, so $(k, x) = (0, -1)$, is instructive. Recall from Theorem 6.8 (b) that $\Delta^{(0)} = \{\pm e_1, \pm e_2, \pm (e_1 + e_2)\}$. The map $\pi_2 \circ \pi_2^{(0)} : (\Delta^{(0)})^2 \to \Gamma^{(0)}$ has the three values $-M$, M^2 and id and the three fibers

$$\begin{aligned}(\pi_2\circ\pi_2^{(0)})^{-1}(-M) &= \{(\pm e_1,\pm e_2),(\pm(e_1+e_2),\pm e_1),(\pm e_2,\pm(e_1+e_2))\}\\ &= \mathcal{B}^{dist},\\ (\pi_2\circ\pi_2^{(0)})^{-1}(M^2) &= \{(\pm e_2,\pm e_1),(\pm(e_1+e_2),\pm e_2),(\pm e_1,\pm(e_1+e_2))\}\\ &= \mathrm{Br}_2\ltimes\{\pm 1\}^2(e_2,e_1),\\ (\pi_2\circ\pi_2^{(0)})^{-1}(\mathrm{id}) &= \{(\pm e_1,\pm e_1),(\pm e_2,\pm e_2),(\pm(e_1+e_2),\pm(e_1+e_2))\}.\end{aligned}$$

This gives equality in (3.3) in the case $(k,x)=(0,-1)$.

(ii) Also in the cases $(k=0, x\leq -2)$ a direct elementary proof of equality in (3.3) is instructive. Equality in (3.3) for $(k,x)=(0,-2)$ and Theorem 6.8 (d) (iv) imply equality in (3.3) for $(k=0, x\leq -3)$. Therefore we restrict to the case $(k,x)=(0,-2)$. Recall

$$\mathrm{Rad}\, I^{(0)}=\mathbb{Z}f_1 \quad\text{with}\quad f_1=e_1+e_2.$$

By Theorem 6.8 (c)

$$\Delta^{(0)}=(e_1+\mathbb{Z}f_1)\,\dot\cup\,(-e_1+\mathbb{Z}f_1)=(e_1+\mathbb{Z}f_1)\,\dot\cup\,(e_2+\mathbb{Z}f_1).$$

One easily sees for $b_1,b_2\in\mathbb{Z}$

$$s^{(0)}_{e_1+b_1f_1}s^{(0)}_{e_2+b_2f_1}=-M \iff b_1+b_2=0,$$

thus

$$\{\underline{v}\in(\Delta^{(0)})^2\,|\,s^{(0)}_{v_1}s^{(0)}_{v_2}=-M\}=\{\pm(e_1+bf_1),\pm(e_2-bf_1)\,|\,b\in\mathbb{Z}\}.$$

This set is a single $\mathrm{Br}_2\ltimes\{\pm 1\}^2$ orbit because of

$$\delta_2\sigma_1(e_1+bf_1,e_2-bf_1)=(e_1+(b+1)f_1,e_2-(b+1)f_1). \qquad \square$$

7.2 Distinguished Bases in the Even Rank 3 Cases

In the even cases with $n=3$ we have complete results on the question when the inclusion in (3.3) is an equality. It is one in all cases except the case $\mathcal{H}_{1,2}$.

Theorem 7.3 *Let $(H_{\mathbb{Z}},L,\underline{e})$ be a unimodular bilinear lattice of rank 3 with a triangular basis $\underline{e}$ with matrix $S=S(\underline{x})=L(\underline{e}^t,\underline{e})^t\in T_3(\mathbb{Z})$.*

(a) Suppose $S\notin(\mathrm{Br}_3\ltimes\{\pm 1\}^3)(S(\mathcal{H}_{1,2}))$. Then the inclusion in (3.3) is an equality.

(b) Suppose $S=S(\mathcal{H}_{1,2})=S(-2,2,-2)$. Recall the basis $(f_1,f_2,f_3)=\underline{e}\begin{pmatrix}1&0&1\\1&1&1\\0&1&1\end{pmatrix}$ of $H_{\mathbb{Z}}$ with $\mathrm{Rad}\, I^{(1)}=\mathbb{Z}f_1\oplus\mathbb{Z}f_2$ and $\mathrm{Rad}\, I^{(0)}=\mathbb{Z}f_3$. The set $\{\underline{v}\in(\Delta^{(0)})^3\,|\,(\pi_3\circ$

$\pi_3^{(0)})(\underline{v}) = -M\}$ splits into countably many orbits. The following list gives one representative for each orbit,

$$(f_3 - g_1, -f_3 + g_1 + g_2, f_3 - g_2) \ \textit{with} \ \begin{pmatrix} g_1 \\ g_2 \end{pmatrix} = \begin{pmatrix} 0 & c_2 \\ c_1 & c_3 \end{pmatrix} \begin{pmatrix} f_1 \\ f_2 \end{pmatrix},$$
$$c_1 \in \mathbb{N} \ \textit{odd}, \ c_2 \in \mathbb{Z} \ \textit{odd}, \ c_3 \in \{0, 1, ..., |c_2| - 1\}.$$

The sublattice $\langle f_3 - g_1, -f_3 + g_1 + g_2, f_3 - g_2\rangle = \langle f_3, g_1, g_2\rangle \subset H_{\mathbb{Z}}$ has finite index $c_1 \cdot |c_2|$ in $H_{\mathbb{Z}}$. It is $H_{\mathbb{Z}}$ in the following two cases:

$$\underline{e} = (f_3 - f_2, -f_3 + f_1 + f_2, f_3 - f_1),$$
$$\text{so} \ \ (g_1, g_2, c_1, c_2, c_3) = (f_2, f_1, 1, 1, 0),$$
$$(f_3 + f_2, -f_3 + f_1 - f_2, f_3 - f_1),$$
$$\text{so} \ \ (g_1, g_2, c_1, c_2, c_3) = (-f_2, f_1, 1, -1, 0)$$

(see also Example 3.23 (ii)) for the second case).

Proof (a) We can replace $\underline{e}$ by an arbitrary element $\underline{\widetilde{e}} \in \mathcal{B}^{dist}$. By Theorem 4.6 the following cases exhaust all $\mathrm{Br}_3 \ltimes \{\pm 1\}^3$ orbits except that of $\mathcal{H}_{1,2}$:

(A) $(H_{\mathbb{Z}}, L, \underline{e})$ is irreducible with $\underline{x} \in \mathbb{Z}^3_{\leq 0}$.
(B) $r(\underline{x}) \leq 0$ and $\underline{x} \neq (0, 0, 0)$.
(C) $\underline{x} = (x_1, 0, 0)$ with $x_1 \in \mathbb{Z}_{\leq 0}$, so $(H_{\mathbb{Z}}, L, \underline{e})$ is reducible (this includes the case A_1^3).
(D) $\underline{x} = (-l, 2, -l)$ with $l \geq 3$.

The cases (A): $\Gamma^{(0)}$ is a Coxeter group by Theorem 3.7 (a). Theorem 3.7 (b) applies.

The cases (B): By Theorem 6.11 (g) $\Gamma^{(0)}$ is a free Coxeter group with generators $s_{e_1}^{(0)}, s_{e_2}^{(0)}, s_{e_3}^{(0)}$. Theorem 3.7 (b) or Example 3.23 (iv) can be used.

The cases (C): Consider a triple $\underline{v} \in (\Delta^{(0)})^3$ with $s_{v_1}^{(0)} s_{v_2}^{(0)} s_{v_3}^{(0)} = -M$. The set $\Delta^{(0)}$ splits into the subsets $\Delta^{(0)} \cap (\mathbb{Z}e_1 + \mathbb{Z}e_2)$ and $\{\pm e_3\}$. Compare $-M|_{\mathbb{Z}e_3} = -\mathrm{id}\,|_{\mathbb{Z}e_3}$ with $s_{e_3}^{(0)}|_{\mathbb{Z}e_3} = -\mathrm{id}\,|_{\mathbb{Z}e_3}$ and $s_a^{(0)}|_{\mathbb{Z}e_3} = \mathrm{id}\,|_{\mathbb{Z}e_3}$ for $a \in \Delta^{(0)} \cap (\mathbb{Z}e_1 + \mathbb{Z}e_2)$. All three $v_i \in \{\pm e_3\}$ is impossible because $(s_{e_3}^{(0)})^3 \neq -M$. Therefore there are i, j, k with $\{i, j, k\} = \{1, 2, 3\}$, $i < j$, $v_i, v_j \in \Delta^{(0)} \cap (\mathbb{Z}e_1 + \mathbb{Z}e_2)$ and $v_k \in \{\pm e_3\}$. The reflection $s_{v_k}^{(0)}$ acts trivially on $\mathbb{Z}e_1 + \mathbb{Z}e_2$ and commutes with $s_{v_i}^{(0)}$ and $s_{v_j}^{(0)}$. Therefore

$$s_{v_i}^{(0)} s_{v_j}^{(0)} s_{v_k}^{(0)} = s_{v_1}^{(0)} s_{v_2}^{(0)} s_{v_3}^{(0)} = (\pi_3 \circ \pi_3^{(0)})(\underline{v}) = -M = s_{e_1}^{(0)} s_{e_2}^{(0)} s_{e_3}^{(0)},$$
$$\text{so} \ \ s_{v_i}^{(0)} s_{v_j}^{(0)} = s_{e_1}^{(0)} s_{e_2}^{(0)}.$$

The reflections $s_{v_i}^{(0)}, s_{v_j}^{(0)}, s_{e_1}^{(0)}$ and $s_{e_2}^{(0)}$ act trivially on $\mathbb{Z}e_3$. The inclusion in (3.3) is an equality because of Theorem 7.1 (a) for the rank 2 cases.

The cases (D): The proof of these cases is prepared by Lemma 7.4 and Lemma 7.5. The proof comes after the proof of Lemma 7.5.

(b) The proof of part (b) is prepared by Lemma 7.6 and comes after the proof of Lemma 7.6. (□)

The following lemma is related to t_λ^+ in Lemma 6.17. Recall also $j^{(k)} : H_\mathbb{Z} \to H_\mathbb{Z}^\sharp$, $a \mapsto I^{(k)}(a, .)$, in Definition 6.1.

Lemma 7.4 *Let $(H_\mathbb{Z}, L)$ be a unimodular bilinear lattice of rank $n \in \mathbb{N}$. Fix $k \in \{0, 1\}$. Suppose* $\mathrm{Rad}\, I^{(k)} \neq \{0\}$ *and choose an element $f \in \mathrm{Rad}\, I^{(k)} - \{0\}$. Denote*

$$\mathrm{Hom}_{0,f}(H_\mathbb{Z}, \mathbb{Z}) := \{\lambda : H_\mathbb{Z} \to \mathbb{Z} \mid \lambda \text{ is } \mathbb{Z}\text{-linear}, \lambda(f) = 0\},$$
$$t_\lambda : H_\mathbb{Z} \to H_\mathbb{Z} \text{ with } t_\lambda(a) = a + \lambda(a) f \text{ for } \lambda \in \mathrm{Hom}_{0,f}(H_\mathbb{Z}, \mathbb{Z}).$$

Then $t_\lambda \in O_u^{(k),Rad}$. The map

$$\mathrm{Hom}_{0,f}(H_\mathbb{Z}, \mathbb{Z}) \to O_u^{(k),Rad}, \quad \lambda \mapsto t_\lambda,$$

is an injective group homomorphism. For $b \in R^{(k)}$ (with $R^{(1)} = H_\mathbb{Z}$, see 3.9 (i)) and $a \in \mathbb{Z}$

$$s_{b+af}^{(k)} = s_b^{(k)} \circ t_{-aj^{(k)}(b)} = t_{(-1)^k a j^{(k)}(b)} \circ s_b^{(k)},$$
$$s_b^{(k)} \circ t_\lambda = t_{\lambda - (-1)^k \lambda(b) j^{(k)}(b)} \circ s_b^{(k)}.$$

Proof The proof is straightforward. We skip the details. □

The following lemma studies the Hurwitz action of Br_3 on triples of reflections in $G^{fCox,2}$. It is related to Theorem 3.2 (b).

Lemma 7.5 *As in Definition 3.1, $G^{fCox,2}$ denotes the free Coxeter group with two generators z_1 and z_2, so generating relations are $z_1^2 = z_2^2 = 1$.*

(a) Its set of reflections is

$$\Delta(G^{fCox,2}) = \bigcup_{i=1}^{2} \{w z_i w^{-1} \mid w \in G^{fCox,2}\} = \{(z_1 z_2)^m z_1 \mid m \in \mathbb{Z}\}.$$

The complement of this set is

$$G^{fCox,2} - \Delta(G^{fCox,2}) = \{(z_1 z_2)^m \mid m \in \mathbb{Z}\}.$$

$\Delta(G^{fCox,2})$ respectively its complement consists of the elements which can be written as words of odd respectively even length in z_1 and z_2.

(b) The set

$$\{(w_1, w_2, w_3) \in (\Delta(G^{fCox,2})^3 \mid w_1 w_2 w_3 = z_1 z_2 z_1\}$$

is the disjoint union of the following Br_3 *orbits:*

$$\dot{\bigcup}_{m\in\mathbb{Z}_{\geq 0}} \mathrm{Br}_3\big((z_1z_2z_1, (z_1z_2)^{1-m}z_1, (z_1z_2)^{1-m}z_1)\big).$$

Proof (a) Clear.

(b) The map

$$\begin{aligned}\{(w_1, w_2, w_3) \in (\Delta(G^{fCox,2}))^3 \mid w_1w_2w_3 = z_1z_2z_1\} &\to M_{2\times 1}(\mathbb{Z}),\\ (w_1, w_2, w_3) &\mapsto (m_1, m_2)^t\\ \text{with } w_1w_2 = (z_1z_2)^{m_1},\ w_2w_3 = (z_1z_2)^{m_2},&\end{aligned}$$

is a bijection because

$$\begin{aligned} z_1z_2z_1 &= (w_1w_2)w_2(w_2w_3), \quad \text{so}\\ w_2 &= (w_1w_2)^{-1}z_1z_2z_1(w_2w_3)^{-1},\\ w_1 &= (w_1w_2)w_2,\\ w_3 &= w_2(w_2w_3),\end{aligned}$$

so a given column vector $(m_1, m_2)^t$ has a unique preimage.

The Hurwitz action of Br_3 on the set on the left hand side of the bijection above translates as follows to an action on $M_{2\times 1}(\mathbb{Z})$.

$$\begin{aligned}\sigma_1(w_1, w_2, w_3) &= (w_1w_2w_1, w_1, w_3),\\ w_1w_2w_1 \cdot w_1 &= w_1w_2 = (z_1z_2)^{m_1},\\ w_1w_3 &= (w_1w_2)(w_2w_3) = (z_1z_2)^{m_1+m_2},\\ \sigma_1\begin{pmatrix}m_1\\ m_2\end{pmatrix} &= \begin{pmatrix}m_1\\ m_1+m_2\end{pmatrix} = \begin{pmatrix}1&0\\1&1\end{pmatrix}\begin{pmatrix}m_1\\ m_2\end{pmatrix},\\ \sigma_2(w_1, w_2, w_3) &= (w_1, w_2w_3w_2, w_2),\\ w_1 \cdot w_2w_3w_2 = (w_1w_2)(w_2w_3)^{-1} &= (z_1z_2)^{m_1-m_2},\\ w_2w_3w_2 \cdot w_2 &= w_2w_3 = (z_1z_2)^{m_2},\\ \sigma_2\begin{pmatrix}m_1\\ m_2\end{pmatrix} &= \begin{pmatrix}m_1-m_2\\ m_2\end{pmatrix} = \begin{pmatrix}1&-1\\0&1\end{pmatrix}\begin{pmatrix}m_1\\ m_2\end{pmatrix}.\end{aligned}$$

So Br_3 acts as multiplication with matrices in $SL_2(\mathbb{Z})$ from the left on $M_{2\times 1}(\mathbb{Z})$. Each orbit has a unique element of the shape $\begin{pmatrix}m\\0\end{pmatrix}$ with $m \in \mathbb{Z}_{\geq 0}$. This element corresponds to

$$(z_1z_2z_1, (z_1z_2)^{1-m}z_1, (z_1z_2)^{1-m}z_1).$$

□

Proof of Theorem 7.3 (a) in the cases (D), $\underline{x} = (-l, 2, -l)$ with $l \geq 3$: Recall from Theorem 6.11 (f)

$$\begin{aligned} \text{Rad } I^{(0)} &= \mathbb{Z} f_1 \quad \text{with} \quad f_1 = e_1 - e_3, \\ \Gamma_s^{(0)} &\cong G^{fCox,2} \quad \text{with generators } z_1 = \overline{s_{e_1}^{(0)}} = \overline{s_{e_3}^{(0)}}, \ z_2 = \overline{s_{e_2}^{(0)}}. \end{aligned}$$

Suppose $\underline{v} \in (\Delta^{(0)})^3$ with $s_{v_1}^{(0)} s_{v_2}^{(0)} s_{v_3}^{(0)} = -M$. We want to show $\underline{v} \in \mathcal{B}^{dist}$ or equivalently $(s_{v_1}^{(0)}, s_{v_2}^{(0)}, s_{v_3}^{(0)}) \in \mathcal{R}^{(0),dist}$.

First we look at the images in $\Gamma_s^{(0)}$: $\overline{s_{v_1}^{(0)} s_{v_2}^{(0)} s_{v_3}^{(0)}} = z_1 z_2 z_1$. Because of Lemma 7.5 (b), we can make a suitable braid group action and then suppose

$$(\overline{s_{v_1}^{(0)}}, \overline{s_{v_2}^{(0)}}, \overline{s_{v_3}^{(0)}}) = (z_1 z_2 z_1, r, r) \text{ with } r = (z_1 z_2)^{1-m} z_1 \text{ for some } m \in \mathbb{Z}_{\geq 0}.$$

Write $\widetilde{e_2} := s_{e_1}^{(0)}(e_2) = e_2 + l e_1$ and observe

$$z_1 z_2 z_1 = \overline{s_{e_1}^{(0)} s_{e_2}^{(0)} s_{e_3}^{(0)}} = \overline{s_{\widetilde{e_2}}^{(0)}}.$$

After possibly changing the signs of v_1 and v_3, $\overline{s_{v_1}^{(0)}} = \overline{s_{\widetilde{e_2}}^{(0)}}$ and $\overline{s_{v_2}^{(0)}} = r = \overline{s_{v_3}^{(0)}}$ imply

$$v_1 = \widetilde{e_2} + a_1 f_1 \quad \text{and} \quad v_3 = v_2 + a_2 f_1 \quad \text{for some } a_1, a_2 \in \mathbb{Z}.$$

With Lemma 7.4 and f_1 =(f in Lemma 7.4) we calculate

$$\begin{aligned} -M &= s_{e_1}^{(0)} s_{e_2}^{(0)} s_{e_3}^{(0)} = s_{e_1}^{(0)} s_{e_2}^{(0)} s_{e_1 - f_1}^{(0)} \\ &= s_{e_1}^{(0)} s_{e_2}^{(0)} s_{e_1}^{(0)} t_{j^{(0)}(e_1)} = s_{\widetilde{e_2}}^{(0)} t_{j^{(0)}(e_1)}, \\ -M &= s_{v_1}^{(0)} s_{v_2}^{(0)} s_{v_3}^{(0)} = s_{\widetilde{e_2} + a_1 f_1}^{(0)} s_{v_2}^{(0)} s_{v_2 + a_2 f_1}^{(0)} \\ &= s_{\widetilde{e_2}}^{(0)} t_{-a_1 j^{(0)}(\widetilde{e_2})} s_{v_2}^{(0)} s_{v_2}^{(0)} t_{-a_2 j^{(0)}(v_2)} \\ &= s_{\widetilde{e_2}}^{(0)} t_{-a_1 j^{(0)}(\widetilde{e_2}) - a_2 j^{(0)}(v_2)}, \end{aligned}$$

so

$$j^{(0)}(e_1) = -a_1 j^{(0)}(\widetilde{e_2}) - a_2 j^{(0)}(v_2).$$

Write

$$\overline{v_2}^{(0)} = b_1 \overline{e_1}^{(0)} + b_2 \overline{e_2}^{(0)} \quad \text{with } b_1, b_2 \in \mathbb{Z}.$$

By Theorem 6.14 (f) the tuple $(\overline{H_{\mathbb{Z}}}^{(0)}, \overline{I}^{(0)}, (\overline{e_1}^{(0)}, \overline{e_2}^{(0)}))$ is isomorphic to the corresponding tuple from the 2×2 matrix $S(-l) = \begin{pmatrix} 1 & -l \\ 0 & 1 \end{pmatrix}$. The set of L1-cycles of this tuple is called $R^{(0)}(S(-l))$. It contains $\overline{v_2}^{(0)}, \overline{e_1}^{(0)}, \overline{e_1}^{(0)}$. By Theorem 6.8 (d)(i) the map

$$\begin{aligned} R^{(0)}(S(-l)) &\to \{\text{units in } \mathbb{Z}[\kappa_1] \text{ with norm } 1\} = \{\pm\kappa_1^m \mid m \in \mathbb{Z}\} \\ y_1\overline{e_1}^{(0)} + y_2\overline{e_2}^{(0)} &\mapsto y_1 - \kappa_1 y_2, \end{aligned}$$

is a bijection, where $\kappa_1 = \frac{l}{2} + \frac{1}{2}\sqrt{l^2 - 4}$. The norm of $b_1 - b_2\kappa_1$ is

$$1 = b_1^2 - lb_1b_2 + b_2^2.$$

Now

$$\begin{aligned} (2, -l) &= j^{(0)}(e_1)(e_1, e_2) = (-a_1 j^{(0)}(\widetilde{e_2}) - a_2 j^{(0)}(v_2))(e_1, e_2) \\ &= -a_1((-l, 2) + l(2, -l)) - a_2(b_1(2, -l) + b_2(-l, 2)) \\ &= (-a_1 l - a_2 b_1)(2, -l) + (-a_1 - a_2 b_2)(-l, 2). \end{aligned}$$

so

$$\begin{aligned} a_1 = -a_2 b_2, \quad 1 &= -a_1 l - a_2 b_1 = a_2(b_2 l - b_1), \\ a_2 &= \pm 1 \quad \text{and} \quad b_1 = -a_2 + b_2 l. \end{aligned}$$

Calculate

$$\begin{aligned} 0 &= -1 + b_1^2 - lb_1b_2 + b_2^2 = -1 + (-a_2 + b_2 l)(-a_2) + b_2^2 \\ &= b_2(b_2 - a_2 l). \end{aligned}$$

We obtain the four solutions

$$\begin{aligned} (a_1, a_2, b_1, b_2) \in \{&(0, 1, -1, 0), (0, -1, 1, 0), \\ &(-l, 1, l^2 - 1, l), (-l, -1, 1 - l^2, -l)\}. \end{aligned}$$

In the case of the third solution

$$\begin{aligned} \overline{v_2}^{(0)} &= (l^2 - 1)\overline{e_1}^{(0)} + l\overline{e_2}^{(0)} = \overline{s_{e_1}^{(0)} s_{e_2}^{(0)}(e_1)}, \\ r = \overline{s_{v_2}^{(0)}} &= \overline{s^{(0)}_{s_{e_1}^{(0)} s_{e_2}^{(0)}(e_1)}} = \overline{s_{e_1}^{(0)} s_{e_2}^{(0)} s_{e_1}^{(0)} s_{e_2}^{(0)} s_{e_1}^{(0)}} \\ &= (z_1 z_2)^2 z_1 = (z_1 z_2)^{1-m} z_1 \quad \text{with } m = -1. \end{aligned}$$

As $m = -1$ is not in the set $\mathbb{Z}_{\geq 0}$, we can discard the third solution. In fact, $(z_1z_2z_1, (z_1z_2)^2z_1, (z_1z_2)^2z_1)$ is in the Br_3 orbit of $(z_1z_2z_1, z_1, z_1)$ because $\begin{pmatrix} -1 \\ 0 \end{pmatrix}$

is in the $SL_2(\mathbb{Z})$ orbit of $\begin{pmatrix} 1 \\ 0 \end{pmatrix}$. We can discard also the fourth solution because its vector $\overline{v_2}^{(0)}$ differs from the vector $\overline{v_2}^{(0)}$ in the third solution only by the sign.

Also the vector $\overline{v_2}^{(0)}$ in the first solution differs from the vector $\overline{v_2}^{(0)}$ in the second solution only by the sign.

The second solution gives $\overline{v_2}^{(0)} = \overline{e_1}^{(0)}$ and thus for some $b_3 \in \mathbb{Z}$

$$\underline{v} = (\widetilde{e_2}, e_1 + b_3 f_1, e_1 + b_3 f_1 - f_1) = (\widetilde{e_2}, e_1 + b_3 f_1, e_3 + b_3 f_1).$$

The observation

$$\begin{aligned}\delta_2\sigma_2(\underline{v}) &= \delta_2(\widetilde{e_2}, e_3 + b_3 f_1 - 2(e_1 + b_3 f_1), e_1 + b_3 f_1) \\ &= (\widetilde{e_2}, e_1 + (b_3 + 1) f_1, e_3 + (b_3 + 1) f_1)\end{aligned}$$

shows $\underline{v} \in \mathrm{Br}_3 \ltimes \{\pm 1\}^3(\widetilde{e_2}, e_1, e_3)$. This orbit is $\mathcal{B}^{dist}$ because $(\widetilde{e_2}, e_1, e_3) = \sigma_1(\underline{e})$. □

Lemma 7.6 states some facts which arise in the proof of part (b) of Theorem 7.3 and which are worth to be formulated explicitly.

Lemma 7.6 *Let $(H_{\mathbb{Z}}, L, \underline{e})$ be the unimodular bilinear lattice of rank 3 with triangular basis $\underline{e}$ with matrix $S = S(\mathcal{H}_{1,2}) = S(-2, 2, -2) = L(\underline{e}^t, \underline{e})^t$. Recall* $\mathrm{Rad}\, I^{(0)} = \mathbb{Z} f_1 + \mathbb{Z} f_2$ *and* $R^{(0)} = \pm f_3 + \mathrm{Rad}\, I^{(0)}$ *(Theorem 6.14 (f)).*

(a) For $g_1, g_2, g_3 \in \mathrm{Rad}\, I^{(0)}$

$$s^{(0)}_{f_3 - g_1} s^{(0)}_{f_3 - g_2} s^{(0)}_{f_3 - g_3} = -M \iff g_2 = g_1 + g_3.$$

(b) The map

$$\begin{aligned}\Phi : M_{2\times 2}(\mathbb{Z}) &\to \{\underline{v} \in (R^{(0)})^3 \mid (\pi_3 \circ \pi_3^{(0)})(\underline{v}) = -M\}/\{\pm 1\}^3, \\ A &\mapsto (f_3 - g_1, f_3 - g_1 - g_3, f_3 - g_3)/\{\pm 1\}^3 \\ &\text{with } \begin{pmatrix} g_1 \\ g_3 \end{pmatrix} = A \begin{pmatrix} f_1 \\ f_2 \end{pmatrix},\end{aligned}$$

is a bijection. The action of Br_3 *on the right hand side translates to the following action on the left hand side,*

$$\sigma_1(A) = \begin{pmatrix} 1 & -1 \\ 0 & 1 \end{pmatrix} A, \quad \sigma_2(A) = \begin{pmatrix} 1 & 0 \\ 1 & 1 \end{pmatrix} A.$$

(c) $\underline{v} \in (R^{(0)})^3$ with $(\pi_3 \circ \pi_3^{(0)})(\underline{v}) = -M$ satisfies either (i) or (ii):

(i) There exists a permutation $\sigma \in S_3$ with $v_i \in \Gamma^{(0)}\{e_{\sigma(i)}\}$ for $i \in \{1, 2, 3\}$. (ii) Either $v_1, v_2, v_3 \in \Gamma^{(0)}\{f_3\}$ or there exists a permutation $\sigma \in S_3$ and an $l \in \{1, 2, 3\}$ with $v_{\sigma(1)} \in \Gamma^{(0)}\{f_3\}$ and $v_{\sigma(2)}, v_{\sigma(3)} \in \Gamma^{(0)}\{e_l\}$.

(i) holds if and only if $\Phi^{-1}(\underline{v}/\{\pm 1\}^3)$ *has an odd determinant.*

(d) Let $SL_2(\mathbb{Z})$ *act by multiplication from the left on* $\{A \in M_{2\times 2}(\mathbb{Z}) \mid \det A$ *is odd*$\}$. *Each orbit has a unique representative of the shape*

$$\begin{pmatrix} 0 & c_2 \\ c_1 & c_3 \end{pmatrix} \quad \textit{with } c_1 \in \mathbb{N} \textit{ odd}, c_2 \in \mathbb{Z} \textit{ odd}, c_3 \in \{0, 1, ..., |c_2| - 1\}.$$

Proof (a) For $g_1, g_2, g_3 \in \operatorname{Rad} I^{(0)}$

$$s^{(0)}_{f_3-g_1}|_{\operatorname{Rad} I^{(0)}} = \operatorname{id}, \quad s^{(0)}_{f_3-g_1}(f_3 + g_2) = -(f_3 - g_2 - 2g_1),$$
$$s^{(0)}_{f_3-g_1} s^{(0)}_{f_3-g_2} s^{(0)}_{f_3-g_3}(f_3) = -f_3 + 2(g_1 - g_2 + g_3).$$

Compare $-M|_{\operatorname{Rad} I^{(0)}} = \operatorname{id}$, $-M(f_3) = -f_3$.

(b) Φ is a bijection because of $R^{(0)} = \pm f_3 + \operatorname{Rad} I^{(0)}$ and part (a). The action of Br_3 on the right hand side translates to the claimed action on the left hand side because of the following,

$$\delta_1\sigma_1(f_3 - g_1, f_3 - g_1 - g_3, f_3 - g_3) = (f_3 - g_1 + g_3, f_3 - g_1, f_3 - g_3),$$
$$\begin{pmatrix} g_1 - g_3 \\ g_3 \end{pmatrix} = \begin{pmatrix} 1 & -1 \\ 0 & 1 \end{pmatrix} \begin{pmatrix} g_1 \\ g_3 \end{pmatrix},$$
$$\delta_2\sigma_2(f_3 - g_1, f_3 - g_1 - g_3, f_3 - g_3) = (f_3 - g_1, f_3 - 2g_1 - g_3, f_3 - g_1 - g_3),$$
$$\begin{pmatrix} g_1 \\ g_1 + g_3 \end{pmatrix} = \begin{pmatrix} 1 & 0 \\ 1 & 1 \end{pmatrix} \begin{pmatrix} g_1 \\ g_3 \end{pmatrix}.$$

(c) Recall that by Theorem 6.14 (e)

$$\Gamma^{(0)}\{e_i\} = \pm e_i + 2 \operatorname{Rad} I^{(0)}, \quad \Gamma^{(0)}\{f_3\} = \pm f_3 + 2 \operatorname{Rad} I^{(0)},$$
$$(e_1, e_2, e_3) = (f_3 - f_2, -f_3 + f_1 + f_2, f_3 - f_1),$$
$$\Delta^{(0)} = \Gamma^{(0)}\{e_1\} \,\dot{\cup}\, \Gamma^{(0)}\{e_2\} \,\dot{\cup}\, \Gamma^{(0)}\{e_3\}, \quad R^{(0)} = \Delta^{(0)} \,\dot{\cup}\, \Gamma^{(0)}\{f_3\}.$$

Observe that $g_1, g_2, g_3 \in \operatorname{Rad} I^{(0)}$ with $g_2 = g_1 + g_3$ satisfy either (i)' or (ii)',

(i)' $g_1, g_2, g_3 \notin 2 \operatorname{Rad} I^{(0)}$, (ii)' There exists a permutation $\sigma \in S_3$ with $g_{\sigma(1)} \in 2 \operatorname{Rad} I^{(0)}$ and $g_{\sigma(2)} - g_{\sigma(3)} \in 2 \operatorname{Rad} I^{(0)}$.

$\underline{v} = (f_3 - g_1, f_3 - g_2, f_3 - g_3)$ satisfies (i) if (i)' holds, and it satisfies (ii) if (ii)' holds. If $\begin{pmatrix} g_1 \\ g_2 \end{pmatrix} = A \begin{pmatrix} f_1 \\ f_2 \end{pmatrix}$ then (i)' holds if and only if $\det A$ is odd.

(d) This is elementary. We skip the details. □

Proof of Theorem 7.3 **(b)**: $\underline{v} \in (\Delta^{(0)})^3$ with $(\pi_3 \circ \pi_3^{(0)})(\underline{v}) = -M$ satisfies property (i) in Lemma 7.6 (c) because it does not satisfy property (ii) in Lemma 7.6 (c). The parts (b) and (d) of Lemma 7.6 show that the set $(\Delta^{(0)})^3 \cap (\pi_3 \circ \pi_3^{(0)})^{-1}(-M)$ consists of countably many orbits. The parts (b) and (d) of Lemma 7.6 also give the claimed representative in each orbit. The rest is obvious. □

7.3 Distinguished Bases in the Odd Rank 3 Cases

Also in the odd cases with $n = 3$ we have complete results on the question when the inclusion (3.4) is an equality. It is one if and only if $\underline{x} \in B_1 \cup B_2$ where

$$\begin{aligned}
B_1 &= \{\underline{x} \in \mathbb{Z}^3 - \{(0,0,0)\} \mid \\
&\qquad ((G^{phi} \ltimes \widetilde{G}^{sign}) \rtimes \langle \gamma \rangle)(\underline{x}) \cap r^{-1}(\mathbb{Z}_{\le 0}) \neq \emptyset\}, \\
B_2 &= \{\underline{x} \in \mathbb{Z}^3 - \{(0,0,0)\} \mid S(\underline{x}) \text{ is reducible, i.e. there are } i, j, k \\
&\qquad \text{with } \{i,j,k\} = \{1,2,3\} \text{ and } x_i \neq 0 = x_j = x_k\}, \\
B_3 &:= \{(0,0,0)\}, \\
B_4 &:= \{\underline{x} \in \mathbb{Z}^3 \mid S(\underline{x}) \in (\mathrm{Br}_3 \ltimes \{\pm 1\}^3)\Big(\{S(A_3), S(\widehat{A}_2), S(\mathcal{H}_{1,2})\} \\
&\qquad \cup \{S(-l, 2, -l) \mid l \ge 3\}\Big)\}, \\
B_5 &:= \mathbb{Z}^3 - (B_1 \cup B_2 \cup B_3 \cup B_4).
\end{aligned}$$

B_2 is the set of $\underline{x} \neq (0,0,0)$ which give reducible cases. B_1 contains $r^{-1}(\mathbb{Z}_{\le 0}) - \{(0,0,0)\}$, but is bigger. $\underline{x} \in B_1$ if and only if the $(G^{phi} \ltimes \widetilde{G}^{sign}) \rtimes \langle \gamma \rangle$ orbit of $\underline{x}$ contains a triple $\underline{\widetilde{x}} \in \mathbb{Z}^3$ as in Lemma 4.18 (a), so with $\underline{\widetilde{x}} \in \mathbb{Z}^3_{\ge 3}$ and $2\widetilde{x}_i \le \widetilde{x}_j \widetilde{x}_k$ for $\{i,j,k\} = \{1,2,3\}$. The Examples 4.20 show that it is not so easy to describe B_1 more explicitly.

In Theorem 7.7 we show $B_4 \subset \mathbb{Z}^3 - (B_1 \cup B_2 \cup B_3)$. For $\underline{x} \in B_3 \cup B_4$ the inclusion in (3.4) is not an equality, but we can add the constraint $\sum_{i=1}^3 \mathbb{Z} v_i = H_{\mathbb{Z}}$ to (3.4) and obtain an equality. For $\underline{x} \in B_5$ we do not know whether (3.4) with the additional constraint $\sum_{i=1}^3 \mathbb{Z} v_i = H_{\mathbb{Z}}$ becomes an equality.

Theorem 7.7 *Let $(H_{\mathbb{Z}}, L, \underline{e})$ be a unimodular bilinear lattice of rank 3 with a triangular basis $\underline{e}$ with matrix $L(\underline{e}^t, \underline{e})^t = S(\underline{x}) \in T_3(\mathbb{Z})$ for some $\underline{x} \in \mathbb{Z}^3$.*

(a) $(H_{\mathbb{Z}}, L, \underline{e})$ is reducible if and only if $\underline{x} \in B_2 \cup B_3$. Then $\Gamma_u^{(1)} = \{\mathrm{id}\}$.

(b) The following conditions are equivalent:

(i) $\underline{x} \in B_1$.

(ii) $\Gamma^{(1)} \cong G^{free,3}$.

(iii) $(H_{\mathbb{Z}}, L, \underline{e})$ is irreducible and $\Gamma_u^{(1)} = \{\mathrm{id}\}$.

(c) $\mathbb{Z}^3 = \dot{\bigcup}_{i \in \{1,2,3,4,5\}} B_i$.

(d) The inclusion in (3.4) is an equality $\iff \underline{x} \in B_1 \cup B_2$.

(e) Consider $\underline{x} = (0,0,0)$. The set $\{\underline{v} \in (\Delta^{(1)})^3 \mid (\pi_3 \circ \pi_3^{(1)})(\underline{v}) = M\}$ is $(\Delta^{(1)})^3$. It consists of ten $\mathrm{Br}_3 \ltimes \{\pm 1\}^3$ orbits, the orbit $\mathcal{B}^{dist}$ of $\underline{e} = (e_1, e_2, e_3)$ and the orbits of the nine triples

$$\begin{aligned}
&(e_1,e_1,e_1), (e_1,e_1,e_2), (e_1,e_2,e_2), (e_2,e_2,e_2), \\
&(e_1,e_1,e_3), (e_1,e_3,e_3), (e_3,e_3,e_3), (e_2,e_2,e_3), (e_2,e_3,e_3).
\end{aligned}$$

(f) Consider $\underline{x} \in B_4 \cup B_5$. Then $\Gamma_u^{(1)} \cong \mathbb{Z}^2$. The map

$$\Psi : \{\underline{v} \in (\Delta^{(1)})^3 \mid (\pi_3 \circ \pi_3^{(1)})(\underline{v}) = M\} \to \mathbb{N} \cup \{\infty\},$$
$$\underline{v} \mapsto \Big(\textit{index of } \sum_{i=1}^{3} \mathbb{Z}v_i \textit{ in } H_{\mathbb{Z}}\Big),$$

has infinitely many values. The set $\{\underline{v} \in (\Delta^{(1)})^3 \mid (\pi_3 \circ \pi_3^{(1)})(\underline{v}) = M\}$ *contains besides* $\mathcal{B}^{dist}$ *infinitely many* $\mathrm{Br}_3 \ltimes \{\pm 1\}^3$ *orbits.*

(g) For $\underline{x} \in B_3 \cup B_4$

$$\mathcal{B}^{dist} = \{\underline{v} \in (\Delta^{(1)})^3 \mid (\pi_3 \circ \pi_3^{(1)})(\underline{v}) = M, \sum_{i=1}^{3} \mathbb{Z}v_i = H_{\mathbb{Z}}\}.$$

Proof (a) Compare Definition 2.10 in the case $n = 3$. For $\Gamma_u^{(1)} = \{\mathrm{id}\}$ see Theorem 6.18 (b).

(b) By the Remarks 4.17 the tuple $(H_{\mathbb{Z}}, \pm I^{(1)}, \Gamma^{(1)}, \Delta^{(1)})$ depends up to isomorphism only on the $(G^{phi} \ltimes \widetilde{G}^{sign}) \rtimes \langle\gamma\rangle$ orbit of $\underline{x}$. Lemma 4.18 gives representatives of each such orbit. Theorem 6.18 studies their groups $\Gamma^{(1)}$. Theorem 6.18 (b) treats $\underline{x} \in B_2 \cup B_3$. Theorem 6.18 (g) treats $\underline{x} \in B_1$. Theorem 6.18 (c)–(f) treats $\underline{x} \in \mathbb{Z}^3 - (B_1 \cup B_2 \cup B_3)$.

One sees

$$\begin{aligned} \Gamma^{(1)} \cong G^{free,3} &\iff \underline{x} \in B_1, \\ \Gamma_u^{(1)} = \{\mathrm{id}\} &\iff \underline{x} \in B_1 \cup B_2 \cup B_3, \\ \Gamma_u^{(1)} \cong \mathbb{Z}^2 &\iff \underline{x} \in \mathbb{Z}^3 - (B_1 \cup B_2 \cup B_3). \end{aligned}$$

(c) $B_1 \cap B_2 = \emptyset$, $B_2 \cap B_4 = \emptyset$ and $(0, 0, 0) \notin B_1 \cup B_2 \cup B_4$ are clear. $B_1 \cap B_4 = \emptyset$ follows from $\Gamma^{(1)}(\underline{x}) \not\cong G^{free,3}$ for $\underline{x} \in B_4$.

(d) The parts (e) and (f) will give $\Longrightarrow$. Here we show $\Longleftarrow$, first for $\underline{x} \in B_1$, then for $\underline{x} \in B_2$.

Let $\underline{x} \in B_1$. Then by the Remarks 4.17 and Theorem 6.18 (g) $\Gamma^{(1)}$ is a free group with generators $s_{e_1}^{(1)}$, $s_{e_2}^{(1)}$ and $s_{e_3}^{(1)}$. Example 3.23 (iv) applies.

Let $\underline{x} \in B_2$. Because of the actions of γ and $\widetilde{G}^{sign}$ (here G^{sign} is sufficient) on B_2 we can suppose $\underline{x} = (x_1, 0, 0)$ with $x_1 \in \mathbb{Z}_{<0}$. Then $e_3 \in \mathrm{Rad}\, I^{(1)}$, $s_{e_3}^{(1)} = \mathrm{id}$, $\Gamma^{(1)} = \langle s_{e_1}^{(1)}, s_{e_2}^{(1)}\rangle$ and by Theorem 6.21 (b) $\Delta^{(1)} = \Delta^{(1)} \cap (\mathbb{Z}e_1 + \mathbb{Z}e_2) \ \cup \ \{\pm e_3\}$. The monodromy M has the characteristic polynomial $(t-1)(t^2 - (2 - r(\underline{x}))t + 1) = (t-1)(t^2 - (2 - x_1^2)t + 1)$, so three different eigenvalues.

Consider $\underline{v} \in (\Delta^{(1)})^3$ with $s_{v_1}^{(1)} \circ s_{v_2}^{(1)} \circ s_{v_3}^{(1)} = M$. Now $v_1, v_2, v_3 \in \{\pm e_3\}$ is impossible because $M \neq \mathrm{id}$. Two of v_1, v_2, v_3 in $\{\pm e_3\}$ cannot be because then M would have the eigenvalue 1 with multiplicity 3.

Claim: All three $v_1, v_2, v_3 \in \Delta^{(1)} \cap (\mathbb{Z}e_1 + \mathbb{Z}e_2)$ is impossible.

Proof of the Claim: Suppose $v_1, v_2, v_3 \in \Delta^{(1)} \cap (\mathbb{Z}e_1 + \mathbb{Z}e_2)$. First we consider a case with $x_1 \leq -2$. By Theorem 6.18 (b) and Theorem 6.10 (c)+(d) $\Gamma^{(1)} \cong G^{free,2}$ with generators $s_{e_1}^{(1)}$ and $s_{e_2}^{(1)}$. There is a unique group homomorphism

$$\Gamma^{(1)} \to \{\pm 1\} \quad \text{with} \quad s^{(1)}_{e_1} \mapsto -1,\ s^{(1)}_{e_2} \mapsto -1.$$

Each $s^{(1)}_{v_i}$ is conjugate to $s^{(1)}_{e_1}$ or $s^{(1)}_{e_2}$ and thus has image -1. Also their product $s^{(1)}_{v_1} \circ s^{(1)}_{v_2} \circ s^{(1)}_{v_3}$ has image -1. But $M = s^{(1)}_{e_1} \circ s^{(1)}_{e_2}$ has image 1, a contradiction.

Now consider the case $x_1 = -1$. By Theorem 6.18 (b) and Theorem 6.10 (a)+(b) $\Gamma^{(1)} \cong SL_2(\mathbb{Z})$ with $s^{(1)}_{e_1} \sim \begin{pmatrix} 1 & 1 \\ 0 & 1 \end{pmatrix}$ and $s^{(1)}_{e_2} \sim \begin{pmatrix} 1 & 0 \\ -1 & 1 \end{pmatrix}$. It is well known that the group $SL_2(\mathbb{Z})$ is isomorphic to the group with the presentation

$$\langle x_1, x_2 \,|\, x_1x_2x_1 = x_2x_1x_2,\ 1 = (x_1x_2)^6 \rangle$$

where $x_1 \mapsto \begin{pmatrix} 1 & 1 \\ 0 & 1 \end{pmatrix}$ and $x_2 \mapsto \begin{pmatrix} 1 & 0 \\ -1 & 1 \end{pmatrix}$. The differences of the lengths of the words in $x_1^{\pm 1}$ and $x_2^{\pm 1}$ which are connected by these relations are $3-3$ and $12-0$, so even. Therefore also in this situation there is a unique group homomorphism

$$\Gamma^{(1)} \to \{\pm 1\} \quad \text{with} \quad s^{(1)}_{e_1} \to -1,\ s^{(1)}_{e_2} \to -1.$$

The argument in the case $\Gamma^{(1)} \cong G^{free,2}$ goes here through, too. The Claim is proved. ($\Box$)

Therefore a permutation $\sigma \in S_3$ with $v_{\sigma(1)}, v_{\sigma(2)} \in \Delta^{(1)} \cap (\mathbb{Z}e_1 + \mathbb{Z}e_2)$, $\sigma(1) < \sigma(2)$ and $v_{\sigma(3)} \in \{\pm e_3\}$ exists. Then $s^{(1)}_{v_{\sigma(3)}} = \mathrm{id}$ and

$$s^{(1)}_{e_1} s^{(1)}_{e_2} = M = s^{(1)}_{v_1} s^{(1)}_{v_2} s^{(1)}_{v_3} = s^{(1)}_{v_{\sigma(1)}} s^{(1)}_{v_{\sigma(2)}}.$$

Because of Theorem 7.1 $(v_{\sigma(1)}, v_{\sigma(2)})$ is in the $\mathrm{Br}_2 \ltimes \{\pm 1\}^2$ orbit of (e_1, e_2). Therefore $\underline{v}$ is in $(\mathrm{Br}_3 \ltimes \{\pm 1\}^3)(\underline{e}) = \mathcal{B}^{dist}$.

(e) See the Examples 3.23 (iii).

(f) $\Gamma^{(1)}_u \cong \mathbb{Z}^2$ for $\underline{x} \in B_4 \cup B_5$ follows from Theorem 6.21 (c)–(f), the Remarks 4.17, Lemma 4.18 and the definition of B_4 and B_5. The last statement in (f) follows from the middle statement because the sublattice $\sum_{i=1}^3 \mathbb{Z}v_i \subset H_{\mathbb{Z}}$ and its index in $H_{\mathbb{Z}}$ are invariants of the $\mathrm{Br}_3 \ltimes \{\pm 1\}^3$ orbit of $\underline{v}$.

For the middle statement we consider

$$\underline{v} = (e_1 + a(-\widetilde{x}_3)f_3, e_2 + a(\widetilde{x}_2 - \widetilde{x}_1x_3)f_3, e_3 + a(-\widetilde{x}_1)f_3) \quad \text{with } a \in \mathbb{Z}.$$

The next Lemma 7.8 implies

$$(\pi_3 \circ \pi_3^{(1)})(\underline{v}) = M,$$

$$\Big(\text{index of } \sum_{i=1}^{3} \mathbb{Z}v_i \text{ in } H_{\mathbb{Z}}\Big) = \left| 1 + a\frac{r(\underline{x})}{\gcd(x_1, x_2, x_3)^2} \right|$$

and that for a suitable $a_0 \in \mathbb{N}$ and any $a \in \mathbb{Z}a_0$ $\underline{v} \in (\Delta^{(1)})^3$. As $r(\underline{x}) \neq 0$ for $\underline{x} \in B_4 \cup B_5$, this shows that the map Ψ has countably many values.

(g) Part (g) will be prepared by Lemma 7.10 and will be proved after the proof of Lemma 7.10. □

Lemma 7.8 *Let $(H_\mathbb{Z}, L, \underline{e})$ be a unimodular bilinear lattice of rank 3 with a triangular basis $\underline{e}$ with matrix $L(\underline{e}^t, \underline{e})^t = S(\underline{x})$ for some $\underline{x} \in \mathbb{Z}^3 - \{(0,0,0)\}$. Recall* $\operatorname{Rad} I^{(1)} = \mathbb{Z} f_3$ *with* $f_3 = -\widetilde{x}_3 e_1 + \widetilde{x}_2 e_2 - \widetilde{x}_1 e_3$ *and* $(\widetilde{x}_1, \widetilde{x}_2, \widetilde{x}_3) = \gcd(x_1, x_2, x_3)^{-1}$ (x_1, x_2, x_3).

(a) For $\underline{a} = (a_1, a_2, a_3) \in \mathbb{Z}^3$

$$s^{(1)}_{e_1+a_1 f_3} \circ s^{(1)}_{e_2+a_2 f_3} \circ s^{(1)}_{e_3+a_3 f_3} = M$$
$$\iff (a_1, a_2, a_3) \in \mathbb{Z}(-\widetilde{x}_3, \widetilde{x}_2 - \widetilde{x}_1 x_3, -\widetilde{x}_1).$$

(b) For $\underline{a} = (a_1, a_2, a_3) = a(-\widetilde{x}_3, \widetilde{x}_2 - \widetilde{x}_1 x_3, -\widetilde{x}_1)$ with $a \in \mathbb{Z}$, the index of $\sum_{i=1}^3 \mathbb{Z}(e_i + a_i f_3)$ in $H_\mathbb{Z}$ is $|1 + a\frac{r(\underline{x})}{\gcd(x_1,x_2,x_3)^2}|$.

(c) If $\Gamma_u^{(1)} \cong \mathbb{Z}^2$ then there is a number $a_0 \in \mathbb{N}$ with

$$(e_1 + a(-\widetilde{x}_3) f_3, e_2 + a(\widetilde{x}_2 - \widetilde{x}_1 x_3) f_3, e_3 + a(-\widetilde{x}_1) f_3) \in (\Delta^{(1)})^3 \text{ for } a \in \mathbb{Z}a_0.$$

Proof (a) With Lemma 7.4 and $f_3 = (f$ in Lemma7.4) one calculates

$$\begin{aligned}
& s^{(1)}_{e_1+a_1 f_3} \circ s^{(1)}_{e_2+a_2 f_3} \circ s^{(1)}_{e_3+a_3 f_3} \circ M^{-1} \\
&= t_{-a_1 j^{(1)}(e_1)} \circ s^{(1)}_{e_1} \circ t_{-a_2 j^{(1)}(e_2)} \circ s^{(1)}_{e_2} \circ t_{-a_3 j^{(1)}(e_3)} \circ s^{(1)}_{e_3} \circ M^{-1} \\
&= t_{-a_1 j^{(1)}(e_1)} \circ t_{-a_2 j^{(1)}(e_2) - a_2 j^{(1)}(e_2)(e_1) j^{(1)}(e_1)} \circ s^{(1)}_{e_1} \\
&\quad \circ t_{-a_3 j^{(1)}(e_3) - a_3 j^{(1)}(e_3)(e_2) j^{(1)}(e_2)} \circ s^{(1)}_{e_2} \circ s^{(1)}_{e_3} \circ M^{-1} \\
&= t_{-A}
\end{aligned}$$

with

$$\begin{aligned}
A &= a_1 j^{(1)}(e_1) + a_2 j^{(1)}(e_2) + a_2 I^{(1)}(e_2, e_1) j^{(1)}(e_1) \\
&\quad + a_3 j^{(1)}(e_3) + a_3 I^{(1)}(e_3, e_2) j^{(1)}(e_2) + a_3 I^{(1)}(e_3, e_1) j^{(1)}(e_1) \\
&\quad + a_3 I^{(1)}(e_3, e_2) I^{(1)}(e_2, e_1) j^{(1)}(e_1) \\
&= j^{(1)}\Big((a_1 - a_2 x_1 - a_3 x_2 + a_3 x_1 x_3) e_1 + (a_2 - a_3 x_3) e_2 + a_3 e_3\Big).
\end{aligned}$$

$t_{-A} = \mathrm{id}$ holds if and only if $A = 0$, so if and only if

$$(a_1 - a_2 x_1 - a_3 x_2 + a_3 x_1 x_3) e_1 + (a_2 - a_3 x_3) e_2 + a_3 e_3 \in \operatorname{Rad} I^{(1)} = \mathbb{Z} f_3.$$

The ansatz that it is $a f_3 = a(-\widetilde{x}_3 e_1 + \widetilde{x}_2 e_2 - \widetilde{x}_1 e_3)$ with $a \in \mathbb{Z}$ gives

$$\begin{aligned}
-a\widetilde{x}_1 = a_3,\ a\widetilde{x}_2 = a_2 - a_3 x_3,\ -a\widetilde{x}_3 = a_1 - a_2 x_1 - a_3 x_2 + a_3 x_1 x_3, \\
\text{so}\quad (a_1, a_2, a_3) = a(-\widetilde{x}_3, \widetilde{x}_2 - \widetilde{x}_1 x_3, -\widetilde{x}_1).
\end{aligned}$$

(b) Write $\underline{a} = \widetilde{a}\,\gcd(x_1, x_2, x_3)(-x_3, x_2 - x_1x_3, -x_1)$ with $\widetilde{a} = \gcd(x_1, x_2, x_3)^{-2}$ $a \in \gcd(x_1, x_2, x_3)^{-2}\mathbb{Z}$. Then

$$
\begin{aligned}
&(e_1 + a_1 f_3, e_2 + a_2 f_3, e_3 + a_3 f_3) \\
&= \underline{e}\begin{pmatrix} 1+\widetilde{a}(-x_3)(-x_3) & \widetilde{a}(x_2 - x_1x_3)(-x_3) & \widetilde{a}(-x_1)(-x_3) \\ \widetilde{a}(-x_3)x_2 & 1+\widetilde{a}(x_2 - x_1x_3)x_2 & \widetilde{a}(-x_1)x_2 \\ \widetilde{a}(-x_3)(-x_1) & \widetilde{a}(x_2 - x_1x_3)(-x_1) & 1+\widetilde{a}(-x_1)(-x_1) \end{pmatrix}.
\end{aligned}
$$

The determinant of this matrix is $1 + \widetilde{a}r(\underline{x})$. The index of the lattice $\sum_{i=1}^3 \mathbb{Z}(e_i + a_i f_3)$ in $H_\mathbb{Z}$ is the absolute value of this determinant.

(c) Suppose $\Gamma_u^{(1)} \cong \mathbb{Z}^2$. Compare Lemma 6.17. The set

$$\Lambda := \{\lambda \in \mathrm{Hom}_0(H_\mathbb{Z}, \mathbb{Z}) \mid t_\lambda^+ \in \Gamma_u^{(1)}\}$$

is a sublattice of rank 2 in the lattice $\mathrm{Hom}_0(H_\mathbb{Z}, \mathbb{Z})$ of rank 2. For $i \in \{1, 2, 3\}$

$$\Gamma_u^{(1)}\{e_i\} = \{e_i + \lambda(e_i) f_3 \mid \lambda \in \Lambda\} \subset (e_i + \mathbb{Z}f_3) \cap \Delta^{(1)}.$$

The triple $(H_\mathbb{Z}, L, \underline{e})$ is irreducible because of $\Gamma_u^{(1)} \cong \mathbb{Z}^2$ and Theorem 7.7 (a). Therefore it is not reducible with a summand of type A_1, and thus $\{e_1, e_2, e_3\} \cap \mathrm{Rad}\, I^{(1)} = \emptyset$. Because of this and because Λ has finite index in $\mathrm{Hom}_0(H_\mathbb{Z}, \mathbb{Z})$, there is a number $b_i \in \mathbb{N}$ with

$$\mathbb{Z}b_i = \{b \in \mathbb{Z} \mid e_i + bf_3 \in \Gamma_u^{(1)}\{e_i\}\}.$$

For each $\underline{a} = (a_1, a_2, a_3)$ with $a_i \in \mathbb{Z}b_i$ $e_i + a_i f_3 \in \Delta^{(1)}$. Any number $a_0 \in \mathbb{N}$ (for example the smallest one) with

$$a_0\widetilde{x}_3 \in \mathbb{Z}b_1,\ a_0(\widetilde{x}_2 - \widetilde{x}_1x_3) \in \mathbb{Z}b_2,\ a_0\widetilde{x}_1 \in \mathbb{Z}b_3$$

works. □

Remarks 7.9 If $\underline{x} \in B_1 \cup B_2$ then the map $\Delta^{(1)} \to \overline{\Delta^{(1)}}$ is a bijection by Theorem 6.18 (b)+(g). Therefore then

$$\underline{v} = (e_1 + a(-\widetilde{x}_3) f_3, e_2 + a(\widetilde{x}_2 - \widetilde{x}_1x_3) f_3, e_3 + a(-\widetilde{x}_1) f_3) \in H_\mathbb{Z}^3$$

for $a \in \mathbb{Z} - \{0\}$ satisfies $(\pi_3 \circ \pi_3^{(1)})(\underline{v}) = M$, but $\underline{v} \notin (\Delta^{(1)})^3$. This fits to Theorem 7.7 (d).

Lemma 7.10 *The* $\mathrm{Br}_3 \ltimes \{\pm 1\}^3$ *orbits in* $r^{-1}(4) \subset \mathbb{Z}^3$ *are classified in Theorem 4.6 (e). They are separated by the isomorphism classes of the pairs* $(\overline{H_\mathbb{Z}}^{(1)}, \overline{M})$ *for corresponding unimodular bilinear lattices* $(H_\mathbb{Z}, L, \underline{e})$ *with triangular bases* $\underline{e}$ *with* $L(\underline{e}^t, \underline{e})^t = S(\underline{x})$ *and* $r(\underline{x}) = 4$. *More precisely,* $\overline{H_\mathbb{Z}}^{(1)}$ *has a* $\mathbb{Z}$*-basis* (c_1, c_2) *with* $\overline{M}(c_1, c_2) = (c_1, c_2)\begin{pmatrix} -1 & \gamma \\ 0 & -1 \end{pmatrix}$ *with a unique* $\gamma \in \mathbb{Z}_{\geq 0}$*, which is as follows:*

$S(\underline{x})$	$S(\mathcal{H}_{1,2})$	$S(\mathbb{P}^1 A_1)$	$S(\widehat{A}_2)$	$S(-l,2,-l)$ with $l\equiv 0(2)$	$S(-l,2,-l)$ with $l\equiv 1(2)$
γ	0	2	3	$\frac{l^2}{2}-2$	l^2-4

The numbers γ in this table are pairwise different.

Proof $S(\mathcal{H}_{1,2})$: See Theorem 5.14 (a) (i).

$S(\mathbb{P}^1 A_1)$: By Theorem 5.13 $(\overline{H_{\mathbb{Z}}}^{(1)}, \overline{M}) \cong (H_{\mathbb{Z},1}, M_1)$ which comes from $S(\mathbb{P}^1) = \begin{pmatrix} 1 & -2 \\ 0 & 1 \end{pmatrix}$ with $S(\mathbb{P}^1)^{-1}S(\mathbb{P}^1)^t = \begin{pmatrix} -3 & 2 \\ -2 & 1 \end{pmatrix}$. This monodromy matrix is conjugate to $\begin{pmatrix} -1 & 2 \\ 0 & -1 \end{pmatrix}$ with respect to $GL_2(\mathbb{Z})$.

$S(\widehat{A}_2)$: Compare Theorem 5.14 (b) (iii) and its proof:

$$\begin{aligned} f_3 &= e_1 - e_2 - e_3, \\ \overline{H_{\mathbb{Z}}}^{(1)} &= \mathbb{Z}\overline{e_1}^{(1)} + \mathbb{Z}\overline{e_2}^{(1)}, \\ M\underline{e} &= \underline{e}\begin{pmatrix} -2 & -1 & 2 \\ -2 & 0 & 1 \\ -1 & -1 & 1 \end{pmatrix}, \\ \overline{M}(\overline{e_1}^{(1)}, \overline{e_2}^{(1)}) &= (\overline{e_1}^{(1)}, \overline{e_2}^{(1)})\begin{pmatrix} -1 & 0 \\ -3 & -1 \end{pmatrix}. \end{aligned}$$

This monodromy matrix is conjugate to $\begin{pmatrix} -1 & 3 \\ 0 & -1 \end{pmatrix}$ with respect to $GL_2(\mathbb{Z})$.

$S(-l,2,-l)$ with $l\geq 4$, $l\equiv 0(2)$: See Theorem 5.14 (a) (ii). Here $(\overline{H_{\mathbb{Z}}}^{(1)}, \overline{M}) \cong (H_{\mathbb{Z},1}, M_1)$.

$S(-l,2,-l)$ with $l\geq 3$, $l\equiv 1(2)$: Compare Theorem 5.14 (b) (iv) and its proof. Define elements $a_1, a_2 \in H_{\mathbb{Z}}$,

$$\begin{aligned} a_1 &:= \frac{l+1}{2}e_1 + e_2 + \frac{l+1}{2}e_3 = \frac{1}{2}f_1 + \frac{1}{2}f_3, \\ a_2 &:= -e_1 = \frac{1}{2}\widetilde{f_2} - \frac{l^2}{4}f_1 - \frac{l}{4}f_3. \end{aligned}$$

The triple (a_1, a_2, f_3) is a $\mathbb{Z}$-basis of $H_{\mathbb{Z}}$. The equality in the proof of Theorem 5.14 (b) (iv),

$$M(f_1, \widetilde{f_2}) = (f_1, \widetilde{f_2})\begin{pmatrix} -1 & l^2-4 \\ 0 & -1 \end{pmatrix},$$

implies

$$M(\overline{a_1}^{(1)}, \overline{a_2}^{(1)}) = (\overline{a_2}^{(1)}, \overline{a_2}^{(1)})\begin{pmatrix} -1 & l^2-4 \\ 0 & -1 \end{pmatrix}.$$

□

Proof of Theorem 7.7 **(g)**: The case $(0, 0, 0)$ is treated first and separately. Compare part (e). Of the ten triples listed there, only the triple $\underline{v} = (e_1, e_2, e_3)$ satisfies $\sum_{i=1}^{3} \mathbb{Z}v_i = H_\mathbb{Z}$. This shows part (g) in the case $\underline{x} = (0, 0, 0)$.

Now consider $\underline{x} \in B_4$. We can suppose

$$\underline{x} \in \{(-1, 0, -1), (-1, -1, -1), (-2, 2, -2)\} \cup \{(-l, 2, -l) \mid l \geq 3\},$$

which are the cases $S(\underline{x}) \in \{S(A_3), S(\widehat{A}_2), S(\mathcal{H}_{1,2})\} \cup \{S(-l, 2, -l) \mid l \geq 3\}$. Consider $\underline{v} \in (\Delta^{(1)})^3$ with $(\pi_3 \circ \pi_3^{(1)})(\underline{v}) = M$ and $\underline{v}$ a $\mathbb{Z}$-basis of $H_\mathbb{Z}$. We want to show $\underline{v} \in \mathcal{B}^{dist}$. We have

$$I^{(1)}(\underline{v}^t, \underline{v}) = \begin{pmatrix} 0 & y_1 & y_2 \\ -y_1 & 0 & y_3 \\ -y_2 & -y_3 & 0 \end{pmatrix} = S(\underline{y}) - S(\underline{y})^t \quad \text{for some } \underline{y} \in \mathbb{Z}^3.$$

Define a new Seifert form $\widetilde{L} : H_\mathbb{Z} \times H_\mathbb{Z} \to \mathbb{Z}$ by $\widetilde{L}(\underline{v}^t, \underline{v})^t = S(\underline{y})$ (only at the end of the proof it will turn out that $\widetilde{L} = L$). Then

$$\widetilde{L}^t - \widetilde{L} = I^{(1)} = L^t - L.$$

$\underline{v}$ is a triangular basis with respect to $(H_\mathbb{Z}, \widetilde{L})$. By Theorem 2.7 for $(H_\mathbb{Z}, \widetilde{L})$ (alternatively, one can calculate the product of the matrices of $s_{v_1}^{(1)}$, $s_{v_2}^{(1)}$ and $s_{v_3}^{(1)}$ with respect to $\underline{v}$)

$$M\underline{v} = (\pi_3 \circ \pi_3^{(1)})(\underline{v}) = (s_{v_1}^{(1)} \circ s_{v_2}^{(1)} \circ s_{v_3}^{(1)})(\underline{v}) = \underline{v}S(\underline{y})^{-1}S(\underline{y})^t.$$

Then

$$3 - r(\underline{x}) = \operatorname{tr}(M) = \operatorname{tr}(S(\underline{y})^{-1}S(\underline{y})^t) = 3 - r(\underline{y}),$$

so $r(\underline{x}) = r(\underline{y})$. In the case of A_3, $r^{-1}(2)$ is a unique $\mathrm{Br}_3 \ltimes \{\pm 1\}^3$ orbit. In the cases of $\widehat{A}_2$, $\mathcal{H}_{1,2}$ and $\underline{x} \in \{(-l, 2, -l) \mid l \geq 3\}$, Lemma 7.10 and $M\underline{v} = \underline{v}S(\underline{y})^{-1}S(\underline{y})^t$ show that $\underline{y}$ is in the same $\mathrm{Br}_3 \ltimes \{\pm 1\}^3$ orbit as $\underline{x}$. Therefore in any case there is an element of $\mathrm{Br}_3 \ltimes \{\pm 1\}^3$ which maps $\underline{v}$ to a $\mathbb{Z}$-basis $\underline{w}$ of $H_\mathbb{Z}$ with

$$\widetilde{L}(\underline{w}^t, \underline{w})^t = S(\underline{x}).$$

Then

$$I^{(1)}(\underline{w}^t, \underline{w}) = S(\underline{x}) - S(\underline{x})^t = I^{(1)}(\underline{e}^t, \underline{e}).$$

Define an automorphism $g \in O^{(1)}$ by $g(\underline{e}) = \underline{w}$. Because of

$$\begin{aligned} M &= s^{(1)}_{v_1} s^{(1)}_{v_2} s^{(1)}_{v_2} = s^{(1)}_{w_1} s^{(1)}_{w_2} s^{(1)}_{w_3} \\ &= s^{(1)}_{g(e_1)} s^{(1)}_{g(e_2)} s^{(1)}_{g(e_3)} = g s^{(1)}_{e_1} s^{(1)}_{e_2} s^{(1)}_{e_3} g^{-1} = g M g^{-1} \end{aligned}$$

g is in $G^{(1)}_{\mathbb{Z}}$. But for the considered cases of $\underline{x}$

$$G^{(1)}_{\mathbb{Z}} \stackrel{\text{Theorem 5.14}}{=} G_{\mathbb{Z}} \stackrel{\text{Theorem 3.28}}{=} Z(\mathrm{Br}_3 \ltimes \{\pm 1\}^3).$$

Therefore there is an element of $\mathrm{Br}_3 \ltimes \{\pm 1\}^3$ which maps $\underline{e}$ to $\underline{w}$. Altogether $\underline{v} \in \mathrm{Br}_3 \ltimes \{\pm 1\}^3(\underline{e}) = \mathcal{B}^{dist}$. (Now also $L(\underline{v}^t, \underline{v})^t \in T_3(\mathbb{Z})$ and thus $L = \widetilde{L}$ are clear).□

7.4 The Stabilizers of Distinguished Bases in the Rank 3 Cases

Let $(H_{\mathbb{Z}}, L, \underline{e})$ be a unimodular bilinear lattice of rank 3 with a triangular basis $\underline{e}$ with $L(\underline{e}^t, \underline{e})^t = S(\underline{x}) \in T_3(\mathbb{Z})$ for some $x \in \mathbb{Z}^3$. We are interested in the stabilizer $(\mathrm{Br}_3)_{\underline{e}/\{\pm 1\}^3}$. The surjective map

$$\begin{aligned} \mathcal{B}^{dist} = (\mathrm{Br}_3 \ltimes \{\pm 1\}^3)(\underline{e}) &\to (\mathrm{Br}_3 \ltimes \{\pm 1\}^3)(\underline{x}) \\ \underline{\widetilde{e}} &\mapsto \underline{\widetilde{x}} \text{ with } L(\underline{\widetilde{e}}^t, \underline{\widetilde{e}})^t = S(\underline{\widetilde{x}}), \end{aligned}$$

is $\mathrm{Br}_3 \ltimes \{\pm 1\}^3$ equivariant. By Theorem 4.13 (a)

$$\mathbb{Z}^3 = \dot{\bigcup}_{\underline{x} \in \bigcup_{i=1}^{24} C_i} (\mathrm{Br}_3 \ltimes \{\pm 1\}^3)(\underline{x}).$$

Therefore we can and will restrict to $\underline{x} \in \bigcup_{i=1}^{24} C_i$.

The stabilizer $(\mathrm{Br}_3)_{\underline{e}/\{\pm 1\}^3}$ is by Lemma 3.25 (e) the kernel of the group antihomomorphism

$$\overline{Z} : (\mathrm{Br}_3)_{\underline{x}/\{\pm 1\}^3} \to G^{\mathcal{B}}_{\mathbb{Z}} / Z((\{\pm 1\}^3)_{\underline{x}}).$$

Here this simplifies to

$$\overline{Z} : (\mathrm{Br}_3)_{\underline{x}/\{\pm 1\}^3} \to G_{\mathbb{Z}} / Z((\{\pm 1\}^3)_{\underline{x}})$$

because $G^{\mathcal{B}}_{\mathbb{Z}} = G_{\mathbb{Z}}$ in the reducible cases (and also in most irreducible cases) and $Z((\{\pm 1\}^3)_{\underline{x}}) = \{\pm \mathrm{id}\}$, which is a normal subgroup of $G_{\mathbb{Z}}$, in the irreducible cases by Lemma 3.25 (f).

Theorem 4.16 gives the stabilizer $(\mathrm{Br}_3)_{\underline{x}/\{\pm 1\}^3}$ in all cases. The following Theorem 7.11 gives the stabilizer $(\mathrm{Br}_3)_{\underline{e}/\{\pm 1\}^3}$ in all cases.

Theorem 7.11 *Consider a local minimum* $\underline{x} \in C_i \subset \mathbb{Z}^3$ *for some* $i \in \{1, ..., 24\}$ *and the pseudo-graph* $\mathcal{G}_j$ *with* $\mathcal{G}_j = \mathcal{G}(\underline{x})$*. In the following table, the entry in the fourth column and in the line of* C_i *is the stabilizer* $(\mathrm{Br}_3)_{\underline{e}/\{\pm 1\}^3}$*. The first, second and third column are copied from the table in Theorem 4.13.*

	sets	$(\mathrm{Br}_3)_{\underline{x}/\{\pm 1\}^3}$	$(\mathrm{Br}_3)_{\underline{e}/\{\pm 1\}^3}$
$\mathcal{G}_1$	$C_1\ (A_1^3)$	Br_3	Br_3^{pure}
$\mathcal{G}_1$	$C_2\ (\mathcal{H}_{1,2})$	Br_3	$\langle (\sigma^{mon})^2 \rangle$
$\mathcal{G}_2$	$C_3\ (A_2A_1)$	$\langle \sigma_1, \sigma_2^2 \rangle$	$\langle \sigma_2^2, (\sigma^{mon})^{-1}\sigma_1^2, \sigma^{mon}\sigma_1 \rangle$
			$= \langle \sigma_2^2, \sigma_1\sigma_2^2\sigma_1^{-1}, \sigma_1^3 \rangle$
$\mathcal{G}_2$	$C_4\ (\mathbb{P}^1 A_1)$, C_5	$\langle \sigma_1, \sigma_2^2 \rangle$	$\langle \sigma_2^2, (\sigma^{mon})^{-1}\sigma_1^2 \rangle$
			$= \langle \sigma_2^2, \sigma_1\sigma_2^2\sigma_1^{-1} \rangle$
$\mathcal{G}_3$	$C_6\ (A_3)$	$\langle \sigma_1\sigma_2, \sigma_1^3 \rangle$	$\langle (\sigma_1\sigma_2)^4, \sigma_1^3 \rangle$
$\mathcal{G}_4$	$C_7\ (\widehat{A}_2)$	$\langle \sigma_2\sigma_1, \sigma_1^3 \rangle$	$\langle \sigma_1^3, \sigma_2^3, \sigma_2\sigma_1^3\sigma_2^{-1} \rangle$
$\mathcal{G}_5$	C_8, $C_9\ ((-l, 2, -l))$	$\langle \sigma^{mon}, \sigma_1^{-1}\sigma_2^{-1}\sigma_1 \rangle$	$\langle (\sigma^{mon})^2\sigma_1^{-1}\sigma_2^{l^2-4}\sigma_1 \rangle$
$\mathcal{G}_6$	$C_{10}\ (\mathbb{P}^2)$, C_{11}, C_{12}	$\langle \sigma_2\sigma_1 \rangle$	$\langle \mathrm{id} \rangle$
$\mathcal{G}_7$	C_{13} (e.g. $(4, 4, 8)$)	$\langle \sigma_2\sigma_1^2 \rangle$	$\langle \mathrm{id} \rangle$
$\mathcal{G}_8$	C_{14} (e.g. $(3, 4, 6)$)	$\langle \sigma^{mon} \rangle$	$\langle \mathrm{id} \rangle$
$\mathcal{G}_9$	C_{15}, C_{16}, C_{23}, C_{24}	$\langle \sigma^{mon} \rangle$	$\langle \mathrm{id} \rangle$
$\mathcal{G}_{10}$	C_{17} (e.g. $(-2, -2, 0)$)	$\langle \sigma^{mon}, \sigma_2 \rangle$	$\langle \sigma_2^2 \rangle$
$\mathcal{G}_{11}$	C_{18} (e.g. $(-3, -2, 0)$)	$\langle \sigma^{mon}, \sigma_2^2 \rangle$	$\langle \sigma_2^2 \rangle$
$\mathcal{G}_{12}$	C_{19} (e.g. $(-2, -1, 0)$)	$\langle \sigma^{mon}, \sigma_2^2, \sigma_2\sigma_1^3\sigma_2^{-1} \rangle$	$\langle \sigma_2^2, \sigma_2\sigma_1^3\sigma_2^{-1} \rangle$
$\mathcal{G}_{13}$	C_{20} (e.g. $(-2, -1, -1)$)	$\langle \sigma^{mon}, \sigma_2^3, \sigma_2\sigma_1^3\sigma_2^{-1} \rangle$	$\langle \sigma_2^3, \sigma_2\sigma_1^3\sigma_2^{-1} \rangle$
$\mathcal{G}_{14}$	C_{21}, C_{22}	$\langle \sigma^{mon}, \sigma_2^3 \rangle$	$\langle \sigma_2^3 \rangle$

Proof **The reducible case** $\mathcal{G}_1$ & $C_1\ (A_1^3)$: Here $\underline{x} = (0, 0, 0)$ and $(\mathrm{Br}_3)_{\underline{x}/\{\pm 1\}^3} = \mathrm{Br}_3$.

$$\mathcal{B}^{dist} = \{(\varepsilon_1 e_{\sigma(1)}, \varepsilon_2 e_{\sigma(2)}, \varepsilon_3 e_{\sigma(3)}) \mid \varepsilon_1, \varepsilon_2, \varepsilon_3 \in \{\pm 1\}, \sigma \in S_3\},$$

and Br_3 acts by permutation of the entries of triples on $\mathcal{B}^{dist}$. Therefore $(\mathrm{Br}_3)_{\underline{e}/\{\pm 1\}^3}$ is the kernel of the natural group homomorphism $\mathrm{Br}_3 \to S_3$, so it is the subgroup Br_3^{pure} of pure braids (see Remark 8.2 (vii) for this group).

The case $\mathcal{G}_1$ & $C_2\ (\mathcal{H}_{1,2})$: Here $\underline{x} = (2, 2, 2) \sim (-2, 2, -2)$ and $(\mathrm{Br}_3)_{\underline{x}/\{\pm 1\}^3} = \mathrm{Br}_3$. Recall the case $\mathcal{H}_{1,2}$ in the proof of Theorem 5.14, recall the $\mathbb{Z}$-basis $\widetilde{f}$ of $H_\mathbb{Z} = H_{\mathbb{Z},1} \oplus H_{\mathbb{Z},2}$, and recall

$$G_\mathbb{Z} = \{g \in \mathrm{Aut}(H_\mathbb{Z}, 1) \mid \det g = 1\} \times \mathrm{Aut}(H_{\mathbb{Z},2}) \cong SL_2(\mathbb{Z}) \times \{\pm 1\}.$$

We found in the proof of Theorem 5.14 (c)

$$Z(\delta_2\sigma_1) = (\widetilde{f} \mapsto \widetilde{f} \begin{pmatrix} 1 & -1 & 0 \\ 0 & 1 & 0 \\ 0 & 0 & 1 \end{pmatrix}), \quad Z(\delta_3\sigma_2) = (\widetilde{f} \mapsto \widetilde{f} \begin{pmatrix} 1 & 0 & 0 \\ 1 & 1 & 0 \\ 0 & 0 & 1 \end{pmatrix}).$$

The group antihomomorphism $\overline{Z} : (\mathrm{Br}_3)_{\underline{x}/\{\pm 1\}^3} = \mathrm{Br}_3 \to G_{\mathbb{Z}}/\{\pm\,\mathrm{id}\} \cong SL_2(\mathbb{Z})$ is surjective with $\overline{Z}(\sigma_1) \equiv A_1$ and $\overline{Z}(\sigma_2) \equiv A_2$. It almost coincides with the group homomorphism $\mathrm{Br}_3 \to SL_2(\mathbb{Z})$ in Remark 4.15 (i). It has the same kernel $\langle(\sigma^{mon})^2\rangle$.
The reducible cases $\mathcal{G}_2$ & C_3 (A_2A_1), C_4 $(\mathbb{P}^1 A_1)$, C_5: Here $\underline{x} = (x_1, 0, 0)$ with $x_1 \le -1$ and $(\mathrm{Br}_3)_{\underline{x}/\{\pm 1\}^3} = \langle \sigma_1, \sigma_2^2\rangle$. The quotient group $(\mathrm{Br}_3)_{\underline{x}/\{\pm 1\}^3}/(\mathrm{Br}_3)_{\underline{e}/\{\pm 1\}^3}$ is by Theorem 3.28 (c) and Lemma 3.25 (e) isomorphic to the quotient group $G_{\mathbb{Z}}/Z((\{\pm 1\}^3)_{\underline{x}})$. Here $Z((\{\pm 1\}^3)_{\underline{x}}) = \langle(-1,-1,-1),(-1,-1,1)\rangle$ with

$$Z((-1,-1,-1)) = -\,\mathrm{id},$$
$$Z((-1,-1,1)) = (\underline{e} \mapsto (-e_1, -e_2, e_3)) = Q.$$

Define

$$M^{root} := Z(\delta_2\sigma_1) = (\underline{e} \mapsto \underline{e}\begin{pmatrix} -x_1 & -1 & 0 \\ 1 & 0 & 0 \\ 0 & 0 & 1 \end{pmatrix}),$$

and recall from Theorem 5.13 and Theorem 5.5

$$G_{\mathbb{Z}} = \{\pm (M^{root})^l \mid l \in \mathbb{Z}\} \times \{\mathrm{id}, Q\}.$$

Therefore

$$(\mathrm{Br}_3)_{\underline{x}/\{\pm 1\}^3}/(\mathrm{Br}_3)_{\underline{e}/\{\pm 1\}^3} \cong G_{\mathbb{Z}}/Z((\{\pm 1\}^3)_{\underline{x}}) \cong \{(M^{root})^l \mid l \in \mathbb{Z}\}.$$

In the case C_3 (A_2A_1) $x_1 = -1$ and M^{root} has order three, so the quotient group $(\mathrm{Br}_3)_{\underline{x}/\{\pm 1\}^3}/(\mathrm{Br}_3)_{\underline{e}/\{\pm 1\}^3}$ is cyclic of order three with generator the class $[\sigma_1]$ of σ_1. In the cases C_4 and C_5 $x_1 \le -2$ and M^{root} has infinite order, so the quotient group $(\mathrm{Br}_3)_{\underline{x}/\{\pm 1\}^3}/(\mathrm{Br}_3)_{\underline{e}/\{\pm 1\}^3}$ is cyclic of infinite order with generator the class $[\sigma_1]$ of σ_1.

The cases C_4 $(\mathbb{P}^1 A_1)$, C_5: Theorem 7.1 (b) can be applied to the subbasis (e_2, e_3) with $x_3 = 0$. It shows $\sigma_2^2 \in (\mathrm{Br}_3)_{\underline{e}/\{\pm 1\}^3}$. Therefore $(\mathrm{Br}_3)_{\underline{e}/\{\pm 1\}^3}$ contains the normal closure of σ_2^2 in $(\mathrm{Br}_3)_{\underline{x}/\{\pm 1\}^3} = \langle\sigma_1, \sigma_2^2\rangle$. This normal subgroup is obviously

$$\langle \sigma_1^l \sigma_2^2 \sigma_1^{-l} \mid l \in \mathbb{Z}\rangle.$$

It can also be written with two generators, namely it is

$$\langle \sigma_2^2, \sigma_1\sigma_2^2\sigma_1^{-1}\rangle = \langle \sigma_2^2, (\sigma^{mon})^{-1}\sigma_1^2\rangle.$$

The equality of left and right side follows from

$$\sigma_2^2 \cdot \sigma_1\sigma_2^2\sigma_1^{-1} = \sigma_2^2\sigma_1\sigma_2^2\sigma_1 \cdot \sigma_1^{-2} = \sigma^{mon}\sigma_1^{-2}.$$

The equality of this group with $\langle \sigma_1^l \sigma_2^2 \sigma_1^{-l} \mid l \in \mathbb{Z}\rangle$ follows from the fact that σ^{mon} is in the center of Br_3. The quotient group $\langle\sigma_1, \sigma_2^2\rangle/\langle \sigma_1^l \sigma_2^2 \sigma_1^{-l} \mid l \in \mathbb{Z}\rangle$ is cyclic of infinite order with generator the class $[\sigma_1]$ of σ_1. Therefore

$$\begin{aligned}(\mathrm{Br}_3)_{\underline{e}/\{\pm 1\}^3} &= \langle \sigma_2^2, (\sigma^{mon})^{-1}\sigma_1^2\rangle = \langle \sigma_2^2, \sigma_1\sigma_2^2\sigma_1^{-1}\rangle \\ &= \langle \sigma_1^l\sigma_2^2\sigma_1^{-l}\,|\, l\in\mathbb{Z}\rangle \\ &= (\text{the normal closure of } \sigma_2^2 \text{ in } \langle\sigma_1,\sigma_2^2\rangle).\end{aligned}$$

The case C_3 (A_2A_1): Theorem 7.1 (b) can be applied to the subbasis (e_1, e_2) with $x_1 = -1$ and to the subbasis (e_2, e_3) with $x_3 = 0$. It shows σ_1^3 and $\sigma_2^2 \in (\mathrm{Br}_3)_{\underline{e}/\{\pm 1\}^3}$. Therefore $(\mathrm{Br}_3)_{\underline{e}/\{\pm 1\}^3}$ contains the normal closure of σ_1^3 and σ_2^2 in $(Br_3)_{\underline{x}/\{\pm 1\}^3} = \langle\sigma_1, \sigma_2^2\rangle$. The quotient group

$$\langle\sigma_1,\sigma_2^2\rangle/(\text{the normal closure of } \sigma_1^3 \text{ and } \sigma_2^2 \text{ in } \langle\sigma_1,\sigma_2^2\rangle)$$

is cyclic of order three with generator the class $[\sigma_1]$ of σ_1. Therefore

$$(\mathrm{Br}_3)_{\underline{e}/\{\pm 1\}^3} = (\text{the normal closure of } \sigma_1^3 \text{ and } \sigma_2^2 \text{ in } \langle\sigma_1,\sigma_2^2\rangle).$$

It coincides with the subgroup generated by σ_1^3 and by the normal closure $\langle\sigma_2^2, (\sigma^{mon})^{-1}\sigma_1^2\rangle$ of σ_2^2 in $\langle\sigma_1,\sigma_2^2\rangle$. Therefore

$$(\mathrm{Br}_3)_{\underline{e}/\{\pm 1\}^3} = \langle \sigma_2^2, (\sigma^{mon})^{-1}\sigma_1^2, \sigma_1^3\rangle = \langle \sigma_2^2, (\sigma^{mon})^{-1}\sigma_1^2, \sigma^{mon}\sigma_1\rangle.$$

The case $\mathcal{G}_3$ & C_6 (A_3): Here $\underline{x} = (-1, 0, -1)$ and $(\mathrm{Br}_3)_{\underline{x}/\{\pm 1\}^3} = \langle\sigma_1\sigma_2, \sigma_1^3\rangle$. By Theorem 5.14 (b) $G_{\mathbb{Z}} = \{\pm M^l\,|\, l\in\{0,1,2,3\}\}$, and M has order four. By Theorem 3.28 and Lemma 3.25 (f), the antihomomorphism

$$\overline{Z} : (\mathrm{Br}_3)_{\underline{x}/\{\pm 1\}^3}/(\mathrm{Br}_3)_{\underline{e}/\{\pm 1\}^3} \to G_{\mathbb{Z}}/\{\pm\,\mathrm{id}\}$$

is an antiisomorphism. Therefore the quotient group $(\mathrm{Br}_3)_{\underline{x}/\{\pm 1\}^3}/(\mathrm{Br}_3)_{\underline{e}/\{\pm 1\}^3}$ is cyclic of order four.

Theorem 7.1 (b) can be applied to the subbasis (e_1, e_2) with $x_1 = -1$. It shows $\sigma_1^3 \in (\mathrm{Br}_3)_{\underline{e}/\{\pm 1\}^3}$. Observe also

$$\begin{aligned}\delta_1\sigma_1\sigma_2(\underline{e}) &= \delta_1\sigma_1(e_1, e_3+e_2, e_2) = \delta_1(e_1+e_2+e_3, e_1, e_2) \\ &= (-e_1-e_2-e_3, e_1, e_2) = -M^{-1}(\underline{e}),\end{aligned}$$

$$\text{so } Z(\delta_1\sigma_1\sigma_2) = -M^{-1}.$$

M and $-M^{-1}$ have order four. Therefore $(\sigma_1\sigma_2)^4 \in (\mathrm{Br}_3)_{\underline{e}/\{\pm 1\}^3}$. Thus $(\mathrm{Br}_3)_{\underline{e}/\{\pm 1\}^3} \supset \langle(\sigma_1\sigma_2)^4, \sigma_1^3\rangle$.

We will show first that $\langle(\sigma_1\sigma_2)^4, \sigma_1^3\rangle$ is a normal subgroup of $\langle\sigma_1\sigma_2, \sigma_1^3\rangle$ and then that the quotient group is cyclic of order four. This will imply $(\mathrm{Br}_3)_{\underline{e}/\{\pm 1\}^3} = \langle(\sigma_1\sigma_2)^4, \sigma_1^3\rangle$.

Recall that $\sigma^{mon} = (\sigma_2\sigma_1)^3 = (\sigma_1\sigma_2)^3$ generates the center of Br_3. Therefore

$$(\sigma_1\sigma_2)^l\sigma_1^3(\sigma_1\sigma_2)^{-l} = (\sigma_1\sigma_2)^{4l}\sigma_1^3(\sigma_1\sigma_2)^{-4l} \in \langle(\sigma_1\sigma_2)^4, \sigma_1^3\rangle \text{ for any } l \in \mathbb{Z}.$$

Thus $\langle(\sigma_1\sigma_2)^4, \sigma_1^3\rangle$ is a normal subgroup of $\langle\sigma_1\sigma_2, \sigma_1^3\rangle$. This also shows that the quotient group $\langle\sigma_1\sigma_1, \sigma_1^3\rangle/\langle(\sigma_1\sigma_2)^4, \sigma_1^3\rangle$ is cyclic of order four. Therefore $\langle(\sigma_1\sigma_2)^4, \sigma_1^3\rangle = (\mathrm{Br}_3)_{\underline{e}/\{\pm 1\}^3}$.

The case $\mathcal{G}_4$ & C_7 ($\widehat{A}_2$): Here $\underline{x} = (-1, -1, -1)$ and $(\mathrm{Br}_3)_{\underline{x}/\{\pm 1\}^3} = \langle\sigma_2\sigma_1, \sigma_1^3\rangle$. By Theorem 5.14 (b) $G_{\mathbb{Z}} = \{\pm(M^{root})^l \mid l \in \mathbb{Z}\}$, and M^{root} has infinite order. By Theorem 3.28 and Lemma 3.25 (f), the antihomomorphism

$$\overline{Z} : (\mathrm{Br}_3)_{\underline{x}/\{\pm 1\}^3}/(\mathrm{Br}_3)_{\underline{e}/\{\pm 1\}^3} \to G_{\mathbb{Z}}/\{\pm \mathrm{id}\}$$

is an antiisomorphism. By Theorem 3.26 (c) and Theorem 5.14 (b) $Z(\delta_3\sigma_2\sigma_1) = M^{root}$. Therefore the quotient group $(\mathrm{Br}_3)_{\underline{x}/\{\pm 1\}^3}/(\mathrm{Br}_3)_{\underline{e}/\{\pm 1\}^3}$ is cyclic of infinite order with generator the class $[\sigma_2\sigma_1]$ of $\sigma_2\sigma_1$.

Theorem 7.1 (b) can be applied to the subbasis (e_1, e_2) with $x_1 = -1$. It shows $\sigma_1^3 \in (\mathrm{Br}_3)_{\underline{e}/\{\pm 1\}^3}$. Therefore $(\mathrm{Br}_3)_{\underline{e}/\{\pm 1\}^3}$ contains the normal closure of σ_1^3 in $\langle\sigma_2\sigma_1, \sigma_1^3\rangle$. We will first determine this normal closure and then show that it equals $(\mathrm{Br}_3)_{\underline{e}/\{\pm 1\}^3}$.

As $\sigma^{mon} = (\sigma_2\sigma_1)^3$ generates the center of Br_3,

$$(\sigma_2\sigma_1)^{\varepsilon+3l}\sigma_1^3(\sigma_2\sigma_1)^{-\varepsilon-3l} = (\sigma_2\sigma_1)^{\varepsilon}\sigma_1^3(\sigma_2\sigma_1)^{-\varepsilon} \quad \text{for } \varepsilon \in \{0; \pm 1\}, l \in \mathbb{Z}.$$

One sees

$$\begin{aligned}
&(\sigma_2\sigma_1)\sigma_1(\sigma_2\sigma_1)^{-1} = \sigma_2\sigma_1\sigma_2^{-1}, \quad \text{so} \\
&(\sigma_2\sigma_1)\sigma_1^3(\sigma_2\sigma_1)^{-1} = \sigma_2\sigma_1^3\sigma_2^{-1}, \\
&(\sigma_2\sigma_1)^{-1}\sigma_1(\sigma_2\sigma_1) = \sigma_1^{-1}(\sigma_2^{-1}\sigma_1\sigma_2)\sigma_1 \overset{(4.14)}{=} \sigma_1^{-1}(\sigma_1\sigma_2\sigma_1^{-1})\sigma_1 = \sigma_2, \quad \text{so} \\
&(\sigma_2\sigma_1)^{-1}\sigma_1^3(\sigma_2\sigma_1) = \sigma_2^3.
\end{aligned}$$

Therefore the normal closure of σ_1^3 in $\langle\sigma_2\sigma_1, \sigma_1^3\rangle$ is $\langle\sigma_1^3, \sigma_2^3, \sigma_2\sigma_1^3\sigma_2^{-1}\rangle$. The quotient group is an infinite cyclic group with generator the class $[\sigma_2\sigma_1]$ of $\sigma_2\sigma_1$. Therefore

$$\begin{aligned}
(\mathrm{Br}_3)_{\underline{e}/\{\pm 1\}^3} &= \langle\sigma_1^3, \sigma_2^3, \sigma_2\sigma_1^3\sigma_2^{-1}\rangle \\
&= (\text{the normal closure of } \sigma_1^3 \text{ in } \langle\sigma_2\sigma_1, \sigma_1^3\rangle).
\end{aligned}$$

The cases $\mathcal{G}_5$ & C_8: Here $\underline{x} = (-l, 2, -l)$ with $l \geq 3$ odd and $(\mathrm{Br}_3)_{\underline{x}/\{\pm 1\}^3} = \langle\sigma^{mon}, \sigma_1^{-1}\sigma_2^{-1}\sigma_1\rangle$.

By Theorem 5.14 (b) $G_{\mathbb{Z}} = \{\pm(M^{root})^l \mid l \in \mathbb{Z}\}$ with M^{root} as in Theorem 5.14 (b) with $(M^{root})^{l^2-4} = -M$. Because l is odd, the cyclic group $G_{\mathbb{Z}}/\{\pm \mathrm{id}\}$ with generator $[M^{root}]$ can also be written as

$$G_{\mathbb{Z}}/\{\pm \mathrm{id}\} = \langle[M], [(M^{root})^2]\rangle \quad \text{with } [M] = [M^{root}]^{l^2-4}.$$

By Theorem 3.28 and Lemma 3.25 (f), the antihomomorphism

$$\overline{Z} : (\mathrm{Br}_3)_{\underline{x}/\{\pm 1\}^3} \to G_{\mathbb{Z}}/\{\pm \mathrm{id}\}$$

is surjective with kernel $(\mathrm{Br}_3)_{\underline{e}/\{\pm 1\}^3}$. By the proof of Theorem 5.14 (b)(iv) $[M] = \overline{Z}(\sigma^{mon})$ and $[(M^{root})^2] = \overline{Z}(\sigma_1^{-1}\sigma_2^{-1}\sigma_1)$. The single relation between $[M]$ and $[(M^{root})^2]$ is $[\mathrm{id}] = [M]^2([(M^{root})^2])^{4-l^2}$. Therefore the kernel $(Br_3)_{\underline{e}/\{\pm 1\}^3}$ of the group antihomomorphism $\overline{Z}$ above is generated by $(\sigma^{mon})^2(\sigma_1^{-1}\sigma_2^{-1}\sigma_1)^{4-l^2} = (\sigma^{mon})^2\sigma_1^{-1}\sigma_2^{l^2-4}\sigma_1$.

The cases $\mathcal{G}_6$ & C_9: Here $\underline{x} = (-l, 2, -l)$ with $l \geq 4$ even and $(\mathrm{Br}_3)_{\underline{x}/\{\pm 1\}^3} = \langle \sigma^{mon}, \sigma_1^{-1}\sigma_2^{-1}\sigma_1\rangle$.

By Theorem 3.28 and Lemma 3.25 (f), the antihomomorphism

$$\overline{Z} : (\mathrm{Br}_3)_{\underline{x}/\{\pm 1\}^3} \to G_{\mathbb{Z}}/\{\pm \mathrm{id}\}$$

is surjective with kernel $(\mathrm{Br}_3)_{\underline{e}/\{\pm 1\}^3}$.

By Theorem 5.14 (a) and the proof of Theorem 5.14 (c)

$$\begin{aligned} G_{\mathbb{Z}} &= \langle -\mathrm{id}, \widetilde{M}, Q\rangle \\ \text{with } \widetilde{M} &= Z(\delta_3\sigma_1^{-1}\sigma_2^{-1}\sigma_1),\ Q = Z(\sigma^{mon})Z(\delta_3\sigma_1^{-1}\sigma_2^{-1}\sigma_1)^{2-l^2/2}, \end{aligned}$$

$\widetilde{M}$ and Q commute, $\widetilde{M}$ has infinite order, Q has order two. Therefore the kernel $(\mathrm{Br}_3)_{\underline{e}/\{\pm 1\}^3}$ of the group antihomomorphism $\overline{Z}$ above is generated by $(\sigma^{mon}(\sigma_1^{-1}\sigma_2^{-1}\sigma_1)^{2-l^2/2})^2 = (\sigma^{mon})^2\sigma_1^{-1}\sigma_2^{l^2-4}\sigma_1$.

The cases $\mathcal{G}_6$ & C_{10} ($\mathbb{P}^2$), C_{11}, C_{12}: Here $(\mathrm{Br}_3)_{\underline{x}/\{\pm 1\}^3} = \langle \sigma_2\sigma_1\rangle$. Here $Z(\sigma_2\sigma_1) = M^{root}$ is a third root of the monondromy. The monodromy M and M^{root} have infinite order. Therefore the kernel $(\mathrm{Br}_3)_{\underline{e}/\{\pm 1\}^3}$ of the group antihomomorphism

$$\overline{Z} : (\mathrm{Br}_3)_{\underline{x}/\{\pm 1\}^3} = \langle \sigma_2\sigma_1\rangle \to G_{\mathbb{Z}}/\{\pm \mathrm{id}\}$$

is $\langle \mathrm{id}\rangle$.

The cases $\mathcal{G}_7$ & C_{13} (e.g. $(4, 4, 8)$): Here $(\mathrm{Br}_3)_{\underline{x}/\{\pm 1\}^3} = \langle \sigma_2\sigma_1^2\rangle$. Here $Z(\sigma_2\sigma_1) = M^{root}$ is a root of the monondromy. The monodromy M and M^{root} have infinite order. Therefore the kernel $(\mathrm{Br}_3)_{\underline{e}/\{\pm 1\}^3}$ of the group antihomomorphism

$$\overline{Z} : (\mathrm{Br}_3)_{\underline{x}/\{\pm 1\}^3} = \langle \sigma_2\sigma_1^2\rangle \to G_{\mathbb{Z}}/\{\pm \mathrm{id}\}$$

is $\langle \mathrm{id}\rangle$.

The cases $\mathcal{G}_8$ & C_{14}, $\mathcal{G}_9$ & C_{15}, C_{16}, C_{23}, C_{24}: Here $(\mathrm{Br}_3)_{\underline{x}/\{\pm 1\}^3} = \langle \sigma^{mon}\rangle$. The monodromy $M = Z(\sigma^{mon})$ has infinite order. Therefore the kernel $(\mathrm{Br}_3)_{\underline{e}/\{\pm 1\}^3}$ of the group antihomomorphism

$$\overline{Z} : (\mathrm{Br}_3)_{\underline{x}/\{\pm 1\}^3} = \langle \sigma^{mon}\rangle \to G_{\mathbb{Z}}/\{\pm \mathrm{id}\}$$

is $\langle \mathrm{id} \rangle$.

The cases $\mathcal{G}_{10}$ & C_{17} (e.g. $(-2,-2,0)$): Here $(\mathrm{Br}_3)_{\underline{x}/\{\pm 1\}^3} = \langle \sigma^{mon}, \sigma_2 \rangle$. Recall from Theorem 5.16 (c) that

$$G_{\mathbb{Z}} = \{\mathrm{id}, Q\} \times \{\pm M^l \mid l \in \mathbb{Z}\} = \langle -\mathrm{id}, Q, M \rangle,$$

Q and M commute, Q has order two, M has infinite order, $-Q = Z(\sigma_2)$ (see the proof of Theorem 5.17 (e)), $M = Z(\sigma^{mon})$. Therefore the kernel $(\mathrm{Br}_3)_{\underline{e}/\{\pm 1\}^3}$ of the group antihomomorphism

$$\overline{Z} : (\mathrm{Br}_3)_{\underline{x}/\{\pm 1\}^3} = \langle \sigma^{mon}, \sigma_2 \rangle \to G_{\mathbb{Z}}/\{\pm \mathrm{id}\}$$

is $\langle \sigma_2^2 \rangle$.

The cases $\mathcal{G}_{11}$ & C_{18}, $\mathcal{G}_{12}$ & C_{19}, $\mathcal{G}_{13}$ & C_{20}, $\mathcal{G}_{14}$ & C_{21}, C_{22}: By Theorem 4.16 in all cases $(\mathrm{Br}_3)_{\underline{x}/\{\pm 1\}^3}$ is generated by σ^{mon} and some other generators. We claim that the other generators are all in $(\mathrm{Br}_3)_{\underline{e}/\{\pm 1\}^3}$. Application of Theorem 7.1 (b) to the subbasis (e_2, e_3) shows this for the following generators:

σ_2^2 in the cases $\mathcal{G}_{11}$&C_{18} and G_{12}&C_{19} because there $x_3 = 0$;

σ_2^3 in the cases $\mathcal{G}_{13}$&C_{20} and G_{14}&C_{21}, C_{22} because there $x_3 = -1$.

σ_2^{-1} maps $\underline{e}$ to the basis $(e_1, e_3, s_{e_3}^{(0)}(e_2))$. Therefore application of Theorem 7.1 (b) to the subbasis (e_1, e_3) with $x_2 = -1$ in the cases $\mathcal{G}_{12}$&C_{19} and $\mathcal{G}_{13}$&C_{20} shows $\sigma_2\sigma_1^3\sigma_2^{-1} \in (\mathrm{Br}_3)_{\underline{e}/\{\pm 1\}^3}$. The claim $\langle$other generators$\rangle \subset (\mathrm{Br}_3)_{\underline{e}/\{\pm 1\}^3}$ is proved.

The monodromy $M = Z(\sigma^{mon})$ has infinite order. Therefore the kernel $(\mathrm{Br}_3)_{\underline{e}/\{\pm 1\}^3}$ of the group antihomomorphism

$$\overline{Z} : (\mathrm{Br}_3)_{\underline{x}/\{\pm 1\}^3} = \langle \sigma^{mon}, \text{other generators} \rangle \to G_{\mathbb{Z}}/\{\pm \mathrm{id}\}$$

is $\langle$other generators$\rangle$. □

Remarks 7.12 (i) In the cases $\mathcal{G}_6$ & C_{10}, C_{11}, C_{12}, $\mathcal{G}_7$ & C_{13}, $\mathcal{G}_8$ & C_{14} and $\mathcal{G}_9$ & C_{15}, C_{16}, C_{23}, C_{24}, the even monodromy group $\Gamma^{(0)}$ is a free Coxeter group with three generators by Theorem 6.11 (b) and (g). Example 3.23 (iv), which builds on Theorem 3.2, shows $(\mathrm{Br}_3)_{\underline{e}/\{\pm 1\}^3} = \langle \mathrm{id} \rangle$.

Using this fact, one derives also $(\mathrm{Br}_3)_{\underline{x}/\{\pm 1\}^3}$ in the following way. In all cases $(H_{\mathbb{Z}}, L, \underline{e})$ is irreducible. The group antihomomorphism

$$\overline{Z} : (\mathrm{Br}_3)_{\underline{x}/\{\pm 1\}^3} \to G_{\mathbb{Z}}/\{\pm \mathrm{id}\}$$

is injective because the kernel is $(\mathrm{Br}_3)_{\underline{e}/\{\pm 1\}^3} = \langle \mathrm{id} \rangle$. By Theorem 3.28 $\overline{Z}$ is surjective in almost all cases. The proof of Theorem 3.28 provides in all cases preimages of generators of $\mathrm{Im}(\overline{Z}) \subset G_{\mathbb{Z}}/\{\pm \mathrm{id}\}$. These preimages generate $(\mathrm{Br}_3)_{\underline{x}/\{\pm 1\}^3}$.

So, the arguments here on one side and the Theorems 4.13, 4.16 and 7.11 on the other side offer two independent ways to derive the stabilizers $(\mathrm{Br}_3)_{\underline{x}/\{\pm 1\}^3}$ and $(\mathrm{Br}_3)_{\underline{e}/\{\pm 1\}^3}$ in the considered cases.

(ii) But the arguments in (i) cannot easily be adapted to the other cases. In the cases $\mathcal{G}_{10}$ & C_{17}, $\mathcal{G}_{11}$ & C_{18}, $\mathcal{G}_{12}$ & C_{19}, $\mathcal{G}_{13}$ & C_{20} and $\mathcal{G}_{14}$ & C_{21}, C_{22}, the even monodromy group $\Gamma^{(0)}$ is a non-free Coxeter group.
Theorem 3.2 (b) generalizes in Theorem 3.7 (b) to a statement on the size of $\mathcal{B}^{dist}$, but not to a statement on the stabilizer $(\text{Br}_3)_{\underline{e}/\{\pm 1\}^3}$.

The cases $\mathcal{G}_1$, $\mathcal{G}_2$, $\mathcal{G}_3$, $\mathcal{G}_4$ and $\mathcal{G}_5$ are with the exception of the reducible cases C_5 and the case C_{10} the cases with $r(\underline{x}) \in \{0, 1, 3, 4\}$. For them it looks possible, but difficult, to generalize the arguments in (i).

The conceptual derivation of the stabilizer groups for all cases with the Theorems 4.13, 4.16 and 7.11 is more elegant.

Remarks 7.13 (i) The table in Theorem 7.11 describes the stabilizer $(\text{Br}_3)_{\underline{e}/\{\pm 1\}^3}$ by generators, except for the case $\mathcal{G}_1$ & $C_1(A_1^3)$ where $(\text{Br}_3)_{\underline{e}/\{\pm 1\}^3} = \text{Br}_3^{pure}$. In fact

$$\begin{aligned} \text{Br}_3^{pure} &= \langle \sigma_1^2, \sigma_2^2, \sigma_2\sigma_1^2\sigma_2^{-1}, \sigma_2^{-1}\sigma_1^2\sigma_2 \rangle \\ &= (\text{the normal closure of } \sigma_1^2 \text{ in } \text{Br}_3), \end{aligned}$$

because $\sigma_2\sigma_1^2\sigma_2^{-1} = \sigma_1^{-1}\sigma_1^2\sigma_2$, $\sigma_2^{-1}\sigma_1^2\sigma_2 = \sigma_1\sigma_2^2\sigma_1^{-1}$ by (4.14).

(ii) In some cases the proof of Theorem 7.11 provides elements so that $(\text{Br}_3)_{\underline{e}/\{\pm 1\}^3}$ is the normal closure of these elements in $(\text{Br}_3)_{\underline{x}/\{\pm 1\}^3}$:

	elements	$(\text{Br}_3)_{\underline{x}/\{\pm 1\}^3}$
$\mathcal{G}_2$ & C_3 (A_2A_1)	σ_1^3, σ_2^2	$\langle \sigma_1, \sigma_2^2 \rangle$
$\mathcal{G}_2$ & C_4 $(\mathbb{P}^1 A_1)$, C_5	σ_2^2	$\langle \sigma_1, \sigma_2^2 \rangle$
$\mathcal{G}_4$ & C_7 $(\widehat{A}_2)$	σ_1^3	$\langle \sigma_2\sigma_1, \sigma_1^3 \rangle$

(iii) In the case $\mathcal{G}_3$ & C_6 (A_3), the stabilizer $(\text{Br}_3)_{\underline{e}/\{\pm 1\}^3} = \langle (\sigma_1\sigma_2)^4, \sigma_1^3 \rangle$ was determined already in [Yu90, Satz 7.3].

The pseudo-graph $\mathcal{G}(\underline{x})$ for $\underline{x} \in \bigcup_{i=1}^{24} C_i$ with vertex set $\mathcal{V} = \text{Br}_3(\underline{x}/\{\pm 1\}^3)$ in Definition 4.9 (f), Lemma 4.10 and the Examples 4.11 had been very useful. All except two edges came from the generators $\varphi_1, \varphi_2, \varphi_3$ of the free Coxeter group G^{phi}, and two edges came from $\gamma(v_0)$ and $\gamma^{-1}(v_0)$. An a priori more natural choice of edges comes from the elementary braids σ_1 and σ_2. It is less useful, but also interesting.

Definition 7.14 Let $\mathcal{V}$ be a non-empty finite or countably infinite set on which Br_3 acts. The triple $\mathcal{G}_\sigma(\mathcal{V}) := (\mathcal{V}, \mathcal{E}_1, \mathcal{E}_2)$ with $\mathcal{E}_1 := \{(v, \sigma_1(v)) \mid v \in \mathcal{V}\}$ and $\mathcal{E}_2 := \{(v, \sigma_2(v)) \mid v \in \mathcal{V}\}$ is called *σ-pseudo-graph of $\mathcal{V}$*. Here $\mathcal{E}_1$ and $\mathcal{E}_2$ are two families of directed edges. A *loop* in $\mathcal{E}_i$ is an edge $(v, \sigma_i(v)) = (v, v)$.

Remarks 7.15 (i) In a picture of a σ-pseudo-graph, edges in $\mathcal{E}_1$ and in $\mathcal{E}_2$ are denoted as follows.

$\longrightarrow$ an edge in $\mathcal{E}_1$,

$\longrightarrow$ an edge in $\mathcal{E}_2$.

(ii) Consider a σ-pseudo-graph $\mathcal{G}_\sigma(\mathcal{V})$. Because $\sigma_1 : \mathcal{V} \to \mathcal{V}$ and $\sigma_2 : \mathcal{V} \to \mathcal{V}$ are bijections, each vertex $v \in \mathcal{V}$ is starting point of one edge in $\mathcal{E}_1$ and one edge in $\mathcal{E}_2$ and end point of one edge in $\mathcal{E}_1$ and one edge in $\mathcal{E}_2$. The σ-pseudo-graph is connected if and only if $\mathcal{V}$ is a single Br_3 orbit.

(iii) Let $(H_\mathbb{Z}, L, \underline{e})$ be a unimodular bilinear lattice with a triangular basis $\underline{e}$ with $L(\underline{e}^t, \underline{e})^t = S(\underline{x})$ for some $\underline{x} \in \mathbb{Z}^3$. Two σ-pseudo-graphs are associated to it, $\mathcal{G}_\sigma(\mathcal{B}^{dist}/\{\pm 1\}^3)$ and $\mathcal{G}_\sigma(\mathrm{Br}_3(\underline{x}/\{\pm 1\}^3))$. The natural map

$$\begin{aligned} \mathcal{B}^{dist}/\{\pm 1\}^3 &\to \mathrm{Br}_3(\underline{x}/\{\pm 1\}^3) \\ \underline{\widetilde{e}}/\{\pm 1\}^3 &\mapsto \underline{\widetilde{x}}/\{\pm 1\}^3 \text{ with } L(\underline{\widetilde{e}}^t, \underline{\widetilde{e}})^t = S(\underline{\widetilde{x}}), \end{aligned}$$

is Br_3 equivariant and surjective. It induces a *covering*

$$\mathcal{G}_\sigma(\mathcal{B}^{dist}/\{\pm 1\}^3) \to \mathcal{G}_\sigma(\mathrm{Br}_3(\underline{x}/\{\pm 1\}^3))$$

of σ-pseudo-graphs. This is even a *normal covering* with group of deck transformations $G_\mathbb{Z}^\mathcal{B}/Z((\{\pm 1\}^3)_{\underline{x}})$ where $G_\mathbb{Z}^\mathcal{B} = Z((\mathrm{Br}_3 \ltimes \{\pm 1\}^3)_{\underline{x}}) \subset G_\mathbb{Z}$ is as in Lemma 3.25 (e). Now we explain what this means and why it holds.
The group $G_\mathbb{Z}^\mathcal{B}$ acts transitively on the fiber over $\underline{x}$ of the map $\mathcal{B}^{dist} \to (\mathrm{Br}_3 \ltimes \{\pm 1\}^3)(\underline{x})$. By Lemma 3.22 (a) the action of this group $G_\mathbb{Z}^\mathcal{B}$ and the action of the group $\mathrm{Br}_3 \ltimes \{\pm 1\}^3$ on $\mathcal{B}^{dist}$ commute, so that $G_\mathbb{Z}^\mathcal{B}$ acts transitively on each fiber of the map $\mathcal{B}^{dist} \to (\mathrm{Br}_3 \ltimes \{\pm 1\}^3)(\underline{x})$. Therefore the group $G_\mathbb{Z}^\mathcal{B}/Z((\{\pm 1\}^3)_{\underline{x}})$ acts simply transitively on each fiber of the covering $\mathcal{G}_\sigma(\mathcal{B}^{dist}/\{\pm 1\}^3) \to \mathcal{G}_\sigma(\mathrm{Br}_3(\underline{x}/\{\pm 1\}^3))$ and is a group of automorphisms of the σ-pseudo-graph $\mathcal{G}_\sigma(\mathcal{B}^{dist}/\{\pm 1\}^3)$. The quotient by this group is the σ-pseudo-graph $\mathcal{G}_\sigma(\mathrm{Br}_3(\underline{x}/\{\pm 1\}^3))$. These statements are the meaning of the *normal covering* $\mathcal{G}_\sigma(\mathcal{B}^{dist}/\{\pm 1\}^3) \to \mathcal{G}_\sigma(\mathrm{Br}_3(\underline{x}/\{\pm 1\}^3))$.

(iv) In part (iii) the σ-pseudo-graph $\mathcal{G}_\sigma(\mathcal{B}^{dist}/\{\pm 1\}^3)$ contains no loops, and for any $v \in \mathcal{B}^{dist}/\{\pm 1\}^3$ $\sigma_1(v) \neq \sigma_2(v)$. The σ-pseudo-graph $\mathcal{G}_\sigma(\mathrm{Br}_3(\underline{x}/\{\pm 1\}^3))$ contains loops in a few cases. It contains a vertex v with $\sigma_1(v) = \sigma_2(v)$ only in the cases $\underline{x} = (0, 0, 0)$ (A_1^3) and $\underline{x} = (2, 2, 2)$ $(\mathcal{H}_{1,2})$ where $\mathrm{Br}_3(\underline{x}/\{\pm 1\}^3)$ has only one vertex anyway.

Example 7.16 (i) By Theorem 4.13 (a) $\mathbb{Z}^3/\{\pm 1\}^3$ consists of the Br_3 orbits $\mathrm{Br}_3(\underline{x}/\{\pm 1\}^3)$ for $\underline{x} \in \bigcup_{i=1}^{24} C_i$. Precisely for $\underline{x} \in \bigcup_{i=1}^{7} C_i$ such an orbit is finite. This led to the four pseudo-graphs $\mathcal{G}_1, \mathcal{G}_2, \mathcal{G}_3, \mathcal{G}_4$ in the Examples 4.11. The four corresponding σ-pseudo-graphs $\mathcal{G}_\sigma(\mathrm{Br}_3(\underline{x}/\{\pm 1\}^3))$ are listed in Fig. 7.1. A vertex $\underline{\widetilde{x}}/\{\pm 1\}^3 \in \mathbb{Z}^3/\{\pm 1\}^3$ is denoted by a representative $\underline{\widetilde{x}} \in \mathbb{Z}^3_{>0} \cup \mathbb{Z}^3_{\leq 0}$. The vertices are positioned at the same places as in the pictures in the Examples 4.11 for $\mathcal{G}_1, \mathcal{G}_2, \mathcal{G}_3, \mathcal{G}_4$.

(ii) The case $\mathcal{H}_{1,2}$, $\underline{x} = (2, 2, 2)$: Here $\mathrm{Br}_3(\underline{x}/\{\pm 1\}^3) = \{\underline{x}/\{\pm 1\}^3\}$ has only one vertex, but the group

$$G_\mathbb{Z}^\mathcal{B}/Z((\{\pm 1\}^3)_{\underline{x}}) = G_\mathbb{Z}/\{\pm \mathrm{id}\} \cong SL_2(\mathbb{Z})$$

is big. There is a natural bijection $\mathcal{B}^{dist}/\{\pm 1\}^3 \to SL_2(\mathbb{Z})$, and the elementary braids σ_1 and σ_2 act by multiplication from the left with the matrices $A_1 = \begin{pmatrix} 1 & -1 \\ 0 & 1 \end{pmatrix}$ and $A_2 = \begin{pmatrix} 1 & 0 \\ 1 & 1 \end{pmatrix}$ on $SL_2(\mathbb{Z})$. This gives a clear description of the σ-pseudo-graph $\mathcal{G}_\sigma(\mathcal{B}^{dist}/\{\pm 1\}^3)$. We do not attempt a picture.

(iii) The reducible case A_1^3, $\underline{x} = (0, 0, 0)$: Also here $\mathrm{Br}_3(\underline{x}/\{\pm 1\}^3) = \{\underline{x}/\{\pm 1\}^3\}$ has only one vertex. The group

$$G_{\mathbb{Z}}^{\mathcal{B}}/Z((\{\pm 1\}^3)_{\underline{x}}) = G_{\mathbb{Z}}/\{\pm 1\}^3 \cong O_3(\mathbb{Z})/\{\pm 1\}^3 \cong S_3$$

has six elements. Therefore $\mathcal{B}^{dist}/\{\pm 1\}^3$ has six elements, and σ_1 and σ_2 act as involutions. The right hand side of the first line in Fig. 7.3 gives the σ-pseudo-graph $\mathcal{G}_\sigma(\mathcal{B}^{dist}/\{\pm 1\}^3)$. Part (iv) offers a different description which applies also to A_1^3 if one sees it as $A_1^2 A_1$.

(iv) The reducible cases, $\underline{x} \in \bigcup_{i \in \{3,4,5\}} C_i$ ($A_2 A_1$, $\mathbb{P}^1 A_1$, other reducible cases): Here $(H_{\mathbb{Z}}, L, \underline{e}) = (H_{\mathbb{Z},1}, L_1, (e_1, e_2)) \oplus (H_{\mathbb{Z},2}, L_2, e_3)$ with $H_{\mathbb{Z},1} = \mathbb{Z}e_1 \oplus \mathbb{Z}e_2$ and $H_{\mathbb{Z},2} = \mathbb{Z}e_3$.

The group of deck transformations of the normal covering $\mathcal{G}_\sigma(\mathcal{B}^{dist}/\{\pm 1\}^3) \to \mathcal{G}_\sigma(\mathrm{Br}_3(\underline{x}/\{\pm 1\}^3))$ is

$$G_{\mathbb{Z}}^{\mathcal{B}}/Z((\{\pm 1\}^3)_{\underline{x}}) = G_{\mathbb{Z}}/\{\pm \mathrm{id}, \pm Q\} \cong \{(M^{root})^l \mid l \in \mathbb{Z}\}.$$

Here M^{root} has order 3 in the case $A_2 A_1$ and infinite order in the other cases. Therefore the σ-pseudo-graph $\mathcal{G}_\sigma(\mathcal{B}^{dist}/\{\pm 1\}^3)$ can be obtained by a triple or infinite covering of the σ-pseudo-graph $\mathcal{G}_\sigma(\mathrm{Br}_3(\underline{x}/\{\pm 1\}^3)$.

More concretely, the type of the covering is determined by the Br_1 orbit of distinguished bases up to signs of $(H_{\mathbb{Z},1}, L_1, (e_1, e_2))$. One such distinguished basis modulo signs $(\widetilde{e}_1, \widetilde{e}_2)/\{\pm 1\}^2$ gives rise to one sheet in the covering $\mathcal{G}_\sigma(\mathcal{B}^{dist}/\{\pm 1\}^3) \to \mathcal{G}_\sigma(\mathrm{Br}_3(\underline{x}/\{\pm 1\}^3))$. Figure 7.2 shows the part of a σ-pseudo-graph which corresponds to one such sheet.

The six pictures in Fig. 7.3 show on the left hand side analogous σ_1-pseudo-graphs for the distinguished bases modulo signs of the rank 2 cases A_1^2, A_2 and $\mathbb{P}^1$, and on the right hand side the σ-pseudo-graphs $\mathcal{G}_\sigma(\mathcal{B}^{dist}/\{\pm 1\}^3)$ for A_1^3, $A_2 A_1$ and $\mathbb{P}^1 A_1$ (respectively only a part of the σ-pseudo-graph in the case of $\mathbb{P}^1 A_1$). The σ-pseudo-graph $\mathcal{G}_\sigma(\mathcal{B}^{dist}/\{\pm 1\}^3)$ for $\underline{x} = (x, 0, 0)$ with $x < -2$ looks the same as the one for $\mathbb{P}^1 A_1$, though of course the distinguished bases are different.

(v) The case A_3, $\underline{x} = (-1, 0, -1)$: The σ-pseudo-graph $\mathcal{G}_\sigma(\mathcal{B}^{dist}/\{\pm 1\}^3)$ was first given in [Yu90, page 40, Fig. 6]. We recall and explain it in our words. The group of deck transformations of the normal covering $\mathcal{G}_\sigma(\mathcal{B}^{dist}/\{\pm 1\}^3) \to \mathcal{G}_\sigma(\mathrm{Br}_3(\underline{x}/\{\pm 1\}^3))$ is

$$G_{\mathbb{Z}}^{\mathcal{B}}/Z((\{\pm 1\}^3)_{\underline{x}}) = G_{\mathbb{Z}}/\{\pm \mathrm{id}\} \cong \{M^l \mid l \in \{0, 1, 2, 3\}\}.$$

Here the monodromy M acts in the natural way,

$\underline{x} \in C_1 \cup C_2 = \{(0,0,0),(2,2,2)\},\ A_1^3, \mathcal{H}_{1,2}$

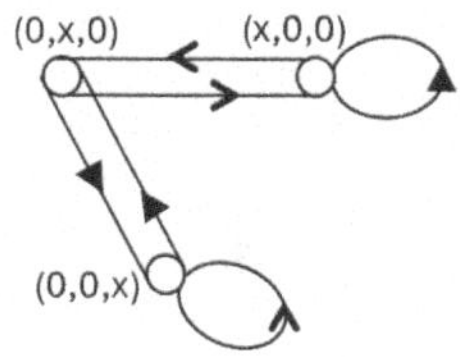

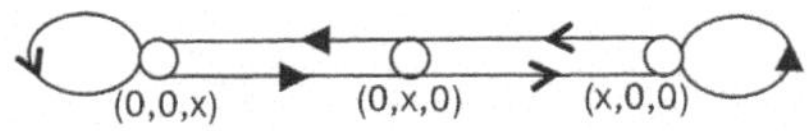

Two equivalent pictures for the cases
$\underline{x} = (x,0,0) \in C_3 \cup C_4 \cup C_5 = \{(\widetilde{x},0,0) \mid \widetilde{x} < 0\}$
$(A_2A_1, \mathbb{P}^1A_1,$ other reducible cases without $A_1^3)$

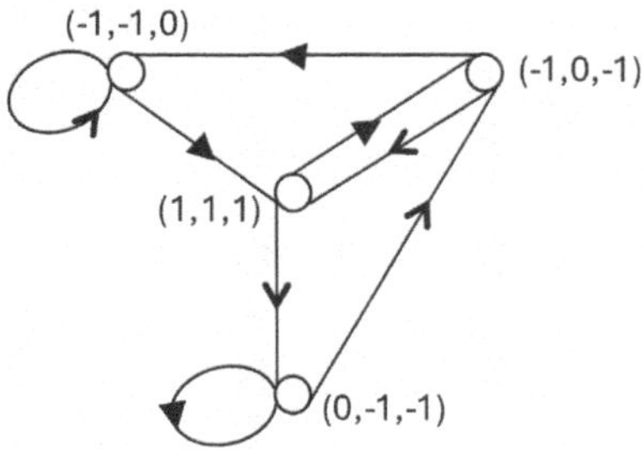

$\underline{x} = (-1,0,-1) \in C_6,\ A_3$

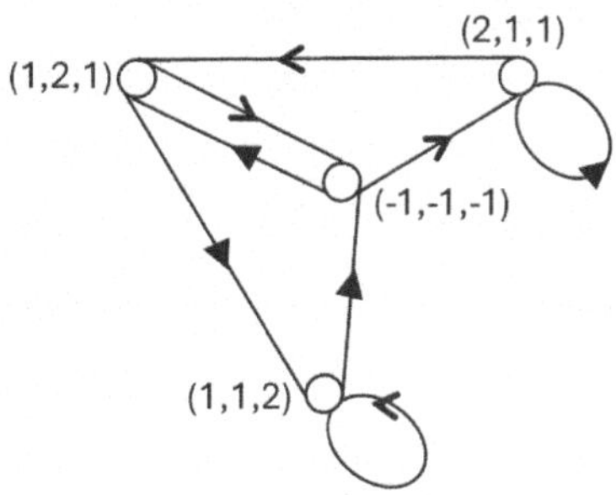

$\underline{x} = (-1,-1,-1) \in C_7,\ \widehat{A}_2$

Fig. 7.1 Examples 7.16 (i): the σ-pseudo-graphs for the finite Br_3 orbits in $\mathbb{Z}^3/\{\pm 1\}^3$

Fig. 7.2 In the reducible cases (without A_1^3) one sheet of the covering $\mathcal{G}_\sigma(\mathcal{B}^{dist}/\{\pm 1\}^3) \to \mathcal{G}_\sigma(\mathrm{Br}_3(\underline{x}/\{\pm 1\}^3))$

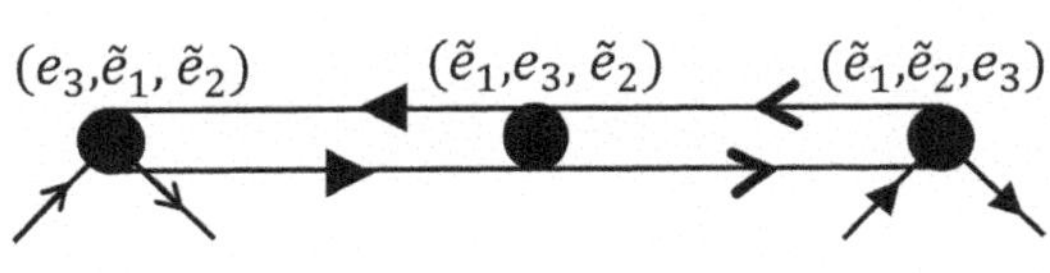

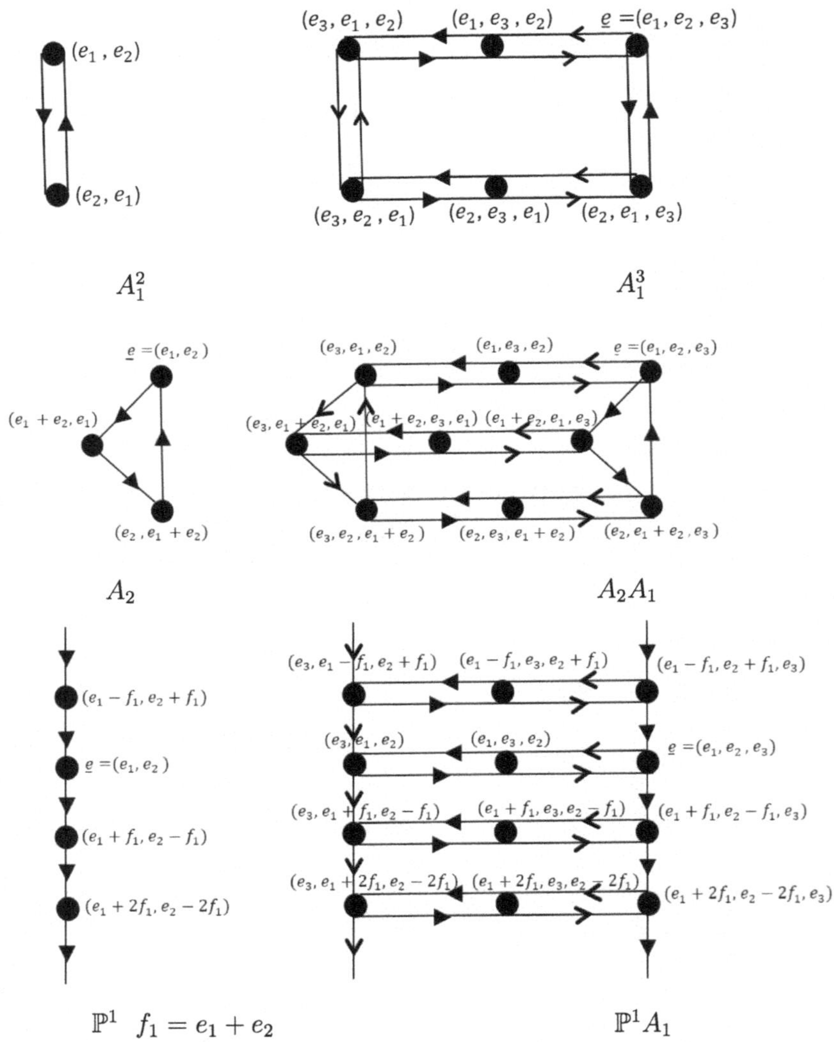

Fig. 7.3 Examples 7.16 (iii) and (iv): the σ-pseudo-graphs for distinguished bases modulo signs in the reducible cases

$$M((\widetilde{e}_1, \widetilde{e}_2, \widetilde{e}_3)/\{\pm 1\}^3) = (M(\widetilde{e}_1), M(\widetilde{e}_2), M(\widetilde{e}_3))/\{\pm 1\}^3),$$

on $\mathcal{B}^{dist}/\{\pm 1\}^3$ and has order four, $M^4 = \text{id}$. Here M and its powers are

$$M(\underline{e}) = \underline{e}\begin{pmatrix} 0 & 0 & 1 \\ -1 & 0 & 1 \\ 0 & -1 & 1 \end{pmatrix}, M^2(\underline{e}) = \underline{e}\begin{pmatrix} 0 & -1 & 1 \\ 0 & -1 & 0 \\ 1 & -1 & 0 \end{pmatrix},$$
$$M^3(\underline{e}) = \underline{e}\begin{pmatrix} 1 & -1 & 0 \\ 1 & 0 & -1 \\ 1 & 0 & 0 \end{pmatrix}.$$

Because of the shape of $\mathcal{G}_\sigma(\mathrm{Br}_3(\underline{x}/\{\pm 1\}^3))$,

$$b^{1,0} := \underline{e}/\{\pm 1\}^3, \quad b^{2,0} := \sigma_1 b^{1,0}, \quad b^{3,0} := \sigma_1^2 b^{1,0}, \quad b^{4,0} := \sigma_2^{-1} b^{1,0}$$

form one sheet of the fourfold cyclic covering $\mathcal{G}_\sigma(\mathcal{B}^{dist}/\{\pm 1\}^3) \to \mathcal{G}_\sigma(\mathrm{Br}_3(\underline{x}/\{\pm 1\}^3))$. Define

$$b^{i,l} := M^l(b^{i,0}) \quad \text{for} \quad l \in \{1, 2, 3\}.$$

Then

$$\mathcal{B}^{dist}/\{\pm 1\}^3 = \{b^{i,l} \mid i \in \{1, 2, 3, 4\}, l \in \{0, 1, 2, 3\}\}$$

has sixteen elements. We claim for $l \in \{0, 1, 2, 3\}$

$$\begin{aligned} \sigma_1 b^{1,l} &= b^{2,l}, & \sigma_2 b^{1,l} &= b^{3,l+3(\mathrm{mod}\,4)}, \\ \sigma_1 b^{2,l} &= b^{3,l}, & \sigma_2 b^{2,l} &= b^{2,l+2(\mathrm{mod}\,4)}, \\ \sigma_1 b^{3,l} &= b^{1,l}, & \sigma_2 b^{3,l} &= b^{4,l+1(\mathrm{mod}\,4)}, \\ \sigma_1 b^{4,l} &= b^{4,l+2(\mathrm{mod}\,4)}, & \sigma_2 b^{4,l} &= b^{1,l}. \end{aligned}$$

It is sufficient to prove the claim for $l = 0$. The equations $\sigma_1 b^{1,0} = b^{2,0}$, $\sigma_1 b^{2,0} = b^{3,0}$, $\sigma_2 b^{4,0} = b^{1,0}$ follow from the definitions of $b^{2,0}, b^{3,0}, b^{4,0}$. The inclusion $\sigma_1^3 \in (\mathrm{Br}_3)_{\underline{e}/\{\pm 1\}^3}$ gives $\sigma_1 b^{3,0} = b^{1,0}$. It remains to show

$$\sigma_1 b^{4,0} = b^{4,2}, \sigma_2 b^{1,0} = b^{3,3}, \ \sigma_2 b^{2,0} = b^{2,2}, \ \sigma_2 b^{3,0} = b^{4,1}.$$

One sees

i	$b^{i,0}$	$\widetilde{\underline{e}}$	$\widetilde{\underline{x}}$ with $L(\widetilde{\underline{e}}^t, \widetilde{\underline{e}}) = S(\widetilde{\underline{x}})$
1	$\underline{e}/\{\pm\}^3$	$\underline{e}$	$(-1, 0, -1)$
2	$\sigma_1(\underline{e})/\{\pm 1\}^3$	$\sigma_1(\underline{e}) = (e_1 + e_2, e_1, e_3)$	$(1, -1, 0)$
3	$\sigma_1^2(\underline{e})/\{\pm 1\}^3$	$\sigma_1^2(\underline{e}) = (-e_2, e_1 + e_2, e_3)$	$(-1, 1, -1)$
4	$\sigma_2^{-1}(\underline{e})/\{\pm 1\}^3$	$\sigma_2^{-1}(\underline{e}) = (e_1, e_3, e_2 + e_3)$	$(0, -1, 1)$

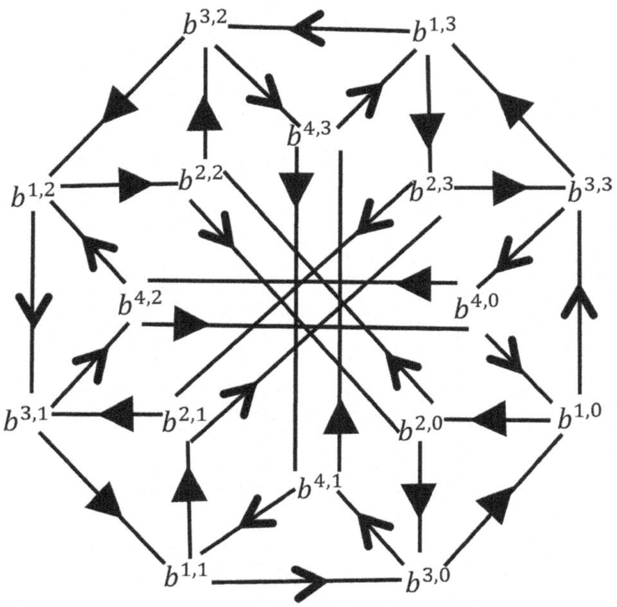

Fig. 7.4 Example 7.16 (v): the σ-pseudo-graph for distinguished bases modulo signs in the case A_3

$$\sigma_1 b^{4,0} = (e_3, e_1, e_2 + e_3)/\{\pm 1\}^3 = M^2 b^{4,0} = b^{4,2},$$
$$\sigma_2 b^{1,0} = (e_1, e_2 + e_3, e_2)/\{\pm 1\}^3 = M^3 b^{3,0} = b^{3,3},$$
$$\sigma_2 b^{2,0} = (e_1 + e_2, e_3, e_1)/\{\pm 1\}^3 = M^2 b^{2,0} = b^{2,2}.$$

$\sigma_2 b^{1,0} = b^{3,3}$, $\sigma_2 b^{4,0} = b^{1,0}$ and $\sigma_2^3 \in (\mathrm{Br}_3)_{\underline{e}/\{\pm 1\}^3}$ show $\sigma_2 b^{3,3} = b^{4,0}$. This implies $\sigma_2 b^{3,0} = b^{4,1}$. The claim is proved. The σ-pseudo-graph $\mathcal{G}_\sigma(\mathcal{B}^{dist}/\{\pm 1\}^3)$ is given in Fig. 7.4.

(vi) The case $\widehat{A}_2$, $\underline{x} = (-1, -1, -1)$: The group of deck transformations of the normal covering $\mathcal{G}_\sigma(\mathcal{B}^{dist}/\{\pm 1\}^3) \to \mathcal{G}_\sigma(\mathrm{Br}_3(\underline{x}/\{\pm 1\}^3))$ is

$$G_{\mathbb{Z}}^{\mathcal{B}}/Z((\{\pm 1\}^3)_{\underline{x}}) = G_{\mathbb{Z}}/\{\pm \mathrm{id}\} \cong \{(M^{root})^l \mid l \in \mathbb{Z}\}.$$

Here M^{root} acts in the natural way on $\mathcal{B}^{dist}/\{\pm 1\}^3$. M^{root} has infinite order and satisfies $(M^{root})^3 = -M$. Recall $f_1 = e_1 + e_2 + e_3$, $\mathbb{Z} f_1 = \mathrm{Rad}\, I^{(0)}$,

$$M^{root}(\underline{e}) = \underline{e} \begin{pmatrix} 1 & 1 & -1 \\ 1 & 0 & 0 \\ 0 & 1 & 0 \end{pmatrix}, \quad M^{root}(f_1) = f_1,$$
$$(M^{root})^2(\underline{e}) = \underline{e} + f_1(1, 0, -1).$$

Because of the shape of $\mathcal{G}_\sigma(\mathrm{Br}_3(\underline{x}/\{\pm 1\}^3))$,

$$b^{1,0} := \underline{e}/\{\pm 1\}^3, \quad b^{2,0} := \sigma_1 b^{1,0}, \quad b^{3,0} := \sigma_1^2 b^{1,0}, \quad b^{4,0} := \sigma_2 b^{1,0}$$

form one sheet of the infinite cyclic covering $\mathcal{G}_\sigma(\mathcal{B}^{dist}/\{\pm 1\}^3) \to \mathcal{G}_\sigma(\mathrm{Br}_3(\underline{x}/\{\pm 1\}^3))$. Define

$$b^{i,l} := (M^{root})^l(b^{i,0}) \quad \text{for} \quad i \in \{1.2,3,4\},\ l \in \mathbb{Z} - \{0\}.$$

Then

$$\mathcal{B}^{dist}/\{\pm 1\}^3 = \{b^{i,l} \mid i \in \{1,2,3,4\}, l \in \mathbb{Z}\}.$$

We claim for $l \in \mathbb{Z}$

$$\begin{aligned} \sigma_1 b^{1,l} &= b^{2,l}, & \sigma_2 b^{1,l} &= b^{4,l}, \\ \sigma_1 b^{2,l} &= b^{3,l}, & \sigma_2 b^{2,l} &= b^{1,l+1}, \\ \sigma_1 b^{3,l} &= b^{1,l}, & \sigma_2 b^{3,l} &= b^{3,l+2}, \\ \sigma_1 b^{4,l} &= b^{4,l+2}, & \sigma_2 b^{4,l} &= b^{2,l-1}. \end{aligned}$$

It is sufficient to prove the claim for $l = 0$. The equations $\sigma_1 b^{1,0} = b^{2,0}$, $\sigma_1 b^{2,0} = b^{3,0}$, $\sigma_2 b^{1,0} = b^{4,0}$ follow from the definitions of $b^{2,0}, b^{3,0}, b^{4,0}$. The inclusion $\sigma_1^3 \in (\mathrm{Br}_3)_{\underline{e}/\{\pm 1\}^3}$ gives $\sigma_1 b^{3,0} = b^{1,0}$. It remains to show

$$\sigma_1 b^{4,0} = b^{4,2}, \sigma_2 b^{2,0} = b^{1,1}, \ \sigma_2 b^{3,0} = b^{3,2}, \ \sigma_2 b^{4,0} = b^{2,-1}.$$

One sees

i	$b^{i,0}$	$\underline{\widetilde{e}}$	$\underline{\widetilde{x}}$ with $L(\underline{\widetilde{e}}^t, \underline{\widetilde{e}}) = S(\underline{\widetilde{x}})$
1	$\underline{e}/\{\pm\}^3$	$\underline{e}$	$(-1,-1,-1)$
2	$\sigma_1(\underline{e})/\{\pm 1\}^3$	$\sigma_1(\underline{e}) = (e_1+e_2, e_1, e_3)$	$(1,-2,-1)$
3	$\sigma_1^2(\underline{e})/\{\pm 1\}^3$	$\sigma_1^2(\underline{e}) = (-e_2, e_1+e_2, e_3)$	$(-1,1,-2)$
4	$\sigma_2(\underline{e})/\{\pm 1\}^3$	$\sigma_2(\underline{e}) = (e_1, e_2+e_3, e_2)$	$(-2,-1,1)$

$$\begin{aligned} \sigma_1 b^{4,0} &= (2e_1+e_2+e_3, e_1, e_2)/\{\pm 1\}^3 \\ &= (e_1+f_1, e_2+e_3-f_1, e_2)/\{\pm 1\}^3 = (M^{root})^2 b^{4,0} = b^{4,2}, \\ \sigma_2 b^{2,0} &= (e_1+e_2, e_1+e_3, e_1)/\{\pm 1\}^3 = M^{root} b^{1,0} = b^{1,1}, \\ \sigma_2 b^{3,0} &= (-e_2, 2e_1+2e_2+e_3, e_1+e_2)/\{\pm 1\}^3 \\ &= (-e_2, e_1+e_2+f_1, -e_3+f_1)/\{\pm 1\}^3 = M^2 b^{3,0} = b^{3,2}. \end{aligned}$$

$\sigma_2 b^{2,0} = b^{1,1}$ implies $\sigma_2 b^{2,-1} = b^{1,0}$. This, $\sigma_2 b^{1,0} = b^{4,0}$ and $\sigma_2^3 \in (\mathrm{Br}_3)_{\underline{e}/\{\pm 1\}^3}$ show $\sigma_2 b^{4,0} = b^{2,-1}$. The claim is proved. A part of the σ-pseudo-graph $\mathcal{G}_\sigma(\mathcal{B}^{dist}/\{\pm 1\}^3)$ is given in Fig. 7.5.

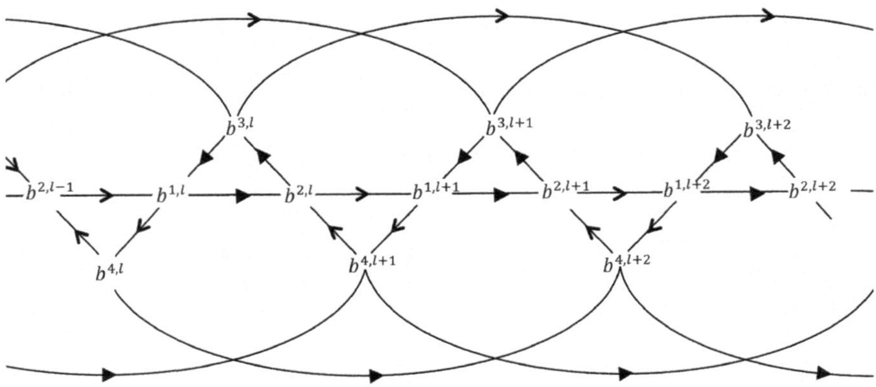

Fig. 7.5 Example 7.16 (vi): the part of the σ-pseudo-graph for distinguished bases modulo signs in the case $\widehat{A}_2$

Chapter 8
Manifolds Induced by the Orbit of Distinguished Bases and the Orbit of Distinguished Matrices

A single matrix $S \in T_n(\mathbb{Z})$ induces a unimodular bilinear lattice $(H_{\mathbb{Z}}, L, \underline{e})$ with triangular basis $\underline{e}$. This triple is unique up to isomorphism. The Chaps. 2–7 studied this structure and the associated structures $\Gamma^{(k)}$, $\Delta^{(k)}$ and $\mathcal{B}^{dist}$ in general, and also systematically in the cases of rank 2 and 3. These chapters are all of an algebraic/combinatorial type.

This chapter comes to geometry. It will present two complex n-dimensional manifolds $C_n^{\underline{e}/\{\pm 1\}^n}$ and $C_n^{S/\{\pm 1\}^n}$ which are induced by S respectively by $(H_{\mathbb{Z}}, L, \underline{e})$. If $n \geq 2$ they are certain coverings of the configuration space of the braid group Br_n. They can also be seen as the results of gluing many *Stokes regions*, one for each element in $\mathcal{B}^{dist}/\{\pm 1\}^n$ respectively in $\mathcal{S}^{dist}/\{\pm 1\}^n$. Their definition is not difficult, but worth to be discussed in detail. This is done in Sect. 8.1.

Many of them are important in algebraic geometry and singularity theory. See Chap. 10 for statements on the cases in singularity theory. They all are carriers of much richer structures. One version of these richer structures is presented in Sect. 8.5, certain *$\mathbb{Z}$-lattice bundles* on them. Another rather elementary differential geometric structure on them is given in Sect. 8.2.

Section 8.2 recalls the notion of an F-manifold and states that $C_n^{\underline{e}/\{\pm 1\}^n}$ and $C_n^{S/\{\pm 1\}^n}$ are semisimple F-manifolds with Euler field and empty Maxwell stratum. As such, often they have partical compactifications to which the F-manifold structure extends.

Section 8.3 considers first the case when $(H_{\mathbb{Z}}, L, \underline{e})$ is reducible. It states that then $C_n^{\underline{e}/\{\pm 1\}^n}$ embeds into the product of the corresponding manifolds for the summands of $(H_{\mathbb{Z}}, L, \underline{e})$. Afterwards it discusses the rank 2 cases.

Section 8.4 recalls a notion which was developed within singularity theory, the *distinguished systems of n paths*. There are a priori two actions of the braid group Br_n on sets of (homotopy classes) of distinguished systems of paths. They are discussed and compared in Sect. 8.4.

They are needed in Sect. 8.5. It constructs from $(H_{\mathbb{Z}}, L, \underline{e})$ natural families of $\mathbb{Z}$-lattices structures over the manifolds C_n^{univ} and $C_n^{\underline{e}/\{\pm 1\}^n}$. Before, it constructs a single

C. Hertling and K. Larabi, *Triangular Bases of Unimodular Bilinear Lattices and Induced Structures*, Lecture Notes in Mathematics 2398,
https://doi.org/10.1007/978-3-032-25997-4_8

$\mathbb{Z}$-lattice structure on $\mathbb{C}-\{u_1,...,u_n\}$ from $(H_\mathbb{Z}, L, \underline{e})$ and the additional choices of a pair $(\underline{u}, r) \in C_n^{pure} \times \mathbb{R}$ with r big enough and of a distinguished system of paths with starting point r and endpoints in $\{u_1, ..., u_n\}$.

The $\mathbb{Z}$-lattice bundle over $C_n^{\underline{e}/\{\pm 1\}^n}$ leads in general (with an additional choice) to a *Dubrovin-Frobenius manifold structure* on a Zariski open subset of $C_n^{\underline{e}/\{\pm 1\}^n}$. Remark 8.9 (ii) offers one definition of a Dubrovin-Frobenius manifold and references. We do not go more into this big story.

8.1 Natural Coverings of the Configuration Space of the Braid Group Br_n

First we fix some notations and recall some facts from the theory of covering spaces and some facts around the configuration space of the braid group Br_n. After this the Remarks 8.4 start the discussion of the manifolds $C_n^{\underline{e}/\{\pm 1\}^n}$ and $C_n^{S/\{\pm 1\}^n}$.

Definition 8.1 Let $n \in \mathbb{Z}_{\geq 2}$. Define

$$\begin{aligned}
D_n^{pure} &:= \bigcup_{1\leq i<j\leq n} \{\underline{u} \in \mathbb{C}^n \mid u_i = u_j\} \subset \mathbb{C}^n, \\
&\quad\ \text{the union of the partial diagonals in } \mathbb{C}^n, \\
C_n^{pure} &:= \mathbb{C}^n - D_n^{pure}, \\
C_n^{conf} &:= C_n^{pure}/S_n \subset \mathbb{C}^n/S_n \\
D_n^{conf} &:= D_n^{pure}/S_n \subset \mathbb{C}^n/S_n, \quad \text{so } \mathbb{C}^n/S_n = C_n^{conf} \dot\cup D_n^{conf}, \\
C_n^{univ} &:= (\text{the universal covering of } C_n^{pure} \text{ and } C_n^{conf}), \\
b_n^{pure} &:= (i, 2i, ..., ni) \in C_n^{pure}, \\
b_n^{conf} &:= [b_n^{pure}] = b_n^{pure}/S_n, \\
b_n^{univ} &:= (\text{the preimage of } b_n^{pure} \text{ in } C_n^{univ} \text{ which corresponds} \\
&\quad\ \text{to the trivial path in } C_n^{pure}).
\end{aligned}$$

C_n^{conf} is called *configuration space* of the braid group Br_n. The three covering maps are denoted as follows,

$$\begin{aligned}
\mathrm{pr}_n^{p,c} &: C_n^{pure} \to C_n^{conf}, \\
\mathrm{pr}_n^{u,p} &: C_n^{univ} \to C_n^{pure}, \\
\mathrm{pr}_n^{u,c} = \mathrm{pr}_n^{p,c} \circ \mathrm{pr}_n^{u,p} &: C_n^{univ} \to C_n^{conf}.
\end{aligned}$$

Remarks 8.2 (i) The permutation group S_n acts from the left on $\mathbb{C}^n$ by the following action for $\sigma \in S_n$,

$$\sigma(\underline{u}) = \sigma((u_1, u_2, ..., u_n)) := (u_{\sigma^{-1}(1)}, u_{\sigma^{-1}(2)}, ..., u_{\sigma^{-1}(n)}).$$

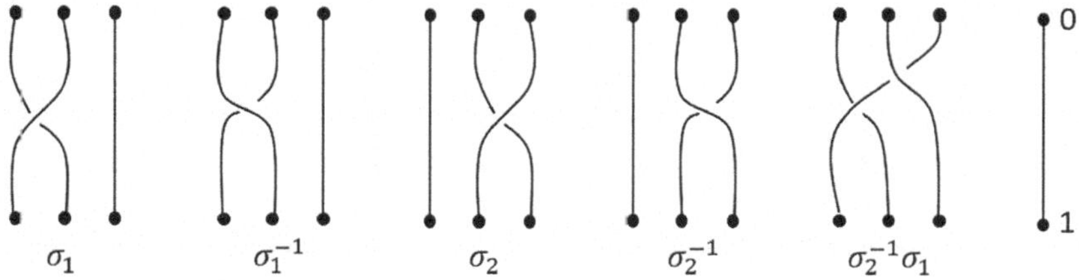

Fig. 8.1 Five braids.

So the entry of $\underline{u}$ at the $\sigma^{-1}(j)$-th place is put to the j-th place. If $\tau \in S_n$, then $\sigma(\tau(\underline{u}))$ has at the j-th place the entry $u_{\tau^{-1}(\sigma^{-1}(j))}$ of $\tau(\underline{u})$ at the $\sigma^{-1}(j)$-th place, so the entry $u_{(\sigma\circ\tau)^{-1}(j)}$. Therefore this is an action from the left of S_n on $\mathbb{C}^n$. Although this is an action from the left, we write in the quotient $\mathbb{C}^n/S_n$ the group on the right.

(ii) The quotient $\mathbb{C}^n/S_n$ is again isomorphic to $\mathbb{C}^n$. It can be identified with the set

$$\mathbb{C}[x]_n := \{f(x) = x^n + \sum_{j=1}^{n} f_j x^{n-j} \in \mathbb{C}[x] \mid f = (f_1, ..., f_n) \in \mathbb{C}^n\}$$

of complex unitary polynomials of degree n. The identification is given by the map $[\underline{u}] = [(u_1, ..., u_n)] \mapsto \prod_{i=1}^{n}(x - u_i)$.

The covering $C_n^{pure} \to C_n^{conf}$ extends to a branched covering $\mathbb{C}^n \to \mathbb{C}^n/S_n$. It maps each partial diagonal $\{\underline{u} \in \mathbb{C}^n \mid u_i = u_j\}$ for $1 \le i < j \le n$ to the same irreducible algebraic hypersurface $D_n^{conf} = D_n^{pure}/S_n \subset \mathbb{C}^n/S_n$. This hypersurface D_n^{conf} and its image

$$\mathbb{C}[x]_n^{mult} := \{f(x) \in \mathbb{C}[x]_n \mid f(x)\text{has multiple roots}\}$$

in $\mathbb{C}[x]_n$ are called *discriminants*. The configuration space C_n^{conf} is identified with the complement

$$\mathbb{C}[x]_n^{reg} := \{f(x) \in \mathbb{C}[x]_n \mid f(x) \text{ has no multiple roots}\}.$$

(iii) The braid group Br_n is isomorphic to the fundamental group $\pi_1(C_n^{conf}, b_n^{conf})$ (e.g. [Bi75, Theorem 1.8], [KT08, 1.4.3]). A closed loop $\gamma : [0, 1] \to C_n^{conf}$ with $\gamma(0) = \gamma(1) = b_n^{conf}$ can be visualized by a braid with n strings. Figure 8.1 shows five braids with three strings which represent the braids $\sigma_1, \sigma_1^{-1}, \sigma_2, \sigma_2^{-1}, \sigma_2^{-1}\sigma_1$. The interval $[0, 1]$ is going top down. The starting points and the end points of the strings are $i, 2i, ..., ni \in \mathbb{C}$ where $\mathbb{C}$ is the complex line in which the entries of the loop live. Recall that in a fundamental group a product of elements is gone through from left to right. Therefore in the braid $\sigma_2^{-1}\sigma_1$ in Fig. 8.1 σ_2^{-1} comes first and σ_1 comes second.

The relations

$$\begin{aligned} \sigma_j\sigma_{j+1}\sigma_j &= \sigma_{j+1}\sigma_j\sigma_{j+1} \quad \text{for } j\in\{,...,n-1\} \\ \text{and} \quad \sigma_j\sigma_k &= \sigma_k\sigma_j \quad \text{for } |j-k|>1 \end{aligned}$$

are now obvious. It is true, but not obvious that they are generating relations in Br_n for the generators $\sigma_1, ..., \sigma_{n-1}$.

(iv) C_n^{conf} is a connected complex manifold and especially locally simply connected. Therefore the main results of the theory of covering spaces apply. The *Galois correspondence* for connected coverings tells the following [Ha01, Theorem 1.38].

Each connected covering $p:(Y,y_0)\to(C_n^{conf},b_n^{conf})$ with base point $y_0\in Y$ with $p(y_0)=b_n^{conf}$ gives rise to an injective group homomorphism $p_*:\pi_1(Y,y_0)\to\pi_1(C_n^{conf},b_n^{conf})$. Vice versa, for each subgroup $U\subset\mathrm{Br}_n$, there is such a connected covering $p:(Y,y_0)\to(C_n^{conf},b_n^{conf})$ with $p_*(\pi_1(Y,y_0))=U$, and it is unique up to isomorphism of connected coverings with base points. This gives a 1–1 correspondence between connected coverings of C_n^{conf} with base points up to isomorphism and subgroups of Br_n. If one forgets the base points, one obtains a 1–1 correspondence between connected coverings and conjugacy classes of subgroups of Br_n.

(v) The group of deck transformations of a connected covering $p:(Y,y_0)\to(C_n^{conf},b_n^{conf})$ is the group of automorphisms of Y which map each fiber of p to itself. A deck transformation $f:Y\to Y$ is determined by the one value $f(y_0)\in p^{-1}(b_n^{conf})$. The group of deck transformations is isomorphic to the quotient

$$N(p_*(\pi_1(Y,y_0)))/p_*(\pi_1(Y,y_0)),$$

where $N(p_*(\pi_1(Y,y_0)))\subset\mathrm{Br}_n$ is the largest subgroup of Br_n in which $p_*(\pi_1(Y,y_0))$ is a normal subgroup.

The covering is normal if $N(p_*(\pi_1(Y,y_0)))=\mathrm{Br}_n$. This holds if and only if the group of deck transformations acts transitively on the fiber $p^{-1}(b_n^{conf})$ (and then transitively on each fiber of p).

For a given subgroup $U\subset\mathrm{Br}_n$, the covering $p:(Y,y_0)\to(C_n^{conf},b_n^{conf})$ can be constructed as the quotient C_n^{univ}/U where U acts from the left as subgroup of the group Br_n, which acts as group of deck transformations of C_n^{univ}.

(vi) The universal covering $\mathrm{pr}_n^{u,c}:C_n^{univ}\to C_n^{conf}$ has trivial fundamental group $\{\mathrm{id}\}\subset\mathrm{Br}_n$ which is normal in Br_n, so it is a normal covering with Br_n as group of deck transformations. One can identify C_n^{univ} as a set with the following set,

$$\bigcup_{b\in C_n^{conf}}\{\text{homotopy classes of paths in } C_n^{conf} \text{ from } b_n^{conf} \text{ to } b\}.$$

Br_n acts on this set from the left as follows: a braid α maps a homotopy class β to the homotopy class $\alpha\beta$, where one first goes along a loop which represents α and then along a path which represents β. Remarkably, this gives an action *from the left* of Br_n on C_n^{univ}, although in Br_n in a product of two loops one walks first along the left loop, whereas in a product of two automorphisms of C_n^{univ} the right automorphism is carried out first.

(vii) The covering $\mathrm{pr}_n^{p,c} : C_n^{pure} \to C_n^{conf}$ is normal of degree $n!$ with group of deck transformations isomorphic to S_n. The fundamental group $\pi_1(C_n^{pure}, b_n^{pure})$ embeds into Br_n as group Br_n^{pure} of *pure braids*. A pure braid is represented by a closed loop in C_n^{pure} which starts and ends at b_n^{pure}. So in a pure braid with n strings, each string ends at the point at which it starts. There is an exact sequence

$$\{1\} \to \mathrm{Br}_n^{pure} \to \mathrm{Br}_n \to S_n \to \{1\}$$

of group homomorphisms. Br_n^{pure} is the kernel of the map $\mathrm{Br}_n \to S_n$, so normal in Br_n, so $N(\mathrm{Br}_n^{pure}) = \mathrm{Br}_n$. Here given a braid $\alpha \in \mathrm{Br}_n$, that permutation $\sigma \in S_n$ is associated to it, such that for a closed loop in C_n^{conf} which represents α, the lift to C_n^{pure} which starts at $b_n^{pure} = (i, 2i, ..., ni)$ ends at $(\sigma^{-1}(1)i, \sigma^{-1}(2)i, ..., \sigma^{-1}(n)i)$. This fits to the action of S_n on $\mathbb{C}^n$ in part (i) of these remarks.

(viii) Arnold [Ar68] showed that the universal covering C_n^{univ} is homeomorphic to an open ball in $\mathbb{R}^{2n}$. Kaliman [Ka75, Ka93] showed that C_n^{univ} is biholomorphic to $\mathbb{C}^2 \times M_{n-2}^{univ}$ where M_{n-2}^{univ} is a bounded domain in $\mathbb{C}^{n-2}$ and a Teichmüller space.

We will need explicitly only the cases $C_2^{univ} \cong \mathbb{C}^2$ and $C_3^{univ} \cong \mathbb{C}^2 \times \mathbb{H}$. They are treated together with the actions of Br_2 respectively Br_3 as groups of deck transformations in Theorem 8.12 ($n = 2$) and in Theorem 9.1 and Remark 9.2 ($n = 3$).

(ix) As C_n^{conf} is a connected complex manifold, each connected covering is a connected complex manifold. As C_n^{conf} is an affine algebraic manifold, each finite connected covering is an affine algebraic manifold. This follows from Grothendieck's version of the Riemann existence theorem, relating especially finite etale coverings of an affine algebraic manifold to finite coverings of the underlying complex manifold [Gr61, XII Théorème 5.1].

Definition 8.3 Fix $n \in \mathbb{Z}_{\geq 2}$. Let $X \neq \emptyset$ be a set on which Br_n acts transitively from the left, and let $x_0 \in X$. The covering of C_n^{conf} with fundamental group the stabilizer $(\mathrm{Br}_n)_{x_0}$ and with a base point $b_n^{x_0} \in C_n^{x_0}$ is denoted by

$$\mathrm{pr}_n^{x_0,c} : (C_n^{x_0}, b_n^{x_0}) \to (C_n^{conf}, b_n^{conf}).$$

It is isomorphic to the quotient manifold $C_n^{univ}/(\mathrm{Br}_n)_{x_0}$ (in the quotient, we write the group on the right although the action on C_n^{univ} is from the left).

Remarks 8.4 Let $(H_\mathbb{Z}, L, \underline{e})$ be a unimodular bilinear lattice of rank $n \geq 2$ with a triangular basis $\underline{e}$ and matrix $S = L(\underline{e}^t, \underline{e})^t \in T_n(\mathbb{Z})$. Consider the coverings

$$\begin{aligned}
\mathrm{pr}_n^{\underline{e},c} &: (C_n^{\underline{e}/\{\pm 1\}^n}, b_n^{\underline{e}/\{\pm 1\}^n}) \to (C_n^{conf}, b_n^{conf}),\\
\mathrm{pr}_n^{S,c} &: (C_n^{S/\{\pm 1\}^n}, b_n^{S/\{\pm 1\}^n}) \to (C_n^{conf}, b_n^{conf}).
\end{aligned}$$

We claim that especially the covering $C_n^{\underline{e}/\{\pm 1\}^n}$ is important.

(i) They are the following quotients of the universal covering,

$$C_n^{\underline{e}/\{\pm 1\}^n} \cong C_n^{univ}/(\mathrm{Br}_n)_{\underline{e}/\{\pm 1\}^n}, \quad \text{with } b_n^{\underline{e}/\{\pm 1\}^n} = [b_n^{univ}],$$
$$C_n^{S/\{\pm 1\}^n} \cong C_n^{univ}/(\mathrm{Br}_n)_{S/\{\pm 1\}^n}, \quad \text{with } b_n^{S/\{\pm 1\}^n} = [b_n^{univ}].$$

(ii) Because of Lemma 3.25 (e) $(\mathrm{Br}_n)_{\underline{e}/\{\pm 1\}^n}$ is a normal subgroup of $(\mathrm{Br}_n)_{S/\{\pm 1\}^n}$. Therefore there is a normal covering $C_n^{\underline{e}/\{\pm 1\}^n} \to C_n^{S/\{\pm 1\}^n}$ with group of deck transformations the quotient group $(\mathrm{Br}_n)_{S/\{\pm 1\}^n}/(\mathrm{Br}_n)_{\underline{e}/\{\pm 1\}^n}$, and the covering space $C_n^{S/\{\pm 1\}^n}$ is canonically isomorphic to the quotient

$$C_n^{\underline{e}/\{\pm 1\}^n} / \frac{(\mathrm{Br}_n)_{S/\{\pm 1\}^n}}{(\mathrm{Br}_n)_{\underline{e}/\{\pm 1\}^n}}$$

(we write the group on the right although it acts from the left).

(iii) Recall from Lemma 3.25 (e) that the quotient group $(\mathrm{Br}_n)_{S/\{\pm 1\}^n}/(\mathrm{Br}_n)_{\underline{e}/\{\pm 1\}^n}$ is isomorphic to the group $G_{\mathbb{Z}}^{\mathcal{B}}/Z((\{\pm 1\}^n)_S)$. Often $G_{\mathbb{Z}}^{\mathcal{B}} = G_{\mathbb{Z}}$.

Recall from Lemma 3.25 (f) that in the case of $(H_{\mathbb{Z}}, L, \underline{e})$ irreducible $Z((\{\pm 1\}^n)_S) = \{\pm \mathrm{id}\} \subset G_{\mathbb{Z}}$, so the quotient group above is then isomorphic to $G_{\mathbb{Z}}^{\mathcal{B}}/\{\pm \mathrm{id}\} \subset G_{\mathbb{Z}}/\{\pm \mathrm{id}\}$. Often equality holds.

(iv) If $I^{(0)}$ is positive definite, then $\widetilde{S} + \widetilde{S}^t$ for $\widetilde{S} \in \mathcal{S}^{dist}$ is positive definite with $(\widetilde{S} + \widetilde{S}^t)_{jj} = 2$, so $\widetilde{S}_{ij} \in \{0, \pm 1\}$ for $i < j$. Then the sets $R^{(0)}$, $\mathcal{B}^{dist}$ and $\mathcal{S}^{dist}$ are finite, the covering $C_n^{\underline{e}/\{\pm 1\}^n} \to C_n^{conf}$ is finite of degree $|\mathcal{B}^{dist}/\{\pm 1\}^n|$, the covering $C_n^{S/\{\pm 1\}^n} \to C_n^{conf}$ is finite of degree $|\mathcal{S}^{dist}/\{\pm 1\}^n|$, and the manifolds $C_n^{\underline{e}/\{\pm 1\}^n}$ and $C_n^{S/\{\pm 1\}^n}$ are affine algebraic.

(v) If $I^{(0)}$ is positive semidefinite, then $\widetilde{S} + \widetilde{S}^t$ for $\widetilde{S} \in \mathcal{S}^{dist}$ is positive semidefinite with $(\widetilde{S} + \widetilde{S}^t)_{jj} = 2$, so $\widetilde{S}_{ij} \in \{0, \pm 1, \pm 2\}$ for $i < j$. Then the set $\mathcal{S}^{dist}$ is finite, the covering $C_n^{S/\{\pm 1\}^n} \to C_n^{conf}$ is finite of degree $|\mathcal{S}^{dist}/\{\pm 1\}^n|$, and the manifold $C_n^{S/\{\pm 1\}^n}$ is affine algebraic.

(vi) Often the covering space $C_n^{\underline{e}/\{\pm 1\}^n}$ has a natural smooth partial compactification $\overline{C_n^{\underline{e}/\{\pm 1\}^n}}^p$, and the covering map $\mathrm{pr}_n^{\underline{e},c} : C_n^{\underline{e}/\{\pm 1\}^n} \to C_n^{conf}$ extends to a holomorphic map $\overline{C_n^{\underline{e}/\{\pm 1\}^n}}^p \to \mathbb{C}^n/S_n$.

(vii) If $H_{\mathbb{Z}}$ has rank $n = 1$, we have $S = (1)$, and we set

$$C_1^{univ} = C_1^{\underline{e_1}/\{\pm 1\}} = C_1^{S/\{\pm 1\}} = C_1^{pure} = C_1^{conf} = \mathbb{C},$$
$$\mathrm{Br}_1 = (\mathrm{Br}_1)_{\underline{e_1}/\{\pm 1\}} = (\mathrm{Br}_1)_{S/\{\pm 1\}} = \{1\}.$$

Then the statements in the parts (i)–(iv) still hold, though in a rather trivial way. The coordinate on $\mathbb{C}$ is here called u_1.

The covering $C_n^{x_0} \to C_n^{conf}$ in Definition 8.3 can and will be described in Lemma 8.6 in a more explicit way, by gluing manifolds with boundaries, one for each element of the set X. The manifolds with boundaries are all copies of $\overline{F_n}$, which is given in Definition 8.5.

Definition 8.5 Fix $n \in \mathbb{Z}_{\geq 2}$. Define for $j \in \{1, ..., n-1\}$

$$
\begin{aligned}
W_n^{j,+} &:= \{\underline{u} \in C_n^{pure} \mid \mathrm{Im}(u_1) \leq ... \leq \mathrm{Im}(u_n), \\
&\qquad \mathrm{Im}(u_j) = \mathrm{Im}(u_{j+1}),\ \mathrm{Re}(u_j) < \mathrm{Re}(u_{j+1})\}, \\
W_n^{j,-} &:= \{\underline{u} \in C_n^{pure} \mid \mathrm{Im}(u_1) \leq ... \leq \mathrm{Im}(u_n), \\
&\qquad \mathrm{Im}(u_j) = \mathrm{Im}(u_{j+1}),\ \mathrm{Re}(u_j) > \mathrm{Re}(u_{j+1})\}, \\
W_n &:= \bigcup_{j=1}^{n-1} (W_n^{j,+} \cup W_n^{j,-}), \\
F_n &:= \{\underline{u} \in C_n^{pure} \mid \mathrm{Im}(u_1) < ... < \mathrm{Im}(u_n)\}.
\end{aligned}
$$

Lemma 8.6 *Fix $n \in \mathbb{Z}_{\geq 2}$.*

(a) The set $W_n^{j,\varepsilon} \subset C_n^{pure}$ for $\varepsilon \in \{\pm 1\}$ is called wall *or* σ_j^{ε}-wall. *It is closed in C_n^{pure}, it has real codimension 1 in C_n^{pure}, it has boundary in C_n^{pure} if $n \geq 3$, it is (outside its boundary if $n \geq 3$) smooth and affine linear, it is contractible. If $n \geq 4$ its boundary in C_n^{pure} is only piecewise smooth and affine linear.*

The set F_n is an open convex polyhedron in $\mathbb{C}^n \supset C_n^{pure}$ and therefore contractible. Its boundary consists of the union W_n of the $2(n-1)$ walls. Two walls $W_n^{j_1,\varepsilon_1}$ and $W_n^{j_2,\varepsilon_2}$ with $j_1 \neq j_2$ intersect in real codimension 1. Two walls $W_n^{j,+}$ and $W_n^{j,-}$ do not intersect.

(b) The covering $\mathrm{pr}_n^{p,c} : C_n^{pure} \to C_n^{conf}$ restricts to a homeomorphism from F_n to its image in C_n^{conf}. The complement of $\mathrm{pr}_n^{p,c}(F_n)$ in C_n^{conf} is $\mathrm{pr}_n^{p,c}(W_n)$. It has real codimension 1 and is the boundary of $\mathrm{pr}_n^{p,c}(F_n)$. The covering $\mathrm{pr}_n^{p,c}$ maps $W_n^{j,+}$ and $W_n^{j,-}$ to the same set. The covering $\mathrm{pr}_n^{p,c}$ is not injective on $W_n^{j,\varepsilon}$ because it maps the disjoint sets $W_n^{j,\varepsilon} \cap W_n^{j_1,+}$ and $W_n^{j,\varepsilon} \cap W_n^{j_1,-}$ for $j_1 \neq j$ to the same set.

(c) Let $X \neq \emptyset$ be a set on which Br_n acts transitively from the left, and let $x_0 \in X$. The following are basic general facts from covering theory. The group Br_n acts from the left on the fiber $(\mathrm{pr}_n^{x_0,c})^{-1}(b_n^{conf})$ in the following way. For $\alpha \in Br_n = \pi_1(C_n^{conf}, b_n^{conf})$ and $b \in (\mathrm{pr}_n^{x_0,c})^{-1}(b_n^{conf})$ let $\widetilde{\alpha}$ be the lift to $C_n^{x_0}$ with end point $\widetilde{\alpha}(1) = b$ of a loop in C_n^{conf} which represents α. Then $\alpha(b) := \widetilde{\alpha}(0)$.

Write $U := (\mathrm{Br}_n)_{x_0}$. This action induces the two bijections

$$
\begin{array}{ccccc}
X & \stackrel{1:1}{\longleftarrow} & \mathrm{Br}_n/U & \stackrel{1:1}{\longleftarrow} & (\mathrm{pr}_n^{x_0,c})^{-1}(b_n^{conf}) \\
x = \alpha(x_0) & \longleftarrow & \alpha U & \longmapsto & \alpha(b_n^{x_0})
\end{array}
$$

For $x = \alpha(x_0)$ write $b_n^{x_0,x} := \alpha(b_n^{x_0}) \in (\mathrm{pr}_n^{x_0,c})^{-1}(b_n^{conf})$ (especially $b_n^{x_0} = b_n^{x_0,x_0}$). So there is a natural bijection between X and the fiber in $C_n^{x_0}$ over b_n^{conf}, and x corresponds to $b_n^{x_0,x}$.

(d) Let X and x_0 be as in part (c). The covering $\mathrm{pr}_n^{x_0,c} : (C_n^{x_0}, b_n^{x_0}) \to (C_n^{conf}, b_n^{conf})$ restricts over the open subset $\mathrm{pr}_n^{p,c}(F_n)$ of C_n^{conf} to an even covering, that means that $(\mathrm{pr}_n^{x_0,c})^{-1}(\mathrm{pr}_n^{p,c}(F_n))$ is a union of disjoint open subsets of $C_n^{x_0}$ each of which is mapped homeomorphically to $\mathrm{pr}_n^{p,c}(F_n)$.

The components of $(\mathrm{pr}_n^{x_0,c})^{-1}(\mathrm{pr}_n^{p,c}(F_n))$ *are called* Stokes regions. *Each contains one element of the fiber* $(\mathrm{pr}_n^{x_0,c})^{-1}(b_n^{conf})$. *For* $x \in X$ *let* $F_n^{x_0,x}$ *be the component which contains the point* $b_n^{x_0,x}$.

The complement $C_n^{x_0} - (\mathrm{pr}_n^{x_0,c})^{-1}(\mathrm{pr}_n^{p,c}(F_n))$ *is the preimage of the walls* $\mathrm{pr}_n^{p,c}(W_n)$ *in* C_n^{conf}, *it has everywhere real codimension 1. It is the boundary of* $(\mathrm{pr}_n^{x_0,c})^{-1}(\mathrm{pr}_n^{p,c}(F_n))$.

(e) In C_n^{conf} *the images of the walls* $W_n^{j,+}$ *and* $W_n^{j,-}$ *coincide, but this image is a smooth oriented real hypersurface and has two sides, the* $W_n^{j,+}$*-side and the* $W_n^{j,-}$*-side. The braid* σ_j *has a representative by a loop which goes once through this hypersurface, from the* $W_n^{j,+}$*-side to the* $W_n^{j,-}$*-side.*

(f) Let X *and* x_0 *be as in the parts (c) and (d). The covering space* $C_n^{x_0}$ *can be obtained by glueing copies of the closure* $\overline{F_n}$ *in* C_n^{pure} *in the following way along their boundaries. More precisely, it is isomorphic to the quotient of the disjoint union* $\bigcup_{x\in X}(\overline{F_n}\times\{x\})$ *by the equivalence relation which is generated by the following equivalences.*

$$(\underline{u}, x) \sim ((u_1, ..., u_{j-1}, u_{j+1}, u_j, u_{j+2}, ..., u_n)), \sigma_j^{-1}(x)) \quad \text{for } x \in X$$
$$\text{and } \underline{u} \in W_n^{j,+} (\text{and thus } (u_1, ..., u_{j+1}, u_j, ..., u_n) \in W_n^{j,-}.$$

So the walls $W_n^{j,+} \times \{x\}$ *and* $W_n^{j,-} \times \{\sigma_j^{-1}(x)\}$ *are identified. The isomorphism maps* $F_n \times \{x\}$ *to* $F_n^{x_0,x}$.

Proof (a) Trivial.

(b) Trivial.

(c) The bijection $\mathrm{Br}_n/U \to X$, $\alpha U \to \alpha(x_0)$, is elementary group theory. The group action from the left of Br_n on the fiber $(\mathrm{pr}_n^{x_0,c})^{-1}(b_n^{conf})$ is described for example in [Ha01, page 65, Sect. 1.3]. Here the image $\alpha(b)$ of the *end* point $b = \widetilde{\alpha}(1)$ under α is the *starting* point $\widetilde{\alpha}(0)$, because composition of paths is written from left to right, whereas the action of Br_n on the fiber $(\mathrm{pr}_n^{x_0,c})^{-1}(b_n^{conf})$ shall be an action from the left.

The stabilizer of the base point $b_n^{x_0}$ is by construction of the covering the group $U = (\mathrm{Br}_n)_{x_0}$. This gives the second bijection $\mathrm{Br}_n/U \to (\mathrm{pr}_n^{x_0,c})^{-1}(b_n^{conf})$, $\alpha U \mapsto \alpha(b_n^{x_0})$.

(d) This follows from the shape of the subset $\mathrm{pr}_n^{p,c}(F_n)$ in C_n^{conf}, from the definition of a covering of C_n^{conf} and from part (c).

(e) Compare the description of the elementary braid σ_j in Remark 8.2 (iii) and the two pictures in Fig. 8.2.

The path $[0, 1] \to C_n^{pure}$, $t \mapsto \underline{u}(t)$, which represents σ_j is chosen such that $u_1(t), ..., u_{j-1}(t), u_{j+2}(t), ..., u_n(t)$ are constant and $u_j(t)$ and $u_{j+1}(t)$ have a constant center $\frac{1}{2}(u_j(t) + u_{j+1}(t))$ and turn around halfway clockwise.

(f) This follows from the parts (c), (d) and (e). □

Remarks 8.7 Let X and x_0 be as in the parts (c), (d) and (f) in Lemma 8.6.

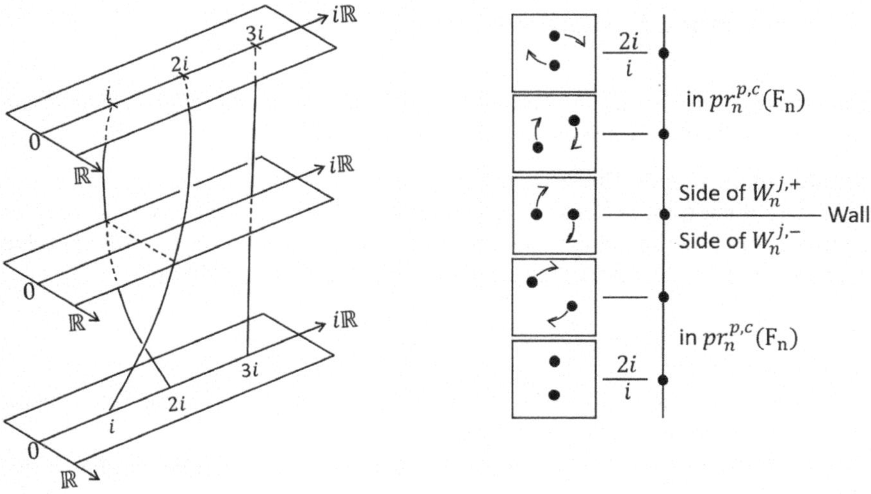

Fig. 8.2 Crossing a wall with an elementary braid

(i) The union

$$W_n^{sing} := \bigcup_{j_1,j_2\in\{1,\dots,n-1\},\, j_1\neq j_2}\ \bigcup_{\varepsilon_1,\varepsilon_2\in\{\pm 1\}} W_n^{j_1,\varepsilon_1}\cap W_n^{j_2,\varepsilon_2} \subset W_n \subset C_n^{pure}$$

has real codimension 2 in C_n^{pure}. Also the image $\mathrm{pr}_n^{p,c}(W_n^{sing}) \subset C_n^{conf}$ has real codimension 2 in C_n^{conf}.

(ii) Choose a sequence $(\sigma_{j_a}^{\varepsilon_a})_{a\in\{1,\dots,l\}}$ of elementary braids for some $l\in\mathbb{N}$, $j_a\in\{1,\dots,n-1\}$, $\varepsilon_a\in\{\pm 1\}$. Choose a closed path p in C_n^{conf} with starting point and end point b_n^{conf} which avoids $\mathrm{pr}_n^{p,c}(W_n^{sing})$ and which passes l times through $\mathrm{pr}_n^{p,c}(W_n)$, first from the $W_n^{j_1,\varepsilon_1}$-side through $\mathrm{pr}_n^{p,c}(W_n^{j_1,\varepsilon_1})$, then from the $W_n^{j_2,\varepsilon_2}$-side through $\mathrm{pr}_n^{p,c}(W_n^{j_2,\varepsilon_2})$, and so on. The path p represents the braid $\alpha = \sigma_{j_1}^{\varepsilon_1}\sigma_{j_2}^{\varepsilon_2}\dots\sigma_{j_l}^{\varepsilon_l}$. The lift of p to $C_n^{x_0}$ which starts at $b_n^{x_0,x_0}$ starts in the Stokes region $F_n^{x_0,x_0}$ and goes through the wall $W_n^{j_1,\varepsilon_1}$ of it to the Stokes region $F_n^{x_0,\sigma_{j_1}^{-\varepsilon_1}(x_0)}$, then through the wall $W_n^{j_2,\varepsilon_2}$ of this Stokes region to the Stokes region $F_n^{x_0,\sigma_{j_2}^{-\varepsilon_2}\sigma_{j_1}^{-\varepsilon_1}(x_0)}$, and so on. It ends in the Stokes region $F_n^{x_0,\alpha^{-1}(x_0)}$ at the point $b_n^{x_0,\alpha^{-1}(x_0)}$.

(iii) A homotopy from a path as in part (ii) for $\sigma_j\sigma_{j+1}\sigma_j$ to a path as in part (ii) for $\sigma_{j+1}\sigma_j\sigma_{j+1}$ contains at least one path which contains a point in $\mathrm{pr}_n^{p,c}(W_n^{j,+}\cap W_n^{j+1,+})$. A preimage in C_n^{pure} of this point satisfies $\mathrm{Im}(u_j) = \mathrm{Im}(u_{j+1}) = \mathrm{Im}(u_{j+2})$.

(iv) If the covering $\mathrm{pr}_n^{x_0,c} : C_n^{x_0} \to C_n^{conf}$ is normal, the deck transformation group of it acts transitively on the set of Stokes regions $\{F_n^{x_0,x}\,|\,x\in X\}$. Then $F_n^{x_0,x_0}$ is a fundamental domain in $C_n^{x_0}$ for the action of the deck transformation group.

8.2 Semisimple F-Manifolds

Each covering space of C_n^{conf} is a semisimple F-manifold with Euler field and with empty Maxwell stratum. We explain these notions here.

Definition 8.8 [HM99, He02, Definition 2.8]

(a) A (holomorphic) *F-manifold* is a tuple $(\mathcal{M}, \circ, e)$ where $\mathcal{M}$ is a complex manifold, $\circ$ is a commutative and associative multiplication on the holomorphic tangent bundle $T\mathcal{M}$ of $\mathcal{M}$ which satisfies the integrability condition

$$\mathrm{Lie}_{X\circ Y}(\circ) = Y \circ \mathrm{Lie}_X(\circ) + X \circ \mathrm{Lie}_Y(\circ)$$

for local holomorphic vector fields X and Y, and e is a global holomorphic vector field with $e\circ = \mathrm{id}$. It is called *unit field.*

(b) An *Euler field* E in an F-manifold $(\mathcal{M}, \circ, e)$ is a global holomorphic vector field with $\mathrm{Lie}_E(\circ) = \circ$.

(c) An F-manifold is semisimple at a point $t \in \mathcal{M}$ if the algebra $(T_t\mathcal{M}, \circ_t, e_t)$ is semisimple, that means, it splits canonically into a sum of 1-dimensional algebras, so $T_t\mathcal{M} = \bigoplus_{j=1}^n \mathbb{C}\cdot e_{t,j}$ with $e_{t,i}\circ e_{t,j} = \delta_{ij}e_{t,i}$ (and automatically $e = \sum_{j=1}^n e_j$).

Remarks 8.9 (i) The theory of F-manifolds is developed in [He02, Chaps. 1–5].

(ii) The condition to be semisimple is an open condition. If an F-manifold is semisimple at one point, it is generically semisimple. More precisely, then the set $\mathcal{K}_3 \subset \mathcal{M}$ of points where it is not semisimple is either empty or an analytic hypersurface [He02, Proposition 2.6]. The set $\mathcal{K}_3$ is called *caustic.*

(iii) A complex manifold with a semisimple (commutative and associative) multiplication on the holomorphic tangent bundle has locally a basis $e_1, ..., e_n$ of holomorphic vector fields which are *the partial units,* so $e_i \circ e_j = \delta_{ij}e_i$. It is unique up to indexing. Here the integrability condition of an F-manifold says $[e_i, e_j] = 0$, so the partial units are locally coordinate vector fields [He02, Theorem 3.2]. The corresponding coordinates are unique up to addition of constants.

The choice of an Euler field E makes them unique, because the eigenvalues $u_1, ..., u_n$ of the endomorphism $E\circ$ on the holomorphic tangent bundle are such local coordinates with $e_1 = \frac{\partial}{\partial u_1}, ..., e_n = \frac{\partial}{\partial u_n}$ and $E = \sum_{j=1}^n u_j e_j$. These local coordinates are unique up to indexing in the case of a semisimple F-manifold with Euler field and are called *canonical coordinates.*

(iv) Let $(\mathcal{M}, \circ, e, E)$ be a generically semisimple F-manifold with Euler field. Besides the caustic $\mathcal{K}_3$ there is often a second analytic hypersurface $\mathcal{K}_2$, the *Maxwell stratum.* It is the closure of the set of points $t \in \mathcal{M}$ such that $(T_t\mathcal{M}, \circ_t, e_t)$ is semisimple (so locally canonical coordinates exist), but at least two eigenvalues of $E\circ$ coincide. Usually $\mathcal{K}_3$ and $\mathcal{K}_2$ intersect. The map

$$\mathcal{L} : \mathcal{M} \to \mathbb{C}[x]_n, \quad t \mapsto (\text{the characteristic polynomial of } E_t\circ),$$

is holomorphic. In singularity theory it is called *Lyashko-Looijenga map.* $\mathcal{K}_3 \cup \mathcal{K}_2$ is mapped to the discriminant $\mathbb{C}[x]_n^{mult}$, the complement $\mathcal{M} - (\mathcal{K}_3 \cup \mathcal{K}_2)$ is mapped locally biholomorphically to the complement $\mathbb{C}[x]_n^{reg} \cong C_n^{conf}$. Near a generic point of $\mathcal{K}_2$ there are local coordinates such that two of them take the same value at points of $\mathcal{K}_2$. Therefore $\mathcal{L}$ is a branched covering of degree two near a generic point of $\mathcal{K}_2$. The index 3 in $\mathcal{K}_3$ stems from the fact that in a family of important cases $\mathcal{L}$ is a branched covering of order 3 near generic points of $\mathcal{K}_3$.

(v) Vice versa, a locally biholomorphic map $\mathcal{L} : \mathcal{M} \to \mathbb{C}[x]_n^{reg} \cong C_n^{conf}$ induces on $\mathcal{M}$ the structure of a semisimple F-manifold with Euler field and empty Maxwell stratum.

(vi) Define

$$\mathbb{C}[x]_n^{mult,reg} := \{f \in \mathbb{C}[x]^{mult} \mid f \text{ has precisely } n-1 \text{ different roots}\}.$$

It is an open part of $\mathbb{C}[x]_n^{mult}$. The complement is a complex hypersurface in $\mathbb{C}[x]_n^{mult}$ (which itself is a complex hypersurface in $\mathbb{C}[x]_n$).

Consider the following situation: A locally biholomorphic map $\mathcal{L} : \mathcal{M} \to \mathbb{C}[x]_n^{reg}$, a point $p \in \mathbb{C}[x]_n^{mult,reg}$, a neighborhood $U \subset \mathbb{C}[x]_n$ of p with $U \cap \mathbb{C}[x]_n^{mult} \subset \mathbb{C}[x]_n^{mult,reg}$, $U' := U - \mathbb{C}[x]_n^{mult}$, a component $\widetilde{U}'$ of $\mathcal{L}^{-1}(U')$ such that the restriction $\mathcal{L} : \widetilde{U}' \to U'$ is an m-fold covering for some $m \in \mathbb{N}$. Then $\widetilde{U}'$ has a partial compactification $\widetilde{U}$ such that $\mathcal{L}$ extends to a branched covering $\widetilde{\mathcal{L}} : \widetilde{U} \to U$, which is branched of order m over $U - U' = U \cap \mathbb{C}[x]_n^{mult}$. If $m \geq 2$ the F-manifold structure extends from $\widetilde{U}'$ to $\widetilde{U}$. If $m = 2$ $\widetilde{U} - \widetilde{U}'$ is the Maxwell stratum in $\widetilde{U}$. If $m \geq 3$ $\widetilde{U} - \widetilde{U}'$ is the caustic in $\widetilde{U}$. See the Examples 8.10.

(vii) A semisimple F-manifold with Euler field is locally a rather trivial object. More interesting is a generically semisimple F-manifold near points of the caustic $\mathcal{K}_3$.

(viii) The notion of a *Dubrovin-Frobenius manifold* is much richer. It is an F-manifold $(\mathcal{M}, \circ, e, E)$ with Euler field together with a flat holomorphic metric g on the holomorphic tangent bundle such that e is flat (with respect to the Levi-Civita connection of g), g is *multiplication invariant*, that means

$$g(X \circ Y, Z) = g(X, Y \circ Z)$$

for local holomorphic vector fields X, Y, Z, and

$$\mathrm{Lie}_E(g) = (2-d)g \quad \text{for some} \quad d \in \mathbb{C}.$$

It was first defined by Dubrovin [Du92][Du96]. To see the equivalence of the definition above with the definition in [Du92][Du96] one needs [He02, Theorem 2.15 and Lemma 2.16].

Dubrovin-Frobenius manifolds arise in singularity theory, in quantum cohomology, in mirror symmetry and in integrable systems. They are all highly transcendental and interesting objects. The construction in singularity theory of a metric g which makes a semisimple F-manifold with Euler field into a Dubrovin-Frobenius manifold uses additional structure above the F-manifold. One version of this additional structure is presented in Sect. 8.5.

Examples 8.10 (i) Of course, $\mathbb{C}^n$ with coordinates $u_1, ..., u_n$ is a semisimple F-manifold with partial units $e_i = \frac{\partial}{\partial u_i}$ and Euler field $E = \sum_{i=1}^n u_i e_i$. It is called of type A_1^n. The Maxwell stratum is the union D_n^{pure} of partial diagonals.

(ii) For $m \in \mathbb{Z}_{\geq 2}$ the manifold $\mathcal{M} = \mathbb{C}^2$ with coordinates z_1, z_2 becomes a generically semisimple F-manifold with Euler field in the following way [He02, Theorem 4.7]. It is called of type $I_2(m)$. Denote by $\partial_1 := \frac{\partial}{\partial z_1}$ and $\partial_2 := \frac{\partial}{\partial z_2}$ the coordinate vector fields and define the unit field, the multiplication $\circ$ and the Euler field E as follows,

$$\begin{aligned} e &:= \partial_1, \quad \text{so } \partial_1 \circ = \text{id on the holomorphic tangent bundle}, \\ \partial_2 \circ \partial_2 &:= z_2^{m-2}\partial_1, \\ E &:= z_1\partial_1 + \frac{2}{m} z_2 \partial_2. \end{aligned}$$

If $m = 2$ it is everywhere semisimple, and the Maxwell stratum is $\mathcal{K}_2 = \{(z_1, z_2) \in \mathcal{M} \mid z_2 = 0\}$. If $m \geq 3$ it is generically semisimple, and the caustic is $\mathcal{K}_3 = \{(z_1, z_2) \in \mathcal{M} \mid z_2 = 0\}$.

For $m = 2$ there are global canonical coordinates, for even $m \geq 3$ there are global canonical coordinates on $\mathcal{M} - \mathcal{K}_3$, for odd $m \geq 3$ there are local canonical coordinates on $\mathcal{M} - \mathcal{K}_3$, which in fact are globally 2-valued. In all cases they are

$$u_{1/2} = z_1 \pm \frac{2}{m} z_2^{m/2}.$$

The partial units are

$$e_{1/2} = \frac{1}{2} e \pm \frac{1}{2} z_2^{1-\frac{m}{2}} \partial_2.$$

The Lyashko-Looijenga map

$$\begin{aligned} \mathcal{L} : \mathcal{M} \to \mathbb{C}[x]_2, \ (z_1, z_2) &\mapsto (-u_1 - u_2, u_1 u_2) = (-2z_1, z_1^2 - \frac{4}{m^2} z_2^m) \\ &\cong x^2 + (-u_1 - u_2)x + u_1 u_2 = (x - u_1)(x - u_2), \end{aligned}$$

restricts to an m-fold covering $\mathcal{L} : \mathcal{M} - \{z_2 = 0\} \to \mathbb{C}[x]^{reg}$, it maps $\{z_2 = 0\}$ to $\mathbb{C}[x]^{mult}$, and it is branched of order m along $\{z_2 = 0\}$.

(iii) The type $I_2(2)$ is the type A_1^2, the type $I_2(3)$ is also called type A_2.

(iv) Theorem 4.7 in [He02] says also that each 2-dimensional irreducible germ of a generically semisimple F-manifold with Euler field is isomorphic to the germ $((\mathcal{M}, t^0), \circ, e, E)$ in the F-manifold $\mathcal{M} = \mathbb{C}^2$ of type $I_2(m)$ for some $t^0 = (0, z_2^0) \in \mathcal{K}_3$ and some $m \in \mathbb{Z}_{\geq 3}$.

(v) If $(\mathcal{M}^{(i)}, \circ^{(i)}, e^{(i)}, E^{(i)})$ for $i \in \{1, 2\}$ are F-manifolds with Euler fields, their product becomes an F-manifold $(\mathcal{M}^{(1)} \times \mathcal{M}^{(2)}, \circ^{(1)} \times \circ^{(2)}, e^{(1)} + e^{(2)}, E^{(1)} + E^{(2)})$

in the natural way, with Maxwell stratum $\mathcal{K}_2^{(1)} \times \mathcal{M}^{(2)} \cup \mathcal{M}^{(1)} \times \mathcal{K}_2^{(2)}$ and caustik $\mathcal{K}_3^{(1)} \times \mathcal{M}^{(2)} \cup \mathcal{M}^{(1)} \times \mathcal{K}_3^{(2)}$. For example, one obtains for $m \geq 2$ and $n \geq 3$ the generically semisimple F-manifold of type $I_2(m)A_1^{n-2}$ with Euler field on $\mathbb{C}^2 \times \mathbb{C}^{n-2}$.

8.3 Reducible Cases and Rank 2 Cases

Let $(H_{\mathbb{Z}}, L, \underline{e})$ be a unimodular bilinear lattice of rank $n \geq 2$ with a triangular basis $\underline{e}$ and matrix $S = L(\underline{e}^t, \underline{e})^t \in T_n(\mathbb{Z})$. First we state a result on $C_n^{\underline{e}/\{\pm 1\}^n}$ in the case when $(H_{\mathbb{Z}}, L, \underline{e})$ is reducible. Then we treat the rank 2 cases.

Theorem 8.11 *Let* $(H_{\mathbb{Z}}, L, \underline{e}) = (H_{\mathbb{Z},1}, L_1, \underline{e}^{(1)}) \oplus (H_{\mathbb{Z},2}, L_2, \underline{e}^{(2)})$ *be reducible with* $\underline{e}^{(1)} = (e_1, e_2, ..., e_{n_1})$ *and* $\underline{e}^{(2)} = (e_{n_1+1}, ..., e_n)$, *so* $S_{ij} = 0$ *for* $i \leq n_1 < j$. *Write* $n_2 := n - n_1$.

There is a natural embedding

$$C_n^{\underline{e}/\{\pm 1\}^n} \hookrightarrow C_{n_1}^{\underline{e}^{(1)}/\{\pm 1\}^{n_1}} \times C_{n_2}^{\underline{e}^{(2)}/\{\pm 1\}^{n_2}}.$$

It is an embedding of semisimple F-manifolds with Euler fields. The complement is a complex hypersurface. This hypersurface is the Maxwell stratum of the semisimple F-manifold with Euler field on the right hand side.

Proof In order to write down the embedding concretely, we will make use of the construction of the three manifolds by the glueing of Stokes regions in Lemma 8.6 (f). Here

$$\begin{aligned} x_0 &:= \underline{e}/\{\pm 1\}^n \in X := \mathrm{Br}_n(x_0) = \mathcal{B}^{dist}/\{\pm 1\}^n, \\ x_0^{(1)} &:= \underline{e}^{(1)}/\{\pm 1\}^{n_1} \in X^{(1)} := \mathrm{Br}_{n_1}(x_0^{(1)}) = \mathcal{B}^{dist,(1)}/\{\pm 1\}^{n_1}, \\ x_0^{(2)} &:= \underline{e}^{(2)}/\{\pm 1\}^{n_2} \in X^{(2)} := \mathrm{Br}_{n_2}(x_0^{(2)}) = \mathcal{B}^{dist,(2)}/\{\pm 1\}^{n_2}, \end{aligned}$$

$$\begin{aligned} C_n^{x_0} &= \Big(\bigcup_{x \in X} \overline{F_n} \times \{x\}\Big)_{\sim}, \\ C_{n_1}^{x_0^{(1)}} &= \Big(\bigcup_{x^{(1)} \in X^{(1)}} \overline{F_{n_1}} \times \{x^{(1)}\}\Big)_{\sim(1)}, \\ C_{n_2}^{x_0^{(2)}} &= \Big(\bigcup_{x^{(2)} \in X^{(2)}} \overline{F_{n_2}} \times \{x^{(2)}\}\Big)_{\sim(2)}. \end{aligned}$$

For any $x = (x_1, ..., x_n) \in X$ there are unique indices $i_1, ..., i_{n_1}$ and $j_1, ..., j_{n_2}$ with

$$1 \leq i_1 < ... < i_{n_1} \leq n, \quad 1 \leq j_1 < ... < j_{n_2} \leq n,$$

$$\{i_1, ..., i_{n_1}, j_1, ..., j_{n_2}\} = \{1, 2, ..., n\},$$
$$x^{(1)} := (x_{i_1}, x_{i_2}, ..., x_{i_{n_1}}) \in X^{(1)}, \ x^{(2)} := (x_{j_1}, x_{j_2}, ..., x_{j_{n_1}}) \in X^{(2)}.$$

In words, the distinguished basis x up to signs for $(H_{\mathbb{Z}}, L, \underline{e})$ is a shuffling of the distinguished basis $x^{(1)}$ up to signs for $(H_{\mathbb{Z},1}, L_1, \underline{e}^{(1)})$ and the distinguished basis $x^{(2)}$ up to signs for $(H_{\mathbb{Z},2}, L_2, \underline{e}^{(2)})$. Here $x^{(1)}$ and $x^{(2)}$ are unique.

A point $(\underline{u}, x) \in \overline{F_n} \times \{x\}$ is mapped to the point

$$(\underline{u}^{(1)}, x^{(1)}; \underline{u}^{(2)}, x^{(2)}) := ((u_{i_1}, u_{i_2}, ..., u_{i_{n_1}}), x^{(1)}; (u_{j_1}, u_{j_2}, ..., u_{j_{n_2}}), x^{(2)})$$
$$\in (\overline{F_{n_1}} \times \{x^{(1)}\}) \times (\overline{F_{n_2}} \times \{x^{(2)}\}).$$

The following observations (1), (2) and (3) are crucial.

(1) Given $x^{(1)} \in X^{(1)}$ and $x^{(2)} \in X^{(2)}$, there are $\binom{n}{n_1}$ distinguished bases $y \in X$ up to signs with $y^{(1)} = x^{(1)}$ and $y^{(2)} = x^{(2)}$. They are all the tuples obtained by shuffling of $x^{(1)}$ and $x^{(2)}$.

(2) For $x^{(1)} \in X^{(1)}$ and $x^{(2)} \in X^{(2)}$ the map

$$\dot{\bigcup}_{y \in X:\, y^{(1)}=x^{(1)}, y^{(2)}=x^{(2)}} \overline{F_n} \times \{y\} \to (\overline{F_{n_1}} \times \{x^{(1)}\}) \times (\overline{F_{n_2}} \times \{x^{(2)}\})$$

is almost surjective and almost injective. More precisely, the following two points (i) and (ii) hold.

(i) The map

$$\Big(\dot{\bigcup}_{y \in X:\, y^{(1)}=x^{(1)}, y^{(2)}=x^{(2)}} \overline{F_n} \times \{y\}\Big)_{\sim} \to (\overline{F_{n_1}} \times \{x^{(1)}\}) \times (\overline{F_{n_2}} \times \{x^{(2)}\})$$

is well defined and injective because of the following. The restriction to $\dot{\bigcup}_{y \in X:\, y^{(1)}=x^{(1)}, y^{(2)}=x^{(2)}} F_n \times \{y\}$ is injective, anyway. Consider $\underline{u} \in W_n^{j,+}$ and $y \in X$ with $x^{(1)} = y^{(1)} = (\sigma_j^{-1} y)^{(1)}$, $x^{(2)} = y^{(2)} = (\sigma_j^{-1} y)^{(2)}$. Then $\mathrm{Im}(u_j) = \mathrm{Im}(u_{j+1})$, and $(\underline{u}, x)$ and $((u_1, ..., u_{j+1}, u_j, ..., u_n), \sigma_j^{-1}(x))$ have the same image.

(ii) Consider any two points $\underline{u}^{(1)} \in \overline{F_{n_1}}$ and $\underline{u}^{(2)} \in \overline{F_{n_2}}$ such that the entries of $\underline{u}^{(1)}$ are pairwise different from the entries of $\underline{u}^{(2)}$. There are indices $\widetilde{i}_1, ..., \widetilde{i}_{n_1}$ and $\widetilde{j}_1, ..., \widetilde{j}_{n_2}$ with

$$1 \le \widetilde{i}_1 < ... < \widetilde{i}_{n_1} \le n, \quad 1 \le \widetilde{j}_1 < ... < \widetilde{j}_{n_2} \le n,$$
$$\{\widetilde{i}_1, ..., \widetilde{i}_{n_1}, \widetilde{j}_1, ..., \widetilde{j}_{n_2}\} = \{1, ..., n\},$$

such that the tuple $\underline{u} \in \mathbb{C}^n$ which is defined by

$$\underline{u}^{(1)} = (u_{\widetilde{i}_1}, ..., u_{\widetilde{i}_{n_1}}), \quad \underline{u}^{(2)} = (u_{\widetilde{j}_1}, ..., u_{\widetilde{j}_{n_2}}),$$

satisfies

$$\mathrm{Im}(u_1) \leq \mathrm{Im}(u_2) \leq \ldots \leq \mathrm{Im}(u_n).$$

The tuple $y = (y_1, \ldots, y_n)$ which is defined by

$$x^{(1)} = (y_{\widetilde{i}_1}, y_{\widetilde{i}_2}, \ldots, y_{\widetilde{i}_{n_1}}), \quad x^{(2)} = (y_{\widetilde{j}_1}, y_{\widetilde{j}_2}, \ldots, y_{\widetilde{j}_{n_2}}),$$

is in X and is obtained by shuffling of $x^{(1)}$ and $x^{(2)}$. The pair $(\underline{u}, y)$ is mapped to the tuple $(\underline{u}^{(1)}, x^{(1)}; \underline{u}^{(2)}, x^{(2)})$. Therefore the only points $(\underline{u}^{(1)}, x^{(1)}; \underline{u}^{(2)}, x^{(2)}) \in (\overline{F_{n_1}} \times \{x^{(1)}\}) \times (\overline{F_{n_2}} \times \{x^{(2)}\})$ which are not met by the map in part (i) are those points where some entry of $\underline{u}^{(1)}$ coincides with some entry of $\underline{u}^{(2)}$. In the semisimple F-manifold $(F_{n_1} \times \{x^{(1)}\}) \times (F_{n_2} \times \{x^{(2)}\})$ with Euler field these points form the Maxwell stratum.

(3) (i) The map

$$\dot{\bigcup}_{x \in X} \overline{F_n} \times \{x\} \to \Big(\dot{\bigcup}_{x^{(1)} \in X^{(1)}} \overline{F_{n_1}} \times \{x^{(1)}\}\Big) \times \Big(\dot{\bigcup}_{x^{(2)} \in X^{(2)}} \overline{F_{n_2}} \times \{x^{(2)}\}\Big)$$

is well defined (and almost surjective and almost injective). This follows from (2)(i) and (ii).

(ii) **Claim**: The map in (3)(i) induces a well defined injective map on the quotients

$$\Big(\dot{\bigcup}_{x \in X} \overline{F_n} \times \{x\}\Big)_{\sim} \to$$
$$\Big(\dot{\bigcup}_{x^{(1)} \in X^{(1)}} \overline{F_{n_1}} \times \{x^{(1)}\}\Big)_{\sim^{(1)}} \times \Big(\dot{\bigcup}_{x^{(2)} \in X^{(2)}} \overline{F_{n_2}} \times \{x^{(2)}\}\Big)_{\sim^{(2)}}.$$

Proof of the Claim: The restriction to $\dot{\bigcup}_{x \in X} F_n \times \{x\}$ is well defined and injective, anyway. Consider $\underline{u} \in W_n^{j,+}$ and $x \in X$. If $(\sigma_j^{-1}(x))^{(1)} = x^{(1)}$ and $(\sigma_j^{-1}(x))^{(2)} = x^{(2)}$, we are in the case treated in (2)(ii). If not, then either x_j and x_{j+1} are both in $H_{\mathbb{Z},1}/\{\pm 1\}^{n_1}$ or x_j and x_{j+1} are both in $H_{\mathbb{Z},2}/\{\pm 1\}^{n_2}$. In the first case $(\underline{u}^{(1)}, x^{(1)}) \sim^{(1)} ((u_1, \ldots, u_{j+1}, u_j, \ldots, u_n)^{(1)}, (\sigma_j^{-1}(x))^{(1)})$, in the second case $(\underline{u}^{(2)}, x^{(2)}) \sim^{(2)} ((u_1, \ldots, u_{j+1}, u_j, \ldots, u_n)^{(2)}, (\sigma_j^{-1}(x))^{(2)})$. This proves the Claim in (3)(ii).

The observations (1)–(3) show that there is a natural open embedding

$$C_n^{\underline{x}_0} \hookrightarrow C_{n_1}^{x_0^{(1)}} \times C_{n_2}^{x_0^{(2)}}$$

and that the complement is a hypersurface, which is the Maxwell stratum of $C_{n_1}^{x_0^{(1)}} \times C_{n_2}^{x_0^{(2)}}$ as a semisimple F-manifold with Euler field. □

Now we come to the rank 2 cases. The parts (a)–(c) of the following theorem treat C_2^{pure}, C_2^{conf} and C_2^{univ} as semisimple F-manifolds with Euler fields, part (d) considers $C_2^{\underline{e}/\{\pm 1\}^2}$ for a unimodular bilinear lattice of rank 2 with a triangular basis.

Theorem 8.12 *(a) C_2^{pure} is isomorphic to the semisimple F-manifold with Euler field $I_2(2)$ in the Examples 8.10, minus its Maxwell stratum. In formulas:*

$$C_2^{pure} = \{(u_1, u_2) \in \mathbb{C}^2 \mid u_1 \neq u_2\} \xrightarrow{\cong} \mathbb{C} \times \mathbb{C}^*,$$
$$(u_1, u_2) \mapsto (z_1, z_2) = (\frac{u_1 + u_2}{2}, \frac{u_1 - u_2}{2}).$$

Therefore C_2^{pure} extends to an F-manifold on $\mathbb{C}^2$ with Maxwell stratum $\{(u_1, u_2) \in \mathbb{C}^2 \mid u_1 = u_2\}$. Multiplication $\circ$, unit field e, Euler field E and partial units e_1 and e_2 are explicitly as follows,

$$u_{1/2} = z_1 \pm z_2, \quad e_i = \frac{\partial}{\partial u_i}, \quad e_i \circ e_j = \delta_{ij} e_i,$$
$$e = e_1 + e_2 = \frac{\partial}{\partial z_1}, \quad \text{so } e \circ = \text{id},$$
$$\frac{\partial}{\partial z_2} = e_1 - e_2, \quad \frac{\partial}{\partial z_2} \circ \frac{\partial}{\partial z_2} = e,$$
$$E = u_1 e_1 + u_2 e_2 = z_1 \frac{\partial}{\partial z_1} + z_2 \frac{\partial}{\partial z_2}.$$

(b) The covering $C_2^{pure} \to C_2^{conf}$ has degree 2. In the coordinates (z_1, z_2) on C_2^{pure} it extends to the branched covering

$$\mathbb{C}^2 \to \mathbb{C}^2, \quad (z_1, z_2) \mapsto (\widetilde{z}_1, \widetilde{z}_2) = (z_1, z_2^2).$$

On $C_2^{conf} \cong \mathbb{C} \times \mathbb{C}^*$ with coordinates $(\widetilde{z}_1, \widetilde{z}_2)$, unit field e, Euler field E, multiplication $\circ$ and partial units e_1 and e_2 are as follows,

$$e = e_1 + e_2 = \frac{\partial}{\partial \widetilde{z}_1}, \quad E = \widetilde{z}_1 e + 2\widetilde{z}_2 \frac{\partial}{\partial \widetilde{z}_2},$$
$$\frac{\partial}{\partial \widetilde{z}_2} \circ \frac{\partial}{\partial \widetilde{z}_2} = \frac{1}{4\widetilde{z}_2} e,$$
$$u_{1/2} = \widetilde{z}_1 \pm \sqrt{\widetilde{z}_2}, \quad e_{1/2} = \frac{1}{2} e \pm \sqrt{\widetilde{z}_2} \frac{\partial}{\partial \widetilde{z}_2}.$$

The canonical coordinates and the partial units are 2-valued. The multiplication does not extend to $\mathbb{C} \times \{0\}$.

(c) The universal covering $C_2^{univ} \to C_2^{pure}$ can be written as follows,

$$C_2^{univ} \cong \mathbb{C} \times \mathbb{C} \to \mathbb{C} \times \mathbb{C}^* \cong C_2^{pure},$$
$$(z_1, \zeta) \mapsto (z_1, \exp(\zeta)) = (z_1, z_2).$$

In the semisimple F-manifold $C_2^{univ} \cong \mathbb{C}^2$ with coordinates (z_1, ζ), unit field e, Euler field E, multiplication $\circ$ and partial units e_1 and e_2 are as follows,

$$e = e_1 + e_2 = \frac{\partial}{\partial z_1}, \quad E = z_1 e + \frac{\partial}{\partial \zeta},$$
$$\frac{\partial}{\partial \zeta} \circ \frac{\partial}{\partial \zeta} = \exp(2\zeta)e,$$
$$u_{1/2} = z_1 \pm \exp(\zeta), \quad e_{1/2} = \frac{1}{2}e \pm \frac{1}{2}\exp(-\zeta)\frac{\partial}{\partial \zeta},$$

(d) Let $(H_{\mathbb{Z}}, L, \underline{e})$ be a unimodular bilinear lattice of rank 2 with triangular basis $\underline{e}$ and matrix $S = L(\underline{e}^t, \underline{e})^t = \begin{pmatrix} 1 & x \\ 0 & 1 \end{pmatrix} \in T_2(\mathbb{Z})$ for some $x \in \mathbb{Z}_{\leq 0}$. Then

$$C_2^{S/\{\pm 1\}^2} = C_2^{conf},$$

and

$$C_2^{\underline{e}/\{\pm 1\}^2} = \begin{cases} C_2^{pure} & \text{if } x = 0, \quad \text{the case } A_1^2, \\ C_2^{A_2} & \text{if } x = -1, \quad \text{the case } A_2, \\ C_2^{univ} & \text{if } x \leq -2, \quad \text{the cases } \mathbb{P}^1 \text{ and beyond.} \end{cases}$$

Here $C_2^{A_2} \cong \mathbb{C} \times \mathbb{C}^*$ with coordinates (z_1, z_2) is the semisimple F-manifold of type A_2 in the Examples 8.10 (ii), minus its Maxwell stratum $\mathbb{C} \times \{0\}$. Explicitly, unit field e, Euler field E, multiplication $\circ$ and partial units are as follows,

$$e = e_1 + e_2 = \frac{\partial}{\partial z_1}, \quad E = z_1 e + \frac{2}{3} z_2 \frac{\partial}{\partial z_2},$$
$$\frac{\partial}{\partial z_2} \circ \frac{\partial}{\partial z_2} = z_2 e,$$
$$u_{1/2} = z_1 \pm \frac{2}{3} z_2^{3/2}, \quad e_{1/2} = \frac{1}{2} e \pm \frac{1}{2} z_2^{-1/2} \frac{\partial}{\partial z_2}.$$

The canonical coordinates and the partial units are 2-valued. The semisimple F-manifold $C_2^{A_2}$ extends to $\mathbb{C} \times \{0\}$ with $\mathbb{C} \times \{0\}$ as caustic.

Proof (a) Trivial.

(b) Locally on C_2^{pure} $(\tilde{z}_1, \tilde{z}_2) = (z_1, z_2^2)$ is a coordinate system with $d\tilde{z}_2 = 2z_2 dz_2$ and $\frac{\partial}{\partial \tilde{z}_2} = \frac{1}{2z_2}\frac{\partial}{\partial z_2}$. One applies part (a).

(c) Locally on $C_2^{univ} \cong \mathbb{C} \times \mathbb{C}$ $(z_1, z_2) = (z_1, e^{\zeta})$ is a coordinate system with $d\zeta = e^{-\zeta} d(e^{\zeta}) = z_2^{-1} dz_2$ and $\frac{\partial}{\partial \zeta} = z_2 \frac{\partial}{\partial z_2}$. One applies part (a).

(d) Recall from Theorem 7.1 (b)

$$(\mathrm{Br}_2)_{S/\{\pm 1\}^2} = \mathrm{Br}_2 \quad \text{for any } x,$$

$$(\mathrm{Br}_2)_{\underline{e}/\{\pm 1\}^2} = \begin{cases} \langle \sigma_1^2 \rangle & \text{for } x = 0, \text{ the case } A_1^2, \\ \langle \sigma_1^3 \rangle & \text{for } x = -1, \text{ the case } A_2, \\ \{\mathrm{id}\} & \text{for } x \leq -2, \text{ the cases } \mathbb{P}^1 \text{ and beyond.} \end{cases}$$

Therefore

$$C_2^{S/\{\pm 1\}^2} = C_2^{conf},$$

and $C_2^{\underline{e}/\{\pm 1\}^2}$ is a cyclic covering of C_2^{conf} of degree 2 for $x = 0$, of degree 3 for $x = -1$ and of infinite degree for $x \leq -2$. With the parts (a)–(c) and the case of type A_2 in Example 8.10 (ii) this shows part (d). □

8.4 Distinguished Systems of Paths, Two Braid Group Actions

For the $\mathbb{Z}$-lattice bundles in Sect. 8.5, *distinguished systems of paths* are needed. This is a notion which was developed within the theory of isolated hypersurfac singularities. A good source is [Eb01, 5.2 and 5.7]. Definition 8.13 and Theorem 8.14 define them and describe one action of the braid group Br_n on the set of distinguished systems of n paths with fixed starting point and fixed set of endpoints.

Though there is also an action of the groupoid of homotopy classes of paths in C_n^{pure} on the set of distinguished systems of n paths with fixed starting point, but variable set of endpoints (Remarks 8.15 and Definition/Lemma 8.16). It leads to a second action of the braid group Br_n on the set of distinguished systems of n paths with fixed starting point and fixed set of endpoints (Remark 8.17 (ii)). This second action is not discussed in [Eb01]. Lemma 8.18 clarifies the relationship with the first action. The two actions commute, and one can be expressed through the other. Example 8.19 gives an example.

Definition 8.13 Let $\underline{u} = (u_1, u_2, ..., u_n) \in C_n^{pure}$ and $r \in \mathbb{R}$ with $r > \max(\mathrm{Re}\, u_1, ..., \mathrm{Re}\, u_n)$.

(a) A *$(\underline{u}, r)$-path* γ is a continuous embedding $\gamma : [0, 1] \to \mathbb{C}$ with $\gamma(0) = r$, $\gamma(1) \in \{u_1, ..., u_n\}$ and $\gamma((0, 1)) \subset \{z \in \mathbb{C} \mid \mathrm{Re}\, z < r\} - \{u_1, ..., u_n\}$.

(b) [Eb01, 5.2] A *distinguished system of paths* is a tuple $(\gamma_1, ..., \gamma_n)$ of $(\underline{u}, r)$-paths together with a permutation $\sigma \in S_n$ such that $\gamma_j(1) = u_{\sigma(j)}$, $\gamma_i((0, 1)) \cap \gamma_j((0, 1)) = \emptyset$ for $i \neq j$, and $\gamma_1, ..., \gamma_n$ leave their common starting point $r = \gamma_1(0) = ... = \gamma_n(0)$ in clockwise order.

(c) [Eb01, 5.7] Two distinguished systems of $(\underline{u}, r)$-paths $(\gamma_1, ..., \gamma_n; \sigma)$ and $(\widetilde{\gamma}_1, ..., \widetilde{\gamma}_n; \sigma)$ are *homotopic*, if continuous maps $\Gamma_1, ..., \Gamma_n : [0, 1] \times [0, 1] \to \mathbb{C}$ exist such that $(\Gamma_1(., s), ..., \Gamma_n(., s); \sigma)$ is for each $s \in [0, 1]$ a distinguished system of $(\underline{u}, r)$-paths and $(\Gamma_1(., 0), ..., \Gamma_n(., 0)) = (\gamma_1, ..., \gamma_n)$, $(\Gamma_1(., 1), ..., \Gamma_n(., 1)) = (\widetilde{\gamma}_1, ..., \widetilde{\gamma}_n)$. Denote by $\mathcal{P}(\underline{u}, r)$ the set of all homotopy classes of distinguished systems of $(\underline{u}, r)$-paths.

(d) A *standard system of paths* is a distinguished system of paths $(\gamma_1, ..., \gamma_n)$ with the following properties (1)–(4).
(1) $\sigma \in S_n$ is determined by

(i) $\text{Im}\, u_{\sigma(1)} \leq ... \leq \text{Im}\, u_{\sigma(n)}$.
(ii) If $\text{Im}\, u_{\sigma(j)} = \text{Im}\, u_{\sigma(j+1)}$ then $\text{Re}\, u_{\sigma(j)} < \text{Re}\, u_{\sigma(j+1)}$.

(2) Choose $r_1 \in \mathbb{R}$ with $\max(\text{Re}\, u_1, ..., \text{Re}\, u_n) < r_1 < r$. Consider $j \in \{1, ..., n\}$ and $n_1 \in \mathbb{N}$ with

$$\text{Im}\, u_{\sigma(j-n_1)} < \text{Im}\, u_{\sigma(j-n_1+1)} = ... = \text{Im}\, u_{\sigma(j)} < \text{Im}\, u_{\sigma(j+1)}.$$

(3) Choose $\varepsilon_j > 0$ so small that $n_1 \varepsilon_j < \text{Im}\, u_{\sigma(j)} - \text{Im}\, u_{\sigma(j-n_1)}$.
(4) Choose the path γ_{j-k} for $k \in \{0, 1, ..., n_1 - 1\}$ as follows:
First go straight from r to $r_1 + i(\text{Im}\, u_{\sigma(j)} - k\varepsilon_j)$.
Then go straight (horizontally to the left) to $\text{Re}(u_{\sigma(j-k)}) + i(\text{Im}\, u_{\sigma(j)} - k\varepsilon_j)$.
Then go straight (vertically upwards) to $u_{\sigma(j-k)}$.

Obviously any two standard systems of paths are homotopic. Figure 8.3 shows three distinguished systems of paths. The lower one is a standard system of paths, the upper left one is homotopic to a standard system of paths, the upper right one not.

Theorem 8.14 *[Eb01, 5.7] Let $\underline{u} = (u_1, ..., u_n) \in C_n^{pure}$ and let $r \in \mathbb{R}$ with $r >$* $\max(\text{Re}\, u_1, ..., \text{Re}\, u_n)$.

(a) (Definition) Define actions of the elementary braids σ_j and σ_j^{-1} in Br_n on the set $\mathcal{P}(\underline{u}, r)$ as follows. σ_j maps the homotopy class of a distinguished system of paths $(\gamma_1, ..., \gamma_n)$ to the homotopy class of a distinguished system of paths $(\gamma_1', ..., \gamma_n')$ where

$$\begin{aligned} \gamma_i' &:= \gamma_i \text{ for } i \in \{1, ..., n\} - \{j, j+1\}, \\ \gamma_{j+1}' &:= \gamma_j, \\ \gamma_j' &: \textit{go from } r \textit{ along } \gamma_j \textit{ almost to } u_{\sigma(j)}, \\ &\quad \textit{turn around } u_{\sigma(j)} \textit{ once clockwise}, \\ &\quad \textit{go along } \gamma_j^{-1} \textit{ back to } r, \\ &\quad \textit{go from } r \textit{ along } \gamma_{j+1} \textit{ to } u_{\sigma(j+1)}, \\ \sigma' &:= \sigma \circ (j\ j+1). \end{aligned}$$

σ_j^{-1} maps the homotopy class of a distinguished system of paths $(\gamma_1, ..., \gamma_n)$ to the homotopy class of a distinguished system of paths $(\gamma_1'', ..., \gamma_n'')$ where

$$\begin{aligned} \gamma_i'' &:= \gamma_i \text{ for } i \in \{1, ..., n\} - \{j, j+1\}, \\ \gamma_j'' &:= \gamma_{j+1}, \\ \gamma_{j+1}'' &: \textit{go from } r \textit{ along } \gamma_{j+1} \textit{ almost to } u_{\sigma(j+1)}, \\ &\quad \textit{turn around } u_{\sigma(j+1)} \textit{ once counter clockwise}, \end{aligned}$$

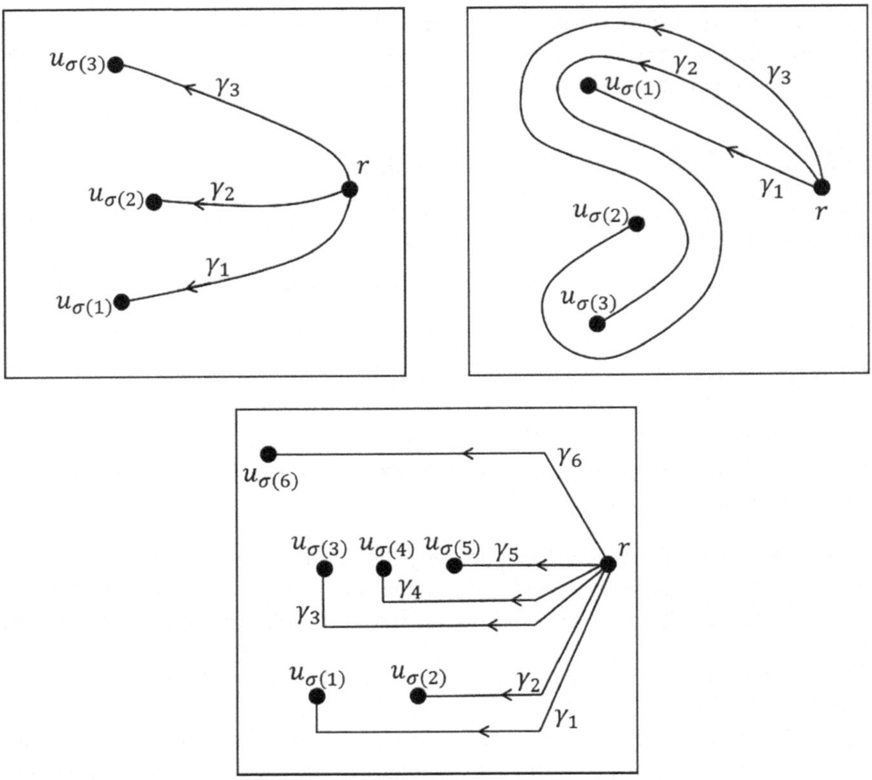

Fig. 8.3 Three distinguished systems of paths

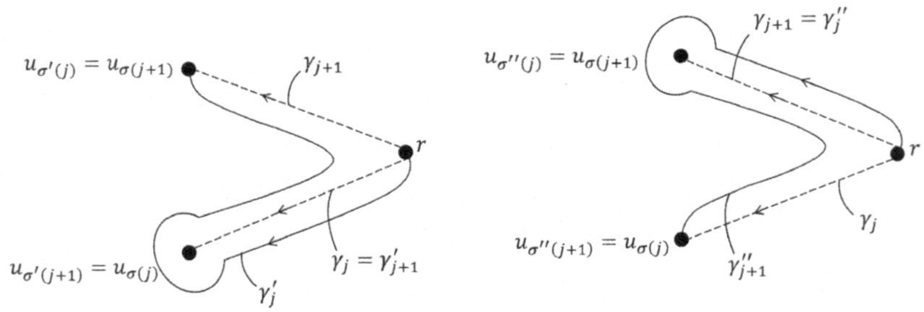

Fig. 8.4 Actions of σ_j and σ_j^{-1} on distinguished systems of paths

$$
\begin{aligned}
&\textit{go along } \gamma_{j+1}^{-1} \textit{ back to } r, \\
&\textit{go from } r \textit{ along } \gamma_j \textit{ to } u_{\sigma(j)}, \\
\sigma'' &:= \sigma \circ (j\ j+1).
\end{aligned}
$$

Fig. 8.4 shows the old and new paths under both operations.

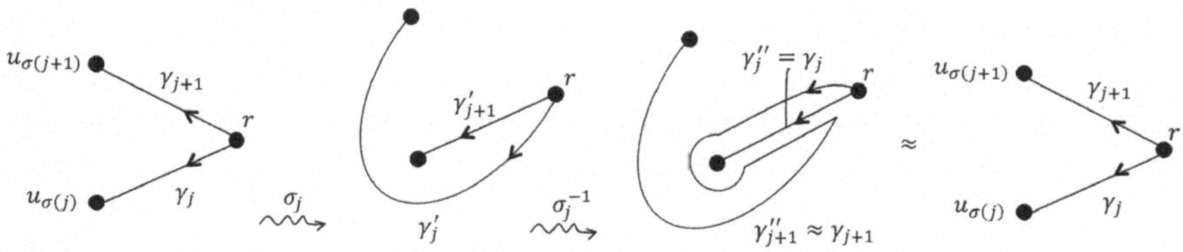

Fig. 8.5 The action of $\sigma_j^{-1}\sigma_j$ on distinguished systems of paths is trivial

(b) The actions of σ_j and σ_j^{-1} extend to an action of the braid group Br_n *from the left on the set $\mathcal{P}(\underline{u}, r)$ of homotopy classes of distinguished systems of $(\underline{u}, r)$-paths.*
(c) The action of Br_n *on the set $\mathcal{P}(\underline{u}, r)$ is transitive.*

The sequence of pictures in Fig. 8.5 illustrates that the action of $\sigma_j^{-1}\sigma_j$ on $\mathcal{P}(\underline{u}, r)$ is trivial. The pictures 5.27 and 5.28 in [Eb01] illustrate that the braids $\sigma_j\sigma_{j+1}\sigma_j$ and $\sigma_{j+1}\sigma_j\sigma_{j+1}$ act in the same way on $\mathcal{P}(\underline{u}, r)$.

In the case of $\underline{u} = b_n^{pure} = (i, 2i, ..., ni)$ and $r = 1$ ($r = 1$ is here an example, any $r > 0$ works here) there is also an action of Br_n from the right on the set $\mathcal{P}(b_n^{pure}, 1)$. It comes from an action of the fundamental groupoid of C_n^{pure} on the union $\bigcup_{(\underline{u},r)} \mathcal{P}(\underline{u}, r)$. This action and the compatibility with the action of Br_n from the left on $\mathcal{P}(b_n^{conf}, 1)$ in Theorem 8.14 are not discussed in [Eb01]. They are given in the points 8.15 to 8.19.

Remarks 8.15 [Sp66, 1.7] For $\underline{u}^0$ and $\underline{u}^1 \in C_n^{pure}$ denote by $H(\underline{u}^0, \underline{u}^1)$ the set of homotopy classes of (continuous maps =) paths $\alpha : [0, 1] \to C_n^{pure}$ with $\alpha(0) = \underline{u}^0$ and $\alpha(1) = \underline{u}^1$. The tuple $(H(\underline{u}^0, \underline{u}^1))_{\underline{u}^0,\underline{u}^1 \in C_n^{pure}}$ is a *groupoid* with $[\alpha_0][\alpha_1] = [\alpha_0\alpha_1] \in H(\underline{u}^0, \underline{u}^2)$ for $[\alpha_0] \in H(\underline{u}^0, \underline{u}^1)$, $[\alpha_1] \in H(\underline{u}^1, \underline{u}^2)$. The multiplication is associative. $[\alpha][\alpha^{-1}] = [\text{trivial path}] \in H(\underline{u}^0, \underline{u}^0)$ for $[\alpha] \in H(\underline{u}^0, \underline{u}^1)$. See [Sp66, 1.7] for more details on this groupoid.

Definition/Lemma 8.16 *(a) Let $\alpha : [0, 1] \to C_n^{pure}$ be a path with $[\alpha] \in H(\underline{u}^0, \underline{u}^1)$. We will construct a natural bijection $\Phi(\alpha) : \mathcal{P}(\underline{u}^0, r_0) \to \mathcal{P}(\underline{u}^1, r_1)$ (in the notation $\Phi(\alpha)$ we should include r_0 and r_1, but for simplicity we drop them).*

Choose $r \in \mathbb{R}$ with

$$r \geq \max(r_0, r_1), \quad r > \max_{i\{1,...,n\}} \max_{t\in[0,1]} \mathrm{Re}\, \alpha_i(t).$$

Because there are natural bijections $\mathcal{P}(\underline{u}^0, r_0) \to \mathcal{P}(\underline{u}^0, r)$ and $\mathcal{P}(\underline{u}^1, r_1) \to \mathcal{P}(\underline{u}^1, r)$, we can replace r_0 and r_1 by r. Choose a continuous family in $s \in [0, 1]$ of homeomorphisms $\Psi_s : \mathbb{C} \to \mathbb{C}$ with the properties

$$\Psi_s|_{\{z\in\mathbb{C} \mid \mathrm{Re}\, z\geq r\}} = \mathrm{id},$$
$$\Psi_s(\{u_1^0, ..., u_n^0\}) = \{\alpha_1(s), ..., \alpha_n(s)\}.$$

For a distinguished system of paths $(\underline{\gamma};\sigma) = (\gamma_1, ..., \gamma_n; \sigma)$ *with* $[(\underline{\gamma};\sigma)] \in \mathcal{P}(\underline{u}^0, r)$*, the composition* $(\Psi_s \circ \underline{\gamma}; \sigma) = (\Psi_s \circ \gamma_1, ..., \Psi_s \circ \gamma_n; \sigma)$ *is a distinguished system of paths with* $[(\Psi_s \circ \underline{\gamma}; \sigma)] \in \mathcal{P}(\alpha(s), r)$*. The class* $[(\Psi_s \circ \underline{\gamma}; \sigma)]$ *depends only on the class* $[(\underline{\gamma};\sigma)]$ *and on the path* α*. Define*

$$\Phi(\alpha)([(\underline{\gamma};\sigma)]) := [(\Psi_1 \circ \underline{\gamma}; \sigma)] \in \mathcal{P}(\underline{u}^1, r).$$

(b) The map $\Phi(\alpha) : \mathcal{P}(\underline{u}^0, r) \to \mathcal{P}(\underline{u}^1, r)$ *depends only on the homotopy class* $[\alpha] \in H(\underline{u}^0, \underline{u}^1)$*.*

(c) If $[\alpha_1] \in H(\underline{u}^0, \underline{u}^1)$ *and* $[\alpha_2] \in H(\underline{u}^1, \underline{u}^2)$ *then* $[\alpha_1\alpha_2] \in H(\underline{u}^0, \underline{u}^2)$ *and*

$$\Phi(\alpha_1\alpha_2) = \Phi(\alpha_2)\Phi(\alpha_1).$$

The groupoid $(H(\underline{u}^0, \underline{u}^1))_{\underline{u}^0, \underline{u}^1 \in C_n^{pure}}$ *acts from the right on the tuple* $(\mathcal{P}(\underline{u}, r))_{\underline{u} \in C_n^{pure}, r > \max(\mathrm{Re}\, u_1, ..., \mathrm{Re}\, u_n)}$*.*

Remarks 8.17 (i) Let $\underline{u} \in C_n^{pure}$ and $r > \max(\mathrm{Re}\, u_1, ..., \mathrm{Re}\, u_n)$. Let $\tau \in S_n$. Then τ as deck transformation of the covering $\mathrm{pr}_n^{p,c} : C_n^{pure} \to C_n^{conf}$ maps $\underline{u} = (u_1, ..., u_n)$ to $\tau(\underline{u}) = (u_{\tau^{-1}(1)}, ..., u_{\tau^{-1}(n)})$, see the Remarks 8.2 (i) and (vii). A distinguished basis $(\underline{\gamma};\sigma)$ for $\underline{u}$ and r is almost the same as the distinguished basis $(\underline{\gamma}, \tau\sigma)$ for $\tau(\underline{u})$ and r. They just differ by the indexing of the entries of $\underline{u}$. This gives a natural bijection $\mathcal{P}(\underline{u}, r) \to \mathcal{P}(\tau(\underline{u}), r)$, $[(\underline{\gamma};\sigma)] \mapsto [(\underline{\gamma};\tau\sigma)]$.

(ii) A closed path β in C_n^{conf} with starting point and end point b_n^{conf} represents a braid $[\beta] \in \mathrm{Br}_n$. It lifts to a path $\widetilde{\beta}$ in C_n^{pure} from $b_n^{pure} = (i, 2i, ..., ni)$ to $\sigma(b_n^{pure}) = (\sigma^{-1}(1)i, \sigma^{-1}(2)i, ..., \sigma^{-1}(n)i)$ for some $\sigma \in S_n$ with homotopy class $[\widetilde{\beta}] \in H(b_n^{pure}, \sigma(b_n^{pure}))$. Now $\Phi(\widetilde{\beta})$ maps $\mathcal{P}(b_n^{pure}, 1)$ to $\mathcal{P}(\sigma(b_n^{pure}), 1)$ which can be identified with $\mathcal{P}(b_n^{pure}, 1)$ because of (i). Therefore $[\beta]$ acts on $\mathcal{P}(b_n^{pure}, 1)$. We obtain an action of Br_n on $\mathcal{P}(b_n^{pure}, 1)$, which is an action from the right because of Lemma 8.16(c).

Remark 8.17 (ii) gives an intuitive geometric action of Br_n on $\mathcal{P}(b_n^{pure}, 1)$ from the right. It comes from moving the endpoints of a distinguished system of paths along a braid and deforming the distinguished system of paths accordingly. Lemma 8.18 connects it to the more algebraic action of Br_n on $\mathcal{P}(b_n^{pure}, 1)$ from the left in Theorem 8.14 (a)+(b). Lemma 8.18 shows first that these actions commute and second that one action can be expressed through the other action.

Lemma 8.18 *The action of* Br_n *on* $\mathcal{P}(b_n^{pure}, 1)$ *from the left in Theorem 8.14 and the action of* Br_n *on* $\mathcal{P}(b_n^{pure}, 1)$ *from the right in Remark 8.17 (ii) commute,*

$$\beta_1([(\underline{\gamma};\sigma)].\beta_2) = (\beta_1[(\underline{\gamma};\sigma)]).\beta_2 \quad \text{for } \beta_1, \beta_2 \in \mathrm{Br}_n, [(\underline{\gamma};\sigma)] \in \mathcal{P}(b_n^{pure}, 1),$$

and satisfy

$$\beta[(\underline{\gamma};\sigma)] = [(\underline{\gamma};\sigma)].\beta \quad \text{for } \beta \in \mathrm{Br}_n, [(\underline{\gamma};\sigma)] \in \mathcal{P}(b_n^{pure}, 1).$$

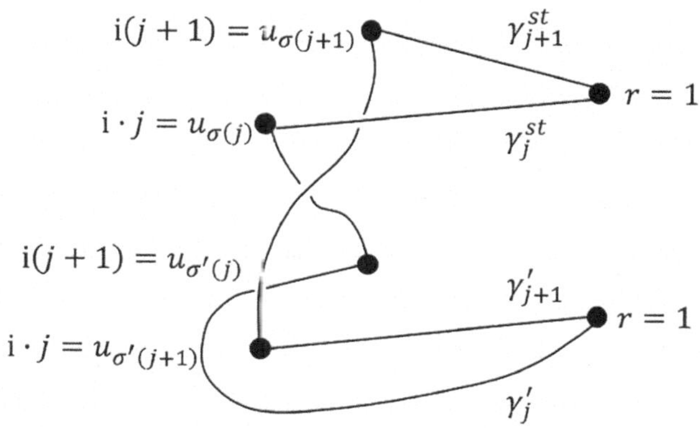

Fig. 8.6 Moving along a braid in C_n^{pure} changes a distinguished system of paths

Proof The action on the right comes from a continuous family of homeomorphisms as in Lemma 8.16 (a) which deforms the distinguished system of paths. The action on the left comes from a construction of new paths which follow closely the old paths. The new paths are deformed together with the old paths. A deformation of the old and new paths does not change this construction of new paths from old paths. Therefore the actions commute.

The action of Br_n from the left on $\mathcal{P}(b_n^{pure}, 1)$ is transitive by Theorem 8.14 (c). Therefore it is sufficient to show for an elementary braid σ_j and a standard system of paths $(\underline{\gamma}^{st}; \sigma)$ with $[(\underline{\gamma}^{st}; \sigma)] \in \mathcal{P}(b_n^{conf}, 1)$

$$\sigma_j[(\underline{\gamma}^{st}; \sigma)] = [(\underline{\gamma}^{st}; \sigma)].\sigma_j.$$

This is obvious respectively shown in the picture in Fig. 8.6. □

Example 8.19 Fig. 8.7 shows for $n = 3$ on the left the action of $\sigma_2^{-1}\sigma_1$ on $[(\underline{\gamma}^{st}; \sigma)]$ from the left and on the right the action of $\sigma_2^{-1}\sigma_1$ on $[(\underline{\gamma}^{st}; \sigma)]$ from the right.

8.5 Two families of $\mathbb{Z}$-lattice structures over the manifold $C_n^{\underline{e}/\{\pm1\}^n}$

Let $(H_{\mathbb{Z}}, L, \underline{e})$ be a unimodular bilinear lattice of rank $n \in \mathbb{N}$ with a triangular basis $\underline{e}$ and matrix $S = L(\underline{e}^t, \underline{e})^t \in T_n(\mathbb{Z})$. The manifold $C_n^{\underline{e}/\{\pm1\}^n}$ is not just a semisimple F-manifold with Euler field. It is the carrier of much richer structures, which we call *$\mathbb{Z}$-lattice structures* and which we explain in this section.

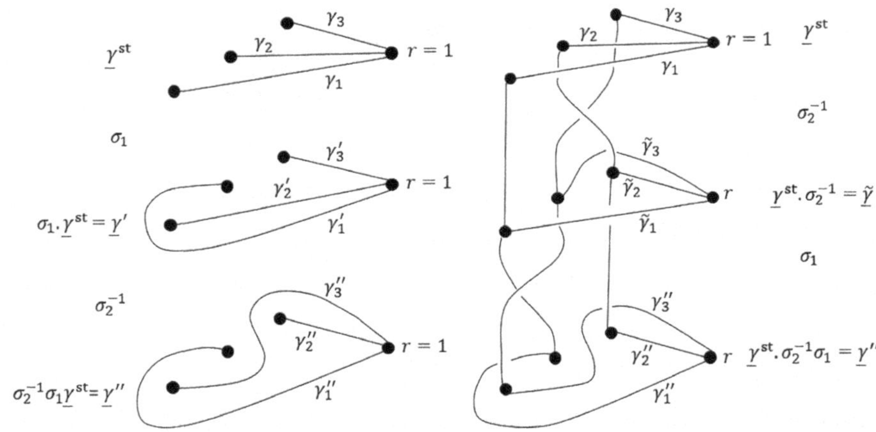

Fig. 8.7 Actions of $\sigma_2^{-1}\sigma_1$ from the left and from the right: same result

Remarks 8.20 A $\mathbb{Z}$-lattice structure leads often in several steps to a strong enrichment of the F-manifold structure, namely a Dubrovin-Frobenius manifold. We will not go through these steps and not treat Dubrovin-Frobenius manifolds here, but we will say a few words in this remark. Details on the steps are given in [Sa02, He02, He03].

The $\mathbb{Z}$-lattice structure can be extended to a holomorphic vector bundle on $\mathbb{C} \times C_n^{\underline{e}/\{\pm 1\}^n}$ with a meromorphic connection with a logarithmic pole along a certain discriminant which will be defined below and a $\mathbb{Z}$-lattice bundle on the complement. A Fourier-Laplace transformation of this bundle along the first factor $\mathbb{C}$ of the base space leads to another holomorphic vector bundle on $\mathbb{C} \times C_n^{\underline{e}/\{\pm 1\}^n}$ with a meromorphic connection with a pole of Poincaré rank 1 along the hypersurface $\{0\} \times C_n^{\underline{e}/\{\pm 1\}^n}$, a $\mathbb{Z}$-lattice bundle on the complement and a certain pairing. In the notation of [He03] it is a *TEZP-structure* (Twistor Extension $\mathbb{Z}$-lattice Pairing). From this and from some additional choice one can usually construct a Dubrovin-Frobenius manifold structure on $C_n^{\underline{e}/\{\pm 1\}^n}$ or at least on the complement of a hypersurface in it. See [Sa02, He02, He03] for this construction. Dubrovin-Frobenius manifolds were first defined by Dubrovin [Du92, Du96]. A definition is given above in Remark 8.9 (viii). Here we will restrict to explain two natural families of $\mathbb{Z}$-lattice structures over $C_n^{\underline{e}/\{\pm 1\}^n}$.

Definition/Lemma 8.21 (c) presents the notion of a single $\mathbb{Z}$-lattice structure. The construction of natural families of $\mathbb{Z}$-lattice structures over C_n^{univ} and $C_n^{\underline{e}/\{\pm 1\}^n}$ comes in Definition 8.22 (c) and Theorem 8.23 (d). The unimodular lattice $(H_{\mathbb{Z}}, L, \underline{e})$ with distinguished basis $\underline{e}$ induces such families. The construction uses distinguished systems of paths. Theorem 8.23 (a)–(c) discusses how $\mathbb{Z}$-lattice structures over different points in C_n^{univ} in such a family of $\mathbb{Z}$-lattice structures are related by the actions of braids.

Definition/Lemma 8.21 *Let $(H_\mathbb{Z}, L, \underline{e})$ and S be as above.*

(a) (Definition) The discriminant $D_{1,n}^{univ} \subset \mathbb{C} \times C_n^{univ}$ *is the smooth complex hypersurface*

$$\{(\tau, b) \in \mathbb{C} \times C_n^{univ} \mid \mathcal{L}(b)(\tau) = 0\}.$$

Recall the map $\mathcal{L} : C_n^{univ} \to \mathbb{C}[x]_n^{reg}$ in Remark 8.9 (iv). The polynomial $\mathcal{L}(b) \in \mathbb{C}[x]_n^{reg} \cong C_n^{conf}$ corresponds to $\mathrm{pr}_n^{u,c}(b) \in C_n^{conf}$.

(Trivial lemma) The projection $\mathrm{pr}_2 : \mathbb{C} \times C_n^{univ} \to C_n^{univ}$, $(\tau, b) \mapsto b$, restricts to a covering $D_{1,n}^{univ} \to C_n^{univ}$ of degree n.

(b) (Trivial lemma) The group Br_n of deck transformations of the universal covering $\mathrm{pr}_n^{u,c} : C_n^{univ} \to C_n^{conf}$ extends with $\mathrm{id}_\mathbb{C}$ on the first factor of $\mathbb{C} \times C_n^{univ}$ to a group of automorphisms of $\mathbb{C} \times C_n^{univ}$. It leaves the discriminant $D_{1,n}^{univ}$ invariant.

(Definition) The image of $D_{1,n}^{univ}$ under the group $\mathrm{id}_\mathbb{C} \times (\mathrm{Br}_n)_{\underline{e}/\{\pm 1\}^n}$ is the discriminant $D_{1,n}^{\underline{e}/\{\pm 1\}^n} \subset \mathbb{C} \times C_n^{\underline{e}/\{\pm 1\}^n}$,

$$D_{1,n}^{\underline{e}/\{\pm 1\}^n} = \{(\tau, b) \in \mathbb{C} \times C_n^{\underline{e}/\{\pm 1\}^n} \mid \mathcal{L}(b)(\tau) = 0\},$$

where now $\mathcal{L}$ is the Lyashko-Looijenga map $\mathcal{L} : C_n^{\underline{e}/\{\pm 1\}^n} \to \mathbb{C}[x]_n^{reg}$.

(c) (Definition) A $\mathbb{Z}$-lattice structure means an (automatically flat) $\mathbb{Z}$-lattice bundle over $\mathbb{C} - \{n$ points$\}$.

(Remark) Therefore a $\mathbb{Z}$-lattice bundle over $\mathbb{C} \times C_n^{univ} - D_{1,n}^{univ}$ is considered as a (flat *or* isomonodromic*) family of $\mathbb{Z}$-lattice structures over C_n^{univ}. A $\mathbb{Z}$-lattice bundle over $\mathbb{C} \times C_n^{\underline{e}/\{\pm 1\}^n} - D_{1,n}^{\underline{e}/\{\pm 1\}^n}$ is considered as a* (flat *or* isomonodromic*) family of $\mathbb{Z}$-lattice structures over $C_n^{\underline{e}/\{\pm 1\}^n}$.*

(Remark) The $\mathbb{Z}$-lattice structures which we will construct will come equipped with an (automatically flat) even or odd intersection form on each fiber.

Recall that a unimodular bilinear lattice $(H_\mathbb{Z}, L)$ with triangular basis $\underline{e}$ and matrix $S = L(\underline{e}^t, \underline{e})^t \in T_n(\mathbb{Z})$ was fixed at the beginning of this section. Together with the choice of a pair $(\underline{u}, r) \in C_n^{pure} \times \mathbb{R}$ with $r > \max(\mathrm{Re}\, u_1, ..., \mathrm{Re}\, u_n)$ and the choice of a homotopy class $[(\underline{\gamma}; \sigma)] \in \mathcal{P}(\underline{u}, r)$ of a distinguished system of paths $(\underline{\gamma}; \sigma)$, it induces a $\mathbb{Z}$-lattice structure on $\mathbb{C} - \{u_1, ..., u_n)$. It also induces (without any additional choice) a natural $\mathbb{Z}$-lattice bundle over C_n^{univ}. The two constructions are given in Definition 8.22(b) and (c).

Definition 8.22 (a) Let $\underline{u} \in C_n^{pure}$ and $r \in \mathbb{R}$ with $r > \max(\mathrm{Re}\, u_1, ..., \mathrm{Re}\, u_n)$. Let $\gamma : [0, 1] \to \mathbb{C}$ be a $(\underline{u}, r)$-path (Definition 8.13 (a)). An *associated loop* $\delta : [0, 1] \to \mathbb{C}$ is a closed path which starts and ends at r, which first goes along γ almost to $\gamma(1)$, which then moves on a small circle once counterclockwise around $\gamma(1)$ and which finally goes along γ^{-1} back to r.

Figure 8.8 illustrates in a picture three loops which are associated to three paths.

(b) Let $\underline{u} \in C_n^{pure}$ and $r \in \mathbb{R}$ with $r > \max(\mathrm{Re}\, u_1, ..., \mathrm{Re}\, u_n)$. Let $(\underline{\gamma}; \sigma)$ be a distingushed system of paths with $[(\underline{\gamma}; \sigma)] \in \mathcal{P}(\underline{u}, r)$. Let $\underline{f} \in \mathcal{B}^{dist}$ be a distinguished

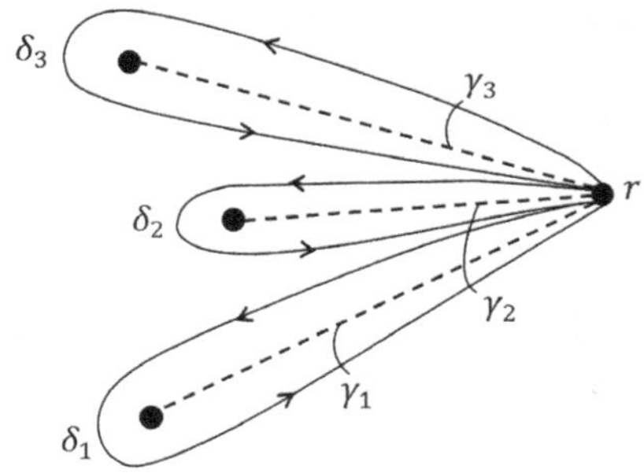

Fig. 8.8 Three loops associated to three paths

basis of $H_{\mathbb{Z}}$. Let $k \in \{0, 1\}$. Then $V_{\mathbb{Z}}^{(k)}((\underline{u}, r), [(\underline{\gamma}; \sigma)], \underline{f})$ is the $\mathbb{Z}$-lattice bundle on $\mathbb{C} - \{u_1, ..., u_n\}$ which is determined by the following conditions (i) and (ii):

(i) The restriction of the bundle to $\{z \in \mathbb{C} \mid \operatorname{Re} z \geq r\}$ is trivial, and each fiber over a point in this set is identified with $H_{\mathbb{Z}}$.
(ii) Let $\delta_1, ..., \delta_n$ be loops which are associated to $\gamma_1, ..., \gamma_n$. The monodromy along the loop δ_j is with the identification of the fiber $V_{\mathbb{Z}}^{(k)}((\underline{u}, r), [(\underline{\gamma}; \sigma)], \underline{f})_r$ with $H_{\mathbb{Z}}$ the automorphism $s_{f_j}^{(k)} : H_{\mathbb{Z}} \to H_{\mathbb{Z}}$.

It is well known that (i) and (ii) determine uniquely a $\mathbb{Z}$-lattice bundle on $\mathbb{C} - \{u_1, ..., u_n\}$. This bundle $V_{\mathbb{Z}}^{(k)}((\underline{u}, r), [(\underline{\gamma}; \sigma)], \underline{f})$ is a $\mathbb{Z}$-lattice structure in the sense of Definition 8.21 (c).

(c) Let $k \in \{0, 1\}$. $V_{\mathbb{Z}}^{(k),univ}$ is the $\mathbb{Z}$-lattice bundle on $\mathbb{C} \times C_n^{univ} - D_{1,n}^{univ}$ which is determined by the following conditions (i) and (ii):

(i) The restriction of the bundle to $\{(z, b) \in \mathbb{C} \times C_n^{univ} \mid \operatorname{Re} z > \max(\operatorname{Re} u \mid (u, b) \in D_{1,n}^{univ})\}$ is trivial, and each fiber over a point in this set is identified with $H_{\mathbb{Z}}$.
(ii) The restriction of $V_{\mathbb{Z}}^{(k),univ}$ to $(\mathbb{C} - \{i, 2i, ..., ni\}) \times \{b_n^{univ}\}$ is equal to the $\mathbb{Z}$-lattice bundle $V_{\mathbb{Z}}^{(k)}((b_n^{pure}, 1), [(\underline{\gamma}^{st}; \mathrm{id})], \underline{e})$. Here $(\underline{\gamma}^{st}, \mathrm{id})$ is a standard system of paths which start at $r = 1$ and end at the entries of $b_n^{pure} = (i, 2i, ..., ni)$.

Because $\mathrm{pr}_2 : \mathbb{C} \times C_n^{univ}$ restricts to a trivial n-sheeted covering $\mathrm{pr}_2 : D_{1,n}^{univ} \to C_n^{univ}$ over the simply connected manifold C_n^{univ}, (i) and (ii) determine uniquely a $\mathbb{Z}$-lattice bundle on $\mathbb{C} \times C_n^{univ} - D_{1,n}^{univ}$. The bundle $V_{\mathbb{Z}}^{(k),univ}$ is a flat family of $\mathbb{Z}$-lattice structures over the manifold C_n^{univ} in the sense of Definition 8.21 (c).

The next theorem is central in this section. It will construct in part (d) a $\mathbb{Z}$-lattice bundle over $\mathbb{C} \times C_n^{\underline{e}/\{\pm 1\}^n} - D_{1,n}^{\underline{e}/\{\pm 1\}^n}$. This is a flat family of $\mathbb{Z}$-lattice structures over $C_n^{\underline{e}/\{\pm 1\}^n}$ in the sense of Definition 8.21 (c). The parts (a), (b) and (c) prepare this.

Part (a) starts with a $\mathbb{Z}$-lattice bundle $V_{\mathbb{Z}}^{(k)}((\underline{u}, r), [(\underline{\gamma}; \sigma)], \underline{f})$ on $\mathbb{C} - \{u_1, ..., u_n\}$ as in Definition 8.22 (b). It answers the question how one has to change $\underline{f}$ in order to obtain the same $\mathbb{Z}$-lattice bundle if one changes the distinguished system of paths from $[(\underline{\gamma}; \sigma)]$ to $\beta[(\underline{\gamma}; \sigma)]$ with some braid $\beta \in \mathrm{Br}_n$.

Part (b) starts with the $\mathbb{Z}$-lattice bundle $V_{\mathbb{Z}}^{(k),univ}$ on $\mathbb{C}\times C_n^{univ}-D_{1,n}^{univ}$ as in Definition 8.22 (c). It considers for a point $\widetilde{\underline{u}}\in C_n^{univ}$ and its image $\underline{u}^1=\mathrm{pr}^{u,p}(\widetilde{\underline{u}})\in C_n^{pure}$ the restriction of $V_{\mathbb{Z}}^{(k),univ}$ to a $\mathbb{Z}$-lattice bundle on $(\mathbb{C}-\{u_1^1,...,u_n^1\})\times\{\widetilde{\underline{u}}\}$. It studies how this restriction is obtained from the restriction to $(\mathbb{C}-\{i,2i,...,ni\})\times\{b_n^{univ}\}$ if one moves from b_n^{univ} to $\widetilde{\underline{u}}$. The groupoid action in Lemma 8.18 is used.

Part (c) specializes the situation in part (b) to the case where $\mathrm{pr}^{u,c}(\widetilde{\underline{u}})=b_n^{conf}$. Then a path in C_n^{univ} from b_n^{univ} to $\widetilde{\underline{u}}$ corresponds to a braid $\beta\in\mathrm{Br}_n$. Part (c) shows with the help of the parts (a) and (b) and Lemma 8.18 that then the restriction of $V_{\mathbb{Z}}^{(k),univ}$ to $(\mathbb{C}-\{i,2i,...,ni\})\times\{\widetilde{\underline{u}}\}$ is simply $V_{\mathbb{Z}}^{(k)}((b_n^{pure},1),[(\underline{\gamma}^{st};\mathrm{id})],\beta^{-1}(\underline{e}))$.

Theorem 8.23 *(a) Let $(\underline{\gamma};\sigma)$ be a distinguished system of paths with $[(\underline{\gamma};\sigma)]\in\mathcal{P}(b_n^{pure},1)$. Let $\underline{f}$ be a distinguished basis of $H_{\mathbb{Z}}$. Let $k\in\{0,1\}$. Let $\beta\in\mathrm{Br}_n$. Then*

$$V_{\mathbb{Z}}^{(k)}((b_n^{pure},1),[(\underline{\gamma};\sigma)],\underline{f})=V_{\mathbb{Z}}^{(k)}((b_n^{pure},1),\beta[(\underline{\gamma};\sigma)],\beta(\underline{f})).$$

(b) Let $\alpha:[0,1]\to C_n^{pure}$ be a path with $[\alpha]\in H(b_n^{pure},\underline{u}^1)$ for some $\underline{u}^1\in C_n^{pure}$. Let $\widetilde{\alpha}:[0,1]\to C_n^{univ}$ be the lift to C_n^{univ} of α with starting point $\widetilde{\alpha}(0)=b_n^{univ}$. Let $k\in\{0,1\}$. The restriction of $V_{\mathbb{Z}}^{(k),univ}$ to $(\mathbb{C}-\{u_1^1,...,u_n^1\})\times\{\widetilde{\alpha}(1)\}$ is equal to $V_{\mathbb{Z}}^{(k)}((\underline{u}^1,r),\Phi(\alpha)[(\underline{\gamma}^{st};\mathrm{id})],\underline{e})$ for some $r\in\mathbb{R}_{\geq 1}$ with $r>\max(\mathrm{Re}\,u_1^1,...,\mathrm{Re}\,u_n^1)$.

(c) Let $\beta\in\mathrm{Br}_n$. Let $\widetilde{\beta}:[0,1]\to C_n^{univ}$ be the lift to C_n^{univ} with $\widetilde{\beta}(0)=b_n^{univ}$ of a loop in C_n^{conf} which represents β. Let $k\in\{0,1\}$. Then the restriction of $V_{\mathbb{Z}}^{(k),univ}$ to $(\mathbb{C}-\{i,2i,...,ni\})\times\{\widetilde{\beta}(1)\}$ is equal to $V_{\mathbb{Z}}^{(k)}((b_n^{pure},1),[(\underline{\gamma}^{st},\mathrm{id})],\beta^{-1}(\underline{e}))$.

(d) For $\beta\in(\mathrm{Br}_n)_{\underline{e}/\{\pm 1\}^n}$ denote the deck transformation of C_n^{univ} which it induces by $\beta^{deck}:C_n^{univ}\to C_n^{univ}$. It extends to an automorphism $\beta^{deck,V,(k)}$ of $V_{\mathbb{Z}}^{(k),univ}$ which is $(\mathrm{id}_{H_{\mathbb{Z}}},\mathrm{id}_{\mathbb{C}}\times\beta^{deck})$ over $\{(z,b)\in\mathbb{C}\times C_n^{univ}\,|\,\mathrm{Re}\,z>\max(\mathrm{Re}\,u\,|\,(u,b)\in D_{1,n}^{univ})\}$. Therefore the quotient $V_{\mathbb{Z}}^{(k),univ}/(\mathrm{Br}_n)_{\underline{e}/\{\pm 1\}^n}^{deck,V,(k)}$ is a $\mathbb{Z}$-lattice bundle over $\mathbb{C}\times C_n^{\underline{e}/\{\pm 1\}^n}-D_{1,n}^{\underline{e}/\{\pm 1\}^n}$ whose restriction to

$$(\{(z,b)\in\mathbb{C}\times C_n^{\underline{e}/\{\pm 1\}^n}\,|\,\mathrm{Re}\,z>\max(\mathrm{Re}\,u\,|\,(u,b)\in D_{1,n}^{\underline{e}/\{\pm 1\}^n})\}$$

is trivial. We call this bundle $V_{\mathbb{Z}}^{(k),\underline{e}/\{\pm 1\}^n}$. It is a family of $\mathbb{Z}$-lattice structures over $C_n^{\underline{e}/\{\pm 1\}^n}$.

Proof (a) Let $\delta_1,...,\delta_n$ be loops associated to $\gamma_1,...,\gamma_n$. By definition, in the bundle $V_{\mathbb{Z}}^{(k)}((b_n^{pure}),[(\underline{\gamma};\sigma)],\underline{f})$ the local monodromy along the loop δ_j is given by $s_{f_j}^{(k)}$. Write $[(\underline{\gamma}';\sigma')]:=\beta[(\underline{\gamma};\sigma)]$ and $\underline{f}':=\beta(\underline{f})$.

We have to show that in the bundle $V_{\mathbb{Z}}^{(k)}((b_n^{pure}),[(\underline{\gamma};\sigma)],\underline{f})$ the local monodromy around the loop δ_j' is given by $s_{f_j'}^{(k)}$.

First we consider $\beta=\sigma_j$. Then

$$\gamma_i'=\gamma_i\text{ for }i\in\{1,...,\}-\{j,j+1\},$$

$$\begin{aligned}
&\gamma'_{j+1} = \gamma_j, \\
&\gamma'_j \text{ is homotopic to } \delta_j^{-1}\gamma_{j+1}, \\
\text{so } &\delta'_i = \delta_i \text{ for } i \in \{1, ..., n\} - \{j, j+1\}, \\
&\delta'_{j+1} = \delta_j, \\
&\delta'_j \text{ is homotopic to } \delta_j^{-1}\delta_{j+1}\delta_j.
\end{aligned}$$

Also

$$\begin{aligned}
&f'_i = f_i \text{ for } i \in \{1, ..., n\} - \{j, j+1\}, \\
&f'_{j+1} = f_j, \\
&f'_j = s^{(k)}_{f_j}(f_{j+1}).
\end{aligned}$$

Because of

$$s^{(k)}_{f'_j} = s^{(k)}_{s^{(k)}_{f_j}(f_{j+1})} = s^{(k)}_{f_j} s^{(k)}_{f_{j+1}} (s^{(k)}_{f_j})^{-1}$$

the local monodromy along δ'_i is indeed given by $s^{(k)}_{f'_i}$ for $i \in \{1, ..., n\}$ (recall that a composition of paths is read from the left, a composition of automorphisms is read from the right). The case $\beta = \sigma_j^{-1}$ is treated analogously. The general case $\beta \in \mathrm{Br}_n$ follows.

(b) In the restriction of $V_{\mathbb{Z}}^{(k),univ}$ to $(\mathbb{C} - \{i, 2i, ..., ni\}) \times \{b_n^{univ}\}$ the local monodromies along loops associated to a standard system of paths $(\underline{\gamma}^{st}; \mathrm{id})$ are $s^{(k)}_{e_1}, ..., s^{(k)}_{e_n}$.

Moving in the base space C_n^{univ} from b_n^{univ} to $\widetilde{\alpha}(1)$, the standard system of paths is deformed to a representative of $\Phi(\alpha)[(\underline{\gamma}^{st}; \mathrm{id})]$, associated loops are deformed accordingly, and the local monodromies along the deformed associated loops are still $s^{(k)}_{e_1}, ..., s^{(k)}_{e_n}$. This shows the claim.

(c) One has to apply part (b), Remark 8.17 (ii), Lemma 8.18 and part (a), as follows. Let $\overline{\beta} : [0, 1] \to C_n^{pure}$ be the lift of a representative of β with $\overline{\beta}(0) = b_n^{pure}$. Then $\overline{\beta}(1) = \sigma(b_n^{pure})$ for some $\sigma \in S_n$. We have the following equalities

$$\begin{aligned}
&\text{the restriction of } V_{\mathbb{Z}}^{(k),univ} \text{ to } (\mathbb{C} - \{i, 2i, ..., ni\}) \times \{\widetilde{\beta}(1)\} \\
= &V_{\mathbb{Z}}^{(k)}((\sigma(b_n^{pure}), 1), \Phi(\overline{\beta})[(\underline{\gamma}^{st}; \mathrm{id})], \underline{e}) \quad \text{(by part (b))} \\
= &V_{\mathbb{Z}}^{(k)}((b_n^{pure}, 1), [(\underline{\gamma}^{st}; \mathrm{id})].\beta, \underline{e}) \quad \text{(by Remark8.17 (ii))} \\
= &V_{\mathbb{Z}}^{(k)}((b_n^{pure}, 1), \beta[(\underline{\gamma}^{st}; \mathrm{id})], \underline{e}) \quad \text{(by Lemma8.18)} \\
= &V_{\mathbb{Z}}^{(k)}((b_n^{pure}, 1), [(\underline{\gamma}^{st}; \mathrm{id})], \beta^{-1}(\underline{e})) \quad \text{(by part (a))}.
\end{aligned}$$

(d) For $\beta \in (\mathrm{Br}_n)_{\underline{e}/\{\pm 1\}^n}$ $\beta^{-1}(\underline{e})$ coincides with $\underline{e}$ up to signs, $\beta^{-1}(\underline{e}/\{\pm 1\}^n) = \underline{e}/\{\pm 1\}^n$. Because of $s^{(k)}_{-e_i} = s^{(k)}_{e_i}$ and part (c), the restriction of $V_{\mathbb{Z}}^{(k),univ}$ to $(\mathbb{C} -$

$\{i, 2i, ..., ni\}) \times \{\widetilde{\beta}(1)\}$ is canonically isomorphicm to the restriction of $V_{\mathbb{Z}}^{(k),univ}$ to $(\mathbb{C} - \{i, 2i, ..., ni\}) \times \{b^{univ}\}$. Therefore β^{deck} extends to an automorphism of $V_{\mathbb{Z}}^{(k),univ}$ as claimed. The quotient bundle $V_{\mathbb{Z}}^{(k),\underline{e}/\{\pm 1\}^n}$ is well defined. □

Part (a) of the next lemma says that in all $\mathbb{Z}$-lattice structures and $\mathbb{Z}$-lattice bundles which are constructed from $(H_{\mathbb{Z}}, L, \underline{e})$, the bilinear form $I^{(k)}$ becomes a flat bilinear form on the bundle. Part (b) is a nice observation of K. Saito [Sa82, Lemma 2]. It says that $I^{(k)}$ is essentially the unique flat bilinear form on these bundles.

Lemma 8.24 *Let $k \in \{0, 1\}$.*

(a) Each of the following $\mathbb{Z}$-lattice bundles comes with a $(-1)^k$-symmetric bilinear form on its fibers, which is induced by $I^{(k)}$ on $H_{\mathbb{Z}}$ and which is also called $I^{(k)}$: the bundle $V_{\mathbb{Z}}^{(k)}((\underline{u}, r), [(\gamma; \sigma)], \underline{f})$ over $\mathbb{C} - \{u_1, ..., u_n\}$ in Definition 8.22 (b), the bundle $V_{\mathbb{Z}}^{(k),univ}$ over $\mathbb{C} \times C_n^{univ} - D_{1,n}^{univ}$ in Definition 8.22 (c), the bundle $V_{\mathbb{Z}}^{(k),\underline{e}/\{\pm 1\}^n}$ over $\mathbb{C} \times C_n^{\underline{e}/\{\pm 1\}^n} - D_{1,n}^{\underline{e}/\{\pm 1\}^n}$ in Theorem 8.23 (d).

(b) [Sa82, Lemma 2] Suppose that $(H_{\mathbb{Z}}, L, \underline{e})$ is irreducible and $(n, k) \neq (1, 1)$. In each of the bundles in part (a), the only flat bilinear forms on its fibers are the bilinear bilinear forms $a \cdot I^{(k)}$ with $a \in \mathbb{Z}$, so the scalar multiples of $I^{(k)}$.

Proof (a) This follows in all cases from $s_{f_i}^{(k)} \in O^{(k)}$, which says that $s_{f_i}^{(k)} : H_{\mathbb{Z}} \to H_{\mathbb{Z}}$ respects $I^{(k)}$.

(b) In each of the bundles in part (a), the monodromy group, which is the image of the group antihomomorphism

$$\pi_1(\text{base space, base point}) \to O^{(k)}$$

is $\Gamma^{(k)}$. This is clear for the bundle $V_{\mathbb{Z}}^{(k)}((\underline{u}, r), [(\gamma; \sigma)], \underline{f})$ in Definition 8.22 (b) and for $V_{\mathbb{Z}}^{(k),univ}$. It follows for the bundle $V_{\mathbb{Z}}^{\underline{e}/\{\pm 1\}^n}$ with the fact that the restriction of this bundle to $\{(z, b) \in \mathbb{C} \times C_n^{\underline{e}/\{\pm 1\}^n} \mid \operatorname{Re} z > \max(\operatorname{Re} u \mid (u, b) \in D_{1,n}^{\underline{e}/\{\pm 1\}^n})\}$ is trivial (without this fact, one might have additional *transversal* monodromy).

The proof of Lemma 2 in [Sa82] shows that the only $\Gamma^{(k)}$ invariant bilinear forms on $H_{\mathbb{Z}}$ are the forms $a \cdot I^{(k)}$ with $a \in \mathbb{Z}$. □

Example 8.25 The case A_2, $(H_{\mathbb{Z}}, L, \underline{e})$ of rank 2 with the matrix $S = L(\underline{e}^t, \underline{e})^t = \begin{pmatrix} 1 & -1 \\ 0 & 1 \end{pmatrix}$. We saw in Theorem 8.12

$$C_2^{\underline{e}/\{\pm 1\}^2} = C_2^{A_2} \cong \mathbb{C} \times \mathbb{C}^* \quad \text{with coordinates } (z_1, z_2)$$
$$\text{with } u_{1/2} = z_1 \pm \frac{2}{3} z_2^{3/2}, \quad z_1 = \frac{1}{2}(u_1 + u_2), z_2 = (\frac{3}{4}(u_1 - u_2))^{2/3}.$$

The base point $b_2^{\underline{e}/\{\pm 1\}^2} \in C_2^{\underline{e}/\{\pm 1\}^2}$ is a point with $(u_1, u_2) = (i, 2i)$. We choose $b_2^{\underline{e}/\{\pm 1\}^2}$ with $(z_1^0, z_2^0) := (\frac{3}{2}i, (\frac{3}{4})^{2/3} e^{2\pi i/6})$ in the coordinates (z_1, z_2). Recall

$$(e_1, e_2) \stackrel{\sigma_1}{\longmapsto} (e_1 + e_2, e_1) \stackrel{\sigma_1}{\longmapsto} (-e_2, e_1 + e_2) \stackrel{\sigma_1}{\longmapsto} (e_1, -e_2).$$

The set $\mathcal{B}^{dist}/\{\pm 1\}^2$ has the three elements

$$\begin{aligned} x_0 &= (e_1, e_2)/\{\pm 1\}^2 = \sigma_1^{-1} x_2, \\ x_1 &= (-e_2, e_1 + e_2)/\{\pm 1\}^2 = \sigma_1^{-1} x_0, \\ x_2 &= (e_1 + e_2, e_1)/\{\pm 1\}^2 = \sigma_1^{-1} x_1. \end{aligned}$$

The walls in $C_2^{\underline{e}/\{\pm 1\}^2}$ are the real hypersurfaces of points (z_1, z_2) with $\operatorname{Im} u_1 = \operatorname{Im} u_2$. They are in the coordinates (z_1, z_2)

$$\mathbb{C} \times \{z_2 \in \mathbb{C}^* \mid \operatorname{Im} z_2^{3/2} = 0\} = \mathbb{C} \times (\mathbb{R}_{>0} \dot\cup \mathbb{R}_{>0} \cdot e^{2\pi i/3} \dot\cup \mathbb{R}_{>0} \cdot e^{2\pi i 2/3}).$$

Between them are the three Stokes regions $F_2^{x_0,x_0}$, $F_2^{x_0,x_1}$, $F_2^{x_0,x_2}$.

The picture in Fig. 8.9 shows in the middle a part of the slice $\{z_1^0\} \times \mathbb{C} \subset C_2^{\underline{e}/\{\pm 1\}^2}$ and the intersection of this part with the three walls and the three Stokes regions. The three lines with arrows are the three lifts of $\sigma_1 \in \pi_1(C_2^{conf}, b_2^{conf})$ to paths in $C_2^{\underline{e}/\{\pm 1\}^2}$. In the outer part, the picture shows for different values of (z_1^0, z_2) standard systems of paths (recall $u_{1/2} = z_1 \pm \frac{2}{3} z_2^{3/2}$). The small circle between the points u_1 and u_2 means the value 0.

We have chosen the indexing of u_1 and u_2 in each Stokes region so that $\operatorname{Im} u_1 < \operatorname{Im} u_2$. Therefore there is a discontinuity of indexing along each wall.

Remarks 8.26 (i) Part (c) and part (a) of Theorem 8.23 give the following. Let $k \in \{0, 1\}$ and let $\beta \in \mathrm{Br}_n$. Let $\widetilde{\beta} : [0, 1] \to C_n^{\underline{e}/\{\pm 1\}^n}$ be the lift to $C_n^{\underline{e}/\{\pm 1\}^n}$ with $\widetilde{\beta}(0) = b_n^{\underline{e}/\{\pm 1\}^n}$ of a loop in C_n^{conf} which represents β. Then the restriction of $V_{\mathbb{Z}}^{(k),\underline{e}/\{\pm 1\}^n}$ to $(\mathbb{C} - \{i, 2i, ..., ni\}) \times \{\widetilde{\beta}(1)\}$ is equal to $V_{\mathbb{Z}}^{(k)}((b_n^{pure}, 1), [(\underline{\gamma}^{st}, \mathrm{id})], \beta^{-1}(\underline{e}))$.

(ii) This fits well to the indexing of the Stokes regions in Lemma 8.6 (d) in the case $X = \mathcal{B}^{dist}/\{\pm 1\}^n$. Then $x_0 := \underline{e}/\{\pm 1\}^n$. The Stokes region which contains $\widetilde{\beta}(1)$ is $F_n^{x_0,x}$ with $x = \beta^{-1}(\underline{e})/\{\pm 1\}^n \in X$. See also Remark 8.7 (ii).

(iii) One can modify the construction of the bundle $V_{\mathbb{Z}}^{(k),\underline{e}/\{\pm 1\}^n}$ in Theorem 8.23 (d). Suppose that a group antihomomorphism

$$\pi_1(C_n^{\underline{e}/\{\pm 1\}^n}, b_n^{\underline{e}/\{\pm 1\}^n}) \to (\{\pm 1\}^n)_S \subset G_{\mathbb{Z}}$$

is given. Here an element of $\{\pm 1\}^n$ means an automorphism of $H_{\mathbb{Z}}$ which changes some signs in the basis $\underline{e}$. Then $(\{\pm 1\}^n)_S \subset G_{\mathbb{Z}}$.

One can twist the automorphism $\beta^{deck,V,(k)}$ in Theorem 8.23 (d) with the correct element of $(\{\pm 1\}^n)_S$ and obtain a twisted quotient bundle over $\mathbb{C} \times C_n^{\underline{e}/\{\pm 1\}^n} - D_{1,n}^{\underline{e}/\{\pm 1\}^n}$.

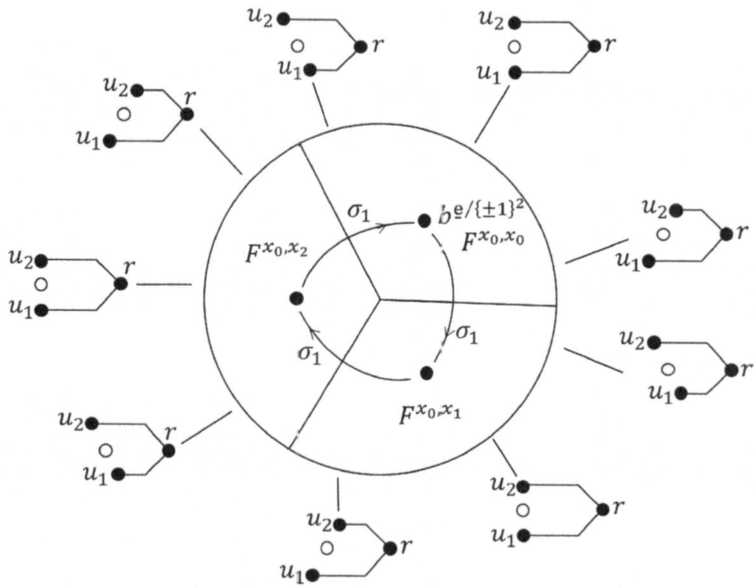

Fig. 8.9 Moving with powers of σ_1 through the three Stokes regions of $C_2^{A_2}$

But one looses the property of $V_{\mathbb{Z}}^{(k),\underline{e}/\{\pm 1\}^n}$ to be a trivial bundle over the open set

$$\{(z, b) \in \mathbb{C} \times C_n^{\underline{e}/\{\pm 1\}^n} \mid \operatorname{Re} z > \max(\operatorname{Re} u \mid (u, b) \in D_{1,n}^{\underline{e}/\{\pm 1\}^n})\}.$$

This property is needed for a possible extension of such a bundle over a partial compactification of $C_n^{\underline{e}/\{\pm 1\}^n}$. Therefore the bundle $V_{\mathbb{Z}}^{(k),\underline{e}/\{\pm 1\}^n}$ is more natural than the twisted bundles.

(iv) Consider a bundle $V_{\mathbb{Z}}^{(k)}((\underline{u}, r), [(\underline{\gamma}; \sigma)], \underline{f})$ as in Definition 8.22(a). In most cases one can recover $\underline{f}/\{\pm 1\}^n$ from the bundle and from $[(\underline{\gamma}; \sigma)]$, namely in all cases with $k = 0$ and in all cases with $k = 1$ where is $(H_{\mathbb{Z}}, L, \underline{e})$ is not reducible with at least two summands of type A_1, so especially in all irreducible cases.

We explain this. In any case, one recovers from the bundle and the distinguished system of paths $(\underline{\gamma}; \sigma)$ the local monodromies $s_{f_1}^{(k)}, \ldots, s_{f_n}^{(k)}$ along the associated loops. With Lemma 3.15 (b) and (c) one obtains from the tuple $(s_{f_1}^{(k)}, \ldots, s_{f_n}^{(k)})$ of reflections or transvections the distinguished basis $\underline{f}/\{\pm 1\}^n$ up to signs.

Chapter 9
The Manifolds in the Rank 3 Cases

The manifolds $C_n^{\underline{e}/\{\pm 1\}^n}$ and $C_n^{S/\{\pm 1\}^n}$ from Chap. 8 had been discussed in the rank 2 cases in Sect. 8.3. There only three different manifolds $C_2^{\underline{e}/\{\pm 1\}^n}$ arose, namely C_2^{pure}, $C_2^{A_2}$ and C_2^{univ}, and only one manifold $C_2^{S/\{\pm 1\}^2} = C_2^{conf}$. In rank 3 the classification is already much richer. The rank 3 cases are treated in this chapter.

Section 9.1 makes the deck transformations of C_3^{univ} as universal covering of C_3^{conf} and of C_3^{pure} explicit. For that it introduces new coordinates on C_3^{pure}. In fact, the new coordinates come from the restriction to C_3^{pure} of a blowing up of the complex line $\{u_1 = u_2 = u_3\}$ in $\mathbb{C}^3 \supset C_3^{pure}$. For the deck transformations the Schwarzian triangle function $T : \mathbb{H} \to \mathbb{C} - \{0, 1\} \cong \mathbb{H}/\overline{\Gamma(2)}$ and a lift $\kappa : \mathbb{H} \to \mathbb{C}$ with T of the logarithm $\ln : \mathbb{C} - \{0\} \dashrightarrow \mathbb{C}$ are needed. Both are treated in Appendix C.

Theorem 9.3 in Sect. 9.2 gives the main result, a description of all possible manifolds $C_3^{\underline{e}/\{\pm 1\}^3}$ and $C_3^{S/\{\pm 1\}^3}$ in rank 3. It starts with a unimodular bilinear lattice $(H_{\mathbb{Z}}, L, \underline{e})$ with a triangular basis $\underline{e}$ and matrix $S = S(\underline{x}) = L(\underline{e}^t, \underline{e})^t \in T_3(\mathbb{Z})$ for some $\underline{x} \in \mathbb{Z}^3$. Thanks to Theorem 4.14 we can and will restrict to a local minimum $\underline{x} \in C_i$ for some $i \in \{1, 2, ..., 24\}$. Theorem 7.11 lists all 16 possible pairs $((\mathrm{Br}_3)_{\underline{e}/\{\pm 1\}^3}, (\mathrm{Br}_3)_{S/\{\pm 1\}^3})$ of stabilizers, the case $S(-l, 2, -l)$ with parameter $l \in \mathbb{Z}_{\geq 3}$.

Theorem 9.3 gives in all 16 cases the manifolds $C_3^{\underline{e}/\{\pm 1\}^3}$ and $C_3^{S/\{\pm 1\}^3}$. Though this is coarse information. The covering map of the normal covering $C_3^{\underline{e}/\{\pm 1\}^3} \to C_3^{S/\{\pm 1\}^3}$ is also important. But its description requires the whole discussion in the proof, and therefore it is given only in the proof of Theorem 9.3.

Also the F-manifold structure and the Euler field are important, as well as possible partial compactifications of $C_3^{\underline{e}/\{\pm 1\}^3}$ such that the F-manifold structure extends. They are treated in Sect. 9.3 in Lemma 9.7 and Corollary 9.8. The cases A_3 and A_2A_1 have partial compactifications which are well known from singularity theory. Seeing them here is a bit involved, especially in the case A_3.

C. Hertling and K. Larabi, *Triangular Bases of Unimodular Bilinear Lattices and Induced Structures*, Lecture Notes in Mathematics 2398,
https://doi.org/10.1007/978-3-032-25997-4_9

9.1 The Deck Transformations on C_3^{univ}

Theorem 9.1 first gives new coordinates on C_3^{pure}. They are better suited to treat the deck transformations of C_3^{univ} as universal covering $\mathrm{pr}_3^{u,c}: C_3^{univ} \to C_3^{conf}$ and as universal covering $\mathrm{pr}_3^{u,p}: C_3^{univ} \to C_3^{pure}$. It writes the F-manifold structure in these new coordinates. It gives in these coordinates explicitly the six deck transformations of C_3^{pure} as normal covering of C_3^{conf}. Finally, it gives explicitly several deck transformations of C_3^{univ} as universal covering of C_3^{conf}. Here the Schwarzian triangle function $T: \mathbb{H} \to \mathbb{C} - \{0, 1\}$ and the lift $\kappa: \mathbb{H} \to \mathbb{C}$ with T of the logarithm $\ln: \mathbb{C} - \{0\} \dashrightarrow \mathbb{C}$ from Appendix C are used.

Theorem 9.1 *(a) The map*

$$f: C_3^{pure} \to \mathbb{C} \times \mathbb{C}^* \times (\mathbb{C} - \{0, 1\}),$$
$$(u_1, u_2, u_3) \mapsto (z_1, z_2, z_3) = (u_1 + u_2 + u_3, u_2 - u_1, \frac{u_3 - u_1}{u_2 - u_1}),$$

is an isomorphism of complex manifolds. The inverse map is

$$f^{-1}: \mathbb{C} \times \mathbb{C}^* \times (\mathbb{C} - \{0, 1\}) \to C_3^{pure}$$
$$(z_1, z_2, z_3) \mapsto (u_1, u_2, u_3) = \frac{1}{3}(z_1 - z_2(1 + z_3), z_1 + z_2(2 - z_3), z_1 + z_2(-1 + 2z_3)).$$

Let $\partial_j := \frac{\partial}{\partial z_j}$ be the coordinate vector fields of the coordinates (z_1, z_2, z_3). The base changes between them and the partial units $e_i = \frac{\partial}{\partial u_i}$ are as follows,

$$(\partial_1, \partial_2, \partial_3) = (e_1, e_2, e_3)\frac{1}{3}\begin{pmatrix} 1 & -1 - z_3 & -z_2 \\ 1 & 2 - z_3 & -z_2 \\ 1 & -1 + 2z_3 & 2z_2 \end{pmatrix},$$
$$(e_1, e_2, e_3) = (\partial_1, \partial_2, \partial_3)\begin{pmatrix} 1 & 1 & 1 \\ -1 & 1 & 0 \\ \frac{z_3 - 1}{z_2} & \frac{-z_3}{z_2} & \frac{1}{z_2} \end{pmatrix}.$$

Unit field e, Euler field E and multiplication $\circ$ are in canonical coordinates given by

$$e = \sum_{j=1}^{3} e_j, \quad E = \sum_{j=1}^{3} u_j e_j, \quad e_i \circ e_j = \delta_{ij} e_i.$$

In the coordinates (z_1, z_2, z_3) they are given by

$$
\begin{aligned}
e &= 3\partial_1, \quad E = z_1\partial_1 + z_2\partial_2, \\
\partial_2 \circ \partial_2 &= \frac{2 - 2z_3 + 2z_3^2}{3}\partial_1 + \frac{1 - 2z_3}{3}\partial_2 + \frac{-z_3 + z_3^2}{z_2}\partial_3, \\
\partial_2 \circ \partial_3 &= \frac{-z_3 + 2z_2z_3}{3}\partial_1 + \frac{-z_2}{3}\partial_2 + \frac{-1 + 2z_3}{3}\partial_3, \\
\partial_3 \circ \partial_3 &= \frac{2z_2^2}{3}\partial_1 + \frac{z_2}{3}\partial_3.
\end{aligned}
$$

(b) The map

$$\sigma \mapsto \chi_\sigma := (C_3^{pure} \to C_3^{pure}, \ \underline{u} \mapsto (u_{\sigma^{-1}(1)}, u_{\sigma^{-1}(2)}, u_{\sigma^{-1}(3)}))$$

is an isomorphism from S_3 to the group of deck transformations of the covering $C_3^{pure} \to C_3^{conf}$. The corresponding automorphisms $\phi_\sigma := f \circ \chi_\sigma \circ f^{-1}$ of $\mathbb{C} \times \mathbb{C}^* \times (\mathbb{C} - \{0, 1\})$ are as follows (see Theorem B.1 (d) for g_σ):

$$
\begin{aligned}
\phi_{\mathrm{id}}(\underline{z}) &= \underline{z}, \\
\phi_{(12)}(\underline{z}) &= (z_1, -z_2, g_{(12)}(z_3)), \\
\phi_{(13)}(\underline{z}) &= (z_1, z_2(1 - z_3), g_{(13)}(z_3)), \\
\phi_{(23)}(\underline{z}) &= (z_1, z_2z_3, g_{(23)}(z_3)), \\
\phi_{(123)}(\underline{z}) &= (z_1, -z_2z_3, g_{(123)}(z_3)), \\
\phi_{(132)}(\underline{z}) &= (z_1, z_2(z_3 - 1), g_{(132)}(z_3)).
\end{aligned}
$$

(c) A universal covering of $\mathbb{C} \times \mathbb{C}^* \times (\mathbb{C} - \{0, 1\})$ is given by the map

$$
\begin{aligned}
\widetilde{\mathrm{pr}}_3^{u,p} : \mathbb{C} \times \mathbb{C} \times \mathbb{H} &\to \mathbb{C} \times \mathbb{C}^* \times (\mathbb{C} - \{0, 1\}) \\
(z_1, \zeta, \tau) &\mapsto (z_1, e^\zeta, T(\tau)),
\end{aligned}
$$

with T as in Theorem B.1 (d). The group of deck transformations of the universal covering

$$
\begin{aligned}
\mathrm{pr}_3^{p,c} \circ f^{-1} \circ \widetilde{\mathrm{pr}}_3^{u,p} : \mathbb{C} \times \mathbb{C} \times \mathbb{H} &\longrightarrow \mathbb{C} \times \mathbb{C}^* \times (\mathbb{C} - \{0, 1\}) \\
&\xrightarrow{\cong} C_3^{pure} \\
&\xrightarrow{/S_3} C_3^{conf}
\end{aligned}
$$

is isomorphic to the group Br_3. An isomorphism ψ is determined by the images $\psi(\sigma_1)$ and $\psi(\sigma_2)$ of the elementary braids σ_1 and σ_2. The following images $\psi(\sigma_1)$ and $\psi(\sigma_2)$ work (recall the lift $\kappa : \mathbb{H} \to \mathbb{C}$ of the (multivalued) logarithm $\log : \mathbb{C}^* \dashrightarrow \mathbb{C}$ in Definition B.2 and Lemma B.3 (c)),

$$\psi(\sigma_1) : \mathbb{C} \times \mathbb{C} \times \mathbb{H} \to \mathbb{C} \times \mathbb{C} \times \mathbb{H},$$

$$
\begin{aligned}
&(z_1, \zeta, \tau) \mapsto (z_1, \zeta - \pi i, \tau - 1) \quad (\textit{recall } \tau - 1 = \mu_{(12)}(\tau)), \\
\psi(\sigma_2) &: \mathbb{C} \times \mathbb{C} \times \mathbb{H} \to \mathbb{C} \times \mathbb{C} \times \mathbb{H}, \\
&(z_1, \zeta, \tau) \mapsto (z_1, \zeta + \kappa(\tau + 1), \frac{\tau}{\tau + 1}) \\
&\qquad (\textit{recall } \frac{\tau}{\tau + 1} = \mu_{(13)}(\tau)).
\end{aligned}
$$

The isomorphism ψ satisfies

$$
\begin{aligned}
\psi(\sigma^{mon}) &: (z_1, \zeta, \tau) \mapsto (z_1, \zeta - 2\pi i, \tau), \\
\psi((\sigma^{mon})^{-1}\sigma_1^2) &: (z_1, \zeta, \tau) \mapsto (z_1, \zeta, \tau - 2), \\
\psi(\sigma_2^2) &: (z_1, \zeta, \tau) \mapsto (z_1, \zeta, \frac{\tau}{2\tau + 1}), \\
\psi(\sigma_1\sigma_2^2) &: (z_1, \zeta, \tau) \mapsto (z_1, \zeta - \pi i, \frac{-\tau - 1}{2\tau + 1}), \\
\psi(\sigma_2\sigma_1) &: (z_1, \zeta, \tau) \mapsto (z_1, \zeta - \pi i + \kappa(\tau), \frac{\tau - 1}{\tau}).
\end{aligned}
$$

Here recall

$$
\begin{aligned}
\mu_{(12)}^2 = (\tau \to \tau - 2), \quad & \mu_{(13)}^2 = (\tau \mapsto \frac{\tau}{2\tau + 1}), \\
\mu_{(123)} = (\tau \mapsto \frac{\tau - 1}{\tau}) &= \textit{rotation of order 3 around } e^{2\pi i/6}, \\
\mu(\begin{pmatrix} -1 & -1 \\ 2 & 1 \end{pmatrix}) &= \textit{rotation of order 2 around } \frac{-1 + i}{2}.
\end{aligned}
$$

Proof (a) It is rather obvious that f is an isomorphism of complex manifolds with inverse f^{-1} as claimed. The base changes between the two bases $(\partial_1, \partial_2, \partial_3)$ and (e_1, e_2, e_3) of coordinate vector fields are obtained by derivating $f(\underline{u})$ and $f^{-1}(\underline{z})$ suitably. Unit field e, Euler field E and multiplication $\circ$ in the new coordinates $\underline{z} = (z_1, z_2, z_3)$ can be calculated straightforwardly.

(b) Straightforward calculations.

(c) $\psi(\sigma_1)$ and $\psi(\sigma_2)$) are lifts to $\mathbb{C} \times \mathbb{C} \times \mathbb{H}$ of the automorphisms $\phi_{(12)}$ and $\phi_{(13)}$ of $\mathbb{C} \times \mathbb{C}^* \times (\mathbb{C} - \{0, 1\})$ because

$$
\begin{aligned}
\widetilde{\mathrm{pr}}_3^{u,p}(\psi(\sigma_1)(z_1, \zeta, \tau))) &= \widetilde{\mathrm{pr}}_3^{u,p}(z_1, \zeta - \pi i, \mu_{(12)}(\tau)) \\
&= (z_1, e^{\zeta - \pi i}, T(\mu_{(12)}(\tau))) \\
&= (z_1, -e^{\zeta}, g_{(12)}(T(\tau))) \\
&= (z_1, -z_2, g_{(12)}(z_3)), \\
\widetilde{\mathrm{pr}}_3^{u,p}(\psi(\sigma_2)(z_1, \zeta, \tau))) &= \widetilde{\mathrm{pr}}_3^{u,p}(z_1, \zeta + \kappa(\tau + 1), \mu_{(13)}(\tau)) \\
&= (z_1, e^{\zeta} T(\tau + 1), T(\mu_{(13)}(\tau))) \\
&= (z_1, e^{\zeta}(1 - T(\tau)), g_{(13)}(T(\tau)))
\end{aligned}
$$

$$= (z_1, z_2(1 - z_3), g_{(13)}(z_3)).$$

They satisfy the relation

$$\psi(\sigma_1)\psi(\sigma_2)\psi(\sigma_1) = \psi(\sigma_2)\psi(\sigma_1)\psi(\sigma_2)$$

because

$$\begin{aligned}
\psi(\sigma_1)\psi(\sigma_2)\psi(\sigma_1) : (z_1, \zeta, \tau) &\stackrel{\sigma_1}{\longmapsto} (z_1, \zeta - \pi i, \tau - 1) \\
&\stackrel{\sigma_2}{\longmapsto} (z_1, \zeta - \pi i + \kappa(\tau), \frac{\tau - 1}{\tau}) \\
&\stackrel{\sigma_1}{\longmapsto} (z_1, \zeta - 2\pi i + \kappa(\tau), \frac{-1}{\tau}), \\
\psi(\sigma_2)\psi(\sigma_1)\psi(\sigma_2) : (z_1, \zeta, \tau) &\stackrel{\sigma_2}{\longmapsto} (z_1, \zeta + \kappa(\tau + 1)), \frac{\tau}{\tau + 1}) \\
&\stackrel{\sigma_1}{\longmapsto} (z_1, \zeta - \pi i + \kappa(\tau + 1), \frac{-1}{\tau + 1}) \\
&\stackrel{\sigma_2}{\longmapsto} (z_1, \zeta - \pi i + \kappa(\tau + 1) + \kappa(\frac{\tau}{\tau + 1}), \frac{-1}{\tau})
\end{aligned}$$

and $-2\pi i + \kappa(\tau) = -\pi i + \kappa(\tau + 1) + \kappa(\frac{\tau}{\tau+1})$ by Theorem B.3 (c) (iv).

Therefore ψ is a group homomorphism from Br_3 to the group of deck transformations of the universal covering $\mathbb{C} \times \mathbb{C} \times \mathbb{H} \to C_3^{conf}$. It is surjective because $\psi(\sigma^{mon})$ is

$$\begin{aligned}
\psi(\sigma^{mon}) &= (\psi(\sigma_1)\psi(\sigma_2)\psi(\sigma_1))^2 : \\
(z_1, \zeta, \tau) &\mapsto (z_1, \zeta - 4\pi i + \kappa(\tau) + \kappa\left(\frac{-1}{\tau}\right), \tau) \\
&= (z_1, \zeta - 2\pi i, \tau)
\end{aligned}$$

because

$$\begin{aligned}
\kappa(\tau) + \kappa\left(\frac{-1}{\tau}\right) &= \kappa(\tau) + \kappa\left(\frac{-1}{\tau} + 2\right) + 2\pi i \\
&= \kappa(\tau) + \kappa\left(\frac{2\tau - 1}{\tau}\right) + 2\pi i \\
&= 2\pi i \quad \text{by Theorem B.3(c) (iv).}
\end{aligned}$$

$\psi(\sigma_2^2)$ is as claimed because by Theorem B.3 (c) (iv)

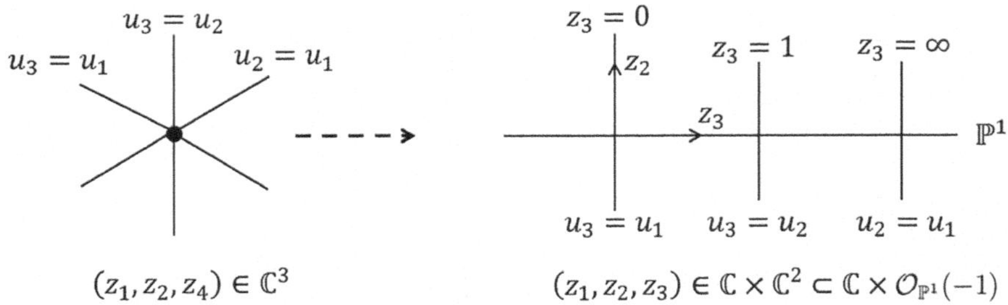

Fig. 9.1 Blowing up of $\mathbb{C} \times \{(0, 0)\} \subset \mathbb{C}^3$

$$\zeta + \kappa(\tau + 1) + \kappa \left(\frac{2\tau + 1}{\tau + 1} \right) = \zeta + \kappa(\tau + 1) + \kappa \left(\frac{2(\tau + 1) - 1}{\tau + 1} \right) = \zeta.$$

The formulas for $\psi((\sigma^{mon})^{-1}\sigma_1^2)$, $\psi(\sigma_1\sigma_2^2)$ and $\psi(\sigma_2\sigma_1)$ are calculated straightforwardly. $\square$

Remark 9.2 (i) The group of deck transformations of the universal covering $\mathbb{C} \times \mathbb{C} \times \mathbb{H} \to C_3^{pure}$ splits into the product

$$\begin{aligned} &\langle \psi(\sigma^{mon}) \rangle \times \langle \psi((\sigma^{mon})^{-1}\sigma_1^2, \psi(\sigma_2^2) \rangle \\ &= \langle ((z_1, \zeta, \tau) \mapsto (z_1, \zeta - 2\pi i, \tau)) \rangle \times \\ &\langle ((z_1, \zeta, \tau) \mapsto (z_1, \zeta, \mu_{(12)}^2(\tau)), (z_1, \zeta, \tau) \mapsto (z_1, \zeta, \mu_{(13)}^2(\tau)) \rangle. \end{aligned}$$

This fits to the splitting of Br_3^{pure},

$$\mathrm{Br}_3^{pure} = \langle \sigma^{mon} \rangle \times \langle (\sigma^{mon})^{-1}\sigma_1^2, \sigma_2^2 \rangle.$$

(ii) The coordinate change

$$\mathbb{C}^3 \to \mathbb{C}^3, \quad (u_1, u_2, u_3) \mapsto (z_1, z_2, z_4) := (u_1 + u_2 + u_3, u_2 - u_1, u_3 - u_1),$$

is linear. f is the restriction to C_3^{pure} of the composition of this coordinate change with the blowing up

$$\mathbb{C}^3 \dashrightarrow \mathbb{C} \times \mathcal{O}_{\mathbb{P}^1}(-1), \quad (z_1, z_2, z_4) \dashrightarrow \left(z_1, z_2, \frac{z_4}{z_2} \right) = (z_1, z_2, z_3),$$

of $\mathbb{C} \times \{(0, 0)\}$ in one of the two natural charts of $\mathbb{C} \times \mathcal{O}_{\mathbb{P}^1}(-1)$ (Fig. 9.1).

Also the multiplication is calculated only in this chart $\mathbb{C}^3$ with coordinates (z_1, z_2, z_3). We refrain from calculating it in the other chart $\mathbb{C}^3$ with coordinates $(z_1, \widetilde{z}_2, \widetilde{z}_3) = (z_1, \frac{z_2}{z_4}, z_4)$ as we do not really need it. Though in the chart considered, the hyperplane in the Maxwell stratum where $u_2 = u_1$ lies above $z_3 = \infty$, so for that hyperplane the other chart would be useful. The hyperplanes where $u_3 = u_1$ respectively $u_3 = u_2$ lie above $z_3 = 0$ respectively $z_3 = 1$.

(iii) The multiplication on $\mathbb{C}\times\mathbb{C}\times\mathbb{H}\cong C_3^{univ}$ involves $\frac{\partial T(\tau)}{\partial\tau}$. Unit field and Euler field on it are $e = 3\partial_1$ and $E = z_1\partial_1 + \partial_\zeta$.

9.2 The Manifolds as Quotients and Their Covering Maps

Let $(H_\mathbb{Z}, L, \underline{e})$ be a unimodular bilinear lattice of rank 3 with triangular basis $\underline{e}$ and matrix $S = S(\underline{x}) = L(\underline{e}^t, \underline{e})^t \in T_3(\mathbb{Z})$ for some $\underline{x}\in\mathbb{Z}^3$. We can and will restrict to a local minimum $\underline{x}\in C_i$ for some $i \in \{1, 2, ..., 24\}$ as in Theorem 7.11. This theorem lists all 16 possible pairs $((\text{Br}_3)_{\underline{e}/\{\pm1\}^3}, (\text{Br}_3)_{S/\{\pm1\}^3})$ of stabilizers, the case $S(-l, 2, -l)$ with parameter $l \in \mathbb{Z}_{\geq 3}$.

Theorem 9.3 below gives in all 16 cases the manifolds $C_3^{\underline{e}/\{\pm1\}^3}$ and $C_3^{S/\{\pm1\}^3}$. Though this is coarse information. The covering map of the normal covering $C_3^{\underline{e}/\{\pm1\}^3} \to C_3^{S/\{\pm1\}^3}$ is also important. But its description requires the whole discussion in the proof, and therefore it is given only in the proof of Theorem 9.3.

The group $(\text{Br}_3)_{\underline{e}/\{\pm1\}^3}$ acts on the third factor $\mathbb{H}$ of $\mathbb{C}\times\mathbb{C}\times\mathbb{H}\cong C_3^{univ}$ via its image in $PSL_2(\mathbb{Z})$ under the homomorphism $\text{Br}_3 \to PSL_2(\mathbb{Z})$ in Remark 4.15 (i). In all cases except A_3 and A_2A_1 this action is free and the quotient of $\mathbb{H}$ by this action is a noncompact complex curve C. Furthermore, in all cases except A_1^3, $\mathcal{H}_{1,2}$, A_2A_1, and A_3 the group $(\text{Br}_3)_{\underline{e}/\{\pm1\}^3}$ does not contain a power of σ^{mon}. Then the quotient $\mathbb{C}\times\mathbb{C}\times\mathbb{H}/(\text{Br}_3)_{\underline{e}/\{\pm1\}^3}\cong C_3^{\underline{e}/\{\pm1\}^3}$ is

$$\mathbb{C}\times(\text{an affine }\mathbb{C}\text{-bundle over }C).$$

Remark 9.4 (i) says what that means, and Remark 9.4 (ii) states that it is isomorphic to $\mathbb{C}\times\mathbb{C}\times C$. Remark 9.4 (iv) states that any $\mathbb{C}^*$-bundle over a noncompact complex curve C is isomorphic to $\mathbb{C}^*\times C$.

Theorem 9.3 *Consider a local minimum $\underline{x}\in C_i\subset\mathbb{Z}^3$ for some $i\in\{1,2,...,24\}$ and the pseudo-graph $\mathcal{G}_j$ with $\mathcal{G}_j = \mathcal{G}(\underline{x})$. In the following table the first and second column are copied from the table in Theorem 7.11. The third and fourth column give the manifolds $C_3^{\underline{e}/\{\pm1\}^3}$ and $C_3^{S/\{\pm1\}^3}$. Here $\mathbb{D} := \{z\in\mathbb{C}\,|\,|z|<1\}$ denotes the unit disk, $\mathbb{D}^* := \mathbb{D}-\{0\}$, $\mathbb{D}^{**} := \mathbb{D}-\{0,\frac{1}{2}\}$ and $\mathbb{C}^{***} := \mathbb{C}-\{z\in\mathbb{C}\,|\,z^3=1\}$.*

	sets	$C_3^{\underline{e}/\{\pm 1\}^3}$	$C_3^{S/\{\pm 1\}^3}$
$\mathcal{G}_1$	C_1 (A_1^3)	C_3^{pure}	C_3^{conf}
$\mathcal{G}_1$	C_2 $(\mathcal{H}_{1,2})$	$\mathbb{C}\times\mathbb{C}^*\times\mathbb{H}$	C_3^{conf}
$\mathcal{G}_2$	C_3 (A_2A_1)	$\frac{\mathbb{C}\times\mathbb{C}^*\times(\mathbb{C}-\{0,1\})}{\langle\phi_{(12)}\rangle}$	$C_3^{pure}/\langle\phi_{(12)}\rangle$
$\mathcal{G}_2$	C_4 $(\mathbb{P}^1A_1)$, C_5	$\mathbb{C}^2\times(\mathbb{C}-\{0,1\})$	$C_3^{pure}/\langle\phi_{(12)}\rangle$
$\mathcal{G}_3$	C_6 (A_3)	$\mathbb{C}\times\frac{\mathbb{C}^*\times\mathbb{C}^{***}}{\textit{group of order 3}}$	$\mathbb{C}\times\frac{\mathbb{C}^*\times\mathbb{C}^{***}}{\textit{group of order 3}}$
$\mathcal{G}_4$	C_7 $(\widehat{A}_2)$	$\mathbb{C}^2\times\mathbb{C}^{***}$	$\mathbb{C}\times\frac{\mathbb{C}^*\times\mathbb{C}^{***}}{\textit{group of order 3}}$
$\mathcal{G}_5$	C_8, C_9 $((-l,2,-l))$	$\mathbb{C}\times\mathbb{C}\times\mathbb{D}^*$	$\mathbb{C}\times\mathbb{C}^*\times\mathbb{D}^*$
$\mathcal{G}_6$	C_{10} $(\mathbb{P}^2)$, C_{11}, C_{12}	C_3^{univ}	$\frac{\mathbb{C}\times\mathbb{C}^*\times\mathbb{H}}{\textit{group of order 3}}$
$\mathcal{G}_7$	C_{13} (e.g. $(4,4,8)$)	C_3^{univ}	$\frac{\mathbb{C}\times\mathbb{C}^*\times\mathbb{H}}{\textit{group of order 2}}$
$\mathcal{G}_8$	C_{14} (e.g. $(3,4,6)$)	C_3^{univ}	$\mathbb{C}\times\mathbb{C}^*\times\mathbb{H}$
$\mathcal{G}_9$	C_{15}, C_{16}, C_{23}, C_{24}	C_3^{univ}	$\mathbb{C}\times\mathbb{C}^*\times\mathbb{H}$
$\mathcal{G}_{10}$	C_{17} (e.g. $(-2,-2,0)$)	$\mathbb{C}\times\mathbb{C}\times\mathbb{D}^*$	$\mathbb{C}\times\mathbb{C}^*\times\mathbb{D}^*$
$\mathcal{G}_{11}$	C_{18} (e.g. $(-3,-2,0)$)	$\mathbb{C}\times\mathbb{C}\times\mathbb{D}^*$	$\mathbb{C}\times\mathbb{C}^*\times\mathbb{D}^*$
$\mathcal{G}_{12}$	C_{19} (e.g. $(-2,-1,0)$)	$\mathbb{C}\times\mathbb{C}\times\mathbb{D}^{**}$	$\mathbb{C}\times\mathbb{C}^*\times\mathbb{D}^{**}$
$\mathcal{G}_{13}$	C_{20} (e.g. $(-2,-1,-1)$)	$\mathbb{C}\times\mathbb{C}\times\mathbb{D}^{**}$	$\mathbb{C}\times\mathbb{C}^*\times\mathbb{D}^{**}$
$\mathcal{G}_{14}$	C_{21}, C_{22}	$\mathbb{C}\times\mathbb{C}\times\mathbb{D}^*$	$\mathbb{C}\times\mathbb{C}^*\times\mathbb{D}^*$

In the case $\mathcal{G}_2$ & C_3 (A_2A_1) the moduli spaces and the covering can also be presented as follows,

$$\begin{array}{ccccc}
C_3^{\underline{e}/\{\pm 1\}^3} & \cong & \mathbb{C}\times\{(z_8,z_9)\in\mathbb{C}^*\times\mathbb{C}\,|\,z_8^3-4z_9^2\neq 0\} & & (z_1,z_8,z_9) \\
\downarrow 3{:}1 & & \downarrow 3{:}1 & & \downarrow 3{:}1 \\
C_3^{S/\{\pm 1\}^3} & \cong & \mathbb{C}\times\{(z_{10},z_9)\in\mathbb{C}^*\times\mathbb{C}\,|\,z_{10}-4z_9^2\neq 0\} & & (z_1,z_8^3,z_9)
\end{array}$$

In the case $\mathcal{G}_3$ & C_6 (A_3) the moduli spaces and the covering can also be presented as follows,

$$\begin{array}{ccccc}
C_3^{\underline{e}/\{\pm 1\}^3} & \cong & \mathbb{C}\times\{(z_{13},z_{14})\in\mathbb{C}^*\times\mathbb{C}\,|\,z_{14}^3-z_{13}^2\neq 0\} & & (z_1,z_{13},z_{14}) \\
\downarrow 4{:}1 & & \downarrow 4{:}1 & & \downarrow 4{:}1 \\
C_3^{S/\{\pm 1\}^3} & \cong & \mathbb{C}\times\frac{\{(z_{15},z_{14})\in\mathbb{C}^*\times\mathbb{C}\,|\,z_{14}^3-z_{15}\neq 0\}}{\langle(z_{15},z_{14})\mapsto(-z_{15},-z_{14})\rangle} & & [(z_1,z_{13}^2,z_{14})]
\end{array}$$

Proof **The reducible case $\mathcal{G}_1$ & C_1 (A_1):**
Here $(\mathrm{Br}_3)_{\underline{e}/\{\pm 1\}^3}=\mathrm{Br}_3^{pure}$ and $(\mathrm{Br}_3)_{\underline{x}/\{\pm 1\}^3}=\mathrm{Br}_3$. Therefore $C_3^{\underline{e}/\{\pm 1\}^3}=C_3^{pure}$ and $C_3^{\underline{x}/\{\pm 1\}^3}=C_3^{conf}$. The normal covering $C_3^{pure}\to C_3^{conf}$ is known.

The case $\mathcal{G}_1$ & C_2 $(\mathcal{H}_{1,2})$:
Here $(\mathrm{Br}_3)_{\underline{e}/\{\pm 1\}^3}=\langle(\sigma^{mon})^2\rangle$ and $(\mathrm{Br}_3)_{\underline{x}/\{\pm 1\}^3}=\mathrm{Br}_3$. Therefore $C_3^{\underline{x}/\{\pm 1\}^3}=C_3^{conf}$. The equality $(\mathrm{Br}_3)_{\underline{e}/\{\pm 1\}^3}=\langle(\sigma^{mon})^2\rangle$ and

$$\psi((\sigma^{mon})^2):\mathbb{C}\times\mathbb{C}\times\mathbb{H}\to\mathbb{C}\times\mathbb{C}\times\mathbb{H},\ (z_1,\zeta,\tau)\mapsto(z_1,\zeta-4\pi i,\tau)$$

show

$$C_3^{univ} \cong \mathbb{C} \times \mathbb{C} \times \mathbb{H} \to \mathbb{C} \times \mathbb{C}^* \times \mathbb{H} \cong C_3^{\underline{e}/\{\pm 1\}^3},$$
$$(z_1, \zeta, \tau) \mapsto (z_1, e^{\zeta/2}, \tau) = (z_1, z_5, \tau).$$

The quotient $(Br_3)_{S/\{\pm1\}^3}/(\mathrm{Br}_3)_{\underline{e}/\{\pm1\}^3} = \mathrm{Br}_3/\langle(\sigma^{mon})^2\rangle$ is by Remark 4.15 (i) isomorphic to $SL_2(\mathbb{Z}) = \langle A_1, A_2\rangle$ where $A_1 = \begin{pmatrix} 1 & -1 \\ 0 & 1 \end{pmatrix}$, $A_2 = \begin{pmatrix} 1 & 0 \\ 1 & 1 \end{pmatrix}$ and $\sigma_1, \sigma_2, \sigma^{mon}$ are mapped to $A_1, A_2, -E_2$. They induce the following deck transformations of $\mathbb{C} \times \mathbb{C}^* \times \mathbb{H} \cong C_3^{\underline{e}/\{\pm1\}^3}$,

$$\begin{aligned} \sigma_1 &: (z_1, z_5, \tau) \mapsto (z_1, (-i)z_5, \tau - 1), \\ \sigma_2 &: (z_1, z_5, \tau) \mapsto (z_1, z_5 \cdot e^{\kappa(\tau+1)/2}, \frac{\tau}{\tau+1}), \\ \sigma^{mon} &: (z_1, z_5, \tau) \mapsto (z_1, -z_5, \tau). \end{aligned}$$

Dividing out $\langle\sigma^{mon}\rangle$ first, one obtains $\mathbb{C} \times \mathbb{C}^* \times \mathbb{H}$ with coordinates $(z_1, z_2, \tau) = (z_1, z_5^2, \tau)$ and an action of $PSL_2(\mathbb{Z})$ on it, which is induced by the following action of the classes $[A_1]$ and $[A_2]$ in $PSL_2(\mathbb{Z})$,

$$\begin{aligned} [A_1] \sim \sigma_1 &: (z_1, z_2, \tau) \mapsto (z_1, -z_2, \tau - 1) = (z_1, -z_2, \mu_{(12)}(\tau)), \\ [A_2] \sim \sigma_2 &: (z_1, z_2, \tau) \mapsto (z_1, z_2 T(\tau + 1), \frac{\tau}{\tau+1}) \\ &\qquad = (z_1, z_2(1 - T(\tau)), \mu_{(13)}(\tau)). \end{aligned}$$

The normal subgroup $\overline{\Gamma(2)} = \langle [A_1^2], [A_2^2]\rangle$ acts nontrivially only on the third factor $\mathbb{H}$ with quotient $\mathbb{C} - \{0, 1\}$, so its quotient is C_3^{pure}. The action of the quotient group $PSL_2(\mathbb{Z})/\overline{\Gamma(2)} \cong S_3$ gives C_3^{conf}.

We treat the cases $\mathcal{G}_2$ & C_4 ($\mathbb{P}^1 A_1$), C_5 before the case $\mathcal{G}_2$ & C_3 ($A_2 A_1$).

The reducible cases $\mathcal{G}_2$ & C_4 ($\mathbb{P}^1 A_1$), C_5:
By the proof of Theorem 7.11, the group $(\mathrm{Br}_3)_{\underline{e}/\{\pm1\}^3} = \langle\sigma_2^2, (\sigma^{mon})^{-1}\sigma_1^2\rangle$ is the normal closure of σ_2^2 in the group $(\mathrm{Br}_3)_{\underline{x}/\{\pm1\}^3} = \langle\sigma_1, \sigma_2^2\rangle$, so especially a normal subgroup.

The deck transformations $\psi(\sigma_2^2)$ and $\psi((\sigma^{mon})^{-1}\sigma_1^2)$ of C_3^{univ} in Theorem 9.1 (c) act nontrivially only on the third factor $\mathbb{H}$, and there they act as the generators $[A_2^2]$ and $[A_1^2]$ of $\overline{\Gamma(2)}$. Therefore

$$C_3^{\underline{e}/\{\pm1\}^3} \cong \mathbb{C} \times \mathbb{C} \times (\mathbb{C} - \{0, 1\}) \quad \text{with coordinates } (z_1, \zeta, z_3).$$

To obtain $C_3^{S/\{\pm1\}^3}$ from $C_3^{\underline{e}/\{\pm1\}^3}$, it is a priori sufficient to divide out the action on $C_3^{\underline{e}/\{\pm1\}^3}$ which the deck transformation $\psi(\sigma_1)$ of C_3^{univ} induces. But it is easier to first divide out the action of $\sigma^{mon} \in \langle\sigma_1, \sigma_2^2\rangle = (\mathrm{Br}_3)_{\underline{x}/\{\pm1\}^3}$ on $C_3^{\underline{e}/\{\pm1\}^3}$. The quotient is

$C_3^{pure} \cong \mathbb{C} \times \mathbb{C}^* \times (\mathbb{C} - \{0, 1\})$. On this space the action of σ_1 is the action of $\phi_{(12)}$, which is a fixed point free involution. Therefore $C_3^{S/\{\pm 1\}^3} = C_3^{pure}/\langle \phi_{(12)} \rangle$.

The reducible case $\mathcal{G}_2$ & C_3 (A_2A_1):
Also here $(\mathrm{Br}_3)_{\underline{x}/\{\pm1\}^3} = \langle \sigma_1, \sigma_2^2 \rangle$, so $C_3^{S/\{\pm 1\}^3} = C_3^{pure}/\langle \phi_{(12)} \rangle$. Here $(Br_3)_{\underline{e}/\{\pm1\}^3} = \langle \sigma_2^2, (\sigma^{mon})^{-1}\sigma_1^2, \sigma^{mon}\sigma_1 \rangle$. First C_3^{univ} is divided by the action of the normal subgroup $\langle \sigma_2^2, (\sigma^{mon})^{-1}\sigma_1^2 \rangle$. The quotient is isomorphic to $\mathbb{C} \times \mathbb{C} \times (\mathbb{C} - \{0, 1\})$ with coordinates (z_1, ζ, z_3) by the discussion in the cases $\mathcal{G}_2$ & C_4, C_5.

The deck transformation $\psi(\sigma^{mon}\sigma_1)$ on C_3^{univ} induces on this manifold the automorphism

$$(z_1, \zeta, z_3) \mapsto (z_1, \zeta - 3\pi i, 1 - z_3).$$

If we divide out first the square of this action, the quotient is isomorphic to $\mathbb{C} \times \mathbb{C}^* \times (\mathbb{C} - \{0, 1\})$ with coordinates (z_1, z_6, z_3) with $z_6 = e^{\zeta/3}$, and $\psi(\sigma^{mon}\sigma_1)$ induces on it the action of $\phi_{(12)} : (z_1, z_6, z_3) \mapsto (z_1, -z_6, 1 - z_3)$. We obtain the following diagram,

$$\begin{array}{ccc}
(z_1, z_6, z_3) & \longrightarrow & (z_1, z_6^3, z_3) = (z_1, z_2, z_3) \\
\mathbb{C} \times \mathbb{C}^* \times (\mathbb{C} - \{0, 1\}) & \xrightarrow[\text{covering}]{3:1} & \mathbb{C} \times \mathbb{C}^* \times (\mathbb{C} - \{0, 1\}) \cong C_3^{pure} \\
\big\downarrow /\langle \phi_{(12)} \rangle & & \big\downarrow /\langle \phi_{(12)} \rangle \\
C_3^{\underline{e}/\{\pm 1\}^3} & \xrightarrow[\text{covering}]{3:1} & C_3^{S/\{\pm 1\}^3} = C_3^{pure}/\langle \phi_{(12)} \rangle
\end{array}$$

Though $C_3^{\underline{e}/\{\pm 1\}^3}$ and $C_3^{S/\{\pm 1\}^3}$ can be presented in a better way. Write $z_7 := z_3 - \frac{1}{2}$. Then $\phi_{(12)} : (z_1, z_6, z_7) \mapsto (z_1, -z_6, -z_7)$ and $\phi_{(12)} : (z_1, z_2, z_7) \mapsto (z_1, -z_2, -z_7)$. The following two diagrams belong together. The upper diagram shows the sets, the lower diagram shows the maps. The horizontal maps are isomorphisms. The sets on the right side in the upper diagram are better presentations of $C_3^{\underline{e}/\{\pm 1\}^3}$ and $C_3^{S/\{\pm 1\}^3}$ than the quotients on the left side. See Corollary 9.8.

$$\begin{array}{ccc}
C_3^{\underline{e}/\{\pm 1\}^3} \cong \frac{\mathbb{C} \times \mathbb{C}^* \times (\mathbb{C} - \{\pm \frac{1}{2}\})}{\langle \phi_{(12)} \rangle} & \xrightarrow{\cong} & \mathbb{C} \times \{(z_8, z_9) \in \mathbb{C}^* \times \mathbb{C} \,|\, z_8^3 - 4z_9^2 = 0\} \\
\big\downarrow 3:1 & & \big\downarrow 3:1 \\
C_3^{S/\{\pm 1\}^3} \cong \frac{\mathbb{C} \times \mathbb{C}^* \times (\mathbb{C} - \{\pm \frac{1}{2}\})}{\langle \phi_{(12)} \rangle} & \xrightarrow{\cong} & \mathbb{C} \times \{(z_{10}, z_9) \in \mathbb{C}^* \times \mathbb{C} \,|\, z_{10} - 4z_9^2 = 0\}
\end{array}$$

$$\begin{array}{ccc}
[(z_1, z_6, z_7)] & \longrightarrow & (z_1, z_6^2, z_6^3 z_7) = (z_1, z_8, z_9) \\
\downarrow 3:1 & & \downarrow 3:1 \\
[(z_1, z_6^3, z_7)] & & (z_1, z_8^3, z_9) \\
\| & & \| \\
[(z_1, z_2, z_7)] & \longrightarrow & (z_1, z_2^2, z_2 z_7) = (z_1, z_{10}, z_9)
\end{array}$$

Geometrically, the horizontal maps are the restrictions of the inversions on the level of $C_3^{\underline{e}/\{\pm 1\}^3}$ and $C_3^{S/\{\pm 1\}^3}$ of the blowing up described in Remark 9.2 (ii).

We treat the case $\mathcal{G}_4$ & C_7 $(\widehat{A}_2)$ before the case $\mathcal{G}_3$ & C_6 (A_3).

The case $\mathcal{G}_4$ & C_7 $(\widehat{A}_2)$:
Here $(\mathrm{Br}_3)_{\underline{e}/\{\pm 1\}^3} = \langle \sigma_1^3, \sigma_2^3, \sigma_2\sigma_1^3\sigma_2^{-1}\rangle$ and $(\mathrm{Br}_3)_{\underline{x}/\{\pm 1\}^3} = \langle \sigma_2\sigma_1, \sigma_1^3\rangle$. The surjective homomorphism $\mathrm{Br}_3 \to PSL_2(\mathbb{Z})$ in Remark 4.15 (i) catches the action of Br_3 on the third factor $\mathbb{H}$ of $\mathbb{C}\times\mathbb{C}\times\mathbb{H} \cong C_3^{univ}$. It maps the subgroup $\langle \sigma_1^3, \sigma_2^3, \sigma_2\sigma_1^3\sigma_2^{-1}\rangle$ to the subgroup $\langle [A_1^3], [A_2^3], [A_2A_1^3A_2^{-1}]\rangle = \overline{\Gamma(3)} \subset PSL_2(\mathbb{Z})$. This group is isomorphic to $G^{free,3}$. The action of $\overline{\Gamma(3)}$ on $\mathbb{H}$ is free.

The quotient $\mathbb{H}/\overline{\Gamma(3)}$ is by Theorem B.1 (c) isomorphic to

$$\mathbb{P}^1 - \{\text{the four vertices of a tetrahedron}\} \cong \mathbb{C} - \{z_{11} \mid z_{11}^3 = 1\} =: \mathbb{C}^{***}.$$

Therefore the quotient $C_3^{\underline{e}/\{\pm 1\}^3}$ of C_3^{univ} by $(\mathrm{Br}_3)_{\underline{e}/\{\pm 1\}^3}$ is isomorphic to

$$\begin{aligned} &\mathbb{C}\times(\text{an affine }\mathbb{C}\text{-bundle over }\mathbb{C}^{***}) \\ &\cong \mathbb{C}\times\mathbb{C}\times\mathbb{C}^{***} \quad \text{with coordinates } (z_1, \zeta, z_{11}) \end{aligned}$$

by Remark 9.4 (ii).

In order to obtain $C_3^{S/\{\pm 1\}^3}$ it is because of $(\mathrm{Br}_3)_{\underline{x}/\{\pm 1\}^3} = \langle \sigma_2\sigma_1, \sigma_1^3\rangle$ a priori sufficient to divide out the action on $C_3^{\underline{e}/\{\pm 1\}^3}$ which the deck transformation $\psi(\sigma_2\sigma_1)$ of C_3^{univ} induces. But it is easier to first divide out the action of $\sigma^{mon} \in \langle \sigma_2\sigma_1, \sigma_1^3\rangle$ on $C_3^{\underline{e}/\{\pm 1\}^3}$. The quotient is isomorphic to $\mathbb{C}\times\mathbb{C}^*\times\mathbb{C}^{***}$ with coordinates (z_1, z_2, z_{11}).

Now $\sigma_2\sigma_1$ acts because of $(\sigma_2\sigma_1)^3 = \sigma^{mon}$ as an automorphism of order three on the $\mathbb{C}^*$-bundle $\mathbb{C}^*\times\mathbb{C}^{***}$. To determine this automorphism, recall the action of $\sigma_2\sigma_1$ on $\mathbb{C}\times\mathbb{C}\times\mathbb{H} \cong C_3^{univ}$ in Theorem 9.1 (c),

$$\psi(\sigma_2\sigma_1) : (z_1, \zeta, \tau) \mapsto \left(z_1, \zeta - \pi i + \kappa(\tau), \frac{\tau - 1}{\tau}\right).$$

The Möbius transformation $\mu_{(123)} = (\tau \mapsto \frac{\tau-1}{\tau})$ is elliptic of order three with fixed point $\tau_0 := e^{2\pi i/6}$. It acts on the tangent space $T_{\tau_0}\mathbb{H}$ by multiplication with $e^{-2\pi i/3}$, because $\mu_{(123)}(\tau_0+\varepsilon) \approx \tau_0 + \varepsilon e^{-2\pi i/3}$ for small $\varepsilon \in \mathbb{C}$. The coordinate z_{11} on $\mathbb{C}^{***}$ can be chosen so that $\sigma_2\sigma_1$ acts on $\mathbb{C}^{***}$ by $z_{11} \mapsto z_{11}e^{-2\pi i/3}$, especially τ_0 maps to $z_{11}=0$. On the $\mathbb{C}^*$-fiber over $z_{11}=0$ $\sigma_2\sigma_1$ acts because of $T(\tau_0)=\tau_0$ by

$$z_2 = e^{\zeta} \mapsto e^{\zeta-\pi i+\kappa(\tau_0)} = z_2(-1)T(\tau_0) = z_2e^{-2\pi i/3}.$$

By Remark 9.4 (iv) we can choose the trivialization of the $\mathbb{C}^*$-bundle $\mathbb{C}^* \times \mathbb{C}^{***}$ over $\mathbb{C}^{***}$ such that the action of $\sigma_2\sigma_1$ on it is the action $(z_2, z_{11}) \mapsto (z_2e^{-2\pi i/3}, z_{11}e^{-2\pi i/3})$. We obtain the following diagram,

$$\begin{array}{ccc}
(z_1,\zeta,z_{11}) & \longrightarrow & (z_1,e^{\zeta},z_{11}) = (z_1,z_2,z_{11}) \\
C_3^{\underline{e}/\{\pm1\}^3} \cong \mathbb{C}\times\mathbb{C}\times\mathbb{C}^{***} & \xrightarrow{/\langle\sigma^{mon}\rangle} & \mathbb{C}\times\mathbb{C}^*\times\mathbb{C}^{***} \\
 & & \big\downarrow /\langle\sigma_2\sigma_1\rangle \\
C_3^{S/\{\pm1\}^3} \cong & & \mathbb{C}\times \dfrac{\mathbb{C}^*\times\mathbb{C}^{***}}{\langle(z_2,z_{11})\mapsto(z_2e^{2\pi i/3},z_{11}e^{2\pi i/3})\rangle}
\end{array}$$

The case $\mathcal{G}_3$ & C_6 (A_3):
Here $(\mathrm{Br}_3)_{\underline{e}/\{\pm1\}^3} = \langle(\sigma_1\sigma_2)^4, \sigma_1^3\rangle$ and $(\mathrm{Br}_3)_{\underline{x}/\{\pm1\}^3} = \langle\sigma_1\sigma_2, \sigma_1^3\rangle$. Especially, $(\sigma_1\sigma_2)^4 = \sigma_1\sigma_2\sigma^{mon}$ and

$$\begin{aligned}(\mathrm{Br}_3)_{\underline{e}/\{\pm1\}^3} &\overset{3:1}{\supset} \langle\sigma_1^3, \sigma_2^{-1}\sigma_1^3\sigma_2 (= \sigma_1\sigma_2^3\sigma_1^{-1}), \sigma_2^3, (\sigma^{mon})^4\rangle \\ &\supset \langle\sigma_1^3, \sigma_2^{-1}\sigma_1^3\sigma_2, \sigma_2^3\rangle,\end{aligned}$$

and the group $\langle\sigma_1^3, \sigma_2^{-1}\sigma_1^3\sigma_2, \sigma_2^3\rangle$ maps under the surjective homomorphism $\mathrm{Br}_3 \to PSL_2(\mathbb{Z})$ in Remark 4.15 (i) to the group

$$\langle[A_1^3], [A_2^{-1}A_1^3A_2], [A_2^3]\rangle = \overline{\Gamma(3)} \subset PSL_2(\mathbb{Z}).$$

The action of $\overline{\Gamma(3)}$ on $\mathbb{H}$ is free, and the quotient is by Theorem B.1 (c) isomorphic to $\mathbb{C}^{***}$. The quotient of C_3^{univ} by $\langle\sigma_1^3, \sigma_2^{-1}\sigma_1^3\sigma_2, \sigma_2^3\rangle$ is isomorphic to

$$\begin{aligned}&\mathbb{C}\times(\text{an affine }\mathbb{C}\text{-bundle over }\mathbb{C}^{***}) \\ &\cong \mathbb{C}\times\mathbb{C}\times\mathbb{C}^{***} \quad \text{with coordinates } (z_1,\zeta,z_{11})\end{aligned}$$

by Remark 9.4 (ii). The quotient of C_3^{univ} by $\langle\sigma_1^3, \sigma_2^{-1}\sigma_1^3\sigma_2, \sigma_2^3, (\sigma^{mon})^4\rangle$ is isomorphic to

$$\mathbb{C}\times\mathbb{C}^*\times\mathbb{C}^{***} \quad \text{with coordinates } (z_1,z_{12},z_{11}) \text{ with } z_{12}=e^{\zeta/4}.$$

On this quotient $(\sigma_1\sigma_2)^4$ acts as an automorphism of order three. We will determine the action now.

First, it acts on $\mathbb{C} \times \mathbb{C}^* \times \mathbb{H}$ with coordinates (z_1, z_{12}, τ) with $z_{12} = e^{\zeta/4}$ by

$$(\sigma_1\sigma_2)^4 : (z_1, z_{12}, \tau) \mapsto (z_1, z_{12}e^{-2\pi i3/8}e^{\kappa(\tau+1)/4}, \frac{-1}{\tau+1}).$$

The Möbius transformation $(\tau \mapsto \frac{-1}{\tau+1})$ is elliptic of order three with fixed point $\tau_1 := e^{2\pi i/3}$. It acts on the tangent space $T_{\tau_1}\mathbb{H}$ by multiplication with $e^{-2\pi i/3}$ because $\frac{-1}{\tau_1+\varepsilon} \equiv \tau_1 + \varepsilon e^{-2\pi i/3}$ for small $\varepsilon \in \mathbb{C}$. The coordinate z_{11} on $\mathbb{C}^{***}$ can be chosen so that $(\sigma_1\sigma_2)^4$ acts on $\mathbb{C}^{***}$ by $z_{11} \mapsto z_{11}e^{-2\pi i/3}$, especially τ_1 maps to $z_{11} = 0$. On the $\mathbb{C}^*$-fiber over $z_{11} = 0$ $(\sigma_1\sigma_2)^4$ acts because of $\kappa(\tau_1 + 1) = \kappa(e^{2\pi i/6}) = \frac{2\pi i}{6}$ by

$$z_{12} = e^{\zeta/4} \mapsto e^{(\zeta-3\pi i+\kappa(\tau_1+1))/4} = z_{12}e^{-2\pi i3/8+2\pi i/24} = z_{12}e^{-2\pi i/3}.$$

By Remark 9.4 (ii) we can choose the trivialization of the $\mathbb{C}^*$-bundle $\mathbb{C}^* \times \mathbb{C}^{***}$ over $\mathbb{C}^{***}$ with coordinates (z_{12}, z_{11}) such that the action of $(\sigma_1\sigma_2)^4$ on it becomes the action $(z_{12}, z_{11}) \mapsto (z_{12}e^{-2\pi i/3}, z_{11}e^{-2\pi i/3})$.

$C_3^{S/\{\pm 1\}^3}$ is obtained by dividing out the action of σ^{mon}, because $(Br_3)_{\underline{x}/\{\pm 1\}^3} = \langle\sigma_1, \sigma_2^2\rangle$ is generated by $(Br_3)_{\underline{e}/\{\pm 1\}^3}$ and by σ^{mon}, with $(\sigma^{mon})^4 \in (Br_3)_{\underline{x}/\{\pm 1\}^3}$.

We obtain the following diagram,

$$\begin{array}{ccc}
(z_1, z_{12}, z_{11}) & \longrightarrow & (z_1, z_{12}^4, z_{11}) = (z_1, z_2, z_{11}) \\
\mathbb{C} \times \mathbb{C}^* \times \mathbb{C}^{***} & \xrightarrow{4:1} & \mathbb{C} \times \mathbb{C}^* \times \mathbb{C}^{***} \\
\downarrow 3:1 & & \downarrow 3:1 \\
\mathbb{C} \times \frac{\mathbb{C}^* \times \mathbb{C}^{***}}{\langle (z_{12},z_{11})\mapsto(z_{12}e^{2\pi i/3},z_{11}e^{2\pi i/3})\rangle} & \xrightarrow{4:1} & \mathbb{C} \times \frac{\mathbb{C}^* \times \mathbb{C}^{***}}{\langle (z_2,z_{11})\mapsto(z_2e^{2\pi i/3},z_{11}e^{2\pi i/3})\rangle} \\
\cong C_3^{\underline{e}/\{\pm 1\}^3} & & \cong C_3^{S/\{\pm 1\}^3}
\end{array}$$

Though $C_3^{\underline{e}/\{\pm 1\}^3}$ and $C_3^{S/\{\pm 1\}^3}$ can be presented in a better way. Write $\xi := e^{2\pi i/3}$. The following two diagrams belong together. The upper diagram shows the sets, the lower diagram shows the maps. The horizontal maps are isomorphisms. In the line of $C_3^{\underline{e}/\{\pm 1\}^3}$ the set on the right side is nicer than the set on the left side. In the line of $C_3^{S/\{\pm 1\}^3}$ both sets are quotients.

$$\begin{array}{ccc}
C_3^{\underline{e}/\{\pm 1\}^3} \cong \frac{\mathbb{C}\times\mathbb{C}^*\times\mathbb{C}^{***}}{\langle (z_1,z_{12},z_{11})\mapsto(z_1,z_{12}\xi,z_{11}\xi)\rangle} & \xrightarrow{\cong} & \mathbb{C} \times \{(z_{13}, z_{14}) \in \mathbb{C}^* \times \mathbb{C} \,|\, z_{14}^3 - z_{13}^2 = 0\} \\
\downarrow 4:1 & & \downarrow 4:1 \\
C_3^{S/\{\pm 1\}^3} \cong \frac{\mathbb{C}\times\mathbb{C}^*\times\mathbb{C}^{***}}{\langle (z_1,z_2,z_{11})\mapsto(z_1,z_2\xi,z_{11}\xi)\rangle} & \xrightarrow{\cong} & \frac{\mathbb{C}\times\{(z_{15},z_{14})\in\mathbb{C}^*\times\mathbb{C}\,|\,z_{14}^3-z_{15}=0\}}{\langle (z_1,z_{15},z_{14})\mapsto(z_1,-z_{15},-z_{14})\rangle}
\end{array}$$

$$\begin{array}{ccc}
[(z_1, z_{12}, z_{11})] & \longrightarrow & (z_1, z_{12}^3, z_{12}^2 z_{11}) \;= (z_1, z_{13}, z_{14}) \\
\downarrow 4:1 & & \downarrow 4:1 \\
[(z_1, z_{12}^4, z_{11})] & & [(z_1, z_{13}^2, z_{14})] \\
\| & & \| \\
[(z_1, z_2, z_{11})] & \longrightarrow & [(z_1, z_2^{3/2}, z_2^{1/2} z_{11})] = [(z_1, z_{15}, z_{14})]
\end{array}$$

We claim that geometrically, the horizontal maps are the restrictions of the inversions on the level of $C_3^{\underline{e}/\{\pm 1\}^3}$ and $C_3^{S/\{\pm 1\}^3}$ of the blowing up described in Remark 9.2 (ii).

The cases $\mathcal{G}_5$ & C_8, C_9 $((-l, 2, -l))$:
Here $(\mathrm{Br}_3)_{\underline{e}/\{\pm 1\}^3} = \langle (\sigma^{mon})^2 \sigma_1^{-1} \sigma_2^{l^2-4} \sigma_1 \rangle$ and $(\mathrm{Br}_3)_{\underline{x}/\{\pm 1\}^3} = \langle \sigma^{mon}, \sigma_1^{-1}\sigma_2\sigma_1 \rangle$. On the last factor $\mathbb{H}$ of $\mathbb{C} \times \mathbb{C} \times \mathbb{H} \cong C_3^{univ}$, $\sigma_1^{-1}\sigma_2\sigma_1$ and $\sigma_1^{-1}\sigma_2^{l^2-4}\sigma_1 = (\sigma_1^{-1}\sigma_2\sigma_1)^{l^2-4}$ act as parabolic elements with fixed point $1 \in \widehat{\mathbb{R}}$, so freely, and σ^{mon} acts trivially. The quotients of $\mathbb{H}$ by the actions of $\sigma_1^{-1}\sigma_2^{l^2-4}\sigma_1$ and $\sigma_1^{-1}\sigma_2\sigma_1$ are both isomorphic to $\mathbb{D}^* = \mathbb{D} - \{0\}$, where $\mathbb{D} = \{z \in \mathbb{C} \mid |z| < 1\}$ is the unit disk, and the covering is as follows,

$$\mathbb{H}/(\text{action of } \sigma_1^{-1}\sigma_2^{l^2-4}\sigma_1) \cong \mathbb{D}^* \to \mathbb{D}^* \cong \mathbb{H}/(\text{action of } \sigma_1^{-1}\sigma_2\sigma_1),$$
$$z_{16} \mapsto z_{16}^{l^2-4}.$$

Recall that $\psi(\sigma_2^2)$ acts trivially on the second factor of $\mathbb{C} \times \mathbb{C} \times \mathbb{H} \cong C_3^{univ}$. The total action of $\psi((\sigma^{mon})^2 \sigma_1^{-1}\sigma_2^{l^2-4}\sigma_1)$ on $\mathbb{C} \times \mathbb{C} \times \mathbb{H} \cong C_3^{univ}$ is as follows, for even l:

$$\begin{aligned}
(z_1, \zeta, \tau) &\overset{\sigma_1}{\mapsto} (z_1, \zeta - \pi i, \tau - 1) \\
&\overset{\sigma_2^{l^2-4}}{\mapsto} (z - 1, \zeta - \pi i, \frac{\tau - 1}{(l^2-4)(\tau - 1) + 1}) \\
&\overset{\sigma_1^{-1}}{\mapsto} (z_1, \zeta, \frac{(\tau - 1) + (l^2-4)(\tau - 1) + 1}{(l^2-4)(\tau - 1) + 1}) \\
&\overset{(\sigma^{mon})^2}{\mapsto} (z_1, \zeta - 4\pi i, \frac{(l^2-3)(\tau - 1) + 1}{(l^2-4)(\tau - 1) + 1}),
\end{aligned}$$

and for odd l:

$$(z_1, \zeta, \tau) \mapsto (z_1, \zeta - 4\pi i + \kappa(\tau), \frac{(l^2-3)(\tau - 1) + 1}{(l^2-4)(\tau - 1) + 1}),$$

Because the action on the third factor $\mathbb{H}$ is free, the action on the second factor $\mathbb{C}$ is not important for the quotient up to isomorphism. The quotient is for even l and for odd l

$$C_3^{\underline{e}/\{\pm 1\}^3} \cong \mathbb{C} \times (\text{an affine } \mathbb{C}\text{-bundle over } \mathbb{D}^*)$$

$$\cong \mathbb{C} \times \mathbb{C} \times \mathbb{D}^* \quad \text{with coordinates } (z_1, \zeta, z_{16})$$

by Remark 9.4 (ii).

Dividing out the action of σ^{mor} gives

$$\begin{aligned} C_3^{\underline{e}/\{\pm 1\}^3} \cong \mathbb{C} \times \mathbb{C} \times \mathbb{D}^* &\to \mathbb{C} \times \mathbb{C}^* \times \mathbb{D}^* \\ (z_1, \zeta, z_{16}) &\mapsto (z_1, e^{\zeta}, z_{16}) = (z_1, z_2, z_{16}). \end{aligned}$$

On this quotient $\sigma_1^{-1}\sigma_2\sigma_1$ acts as a cyclic automorphism of order $l^2 - 4$ because $(\sigma_1^{-1}\sigma_2\sigma_1)^{l^2-4} \in \langle \sigma^{mon}, (\sigma^{mon})^2\sigma_1^{-1}\sigma_2^{l^2-4}\sigma_1\rangle$. The action on the third factor $\mathbb{D}^*$ is free. Therefore we can choose a posteriori the trivialization $\mathbb{C} \times \mathbb{D}^*$ with coordinates (ζ, z_{16}) and $\mathbb{C}^* \times \mathbb{D}^*$ with coordinates (z_2, z_{16}) so that we obtain the following diagram,

$$\begin{array}{ccccc} & (z_1, \zeta, z_{16}) & \longrightarrow & (z_1, e^{\zeta}, z_{16}) & \\ C_3^{\underline{e}/\{\pm 1\}^3} \cong & \mathbb{C} \times \mathbb{C} \times \mathbb{D}^* & \xrightarrow{/\langle \sigma^{mon}\rangle} & \mathbb{C} \times \mathbb{C}^* \times \mathbb{D}^* & (z_1, z_2, z_{16}) \\ & & & \downarrow{\scriptstyle (l^2-4):1} & \downarrow{\scriptstyle /\langle \sigma_1^{-1}\sigma_2\sigma_1\rangle} \\ & C_3^{S/\{\pm 1\}^3} \cong & & \mathbb{C} \times \mathbb{C}^* \times \mathbb{D}^* & (z_1, z_2, z_{16}^{l^2-4}) \end{array}$$

The cases $\mathcal{G}_6$ & C_{10} ($\mathbb{P}^2$), C_{11}, C_{12}:

Here $(\mathrm{Br}_3)_{\underline{e}/\{\pm 1\}^3} = \{\mathrm{id}\}$ and $(\mathrm{Br}_3)_{\underline{x}/\{\pm 1\}^3} = \langle \sigma_2\sigma_1\rangle$. Therefore

$$C_3^{\underline{e}/\{\pm 1\}^3} = C_3^{univ} \cong \mathbb{C} \times \mathbb{C} \times \mathbb{H}.$$

In order to obtain $C_3^{S/\{\pm 1\}^3}$ it is a priori sufficient to divide out the action of $\sigma_2\sigma_1$. But it is easier to first divide out the action of $\sigma^{mon} = (\sigma_2\sigma_1)^3$. On the quotient $C_3^{\underline{e}/\{\pm 1\}^3}/\langle \sigma^{mon}\rangle \cong \mathbb{C} \times \mathbb{C}^* \times \mathbb{H}$ with coordinates (z_1, z_2, τ) $\sigma_2\sigma_1$ acts as automorphism of order three. Concretely,

$$\begin{array}{ccccc} & (z_1, \zeta, \tau) & \longrightarrow & (z_1, \zeta - \pi i + \kappa(\tau), \frac{\tau-1}{\tau}) \\ (z_1, \zeta, \tau) & C_3^{\underline{e}/\{\pm 1\}^3} \cong \mathbb{C}^2 \times \mathbb{H} & \xrightarrow{\sigma_2\sigma_1} & C_3^{\underline{e}/\{\pm 1\}^3} \cong \mathbb{C}^2 \times \mathbb{H} \\ \downarrow & \downarrow{\scriptstyle /\langle \sigma^{mon}\rangle} & & \downarrow{\scriptstyle /\langle \sigma^{mon}\rangle} \\ (z_1, e^{\zeta}, \tau) & \mathbb{C} \times \mathbb{C}^* \times \mathbb{H} & \xrightarrow{\sigma_2\sigma_1} & \mathbb{C} \times \mathbb{C}^* \times \mathbb{H} \\ & (z_1, z_2, \tau) & \longrightarrow & (z_1, z_2(-T(\tau)), \frac{\tau-1}{\tau}) \end{array}$$

The action is fixed point free. One reason is that $C_3^{S/\{\pm 1\}^3}$ is smooth. Another reason is given by the following chain of arguments. The Möbius transformation $\mu_{(123)} = (\tau \mapsto \frac{\tau-1}{\tau})$ is elliptic of order three with fixed point $e^{2\pi i/6}$. The quotient is again isomorphic to $\mathbb{H}$. On the fiber of $\mathbb{C}^* \times \mathbb{H}$ over $e^{2\pi i/6}$, $\sigma_2\sigma_1$ acts by multiplication with $-T(e^{2\pi i/6}) = -e^{2\pi i/6} = e^{-2\pi i/3}$, so fixed point free. Also

$$(-T(\tau))\left(-T\left(\frac{\tau-1}{\tau}\right)\right)\left(-T\left(\frac{-1}{\tau-1}\right)\right) = -T(\tau)\frac{T(\tau)-1}{T(\tau)}\frac{1}{1-T(\tau)} = 1$$

because of Theorem B.1 (d). Therefore $\sigma_2\sigma_1$ acts on $\mathbb{C} \times \mathbb{C}^* \times \mathbb{H}$ fixed point free and has order three.

We obtain the following diagram,

$$\begin{array}{ccc}
(z_1, \zeta, \tau) & \longrightarrow & (z_1, e^{\zeta}, \tau) = (z_1, z_2, \tau) \\
C_3^{\underline{e}/\{\pm 1\}^3} \cong \mathbb{C} \times \mathbb{C} \times \mathbb{H} & \xrightarrow{/\langle \sigma^{mon} \rangle} & \mathbb{C} \times \mathbb{C}^* \times \mathbb{H} \\
 & & \big\downarrow /\langle \sigma_2\sigma_1 \rangle \\
C_3^{S/\{\pm 1\}^3} \cong & & \mathbb{C} \times \frac{\mathbb{C}^* \times \mathbb{H}}{\langle (z_2,\tau) \mapsto (z_2(-T(\tau)), \frac{\tau-1}{\tau}) \rangle}
\end{array}$$

$C_3^{S/\{\pm 1\}^3}$ is $\mathbb{C} \times$ (a $\mathbb{C}^*$-orbibundle over $\mathbb{H}/\langle \mu_{(123)} \rangle$) in the sense of Remark 9.5 with the only exceptional fiber over $[e^{2\pi i/6}] \in \mathbb{H}/\langle \mu_{(123)} \rangle$.

The cases $\mathcal{G}_7$ & C_{13} (e.g. (4, 4, 8)):
Here $(\mathrm{Br}_3)_{\underline{e}/\{\pm 1\}^3} = \{\mathrm{id}\}$ and $(\mathrm{Br}_3)_{\underline{x}/\{\pm 1\}^3} = \langle \sigma_2\sigma_1^2 \rangle$. Therefore

$$C_3^{\underline{e}/\{\pm 1\}^3} = C_3^{univ} \cong \mathbb{C} \times \mathbb{C} \times \mathbb{H}.$$

In order to obtain $C_3^{S/\{\pm 1\}^3}$ it is a priori sufficient to divide out the action of $\sigma_2\sigma_1^2$. But it is easier to first divide out the action of $\sigma^{mon} = (\sigma_2\sigma_1^2)^2$. On the quotient $C_3^{\underline{e}/\{\pm 1\}^3}/\langle \sigma^{mon} \rangle \cong \mathbb{C} \times \mathbb{C}^* \times \mathbb{H}$ with coordinates (z_1, z_2, τ) $\sigma_2\sigma_1^2$ acts as automorphism of order two. Concretely,

$$\begin{array}{cccc}
 & (z_1, \zeta, \tau) & \longrightarrow & (z_1, \zeta - 2\pi i + \kappa(\tau - 1), \frac{\tau-2}{\tau-1}) \\
(z_1, \zeta, \tau) & C_3^{\underline{e}/\{\pm 1\}^3} \cong \mathbb{C}^2 \times \mathbb{H} & \xrightarrow{\sigma_2\sigma_1^2} & C_3^{\underline{e}/\{\pm 1\}^3} \cong \mathbb{C}^2 \times \mathbb{H} \\
\big\downarrow & \big\downarrow /\langle \sigma^{mon} \rangle & & \big\downarrow /\langle \sigma^{mon} \rangle \\
(z_1, e^{\zeta}, \tau) & \mathbb{C} \times \mathbb{C}^* \times \mathbb{H} & \xrightarrow{\sigma_2\sigma_1^2} & \mathbb{C} \times \mathbb{C}^* \times \mathbb{H} \\
 & (z_1, z_2, \tau) & \longrightarrow & (z_1, z_2(1 - T(\tau)), \frac{\tau-2}{\tau-1})
\end{array}$$

The action is fixed point free. One reason is that $C_3^{S/\{\pm 1\}^3}$ is smooth. Another reason is given by the following chain of arguments. The Möbius transformation $(\tau \mapsto \frac{\tau-2}{\tau-1})$ is elliptic of order two with fixed point $1 + i$. The quotient is again isomorphic to $\mathbb{H}$.

On the fiber of $\mathbb{C}^* \times \mathbb{H}$ over $1+i$, $\sigma_2\sigma_1^2$ acts by multiplication with $1 - T(1+i) = 1 - 2 = -1$, so fixed point free. Also

$$\begin{aligned}(1-T(\tau))\left(1-T\left(\frac{\tau-2}{\tau-1}\right)\right) &= (1-T(\tau))(1-T(\mu_{(123)}\mu_{(12)}))\\ &= (1-T(\tau))(1-g_{(123)}g_{(12)}(T(\tau)))\\ &= (1-T(\tau))\left(1-\frac{(1-T(\tau))-1}{1-T(\tau)}\right) = 1\end{aligned}$$

because of Theorem B.1 (d). Therefore $\sigma_2\sigma_1$ acts on $\mathbb{C} \times \mathbb{C}^* \times \mathbb{H}$ fixed point free and has order two.

We obtain the following diagram,

$$\begin{array}{ccc} (z_1, \zeta, \tau) & \longrightarrow & (z_1, e^\zeta, \tau) = (z_1, z_2, \tau) \\ C_3^{e/\{\pm 1\}^3} \cong \mathbb{C} \times \mathbb{C} \times \mathbb{H} & \xrightarrow{/\langle \sigma^{mon}\rangle} & \mathbb{C} \times \mathbb{C}^* \times \mathbb{H} \\ & & \downarrow /\langle \sigma_2\sigma_1^2\rangle \\ C_3^{S/\{\pm 1\}^3} \cong & & \mathbb{C} \times \frac{\mathbb{C}^* \times \mathbb{H}}{\langle (z_2,\tau)\mapsto (z_2(1-T(\tau)), \frac{\tau-2}{\tau-1})\rangle} \end{array}$$

$C_3^{S/\{\pm 1\}^3}$ is $\mathbb{C} \times$ (a $\mathbb{C}^*$-orbibundle over $\mathbb{H}/\langle(\tau \mapsto \frac{\tau-2}{\tau-1})\rangle$) in the sense of Remark 9.5 with the only exceptional fiber over $[1+i] \in \mathbb{H}/\langle(\tau \mapsto \frac{\tau-2}{\tau-1})\rangle$.

The cases $\mathcal{G}_8$ & C_{14} and $\mathcal{G}_9$ & $C_{15}, C_{16}, C_{23}, C_{24}$:
Here $(\mathrm{Br}_3)_{\underline{e}/\{\pm 1\}^3} = \{\mathrm{id}\}$ and $(\mathrm{Br}_3)_{\underline{x}/\{\pm 1\}^3} = \langle \sigma^{mon}\rangle$. Therefore

$$\begin{array}{ccc} C_3^{\underline{e}/\{\pm 1\}^3} = C_3^{univ} \cong \mathbb{C} \times \mathbb{C} \times \mathbb{H} & \xrightarrow{/\langle \sigma^{mon}\rangle} & \mathbb{C} \times \mathbb{C}^* \times \mathbb{H} \cong C_3^{S/\{\pm 1\}^3} \\ (z_1, \zeta, \tau) & \longrightarrow & (z_1, e^\zeta, \tau) = (z_1, z_2, \tau). \end{array}$$

The cases $\mathcal{G}_{10}$ & C_{17} (e.g. $(-2, -2, 0)$)**:**
Here $(\mathrm{Br}_3)_{\underline{e}/\{\pm 1\}^3} = \langle \sigma_2^2\rangle$ and $(\mathrm{Br}_3)_{\underline{x}/\{\pm 1\}^3} = \langle \sigma^{mon}, \sigma_2\rangle$. The deck transformation $\psi(\sigma_2^2)$ of $C_3^{univ} \cong \mathbb{C} \times \mathbb{C} \times \mathbb{H}$ acts nontrivially only on the third factor $\mathbb{H}$, and there it acts as the parabolic transformation $(\tau \mapsto \frac{\tau}{2\tau+1})$ with fixed point $0 \in \widehat{\mathbb{R}}$ and quotient $\mathbb{H}/\langle(\tau \mapsto \frac{\tau}{2\tau+1})\rangle \cong \mathbb{D}^*$. Therefore

$$C_3^{\underline{e}/\{\pm 1\}^3} \cong \mathbb{C} \times \mathbb{C} \times \mathbb{D}^* \quad \text{with coordinates } (z_1, \zeta, z_{16}).$$

On this quotient σ_2 acts as fixed point free automorphism of order two with the action $\mathbb{D}^* \to \mathbb{D}^*$, $z_{16} \mapsto z_{16}^2 = z_{17}$, on the last factor. The quotient is isomorphic to $\mathbb{C} \times \mathbb{C} \times \mathbb{D}^*$ with coordinates (z_1, ζ, z_{17}). σ^{mon} acts on the quotient with nontrivial action $\zeta \mapsto \zeta - 2\pi i$ only on the second factor $\mathbb{C}$. We obtain the following diagram,

$$\begin{array}{ccccc} (z_1, \zeta, z_{16}) & \longrightarrow & (z_1, \zeta + \kappa(\tau+1), z_{16}^2) & \text{with } z_{16} = [\tau] & \\ C_3^{\underline{e}/\{\pm1\}^3} \cong \mathbb{C}\times\mathbb{C}\times\mathbb{D}^* & \xrightarrow{/\langle\sigma_2\rangle} & \mathbb{C}\times\mathbb{C}\times\mathbb{D}^* & & (z_1, \zeta, z_{17}) \\ & & \downarrow /\langle\sigma^{mon}\rangle & & \downarrow \\ C_3^{S/\{\pm1\}^3} \cong & & \mathbb{C}\times\mathbb{C}^*\times\mathbb{D}^* & & (z_1, e^\zeta, z_{17}) \end{array}$$

The cases $\mathcal{G}_{11}$ & C_{18} (e.g. $(-3,-2,0)$):
Here $(\mathrm{Br}_3)_{\underline{e}/\{\pm1\}^3} = \langle\sigma_2^2\rangle$ and $(\mathrm{Br}_3)_{\underline{x}/\{\pm1\}^3} = \langle\sigma^{mon}, \sigma_2^2\rangle$. The manifold $C_3^{\underline{e}/\{\pm1\}^3}$ is the same quotient of C_3^{univ} as in the cases $\mathcal{G}_{10}$ & C_{17}, namely

$$C_3^{\underline{e}/\{\pm1\}^3} \cong \mathbb{C}\times\mathbb{C}\times\mathbb{D}^* \quad \text{with coordinates } (z_1, \zeta, z_{16}).$$

σ^{mon} acts nontrivially only on the second factor $\mathbb{C}$ of this manifold, by $\zeta \mapsto \zeta - 2\pi i$. We obtain the following diagram,

$$\begin{array}{ccc} C_3^{\underline{e}/\{\pm1\}^3} \cong \mathbb{C}\times\mathbb{C}\times\mathbb{D}^* & \xrightarrow{/\langle\sigma^{mon}\rangle} & \mathbb{C}\times\mathbb{C}^*\times\mathbb{D}^* \cong C_3^{S/\{\pm1\}^3} \\ (z_1, \zeta, z_{16}) & \longrightarrow & (z_1, e^\zeta, z_{16}) = (z_1, z_2, z_{16}). \end{array}$$

The cases $\mathcal{G}_{12}$ & C_{19} (e.g. $(-2,-1,0)$):
Here $(\mathrm{Br}_3)_{\underline{e}/\{\pm1\}^3} = \langle\sigma_2^2, \sigma_2\sigma_1^3\sigma_2^{-1}\rangle$ and $(\mathrm{Br}_3)_{\underline{x}/\{\pm1\}^3} = \langle\sigma^{mon}, \sigma_2^2, \sigma_2\sigma_1^3\sigma_2^{-1}\rangle$. The group $\langle\sigma_2^2, \sigma_2\sigma_1^3\sigma_2^{-1}\rangle$ acts on $\mathbb{H}$ as the group $\langle[A_2^2], [A_2A_1^3A_2^{-1}]\rangle \subset PSL_2(\mathbb{Z})$. Both generators are parabolic. Consider the hyperbolic polygon P whose relative boundary consists of the four arcs $A(\infty,0)$, $A(0,\frac{1}{2})$, $A(\frac{2}{3},1)$, $A(1,\infty)$, see the left picture in Fig. 9.2.

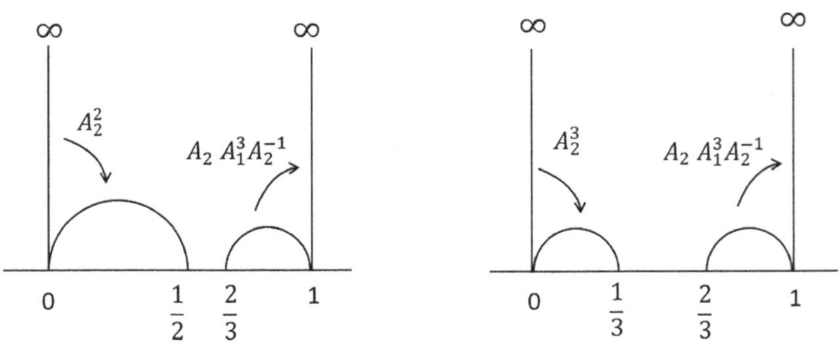

Fig. 9.2 Fundamental domains in $\mathbb{H}$ for the subgroups $\langle[A_2^2], [A_2A_1^3A_2^{-1}]\rangle$ (left) and $\langle[A_2^3], [A_2A_1^3A_2^{-1}]\rangle$ (right) of $PSL_2(\mathbb{Z})$

Recall

$$A_1 = \begin{pmatrix} 1 & -1 \\ 0 & 1 \end{pmatrix}, \ A_2 = \begin{pmatrix} 1 & 0 \\ 1 & 1 \end{pmatrix}, \ A_2^2 = \begin{pmatrix} 1 & 0 \\ 2 & 1 \end{pmatrix},$$
$$A_2 A_1 A_2^{-1} = \begin{pmatrix} 2 & -1 \\ 0 & 1 \end{pmatrix}, \ A_2 A_1^3 A_2^{-1} = (A_2 A_1 A_2^{-1})^3 = \begin{pmatrix} 4 & -3 \\ 3 & -2 \end{pmatrix}.$$

$[A_2^2]$ has the fixed point $0 \in \widehat{\mathbb{R}}$ and maps $A(\infty, 0)$ to $A(0, \frac{1}{2})$, and $[A_2 A_1^3 A_2^{-1}]$ has the fixed point $1 \in \widehat{\mathbb{R}}$ and maps $A(\frac{2}{3}, 1)$ to $A(1, \infty)$. By Theorem A.2 (c) the hyperbolic polygon P is a fundamental domain for the group $\langle [A_2^2], [A_2 A_1^3 A_2^{-1}] \rangle$. Because of the two parabolic fixed points and the euclidean boundary $[\frac{1}{2}, \frac{2}{3}]$ of P, the quotient $\mathbb{H}/\langle [A_2^2], [A_2 A_1^3 A_2^{-1}] \rangle$ is isomorphic to $\mathbb{D} - \{0, \frac{1}{2}\} =: \mathbb{D}^{**}$ where $z_{16} = 0$ is the image of $0 \in \widehat{\mathbb{R}}$ and $z_{16} = \frac{1}{2}$ is the image of $1 \in \widehat{\mathbb{R}}$.

As the action of $\langle \sigma_2^2, \sigma_2 \sigma_1^3 \sigma_2^{-1} \rangle$ on the third factor $\mathbb{H}$ of $\mathbb{C} \times \mathbb{C} \times \mathbb{H} \cong C_3^{univ}$ is free, the quotient of $\mathbb{C} \times \mathbb{H}$ with coordinates (ζ, τ) is an affine $\mathbb{C}$-bundle over $\mathbb{H}/\langle [A_2^2], [A_2 A_1^3 A_2^{-1}] \rangle$, so

$$\begin{aligned} C_3^{\underline{e}/\{\pm 1\}^3} &\cong \mathbb{C} \times (\text{an affine } \mathbb{C}\text{-bundle over } \mathbb{D}^{**}) \\ &\cong \mathbb{C} \times \mathbb{C} \times \mathbb{D}^{**} \quad \text{with coordinates } (z_1, \zeta, z_{16}) \end{aligned}$$

by Remark 9.4 (ii). σ^{mon} acts nontrivially only on the second factor of this product, by the action $\zeta \to \zeta - 2\pi i$. Therefore

$$\begin{array}{ccc} C_3^{\underline{e}/\{\pm 1\}^3} \cong \mathbb{C} \times \mathbb{C} \times \mathbb{D}^{**} & \xrightarrow{/\langle \sigma^{mon} \rangle} & \mathbb{C} \times \mathbb{C}^* \times \mathbb{D}^{**} \cong C_3^{S/\{\pm 1\}^3} \\ (z_1, \zeta, z_{16}) & \longrightarrow & (z_1, e^{\zeta}, z_{16}) = (z_1, z_2, z_{16}). \end{array}$$

The cases $\mathcal{G}_{13}$ & C_{20} (e.g. $(-2, -1, -1)$):
Here $(\mathrm{Br}_3)_{\underline{e}/\{\pm 1\}^3} = \langle \sigma_2^3, \sigma_2 \sigma_1^3 \sigma_2^{-1} \rangle$ and $(\mathrm{Br}_3)_{\underline{x}/\{\pm 1\}^3} = \langle \sigma^{mon}, \sigma_2^3, \sigma_2 \sigma_1^3 \sigma_2^{-1} \rangle$. This is analogous to the cases $\mathcal{G}_{12}$ & C_{19}. The group $\langle \sigma_2^3, \sigma_2 \sigma_1^3 \sigma_2^{-1} \rangle$ acts on $\mathbb{H}$ as the group $\langle [A_2^3], [A_2 A_1^3 A_2^{-1}] \rangle \subset PSL_2(\mathbb{Z})$. Both generators are parabolic. Consider the hyperbolic polygon P whose relative boundary consists of the four arcs $A(\infty, 0)$, $A(0, \frac{1}{3})$, $A(\frac{2}{3}, 1)$, $A(1, \infty)$, see the right picture in Fig. 9.2.

$[A_2^3]$ has the fixed point 0 and maps $A(\infty, 0)$ to $A(0, \frac{1}{3})$, and $[A_2 A_1^3 A_2^{-1}]$ has the fixed point 1 and maps $A(\frac{2}{3}, 1)$ to $A(1, \infty)$. By Theorem A.2 (c) the hyperbolic polygon P is a fundamental domain for the group $\langle [A_2^3], [A_2 A_1^3 A_2^{-1}] \rangle$. As in the cases $\mathcal{G}_{12}$ & C_{19} we obtain

$$\begin{array}{ccc} C_3^{\underline{e}/\{\pm 1\}^3} \cong \mathbb{C} \times \mathbb{C} \times \mathbb{D}^{**} & \xrightarrow{/\langle \sigma^{mon} \rangle} & \mathbb{C} \times \mathbb{C}^* \times \mathbb{D}^{**} \cong C_3^{S/\{\pm 1\}^3} \\ (z_1, \zeta, z_{16}) & \longrightarrow & (z_1, e^{\zeta}, z_{16}) = (z_1, z_2, z_{16}). \end{array}$$

The cases $\mathcal{G}_{14}$ & C_{21}, C_{22}:
Here $(\mathrm{Br}_3)_{\underline{e}/\{\pm 1\}^3} = \langle \sigma_2^3 \rangle$ and $(\mathrm{Br}_3)_{\underline{x}/\{\pm 1\}^3} = \langle \sigma^{mon}, \sigma_2^3 \rangle$. The group $\langle \sigma_2^3 \rangle$ acts on the third factor $\mathbb{H}$ of $\mathbb{C} \times \mathbb{C} \times \mathbb{H} \cong C_3^{univ}$ as the group $\langle [A_2^3] \rangle \subset PSL_2(\mathbb{Z})$. The generator $[A_2^3]$ is parabolic. Therefore the quotient $\mathbb{H}/\langle [A_2^3] \rangle$ is isomorphic to $\mathbb{D}^*$. We obtain

$$\begin{aligned} C_3^{\underline{e}/\{\pm 1\}^3} &\cong \mathbb{C} \times (\text{affine } \mathbb{C}\text{-bundle over } \mathbb{D}^*) \\ &\cong \mathbb{C} \times \mathbb{C} \times \mathbb{D}^* \quad \text{with coordinates } (z_1, \zeta, z_{16}) \end{aligned}$$

by Remark 9.4 (ii) and the diagram

$$\begin{array}{ccc} C_3^{\underline{e}/\{\pm 1\}^3} \cong \mathbb{C} \times \mathbb{C} \times \mathbb{D}^* & \xrightarrow{/\langle \sigma^{mon} \rangle} & \mathbb{C} \times \mathbb{C}^* \times \mathbb{D}^* \cong C_3^{\underline{S}/\{\pm 1\}^3} \\ (z_1, \zeta, z_{16}) & \longrightarrow & (z_1, e^{\zeta}, z_{16}) = (z_1, z_2, z_{16}). \end{array}$$

□

Remark 9.4 Let X be a non-compact smooth complex curve.

(i) An *affine* $\mathbb{C}$*-bundle over* X is a holomorphic bundle $p : Y \to X$ such that the following exist: An open covering $(U_i)_{i \in I}$ of X, isomorphisms $f_i : p^{-1}(U_i) \to \mathbb{C} \times U_i$, and for i and $j \in I$ with $i \neq j$ and $U_i \cap U_j \neq \emptyset$ a holomorphic map $f_{ij} \in \mathcal{O}(U_i \cap U_j)$ with

$$\begin{aligned} \mathbb{C} \times (U_i \cap U_j) \xrightarrow{f_i^{-1}} p^{-1}(U_i \cap U_j) &\xrightarrow{f_j} \mathbb{C} \times (U_i \cap U_j), \\ (z, x) &\mapsto (z + f_{ij}(x), x). \end{aligned}$$

(ii) $H^1(X, \mathcal{O}) = 0$ (e.g. [Fo77, 26.1 Satz]). This sheaf cohomology coincides with the Čech cohomology $H^1((U_i)_{i \in I}, \mathcal{O})$ with values in $\mathcal{O}$ (e.g. [Fo77, 12.8 Satz]). Therefore any affine $\mathbb{C}$-bundle over X is trivial, so isomorphic to the affine $\mathbb{C}$-bundle $\mathbb{C} \times X$.

(iii) Let $a : \mathbb{C} \times X \to \mathbb{C} \times X$ be an automorphism of some finite order of $\mathbb{C} \times X$ as affine $\mathbb{C}$-bundle over X, with a discrete set of fixed points in X, such that a acts as identity on the fiber of each fixed point. Then $(\mathbb{C} \times X)/\langle a \rangle$ is again an affine $\mathbb{C}$-bundle. The pull back of a trivialization of this quotient bundle gives a new trivialization of $\mathbb{C} \times X$ such that with respect to this new trivialization a acts nontrivially only on the base X, $a : (z, x) \mapsto (z, a_2(x))$, where $a_2 : X \to X$ is the induced automorphism of the base X.

(iv) A *line bundle on* X means a holomorphic rank 1 vector bundle. Any line bundle on X is trivial, so isomorphic to $\mathbb{C} \times X$ as vector bundle (e.g. [Fo77, 30.3 Satz]). Therefore any $\mathbb{C}^*$-bundle on X is trivial, so isomorphic to $\mathbb{C}^* \times X$.

9.3 Partial Compactifications of the F-Manifolds

Theorem 9.3 gives all possible manifolds $C_3^{\underline{e}/\{\pm 1\}^3}$ and $C_3^{S/\{\pm 1\}^3}$ as complex manifolds. The (long) proof describes also the coverings $C_3^{\underline{e}/\{\pm 1\}^3} \to C_3^{S/\{\pm 1\}^3}$. But the F-manifold structure and the Euler field as well as possible partial compactifications are not treated in Theorem 9.3. They are subject of this section. Corollary 9.8 gives the Euler fields. We refrain from writing down the multiplications in each case. They can in principle be extracted from the construction as quotients of C_3^{univ} or as coverings of C_3^{conf}.

But we care about possible partial compactifications of $C_3^{\underline{e}/\{\pm 1\}^3}$ such that the F-manifold structure extends. Lemma 9.7 says something conceptually about such partial compactifications. Corollary 9.8 writes them down in all cases and makes the Maxwell strata $\mathcal{K}_2$ and the caustics $\mathcal{K}_3$ in them explicit.

Especially interesting are the cases A_3 and A_2A_1. In both cases $C_3^{\underline{e}/\{\pm 1\}^3}$ has a partial compactification to $\mathbb{C}^3$ which is well known from singularity theory. To recover it starting from $C_3^{\underline{e}/\{\pm 1\}^3}$ is especially in the case A_3 a bit involved. It requires in the case A_3 the notion of a $\mathbb{C}^*$-orbibundle, which is given in the Remarks 9.5, and the Example 9.6.

Remark 9.5 Let $\widetilde{X}$ be a smooth complex curve, compact or not. A *$\mathbb{C}^*$-orbibundle over $\widetilde{X}$* generalizes slightly the notion of a $\mathbb{C}^*$-bundle over $\widetilde{X}$. It is a complex two dimensional manifold Y with a projection $p : Y \to \widetilde{X}$ and a $\mathbb{C}^*$-action which restricts to a transitive $\mathbb{C}^*$-action on each fiber and trivial stabilizer on almost each fiber. There is a discrete subset $\Sigma \subset \widetilde{X}$ such that the stabilizer of the $\mathbb{C}^*$-action is trivial on $p^{-1}(q)$ for $q \in \widetilde{X} - \Sigma$ and nontrivial on $p^{-1}(q)$ for $q \in \Sigma$.

Then $Y|_{\widetilde{X}-\Sigma}$ is a $\mathbb{C}^*$-bundle over $\widetilde{X} - \Sigma$. For any $q \in \Sigma$ the restriction $p^{-1}(\Delta) \subset Y$ for a small disk $\Delta \subset \widetilde{X}$ with center q and $\Delta \cap \Sigma = \{q\}$ is isomorphic to one of the following models,

$$\begin{aligned} Y_{a,m} &:= (\mathbb{C}^* \times \mathbb{D})/\langle((y_1, y_2) \mapsto (e^{2\pi i a/m} y_1, e^{2\pi i/m} y_2))\rangle, \\ \text{with } \mathbb{C}^*\text{-action } &\mathbb{C}^* \times Y_{a,m} \to Y_{a,m}, \quad (t, [(y_1, y_2)]) \mapsto [(ty_1, y_2)], \\ \text{and projection } &p_{a,m} : Y_{a,m} \to \mathbb{D}/\langle(y_2 \mapsto e^{2\pi i/m} y_2)\rangle, \quad [(y_1, y_2)] \mapsto [y_2] \end{aligned}$$

for some $m \in \mathbb{Z}_{\geq 2}$ and some $a \in \{1, 2, ..., m-1\}$ with $\gcd(a, m) = 1$. The fiber over $[0]$ in this model is called *exceptional fiber of type* (a, m). The stabilizer of the $\mathbb{C}^*$-action on it has order m.

If $\widetilde{X}$ is compact there is a notion of an Euler number $\in \mathbb{Q}$ of a $\mathbb{C}^*$-orbibundle over $\widetilde{X}$ which generalizes the Euler number $\in \mathbb{Z}$ of a $\mathbb{C}^*$-bundle over $\widetilde{X}$ [OW71] (see also [Do83]). If one takes out the center of an isolated quasihomogeneous surface singularity one obtains a $\mathbb{C}^*$-orbibundle over a smooth compact curve $\widetilde{X}$ with negative Euler number. Vice versa, any $\mathbb{C}^*$-orbibundle over a smooth compact curve $\widetilde{X}$ with negative Euler number arises in this way [OW71].

We care about $\mathbb{C}^*$-orbibundles here mainly because of the case A_3. The following example will be used in the proof of the case A_3 in Corollary 9.8.

Example 9.6 Consider $Y_1 := \mathbb{C}^2 - \{(0,0)\}$ with coordinates (z_{13}, z_{14}) and the $\mathbb{C}^*$-action

$$\mathbb{C}^* \times Y_1 \to Y_1, \quad (t, z_{13}, z_{14}) \mapsto (t^3 z_{13}, t^2 z_{14}).$$

It is a $\mathbb{C}^*$-orbibundle over $\widetilde{X} = \mathbb{P}^1$ with projection

$$p_1 : Y_1 \to \mathbb{P}^1, \quad (z_{13}, z_{14}) \mapsto \frac{z_{14}^3}{z_{13}^2},$$

and two exceptional fibers, the exceptional fiber $\{z_{14} = 0\}$ over 0 with stabilizer of order three and the exceptional fiber $\{z_{13} = 0\}$ over ∞ of type $(1, 2)$. We claim that the exceptional fiber $\{z_{14} = 0\}$ over 0 is of type $(1, 3)$ and not of type $(2, 3)$. We will show this below after studying the $\mathbb{C}^*$-orbibundle $p_2 : Y_2 \to \mathbb{P}^1$ which is obtained from $p_1 : Y_1 \to \mathbb{P}^1$ via pull back with the $3 : 1$ branched covering $b_2 : \mathbb{P}^1 \to \mathbb{P}^1, z \mapsto z^3$.

The $\mathbb{C}^*$-orbibundle $p_2 : Y_2 \to \mathbb{P}^1$ has only one exceptional fiber, the fiber of type $(1, 2)$ over ∞. This $\mathbb{C}^*$-orbibundle is given by the following two charts and their glueing.
First chart over $\mathbb{C}$:

$$\begin{aligned} p_2^{(1)} : Y_2^{(1)} = \mathbb{C}^* \times \mathbb{C} \to \mathbb{C}, &\quad (z_{12}, z_{11}) \mapsto z_{11}, \\ \mathbb{C}^*\text{-action} \quad \mathbb{C}^* \times Y_2^{(1)} \to Y_2^{(1)}, &\quad (t, z_{12}, z_{11}) \mapsto (t z_{12}, z_{11}), \\ Y_2^{(1)} \xrightarrow{3:1} Y_1^{(1)} = p_1^{-1}(\mathbb{C}) = \mathbb{C}^* \times \mathbb{C}, &\quad (z_{12}, z_{11}) \mapsto (z_{12}^3, z_{12}^2 z_{11}) = (z_{13}, z_{14}). \end{aligned}$$

Second chart over $\mathbb{P}^1 - \{0\}$:

$$\begin{aligned} p_2^{(2)} : Y_2^{(2)} = \mathbb{C} \times \mathbb{C}^* \to \mathbb{P}^1 - \{0\}, &\quad (y_1, y_2) \mapsto \frac{y_2}{y_1^2}, \\ \mathbb{C}^*\text{-action} \quad \mathbb{C}^* \times Y_2^{(2)} \to Y_2^{(2)}, &\quad (t, y_1, y_2) \mapsto (t y_1, t^2 y_2), \\ Y_2^{(2)} \longrightarrow Y_1^{(2)} = p_1^{-1}(\mathbb{P}^1 - \{0\}) = \mathbb{C} \times \mathbb{C}^*, &\quad (y_1, y_2) \mapsto (y_1^3, y_2) = (z_{13}, z_{14}), \end{aligned}$$

Glueing of the two charts:

$$\begin{aligned} Y_2^{(1)} - (p_2^{(1)})^{-1}(0) = \mathbb{C}^* \times \mathbb{C}^* \to Y_2^{(2)} - (p_2^{(2)})^{-1}(\infty) = \mathbb{C}^* \times \mathbb{C}^*, \\ (z_{12}, z_{11}) \mapsto (z_{12}, z_{12}^2 z_{11}) = (y_1, y_2). \end{aligned}$$

The map $Y_2^{(1)} \to Y_1^{(1)}, (z_{12}, z_{11}) \mapsto (z_{12}^3, z_{12}^2 z_{11})$, shows that the chart $Y_1^{(1)}$ of the first $\mathbb{C}^*$-orbibundle $p_1 : Y_1 \to \mathbb{P}^1$ is a $\mathbb{C}^*$-orbibundle over $\mathbb{C}$ with exceptional fiber over 0 and that the $\mathbb{C}^*$-orbibundle is isomorphic to the model of type $(1, 3)$ in Remark 9.4(v), namely

$$Y_1^{(1)} \cong (\mathbb{C}^* \times \mathbb{C})/\langle((z_{12}, z_{11}) \mapsto (e^{2\pi i/3} z_{12}, e^{2\pi i/3} z_{11}))\rangle.$$

Consider a further $2:1$ branched covering $b_3 : \mathbb{P}^1 \to \mathbb{P}^1$, $\widetilde{z}_{11} \to \widetilde{z}_{11}^2 = z_{11}$. The pull back of $p_2 : Y_2 \to \mathbb{P}^1$ via b_3 gives a $\mathbb{C}^*$-bundle $p_3 : Y_3 \to \mathbb{P}^1$ over $\mathbb{P}^1$. We claim that its Euler number is -1, so it is isomorphic to $(\mathcal{O}_{\mathbb{P}^1}(-1) - (\text{zero section}))$. To prove this, we consider two charts for $p_3 : Y_3 \to \mathbb{P}^1$ and their glueing.
First chart over $\mathbb{C}$:

$$\begin{aligned} p_3^{(1)} : Y_3^{(1)} = \mathbb{C}^* \times \mathbb{C} \to \mathbb{C},\ & (z_{12}, \widetilde{z}_{11}) \mapsto \widetilde{z}_{11}, \\ \mathbb{C}^*\text{-action}\ \ \mathbb{C}^* \times Y_3^{(1)} \to Y_3^{(1)},\ & (t, z_{12}, \widetilde{z}_{11}) \mapsto (t z_{12}, \widetilde{z}_{11}), \\ Y_3^{(1)} \xrightarrow{2:1} Y_2^{(1)} = \mathbb{C}^* \times \mathbb{C},\ & (z_{12}, \widetilde{z}_{11}) \mapsto (z_{12}, \widetilde{z}_{11}^2) = (z_{12}, z_{11}). \end{aligned}$$

Second chart over $\mathbb{P}^1 - \{0\}$:

$$\begin{aligned} p_3^{(2)} : Y_3^{(2)} = \mathbb{C}^* \times \mathbb{C} \to \mathbb{P}^1 - \{0\},\ & (y_3, y_4) \mapsto \frac{1}{y_4}, \\ \mathbb{C}^*\text{-action}\ \ \mathbb{C}^* \times Y_3^{(2)} \to Y_3^{(2)},\ & (t, y_3, y_4) \mapsto (t y_3, y_4), \\ Y_3^{(2)} \longrightarrow Y_2^{(2)} = \mathbb{C} \times \mathbb{C}^*,\ & (y_3, y_4) \mapsto (y_3 y_4, y_3^2) = (y_1, y_2), \end{aligned}$$

Glueing of the two charts:

$$\begin{aligned} Y_3^{(1)} - (p_3^{(1)})^{-1}(0) = \mathbb{C}^* \times \mathbb{C}^* \to Y_3^{(2)} - (p_3^{(2)})^{-1}(\infty) = \mathbb{C}^* \times \mathbb{C}^*, \\ (z_{12}, \widetilde{z}_{11}) \mapsto (z_{12}\widetilde{z}_{11}, \frac{1}{\widetilde{z}_{11}}) = (y_3, y_4). \end{aligned}$$

The glueing map shows that $p_3 : Y_3 \to \mathbb{P}^1$ is the $\mathbb{C}^*$-bundle with Euler number -1.

Lemma 9.7 *Fix $m \in \mathbb{Z}_{\geq 2}$.*

(a) The following table defines four elements $\beta_1, \beta_2, \beta_3, \beta_4 \in \mathrm{Br}_3$, it gives their images g_1, g_2, g_3, g_4 in $PSL_2(\mathbb{Z})$ under the homomorphism $\mathrm{Br}_3 \to PSL_2(\mathbb{Z})$ in Remark 4.15 (i), and it defines four points $x_1, x_2, x_3, x_4 \in \widehat{\mathbb{R}}$,

i	1	2	3	4
β_i	σ_2^m	$\sigma_1\sigma_2^m\sigma_1^{-1} = \sigma_2^{-1}\sigma_1^m\sigma_2$	$\sigma_1^{-1}\sigma_2^m\sigma_1 = \sigma_2\sigma_1^m\sigma_2^{-1}$	σ_1^m
g_i	$[A_2^m]$	$[A_1A_2^mA_1^{-1}] = [A_2^{-1}A_1^mA_2]$	$[A_1^{-1}A_2^mA_1] = [A_2A_1^mA_2^{-1}]$	$[A_1^m]$
x_i	0	-1	1	∞

The Möbius transformation g_i is parabolic with fixed point x_i. The action of the braid β_i on $\mathbb{C} \times \mathbb{C} \times \mathbb{H} \cong C_3^{univ}$ restricts to the action of g_i on $\mathbb{H}$. The quotient $\mathbb{H}/\langle g_i\rangle$ of that action is isomorphic to $\mathbb{D}^* = \mathbb{D} - \{0\}$. The quotient $(\mathbb{C} \times \mathbb{C} \times \mathbb{H})/\langle\beta_i\rangle$ is isomorphic to $\mathbb{C} \times$ (an affine $\mathbb{C}$-bundle over $\mathbb{D}^*$).

(b) This quotient extends to $\mathbb{C} \times$ (an affine $\mathbb{C}$-bundle over $\mathbb{D}$) such that this is an F-manifold with Euler field and at each point in the fiber over $0 \in \mathbb{D}$ the germ is of type $I_2(m)A_1$ (compare Example 8.10 (ii)).

(c) (Concretization of a family of cases in part (b)) If m is even then

$$\begin{aligned}(\mathbb{C}\times\mathbb{C}\times\mathbb{H})/\langle\sigma_2^m\rangle &\xrightarrow{\cong} \mathbb{C}\times\mathbb{C}\times\mathbb{D}^*\\ [(z_1,\zeta,\tau)] &\mapsto (z_1,\zeta,z_4)\quad \textit{with } z_4=z_4(\tau),\end{aligned}$$

and the partial compactification in part (b) is $\mathbb{C}\times\mathbb{C}\times\mathbb{D}$.

(d) The semisimple F-manifold with Euler field $\mathbb{C}\times\mathbb{C}^*\times(\mathbb{C}-\{0,1\})\cong C_3^{pure}$ *extends to a semisimple F-manifold with Euler field*

$$\mathbb{C}\times(\textit{the } \mathbb{C}^*\textit{-bundle over } \mathbb{P}^1 \textit{ which is } \mathcal{O}_{\mathbb{P}^1}(-1) \textit{ minus the } 0\textit{-section})$$

with Maxwell stratum the union of the fibers of $0,1,\infty\in\mathbb{P}^1$. *At each point in the Maxwell stratum the germ is of type* $I_2(2)A_1$ *(compare Example 8.10 (ii)).*

(e) Consider $m\in\mathbb{N}$. *The F-manifold with Euler field*

$$\mathbb{C}\times\mathbb{C}^*\times\mathbb{H}\cong(\mathbb{C}\times\mathbb{C}\times\mathbb{H})/\langle(\sigma^{mon})^m\rangle\cong C_3^{univ}/\langle(\sigma^{mon})^m\rangle$$

extends to $\mathbb{C}\times\mathbb{C}\times\mathbb{H}$ *if and only if* $m\geq 2$. *If* $m\geq 2$ *then at each point of* $\mathbb{C}\times\{0\}\times\mathbb{H}$ *the germ of the F-manifold is irreducible.*

Proof (a) The equations $\sigma_2\sigma_1^m\sigma_2^{-1}=\sigma_1^{-1}\sigma_2^m\sigma_1$ and $\sigma_2^{-1}\sigma_1^m\sigma_2=\sigma_1\sigma_2^m\sigma_1^{-1}$ are (4.14). They show that β_1,β_2,β_3 and β_4 are all conjugate in Br_3. Therefore also g_1,g_2,g_3 and g_4 are all conjugate in $PSL_2(\mathbb{Z})$. They are parabolic with the fixed points x_1,x_2,x_3 and x_4. The rest of part (a) is also obvious.

Part (d) is proved before the parts (b) and (c).

(d) $C_3^{pure}=\mathbb{C}^3-D_3^{pure}$ extends into the semisimple F-manifold $\mathbb{C}^3$ of type A_1^3 with Maxwell stratum D_3^{pure}, which consists of three hyperplanes which intersect all three in the complex line $\{\underline{u}\in\mathbb{C}^3\,|\,u_1=u_2=u_3\}$. The isomorphism $C_3^{pure}\cong\mathbb{C}\times\mathbb{C}^*\times(\mathbb{C}-\{0,1\})$ is the restriction to C_3^{pure} of the blowing up of this complex line,

$$\begin{array}{rcl}\mathbb{C}\times\mathcal{O}_{\mathbb{P}^1}(-1) & \longrightarrow & \mathbb{C}^3\\ \cup & & \cup\\ \mathbb{C}\times\mathbb{C}\times\mathbb{C} & \longrightarrow & \mathbb{C}\times(\mathbb{C}\times\mathbb{C}-\{0\}\times\mathbb{C}^*)\\ \cup & & \cup\\ \mathbb{C}\times\mathbb{C}^*\times(\mathbb{C}-\{0,1\}) & \longrightarrow & \{(u_1+u_2+u_3,u_2-u_1,u_3-u_1)\in\mathbb{C}^3\,|\\ & & \quad u_1\neq u_2\neq u_3\neq u_1\}\cong C_3^{pure}\\ (z_1,z_2,z_3) & \mapsto & (z_1,z_2,z_2z_3)\\ & & =(u_1+u_2+u_3,u_2-u_1,u_3-u_1)\end{array}$$

In Theorem 9.1 (a) and here only one of the two standard charts of $\mathbb{C}\times\mathcal{O}_{\mathbb{P}^1}(-1)$ is made explicit. This chart excludes the fiber over $z_3=\infty\in\mathbb{P}^1$. The three hyperplanes in D_3^{pure} give the fibers over $z_3=0,1,\infty\in\mathbb{P}^1$. Compare Remark 9.2 (ii).

(b) and (c) Because $\beta_1, \beta_2, \beta_3$ and β_4 are all conjugate in Br_3, it is for the proof of part (b) sufficient to prove it for $\beta_1 = \sigma_2^m$. The action of σ_2^m on $\mathbb{C} \times \mathbb{C} \times \mathbb{H} \cong C_3^{univ}$ is as follows,

$$\psi(\sigma_2^m): (z_1, \zeta, \tau) \mapsto \left(z_1, \zeta, \frac{\tau}{m\tau + 1}\right) \quad \text{for even } m,$$

$$\psi(\sigma_2^m): (z_1, \zeta, \tau) \mapsto \left(z_1, \zeta + \kappa(\tau + 1), \frac{\tau}{m\tau + 1}\right) \quad \text{for odd } m.$$

So, for even m $\psi(\sigma_2^m)$ acts nontrivially only on the third factor $\mathbb{H}$. Part (d) and this fact give the parts (b) and (c) in the case $m = 2$.

For even m the map $(\mathbb{C} \times \mathbb{C} \times \mathbb{H})/\langle\sigma_2^m\rangle \to (\mathbb{C} \times \mathbb{C} \times \mathbb{H})/\langle\sigma_2^2\rangle$ is a branched covering of degree $\frac{m}{2}$, cyclically branched along the fiber over $[0] \in \mathbb{H}/\langle[A_2^2]\rangle$. With the Examples 8.10 (ii) and 8.10 (v) (the case $I_2(m)A_1$) this shows for even m part (b) for β_1 and part (c).

For odd m observe that $\kappa(\tau + 1)$ has for $\tau \to 0$ along each hyperbolic line in $\mathbb{H}$ a limit, the limit $\kappa(1) = 0$. Therefore for odd m we obtain a 2-valued map $(\mathbb{C} \times \mathbb{C} \times \mathbb{H})/\langle\sigma_2^m\rangle \to (\mathbb{C} \times \mathbb{C} \times \mathbb{H})/\langle\sigma_2^2\rangle$ which can be considered as a branched covering of the half-degree $\frac{m}{2}$, cyclically branched along the fiber over $[0] \in \mathbb{H}/\langle A_2^2\rangle$. Again with the Examples 8.10 (ii) and 8.10 (v) this shows part (b) for $\beta_1 = \sigma_2^m$.

(e) For $m = 1$ the multiplication in $\mathbb{C} \times \mathbb{C}^* \times \mathbb{H} \cong C_3^{univ}/\langle\sigma^{mon}\rangle$ degenerates near $\mathbb{C} \times \{0\} \times \mathbb{H}$ in the same way as the multiplication in $\mathbb{C} \times \mathbb{C}^* \times (\mathbb{C} - \{0, 1\}) \cong C_3^{pure}$ degenerates near $\mathbb{C} \times \{0\} \times (\mathbb{C} - \{0, 1\})$. Because of the denominator z_2 in the third summand on the right hand side of

$$\partial_2 \circ \partial_2 = \frac{2 - 2z_3 + 2z_3^2}{3}\partial_1 + \frac{1 - 2z_3}{3}\partial_2 + \frac{-z_3 + z_3^2}{z_2}\partial_3,$$

the multiplication does not extend holomorphically to $\mathbb{C} \times \{0\} \times (\mathbb{C} - \{0, 1\})$.

For $m \geq 2$ the (vertical) maps

$$\begin{array}{ccc} C_3^{univ}/\langle(\sigma^{mon})^m\rangle \cong \mathbb{C} \times \mathbb{C}^* \times \mathbb{H} & & (z_1, z_{18}, \tau) \\ \downarrow \qquad\qquad \downarrow & & \downarrow \\ C_3^{univ}/\langle\sigma^{mon}\rangle \cong \mathbb{C} \times \mathbb{C}^* \times \mathbb{H} & & (z_1, z_{18}^m, \tau) = (z_1, z_2, \tau) \end{array}$$

are m-fold coverings. Again, we can go from $\mathbb{H}$ to $\mathbb{C} - \{0, 1\}$. The multiplication upstairs is obtained by pull back from the multiplication downstairs via the m-fold covering,

$$\begin{array}{ccc} \mathbb{C} \times \mathbb{C}^* \times (\mathbb{C} - \{0, 1\}) & & (z_1, z_{18}, z_3) \\ \downarrow & & \downarrow \\ C_3^{pure} \cong \mathbb{C} \times \mathbb{C}^* \times (\mathbb{C} - \{0, 1\}) & & (z_1, z_{18}^m, z_3) = (z_1, z_2, z_3) \end{array}$$

Write $\partial_{18} := \frac{\partial}{\partial z_{18}}$. Because of $\partial_{18} \mapsto m z_{18}^{m-1}\partial_2$, the multiplication is given by $e = 3\partial_1$ and

$$\begin{aligned}
\partial_{18} \circ \partial_{18} &= m^2 z_{18}^{2m-2} \frac{2 - 2z_3 + 2z_3^2}{3}\partial_1 + m z_{18}^{m-1}\frac{1 - 2z_3}{3}\partial_{18} \\
&\quad + m^2 z_{18}^{m-2}(-z_3 + z_3^2)\partial_3, \\
\partial_{18} \circ \partial_3 &= m z_{18}^{m-1}\frac{-z_3 + 2z_{18}^m z_3}{3}\partial_1 - z_{18}^m \frac{1}{3}\partial_{18} + m z_{18}^{m-1}\frac{-1 + 2z_3}{3}\partial_3, \\
\partial_3 \circ \partial_3 &= \frac{2z_{18}^{2m}}{3}\partial_1 + \frac{z_{18}^m}{3}\partial_3.
\end{aligned}$$

The multiplication extends holomorphically to $\mathbb{C} \times \{0\} \times (\mathbb{C} - \{0, 1\})$. Moreover $\partial_{18} \circ \partial_3$ and $\partial_3 \circ \partial_3$ vanish there, and also $\partial_{18} \circ \partial_{18}$ if $m \geq 3$. If $m = 2$ then $\partial_{18} \circ \partial_{18} = 4(-z_3 + z_3^2)\partial_3$ on $\mathbb{C} \times \{0\} \times (\mathbb{C} - \{0, 1\})$. Therefore for any $m \geq 2$ the algebra $(T_p, \circ)$ for $p \in \mathbb{C} \times \{0\} \times (\mathbb{C} - \{0, 1\})$ is irreducible, and thus the germ at p of the F-manifold is irreducible. □

Recall that $C_3^{\underline{e}/\{\pm 1\}^3}$ is in all cases a semisimple F-manifold (with Euler field) with empty Maxwell stratum. We consider the manifold in Theorem 9.3 which is isomorphic to $C_3^{\underline{e}/\{\pm 1\}^3}$ with the coordinates in the proof of Theorem 9.3.

Together Theorem 7.11, Theorem 9.1 (a), Theorem 9.3 and its proof, and Lemma 9.7 allow in many cases to embed $C_3^{\underline{e}/\{\pm 1\}^3}$ into a bigger F-manifold $\overline{C_3^{\underline{e}/\{\pm 1\}^3}}^p$ with nonempty Maxwell stratum or nonempty caustic or both. This is subject of Corollary 9.8.

Corollary 9.8 *In many cases in Theorem 9.3, $C_3^{\underline{e}/\{\pm 1\}^3}$ has a partial compactification, which we call $\overline{C_3^{\underline{e}/\{\pm 1\}^3}}^p$, to an F-manifold with nonempty Maxwell stratum or nonempty caustic or both.*

(a) The exceptions are the cases $\mathcal{G}_5, \mathcal{G}_6, \mathcal{G}_7, \mathcal{G}_8$ and $\mathcal{G}_9$. There $E = z_1\partial_1 + \partial_\zeta$ and

$$\begin{aligned}
C_3^{\underline{e}/\{\pm 1\}^3} &= \mathbb{C} \times \mathbb{C} \times \mathbb{D}^* \quad \textit{in the case } \mathcal{G}_5, \\
C_3^{\underline{e}/\{\pm 1\}^3} &= C_3^{univ} \quad \textit{in the cases } \mathcal{G}_6, \mathcal{G}_7, \mathcal{G}_8 \textit{ and } \mathcal{G}_9.
\end{aligned}$$

(b) All cases except the cases in part (a) and the cases A_2A_1 and A_3: The first table below gives the manifold in Theorem 9.3 isomorphic to $C_3^{\underline{e}/\{\pm 1\}^3}$ and the manifold isomorphic to its extension $\overline{C_3^{\underline{e}/\{\pm 1\}^3}}^p$. The second table gives the strata in the Maxwell stratum $\mathcal{K}_2$ and in the caustic $\mathcal{K}_3$ of $\overline{C_3^{\underline{e}/\{\pm 1\}^3}}^p$. In the case $\mathcal{H}_{1,2}$ at each point $p \in \mathcal{K}_3$ the germ of the F-manifold is irreducible. In all other cases with $\mathcal{K}_3 \neq \emptyset$, at each point $p \in \mathcal{K}_3$ the germ of the F-manifold is of type $I_2(3)A_1$.

	sets	$C_3^{\underline{e}/\{\pm 1\}^3}$	$\overline{C_3^{\underline{e}/\{\pm 1\}^3}}^p$
$\mathcal{G}_1$	C_1 (A_1^3)	C_3^{pure}	$\mathbb{C}^3$
$\mathcal{G}_1$	C_2 $(\mathcal{H}_{1,2})$	$\mathbb{C}\times\mathbb{C}^*\times\mathbb{H}$	$\mathbb{C}\times\mathbb{C}\times\mathbb{H}$
$\mathcal{G}_2$	C_4 $(\mathbb{P}^1A_1)$, C_5	$\mathbb{C}^2\times(\mathbb{C}-\{0,1\})$	$\mathbb{C}^3$
$\mathcal{G}_4$	C_7 $(\widehat{A}_2)$	$\mathbb{C}^2\times\mathbb{C}^{***}$	$\mathbb{C}^3$
$\mathcal{G}_{10}$	C_{17}	$\mathbb{C}\times\mathbb{C}\times\mathbb{D}^*$	$\mathbb{C}\times\mathbb{C}\times\mathbb{D}$
$\mathcal{G}_{11}$	C_{18}	$\mathbb{C}\times\mathbb{C}\times\mathbb{D}^*$	$\mathbb{C}\times\mathbb{C}\times\mathbb{D}$
$\mathcal{G}_{12}$	C_{19}	$\mathbb{C}\times\mathbb{C}\times\mathbb{D}^{**}$	$\mathbb{C}\times\mathbb{C}\times\mathbb{D}$
$\mathcal{G}_{13}$	C_{20}	$\mathbb{C}\times\mathbb{C}\times\mathbb{D}^{**}$	$\mathbb{C}\times\mathbb{C}\times\mathbb{D}$
$\mathcal{G}_{14}$	C_{21}, C_{22}	$\mathbb{C}\times\mathbb{C}\times\mathbb{D}^*$	$\mathbb{C}\times\mathbb{C}\times\mathbb{D}$

	sets	$\mathcal{K}_2$	$\mathcal{K}_3$
$\mathcal{G}_1$	C_1 (A_1^3)	D_3^{pure}	–
$\mathcal{G}_1$	C_2 $(\mathcal{H}_{1,2})$	–	$\mathbb{C}\times\{0\}\times\mathbb{H}$
$\mathcal{G}_2$	C_4 $(\mathbb{P}^1A_1)$, C_5	$\mathbb{C}^2\times\{0,1\}$	–
$\mathcal{G}_4$	C_7 $(\widehat{A}_2)$	–	$\mathbb{C}^2\times\{z_5\,\|\,z_5^3=1\}$
$\mathcal{G}_{10}$	C_{17}	$\mathbb{C}\times\mathbb{C}\times\{0\}$	–
$\mathcal{G}_{11}$	C_{18}	$\mathbb{C}\times\mathbb{C}\times\{0\}$	–
$\mathcal{G}_{12}$	C_{19}	$\mathbb{C}\times\mathbb{C}\times\{0\}$	$\mathbb{C}\times\mathbb{C}\times\{\frac{1}{2}\}$
$\mathcal{G}_{13}$	C_{20}	–	$\mathbb{C}\times\mathbb{C}\times\{0,\frac{1}{2}\}$
$\mathcal{G}_{14}$	C_{21}, C_{22}	–	$\mathbb{C}\times\mathbb{C}\times\{0\}$

The Euler field is

	A_1^3	$\mathcal{H}_{1,2}$	$(\mathcal{G}_2$ & $C_4, C_5)$, $\mathcal{G}_4, \mathcal{G}_{10}, \mathcal{G}_{11}, \mathcal{G}_{12}, \mathcal{G}_{13}, \mathcal{G}_{14}$
E	$\sum_{i=1}^3 u_ie_i$	$z_1\partial_1+\frac{1}{2}z_5\partial_5$	$z_1\partial_1+\partial_\varsigma$

(c) The case A_2A_1: We use the second presentation in Theorem 9.1.

$$
\begin{aligned}
C_3^{\underline{e}/\{\pm 1\}^3} &\cong \mathbb{C}\times\{(z_8,z_9)\in\mathbb{C}^*\times\mathbb{C}\,|\,z_8^3-4z_9^2\neq 0\},\\
\overline{C_3^{\underline{e}/\{\pm 1\}^3}}^p &\cong \mathbb{C}^3,\\
\mathcal{K}_2 &\cong \mathbb{C}\times\{(z_8,z_9)\in\mathbb{C}\times\mathbb{C}\,|\,z_8^3-4z_9^2=0\},\\
\mathcal{K}_3 &\cong \mathbb{C}\times\{0\}\times\mathbb{C},\\
E &= z_1\partial_1+\frac{2}{3}z_8\partial_8+z_9\partial_9.
\end{aligned}
$$

At each point $p\in\mathcal{K}_3$ the germ of the F-manifold is of type $I_2(3)A_1$. The left part of Fig. 9.3 shows parts of $\mathcal{K}_2$ and $\mathcal{K}_3$ near 0.

(d) The case A_3: We use the second presentation in Theorem 9.1.

$$
C_3^{\underline{e}/\{\pm 1\}^3} \cong \mathbb{C}\times\{(z_{13},z_{14})\in\mathbb{C}^*\times\mathbb{C}\,|\,z_{14}^3-z_{13}^2\neq 0\},
$$

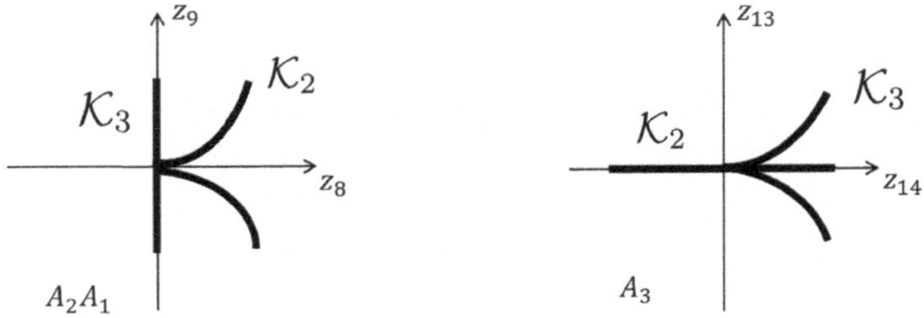

Fig. 9.3 In the cases A_2A_1 (left) and A_3 (right) the restrictions of $\mathcal{K}_2$ and $\mathcal{K}_3$ to a hyperplane $\{z_1 = \text{constant}\}$ in the F-manifold $\mathbb{C}^3$

$$\begin{aligned}
\overline{C_3^{\underline{e}/\{\pm1\}^3}}^p &\cong \mathbb{C}^3, \\
\mathcal{K}_2 &\cong \mathbb{C} \times \{0\} \times \mathbb{C}, \\
\mathcal{K}_3 &\cong \mathbb{C} \times \{(z_{13}, z_{14}) \in \mathbb{C} \times \mathbb{C} \,|\, z_{14}^3 - z_{13}^2 = 0\}, \\
E &= z_1\partial_1 + \frac{3}{4} z_{13}\partial_{13} + \frac{1}{2} z_{14}\partial_{14}.
\end{aligned}$$

At each point $p \in \mathcal{K}_3 - \mathcal{K}_2 = \mathcal{K}_3 - \mathbb{C} \times \{(0,0)\}$ the germ of the F-manifold is of type $I_2(3)A_1$. At each point $p \in \mathbb{C} \times \{(0,0)\} = \mathcal{K}_3 \cap \mathcal{K}_2$ it is irreducible. The right part of Fig. 9.3 shows parts of K_2 and K_3 near 0.

Proof (a) Here nothing has to be proved. Though a remark on the cases $\mathcal{G}_5$ & C_8, C_9: Here $(\mathrm{Br}_3)_{\underline{e}/\{\pm1\}^3} = \langle (\sigma^{mon})^2 \sigma_1^{-1} \sigma_2^{l^2-4} \sigma_1 \rangle$. Without the factor $(\sigma^{mon})^2$ in the generator, Lemma 9.7 (b) would apply with $m = l^2 - 4$, and it would lead to an extension of the F-manifold to $\mathbb{C} \times \mathbb{C} \times \mathbb{D}$ with caustic $\mathbb{C} \times \mathbb{C} \times \{0\}$.

(b) The case $\mathcal{G}_1$ & C_1 (A_1^3): See Example 8.10 (i).

The case $\mathcal{G}_1$ & C_2 ($\mathcal{H}_{1,2}$): Lemma 9.7(e) applies with $m = 2$.

The cases $\mathcal{G}_2$ & C_4 ($\mathbb{P}^1 A_1$), C_5: Recall $(\mathrm{Br}_3)_{\underline{e}/\{\pm1\}^3} = \langle \sigma_2^2, \sigma_1\sigma_2^2\sigma_1^{-1} \rangle$ and consider its action on the third factor $\mathbb{H}$ of $\mathbb{C} \times \mathbb{C} \times \mathbb{H} \cong C_3^{univ}$. The point $z_3 = 0$ in the closure of the quotient $\mathbb{C} - \{0,1\}$ of $\mathbb{H}$ is the image of the fixed point $0 \in \widehat{\mathbb{R}}$ of σ_2^2, the point $z_3 = 1$ is the image of the fixed point $-1 \in \widehat{\mathbb{R}}$ of $\sigma_1\sigma_2^2\sigma_1^{-1}$. By Lemma 9.7 (b) the F-manifold extends to the fibers over 0 and 1 of $\mathbb{C} \times$ (an affine $\mathbb{C}$-bundle over $\mathbb{C}$) and has there its Maxwell stratum. By Remark 9.4 (ii) one can choose the affine $\mathbb{C}$-bundle over $\mathbb{C}$ to be $\mathbb{C} \times \mathbb{C}$.

The case $\mathcal{G}_4$ & C_7 ($\widehat{A}_2$). Recall $(\mathrm{Br}_3)_{\underline{e}/\{\pm1\}^3} = \langle \sigma_1^3, \sigma_2^3, \sigma_2\sigma_1^3\sigma_2^{-1} \rangle$ and consider its action on the third factor $\mathbb{H}$ of $\mathbb{C} \times \mathbb{C} \times \mathbb{H} \cong C_3^{univ}$. The points z_5 with $z_5^3 = 1$ in the closure of the quotient $\mathbb{C}^{***}$ of $\mathbb{H}$ are the images of the fixed points ∞, 0 and $1 \in \widehat{\mathbb{R}}$ of σ_1^3, σ_2^3 and $\sigma_2\sigma_1^3\sigma_2^{-1}$. By Lemma 9.7 (b) the F-manifold extends to the fibers over z_5 with $z_5^3 = 1$ of $\mathbb{C} \times$ (an affine $\mathbb{C}$-bundle over $\mathbb{C}$) and has there its caustic $\mathcal{K}_3$, and at each point $p \in \mathcal{K}_3$ the germ of the F-manifold is of type $I_2(3)A_1$. By Remark 9.4 (ii) one can choose the affine $\mathbb{C}$-bundle over $\mathbb{C}$ to be $\mathbb{C} \times \mathbb{C}$.

A remark on this case: Observe

$$(\sigma_1^3)(\sigma_2\sigma_1^3\sigma_2^{-1})(\sigma_2^3) = (\sigma^{mon})^2 (\sigma_2^{-1}\sigma_1^3\sigma_2)^{-1}.$$

Therefore $(\mathrm{Br}_3)_{\underline{e}/\{\pm 1\}^3}$ contains $(\sigma^{mon})^{-2}\sigma_2^{-1}\sigma_1^3\sigma_2$. The point $z_5=\infty$ is the image of the fixed point of $A_2^{-1}A_1^3A_2$ which is the image in $PSL_2(\mathbb{Z})$ of this braid. But Lemma 9.7 (b) does not apply to $z_5=\infty$, because $(\mathrm{Br}_3)_{\underline{e}/\{\pm 1\}^3}$ does not contain $\sigma_2^{-1}\sigma_1^3\sigma_2$.

The cases $\mathcal{G}_{10},\mathcal{G}_{11},\mathcal{G}_{12},\mathcal{G}_{13},\mathcal{G}_{14}$: Recall from Theorem 7.11

	$\mathcal{G}_{10}$ and $\mathcal{G}_{11}$	$\mathcal{G}_{12}$	$\mathcal{G}_{13}$	$\mathcal{G}_{14}$
$(\mathrm{Br}_3)_{\underline{e}/\{\pm 1\}^3}$	$\langle\sigma_2^2\rangle$	$\langle\sigma_2^2,\sigma_2\sigma_1^3\sigma_2^{-1}\rangle$	$\langle\sigma_2^3,\sigma_2\sigma_1^3\sigma_2^{-1}\rangle$	$\langle\sigma_2^3\rangle$

The point $0\in\mathbb{D}$ is the image of the fixed point $0\in\widehat{\mathbb{R}}$ of σ_2^2 respectively σ_2^3, the point $\frac{1}{2}\in\mathbb{D}$ is in the cases $\mathcal{G}_{12}$ and $\mathcal{G}_{13}$ the image of the fixed point $-1\in\widehat{\mathbb{R}}$ of the action of $\sigma_2\sigma_1^3\sigma_2^{-1}$ on the third factor $\mathbb{H}$ cf $\mathbb{C}\times\mathbb{C}\times\mathbb{H}\cong C_3^{univ}$. Lemma 9.7 (b) applies in all cases and yields an extension to an F-manifold $\mathbb{C}\times$ (an affine $\mathbb{C}$-bundle over $\mathbb{D}$) with the claimed Maxwell stratum and caustic. By Remark 9.4 (ii) one can choose the affine $\mathbb{C}$-bundle over $\mathbb{D}$ to be $\mathbb{C}\times\mathbb{D}$.

On the Euler field: For the case A_1^3 see Example 8.10 (i).

In the case $\mathcal{H}_{1,2}$ the moduli space is $C_3^{\underline{e}/\{\pm 1\}^3}\cong\mathbb{C}\times\mathbb{C}^*\times\mathbb{H}$ with coordinates (z_1,z_5,τ) with $z_5=e^{\zeta/2}$. Therefore the Euler field $z_1\partial_1+\partial_\zeta$ on C_3^{univ} pulls down to $E=z_1\partial_1+\frac{1}{2}z_5\partial_5$.

In all other cases no power of σ^{mon} is in $(\mathrm{Br}_3)_{\underline{e}/\{\pm 1\}^3}$, and

$$C_3^{\underline{e}/\{\pm 1\}^3}\cong\mathbb{C}\times(\text{an affine }\mathbb{C}\text{-bundle over a noncompact curve }C)$$
$$\cong\mathbb{C}\times\mathbb{C}\times C\quad\text{with coordinates }(z_1,\zeta,z_{3/11/16}).$$

Therefore the Euler field $z_1\partial_1+\partial_\zeta$ on C_3^{univ} pulls down to $E=z_1\partial_1+\partial_\zeta$ on $C_3^{\underline{e}/\{\pm 1\}^3}$.

(c) Recall the discussion of the case A_2A_1 in the proof of Theorem 9.3. Because $z_6=e^{\zeta/3}$, the Euler field on $\mathbb{C}\times\mathbb{C}^*\times(\mathbb{C}-\{\pm\frac{1}{2}\})$ with coordinates (z_1,z_6,z_7) is $z_1\partial_1+\frac{1}{3}z_6\partial_6$. Under the 2 : 1 map

$$\mathbb{C}\times\mathbb{C}^*\times(\mathbb{C}-\{\pm\frac{1}{2}\})\to\mathbb{C}\times\{(z_8,z_9)\in\mathbb{C}^*\times\mathbb{C}\,|\,z_8^3-4z_9^2=0\}$$
$$(z_1,z_6,z_7)\mapsto(z_1,z_6^2,z_6^3z_7)=(z_1,z_8,z_9),$$

it becomes

$$E=z_1\partial_1+\frac{1}{3}z_6(2z_6\partial_8+3z_6^2z_7\partial_9)=z_1\partial_1+\frac{2}{3}z_8\partial_8+z_9\partial_9.$$

Consider the isomorphism

$$C_3^{pure}\cong\mathbb{C}\times\mathbb{C}^*\times\left(\mathbb{C}-\{\pm\frac{1}{2}\}\right)\to\mathbb{C}\times\{(z_2,z_9)\in\mathbb{C}^*\times\mathbb{C}\,|\,z_2^3-4z_9^2=0\},$$
$$(u_1,u_2,u_3)\qquad (z_1,z_2,z_7)\mapsto(z_1,z_2,z_2z_7)=(u_1+u_2+u_3,$$
$$u_2-u_1,(u_3-u_1)-\frac{1}{2}(u_2-u_1)).$$

The multiplication in the manifold on the right side extends to $\mathbb{C}^3$ with Maxwell stratum the three hyperplanes

$$\mathbb{C} \times \{0\} \times \mathbb{C} \cup \mathbb{C} \times \{(z_2, z_9) \in \mathbb{C}^2 \mid z_2^2 - 4z_9^2 = 0\}.$$

The two-valued map

$$\mathbb{C}^3 \dashrightarrow \mathbb{C}^3, \quad (z_1, z_8, z_9) \mapsto (z_1, z_8^{3/2}, z_9) = (z_1, z_2, z_9)$$

is outside of the branching locus $\{z_8 = 0\}$ locally an isomorphism of F-manifolds, and along $\{z_8 = 0\}$ it is branched of order 3/2.

Therefore the F-manifold $\mathbb{C}^3$ on the left side with coordinates (z_1, z_8, z_9) has Maxwell stratum $\mathcal{K}_2 = \mathbb{C} \times \{(z_8, z_9) \in \mathbb{C}^2 \mid z_8^3 - 4z_9^2 = 0\}$ and caustic $\mathcal{K}_3 = \mathbb{C} \times \{0\} \times \mathbb{C}$, and at each point of the caustic the germ of the F-manifold is of type $I_2(3)A_1$.

(d) Recall the discussion of the case A_3 in the proof of Theorem 9.3. Because $z_{12} = e^{\zeta/4}$, the Euler field on $\mathbb{C} \times \mathbb{C}^* \times \mathbb{C}^{***}$ with coordinates (z_1, z_{12}, z_{11}) is $z_1\partial_1 + \frac{1}{4}z_{12}\partial_{12}$. Under the 3 : 1 map

$$\begin{aligned} \mathbb{C} \times \mathbb{C}^* \times \mathbb{C}^{***} &\to \mathbb{C} \times \{(z_{13}, z_{14}) \in \mathbb{C}^* \times \mathbb{C} \mid z_{14}^3 - z_{13}^2 \neq 0\} \\ (z_1, z_{12}, z_{11}) &\mapsto (z_1, z_{12}^3, z_{12}^2 z_{11}) = (z_1, z_{13}, z_{14}), \end{aligned}$$

it becomes

$$E = z_1\partial_1 + \frac{1}{4}z_{12}(3z_{12}^2\partial_{13} + 2z_{12}z_{11}\partial_{14}) = z_1\partial_1 + \frac{3}{4}z_{13}\partial_{13} + \frac{1}{2}z_{14}\partial_{14}.$$

In the proof of Theorem 9.3 the quotient of C_3^{univ} by

$$\langle \sigma_1^3, \sigma_2^{-1}\sigma_1^3\sigma_2, \sigma_2^3, (\sigma^{mon})^4 \rangle \overset{1:3}{\subset} (\mathrm{Br}_3)_{\underline{e}/\{\pm 1\}^3} = \langle (\sigma_1\sigma_2)^4, \sigma_1^3 \rangle$$

was constructed as

$$\begin{aligned} \mathbb{C} \times \mathbb{C}^* \times \mathbb{C}^{***} \quad & \text{with coordinates } (z_1, z_{12}, z_{11}) \\ & \text{with } z_{12} = e^{\zeta/4}, z_{11} = z_{11}(\tau). \end{aligned}$$

The three values $z_{11} \in \mathbb{C}$ with $z_{11}^3 = 1$ are the images of the parabolic fixed points $\infty, -1, 0 \in \widehat{\mathbb{R}}$ of the action of $\sigma_1^3, \sigma_2^{-1}\sigma_1^3\sigma_2, \sigma_2^3$ on $\mathbb{H}$.

By Lemma 9.7 (b) and Remark 9.4 (iv) the F-manifold $\mathbb{C} \times \mathbb{C}^* \times \mathbb{C}^{***}$ extends to $\mathbb{C} \times \mathbb{C}^* \times \mathbb{C}$ with caustic $\mathbb{C} \times \mathbb{C}^* \times \{z_{11} \mid z_{11}^3 = 1\}$, and at each point p in the caustic the germ of the F-manifold is of type $I_2(3)A_1$. Under the extension

$$\begin{aligned} \mathbb{C} \times \mathbb{C}^* \times \mathbb{C} &\to \mathbb{C} \times \mathbb{C}^* \times \mathbb{C} \\ (z_1, z_{12}, z_{11}) &\mapsto (z_1, z_{12}^3, z_{12}^2 z_{11}) = (z_1, z_{13}, z_{14}) \end{aligned}$$

of the 3 : 1 map above this caustic maps to the complex surface

$$\mathbb{C} \times \{(z_{13}, z_{14} \in \mathbb{C}^* \times \mathbb{C} \,|\, z_{14}^3 - z_{13}^2 = 0\},$$

and the F-manifold extends to $\mathbb{C} \times \mathbb{C}^* \times \mathbb{C}$ with coordinates (z_1, z_{13}, z_{14}) and with this surface as caustic.

We claim that the F-manifold also extends to a fiber over $z_{14} = \infty$. Though proving this and understanding how this fiber glues in requires more care. The stabilizer $(\mathrm{Br}_3)_{\underline{e}/\{\pm 1\}^3}$ contains $(\sigma^{mon})^4$ and

$$(\sigma_2^3)(\sigma_2^{-1}\sigma_1^3\sigma_2)(\sigma_1^3) = (\sigma^{mon})^2(\sigma_2\sigma_1^3\sigma_2^{-1})^{-1}$$

so also $\sigma_2\sigma_1^6\sigma_2^{-1}$. The parabolic fixed point $1 \in \widehat{\mathbb{R}}$ of the action of $\sigma_2\sigma_1^6\sigma_2^{-1}$ corresponds to $z_{11} = \infty$. By Lemma 9.7 (b) the F-manifold $\mathbb{C} \times \mathbb{C}^* \times \mathbb{C}$ with coordinates (z_1, z_{12}, z_{11}) extends to an F-manifold with fiber over $z_{11} = 0$, namely

$$\mathbb{C} \times (\text{a certain } \mathbb{C}^*\text{-orbibundle over } \mathbb{P}^1) =: \mathbb{C} \times \widetilde{Y}_2,$$

and for a point in the fiber over $z_{11} = \infty$ the germ of the F-manifold is of type $I_2(6)A_1$.

The fiber over $z_{11} = \infty$ is exceptional of type $(1, 2)$ (in the sense of Remark 9.5), because

$$(\sigma^{mon})^4, (\sigma^{mon})^2\sigma_2\sigma_1^3\sigma_2^{-1}, \sigma_2\sigma_1^6\sigma_2^{-1} \in (\mathrm{Br}_3)_{\underline{e}/\{\pm 1\}^3}.$$

The pull back of $\widetilde{p}_2 : \widetilde{Y}_2 \to \mathbb{P}^1$ via the 2 : 1 branched covering $b_3 : \mathbb{P}^1 \to \mathbb{P}^1$, $\widetilde{z}_{11} \to \widetilde{z}_{11}^2 = z_{11}$, is a $\mathbb{C}^*$-bundle $\widetilde{Y}_3$ with projection $\widetilde{p}_3 : \widetilde{Y}_3 \to \mathbb{P}^1$. We claim that its Euler number is -1.

To see this, compare it with C_3^{pure} as $\mathbb{C}^*$-bundle with the standard action of $\mathbb{C}^*$. This is the tautological bundle $\mathcal{O}_{\mathbb{P}^1}(-1)$ minus its zero section, so it has Euler number -1. Because of the 2 : 1 branched covering $b_3 : \mathbb{P}^1 \to \mathbb{P}^1$ and because of $[PSL_2(\mathbb{Z}) : \overline{\Gamma(2)}] = 6$ and $[PSL_2(\mathbb{Z}) : \overline{\Gamma(3)}] = 12$, the base space $\mathbb{P}^1$ of $\widetilde{Y}_3$ is as a quotient of $\mathbb{H} \cup \{\text{cusps}\}$ four times as large as the base space $\mathbb{P}^1$ of C_3^{pure} as $\mathbb{C}^*$-bundle. On the other hand, to obtain C_3^{pure} from C_3^{univ}, we divided out σ^{mon}. To obtain $\widetilde{Y}_3$, we divided out only $(\sigma^{mon})^4$. Therefore the Euler number of $\widetilde{Y}_3$ is also -1.

Now Example 9.6 applies. One sees $\widetilde{Y}_3 = Y_3$ and $\widetilde{Y}_2 = Y_2$. The automorphism of order three in the proof of Theorem 9.3, which $(\sigma_1\sigma_2)^4$ induces on the chart $Y_2^{(1)}$, $(z_{12}, z_{11}) \mapsto (e^{-2\pi i/3}z_{12}, e^{-2\pi i/3}z_{11})$, extends to Y_2. It restricts because of $y_2 = z_{12}^2 z_{11}$ to the identity on the fiber over $z_{11} = \infty$ which is the fiber $\{0\} \times \mathbb{C}$ in the chart $Y_2^{(2)}$. The quotient is Y_1. The quotient map is a 3 : 1 branched covering with branching locus the fiber over $z_{11} = \infty$ and the image $\{0\} \times \mathbb{C}^* \subset Y_1$ of this fiber. Therefore at each point in $\mathbb{C} \times$ (this image) $= \mathbb{C} \times \{0\} \times \mathbb{C}^*$ the germ of the F-manifold is of type $I_2(2)A_1$. Therefore $\mathbb{C} \times \{0\} \times \mathbb{C}^*$ is the Maxwell stratum of $\mathbb{C} \times Y_1$ as F-manifold.

Finally, the complement of $\mathbb{C} \times Y_1$ in $\mathbb{C}^3$ has codimension 2. Therefore the multiplication extends to this complement holomorphically, so the F-manifold extends to $\mathbb{C}^3$ with coordinates (z_1, z_{13}, z_{14}). Because this codimension 2 complement is in the closure of the caustic of the F-manifold $\mathbb{C} \times Y_1$, it is part of the caustic of $\mathbb{C}^3$. Part (d) is proved. □

Chapter 10
Isolated Hypersurface Singularities

An isolated hypersurface singularity gives rise to a unimodular bilinear lattice $(H_{\mathbb{Z}}, L)$ of some rank $n \in \mathbb{N}$ and a $\mathrm{Br}_n \ltimes \{\pm 1\}^n$-orbit $\mathcal{B}^{dist}$ of triangular bases. This chapter reports about these structures and gives references. It does not contain new results.

The theory of isolated hypersurface singularities came to life in the 1960ies. Milnor's book [Mi69] was a crucial early step. Soon several groups around the world joined, especially Arnold and his students, and also the first author's teacher Brieskorn.

Nowadays there are several thorough and extensive presentations of the theory of isolated hypersurface singularities, foremost the two volumes [AGV85][AGV88], but also the more recent books [AGLV98] and [Eb01]. A few years ago Ebeling wrote a survey [Eb20] on distinguished bases for isolated hypersurface singularities. But the sections below offer quite a lot of material which is not covered in that survey.

Section 10.1 introduces basic structures induced by one isolated hypersurface singularity, especially a unimodular bilinear lattice $(H_{\mathbb{Z}}, L)$, but also the notion of a universal unfolding. The base space $\mathcal{M}$ of such an unfolding has to be compared with a space $C_n^{\underline{e}/\{\pm 1\}^n}$. The section finishes with a loose description of Arnold's classification of singularities with modality ≤ 2.

Section 10.2 equips the pair $(H_{\mathbb{Z}}, L)$ with a natural $\mathrm{Br}_n \ltimes \{\pm 1\}^n$ orbit $\mathcal{B}^{dist}$ of distinguished bases. Here the $\mathbb{Z}$-lattice bundles from Sect. 8.5 come from geometry and induce this set $\mathcal{B}^{dist}$. These triples $(H_{\mathbb{Z}}, L, \mathcal{B}^{dist})$ are rather special. The only triples in rank 2 and rank 3 which appear here are those of A_2 and A_3. Section 10.2 points at techniques and cites results on the triples from singularities. It ends with the best studied cases, the simple singularities and the simple elliptic singularities. It remains an interesting open problem to find properties which distinguish the triples $(H_{\mathbb{Z}}, L, \mathcal{B}^{dist})$ for singularities from all other triples.

Section 10.3 discusses results of the first author and coauthors on the groups $G_{\mathbb{Z}}$ for singularities. A crucial notion is that of an *Orlik block*. The most concrete results are those on the singularities with modality ≤ 2.

C. Hertling and K. Larabi, *Triangular Bases of Unimodular Bilinear Lattices and Induced Structures*, Lecture Notes in Mathematics 2398,
https://doi.org/10.1007/978-3-032-25997-4_10

Section 10.4 explains classical results from the 1980ies on the monodromy groups and the vanishing cycles for singularities. The situation for the even monodromy group $\Gamma^{(0)}$ is very satisfying, thanks to results of Ebeling. In almost all cases it is determined by the pair $(H_{\mathbb{Z}}, I^{(0)})$. The odd case is more complicated. There the main work was done by Wajnryb, Chmutov and Janssen. It is also subject of Sect. 10.4.

Section 10.5 comes to the few moduli spaces $C_n^{\underline{e}/\{\pm 1\}^n}$ for singularities, which had been studied, namely those for the simple singularities and the simple elliptic singularities. They turn out to be the complements of Maxwell stratum and caustic in base spaces of certain global unfoldings. In the case of the simple singularities, this is essentially due to Looijenga and Deligne 1974. Hertling and Roucairol generalized it to the simple elliptic singularities. The section offers key ideas of the proofs.

10.1 Topology, Milnor Lattice, Unfolding, Classification

Definition 10.1 (E.g. [Mi69][Lo84][AGV88][Eb01]) An isolated hypersurface singularity is a holomorphic function germ

$$f : (\mathbb{C}^{m+1}, 0) \to (\mathbb{C}, 0)$$

(with $m \in \mathbb{Z}_{\geq 0}$) with $f(0) = 0$ and with an isolated singularity at 0, i.e. the Jacobi ideal $J_f := \left(\frac{\partial f}{\partial z_0}, ..., \frac{\partial f}{\partial z_m}\right) \subset \mathbb{C}\{z_0, ..., z_m\}$ has an isolated zero at 0. Its Milnor number n is

$$n := \dim \mathbb{C}\{z_0, ..., z_m\}/J_f \in \mathbb{N}. \tag{10.1}$$

Usually the Milnor number is called μ, and the number of variables is called n [AGV88] or $n + 1$ [Eb01]. In order to be consistent with the other chapters of this book, here the Milnor number is called n and the number of variables is called $m + 1$.

The first thing to do is to construct *a good representative* of f. One can choose and chooses a small positive number $\varepsilon > 0$ such that the boundary of each ball $B_{\widetilde{\varepsilon}} \subset \mathbb{C}^{m+1}$ with center at 0 and radius $\widetilde{\varepsilon} \in (0, \varepsilon]$ intersects the fiber $f^{-1}(0)$ transversally. Then one chooses a small positive number $\eta > 0$ such that each fiber $f^{-1}(\tau)$ for $\tau \in \Delta_\eta$ intersects the boundary of B_ε transversally. Here $\Delta_\eta := \{\tau \in \mathbb{C} \mid |\tau| < \eta\} \subset \mathbb{C}$ is the disk of radius η around 0. Then

$$f : X \to \Delta_\eta \quad \text{with} \quad X := B_\varepsilon \cap f^{-1}(\Delta_\eta) \tag{10.2}$$

is a good representative of f with fibers $X_\tau = f^{-1}(\tau)$ for $\tau \in \Delta_\eta$. The restriction $f : X - X_0 \to \Delta_\eta - \{0\}$ is a locally trivial C^∞ fiber bundle. Only the 0-fiber X_0 is singular, and its only singularity is at 0. The fibers X_τ for $\tau \in \Delta_\eta - \{0\}$ are called *Milnor fibers*. That all this works and is true can be found in [Mi69], [Lo84], [AGV88] or [Eb01]. See Fig. 10.1 for a schematic picture.

The references [Lo84, (5.11)], [AGV88, ch. 2] or [Eb01] tell also the following. Our signs and conventions are consistent with those in [AGV88, chap. 2].

Theorem 10.2 *Let $f : (\mathbb{C}^{m+1}, 0) \to (\mathbb{C}, 0)$ be an isolated hypersurface singularity with Milnor number $n \in \mathbb{N}$. Let $f : X \to \Delta_\eta$ be a good representative as above.*

(a) Each regular fiber X_τ with $\tau \in \Delta_\eta - \{0\}$ is homotopy equivalent to a bouquet of n m-spheres, so that its only (reduced if $m = 0$) homology is the middle homology and that is

$$H_m^{(red)}(X_\tau, \mathbb{Z}) \cong \mathbb{Z}^n.$$

This $\mathbb{Z}$-module is called a Milnor lattice.

(b) For $\zeta \in S^1$ there is a canonical isomorphism

$$H_m^{(red)}(X_{\zeta\eta}, \mathbb{Z}) \cong H_{m+1}(X, X_{\zeta\eta}, \mathbb{Z}) =: Ml(f, \zeta) \tag{10.3}$$

of the Milnor lattice over a point $\zeta\eta \in \partial\Delta_\eta$ to the relative homology group with boundary in $X_{\zeta\eta}$. The union $\bigcup_{\zeta\in S^1} Ml(f, \zeta)$ is a flat bundle over S^1 of $\mathbb{Z}$-lattices of rank n.

(c) An intersection form for Lefschetz thimbles is well defined on relative homology groups with different boundary parts. It is for any $\zeta \in S^1$ a $(-1)^{m+1}$ symmetric unimodular bilinear form

$$I^{Lef} : Ml(f, \zeta) \times Ml(f, -\zeta) \to \mathbb{Z} \tag{10.4}$$

Let γ_π be the isomorphism $Ml(f, \zeta) \to Ml(f, -\zeta)$ by flat shift in mathematically positive direction. Then the classical Seifert form is given by

$$L^{sing} : Ml(f, \zeta) \times Ml(f, \zeta) \to \mathbb{Z}, \tag{10.5}$$

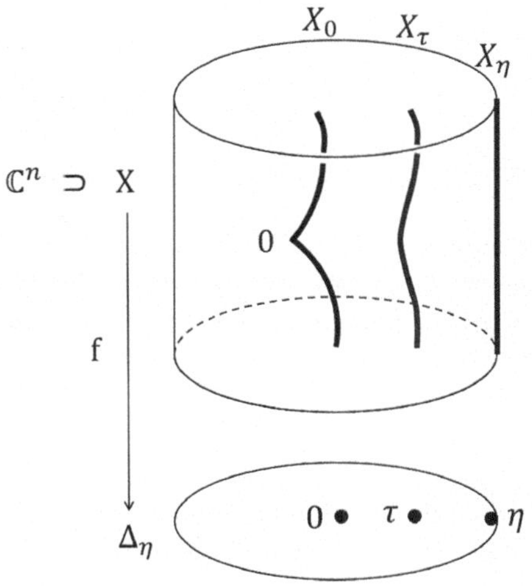

Fig. 10.1 A schematic picture of a representative $f : X \to \Delta_\eta$ of a singularity $f : (\mathbb{C}^{m+1}, 0) \to (\mathbb{C}, 0)$

$$L^{sing}(a,b) := (-1)^{m+1} I^{Lef}(a, \gamma_\pi b).$$

It has determinant $(-1)^{n(m+1)(m+2)/2}$. *The classical monodromy* M^{sing} *and the intersection form* I^{sing} *on* $Ml(f,\zeta)$ *are given by*

$$L^{sing}(M^{sing}a, b) = (-1)^{m+1} L^{sing}(b,a), \tag{10.6}$$

$$\begin{aligned} I^{sing}(a,b) &= -L^{sing}(a,b) + (-1)^{m+1} L^{sing}(b,a) \\ &= L^{sing}((M^{sing} - \mathrm{id})a, b). \end{aligned} \tag{10.7}$$

The monodromy M^{sing} *is the monodromy operator of the flat* $\mathbb{Z}$*-lattice bundle* $\bigcup_{\zeta\in S^1} Ml(f,\zeta) \cong \bigcup_{\zeta\in S^1} H_m(X_{\zeta\eta}, \mathbb{Z})$. *The intersection form* I^{sing} *is the natural intersection form on* $H_m(X_\tau, \mathbb{Z})$. *It is* $(-1)^m$*-symmetric.*

(d) The pair $(Ml(f,1), L^{sing})$ *is up to a canonical isomorphism independent of the choice of a good representative* $f : X \to \Delta_\eta$ *of the function germ* f. *So it is an invariant of the singularity* f.

Definition 10.3 Consider the situation in Theorem 10.2. We define a unimodular bilinear lattice $(H_\mathbb{Z}, L)$ by

$$H_\mathbb{Z}(f) := H_\mathbb{Z} := Ml(f,1), \tag{10.8}$$

$$L(f) := L := (-1)^{(m+1)(m+2)/2} \cdot L^{sing}. \tag{10.9}$$

We call L the *normalized* Seifert form. Define $k \in \{0; 1\}$ by

$$k := m \bmod 2, \quad \text{i.e.} \quad \begin{cases} k := 0 \text{ if } m \text{ is even}, \\ k := 1 \text{ if } m \text{ is odd}. \end{cases} \tag{10.10}$$

Obviously, the monodromy M and the bilinear form $I^{(k)}$ of the unimodular bilinear lattice $(H_\mathbb{Z}, L)$ are related to M^{sing} and I^{sing} by

$$M = (-1)^{m+1} M^{sing}, \tag{10.11}$$

$$I^{(k)} = (-1)^{m(m+1)/2} I^{sing}. \tag{10.12}$$

We call M the *normalized* monodromy.

Remark 10.4 Above the Seifert form L^{sing} is defined with the help of the isomorphism (10.3) and the intersection form I^{Lef} in (10.4). An alternative is to derive it from a *linking number* of cycles in different Milnor fibers in the closely related fibration

$$\partial B_\varepsilon - f^{-1}(0) \to S^1, \quad z \mapsto \frac{f(z)}{|f(z)|}.$$

We prefer I^{Lef} as the Lefschetz thimbles appear implicitly anyway in the construction of distinguished bases in Sect. 10.2.

The notion of an *unfolding* of an isolated hypersurface singularity f goes back to Thom and Mather. An *unfolding* of $f : (\mathbb{C}^{m+1}, 0) \to (\mathbb{C}, 0)$ is a holomorphic function germ $F : (\mathbb{C}^{m+1} \times \mathcal{M}, 0) \to (\mathbb{C}, 0)$ such that $F|_{(\mathbb{C}^{m+1},0)\times\{0\}} = f$ and such that $(\mathcal{M}, 0)$ is the germ of a complex manifold. Its Jacobi ideal is $J_F := (\frac{\partial F}{\partial z_i}) \subset \mathcal{O}_{\mathbb{C}^{m+1}\times\mathcal{M},0}$, its critical space is the germ $(C, 0) \subset (\mathbb{C}^{m+1} \times \mathcal{M}, 0)$ of the zero set of J_F with the canonical complex structure. The projection $(C, 0) \to (\mathcal{M}, 0)$ is finite and flat of degree n. A kind of Kodaira-Spencer map is the $\mathcal{O}_{\mathcal{M},0}$-linear map

$$\mathbf{a}_C : \mathcal{T}_{\mathcal{M},0} \to \mathcal{O}_{C,0}, \quad X \mapsto \widetilde{X}(F)|_{(C,0)} \tag{10.13}$$

where $\widetilde{X}$ is an arbitrary lift of $X \in \mathcal{T}_{\mathcal{M},0}$ to $(\mathbb{C}^{m+1} \times \mathcal{M}, 0)$.

We will use the following notion of morphism between unfoldings. Let $F_i : (\mathbb{C}^{m+1} \times \mathcal{M}_i, 0) \to (\mathbb{C}, 0)$ for $i \in \{1, 2\}$ be two unfoldings of f with projections $\mathrm{pr}_i : (\mathbb{C}^{m+1} \times \mathcal{M}_i, 0) \to (\mathcal{M}_i, 0)$. A *morphism* from F_1 to F_2 is a pair (Φ, φ) of map germs such that the following diagram commutes,

$$\begin{array}{ccc} (\mathbb{C}^{m+1} \times \mathcal{M}_1, 0) & \xrightarrow{\Phi} & (\mathbb{C}^{m+1} \times \mathcal{M}_2, 0) \\ \downarrow{\scriptstyle \mathrm{pr}_1} & & \downarrow{\scriptstyle \mathrm{pr}_2} \\ (\mathcal{M}_1, 0) & \xrightarrow{\varphi} & (\mathcal{M}_2, 0) \end{array}$$

and

$$\begin{aligned} \Phi|_{(\mathbb{C}^{m+1},0)\times\{0\}} &= \mathrm{id}, \\ F_1 &= F_2 \circ \Phi \end{aligned}$$

hold. Then one says that F_1 *is induced* by (Φ, φ) from F_2. An unfolding is *versal* if any unfolding is induced from it by a suitable morphism. A versal unfolding $F : (\mathbb{C}^{n+1} \times \mathcal{M}, 0) \to (\mathbb{C}, 0)$ is *universal* if the dimension of the parameter space $(\mathcal{M}, 0)$ is minimal. Universal unfoldings exist by work of Thom and Mather. More precisely, an unfolding is versal if and only if the map $\mathbf{a}_C$ is surjective, and it is universal if and only if the map $\mathbf{a}_C$ is an isomorphism (see e.g. [AGV85, chap. 8] for a proof). Observe that $\mathbf{a}_C$ is surjective/an isomorphism if and only if its restriction to $0 \in \mathcal{M}$, the map

$$\mathbf{a}_{C,0} : T_0\mathcal{M} \to \mathcal{O}_{\mathbb{C}^{m+1},0}/J_f \tag{10.14}$$

is surjective/an isomorphism. Therefore an unfolding

$$F(z_0, ..., z_m, t_1, ..., t_n) = F(x, t) = F_t(z) = f(z) + \sum_{j=1}^{n} m_j t_j, \tag{10.15}$$

with $(\mathcal{M},0)=(\mathbb{C}^n,0)$ with coordinates $t=(t_1,...,t_n)$ where $m_1,...,m_n\in\mathcal{O}_{\mathbb{C}^{m+1},0}$ represent a basis of the Jacobi algebra $\mathcal{O}_{\mathbb{C}^{m+1},0}/J_f$, is universal. In a versal unfolding the critical space C is smooth.

Just as for f, one can also choose a good representative for an unfolding F. In the following we suppose that such a representative is chosen. Then $\mathcal{M}$ is an open subset of $\mathbb{C}^n$.

Theorem 10.5 *[He02, Theorem 5.3] Let $f:(\mathbb{C}^{m+1},0)\to(\mathbb{C},0)$ be an isolated hypersurface singularity with Milnor number n. The base space $\mathcal{M}$ of a good representative of a universal unfolding is a generically semisimple F-manifold with Euler field. The multiplication comes from the multiplication on the right side of the isomorphism $\mathbf{a}_C:\mathcal{T}_{\mathcal{M},0}\to\mathcal{O}_{C,0}$ in (10.13), the unit field is $e=\mathbf{a}_C^{-1}([1])$, the Euler field is $E=\mathbf{a}_C^{-1}([F])$. For each $t\in\mathcal{M}$, the eigenvalues of $E\circ$ are the values of the critical points of F_t.*

The algebra $(T_0\mathcal{M},\circ)\cong(\mathcal{O}_{\mathbb{C}^{m+1},0}/J_f)$ is irreducible. Therefore the caustic $\mathcal{K}_3\subset\mathcal{M}$ is not empty, so it is a hypersurface. Also the Maxwell stratum $\mathcal{K}_2\subset\mathcal{M}$ is not empty, but a hypersurface. Within the caustic there is the *μ-constant stratum*

$$\mathcal{M}_\mu:=\{t\in\mathcal{M}\mid F_t \text{ has only one singularity } z^0,\ F_t(z^0)=0\}. \qquad (10.16)$$

The only singularity which F_t for $t\in\mathcal{M}_\mu$ has, has automatically Milnor number n (because $\mathbf{a}_C$ is finite and flat of degree n). The μ-constant stratum is a complex subvariety of $\mathcal{M}$. Its dimension is called the *modality* of the singularity f.

It is nontrivial, but true (see e.g. [He11, Theorem 2.2]) that the lattices $(H_\mathbb{Z},L)(F_t)$ for $t\in\mathcal{M}_\mu$ glue to a local system of $\mathbb{Z}$-lattices with Seifert forms (and monodromy M and intersection forms $I^{(0)}$ and $I^{(1)}$) over $\mathcal{M}_\mu$.

Arnold classified all singularities with modality ≤ 2 up to coordinate changes. We will describe them loosely in the following theorem. We give normal forms for the simple singularities, the simple elliptic singularities and the hyperbolic singularities. Normal forms of the other singularities with modality ≤ 2 can be found in [AGV85, II chap. 15.1]. Wall classified also all singularities with modality 3 [Wa99].

From a normal form $F_t(z)$ in $m+1$ variables $z_0,...,z_m$ and some parameters t one obtains also a normal form $F_t(z)+\sum_{j=m+1}^{m_2+1}z_j^2$ in m_2+1 variables. In the following theorem, $(m\ge 2)$ after the symbol of a family means that the smallest number of variables for which this family exists is $m=2$. The meaning of $(m\ge 0)$ and $(m\ge 1)$ is analogous. See also Remark 10.11.

Theorem 10.6 *(Arnold, see [AGV85, II 15.1])*

(a) The singularities with modality 0 are called simple singularities. *There are two series and three exceptional cases (the index is the Milnor number):*

$A_n,\ n\ge 1$ $(m\ge 0)$	$D_n,\ n\ge 4$ $(m\ge 1)$	E_6 $(m\ge 1)$	E_7 $(m\ge 1)$	E_8 $(m\ge 1)$
z_0^{n+1}	$z_0^{n-1}+z_0z_1^2$	$z_0^4+z_1^3$	$z_0^3z_1+z_1^3$	$z_0^5+z_1^3$

(b) The singularities with modality 1 are called unimodal singularities. *They fall into three groups of 1-parameter families:*

(i) The three (1-parameter families of) simple elliptic singularities $\widetilde{E}_6$, $\widetilde{E}_7$ *and* $\widetilde{E}_8$, *here the parameter is* $\lambda \in \mathbb{C} - \{0, 1\}$,

name	n	normalform
$\widetilde{E}_6 (m \geq 2)$	8	$f_\lambda = z_1(z_1 - z_0)(z_1 - \lambda z_0) - z_0 z_2^2$
$\widetilde{E}_7\ (m \geq 1)$	9	$f_\lambda = z_0 z_1 (z_1 - z_0)(z_1 - \lambda z_0)$
$\widetilde{E}_8\ (m \geq 1)$	10	$f_\lambda = z_1(z_1 - z_0^2)(z_1 - \lambda z_0^2)$

These normal forms are called Legendre normal forms; *they are from [Sa74, 1.9].*

(ii) The (1-parameter families of) hyperbolic singularities T_{pqr} *with three discrete parameters* $p, q, r \in \mathbb{Z}_{\geq 2}$ *with* $p \geq q \geq r$ *and* $\frac{1}{p} + \frac{1}{q} + \frac{1}{r} < 1$, *here the continuous parameter is* $t \in \mathbb{C}^*$,

$$z_0^p + z_1^q + z_2^r + t z_0 z_1 z_2.$$

(iii) 14 (1-parameter families of) exceptional unimodal singularities *with the following names (the index is the Milnor number),*

$$m \geq 1: \ E_{12}, \ E_{13}, \ E_{14}, \ Z_{11}, \ Z_{12}, \ Z_{13}, \ W_{12}, \ W_{13}$$
$$m \geq 2: \ Q_{10}, \ Q_{11}, \ Q_{12}, \ S_{11}, \ S_{12}, \ U_{12}.$$

(c) The singularities with modality 2 are called bimodal singularities. *They fall into three groups of 2-parameter families:*

(i) The six (2-parameter families of) quadrangle singularities *with the following names,*

$m \geq 1$	name	$E_{3,0}$	$Z_{1,0}$	$W_{1,0}$	$m \geq 2$	name	$Q_{2,0}$	$S_{1,0}$	$U_{1,0}$
	n	16	15	15		n	14	14	14

(ii) The eight (2-parameter families of) bimodal series *with one discrete paramter* $p \in \mathbb{N}$ *and with the following names*

$m \geq 1$	name	$E_{3,p}$	$Z_{1,p}$	$W_{1,p}$	$W_{1,p}^\sharp$
	n	$16 + p$	$15 + p$	$15 + p$	$15 + p$
$m \geq 2$	name	$Q_{2,p}$	$S_{1,p}$	$S_{1,p}^\sharp$	$U_{1,p}$
	n	$14 + p$	$14 + p$	$14 + p$	$14 + p$

(iii) The 14 (2-parameter families of) exceptional bimodal singularities *with the following names (the index is the Milnor number),*

$$m \geq 1: \ E_{18}, \ E_{19}, \ E_{20}, \ Z_{17}, \ Z_{18}, \ Z_{19}, \ W_{17}, \ W_{18}$$
$$m \geq 2: \ Q_{16}, \ Q_{17}, \ Q_{18}, \ S_{16}, \ S_{17}, \ U_{16}.$$

The simple singularities and the simple elliptic singularities can be characterized in many different ways. Arnold found the following characterizations.

Theorem 10.7 *(a) (Arnold [Ar73-1]) The simple singularities are the only isolated hypersurface singularities where $I^{(0)}$ on $H_{\mathbb{Z}}$ is positive definite.*

(Classical) Then $(H_{\mathbb{Z}}, I^{(0)})$ is a root lattice of the same type as the name, M is a Coxeter element, $\Gamma^{(0)}$ is the Weyl group, $\Delta^{(0)}$ is the set of roots, so especially $\Delta^{(0)} = R^{(0)}$.

(b) (Arnold [Ar73-2]) The simple elliptic singularities are the only isolated hypersurface singularities where $I^{(0)}$ on $H_{\mathbb{Z}}$ is positive semidefinite, but not positive definite.

10.2 Distinguished Bases

Let $f : (\mathbb{C}^{m+1}, 0) \to (\mathbb{C}, 0)$ be an isolated hypersurface singularity. Theorem 10.2 and Definition 10.3 gave a unimodular bilinear lattice $(H_{\mathbb{Z}}, L)$ which is associated canonically to the singularity f. Here $H_{\mathbb{Z}}$ is a Milnor lattice of a fiber with value in $\mathbb{R}_{>0}$ and L is the normalized Seifert form. In fact, the pair $(H_{\mathbb{Z}}, L)$ comes equipped naturally with a $\mathrm{Br}_n \ltimes \{\pm 1\}^n$ orbit $\mathcal{B}^{dist}$ of triangular bases. They are called *distinguished bases*. Also the set $\mathcal{B}^{dist}$ is an invariant of the singularity f. This section gives the construction of the distinguished bases and lists several results on them. They have special properties. There are many open questions around them.

Let $F : \mathcal{X} \to \Delta_\eta$ be a good representative of a universal unfolding F of f, with $\mathcal{X} = F^{-1}(\Delta_\eta) \cap (B_\varepsilon \times \mathcal{M})$ where $\mathcal{M} \subset \mathbb{C}^n$ is a sufficiently small open ball around 0 in $\mathbb{C}^n$. As above the critical space is $C \subset \mathcal{X}$. Its image under the map

$$(F, \mathrm{pr}_{\mathcal{M}}) : \mathcal{X} \subset B_\varepsilon \times \mathcal{M} \to \Delta_\eta \times \mathcal{M}, \quad (z, t) \mapsto (F(z, t), t)$$

is the *discriminant* $D_{1,n}$, a hypersurface in $\Delta_\eta \times \mathcal{M}$. The projection $D_{1,n} \to \mathcal{M}$ is finite and flat of degree n. Of course, for that $\mathcal{M}$ must have been chosen small enough. For any $t \in \mathcal{M}$ the intersection $D_{1,n} \times \{t\} \subset \Delta_\eta \times \{t\}$ is the set of critical values of the critical points of F_t. See Fig. 10.2.

Recall that the caustic $\mathcal{K}_3 \subset \mathcal{M}$ and the Maxwell stratum $\mathcal{K}_2 \subset \mathcal{M}$ are complex hypersurfaces. Choose any point $t^0 \in \mathcal{M} - (\mathcal{K}_3 \cup \mathcal{K}_2)$, so a generic point. Then $F_{t^0} : \mathcal{X}_{t^0} \to \Delta_\eta$ with $\mathcal{X}_{t^0} := \mathcal{X} \cap (B_\varepsilon \times \{t^0\})$ is a holomorphic function with n A_1-singularities and such that the critical values $u_1, ..., u_n \in \Delta_\eta$ of these A_1-singularities are pairwise different.

The union $\bigcup_{\tau \in \Delta_\eta - \{u_1, ..., u_n\}} H_m(F_t^{-1}(\tau), \mathbb{Z})$ is a flat $\mathbb{Z}$-lattice bundle of rank n. Near each critical value u_j there is a lattice vector, which is unique up to the sign and which comes geometrically from the vanishing cycle of the A_1 singularity in the fiber over u_j. One chooses a distinguished system $(\underline{\gamma}; \sigma)$ of paths $\gamma_1, ..., \gamma_n$ from η to $u_{\sigma(1)}, ..., u_{\sigma(n)}$ and pushes the n vanishing cycles along these paths to $H_m(F_t^{-1}(\eta), \mathbb{Z})$. There is a canonical isomorphism

$$H_m(F_t^{-1}(\eta), \mathbb{Z}) \cong H_m(X_\eta, \mathbb{Z}) \cong Ml(f, 1) = H_{\mathbb{Z}}.$$

One obtains a tuple $\underline{v} = (v_1, ..., v_n)$ of cycles in $H_{\mathbb{Z}}$. Brieskorn [Br70, Appendix] first wrote a proof of part (a) of the following theorem (see also [AGV88], [Eb01, Satz 5.5 in 5.4, and 5.6], [AGLV98, §1 1.5]).

Theorem 10.8 *(a) The resulting tuple* $\underline{v}$ *is a triangular basis of* $(H_{\mathbb{Z}}, L)$ *with* $S := L(\underline{v}^t, \underline{v})^t \in T_n(\mathbb{Z})$

(b) The $\mathrm{Br}_n \ltimes \{\pm 1\}^n$ *orbit* $\mathcal{B}^{dist}$ *of this basis* $\underline{v}$ *is an invariant of the singularity* f*. Its elements are called* distinguished bases.

Remarks 10.9 (i) If one pushes the vanishing cycle above $u_{\sigma(j)}$ along the path γ_j to the fiber $F_{t^0}^{-1}(\eta)$ over η, the union of cycles in the fibers above γ_j forms an $m+1$ dimensional cycle with boundary in $F_{t^0}^{-1}(\eta)$, which gives a homology class in $H_{m+1}(\mathcal{X}_{t^0}, F_{t^0}^{-1}(\eta), \mathbb{Z})$. It is a *Lefschetz thimble*. Its homology class is mapped by the isomorphism

$$H_{m+1}(\mathcal{X}_{t^0}, F_{t^0}^{-1}(\eta), \mathbb{Z}) \cong H_m(F_{t^0}^{-1}(\eta), \mathbb{Z})$$

to the homology class of its boundary, which is the cycle in the fiber $F_{t^0}^{-1}(\eta)$. See Fig. 10.3.

(ii) Recall $k = 0$ if $m \equiv 0(2)$ and $k = 1$ if $m \equiv 1(2)$. The $\mathbb{Z}$-lattice bundle $\bigcup_{\tau \in \Delta_\eta - \{u_1, ..., u_n\}} H_m(F_t^{-1}(\tau), \mathbb{Z})$ is the $\mathbb{Z}$-lattice structure $V_{\mathbb{Z}}^{(k)}((\underline{u}, \eta), [(\underline{\gamma}; \sigma)], \underline{v})$. So, in the case of an isolated hypersurface singularity, this bundle comes from geometry and is the source of the triangular basis $\underline{v}$ in $(H_{\mathbb{Z}}, L)$.

(iii) In [Eb01, 5.6] the bilinear form L^{sing} is called v. Korollar 5.3 (i) there claims that the matrix V is upper triangular. But it is lower triangular.

(iii) In [Br83] Brieskorn considered the larger set $\mathcal{B}^* := \bigcup_{g \in \Gamma^{(k)}} g(\mathcal{B}^{dist})$, possibly because he wanted an invariant of the singularity f and did not see that $\mathcal{B}^{dist}$ is one.

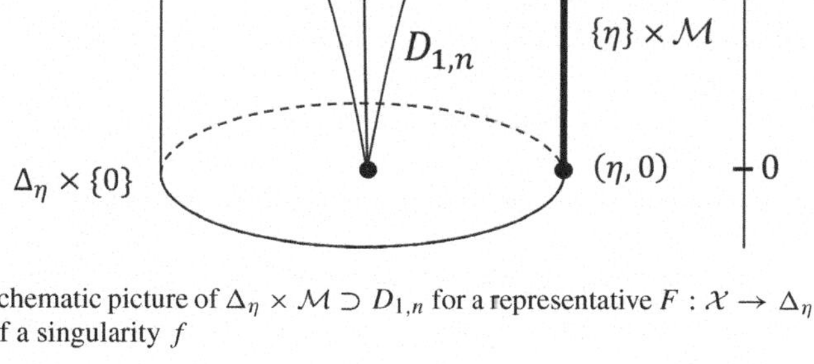

Fig. 10.2 A schematic picture of $\Delta_\eta \times \mathcal{M} \supset D_{1,n}$ for a representative $F : \mathcal{X} \to \Delta_\eta$ of a universal unfolding F of a singularity f

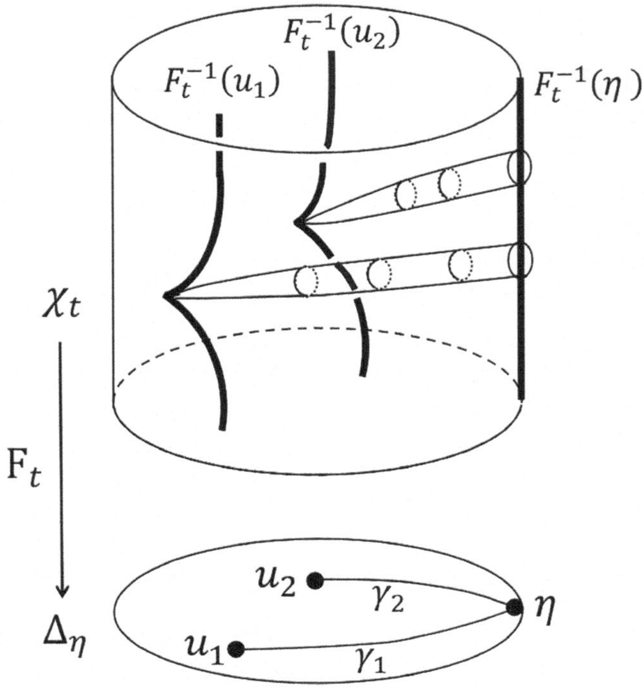

Fig. 10.3 Two Lefschetz thimbles above paths from η to critical values u_1 and u_2

$\mathcal{B}^{dist}$ is one because the set $\{\eta\} \times \mathcal{M}$ does not mix with the discriminant. Therefore the $\mathbb{Z}$-lattice $H_m(F_t^{-1}(\eta), \mathbb{Z})$ is canonically isomorphic to $H_m(X_\eta, \mathbb{Z})$. Moving $t \in \mathcal{M} - (\mathcal{K}_3 \cup \mathcal{K}_2)$ changes the tuple $(u_1, ..., u_n)$ of critical values and deforms the distinguished system of paths. It does not do more. η is fixed.

Now we start a report on specific properties of the unimodular bilinear lattices $(H_\mathbb{Z}, L, \underline{v})$ with distinguished bases which come from isolated hypersurface singularities. We start with the Thom-Sebastiani sum.

Theorem 10.10 *(a) (Definition) The* Thom-Sebastiani sum *of a singularity $f = f(z_0, ..., z_{m_1})$ in $m_1 + 1$ variables and a singularity $g = g(z_{m_1+1}, ..., z_{m_1+m_2+1})$ in $m_2 + 1$ different variables is the singularity $f + g$.*

(b) Thom and Sebastiani [ST71] observed that there is a canonical isomorphism

$$H_\mathbb{Z}(f) \otimes H_\mathbb{Z}(g) \cong H_\mathbb{Z}(f + g), \tag{10.17}$$

which respects the monodromies and the normalized monodromies.

$$M^{sing}(f) \otimes M^{sing}(g) \cong M^{sing}(f + g),$$
$$M(f) \otimes M(g) \cong M(f + g), \tag{10.18}$$

Deligne (1973?, cited in [AGV88]) observed that it respects the Seifert forms (up to a sign) and the normalized Seifert forms,

$$L^{sing}(f) \otimes L^{sing}(g) \cong (-1)^{(m_1+1)(m_2+1)} L^{sing}(f+g).$$
$$L(f) \otimes L(g) \cong L(f+g). \tag{10.19}$$

Any distinguished basis $(\delta_1^f, ..., \delta_{n(f)}^f)$ *of* f *and any distinguished basis* $(\delta_1^g, ..., \delta_{n(g)}^g)$ *of* g *give rise to a basis*

$$(\delta_1^f \otimes \delta_1^g, ..., \delta_1^f \otimes \delta_{n(g)}^g, \delta_2^f \otimes \delta_1^g, ..., \delta_2^f \otimes \delta_{n(g)}^g, ..., \delta_{n(f)}^f \otimes \delta_1^g, ..., \delta_{n(g)}^f \otimes \delta_{n(g)}^g) \tag{10.20}$$

of $H_{\mathbb{Z}}(f) \otimes H_{\mathbb{Z}}(g)$*, with the lexicographic order. Gabrielov [Ga73] observed that its image under the isomorphism (10.17) is a distinguished basis of* $H_{\mathbb{Z}}(f+g)$*.*

Remark 10.11 The Thom-Sebastiani sum $f+g$ of an isolated hypersurface singularity $f = f(z_0, ..., z_{m_1})$ and an A_1-singularity $g = z_{m_1+1}^2 + \cdots + z_{m_1+m_2+1}^2$ is close to the singularity f. Here $f+g$ is called a *suspension* of f. The generator of the Milnor lattice $H_{\mathbb{Z}}(g) \cong \mathbb{Z}$ is unique up to a sign. Therefore here the canonical isomorphism $H_{\mathbb{Z}}(f) \otimes H_{\mathbb{Z}}(g) \cong H_{\mathbb{Z}}(f+g)$ induces an isomorphism $H_{\mathbb{Z}}(f) \cong H_{\mathbb{Z}}(f+g)$ which is unique up to the sign. Therefore any bilinear form and any endomorphism on $H_{\mathbb{Z}}(f)$ are also well-defined on $H_{\mathbb{Z}}(f+g)$. The isomorphism (10.18) tells that the normalized monodromies $M(f)$ and $M(f+g)$ coincide, because $M(g) = \mathrm{id}$. The isomorphism (10.19) tells that the normalized Seifert forms $L(f)$ and $L(f+g)$ coincide, because $L(g)(\delta^g, \delta^g) = 1$ where $H_{\mathbb{Z}}(g) = \mathbb{Z}\delta^g$. Gabrielov's result implies that f and $f+g$ lead to the same unimodular bilinear lattice $(H_{\mathbb{Z}}, L)$ with $\mathrm{Br}_n \ltimes \{\pm 1\}^n$ orbit $\mathcal{B}^{dist}$ of distinguished bases.

The technique which was most successfully applied for the calculation of distinguished bases of isolated hypersurface singularities, is due to A'Campo and Gusein-Zade. It works for plane curve singularities, so functions in two variables. Together with the Thom-Sebastiani result above, it can be applied to many singularities in more than two variables. We will not describe the recipe here. The following theorem will just describe roughly the result. For details see the given references as well as [AGV88]. The recipe builds on resolution of plane curve singularities. In concrete cases it leads to many pictures of curves in the real plane and an interesting way how to construct them and deal with them.

Theorem 10.12 *(A'Campo [AC75-1][AC75-2] and Gusein-Zade [Gu74-1] [Gu74-2], see also [AGLV98, §2 2.1] [BK96] [LSh18]) Let* f *be a plane curve singularity, so an isolated hypersurface singularity with* $m = 1$*. Suppose that* $f = f_1 ... f_r$ *is the decomposition of* f *into its irreducible branches and that* $f_j((\mathbb{R}^2, 0)) \subset (\mathbb{R}, 0)$ *for any* j *(so* f *is a complexification of a* totally real singularity*). Any* totally real morsification *(see e.g. [AGLV98] for its definition) gives rise to three sets* $B^{\ominus}, B^0, B^{\oplus} \subset H_{\mathbb{Z}}$ *with the following properties: If* $\underline{v} = (v_1, ..., v_n)$ *is a list of the* *elements of* $B^{\ominus} \cup B^0 \cup B^{\oplus}$ *which lists first the elements of* $B^{\ominus}$*, then the elements of* B^0 *and finally the elements of* $B^{\oplus}$*, then* $\underline{v}$ *is a distinguished basis, and its matrix* S *takes the following shape,*

$$S = L(\underline{v}^t, \underline{v})^t = E_n + \begin{pmatrix} 0 & - & + \\ 0 & 0 & - \\ 0 & 0 & 0 \end{pmatrix} \in T_n(\mathbb{Z}), \tag{10.21}$$

more precisely, for $i < j$

$$\begin{aligned} & -I^{sing}(v_i, v_j) = I^{(1)}(v_i, v_j) = L(v_j, v_i) = S_{ij} \\ = & \begin{cases} 0 & \textit{if } v_i, v_j \in B^{\ominus} \textit{ or } v_i, v_j \in B^0 \textit{ or } v_i, v_j \in B^{\oplus}, \\ \leq 0 & \textit{if } v_i \in B^{\ominus}, v_j \in B^0, \\ \geq 0 & \textit{if } v_i \in B^{\ominus}, v_j \in B^{\oplus}, \\ \leq 0 & \textit{if } v_i \in B^0, v_j \in B^{\oplus}. \end{cases} \end{aligned} \tag{10.22}$$

Theorem 10.10 and Theorem 10.12 together allow to deal with singularities $f(z_0, ..., z_m)$ with $m \geq 2$ if f is an iterated Thom-Sebastiani sum of plane curve singularities. The following Theorem 10.13 of Gabrielov is stronger. It allows to construct a distinguished basis of any singularity (in principle) from results for singularities with less variables. Again we do not give the precise recipe, but only the rough result of it. For details see the given references.

Theorem 10.13 *(a) [Ga79] (see also [Eb01, 5.10] [AGLV98, §2 2.4]) Given an isolated hypersurface singularity* $f : (\mathbb{C}^{m+1}, 0) \to (\mathbb{C}, 0)$*, there is a recipe of Gabrielov which allows to construct a certain distinguished basis of* f *from the following data: a generic linear function* $g : \mathbb{C}^{m+1} \to \mathbb{C}$ *and a distinguished basis of* $\widetilde{f} := f|_{g^{-1}(0)}$*. For the recipe one needs sufficient information on the* polar curve *of* f *with respect to* g*. Let* S *be the matrix of the distinguished basis of* f*, and let* $\widetilde{S}$ *be the matrix of the distinguished basis of* $\widetilde{f}$*. Then especially*

$$\{S_{ij} \mid i, j \in \{1, .., n(f)\}\} \subset \{0, \pm 1\} \cup \{\pm \widetilde{S}_{ij} \mid i, j \in \{1, ..., n(\widetilde{f})\}\}. \tag{10.23}$$

(b) (Corollary of part (a)) Each isolated hypersurface singularity has a distinguished basis with matrix S *with entries only* $0, \pm 1$*, so*

$$\{S_{ij} \mid i, j \in \{1, .., n(f)\}\} \subset \{0, \pm 1\}.$$

The following theorem lists further properties of the unimodular bilinear lattice $(H_{\mathbb{Z}}, L, \underline{e})$ with a distinguished basis $\underline{e}$, which come from a singularity f.

Theorem 10.14 *Let* $f : (\mathbb{C}^{m+1}, 0) \to (\mathbb{C}, 0)$ *be an isolated hypersurface singularity with Milnor number* $n \in \mathbb{N}$*, and let* $(H_{\mathbb{Z}}, L, \underline{e})$ *be the induced unimodular bilinear lattice with one chosen distinguished basis* $\underline{e}$ *and matrix* $S = L(\underline{e}^t, \underline{e})^t \in T_n(\mathbb{Z})$*. Fix* $k \in \{0; 1\}$*.*

(a) (Classical, e.g. [Ga79]) $\Delta^{(k)} = \Gamma^{(k)}(e_j)$ *for any element* e_j *of the distinguished basis, so* $\Delta^{(k)}$ *is a single orbit.*

(b) [Ga74-1][La73][Le73] The triple $(H_\mathbb{Z}, L, \underline{e})$ *is irreducible.*

(c) If $n \geq 2$ *then there are* $\delta_1, \delta_2 \in \Delta^{(k)}$ *with* $I^{(k)}(\delta_1, \delta_2) = 1$.

(d) (Classical) The monodromy M is quasiunipotent. The sizes of the Jordan blocks are at most $m+1$. *The sizes of the Jordan blocks with eigenvalue 1 are at most m.*

(e) [AC73] $\operatorname{tr}(M) = 1$.

Remarks 10.15 (i) The discriminant $D_{1,n} \subset \Delta_\eta \times M$ of a good representative F of a universal unfolding of f is irreducible because it is the image of the smooth critical set $C \subset \mathcal{X}$. Its irreducibility leads easily to part (a), $\Delta^{(k)} = \Gamma^{(k)}(e_j)$. See e.g. [Eb01, Chap. 5 Satz 5.20].

(ii) Part (a) implies immediately part (b), the irreducibility of the triple $(H_\mathbb{Z}, L, \underline{e})$. This is the proof of part (b) in [Ga74-1]. The proof of part (b) in [Le73] uses part (e).

(iii) Part (c) follows from the fact that any singularity with Milnor number $n \geq 2$ deforms to the A_2 singularity. See e.g. [Eb01].

(iv) The triple $(H_\mathbb{Z}, I^{(k)}, \Delta^{(k)})$ with the properties in the parts (a) and (c) and the property that $\Delta^{(k)}$ generates $H_\mathbb{Z}$ and the further property $\Delta^{(0)} \subset R^{(0)}$ if $k = 0$ is a *vanishing lattice*. This notion is defined in [Ja83] and [Eb84]. Here $\Gamma^{(k)} = \langle s_\delta^{(k)} \mid \delta \in \Delta^{(k)} \rangle$ is determined by the triple.

(v) In rank 2 and rank 3 the only unimodular bilinear lattices whose monodromy has trace 1 are those of A_2 and A_3. So, the condition $\operatorname{tr}(M) = 1$ is very restrictive.

(vi) It is an interesting open question how to characterize those matrices $S \in T_n(\mathbb{Z})$ which come from distinguished bases of some isolated hypersurface singularities. Theorem 10.13 (b) and Theorem 10.14 give a number of necessary conditions. But very probably these conditions are not sufficient.

Finally we come to concrete results for the first singularities in Arnold's classification. Recall that the matrices $S \in \mathcal{S}^{dist} \subset T_n(\mathbb{Z})$ of distinguished bases are called *distinguished matrices*.

Gabrielov gave distinguished matrices for all unimodal singularities in [Ga73] and [Ga74-2]. Ebeling gave distinguished matrices for all bimodal singularities in [Eb83]. Distinguished matrices for the simple singularities are classical.

Figure 10.4 encodes the information of matrices S of distinguished bases for the simple singularities and the simple elliptic singularities in *Coxeter-Dynkin diagrams*. This is more efficient than writing down series of matrices. See [Eb01, Chap. 5 Table 5.3] for the given Coxeter-Dynkin diagrams of the simple elliptic singularities. In the given Coxeter-Dynkin diagrams for the simple singularities one can in fact change the numbering arbitrarily.

Definition 10.16 Let $S = (S_{ij}) \in T_n(\mathbb{Z})$. Its *Coxeter-Dynkin diagram* CDD(S) is a graph with n vertices which are numbered from 1 to n and with weighted edges which are defined as follows: Between the vertices i and j with $i < j$ one has no edge if $S_{ij} = 0$ and an edge with weight S_{ij} if $S_{ij} \neq 0$. Alternatively, one draws $|S_{ij}|$ edges if $S_{ij} < 0$ and S_{ij} dotted edges if $S_{ij} > 0$.

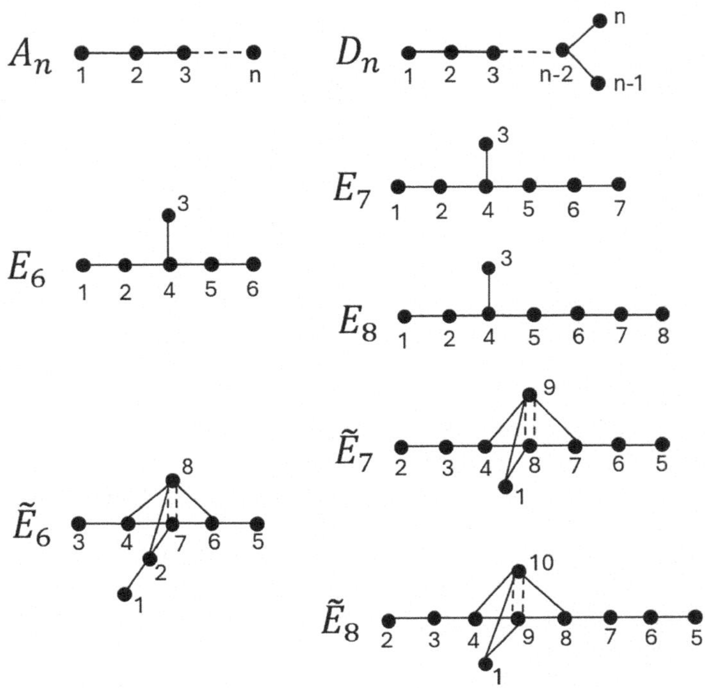

Fig. 10.4 Coxeter-Dynkin diagrams of some distinguished matrices for the simple and the simple elliptic singularities

The matrix S and its Coxeter-Dynkin diagram CDD(S) determine one another. A unimodular bilinear lattice $(H_{\mathbb{Z}}, L)$ with a triangular basis $\underline{e}$ with $S = L(\underline{e}^t, \underline{e})^t$ is reducible if and only if $CDD(S)$ consists of several components.

Recall from Theorem 10.7 that the simple singularities are the only singularities where $I^{(0)}$ on $H_{\mathbb{Z}}$ is positive definite and that the simple elliptic singularities are the only singularities where $I^{(0)}$ on $H_{\mathbb{Z}}$ is positive semidefinite. The following Lemma applies.

Lemma 10.17 *Let $S \in T_n(\mathbb{Z})$ and let $(H_{\mathbb{Z}}, L, \underline{e})$ be a unimodular bilinear lattice with triangular basis $\underline{e}$ with $S = L(\underline{e}^t, \underline{e})^t$.*

(a) Suppose that $I^{(0)}$ is positive definite. Then $S_{ij} \in \{0, \pm 1\}$. The sets $R^{(0)}$, $\mathcal{B}^{dist} = \mathrm{Br}_n \ltimes \{\pm 1\}^n(\underline{e})$ and $\mathcal{S}^{dist} = \mathrm{Br}_n \ltimes \{\pm 1\}^n(S)$ are finite.

(b) Suppose that $I^{(0)}$ is positive semidefinite. Then $S_{ij} \in \{0, \pm 1, \pm 2\}$. The set $\mathcal{S}^{dist} = \mathrm{Br}_n \ltimes \{\pm 1\}^n(S)$ is finite.

Proof (a) $R^{(0)}$ is a discrete subset of the compact set $\{\delta \in H_{\mathbb{R}} \mid I^{(0)}(\delta, \delta) = 2\}$ and therefore finite. Therefore also the sets $\Delta^{(0)}$, $\mathcal{B}^{dist}$ and $\mathcal{S}^{dist}$ are finite. With $S + S^t$ also the submatrix $\begin{pmatrix} 2 & S_{ij} \\ S_{ij} & 2 \end{pmatrix}$ (with $i < j$) is positive definite, so $S_{ij} \in \{0, \pm 1\}$.

(b) With $S + S^t$ also the submatrix $\begin{pmatrix} 2 & S_{ij} \\ S_{ij} & 2 \end{pmatrix}$ (with $i < j$) is positive semidefinite, so $S_{ij} \in \{0, \pm 1, \pm 2\}$. Therefore the set $\mathcal{S}^{dist}$ is finite. □

In Sect. 3.3 we considered the unimodular bilinear lattice $(H_{\mathbb{Z}}, L, \underline{e})$ with triangular basis $\underline{e}$ with arbitrary matrix $S = L(\underline{e}^t, \underline{e})^t \in T_n(\mathbb{Z})$ and posed the questions when the following two inclusions are equalities,

$$\mathcal{B}^{dist} \subset \{\underline{v} \in (\Delta^{(0)})^n \mid s_{v_1}^{(0)} \ldots s_{v_n}^{(0)} = -M\}, \tag{3.3}$$

$$\mathcal{B}^{dist} \subset \{\underline{v} \in (\Delta^{(1)})^n \mid s_{v_1}^{(1)} \ldots s_{v_n}^{(1)} = M\}. \tag{3.4}$$

Chapter 7 was devoted to answer these questions in the cases of rank 2 and 3. General positive answers were given in Theorem 3.7 and the Examples 3.22 (iv) and (v). They embrace the cases of the simple singularities.

References for the results in the next theorem are given after the theorem.

Theorem 10.18 *(a)*

f	S_{ij}	$\lvert\mathcal{B}^{dist}\rvert$	$\lvert\mathcal{S}^{dist}\rvert$
ADE-singularity	*in* $\{0, \pm 1\}$	*finite*	*finite*
simple elliptic sing.	*in* $\{0, \pm 1, \pm 2\}$	*infinite*	*finite*
any other singularity	*unbounded*	*infinite*	*infinite*

More details are given below in (10.33) and (10.47).

(b) The cases of the simple singularities: $\Delta^{(0)} = R^{(0)}$*. The inclusion in (3.3) is an equality.*

(c) The cases of the simple elliptic singularities: $\Delta^{(0)} = R^{(0)}$*. The inclusion in (3.3) becomes an equality if one adds on the right hand side the condition* $\sum_{i=1}^n \mathbb{Z}v_i = H_{\mathbb{Z}}$*.*

(d) The cases of the hyperbolic singularities: The inclusion in (3.3) becomes an equality if one adds on the right hand side the condition $\sum_{i=1}^n \mathbb{Z}v_i = H_{\mathbb{Z}}$*.*

Remarks 10.19 (a) The first two lines in the table in part (a) follow with the exception of $|\mathcal{B}^{dist}| = \infty$ for the simple elliptic singularities from Lemma 10.17 and Theorem 10.7. The fact $|\mathcal{B}^{dist}| = \infty$ for the simple elliptic singularities is proved for example in [Kl87] or in [HR21]. The third line in the table is due to [Eb18].

(b) $\Delta^{(0)} = R^{(0)}$ is classical. The equality in (3.3) was first proved by Deligne in a letter to Looijenga [De74]. It was rediscovered and generalized by Igusa and Schiffler [IS10] to all Coxeter groups, see the Theorems 3.6 and 3.7.

(c) [Kl83, Korollar 4.4] says $\Delta^{(0)} = R^{(0)}$, [Kl87, VI 1.2 Theorem] gives equality in (3.3) with the additional condition $\sum_{i=1}^n \mathbb{Z}v_i = H_{\mathbb{Z}}$ on the right hand side.

(d) [BWY21, Theorem 1.5].

(e) Theorem 7.1 (a) implies that in the case A_2 the inclusion in (3.4) is an equality. Theorem 7.7 (g) implies that in the case A_3 the inclusion in (3.4) is not an equality, but that it becomes an equality if one adds on the right hand side the condition $\sum_{i=1}^n \mathbb{Z}v_i = H_{\mathbb{Z}}$. These two results for (3.4) and the results in Theorem 10.18 for (3.3) are the only results on the interesting question for which singularities the inclusions in (3.3) or (3.4) are equalities or become equalities if one adds on the right hand side the condition $\sum_{i=1}^n \mathbb{Z}v_i = H_\mathbb{Z}$.

10.3 Orlik Blocks and the Groups $G_\mathbb{Z}$ for the Simple, Unimodal and Bimodal Singularities

The groups $G_\mathbb{Z}$ for the unimodular bilinear lattices $(H_\mathbb{Z}, L)$ from isolated hypersurface singularities have been mainly studied by the first author and F. Gauß [He92][He11][GH17][GH18]. This section reports about results, especially for the singularities with modality ≤ 2. But it starts with some rather recent general results on Orlik blocks [He20] [HM22-1] [HM22-2] (which would have simplified part of the work in the references above, if they had been available at that time).

Definition 10.20 Consider a $\mathbb{Z}$-lattice $H_\mathbb{Z}$ of rank $n \in \mathbb{N}$ with an automorphism M which is of finite order and cyclic. Here *cyclic* means that an element $e_1 \in H_\mathbb{Z}$ with $H_\mathbb{Z} = \bigoplus_{i=0}^{n-1} \mathbb{Z}M^{i-1}(e_1)$ exists. The pair $(H_\mathbb{Z}, M)$ is called an *Orlik block*.

Remarks 10.21 Consider an Orlik block $(H_\mathbb{Z}, M)$ with $H_\mathbb{Z}$ a $\mathbb{Z}$-lattice of rank n.

(i) The eigenvalues of M are roots of unity because M is of finite order. Cyclic implies regular. Finite order and regular together imply that each eigenvalue has multiplicity one. Especially, M is semisimple. Denote by $\mathrm{Ord}(M) \subset \mathbb{N}$ the finite set of orders of the eigenvalues of M. Then $p_{ch,M} = \prod_{m\in \mathrm{Ord}(M)} \Phi_m$ is a product of cyclotomic polynomials. The pair $(H_\mathbb{Z}, M)$ is up to isomorphism determined by the set $\mathrm{Ord}(M)$ or by the polynomial $p_{ch,M}$.

(ii) Lemma 5.2 (a) implies $\mathrm{End}(H_\mathbb{Z}, M) = \mathbb{Z}[M]$. Consider the group

$$\begin{aligned}\mathrm{Aut}_{S^1}(H_\mathbb{Z}, M) &:= \{g \in \mathrm{Aut}(H_\mathbb{Z}, M) \mid \text{all eigenvalues of } g \text{ are in } S^1\}\\ &= \{p(M) \mid p(t) = \sum_{i=0}^{n-1} p_i t^i \in \mathbb{Z}[t],\\ &\qquad p(\lambda)p(\lambda^{-1}) = 1 \text{ for all eigenvalues } \lambda \text{ of } M\}.\end{aligned}$$

Here we used $\lambda^{-1} = \overline{\lambda}$ for λ eigenvalues of M.

(iii) If M is the monodromy of a unimodular bilinear lattice $(H_\mathbb{Z}, L)$ then $\mathrm{Aut}_{S^1}(H_\mathbb{Z}, M) = G_\mathbb{Z} = G_\mathbb{Z}^{(0)} = G_\mathbb{Z}^{(1)}$ by Lemma 5.2 (b) (ii).

(iv) $\mathrm{Aut}_{S^1}(H_\mathbb{Z}, M)$ contains the finite group $\{\pm M^l \mid l \in \mathbb{Z}\}$. The main result Theorem 1.2 in [He20] gives necessary and sufficient conditions on the set $\mathrm{Ord}(M)$ or (equivalently) on the polynomial $p_{ch,M}$ for

$$\mathrm{Aut}_{S^1}(H_\mathbb{Z}, M) = \{\pm M^l \mid l \in \mathbb{Z}\}.$$

Orlik [Or72] conjectured that in the case of a *quasihomogeneous singularity* the pair $(H_\mathbb{Z}, M)$ decomposes in a specific way into a sum of Orlik blocks. Part of this conjecture was proved in [HM22-1][HM22-2]. Before stating the conjecture and the results precisely, we define quasihomogeneous singularities and two special families of quasihomogeneous singularities.

Definition 10.22 (a) An isolated hypersurface singularity $f : (\mathbb{C}^{m+1}, 0) \to (\mathbb{C}, 0)$ is a *quasihomogoneous singularity*, if $f \in \mathbb{C}[z_0, ..., z_m]$ is a *quasihomogeneous polynomial*, that means there exists a weight system $\underline{w} = (w_0, ..., w_m) \in (0, \frac{1}{2}] \cap \mathbb{Q}$ such that for any monomial $\prod_{i=0}^m z_i^{e_i}$ with nonvanishing coefficient in f its weighted degree $\deg_{\underline{w}}(\prod_{i=0}^m z_i^{e_i}) := \sum_{i=0}^m w_i e_i$ is equal to 1.

(b) A *chain type singularity* is a quasihomogeneous singularity of the special shape

$$f = f(z_0, ..., z_m) = z_0^{a_0+1} + \sum_{i=1}^{m} z_{i-1} z_i^{a_i} \tag{10.24}$$

for some $m \in \mathbb{N}$ and some $a_0, ..., a_m \in \mathbb{N}$. This quasihomogeneous polynomial has indeed an isolated singularity.

(c) A *cycle type singularity* is a quasihomogeneous singularity of the special shape

$$f = f(z_0, ..., z_m) = \sum_{i=0}^{m-1} z_i^{a_i} z_{i+1} + z_m^{a_m} z_0 \tag{10.25}$$

for some $m \in \mathbb{N}$ and some $a_0, ..., a_m \in \mathbb{N}$ which satisfy in the case of odd m that neither $a_j = 1$ for all even j nor $a_j = 1$ for all odd j. This quasihomogeneous polynomial has indeed an isolated singularity.

The quasihomogeneous singularities form an important and especially well studied subfamily of all isolated hypersurface singularities. Their monodromies are semisimple.

Remarks 10.23 Let $f \in \mathbb{C}[z_0, ..., z_m]$ be a quasihomogeneous singularity.

(i) Milnor and Orlik [MO70] gave a formula for the characteristic polynomial $p_{ch,M^{sing}}$ of its monodromy. Recall $M = (-1)^{m+1} M^{sing}$.

(ii) The polynomial $p_{ch,M}$ has a unique decomposition $p_{ch,M} = p_1 p_2 ... p_l$ with $p_1 \mid p_2 \mid ... \mid p_l$ and p_l the minimal polynomial of M and $l \in \mathbb{N}$ suitable.

(iii) Orlik [Or72, Conjecture 3.1] conjectured that the pair $(H_{\mathbb{Z}}, M)$ has a decomposition into l Orlik blocks with characteristic polynomials $p_1, ..., p_l$.

(iv) His conjecture was proved in [HM22-1, Theorem 1.3] for all cycle type singularities and in [HM22-2, Theorem 1.3] for all iterated Thom-Sebastiani sums of chain type singularities and cycle type singularities.

(v) Furthermore, in [HM22-3, Theorem 1.5] it was proved that each polynomial p_j in (ii) satisfies the sufficient condition (I) in Theorem 1.2 in [He20] for

$$\mathrm{Aut}_{S^1}(\text{Orlik block for } p_j)$$

$$= \{\pm(\text{monodromy of the Orlik block})^l \mid l \in \mathbb{Z}\}.$$

(vi) Therefore if the singularity f satisfies Orlik's conjecture, one knows the automorphisms of the summands of a decomposition of $(H_\mathbb{Z}, M)$ into Orlik blocks. This is a big step towards $G_\mathbb{Z}$. Though in general, the decomposition is not left or right L-orthogonal. Also, automorphisms might mix or exchange Orlik blocks. So, there are still more steps to do towards $G_\mathbb{Z}$.

(vii) Many families of singularities with modality ≤ 2 contain quasihomogeneous singularities, namely all families except the hyperbolic singularities T_{pqr} and the eight bimodal series $E_{3,p}$, $Z_{1,p}$, $W_{1,p}$, $W^\sharp_{1,p}$, $Q_{2,p}$, $S_{1,p}$, $S^\sharp_{1,p}$, $U_{1,p}$. Each family which contains at all a quasihomogeneous singularity, contains also a quasihomogeneous singularity which is an iterated Thom-Sebastiani sum of chain type singularities and cycle type singularities. Therefore Orlik's conjecture is true for all of these families of singularities. In [He92] this was proved in a case-by-case work using Coxeter-Dynkin diagrams.

(viii) But Orlik blocks are not only useful for quasihomogeneous singularities. Also in the families with modality ≤ 2 which do not contain quasihomogeneous singularities, $(H_\mathbb{Z}, M)$ usually contains an M-invariant sublattice of finite index in $H_\mathbb{Z}$, which decomposes into Orlik blocks which each satisfy the sufficient condition (I) in Theorem 1.2 in [He20]. This fact was elaborated and used in many cases in [GH17] and [GH18]. See for example Theorem 10.26 (ii) below.

(ix) Orlik and Randell [OR77] found for the chain type singularities an n-th root of the monodromy, without using Theorem 3.26 (c). The pair ($H_\mathbb{Z}$, this n-th root) is an Orlik block. Though they conjectured that a distinguished basis with matrix S as in Theorem 3.26 (c) exists. This conjecture was proved recently by Varolgunes [Va23].

Now we describe a part of the results in [He11][GH17] and [GH18] on $G_\mathbb{Z}$ in the families of singularities with modality ≤ 2. We start with the table in the proof of Theorem 8.3 in [He11] of the characteristic polynomials $p_{ch,-M}$ for all families of singularities with modality ≤ 2 which contain quasihomogeneous singularities. They can be extracted from the tables of spectral numbers in [AGV88, 13.3.4] or from [He92]. The families consist of all singularities with modality ≤ 2 except the hyperbolic singularities and the eight bimodal series.

A_n	$\frac{t^{n+1}-1}{t-1}$	$\widetilde{E}_6$	$\Phi_3^3\Phi_1^2$	$E_{3,0}$	$\Phi_{18}^2\Phi_6\Phi_2^2$
D_n	$(t^{n-1}+1)\Phi_2$	$\widetilde{E}_7$	$\Phi_4^2\Phi_2^3\Phi_1^2$	$Z_{1,0}$	$\Phi_{14}^2\Phi_2^3$
E_6	$\Phi_{12}\Phi_3$	$\widetilde{E}_8$	$\Phi_6\Phi_3^2\Phi_2^2\Phi_1^2$	$Q_{2,0}$	$\Phi_{12}^2\Phi_4^2\Phi_3$
E_7	$\Phi_{18}\Phi_2$			$W_{1,0}$	$\Phi_{12}^2\Phi_6\Phi_4\Phi_3\Phi_2$
E_8	Φ_{30}			$S_{1,0}$	$\Phi_{10}^2\Phi_5\Phi_2^2$
				$U_{1,0}$	$\Phi_9^2\Phi_3$

E_{12} Φ_{42}	E_{13} $\Phi_{30}\Phi_{10}\Phi_2$	E_{14} $\Phi_{24}\Phi_{12}\Phi_3$
Z_{11} $\Phi_{30}\Phi_6\Phi_2$	Z_{12} $\Phi_{22}\Phi_2^2$	Z_{13} $\Phi_{18}\Phi_9\Phi_2$
Q_{10} $\Phi_{24}\Phi_3$	Q_{11} $\Phi_{18}\Phi_6\Phi_3\Phi_2$	Q_{12} $\Phi_{15}\Phi_3^2$
W_{12} $\Phi_{20}\Phi_5$	W_{13} $\Phi_{16}\Phi_8\Phi_2$	
S_{11} $\Phi_{16}\Phi_4\Phi_2$	S_{12} Φ_{13}	U_{12} $\Phi_{12}\Phi_6\Phi_4^2\Phi_2^2$

E_{18} $\Phi_{30}\Phi_{15}\Phi_3$	E_{19} $\Phi_{42}\Phi_{14}\Phi_2$	E_{20} Φ_{66}
Z_{17} $\Phi_{24}\Phi_{12}\Phi_6\Phi_3\Phi_2$	Z_{18} $\Phi_{34}\Phi_2^2$	Z_{19} $\Phi_{54}\Phi_2$
Q_{16} $\Phi_{21}\Phi_3^2$	Q_{17} $\Phi_{30}\Phi_{10}\Phi_6\Phi_3\Phi_2$	Q_{18} $\Phi_{48}\Phi_3$
W_{17} $\Phi_{20}\Phi_{10}\Phi_5\Phi_2$	W_{18} $\Phi_{28}\Phi_7$	
S_{16} Φ_{17}	S_{17} $\Phi_{24}\Phi_8\Phi_6\Phi_3\Phi_2$	U_{16} $\Phi_{15}\Phi_5^2$

One sees that $p_{ch,-M}$ has no multiple roots in the following cases: The simple singularities A_n, D_{2n+1}, E_6, E_7, E_8, so all simple singularities except the D_{2n}, and 22 of the 28 exceptional unimodal and bimodal singularities, namely all except Z_{12}, Q_{12}, U_{12}, Z_{18}, Q_{16} and U_{16}. Part (a) of the next theorem follows from the Remarks 10.21 (iv) and 10.23 (iv), (v) and (vii).

Theorem 10.24 *(a) [He11, Theorems 8.3 and 8.4] In the cases* A_n, D_{2n+1}, E_6, E_7, E_8 *and the 22 of the 28 exceptional unimodal and bimodal singularities with the exceptions* Z_{12}, Q_{12}, U_{12}, Z_{18}, Q_{16} *and* U_{16}, *the pair* $(H_{\mathbb{Z}}, M)$ *is a single Orlik block and*

$$G_{\mathbb{Z}} = \{\pm M^l \mid l \in \mathbb{Z}\}.$$

(b) [GH17, Theorem 4.1] In the cases D_{2n}, Z_{12}, Q_{12}, U_{12}, Z_{18}, Q_{16} *and* U_{16}, $(H_{\mathbb{Z}}, M)$ *has a decomposition into two Orlik blocks and*

$$G_{\mathbb{Z}} = \{\pm M^l \mid l \in \mathbb{Z}\} \times U$$

with

	D_4	D_{2n} *with* $n \geq 3$	Z_{12}	Q_{12}	U_{12}	Z_{18}	Q_{16}	U_{16}
$U \cong$	S_3	S_2	$\{\text{id}\}$	S_2	S_3	$\{\text{id}\}$	S_2	S_3

(c) [GH17, Theorem 3.1] In the simple elliptic cases $\text{Rad}\, I^{(0)} \subset H_{\mathbb{Z}}$ *has rank 2. The restriction map* $G_{\mathbb{Z}} \to G_{\mathbb{Z}}|_{\text{Rad}\, I^{(0)}}$ *is called* $\text{pr}^{\text{Rad}\, I^{(0)}}$. *There is an exact non-splitting sequence*

$$\{\text{id}\} \to \ker \text{pr}^{\text{Rad}\, I^{(0)}} \to G_{\mathbb{Z}} \stackrel{\text{pr}^{\text{Rad}\, I^{(0)}}}{\to} G_{\mathbb{Z}}|_{\text{Rad}\, I^{(0)}} \cong SL_2(\mathbb{Z}) \to \{\text{id}\}$$

with finite group $\ker \text{pr}^{\text{Rad}\, I^{(0)}}$. *The sublattice* $(H_{\mathbb{C},\neq 1} \cap H_{\mathbb{Z}}, M|_{..})$ *has a left and right L-orthogonal decomposition into three Orlik blocks* $H^{(1)} \oplus H^{(2)} \oplus H^{(3)}$ *with restricted monodromies* $M^{(1)}$, $M^{(2)}$ *and* $M^{(3)}$ *and characteristic polynomials in the following table,*

$$p_{ch,-M^{(j)}} = \frac{t^{p_j}-1}{t-1} \quad \text{with} \quad \begin{array}{c|c|c|c} & \widetilde{E}_6 & \widetilde{E}_7 & \widetilde{E}_8 \\ \hline (p_1,p_2,p_3) & (3,3,3) & (4,4,2) & (6,3,2) \end{array}$$

Then $\ker \mathrm{pr}^{\mathrm{Rad}\, I^{(0)}} = U_1 \rtimes U_2$, *where*
$U_2 \cong S_3$ *permutes the three Orlik blocks in the case of* $\widetilde{E}_6$,
$U_2 \cong S_2$ *permutes* $H^{(1)}$ *and* $H^{(2)}$ *in the case of* $\widetilde{E}_7$,
and $U_2 = \{\mathrm{id}\}$ *in the case of* $\widetilde{E}_8$.
The elements of U_1 *are the extensions to* Rad $I^{(0)}$ *by* id *of the following automorphisms of* $H_{\mathbb{C},\neq 1} \cap H_{\mathbb{Z}}$,

$$\{(M^{(1)})^{\alpha} \times (M^{(2)})^{\beta} \times (M^{(3)})^{\gamma} \mid (\alpha,\beta,\gamma) \in \prod_{j=1}^{3} \{0,1,..,p_j-1\}$$
$$\text{with } \frac{\alpha}{p_1} + \frac{\beta}{p_2} + \frac{\gamma}{p_3} \equiv 0 \bmod \mathbb{Z}\}$$

Remark 10.25 Kluitmann [Kl87, III 2.4–2.6] calculated for the simple elliptic singularities the group $\mathrm{Aut}(H_{\mathbb{Z}}, I^{(0)}, M)$, which contains $G_{\mathbb{Z}}$. Comparison with part (c) above gives equality $\mathrm{Aut}(H_{\mathbb{Z}}, I^{(0)}, M) = G_{\mathbb{Z}}$. Kluitmann described the group by an exact sequence

$$\{\mathrm{id}\} \to \ker \mathrm{pr}^{\neq 1} \to \mathrm{Aut}(H_{\mathbb{Z}}, I^{(0)}, M)$$
$$\stackrel{\mathrm{pr}^{\neq 1}}{\to} \mathrm{Aut}(H_{\mathbb{C},\neq 1} \cap H_{\mathbb{Z}}, I^{(0)}|_{..}, M|_{..}) \to \{\mathrm{id}\}.$$

The group on the right side is finite of order 1296, 768 respectively 864 in the case $\widetilde{E}_6$, $\widetilde{E}_7$ respectively $\widetilde{E}_8$. The kernel $\ker \mathrm{pr}^{\neq 1}$ is isomorphic to $\Gamma(3)$, $\Gamma(4)$ respectively $\Gamma(6) \subset SL_2(\mathbb{Z})$ in the case $\widetilde{E}_6$, $\widetilde{E}_7$ respectively $\widetilde{E}_8$.

In the cases of the other families of singularities with modality ≤ 2, namely the hyperbolic singularities, the six families of quadrangle singularities and the eight bimodal series, we restrict here to the rough information in the following tables and refer to [GH17, Theorem 3.1] and [GH18, Theorem 5.1 and Theorem 6.1] for details. The characteristic polynomial $p_{ch,-M}$ for the family T_{pqr} of hyperbolic singularities with $p \geq q \geq r$ and $\frac{1}{p} + \frac{1}{q} + \frac{1}{r} < 1$ is

$$p_{ch,-M}(t) = (t^p - 1)(t^q - 1)(t^r - 1)(t-1)^{-1}.$$

The next table from [GH18, (5.1)] gives the characteristic polynomial $p_{ch,-M}$ for the eight bimodal series as a product $b_1 b_2$ respectively $b_1 b_2 b_3$ in the case $Z_{1,p}$. The meaning of m and r_I is explained in Theorem 10.26.

series	n	b_1	b_2	b_3	m	r_I
$W^\sharp_{1,p}$	$15+p$	Φ_{12}	$(t^{12+p}-1)/\Phi_1$	$-$	12	1
$S^\sharp_{1,p}$	$14+p$	$\Phi_{10}\Phi_2$	$(t^{10+p}-1)/\Phi_1$	$-$	10	1
$U_{1,p}$	$14+p$	Φ_9	$(t^{9+p}-1)/\Phi_1$	$-$	9	1
$E_{3,p}$	$16+p$	$\Phi_{18}\Phi_2$	$t^{9+p}-1$	$-$	18	2
$Z_{1,p}$	$15+p$	$\Phi_{14}\Phi_2$	$t^{7+p}-1$	Φ_2	14	2
$Q_{2,p}$	$14+p$	$\Phi_{12}\Phi_4\Phi_3$	$t^{6+p}-1$	$-$	12	2
$W_{1,p}$	$15+p$	$\Phi_{12}\Phi_6\Phi_3\Phi_2$	$t^{6+p}+1$	$-$	12	2
$S_{1,p}$	$14+p$	$\Phi_{10}\Phi_5\Phi_2$	$t^{5+p}+1$	$-$	10	2

(10.26)

Theorem 10.26 *[He11][GH17][GH18]*

(a) (i) Within the families of singularities with modality ≤ 2 the monodromy is not finite only in the hyperbolic singularities. It has one 2×2 Jordan block with eigenvalue -1 for the hyperbolic singularities.

(ii) In the bimodal series, $H_{\mathbb{Z}}$ contains with index r_I a direct sum $H^{(1)}\oplus H^{(2)}$ respectively $H^{(1)}\oplus H^{(2)}\oplus H^{(3)}$ for $Z_{1,p}$ of Orlik blocks with characteristic polynomials $b_1(-t)$ and $b_2(-t)$ and in the case $Z_{1,p}$ $b_3(-t)$.

(iii) Within the bimodal series, the eigenvalue $\zeta := e^{2\pi i/m}$ has multiplicity 2 exactly in the eight subseries with $m\mid p$, so the subseries $E_{3,18p}, Z_{1,14p}, W_{1,12p}, W^\sharp_{1,12p}, Q_{2,12p}, S_{1,10p}, S^\sharp_{1,10p}, U_{1,9p}$. In these cases $G_{\mathbb{Z}}$ contains automorphisms which act nontrivially on the 2-dimensional eigenspace $H_{\mathbb{C},\zeta}$ and which do not exist for the other cases.

(b) The quotient $G_{\mathbb{Z}}/\{\pm M^l\mid l\in\mathbb{Z}\}$ looks roughly as follows in the families of singularities with modality ≤ 2.

Singularity family	$G_{\mathbb{Z}}/\{\pm M^l\mid l\in\mathbb{Z}\}$
ADE-singularities	$\{\mathrm{id}\}$ *or* S_2 *or* S_3
simple elliptic sing.	*a finite extension of* $SL(2,\mathbb{Z})$
hyperbolic sing.	*a finite group*
exc. unimodal sing.	$\{\mathrm{id}\}$ *or* S_2 *or* S_3
exc. bimodal sing.	$\{\mathrm{id}\}$ *or* S_2 *or* S_3
quadrangle sing.	*a triangle group*
the 8 series, for $m\not| p$	*a cyclic finite group*
the 8 subseries with $m\mid p$	*an infinite Fuchsian group*

(This table is taken from the introduction in [GH18].)

Finally, we come to the subgroup $G_{\mathbb{Z}}^{\mathcal{B}}\subset G_{\mathbb{Z}}$.

Remarks 10.27 In [He11] a subgroup $G^{mar}\subset G_{\mathbb{Z}}$ is defined for any isolated hypersurface singularity. It is called *μ-constant monodromy group*. It comes from transversal monodromies along μ-constant families and from $\pm$ id. It has the property

$$G^{mar}\subset G_{\mathbb{Z}}^{\mathcal{B}}\subset G_{\mathbb{Z}}$$

(this is stated, though not really explained in [He11, Remark 3.4]). In [He11], [GH17] and [GH18] $G_{\mathbb{Z}}$ and G^{mar} are determined for all singularities with modality ≤ 2. It turns out that $G^{mar} = G_{\mathbb{Z}}^{\mathcal{B}} = G_{\mathbb{Z}}$ for almost all of them, namely all except the subseries with $m \mid p$ of the eight bimodal series. For these subseries $G^{mar} \subsetneqq G_{\mathbb{Z}}$, and it is not clear where in between $G_{\mathbb{Z}}^{\mathcal{B}}$ is.

10.4 Monodromy Groups and Vanishing Cycles

Consider the unimodular bilinear lattice $(H_{\mathbb{Z}}, L, \mathcal{B})$ with set of distinguished bases $\mathcal{B}$ from an isolated hypersurface singularity $f : (\mathbb{C}^{m+1}, 0) \to (\mathbb{C}, 0)$. The distinguished bases induce the even and odd monodromy groups $\Gamma^{(0)}$ and $\Gamma^{(1)}$ and the sets $\Delta^{(0)}$ and $\Delta^{(1)}$ of even and odd vanishing cycles. These two groups and these two sets have been studied a lot and by many people. In the even case the pair $(H_{\mathbb{Z}}, I^{(0)})$ alone determines $\Gamma^{(0)}$ and $\Delta^{(0)}$ in almost all cases. This is satisfying. The precise results below are due to Ebeling. In the odd case, the situation is more complicated. It was worked upon by Wajnryb, Chmutov and Janssen. In the following we report first on the even case and then on the odd case.

The lattice $(H_{\mathbb{Z}}, I^{(0)})$ with even intersection form is an a priori rather rough invariant. Lattices with even intersection forms are well understood, due to work of Durfee, Kneser, Nikulin, Wall and others.

The following theorem of Ebeling was developed by him in several papers, in greater and greater generality, until the final state in [Eb84, (5.5) and (2.5)]. The version here is taken from [Eb01, Theorem 5.9], as [Eb84] does not cover the characterization of $\Delta^{(0)}$ for some special cases.

Theorem 10.28 *[Eb01, Theorem 5.9] Suppose that the singularity f is a hyperbolic singularity of type T_{pqr} with $(p, q, r) \in \{(2, 3, 7), (2, 4, 5), (3, 3, 4)\}$ (so these three triples are allowed, all other triples (p, q, r) with $p \geq q \geq r$ and $\frac{1}{p} + \frac{1}{q} + \frac{1}{r} < 1$ are not allowed) or any not hyperbolic singularity. Then*

$$\begin{aligned}
\Gamma^{(0)} &= O^{(0),*}, \\
\Gamma_u^{(0)} &= (\ker \tau^{(0)})_u \overset{5.2\,(e)}{=} T(\overline{j}^{(0)}(\overline{H_{\mathbb{Z}}}^{(0)}) \otimes \operatorname{Rad} I^{(0)}), \\
\Gamma_s^{(0)} &= \ker \overline{\tau}^{(0)} \cap \ker \overline{\sigma}, \\
[O_s^{(0)} : \Gamma_s^{(0)}] &< \infty, \\
\Delta^{(0)} &= \{a \in R^{(0)} \mid I^{(0)}(a, H_{\mathbb{Z}}) = \mathbb{Z}\},
\end{aligned}$$

(recall Lemma 6.2 (f) for $\overline{\tau}^{(0)}$ and Remark 6.3 (iii) for $\overline{\sigma}$), the exact sequence

$$\{\mathrm{id}\} \to \Gamma_u^{(0)} \to \Gamma^{(0)} \to \Gamma_s^{(0)} \to \{\mathrm{id}\}$$

splits non-canonically.

Remarks 10.29 (i) The characterizations of $\Gamma^{(0)}$ and $\Delta^{(0)}$ in the theorem show that they are determined by $(H_{\mathbb{Z}}, I^{(0)})$ alone.

(ii) $\Gamma_u^{(0)} = (\ker \tau^{(0)})_u$ follows from $\Gamma^{(0)} = O^{(0),*} \stackrel{Def.}{=} \ker \tau^{(0)} \cap \ker \sigma$ and $O_u^{(0),Rad} \subset \ker \sigma$. The simple part $\Gamma_s^{(0)}$ has finite index in $O_s^{(0)}$ because $\overline{H_{\mathbb{Z}}}^{(0),\sharp} / \overline{j}^{(0)}(\overline{H_{\mathbb{Z}}}^{(0)})$ is a finite abelian group and therefore $\ker \overline{\tau}^{(0)}$ has finite index in $O_s^{(0)}$.

(iii) We do not know a singularity where $R^{(0)} \supsetneqq \Delta^{(0)}$. Equivalent: We do not know a singularity where a L1-cycle $a \in R^{(0)}$ with $I^{(0)}(a, H_{\mathbb{Z}}) \subsetneqq \mathbb{Z}$ exists.

(iv) The first idea of a proof of $\Delta^{(0)} = \{a \in R^{(0)} \mid I^{(0)}(a, H_{\mathbb{Z}}) = \mathbb{Z}\}$ is due to Looijenga. It is given in [Br83, Korollar 1 in 4.2]. This paper treats the exceptional unimodal singularities. But it gives also many useful references and general facts.

(v) In the case of a hyperbolic singularity T_{pqr} with $(p, q, r) \notin \{(2, 3, 7), (2, 4, 5), (3, 3, 4)\}$, Gabrielov [Ga74-2] determined $\Gamma^{(0)}$. Then especially $[O_s^{(0)} : \Gamma_s^{(0)}| = \infty$ (see also [Eb84, (5.2)]). For example the singularities $T_{2,7,7}$ and $T_{3,3,10}$ have isomorphic pairs $(H_{\mathbb{Z}}, I^{(0)})$, but the monodromy groups for all hyperbolic singularities are pairwise different [Br83]. So in the case of the hyperbolic singularities, the pair $(H_{\mathbb{Z}}, I^{(0)})$ does not determine the monodromy group $\Gamma^{(0)}$. See [BWY21, 7.2] for a statement on $\Delta^{(0)}$ for the hyperbolic singularities.

(vi) An example [Eb83, Example after Theorem 3.3]: The exceptional bimodal singularities E_{18} and Q_{18} have isomorphic pairs $(H_{\mathbb{Z}}, I^{(0)})$. Both are isomorphic to the orthogonal sum $E_6 \perp E_8 \perp U \perp U$ where E_6 and E_8 mean the root lattices and U means the rank 2 hyperbolic lattice, so with matrix $\begin{pmatrix} 0 & 1 \\ 1 & 0 \end{pmatrix}$ for its even intersection form. Therefore also $\Gamma^{(0)}$ and $\Delta^{(0)}$ are isomorphic for these singularities. But Seifert form L and monodromy M are not isomorphic. The monodromy M has order 30 for E_{18} and order 48 for Q_{18}.

Now we come to the odd case. The lattice $(H_{\mathbb{Z}}, I^{(1)})$ with odd intersection form has a basis $\underline{v}$ whose intersection matrix is

$$I^{(1)}(\underline{v}^t, \underline{v}) = \begin{pmatrix} 0 & d_1 & & & & & & & & \\ -d_1 & 0 & & & & & & & & \\ & & 0 & d_2 & & & & & & \\ & & -d_2 & 0 & & & & & & \\ & & & & \ddots & & & & & \\ & & & & & 0 & d_l & & & \\ & & & & & -d_l & 0 & & & \\ & & & & & & & 0 & & \\ & & & & & & & & \ddots & \\ & & & & & & & & & 0 \end{pmatrix}$$

(at empty places put 0) where $d_1, ..., d_l \in \mathbb{N}$ with $d_l \mid d_{l-1} \mid ... \mid d_1$. This is a linear algebra fact. The numbers l and $d_1, ..., d_l$ are unique. The last rk Rad $I^{(1)}$ vectors of

$\underline{v}$ are a basis of the radikal Rad $I^{(1)}$ of $I^{(1)}$, so $2l + \text{rk Rad } I^{(1)} = n$. The basis $\underline{v}$ is called a *symplectic basis*. Define the sublattice of rank n

$$H_{\mathbb{Z}}^{(d_1)} := \{a \in H_{\mathbb{Z}} \mid I^{(1)}(a, H_{\mathbb{Z}}) \subset d_1\mathbb{Z}\},$$

and the finite abelian quotient group

$$H^{quot} := H_{\mathbb{Z}}^{(d_1)}/2d_1 H_{\mathbb{Z}}.$$

Each automorphism in $O^{(1),Rad}$ respects $H^{(d_1)}$ and $2d_1 H_{\mathbb{Z}}$, so it acts on the quotient H^{quot}. The induced finite group of automorphisms of H^{quot} is called $O^{(1),quot}$, the homomorphism is called $p^{quot} : O^{(1),Rad} \to O^{(1),quot}$. Each element of $O^{(1),quot}$ acts trivially on the image of the radical Rad $I^{(1)}$ in the quotient H^{quot}.

The following theorem cites four results of Chmutov in [Ch82]:
(i) A relative characterization of $\Gamma^{(1)}$,
(ii) a relative characterization of $\Delta^{(1)}$,
(iii) that $\Gamma_s^{(1)}$ has finite index in $O_s^{(1),Rad}$,
(iv) and a characterization of the subgroup ker $p^{quot} \subset \Gamma^{(1)}$.

Theorem 10.30 *[Ch82, Theorem 1, Corollary of Proposition 1, Theorem 2, Proposition 2]*

(i) $\Gamma^{(1)}$ is the full preimage in $O^{(1),Rad}$ of $p^{quot}(\Gamma^{(1)})$ in $O^{(1),quot}$. Equivalent:

$$\ker\left(p^{quot} : O^{(1),rad} \to O^{(1),quot}\right) \subset \Gamma^{(1)}.$$

So if one knowns the image $p^{quot}(\Gamma^{(1)})$, one knows $\Gamma^{(1)}$.

(ii)

$$\Delta^{(1)} = \{a \in H_{\mathbb{Z}} \mid \text{there exists } b \in \Delta^{(1)} \text{ with } a - b \in 2H_{\mathbb{Z}} \text{ and } I^{(1)}(a, H_{\mathbb{Z}}) = \mathbb{Z}\}.$$

So if one knows the image of $\Delta^{(1)}$ in $H_{\mathbb{Z}}/2H_{\mathbb{Z}}$, one knows $\Delta^{(1)}$.

(iii)

$$\{g \in O_s^{(1),Rad} \mid g \text{ acts trivially on the quotient } \overline{H_{\mathbb{Z}}}^{(1)}/2d_1\overline{H_{\mathbb{Z}}}^{(1)}\} \subset \Gamma_s^{(1)}.$$

As the group on the left hand side has finite index in $O_s^{(1),Rad}$, also $\Gamma_s^{(1)}$ has finite index in $O_s^{(1),Rad}$.

(iv)

$$\langle (s_a^{(1)})^2 \mid a \in H_{\mathbb{Z}}\rangle = \ker\left(p^{quot} : O^{(1),rad} \to O^{(1),quot}\right) \overset{(i)}{\subset} \Gamma^{(1)}.$$

Remarks 10.31 (i) In the case of a curve singularity (so $m = 1$) the number of components is rk Rad $I^{(1)} + 1$. So the curve singularity is irreducible if and only if $I^{(1)}$ is nondegenerate.

(ii) In the case of a curve singularity $d_1 = ... = d_l = 1$.

(iii) Wajnryb [Wa80] proved Theorem 10.30 for irreducible curve singularities. Chmutov [Ch81] generalized it to all curve singularities. Here $d_1 = 1$ implies $H^{(d_1)} = H_{\mathbb{Z}}$ and $H^{quot} = H_{\mathbb{Z}}/2H_{\mathbb{Z}}$.

(iv) Wajnryb characterized in the case of irreducible curve singularities also the image $p^{quot}(\Gamma^{(1)}) \subset O^{(1),quot}$ and the image of $\Delta^{(1)}$ in $H_{\mathbb{Z}}/2H_{\mathbb{Z}}$. Also this was generalized by Chmutov first to all curve singularities [Ch81] and then to all singularities [Ch83].

First we discuss the characterization in [Ch83] of the image

$$\Delta^{(1)}_{\mathbb{F}_2} := \text{image of } \Delta^{(1)} \text{ in } H_{\mathbb{Z}}/2H_{\mathbb{Z}}$$

of $\Delta^{(1)}$ in $H_{\mathbb{Z}}/2H_{\mathbb{Z}}$. The following Lemma is elementary. It is essentially formulated in [Ch83, 3.1], extending an observation of Wajnryb [Wa80].

Lemma 10.32 *Let $\widetilde{H}_{\mathbb{F}_2}$ be an $\mathbb{F}_2$-vector space of dimension $\widetilde{n} \in \mathbb{Z}_{\geq 2}$, let $\widetilde{I}^{(1)}$ be an odd bilinear form $\widetilde{I}^{(1)} : \widetilde{H}_{\mathbb{F}_2} \times \widetilde{H}_{\mathbb{F}_2} \to \mathbb{F}_2$ on $\widetilde{H}_{\mathbb{F}_2}$ (*odd* means here only $\widetilde{I}^{(1)}(a, a) = 0$ for $a \in \widetilde{H}_{\mathbb{F}_2}$), and let $\underline{v}$ be a basis of $\widetilde{H}_{\mathbb{F}_2}$.*

(a) (Definition) A quadratic form *$q : \widetilde{H}_{\mathbb{F}_2} \to \mathbb{F}_2$ is a map with*

$$q(x + y) = q(x) + q(y) + \widetilde{I}^{(1)}(x, y) \quad \text{for all } x, y \in \widetilde{H}_{\mathbb{F}_2}.$$

(b) There is a unique quadratic form q with $q(v_j) = 1$ for all $j \in \{1, ..., n\}$. The value $q(x)$ for $x \in \widetilde{H}_{\mathbb{F}_2}$ can be characterized as follows. Write $x = \sum_{j=1}^{\widetilde{n}} a_j v_j$ with $a_j \in \mathbb{F}_2$. Define a graph with vertex set $V(x) := \{j \in \{1, ..., n\} \mid a_j = 1\}$ and edge set $E(x) := \{(i, j) \in \{1, ..., n\}^2 \mid i < j, \widetilde{I}^{(1)}(v_i, v_j) = 1\}$. Then $q(x)$ is the Euler characteristic of the graph mod 2 *so*

$$q(x) \equiv |V(a)| - |E(a)| \mod 2.$$

The next theorem collects results of Chmutov and Janssen. The main point is part (b). It characterizes $\Delta^{(1)}_{\mathbb{F}_2}$.

Theorem 10.33 *Suppose that the singularity f is neither the singularity A_n nor the singularity D_n. Choose a distinguished basis $\underline{v} \in \mathcal{B}^{dist}$. Consider the quadratic form q on $H_{\mathbb{Z}}/2H_{\mathbb{Z}}$ with $q(v_j) = 1$ for any $j \in \{1, ..., n\}$ (it exists and is unique by Lemma 10.32).*

(a) [Ch83, 7.1] The quadratic form q on $H_{\mathbb{Z}}/2H_{\mathbb{Z}}$ is independent of the choice of the distinguished basis $\underline{v} \in \mathcal{B}^{dist}$.

(b) [Ch83, Proposition 1]

$$\Delta^{(1)}_{\mathbb{F}_2} = q^{-1}(1) - (\textit{image of } \operatorname{Rad} I^{(1)} \textit{ in } H_{\mathbb{Z}}/2H_{\mathbb{Z}}).$$

(c) [Ja83, (5.1) and (5.2)] The subset

$$(\mathrm{Rad}\, I^{(1)})^q := \{a \in \mathrm{Rad}\, I^{(1)} \mid q(\textit{image of } a \textit{ in } H_{\mathbb{Z}}/2H_{\mathbb{Z}}) = 0\}$$

of the radical Rad $I^{(1)}$ *is either equal to* Rad $I^{(1)}$ *or a subgroup of index 2 in* Rad $I^{(1)}$. *In any case it is the set* $\{a \in \mathrm{Rad}\, I^{(1)} \mid a + b \in \Delta^{(1)}\}$, *where* $b \in \Delta^{(1)}$ *is an arbitrarily chosen odd vanishing cycle.*

(d) [Ja83, (5.3)]

$$T(\overline{j}^{(1)}(\overline{H_{\mathbb{Z}}}^{(1)}) \otimes (\mathrm{Rad}\, I^{(1)})^q) \subset \Gamma_u^{(1)}.$$

Remarks 10.34 The singularities A_n and D_n play a special role in the odd case, in [Ch83] and in [Ja83][Ja85]. They are the only singularities which have no deformation to the singularity E_6. The groups $\Gamma^{(1)}$ and the sets $\Delta^{(1)}$ for them had been determined by Varchenko. Chmutov offers an account of this at the end of [Ch83].

It remains to characterize the finite group $p^{quot}(\Gamma^{(1)})$. The following holds for curve singularities (irreducible curve singularities without A_n and D_n [Wa80, Theorem 4 (c)], arbitrary curve singularities [Ch83, Theorems 1 and 3]).

Theorem 10.35 *Suppose that* f *is a curve singularity. Recall that then* $d_1 = 1$ *and* $H^{quot} = H_{\mathbb{Z}}/2H_{\mathbb{Z}}$. *The following three properties for an element* $g \in O^{(1),quot}$ *are equivalent:*

(i) $g \in p^{quot}(\Gamma^{(1)})$.
(ii) g *respects the quadratic form* q *in Theorem 10.33.*
(iii) g *maps* $\Delta^{(1)}_{\mathbb{F}_2}$ *to itself.*

Remarks 10.36 (i) So, such an automorphism g maps in the disjoint union

$$H_{\mathbb{Z}}/2H_{\mathbb{Z}} = (\text{image of } \mathrm{Rad}\, I^{(1)}) \,\dot\cup\, \Delta^{(1)}_{\mathbb{F}_2} \,\dot\cup\, (\text{the complement})$$

each of the three sets to itself, and g is the identity on the first set.

(ii) Chmutov found a generalization of this to arbitrary singularities. In Sect. 5 in [Ch83] an equivalence relation for the elements of H^{quot} for an arbitrary singularity is defined. Theorem 1 in [Ch83] says that for an arbitrary singularity an automorphism $g \in O^{(1),quot}$ is in $p^{quot}(\Gamma^{(1)})$ if and only if it respects this equivalence relation. For details see [Ch83, Section 5].

Janssen [Ja83][Ja85] recovered a good part of Chmutov's results. Whereas Chmutov worked mainly within the singularity cases, Janssen took a more abstract point of view. He started with the notion of an *odd vanishing lattice*. He classified odd vanishing lattices over $\mathbb{F}_2$ and over $\mathbb{Z}$.

Definition 10.37 Let $R = \mathbb{Z}$ or $R = \mathbb{F}_2$. An odd vanishing lattice over R is a triple (V, I_V, Δ_V) with the following properties. V is a free R-module of some rank $n_V \in \mathbb{Z}_{\geq 2}$. $I_V : V \times V \to R$ is an odd bilinear form. Δ_V is a subset of V with three specific properties. Define the group $\Gamma_{\Delta_V} := \langle s_a^{[1]} \mid a \in \Delta_V\rangle$ where $s_a^{[1]} \in O(V, I_V)$ is as usual the transvection with $s_a^{[1]}(b) := b - I_V(a, b)a$ for $b \in V$. The three specific properties:

(i) Δ_V is a single Γ_{Δ_V}-orbit.
(ii) Δ_V generates V as R-module.
(iii) There exist a_1 and $a_2 \in \Delta_V$ with $I_V(a_1, a_2) = 1$.

Remarks 10.38 (i) It is true that for any singularity with $n \geq 2$ the triple $(H_{\mathbb{Z}}, I^{(1)}, \Delta^{(1)})$ is an odd vanishing lattice over $\mathbb{Z}$.

(ii) Janssen classified all odd vanishing lattices over $\mathbb{F}_2$ [Ja83, (4.8) Theorem] and over $\mathbb{Z}$ [Ja85, (7.8) Theorem]. There are surprisingly few families, only seven families, for $R = \mathbb{F}_2$ as well as for $R = \mathbb{Z}$

(iii) In the case of $R = \mathbb{Z}$, in each of the seven families, an odd vanishing lattice (V, I_V, Δ_V) is determined up to isomorphism by the invariants $l, d_1, ..., d_l$ and rk Rad I_V of the pair (V, I_V), and additionally in two of the seven families by one more number.

(iv) Except for the A_n and D_n singularities, for any singularity f the triple $(H_{\mathbb{Z}}, I^{(1)}, \Delta^{(1)})$ fits into one of the three families of odd vanishing lattices with symbols $O_1^\sharp(d_1, ..., d_l, \rho)$, $O_0^\sharp(d_1, ..., d_l, \rho)$ and $O^\sharp(d_1, ..., d_l, \rho, k_0)$ where $\rho =$ rk Rad $I^{(1)}$ and k_0 is the additional number [Ja83, (6.6) Theorem][Ja85]

(v) But even if one knows that for a singularity the triple $(H_{\mathbb{Z}}, I^{(1)}, \Delta^{(1)})$ is for example of some type $O_1^\sharp(d_1, ..., d_l, \rho)$ where all invariants are determined by the pair $(H_{\mathbb{Z}}, I^{(1)})$, this does not mean that the pair $(H_{\mathbb{Z}}, I^{(1)})$ determines $\Delta^{(1)}$ uniquely. There might by an automorphism of the pair $(H_{\mathbb{Z}}, I^{(1)})$ which maps $\Delta^{(1)}$ to a different set. Then the pair $(H_{\mathbb{Z}}, I^{(1)})$ determines the triple $(H_{\mathbb{Z}}, I^{(1)}, \Delta^{(1)})$ only up to isomorphism.

(vi) The singularities A_6 and E_6 exist both as irreducible curve singularities, so the pairs $(H_{\mathbb{Z}}, I^{(1)})$ have the same invariants $d_1 = d_2 = d_3 = 1$ and rk Rad $I^{(1)} = 0$, so they are isomorphic. But the sets $\Delta^{(1)}$, the groups $\Gamma^{(1)}$ and the quadratic forms q on $H_{\mathbb{Z}}/2H_{\mathbb{Z}}$ are very different.

Example 10.39 In [DM94, 5.2 and 5.4] several examples of pairs (f_1, f_2) of curve singularities with the following property are given. The pairs $(H_{\mathbb{Z}}, L)$ of Milnor lattice and normalized Seifert form are isomorphic for f_1 and f_2, but f_1 and f_2 have distinct topological type. In all examples f_1 and f_2 are both reducible, with two components each (so rk Rad $I^{(1)} = 1$).

This leads naturally to the following questions for each of these pairs of curve singularities. We do no know their answers.

(A) Are the triples $(H_{\mathbb{Z}}, I^{(1)}, \Delta^{(1)})$ isomorphic?
(B) Are the triples $(H_{\mathbb{Z}}, L, \Delta^{(1)})$ isomorphic?
(C) Are the triples $(H_{\mathbb{Z}}, L, \mathcal{B}^{dist})$ isomorphic?

10.5 Moduli Spaces for the Simple and the Simple Elliptic Singularities

The only cases where $C_n^{\underline{e}/\{\pm 1\}^n}$ and $C_n^{S/\{\pm 1\}^n}$ had been studied in detail up to now are the cases from the simple singularities and the simple elliptic singularities, the cases from the simple singularities by Looijenga [Lo74] and (implicitly) Deligne [De74], all cases by Hertling and Roucairol [HR21]. Results on these cases are recalled and explained here. First we discuss the simple singularities, then the simple elliptic singularities.

The simple singularities in the normal forms in Theorem 10.6 are quasihomogeneous polynomials. They have universal unfoldings which are also quasihomogeneous polynomials. Here the good representatives and the base spaces $\mathcal{M}$ can be chosen as algebraic and global objects. We reproduce the universal unfoldings which are given in [Lo74]:

$$F^{alg} : \mathbb{C}^{m+1} \times \mathcal{M}^{alg} \to \mathbb{C} \quad \text{with} \quad \mathcal{M}^{alg} = \mathbb{C}^n, \tag{10.27}$$
$$F^{alg}(z_0, ..., z_m, t_1, ..., t_n) = F^{alg}(z, t) = F_t^{alg}(z) = f(z) + \sum_{j=1}^{n} t_j m_j$$

with $f = F_0^{alg}$ and $m_1, ..., m_n$ the monomials in the tables (10.28) and (10.29),

name	m_1	m_2	m_3	m_4	...	m_n
A_n	1	z_0	z_0^2	z_0^3	...	z_0^{n-1}
D_n	1	z_1	z_0	z_0^2	...	z_0^{n-2}

(10.28)

name	m_1	m_2	m_3	m_4	m_5	m_6	m_7	m_8
E_6	1	z_0	z_1	z_0^2	z_0z_1	$z_0^2z_1$		
E_7	1	z_0	z_1	z_0^2	z_0z_1	z_0^3	z_0^4	
E_8	1	z_0	z_1	z_0^2	z_0z_1	z_0^3	$z_0^2z_1$	$z_0^3z_1$

(10.29)

One checks easily that the monomials form a basis of the Jacobi algebra $\mathcal{O}_{\mathbb{C}^{m+1},0}/J_f$. Therefore the unfolding F^{alg} is indeed universal. The following tables list the weights $\deg_{\mathbf{w}} t_j$ of the unfolding parameters $t_1, ..., t_n$ which make F into a quasihomogeneous polynomial of weighted degree 1. The weights are all positive. The tables list also the Coxeter number $N_{Coxeter}$ of the corresponding root system.

	$N_{Coxeter}$	z_0	z_1	t_1	t_2	t_3	t_4	...	t_n
A_n	$n+1$	$\frac{1}{n+1}$		1	$\frac{n}{n+1}$	$\frac{n-1}{n+1}$	$\frac{n-2}{n+1}$	...	$\frac{2}{n+1}$
D_n	$2(n-1)$	$\frac{1}{n-1}$	$\frac{n-2}{2(n-1)}$	1	$\frac{n}{2(n-1)}$	$\frac{n-2}{n-1}$	$\frac{n-3}{n-1}$	...	$\frac{1}{n-1}$

	$N_{Coxeter}$	z_0	z_1	t_1	t_2	t_3	t_4	t_5	t_6	t_7	t_8
E_6	12	$\frac{1}{4}$	$\frac{1}{3}$	1	$\frac{3}{4}$	$\frac{2}{3}$	$\frac{1}{2}$	$\frac{5}{12}$	$\frac{1}{6}$		
E_7	18	$\frac{2}{9}$	$\frac{1}{3}$	1	$\frac{7}{9}$	$\frac{2}{3}$	$\frac{5}{9}$	$\frac{4}{9}$	$\frac{1}{3}$	$\frac{1}{9}$	
E_8	30	$\frac{1}{5}$	$\frac{1}{3}$	1	$\frac{4}{5}$	$\frac{2}{3}$	$\frac{3}{5}$	$\frac{7}{15}$	$\frac{2}{5}$	$\frac{4}{15}$	$\frac{1}{15}$

Because the weights are positive, F_t for each t satisfies that the sum of Milnor numbers of its singular points is equal to n. It follows with little work that $\mathcal{M}$ is an F-manifold with

$$\text{unit field } e = \partial_1, \quad \text{Euler field } E = \sum_{j=1}^{n} \deg_{\mathbf{w}}(t_j) t_j \partial_j,$$

and polynomial multiplication [HR21, Theorem 4.3].

The positive weights and general properties of the Lyashko-Looijenga map lead easily to the following result which was found independently by Looijenga and Lyashko (but Lyashko published his version only much later).

Theorem 10.40 *[Lo74][Ly79] The Lyashko-Looijenga map*

$$\mathcal{L}^{alg} : \mathcal{M}^{alg} \to \mathbb{C}[x]_n \cong \mathbb{C}^n \tag{10.30}$$

is a branched covering of degree

$$\deg \mathcal{L}^{alg} = \frac{n!}{\prod_{j=1}^{n} \deg_{\mathbf{w}} t_j}. \tag{10.31}$$

It is branched along the caustic $\mathcal{K}_3^{alg} \subset \mathcal{M}^{alg}$ (at generic points of order 3) and along the Maxwell stratum $\mathcal{K}_2^{alg} \subset \mathcal{M}^{alg}$ (at generic points of order 2). The restriction

$$\mathcal{L}^{alg} : \mathcal{M}^{alg} - (\mathcal{K}_3^{alg} \cup \mathcal{K}_2^{alg}) \to \mathbb{C}[x]_n^{reg} \cong C_n^{conf} \tag{10.32}$$

is a covering.

Remarks 10.41 (i) It is known that $|\Gamma^{(0)}| = (N_{Coxeter})^n \prod_{j=1}^{n} \deg_{\mathbf{w}} t_j$ in the case of an ADE root lattice. Therefore

$$\deg \mathcal{L}^{alg} = \frac{n! \cdot (N_{Coxeter})^n}{|\Gamma^{(0)}|}.$$

(ii) The following table lists the numbers $|G_{\mathbb{Z}}|$ and $|\mathcal{S}^{dist}/\{\pm 1\}^n|$ for the simple singularities. Theorem 10.42 and (10.39) will tell

$$\deg \mathcal{L}^{alg} = |\mathcal{B}^{dist}/\{\pm 1\}^n| = \frac{1}{2} \cdot |G_{\mathbb{Z}}| \cdot |\mathcal{S}^{dist}/\{\pm 1\}^n|.$$

	$\lvert G_{\mathbb{Z}}\rvert$	$\lvert \mathcal{S}^{dist}/\{\pm 1\}^n\rvert$
A_n	$2(n+1)$	$(n+1)^{n-2}$
D_4	36	9
$D_n, n \geq 5$	$4(n-1)$	$(n-1)^{n-1}$
E_6	24	$2^7 \cdot 3^3 = 3456$
E_7	18	$2 \cdot 3^{10} = 118098$
E_8	30	$2 \cdot 3^4 \cdot 5^6 = 2531250$

(10.33)

Recall from Definition 8.5 the open convex polyhedron $F_n \subset C_n^{pure} \subset \mathbb{C}^n$ and its boundary W_n, a union of walls. Also recall

$$C_n^{conf} \subset \mathbb{C}^n = \mathrm{pr}_n^{p,c}(F_n) \,\dot{\cup}\, \mathrm{pr}_n^{p,c}(W_n).$$

The union of walls $(\mathcal{L}^{alg})^{-1}(\mathrm{pr}_n^{p,c}(W_n))$ upstairs in $\mathcal{M}^{alg}$ contains $\mathcal{K}_3^{alg} \cup \mathcal{K}_2^{alg}$. The restriction

$$\mathcal{L}^{alg} : (\mathcal{L}^{alg})^{-1}(\mathrm{pr}_n^{p,c}(F_n)) \to F_n$$

is an even covering. The components of the preimages are called *Stokes regions*. Each of them is mapped isomorphically to F_n. The number of Stokes regions is $\deg \mathcal{L}^{alg}$.

We obtain a map

$$LD : \{\text{Stokes regions in } \mathcal{M}^{alg}\} \to \mathcal{B}^{dist}/\{\pm 1\}^n$$

in the following way. For t in one Stokes region choose a standard system $(\underline{\gamma}; \mathrm{id})$ of paths and obtain as in Theorem 10.8 from F_t a distinguished basis up to signs. This works because $\mathcal{M} = \mathbb{C}^n$ is simply connected and therefore there is a canonical isomorphism

$$H_m(F_t^{-1}(\eta), \mathbb{Z}) \cong H_m(X_\eta, \mathbb{Z}) \cong H_{\mathbb{Z}}$$

for (depending on t) sufficiently large $\eta > 0$. It is independent of the choice of t within a fixed Stokes region. The distinguished basis up to signs is $LD(\text{Stokes region}) \in \mathcal{B}^{dist}/\{\pm 1\}^n$.

Choose one distinguished basis $\underline{e} \in \mathcal{B}^{dist}$ for reference. The map LD is surjective because the restriction of $\mathcal{L}^{alg}$ in (10.32) is a covering and $\mathcal{B}^{dist}/\{\pm 1\}^n$ is a single Br_n orbit. Even more, the covering factorizes through the covering $\mathrm{pr}_n^{e,c} : C_n^{\underline{e}/\{\pm 1\}^n} \to C_n^{conf}$, so it is a composition of two coverings,

$$\mathcal{M}^{alg} - (\mathcal{K}_3^{alg} \cup \mathcal{K}_2^{alg}) \xrightarrow{\mathrm{pr}_n^{alg,e}} C_n^{\underline{e}/\{\pm 1\}^n} \xrightarrow{\mathrm{pr}_n^{e,c}} C_n^{conf}. \tag{10.34}$$

This follows from the construction of the spaces and the compatibility of the braid group actions. The main result in this section for the simple singularities is the following.

Theorem 10.42 *[Lo74][De74] [HR21, Theorem 7.1] Consider a simple singularity. The map*

$$LD : \{\text{Stokes regions in } \mathcal{M}^{alg}\} \to \mathcal{B}^{dist}/\{\pm 1\}^n \tag{10.35}$$

is a bijection. Equivalent: The covering

$$\mathrm{pr}_n^{alg,e} : \mathcal{M}^{alg} - (\mathcal{K}_3^{alg} \cup \mathcal{K}_2^{alg}) \to C_n^{\underline{e}/\{\pm 1\}^n} \tag{10.36}$$

is an isomorphism.

Looijenga [Lo74] asked whether the map LD is a bijection and proved it in the case of the A_n singularity. Deligne [De74] answered the question for all simple singularities positively in the following way. He calculated in a letter to Looijenga for each simple singularity the number $|\mathcal{B}^{dist}/\{\pm 1\}^n|$ and found equality with $\deg \mathcal{L}^{alg}$. The map LD is a surjection between finite sets of same size, so it is a bijection. This is one proof of Theorem 10.42.

[HR21] gives a different proof, which generalizes to the simple elliptic singularities where LD is a map between infinite sets. Here we sketch part of this proof.

We need a general result about symmetries of a singularity from [He02, 13.2].

Let $f : (\mathbb{C}^{m+1}, 0) \to (\mathbb{C}, 0)$ be an isolated hypersurface singularity with Milnor number n. Let $F : (\mathbb{C}^{m+1} \times \mathcal{M}, 0) \to (\mathbb{C}, 0)$ be a universal unfolding with base space a germ $(\mathcal{M}, 0) = (\mathbb{C}^n, 0)$. Denote by

$$\mathcal{R}^f := \{\varphi : (\mathbb{C}^{m+1}, 0) \to (\mathbb{C}^{m+1}, 0) \mid \varphi \text{ is a biholomorphic map germ with } f = f \circ \varphi\}$$

the (very large) group of germs of coordinate changes which leave f invariant. Denote by

$$\mathrm{Aut}_{\mathcal{M}} := \mathrm{Aut}((\mathcal{M}, 0), \circ, e, E)$$

the group of automorphisms of $(\mathcal{M}, 0)$ as a germ of an F-manifold with Euler field. It is a finite group [He02, Theorem 4.14]. Consider a good representative $F : \mathcal{X} \to \Delta_\eta$ of the universal unfolding, whose base space $\mathcal{M} \subset \mathbb{C}^n$ is a small ball in $\mathbb{C}^n$. The Lyashko-Looijenga map $\mathcal{L} : \mathcal{M} \to \mathbb{C}[x]_n$ is holomorphic. The restriction $\mathcal{L} : \mathcal{M} - (\mathcal{K}_3 \cup \mathcal{K}_2) \to C_n^{conf}$ is locally biholomorphic. The components of the even smaller restriction

$$\mathcal{L} : \mathcal{L}^{-1}(\mathrm{pr}_n^{p,c}(F_n)) \to F_n \tag{10.37}$$

are also now called *Stokes regions*, although the restriction $\mathcal{L}$: (one Stokes region) $\to$ F_n is injective, but in general not an isomorphism. Nevertheless, as in the case of the simple singularities, we obtain a natural map

$$LD : \{\text{Stokes regions in } \mathcal{M}\} \to \mathcal{B}^{dist}/\{\pm 1\}^n \tag{10.38}$$

by the construction in Theorem 10.8 applied to F_t for some t in the Stokes region together with a standard system of paths.

Theorem 10.43 *[He02, 13.2]*

(a) Each $\varphi \in \mathcal{R}^f$ can be lifted to an automorphism of the unfolding. It induces an automorphism $(\varphi)_{hom} \in G_{\mathbb{Z}}$ of $H_{\mathbb{Z}}$ and an automorphism $(\varphi)_{\mathcal{M}} \in \mathrm{Aut}_{\mathcal{M}}$. The group homomorphism $()_{\mathcal{M}} : \mathcal{R}^f \to \mathrm{Aut}_{\mathcal{M}}$ is surjective. The group homomorphism $()_{hom} : \mathcal{R}^f \to G_{\mathbb{Z}}$ has finite image. The image is a subgroup of $G_{\mathbb{Z}}^{\mathcal{B}}$.

(b)

$$\begin{array}{llll} -\mathrm{id} \notin (\mathcal{R}^f)_{hom} & \text{and} & \ker()_{\mathcal{M}} = \ker()_{hom} & \text{if } \mathrm{mult}(f) \geq 3. \\ -\mathrm{id} \in (\mathcal{R}^f)_{hom} & \text{and} & \ker()_{\mathcal{M}} = \ker()_{hom} \times \{\pm \mathrm{id}\} & \text{if } \mathrm{mult}(f) = 2. \end{array}$$

(c) One can choose a finite subgroup $\mathcal{R}^{finite} \subset \mathcal{R}^f$ which lifts to a group of automorphisms of a (sufficiently) good representative $F : \mathcal{X} \to \Delta_\eta$ of the universal unfolding and such that $()_{hom} : \mathcal{R}^{finite} \to G_{\mathbb{Z}}$ is injective with image $(\mathcal{R}^f)_{hom}$. Then $(\mathcal{R}^{finite})_{\mathcal{M}} = \mathrm{Aut}_{\mathcal{M}}$. This group acts on the set of Stokes regions of $\mathcal{M}$. The map LD is equivariant, that means, for $\varphi \in \mathcal{R}^{finite}$ the following diagram commutes,

$$\begin{array}{ccc} \{\textit{Stokes regions in } \mathcal{M}\} & \xrightarrow{LD} & \mathcal{B}^{dist}/\{\pm 1\}^n \\ \downarrow (\varphi)_{\mathcal{M}} & & \downarrow (\varphi)_{hom} \\ \{\textit{Stokes regions in } \mathcal{M}\} & \xrightarrow{LD} & \mathcal{B}^{dist}/\{\pm 1\}^n \end{array}$$

The last statement in part (a), $(\mathcal{R}^f)_{hom} \subset G_{\mathbb{Z}}^{\mathcal{B}}$, and the last statement in part (c) that the map LD is equivariant, are not formulated in [He02, 13.2]. But they follow easily from the fact that any automorphism $\varphi \in \mathcal{R}^{finite}$ lifts to an automorphism of the good representative of the universal unfolding.

We return to the simple singularities. Recall that the group $G_{\mathbb{Z}}$ is

$$G_{\mathbb{Z}} = \{\pm M^l \mid l \in \mathbb{Z}\} \times U$$
$$\text{with } U \cong \begin{cases} \{\mathrm{id}\} & \text{in the cases } A_n, D_{2n+1}, E_6, E_7, E_8, \\ S_2 & \text{in the cases } D_{2n} \text{ with } n \geq 3, \\ S_3 & \text{in the case } D_4. \end{cases}$$

In [He11, Chap. 8] as well as in [HR21, 5.1] it is proved that the map $()_{hom} : \mathcal{R}^{finite} \to G_{\mathbb{Z}}$ is an isomorphism if $\mathrm{mult}(f) = 2$ and that its image maps isomorphically to $G_{\mathbb{Z}}/\{\pm \mathrm{id}\}$ if $\mathrm{mult}(f) = 3$. As a conclusion from this discussion of symmetries of the simple singularities, we obtain the following lemma for the simple singularities.

Lemma 10.44 *The elements $(\varphi)_{\mathcal{M}}$ for $\varphi \in \mathcal{R}^{finite}$ form a group of deck transformations for the covering*

$$\mathrm{pr}_n^{alg,S} : \mathcal{M}^{alg} - (\mathcal{K}_3^{alg} \cup \mathcal{K}_2^{alg}) \to C_n^{S/\{\pm 1\}^n}$$

which is isomorphic to $G_{\mathbb{Z}}/\{\pm \mathrm{id}\}$. *They are the restriction to* $\mathcal{M}^{alg} - (\mathcal{K}_3^{alg} \cup \mathcal{K}_2^{alg})$ *of the elements of* $\mathrm{Aut}_{\mathcal{M}}$. *Here* $G_{\mathbb{Z}} = G_{\mathbb{Z}}^{\mathcal{B}}$ *and* $\mathrm{Aut}_{\mathcal{M}} \cong G_{\mathbb{Z}}/\{\pm \mathrm{id}\}$.

On the other hand recall that the covering $\mathrm{pr}_n^{e,S} : C_n^{\underline{e}/\{\pm 1\}^n} \to C_n^{S/\{\pm 1\}^n}$ is normal with group of deck transformations

$$\frac{(\mathrm{Br}_n)_{S/\{\pm 1\}^n}}{(\mathrm{Br}_n)_{\underline{e}/\{\pm 1\}^n}} \cong \frac{G_{\mathbb{Z}}^{\mathcal{B}}}{Z((\{\pm 1\}^n)_{\underline{e}})} \overset{\text{here}}{=} \frac{G_{\mathbb{Z}}}{\{\pm \mathrm{id}\}}. \tag{10.39}$$

In the composition of coverings

$$\mathcal{M}^{alg} - (\mathcal{K}_3^{alg} \cup \mathcal{K}_2^{alg}) \xrightarrow{\mathrm{pr}_n^{alg,e}} C_n^{\underline{e}/\{\pm 1\}^n} \xrightarrow{\mathrm{pr}_n^{e,S}} C_n^{S/\{\pm 1\}^n}$$

the second one is normal, and its group of deck transformations lifts to a group of deck transformations of the composite covering $\mathrm{pr}_n^{alg,S} = \mathrm{pr}_n^{e,S} \circ \mathrm{pr}_n^{alg,e}$. Because of the next lemma, the covering $\mathrm{pr}_n^{alg,e}$ is an isomorphism.

Lemma 10.45 *The covering* $\mathrm{pr}_n^{alg,S}$ *is normal with group of deck transformations the group* $\mathrm{Aut}_{\mathcal{M}}$.

The proof uses the following idea, which had been used first by Jaworski [Ja88, Proposition 2] in the case of the Lyashko-Looijenga map for the simple elliptic singularitiees. Consider two points $t^{(1)}$ and $t^{(2)} \in \mathcal{M}^{alg}$ with the same image $S/\{\pm 1\}^n = \mathrm{pr}_n^{alg,S}(t^{(1)}) = \mathrm{pr}_n^{alg,S}(t^{(2)}) \in \mathcal{S}^{dist}/\{\pm 1\}^n$. Consider in C_n^{conf} a path from $\mathrm{pr}_n^{alg,c}(t^{(1)}) = \mathrm{pr}_n^{alg,c}(t^{(2)})$ to a generic point of D_n^{conf}. Then the lifts of this path to $\mathcal{M}^{alg} - (\mathcal{K}_3^{alg} \cup \mathcal{K}_2^{alg})$ which start at $t^{(1)}$ or $t^{(2)}$ tend either both to $\mathcal{K}_3^{alg}$ (if the relevant entry of the relevant matrix in the orbit of S is ± 1) or both to $\mathcal{K}_2^{alg}$ (if the relevant entry of the relevant matrix in the orbit of S is 0). Therefore the covering $\mathrm{pr}_n^{alg,c}$ looks the same from $t^{(1)}$ and $t^{(2)}$, so there is a deck transformation which maps $t^{(1)}$ to $t^{(2)}$. It lifts to a deck transformation of the covering $\mathrm{pr}_n^{alg,S}$. It extends to an automorphism of $\mathcal{M}$ as F-manifold with Euler field.

Now we turn to the simple elliptic singularities. The ideas above go through and lead to Theorem 10.47, which is analogous to Theorem 10.42. A common point of the simple and simple elliptic singularities, which is not shared by the other singularities, is that they have global algebraic unfoldings.

The 1-parameter families of simple elliptic singularities in the Legendre normal forms in Theorem 10.6 with parameter $\lambda \in \mathbb{C} - \{0, 1\}$ are quasihomogeneous polynomials except for λ which has weight $\deg_{\mathbf{w}}(\lambda) = 0$. They have global unfoldings which are also quasihomogeneous polynomials except for λ.

$$F^{alg} : \mathbb{C}^{m+1} \times \mathcal{M}^{alg} \to \mathbb{C} \quad \text{with} \quad \mathcal{M}^{alg} = \mathbb{C}^{n-1} \times (\mathbb{C} - \{0, 1\}),$$

$$\begin{aligned} F^{alg}(z_0, ..., z_m, t_1, ..., t_{n-1}, \lambda) &= F^{alg}(z, t', \lambda) \\ &= F^{alg}_{t',\lambda}(z) = f_\lambda(z) + \sum_{j=1}^{n-1} t_j m_j \end{aligned} \tag{10.40}$$

with $f_\lambda = F^{alg}_{0,\lambda}$ and $m_1, ..., m_{n-1}$ the monomials in table (10.41),

(10.41)

name	m_1	m_2	m_3	m_4	m_5	m_6	m_7	m_8	m_9
$\widetilde{E}_6$	1	z_0	z_1	z_2	z_0^2	z_0z_1	z_1z_2		
$\widetilde{E}_7$	1	z_0	z_1	z_0^2	z_0z_1	z_1^2	$z_0^2z_1$	$z_0z_1^2$	
$\widetilde{E}_8$	1	z_0	z_0^2	z_1	z_0^3	z_0z_1	$z_0^2z_1$	z_1^2	$z_0z_1^2$

Let $\lambda : \mathbb{H} \to \mathbb{C} - \{0, 1\}, t_n \mapsto \lambda(t_n)$, be the standard universal covering. For each of the three Legendre families of simple elliptic singularities, we will also consider the global family of functions

$$F^{mar} : \mathbb{C}^{m+1} \times \mathcal{M}^{mar} \to \mathbb{C}, (z, t) \mapsto F^{alg}(z, t', \lambda(t_m)) \tag{10.42}$$
$$\text{where} \quad \mathcal{M}^{mar} = \mathbb{C}^{n-1} \times \mathbb{H}.$$

The μ-constant stratum within $\mathcal{M}^{alg}$ is the 1-dimensional set $\mathcal{M}^{alg}_\mu = \{0\} \times (\mathbb{C} - \{0, 1\}) \subset \mathcal{M}^{alg}$. The global unfolding F^{alg} is near any point $(0, \lambda) \in \mathcal{M}^{alg}_\mu$ a universal unfolding of $f_\lambda = F^{alg}_{0,\lambda}$ [HR21, Lemma 4.1]. The weights $\deg_{\mathbf{w}} t_j$ for $j \in \{1, ..., n-1\}$ are all positive. The next table lists them and a rational number which will appear in the degree of a Lyashko-Looijenga map.

	z_0	z_1	z_2	t_1	t_2	t_3	t_4	t_5	t_6	t_7	t_8	t_9	$\frac{1}{2}\sum_{j=2}^{n-1}\frac{1}{\deg_{\mathbf{w}} t_j}$
$\widetilde{E}_6$	$\frac{1}{3}$	$\frac{1}{3}$	$\frac{1}{3}$	1	$\frac{2}{3}$	$\frac{2}{3}$	$\frac{2}{3}$	$\frac{1}{3}$	$\frac{1}{3}$	$\frac{1}{3}$			$\frac{27}{4}$
$\widetilde{E}_7$	$\frac{1}{4}$	$\frac{1}{4}$		1	$\frac{3}{4}$	$\frac{3}{4}$	$\frac{1}{2}$	$\frac{1}{2}$	$\frac{1}{2}$	$\frac{1}{4}$	$\frac{1}{4}$		$\frac{25}{3}$
$\widetilde{E}_8$	$\frac{1}{6}$	$\frac{1}{3}$		1	$\frac{5}{6}$	$\frac{2}{3}$	$\frac{2}{3}$	$\frac{1}{2}$	$\frac{1}{2}$	$\frac{1}{3}$	$\frac{1}{3}$	$\frac{1}{6}$	$\frac{101}{10}$

Because the weights are positive, $F^{alg}_{t',\lambda}$ for each (t', λ) satisfies that the sum of Milnor numbers of its singular points is equal to n. It follows with little work that $\mathcal{M}^{alg}$ is an F-manifold with

$$\text{unit field } e = \partial_1, \quad \text{Euler field } E = \sum_{j=1}^{n-1} \deg_{\mathbf{w}}(t_j) t_j \partial_j$$

and polynomial multiplication [HR21, Theorem 4.3] and also that $\mathcal{M}^{mar}$ is an F-manifold with Euler field.

The Lyashko-Looijenga map for $\mathcal{M}^{alg}$ was first studied by Jaworski [Ja86][Ja88]. He found that the restriction of $\mathcal{L}^{alg} : \mathcal{M}^{alg} \to \mathbb{C}[x]_n$ to $\mathcal{L}^{alg} : \mathcal{M}^{alg} - (\mathcal{K}_3^{alg} \cup \mathcal{K}_2^{alg}) \to C_n^{conf}$ is a finite covering. But contrary to Theorem 10.40 for the simple singularities this is not at all trivial. The map $\mathcal{L}^{alg}$ is holomorphic, but not everywhere finite. The 1-dimensional μ-constant stratum $\mathcal{M}_\mu^{alg}$ is mapped to $0 \in \mathbb{C}[x]_n$.

His methods did not allow Jaworski to calculate the degree $\deg \mathcal{L}^{alg}$. This was done by a huge effort in [HR21, sections 5, 8, 9, 10] by glueing into $\mathcal{M}^{alg}$ suitable fibers over $\lambda \in \{0, 1, \infty\}$ and extending $\mathcal{L}^{alg}$ to these fibers. A simpler way to calculate $\deg \mathcal{L}^{alg}$ was recently found by Takahashi and Zhang [TZ23].

Theorem 10.46 *(a) [Ja86][Ja88] The restriction*

$$\mathcal{L}^{alg} : \mathcal{M}^{alg} - (\mathcal{K}_3^{alg} \cup \mathcal{K}_2^{alg}) \to C_n^{conf}$$

of the Lyashko-Looijenga map $\mathcal{L}^{alg} : \mathcal{M}^{alg} \to \mathbb{C}[x]_n$ *is a finite covering.*

(b) [HR21, Theorem 6.3][TZ23] Its degree is

$$\deg \mathcal{L}^{alg} = \frac{n! \cdot \frac{1}{2} \cdot \sum_{j=2}^{n-1} \frac{1}{\deg_{\mathbf{w}} t_j}}{\prod_{j=2}^{n-1} \deg_{\mathbf{w}} t_j}. \tag{10.43}$$

The Stokes regions in $\mathcal{M}^{alg}$ and in $\mathcal{M}^{mar}$ are the components of the preimages in $\mathcal{M}^{alg}$ and $\mathcal{M}^{mar}$ of $\mathrm{pr}_n^{p,c}(F_n) \subset C_n^{conf}$. The restriction of $\mathcal{L}^{alg}$ respectively $\mathcal{L}^{mar}$ to a Stokes region is an isomorphism to $\mathrm{pr}_n^{p,c}(F_n)$.

We obtain a map

$$LD : \{\text{Stokes regions in } \mathcal{M}^{mar}\} \to \mathcal{B}^{dist}/\{\pm 1\}^n$$

as for the simple singularities. Though here we have to choose a reference point $(0, t_n^{ref}) \in \mathcal{M}_\mu^{mar}$. We choose it with $\lambda(t_n^{ref}) = \frac{1}{2}$. Then $H_{\mathbb{Z}} := H_{\mathbb{Z}}(f_{1/2})$. Because the space $\mathcal{M}^{mar} = \mathbb{C}^{n-1} \times \mathbb{H}$ is simply connected there is a canonical isomorphism

$$H_m(F_t^{-1}(\eta), \mathbb{Z}) \cong H_m(f_{1/2}^{-1}(\eta), \mathbb{Z}) \cong H_{\mathbb{Z}}(f_{1/2})$$

for any $t \in \mathcal{M}^{mar}$ and (depending on t) sufficiently large $\eta > 0$. We do not obtain a map LD for the set of Stokes regions in $\mathcal{M}^{alg}$ because there we do not have such canonical isomorphisms. The $\mathbb{Z}$-lattice bundle $\bigcup_{\lambda \in \mathbb{C}-\{0,1\}} H_{\mathbb{Z}}(f_\lambda)$ has transversal monodromy.

Choose one distinguished basis $\underline{e} \in \mathcal{B}^{dist}$ for reference. The map LD is surjective because the restriction of $\mathcal{L}^{mar}$ to a map

$$\mathcal{L}^{mar} : \mathcal{M}^{mar} - (\mathcal{K}_3^{mar} \cup \mathcal{K}_2^{mar}) \to C_n^{conf}$$

is by Theorem 10.46 (a) a covering. Even more, it factorizes in ways described by the following commutative diagram of coverings,

$$
\begin{array}{ccccc}
\mathcal{M}^{alg} - (\mathcal{K}_3^{alg} \cup \mathcal{K}_2^{alg}) & & \xrightarrow{\text{finite}} & & C_n^{conf} \\
\uparrow & \searrow^{\text{finite}} & & \nearrow^{\text{finite}} & \uparrow \\
\infty & & C_n^{S/\{\pm 1\}^n} & & \infty \\
| & \nearrow^{\infty} & & \nwarrow^{\infty} & | \\
\mathcal{M}^{mar} - (\mathcal{K}_3^{mar} \cup \mathcal{K}_2^{mar}) & & \longrightarrow & & C_n^{\underline{e}/\{\pm 1\}^n}
\end{array}
\tag{10.44}
$$

This follows from the construction of the spaces and the compatibility of the braid group actions. The main result in this section for the simple elliptic singularities is the following.

Theorem 10.47 *[HR21, Theorem 7.1] Consider a 1-parameter family of simple elliptic singularities. The map*

$$LD : \{\textit{Stokes regions in } \mathcal{M}^{mar}\} \to \mathcal{B}^{dist}/\{\pm 1\}^n \tag{10.45}$$

is a bijection. Equivalent: The covering

$$\mathrm{pr}_n^{mar,e} : \mathcal{M}^{mar} - (\mathcal{K}_3^{mar} \cup \mathcal{K}_2^{mar}) \to C_n^{\underline{e}/\{\pm 1\}^n} \tag{10.46}$$

is an isomorphism.

The ideas of the proof are the same as in the proof of Theorem 10.42 for the simple singularities in [HR21]. Recall that the group $G_{\mathbb{Z}} := G_{\mathbb{Z}}(f_{1/2})$ is by Theorem 10.24 (c) the extension of the group $G_{\mathbb{Z}}|_{\mathrm{Rad}\, I^{(0)}} \cong SL_2(\mathbb{Z})$ by the finite group $U_1 \rtimes U_2$. The following results are proved in [HR21] in a similar way as for the simple singularities.

Lemma 10.48 *(a)* $(\mathcal{R}^{f_{1/2}})_{hom} = U_1 \rtimes U_2$ *if* $\mathrm{mult}(f) = 3$, *and* $(\mathcal{R}^{f_{1/2}})_{hom} = U_1 \rtimes U_2 \times \{\pm \mathrm{id}\}$ *if* $\mathrm{mult}(f) = 2$.

(b) $G_{\mathbb{Z}} = G_{\mathbb{Z}}^{\mathcal{B}}$.

(c) The covering $\mathrm{pr}_n^{mar,S} : \mathcal{M}^{mar} - (\mathcal{K}_3^{mar} \cup \mathcal{K}_2^{mar}) \to C_n^{S/\{\pm 1\}^n}$ *is a normal covering. The group of deck transformations is isomorphic to* $\mathrm{Aut}_{\mathcal{M}^{mar}} := \mathrm{Aut}(\mathcal{M}^{mar}, \circ, e, E)$ *of* $\mathcal{M}^{mar}$ *as F-manifold with Euler field and isomorphic to* $G_{\mathbb{Z}}/\{\pm \mathrm{id}\}$.

(d) The covering $\mathrm{pr}_n^{e,S} : C_n^{\underline{e}/\{\pm 1\}^n} \to C_n^{S/\{\pm 1\}^n}$ *is a normal covering with group of deck transformations isomorphic to* $G_{\mathbb{Z}}/\{\pm \mathrm{id}\}$.

The proof of part (c) uses Jaworski's idea which was described after Lemma 10.45. Now a matrix $S \in \mathcal{S}^{dist}$ can have the entries $0, \pm 1, \pm 2$. The lifts of a path in C_n^{conf} which tends to a generic point of D_n^{conf} tend to $\mathcal{K}_2^{mar}$ if the relevant entry is 0, to $\mathcal{K}_3^{mar}$ if the relevant entry is ± 1, and to $\mathbb{C}^n \times \{0, 1, \infty\}$ if the relevant entry is ± 2.

Lemma 10.48 (c)+(d) and the diagram (10.44) of coverings show Theorem 10.47.

Remarks 10.49 (i) The following table gives the degrees of the finite coverings in the diagram (10.44) of coverings.

	$\deg \mathrm{pr}_n^{alg,S} = 6\lvert U_1\rvert\lvert U_2\rvert$	$\deg \mathrm{pr}_n^{S,c} = \lvert \mathcal{S}^{dist}/\{\pm 1\}^n\rvert$
$\widetilde{E}_6$	$6\cdot 2\cdot 3\cdot 3^2 = 324$	$3^7\cdot 5\cdot 7 = 76545$
$\widetilde{E}_7$	$6\cdot 1\cdot 4\cdot 2^2 = 96$	$2^{13}\cdot 5^3\cdot 7 = 7168000$
$\widetilde{E}_8$	$6\cdot 1\cdot 6\cdot 1^2 = 36$	$2^7\cdot 3^8\cdot 7\cdot 101 = 593744256$

(10.47)

The surprising factor 101 in the case of $\widetilde{E}_8$ comes from the term $\sum_{j=2}^{n-1}\frac{1}{\deg_{\mathbf{w}} t_j}$ in formula (10.43) for $\deg \mathcal{L}^{alg}$.

(ii) The number $\deg \mathcal{L}^{alg} = \deg \mathrm{pr}_n^{alg,S}\cdot \lvert \mathcal{S}^{dist}/\{\pm 1\}\rvert$ was calculated in [HR21] and again in [TZ23]. With the easily determined number $\deg \mathrm{pr}_n^{alg,S} = 6\lvert U_1\rvert\lvert U_2\rvert$ it gives $\lvert \mathcal{S}^{dist}/\{\pm 1\}^n\rvert$. Though this number $\lvert \mathcal{S}^{dist}/\{\pm 1\}^n\rvert$ was determined in the cases of $\widetilde{E}_6$ and $\widetilde{E}_7$ much earlier by Kluitmann, for $\widetilde{E}_5$ in [Kl83] and [Kl87, VII], for $\widetilde{E}_7$ in [Kl87, VII], by a long combinatorial and inductive procedure.

Remarks 10.50 In the cases of other singularities than the simple and simple elliptic singularities, there are a priori no natural global unfoldings. A priori the base space of a universal unfolding is a germ $(\mathcal{M}, 0) \cong (\mathbb{C}^n, 0)$.

Consider a given μ-homotopy class of singularities. In [He11] a global moduli space $\mathcal{M}_\mu^{mar}$ for *right equivalence classes* of *marked singularities* is constructed. Locally it is isomorphic to the μ-constant stratum in the base space of a universal unfolding of one singularity. So one can consider it as a global μ-constant stratum for all singularities in one μ-homotopy class.

In order to obtain global moduli spaces of dimension n, it would be good to thicken $\mathcal{M}_\mu^{mar}$ to an n-dimensional manifold which is near each point of $\mathcal{M}_\mu^{mar}$ the base space of a universal unfolding and then glue this manifold to the corresponding manifold $C_n^{\underline{e}/\{\pm 1\}^n}$ where $\underline{e} \in \mathcal{B}^{dist}$ is one chosen distinguished basis.

By Theorem 10.42 and 10.47 this works for the simple singularities and the simple elliptic singularities. It gives the manifold $\mathcal{M}^{alg}$ for the simple singularities and the manifold $\mathcal{M}^{mar}$ for the simple elliptic singularities.

For other singularities this is an open project.

Appendix A
Tools from Hyperbolic Geometry

The upper half plane $\mathbb{H} = \{z \in \mathbb{C} \mid \Im(z) > 0\}$ together with its natural metric (whose explicit form we will not need) is one model of the hyperbolic plane. In the study of the monodromy groups $\Gamma^{(0)}$ and $\Gamma^{(1)}$ for the cases with $n = 2$ or $n = 3$, we will often encounter subgroups of $\mathrm{Isom}(\mathbb{H})$. The theorem of Poincaré-Maskit [Po82][Ma71] allows under some conditions to show for such a group that it is discrete, to find a fundamental domain and to find a presentation. Three special cases of this theorem, which will be sufficient for us, are formulated in Theorem A.2.

Before, we collect basic facts and set up some notations in the following Remarks and Notations A.1.

Subgroups of $\mathrm{Isom}(\mathbb{H})$ arise here in two ways. Either they come from groups of real 2×2 matrices. This is covered by the Remarks and Notations A.1 (v). Or they come from the action of certain groups of real 3×3 matrices on $\mathbb{R}^3$ with an indefinite metric. This will treated in Theorem A.4.

Remarks and Notations A.1 (Some references for the following material are [Fo51][Le66][Be83])

(i) Let $n \in \mathbb{N}$. Recall the notions of the free group $G^{free,n}$ with n generators and of the free Coxeter group $G^{fCox,n}$ with n generators from Definition 3.1.

(ii) $\widehat{\mathbb{C}} := \mathbb{C} \cup \{\infty\}$, $\widehat{\mathbb{R}} := \mathbb{R} \cup \{\infty\}$. The hyperbolic lines in $\mathbb{H}$ are the parts in $\mathbb{H}$ of circles and euclidean lines which meet $\mathbb{R}$ orthogonally. For any $z_1, z_2 \in \mathbb{H} \cup \widehat{\mathbb{R}}$ with $z_1 \neq z_2$, denote by $A(z_1, z_2)$ the part between z_1 and z_2 of the unique hyperbolic line whose closure in $\mathbb{H} \cup \widehat{\mathbb{R}}$ contains z_1 and z_2. Here $z_i \in A(z_1, z_2)$ if $z_i \in \mathbb{H}$, but $z_i \notin A(z_1, z_2)$ if $z_i \in \widehat{\mathbb{R}}$. Such sets are called *arcs*.

(iii) We simplify the definition of a polygon in [Ma71]. A *hyperbolic polygon* P is a contractible open subset $P \subset \mathbb{H}$ whose relative boundary in $\mathbb{H}$ consists of finitely many arcs $A_1 = A(z_{1,1}, z_{1,2})$, ..., $A_m = A(z_{m,1}, z_{m,2})$ (Maskit allows countably many arcs). The arcs and the points are numbered such that one runs through

C. Hertling and K. Larabi, *Triangular Bases of Unimodular Bilinear Lattices and Induced Structures*, Lecture Notes in Mathematics 2398,
https://doi.org/10.1007/978-3-032-25997-4

them in the order $A_1, ..., A_m$ and $z_{1,1}, z_{1,2}, z_{2,1}, z_{2,2}, ..., z_{m,1}, z_{m,2}$ if one runs mathematically positive on the euclidean boundary of P in $\mathbb{H} \cup \widehat{\mathbb{R}}$. For A_i and A_{i+1} (with $A_{m+1} := A_1$) there are three possibilities:

(a) $z_{i,2} = z_{i+1,1} \in \mathbb{H}$; then this point is called a *vertex* of P;
(b) $z_{i,2} = z_{i+1,1} \in \widehat{\mathbb{R}}$;
(c) $z_{i,2} \in \widehat{\mathbb{R}}, z_{i+1,1} \in \widehat{\mathbb{R}}, z_{i,2} \neq z_{i+1,1}$; then the part of $\widehat{\mathbb{R}}$ between $z_{i,2}$ and $z_{i+1,1}$ (moving from smaller to larger values) is in the euclidean boundary of P between A_i and A_{i+1}.

In the second and third case $A_i \cap A_{i+1} = \emptyset$. A polygon has no vertices if and only if all arcs $A_1, ..., A_m$ are hyperbolic lines, and if and only if all points $z_{1,1}, ..., z_{m,2} \in \widehat{\mathbb{R}}$.

(iv) Denote

$$Gl_2^{(-1)}(\mathbb{R}) := \{A \in GL_2(\mathbb{R}) \mid \det A = -1\},$$
$$Gl_2^{(\pm 1)}(\mathbb{R}) := \{A \in GL_2(\mathbb{R}) \mid \det A = \pm 1\} = SL_2(\mathbb{R}) \cup GL_2^{(-1)}(\mathbb{R}),$$

and analogously $Gl_2^{(-1)}(\mathbb{Z}), Gl_2^{(\pm 1)}(\mathbb{Z})$. Recall

$$A^t \begin{pmatrix} 0 & 1 \\ -1 & 0 \end{pmatrix} A = \det A \cdot \begin{pmatrix} 0 & 1 \\ -1 & 0 \end{pmatrix} \quad \text{for } A \in Gl_2^{(\pm 1)}(\mathbb{R}).$$

(v) The following map $\mu : Gl_2^{(\pm 1)}(\mathbb{R}) \to \text{Isom}(\mathbb{H})$ is a surjective group homomorphism with kernel $\ker \mu = \{\pm E_2\}$,

$$\mu(A) = \left(z \mapsto \frac{za+b}{cz+d}\right) \quad \text{if } A = \begin{pmatrix} a & b \\ c & d \end{pmatrix} \in SL_2(\mathbb{R}),$$
$$\mu(A) = \left(z \mapsto \frac{\overline{z}a+b}{c\overline{z}+d}\right) \quad \text{if } A = \begin{pmatrix} a & b \\ c & d \end{pmatrix} \in GL_2^{(-1)}(\mathbb{R}).$$

$\mu(A)$ for $A \in SL_2(\mathbb{R})$ is orientation preserving and is called a *Möbius transformation*. $\mu(A)$ for $A \in GL_2^{(-1)}(\mathbb{R})$ is orientation reversing.

If $A \in SL_2(\mathbb{R}) - \{\pm E_2\}$, there are three possibilities:

(a) $|\text{tr}(A)| < 2$; then A has a fixed point in $\mathbb{H}$ (and the complex conjugate number is a fixed point in $-\mathbb{H}$) and is called *elliptic*.
(b) $|\text{tr}(A)| = 2$; then A has one fixed point in $\widehat{\mathbb{R}}$ and is called *parabolic*.
(c) $|\text{tr}(A)| > 2$; then A has two fixed points in $\widehat{\mathbb{R}}$ and is called *hyperbolic*.

If $A = \begin{pmatrix} a & b \\ c & d \end{pmatrix} \in GL_2^{(-1)}(\mathbb{R})$ with $\text{tr}(A) = 0$ then $\mu(A)$ is a reflection along the hyperbolic line

$$\{z \in \mathbb{H} \mid z = \mu(A)(z)\} = \{z \in \mathbb{H} \mid 0 = cz\overline{z} - 2a\,\text{Re}(z) - b\}$$
$$\left(= \{z \in \mathbb{H} \mid 0 = (z - \frac{a}{c})(\overline{z} - \frac{a}{c}) - \frac{1}{c^2}\} \quad \text{if } c \neq 0\right).$$

The theorem of Poincaré-Maskit starts with a hyperbolic polygon P whose relative boundary in $\mathbb{H}$ consists of arcs $A_1, ..., A_m$, with an involution $\sigma \in S_m$ and with elements $g_1, ..., g_m \in \text{Isom}(\mathbb{H})$ with $g_i(A_i) = A_{\sigma(i)}$ and $g_{\sigma(i)} = g_i^{-1}$. Under some additional conditions, it states that the group $G := \langle g_1, ..., g_m\rangle \subset \text{Isom}(\mathbb{H})$ is discrete, that P is a fundamental domain (i.e. each orbit of G in $\mathbb{H}$ meets the relative closure of P in $\mathbb{H}$, no orbit of G in $\mathbb{H}$ meets P in more than one point), and it gives a complete set of relations with respect to $g_1, ..., g_m$ of G. Poincaré [Po82] had the case when $\mathbb{H}/G$ is compact, Maskit [Ma71] generalized it greatly. In [Ma71] the relative boundary of P in $\mathbb{H}$ may consist of countably many arcs. The following theorem singles out three special cases, which are sufficient for us. Remark A.3 (ii) illustrates them with pictures.

Theorem A.2 *[Ma71] Let $P \subset \mathbb{H}$ be a hyperbolic polygon whose relative boundary in $\mathbb{H}$ consists of arcs $A_1, ..., A_m$ with $A_i = A(z_{i,1}z_{i,2})$ where one runs through these arcs and these points in the order $A_1, ..., A_m$ and $z_{1,1}, z_{1,2}, z_{2,1}, z_{2,2}, ..., z_{m,1}, z_{m,2}$ if one runs mathematically positive on the euclidean boundary of P in $\mathbb{H} \cup \widehat{\mathbb{R}}$.*

(a) Let $I \subset \{1, ..., m\}$ be the set of indices such that $z_{i,2}$ is a vertex, so $z_{i,2} = z_{i+1,1} \in \mathbb{H}$, with $A_{m+1} := A_1$ and $z_{m+1,i} := z_{1,i}$ (I may be empty). Suppose that at a vertex $z_{i,2}$ the arcs A_i and A_{i+1} meet at an angle $\frac{\pi}{n_i}$ for some number $n_i \in \mathbb{Z}_{\geq 2}$. For $i \in \{1, ..., m\}$ let $g_i \in \text{Isom}(\mathbb{H})$ be the reflection along the hyperbolic line which contains A_i.

The group $G := \langle g_1, ..., g_m\rangle \subset \text{Isom}(\mathbb{H})$ is discrete, P is a fundamental domain, and the set of relations

$$g_1^2 = ... = g_m^2 = \text{id}, \quad (g_i g_{i+1})^{n_i} = \text{id} \quad \text{for } i \in I,$$

form a complete set of relations. Especially, if $I = \emptyset$ then G is a free Coxeter group with generators $g_1, ..., g_m$.

(b) Let $P \subset \mathbb{H}$ have no vertices. Choose on each hyperbolic line A_i a point p_i, and let g_i be the elliptic element with fixed point p_i and rotation angle π.

The group $G := \langle g_1, ..., g_m\rangle \subset \text{Isom}(\mathbb{H})$ is discrete, P is a fundamental domain, and the set of relations $g_1^2 = ... = g_m^2 = \text{id}$ a complete set of relation, so G is a free Coxeter group with generators $g_1, ..., g_m$.

(c) Let $P \subset \mathbb{H}$ have no vertices. Suppose that m is even. Suppose $z_{2i-1,2} = z_{2i,1}$ for $i \in \{1, 2, ..., \frac{m}{2}\}$. Let g_i for $i \in \{1, 2, ..., \frac{m}{2}\}$ be the parabolic element with fixed point $z_{2i-1,2}$ which maps A_{2i-1} to A_{2i}.

The group $G := \langle g_1, ..., g_m\rangle \subset \text{Isom}(\mathbb{H})$ is discrete, P is a fundamental domain, and the group G is a free group with generators $g_1, ..., g_m$.

Remarks A.3 (i) The Cayley transformation $\widehat{\mathbb{C}} \to \widehat{\mathbb{C}}, \left(z \mapsto \frac{z-i}{z+i}\right)$ maps the upper half plane $\mathbb{H}$ to the unit disk $\mathbb{D} = \{z \in \mathbb{C} \mid |z| < 1\}$. It leads to the unit disk model of the hyperbolic plane. The hyperbolic lines in this model are the parts in $\mathbb{D}$ of circles and euclidean lines which intersect $\partial\mathbb{D}$ orthogonally.

(ii) The three pictures in Fig. A.1 illustrate Theorem A.2 in the unit disk model instead of the upper half plane model.

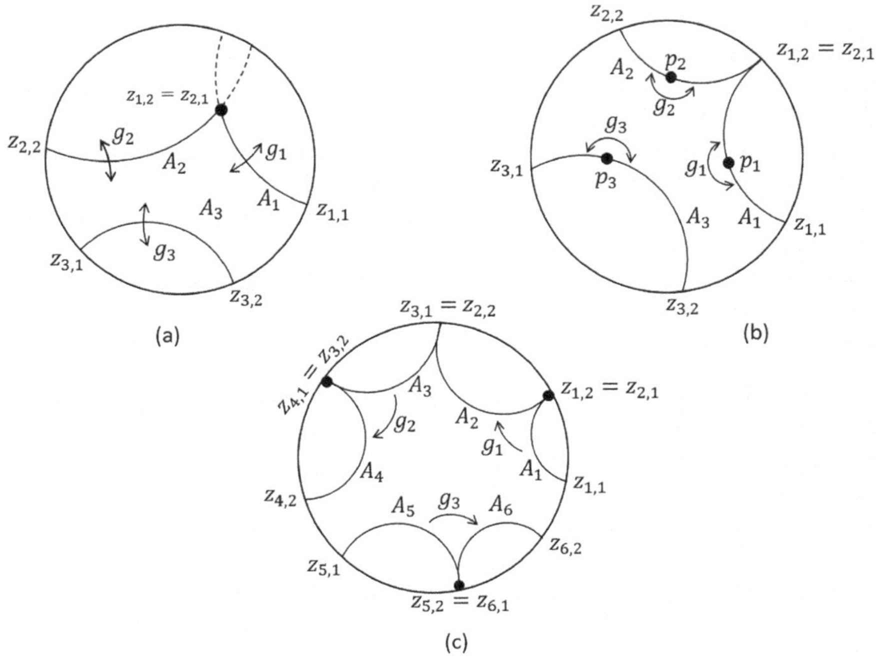

Fig. A.1 Three pictures for Theorem A.2

The surjective group homomorphism $\mu : Gl_2^{(\pm 1)}(\mathbb{R}) \to \text{Isom}(\mathbb{H})$ in Remark A.1 (v) shows how to go from groups of real 2×2 matrices to subgroups of $\text{Isom}(\mathbb{H})$. The next theorem shows how to go from groups of certain real 3×3 matrices to subgroups of $\text{Isom}(\mathbb{H})$. It is classical. But as we need some details, we prefer to explain these details and not refer to some literature.

Theorem A.4 *Let $(H_{\mathbb{R}}, I^{[0]})$ be a 3-dimensional real vector space with a symmetric bilinear form $I^{[0]}$ with signature $(+--)$.*

(a) (Elementary linear algebra) A vector $v \in H_{\mathbb{R}} - \{0\}$ is called positive *if $I^{[0]}(v, v) > 0$,* isotropic *if $I^{[0]}(v, v) = 0$,* negative *if $I^{[0]}(v, v) < 0$. The positive vectors form a (double) cone $\mathcal{K} \subset H_{\mathbb{R}}$, the isotropic vectors and the vector 0 form its boundary, the negative vectors form its complement. The orthogonal hyperplanes $(\mathbb{R} \cdot v)^{\perp}$ satisfy the following:*

(i) $(\mathbb{R} \cdot v)^{\perp} \cap \mathcal{K} \neq \emptyset$ *if v is negative.*
(ii) $(\mathbb{R} \cdot v)^{\perp} \cap \overline{\mathcal{K}} = \mathbb{R} \cdot v$ *if v is isotropic.*
(iii) $(\mathbb{R} \cdot v)^{\perp} \cap \overline{\mathcal{K}} = \{0\}$ *if v is positive.*

$\mathcal{K}/\mathbb{R}^*$ denotes the lines in $\mathcal{K}$, i.e. the 1-dimensional subspaces.

(b) (Basic properties of $\mathrm{Aut}(H_{\mathbb{R}}, I^{[0]})$*) Let* $\sigma : \mathrm{Aut}(H_{\mathbb{R}}, I^{[0]}) \to \{\pm 1\}$ *be the spinor norm map (see Remark 6.3 (iii)). The group* $\mathrm{Aut}(H_{\mathbb{R}}, I^{[0]})$ *is a real 3-dimensional Lie group with four components. The components are the fibers of the group homomorphism*

$$(\det, \sigma) : \mathrm{Aut}(H_{\mathbb{R}}, I^{[0]}) \to \{\pm 1\} \times \{\pm 1\}.$$

$-\mathrm{id} \in \mathrm{Aut}(H_{\mathbb{R}}, I^{[0]})$ *has value* $(\det, \sigma)(-\mathrm{id}) = (-1, 1)$. *If* v *is positive then* $(\det, \sigma)(s_v^{(0)}) = (-1, 1)$. *If* v *is negative then* $(\det, \sigma)(s_v^{(0)}) = (-1, -1)$. *An isometry* $g \in \mathrm{Aut}(H_{\mathbb{R}}, I^{[0]})$ *maps each of the two components of the cone* $\mathcal{K}$ *to itself if and only if* g *is in the two components which together form the kernel of* $\det \cdot \sigma$, *so if* $\det(g)\sigma(g) = 1$.

(c) Choose a basis $\underline{f} = (f_1, f_2, f_3)$ *of* $H_{\mathbb{R}}$ *with*

$$I^{[0]}(\underline{f}^t, \underline{f}) = \begin{pmatrix} 0 & 0 & 1 \\ 0 & -2 & 0 \\ 1 & 0 & 0 \end{pmatrix}.$$

(i) The map

$$\Theta : Gl_2^{(\pm 1)}(\mathbb{R}) \to \mathrm{Aut}(H_{\mathbb{R}}), \quad \begin{pmatrix} a & b \\ c & d \end{pmatrix} \mapsto \left(\underline{f} \mapsto \underline{f} \begin{pmatrix} a^2 & 2ab & b^2 \\ ac & ad + bc & bd \\ c^2 & 2cd & d^2 \end{pmatrix}\right),$$

is a group homomorphism with kernel $\ker \Theta = \{\pm E_2\}$ *and image* $\ker(\det \cdot \sigma : \mathrm{Aut}(H_{\mathbb{R}}, I^{[0]}) \to \{\pm 1\})$.

(ii) The map

$$\vartheta : \mathbb{H} \to (H_{\mathbb{R}} - \{0\})/\mathbb{R}^*, \quad z \mapsto \mathbb{R}^*(z\overline{z} f_1 + \mathrm{Re}(z) f_2 + f_3),$$

is a bijection $\vartheta : \mathbb{H} \to \mathcal{K}/\mathbb{R}^*$.

(iii) For $A \in Gl_2^{(\pm 1)}(\mathbb{R})$, *the automorphism* $\vartheta \circ \mu(A) \circ \vartheta^{-1} : \mathcal{K}/\mathbb{R}^* \to \mathcal{K}/\mathbb{R}^*$ *coincides with the action of* $\Theta(A)$ *on* $\mathcal{K}/\mathbb{R}^*$.

(iv) The natural maps

$$\begin{array}{ccccc} \mathrm{Aut}(H_{\mathbb{R}}, I^{[0]})/\{\pm \mathrm{id}\} & \longleftarrow & \ker(\det \cdot \sigma) & \longrightarrow & \mathrm{Isom}(\mathbb{H}) \\ \{\pm B\} & \longleftarrow & B \;,\; \Theta(A) & \longmapsto & \mu(A) \end{array}$$

are group isomorphisms.

(v) For each hyperbolic line l *there is a negative vector* $v \in H_{\mathbb{R}}$ *with* $\vartheta(l) = ((\mathbb{R} \cdot v)^{\perp} \cap \mathcal{K})/\mathbb{R}^*$.

(vi) Let $\sigma_l \in \mathrm{Isom}(\mathbb{H})$ *be the reflection along a hyperbolic line* l, *and let* v *be as in (v). The action of* $s_v^{(0)}$ *on* $\mathcal{K}/\mathbb{R}^*$ *coincides with* $\vartheta \circ \sigma_l \circ \vartheta^{-1}$.

(vii) Let $\delta_p \in \mathrm{Isom}(\mathbb{H})$ be the elliptic element with fixed point $p \in \mathbb{H}$ and order 2 (so rotation angle π). Let $v \in H_{\mathbb{R}}$ be a positive vector with $\mathbb{R} \cdot v = \vartheta(p)$. The action of $s_v^{(0)}$ on $\mathcal{K}/\mathbb{R}^*$ coincides with $\vartheta \circ \delta_p \circ \vartheta^{-1}$.

Proof (a) and (b) are elementary and classical, their proofs are skipped.

(c) (i) Start with a real 2-dimensional vector space $V_{\mathbb{R}}$ with basis $\underline{e} = (e_1, e_2)$ and a skew-symmetric bilinear form $I^{[1]}$ on $V_{\mathbb{R}}$ with matrix $I^{[1]}(\underline{e}^t, \underline{e}) = \begin{pmatrix} 0 & 1 \\ -1 & 0 \end{pmatrix}$.

The tensor product $V_{\mathbb{R}} \otimes V_{\mathbb{R}}$ comes equipped with an induced symmetric bilinear form $\widetilde{I}^{(0)}$ via (here I and J are finite index sets and $a_i, b_i, c_j, d_j \in V_{\mathbb{R}}$)

$$\widetilde{I}^{(0)}\left(\sum_{i\in I} a_i \otimes b_i, \sum_{j\in J} c_j \otimes d_j\right) = \sum_{i\in I}\sum_{j\in J} I^{[1]}(a_i, c_j) I^{[1]}(b_i, d_j).$$

An element $g \in \mathrm{Aut}(H_{\mathbb{R}})$ with $g\underline{e} = \underline{e}A$ and $A \in Gl_2^{(\pm 1)}(\mathbb{R})$ respects $I^{[1]}$ in the following weak sense:

$$I^{[1]}(g(v_1), g(v_2)) = \det A \cdot I^{[1]}(v_1, v_2),$$
$$\text{because } A^t \begin{pmatrix} 0 & 1 \\ -1 & 0 \end{pmatrix} A = \det A \cdot \begin{pmatrix} 0 & 1 \\ -1 & 0 \end{pmatrix}.$$

It induces an element $\widetilde{\Theta}(g) \in \mathrm{Aut}(V_{\mathbb{R}} \otimes V_{\mathbb{R}}, \widetilde{I}^{(0)})$ via

$$\widetilde{\Theta}(g)(\sum_{i\in I} a_i \otimes b_i) = \sum_{i\in I} g(a_i) \otimes g(b_i).$$

The symmetric part $\widetilde{H}_{\mathbb{R}} \subset V_{\mathbb{R}} \otimes V_{\mathbb{R}}$ of the tensor product has the basis $\underline{\widetilde{f}} = (\widetilde{f}_1, \widetilde{f}_2, \widetilde{f}_3) = (e_1 \otimes e_1, e_1 \otimes e_2 + e_2 \otimes e_1, e_2 \otimes e_2)$. One sees

$$\widetilde{I}^{(0)}(\underline{\widetilde{f}}^t, \underline{\widetilde{f}}) = \begin{pmatrix} 0 & 0 & 1 \\ 0 & -2 & 0 \\ 1 & 0 & 0 \end{pmatrix}.$$

From now on we identify $(\widetilde{H}_{\mathbb{R}}, \widetilde{I}^{(0)}|_{\widetilde{H}_{\mathbb{R}}}, \underline{\widetilde{f}})$ with $(H_{\mathbb{R}}, I^{[0]}, \underline{f})$.

For an element $g \in \mathrm{Aut}(V_{\mathbb{R}})$ with $g\underline{e} = \underline{e}A$ with $A = \begin{pmatrix} a & b \\ c & d \end{pmatrix} \in Gl_2^{(\pm 1)}(\mathbb{R})$, the automorphism $\widetilde{\Theta}(g)$ on $V_{\mathbb{R}} \otimes V_{\mathbb{R}}$ restricts to an automorphism of the symmetric part $H_{\mathbb{R}}$ with matrix

$$\widetilde{\Theta}(g)\underline{f} = \underline{f} \begin{pmatrix} a^2 & 2ab & b^2 \\ ac & ad + bc & bd \\ c^2 & 2cd & d^2 \end{pmatrix}.$$

This fits to Θ. It shows especially $\Theta(A) \in \mathrm{Aut}(H_{\mathbb{R}}, I^{[0]})$.

The kernel of Θ is $\{\pm E_2\}$. The Lie groups $Gl_2^{(\pm 1)}(\mathbb{R})$ and $\mathrm{Aut}(H_{\mathbb{R}}, I^{[0]})$ are real 3-dimensional. The Lie group $Gl_2^{(\pm 1)}(\mathbb{R})$ has two components. $(\det, \sigma)(\Theta(\begin{pmatrix} 1 & 0 \\ 0 & -1 \end{pmatrix})) = (\det, \sigma)(s_{f_2}^{(0)}) = (-1, -1)$. Therefore the image of Θ consists of the two components of $\mathrm{Aut}(H_{\mathbb{R}}, I^{[0]})$ which together form the kernel of $\det \cdot \sigma$. This finishes the proof of part (i).

(ii) Define $\widetilde{\vartheta}(z) := z\overline{z} f_1 + \mathrm{Re}(z) f_2 + f_3$. It is a positive vector because

$$I^{[0]}(\widetilde{\vartheta}(z), \widetilde{\vartheta}(z)) = z\overline{z} \cdot 1 - 2(\mathrm{Re}(z))^2 + 1 \cdot z\overline{z} = 2(\Im(z))^2 > 0.$$

It is easy to see that ϑ is a bijection from $\mathbb{H}$ to $\mathcal{K}/\mathbb{R}^*$.

(iii) In fact, $\widetilde{\vartheta}(z)$ is the symmetric part of

$$(ze_1 + e_2) \otimes (\overline{z}e_1 + e_2) = (\underline{e} \begin{pmatrix} z \\ 1 \end{pmatrix}) \otimes (\underline{e} \begin{pmatrix} \overline{z} \\ 1 \end{pmatrix}) \in V_{\mathbb{C}} \otimes V_{\mathbb{C}}.$$

For $A = \begin{pmatrix} a & b \\ c & d \end{pmatrix} \in Gl_2^{(\pm 1)}(\mathbb{R})$

$$\begin{aligned}
&\widetilde{\vartheta}(\mu(A)(z)) \\
&= \Big(\text{symmetric part of } (\underline{e} \begin{pmatrix} \mu(A)(z) \\ 1 \end{pmatrix}) \otimes (\underline{e} \begin{pmatrix} \mu(A)(\overline{z}) \\ 1 \end{pmatrix}\Big) \\
&= |cz+d|^{-2} \Big(\text{symmetric part of } (\underline{e} A \begin{pmatrix} z \\ 1 \end{pmatrix}) \otimes (\underline{e} A \begin{pmatrix} \overline{z} \\ 1 \end{pmatrix})\Big) \\
&= |cz+d|^{-2} \underline{f} \begin{pmatrix} a^2 & 2ab & b^2 \\ ac & ad+bc & bd \\ c^2 & 2cd & d^2 \end{pmatrix} \begin{pmatrix} z\overline{z} \\ \mathrm{Re}(z) \\ 1 \end{pmatrix} \\
&= |cz+d|^{-2} \Theta(A)(\underline{f}) \begin{pmatrix} z\overline{z} \\ \mathrm{Re}(z) \\ 1 \end{pmatrix} \\
&= |cz+d|^{-2} \Theta(A)(\widetilde{\vartheta}(z)).
\end{aligned}$$

This shows part (iii). Part (iv) follows from part (iii).

(v) A hyperbolic line l is the fixed point set of a reflection $\mu(A)$ for a matrix $A = \begin{pmatrix} a & b \\ c & -a \end{pmatrix} \in Gl_2^{(-1)}(\mathbb{R})$, so

$$l = \{z \in \mathbb{H} \mid z = \mu(A)(z)\} = \{z \in \mathcal{H} \mid 0 = cz\overline{z} - 2a\,\mathrm{Re}(z) - b\}.$$

Fig. A.2 Two disk models of the hyperbolic plane

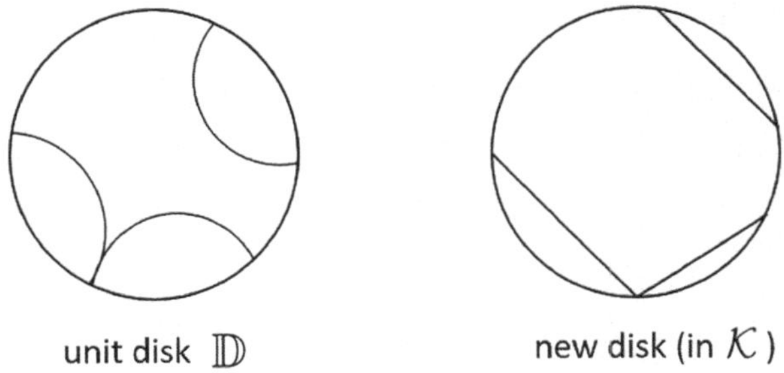

Observe

$$cz\overline{z} - 2a\,\text{Re}(z) - b = I^{[0]}\left(\underline{f}\begin{pmatrix} -b \\ a \\ c \end{pmatrix}, \underline{f}\begin{pmatrix} z\overline{z} \\ \text{Re}(z) \\ 1 \end{pmatrix}\right).$$

Therefore

$$\vartheta(l) = \Big(\mathbb{R}(-bf_1 + af_2 + cf_3)^{\perp} \cap \mathcal{K}\Big)/\mathbb{R}^*.$$

(vi) and (vii) are clear now. □

Remarks A.5 (i) In the model $\mathcal{K}/\mathbb{R}^*$ of the hyperbolic plane, the isometries of $\mathbb{H}$ are transformed to linear isometries of $(H_{\mathbb{R}}, I^{[0]})$. The hyperbolic lines in $\mathbb{H}$ are transformed to linear hyperplanes in $H_{\mathbb{R}}$ (modulo $\mathbb{R}^*$) which intersect $\mathcal{K}$.

(ii) If one chooses an affine hyperplane in $H_{\mathbb{R}}$ which intersects one component of $\mathcal{K}$ in a disk, this disk gives a new disk model of the hyperbolic plane, which is not conformal to $\mathbb{H}$ and $\mathbb{D}$ (angles are not preserved), but where the hyperbolic lines in $\mathbb{H}$ are transformed to the segments in the new disk of euclidean lines in the affine hyperplane which intersect the disk. Figure A.2 sketches three hyperbolic lines in the unit disk $\mathbb{D}$ and in the new disk which is part of the cone $\mathcal{K}$.

Appendix B
The First Congruence Subgroups

For the 3-dimensional covering spaces in Sect. 9.2, we need some classical facts on the quotient map $\mathbb{H} \to \mathbb{H}/\Gamma(n)$ for $n \in \{2, 3\}$. In the case $n = 2$ we need additionally a lift of the multivalued logarithm on $\mathbb{C} - \{0, 1\} \cong \mathbb{H}/\Gamma(2)$ to a univalued holomorphic function on $\mathbb{H}$.

For $n \in \mathbb{Z}_{\geq 2}$ the congruence subgroup $\Gamma(n)$ is the group

$$\Gamma(n) := \{A \in SL_2(\mathbb{Z}) \mid A \equiv E_2 \bmod n\}.$$

For $n = 2$ we have $-E_2 \in \Gamma(2)$, and $\Gamma(2)/\{\pm E_2\}$ embeds into $PSL_2(\mathbb{Z})$. For $n \geq 3$ we have $-E_2 \notin \Gamma(n)$, and $\Gamma(n)$ embeds into $PSL_2(\mathbb{Z})$. In any case we write $\overline{\Gamma(n)} \subset PSL_2(\mathbb{Z})$ for the image in $PSL_2(\mathbb{Z})$. The following theorem collects classical facts (see e.g. [La09, 5.4 and 5.7.3]).

Theorem B.1 *(a) Fix $n \in \mathbb{Z}_{\geq 2}$. The group $\overline{\Gamma(n)}$ is a normal subgroup of finite index in $PSL_2(\mathbb{Z})$. It does not contain elliptic elements. It acts properly discontinuously and freely on $\mathbb{H}$. The quotient $\mathbb{H}/\Gamma(n)$ is the complement of finitely many points in a smooth compact complex curve. The finitely many missing points correspond to the $\overline{\Gamma(n)}$ orbits in $\mathbb{Q} \cup \{\infty\}$, which is the set of fixed points of the parabolic elements in $\overline{\Gamma(n)}$. The number of the missing points is $[PSL_2(\mathbb{Z}) : \overline{\Gamma(n)}] \cdot n^{-1}$.*

(b) Fix $n \in \{2, 3, 4, 5\}$. Then $\mathbb{H}/\Gamma \cong \mathbb{P}^1\mathbb{C} - \{$finitely many points$\}$.

n	$PSL_2(\mathbb{Z})/\overline{\Gamma(n)}$	$[PSL_2(\mathbb{Z}) : \overline{\Gamma(n)}]$	$\|\{$*missing points*$\}\|$
2	$\cong S_3$	6	3
3	$\cong A_4$	12	4
4	$\cong S_4$	24	6
5	$\cong A_5$	60	12

There are a natural homeomorphism $h : \mathbb{P}^1\mathbb{C} \to S^2 \subset \mathbb{R}^3$ and embeddings of the vertices of a tetrahedron, an octahedron and an icosahedron into S^2 such that the image $h(\{$missing points$\})$ is the set of vertices of the tetrahedron in the case $n = 3$,

C. Hertling and K. Larabi, *Triangular Bases of Unimodular Bilinear Lattices and Induced Structures*, Lecture Notes in Mathematics 2398,
https://doi.org/10.1007/978-3-032-25997-4

the set of vertices of the octahedron in the case $n = 4$ and the set of vertices of the icosahedron in the case $n = 5$. The group $PSL_2(\mathbb{Z})/\overline{\Gamma(n)}$ for $n \in \{3, 4, 5\}$ acts on S^2 as group of rotations which fix the corresponding platonic solid. For $n = 2$ $\mathbb{H}/\overline{\Gamma(2)} \cong \mathbb{C} - \{0, 1\}$ (for the group $PSL_2(\mathbb{Z})/\overline{\Gamma(2)}$ see part (d)).

For $n \in \{2, 3, 4, 5\}$ the composition

$$T : \mathbb{H} \to \mathbb{H}/\overline{\Gamma(n)} \xrightarrow{\cong} \mathbb{P}^1\mathbb{C} - \{\textit{missing points}\} \hookrightarrow \mathbb{C}$$

is a Schwarzian triangle function. It maps the degenerate hyperbolic triangle with vertices $0, 1, \infty$ and angles $0, 0, 0$, to a spherical triangle with angles $\frac{2\pi}{n}, \frac{2\pi}{n}, \frac{2\pi}{n}$. It maps the degenerate hyperbolic triangle with vertices $i, e^{2\pi i/6}, \infty$ with angles $\frac{\pi}{2}, \frac{\pi}{3}, 0$ to a spherical triangle with angles $\frac{\pi}{2}, \frac{\pi}{3}, \frac{\pi}{n}$.

(c) In the case $n = 3$, there is an isomorphism $\mathbb{P}^1\mathbb{C} \cong \mathbb{C} \cup \{\infty\}$ such that the 24 spherical triangles with angles $\frac{\pi}{2}, \frac{\pi}{3}, \frac{\pi}{3}$ in $\mathbb{P}^1\mathbb{C}$ have the following images in $\mathbb{C} - \{0, 1\} \subset \mathbb{C} \cup \{\infty\}$. Figure B.1 shows a subdivision of each of the two visible faces of a tetrahedron into 6 triangles. Projecting the tetrahedron to S^2 gives 24 spherical triangles. Their boundaries consist of altogether six great circles in S^2. Figure B.2 shows the images of the six great circles in $\mathbb{C}$. Three are circles, three are lines (where the point ∞ is missing). Figure B.3 restricts to showing the preimages of the edges of the tetrahedron. The preimages in $\mathbb{C} \cup \{\infty\}$ of the vertices of the tetrahedron are the points $\{\infty, 1, e^{2\pi i/3}, e^{-2\pi i/3}\}$. Therefore

$$\mathbb{H}/\overline{\Gamma(3)} \cong \mathbb{C} - \{z \mid z^3 = 1\}.$$

(d) The case $n = 2$. The group of Möbius transformations of $\mathbb{C} \cup \{\infty\}$ which fix the set $\{0, 1, \infty\}$ is called anharmonic group G^{anh}. *The first four entries of the following table fix an isomorphism $g_\bullet : S_3 \to G^{anh}$ with $g_{\sigma_1} g_{\sigma_2} = g_{\sigma_1 \sigma_2}$.*

σ	g_σ	$g_\sigma(z)$	$(g_\sigma(1), g_\sigma(0), g_\sigma(\infty))$	μ_σ	$\mu_\sigma(\tau)$	μ_σ	*fp*
id	g_{id}	z	$(1, 0, \infty)$	μ_{id}	τ	*identity*	$\mathbb{H}$
$(1\,2)$	$g_{(12)}$	$1 - z$	$(0, 1, \infty)$	$\mu_{(12)}$	$\tau - 1$	*parabolic*	∞
$(1\,3)$	$g_{(13)}$	$\frac{z}{z-1}$	$(\infty, 0, 1)$	$\mu_{(13)}$	$\frac{\tau}{\tau+1}$	*parabolic*	0
$(2\,3)$	$g_{(23)}$	$\frac{1}{z}$	$(1, \infty, 0)$	$\mu_{(23)}$	$\frac{-1}{\tau}$	*rotation*	i
$(1\,2\,3)$	$g_{(123)}$	$\frac{z-1}{z}$	$(0, \infty, 1)$	$\mu_{(123)}$	$\frac{\tau-1}{\tau}$	*rotation*	$e^{2\pi i/6}$
$(1\,3\,2)$	$g_{(132)}$	$\frac{1}{1-z}$	$(\infty, 1, 0)$	$\mu_{(132)}$	$\frac{-1}{\tau-1}$	*rotation*	$e^{2\pi i/6}$

The Schwarzian triangle function $T : \mathbb{H} \to \mathbb{H}/\overline{\Gamma(2)} \cong \mathbb{P}^1\mathbb{C} - \{\textit{missing points}\}$ can here be chosen as

$$T = g_{(12)} \circ (\textit{the classical } \lambda\textit{-function}) : \mathbb{H} \to \mathbb{C} - \{0, 1\}$$

(also the λ-function itself works, but we prefer T as it maps $0, 1, \infty$ to $0, 1, \infty$). The following table gives the values under T of some points or sets.

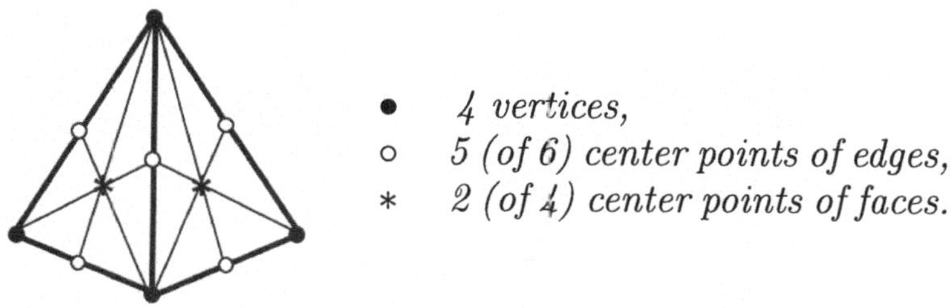

Fig. B.1 A tetrahedron and a subdivision of each face into six triangles, only two faces shown

point	T(*point*)	*set*	T(*set*)
0	0	$i\mathbb{R}_{>0} = A(0,\infty)$	$\mathbb{R}_{<0}$
1	1	$1 + i\mathbb{R}_{>0} = A(\infty, 1)$	$\mathbb{R}_{>1}$
∞	∞	$A(1,0)$	$(0, 1)$
i	-1	$\frac{1}{2} + i\mathbb{R}_{>0} = A(\frac{1}{2}, \infty)$	$\frac{1}{2} + i\mathbb{R}$
$\frac{1}{2}(1+i)$	$\frac{1}{2}$	$A(1,-1)$	$S^1 - \{1\}$
$e^{2\pi i/6}$	$e^{2\pi i/6}$	$A(2,0)$	$1 + (S^1 - \{-1\})$

The hyperbolic polygon with boundary $A(1,0) \cup A(0,-1) \cup A(-1,\infty) \cup A(\infty,1)$ *is a fundamental domain for the action of* $\overline{\Gamma(2)}$ *on* $\mathbb{H}$. *It consists of 12 degenerate hyperbolic triangles with angles* $\frac{\pi}{2}, \frac{\pi}{3}, 0$. *They are shown in the left picture in Fig. B.4. The right picture in Fig. B.4 shows their images in* $\mathbb{C} - \{0, 1\}$. *The images are the connected components of the complement of the two circles and the lines* $\mathbb{R}$ *and* $i\mathbb{R}$.

The six Möbius transformations in the fifth and sixth column of the table above are representatives of the classes in the quotient group $\mu(SL_2(\mathbb{Z}))/\mu(\Gamma(2)) \cong S_3$, *with*

$$T(\mu(A)\mu_\sigma(\tau)) = T(\mu_\sigma\mu(A)(\tau)) = g_\sigma(T(\tau)) \quad \textit{for } \sigma \in S_3,\ A \in \Gamma(2).$$

*The seventh and eighth column give properties of them (*fixed point *is abbreviated* fp*).*

Proof See e.g. [La09, 5.4 and 5.7.3]. □

Definition B.2 The logarithm $\ln : \mathbb{C} - \{0\} \dashrightarrow \mathbb{C}$ is multivalued. As $T : \mathbb{H} \to \mathbb{C} - \{0, 1\}$ is a universal covering of $\mathbb{C} - \{0, 1\}$, the logarithm has univalued lifts $\mathbb{H} \to \mathbb{C}$. We denote by $\kappa : \mathbb{H} \to \mathbb{C}$ the lift with values in $\mathbb{R}_{>0}$ on $1 + i\mathbb{R}_{>0}$ (this set maps under T to $\mathbb{R}_{>1}$).

The geometry of this lift is described in part (c) of Theorem B.3. The parts (a) and (b) help to make part (c) more transparent. They contain basic well known facts.

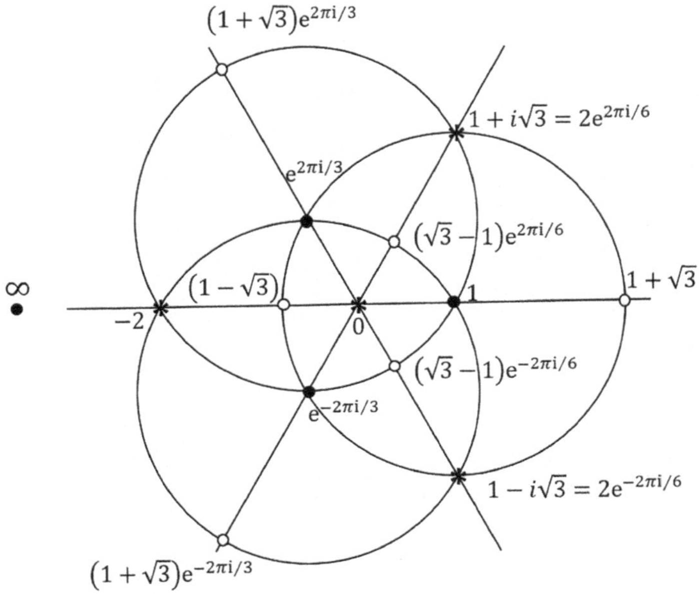

Fig. B.2 Images in $\mathbb{C}\cup\{\infty\}$ of 24 spherical triangles

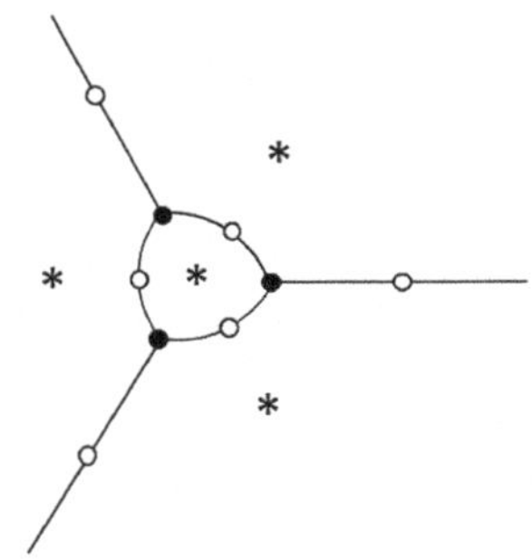

Fig. B.3 The edges of the tetrahedron in Fig. B.2

Theorem B.3 *(a) For each cusp $c\in\mathbb{Q}\cup\{\infty\}$ of $PSL_2(\mathbb{Z})$, the stabilizer $(\mu(SL_2(\mathbb{Z})))_c$ of c in $\mu(SL_2(\mathbb{Z}))$ is infinite cyclic with a parabolic generator. For example, for $c\in\{\infty,0,1\}$ generators are as follows,*

c	*generator of* $(\mu(SL_2(\mathbb{Z}))_c$
∞	$\mu_{(12)}=\mu(\begin{pmatrix}1&-1\\0&1\end{pmatrix})$
0	$\mu_{(13)}=\mu(\begin{pmatrix}1&0\\1&1\end{pmatrix})$
1	$\mu(\begin{pmatrix}2&-1\\1&0\end{pmatrix})$

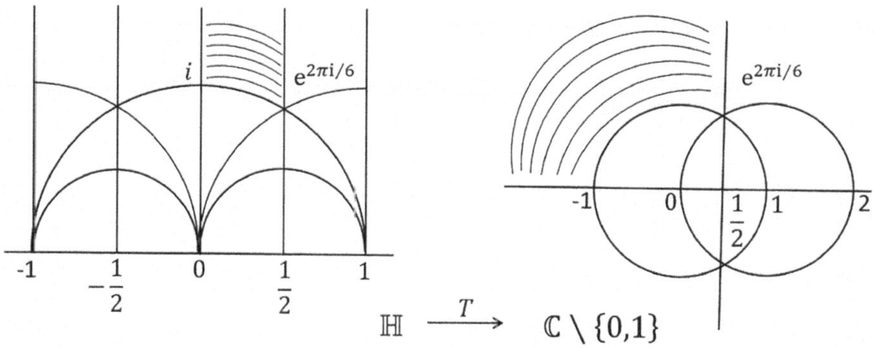

Fig. B.4 *T* maps 12 hyperbolic triangles to 12 spherical triangles

The cusps form a single $\mu(SL_2(\mathbb{Z}))$ *orbit. If* $c = \mu(B)(\infty)$ *then* $(\mu(SL_2(\mathbb{Z})))_c = \langle \mu(B)\mu_{(12)}\mu(B)^{-1} \rangle$.

(b) For each cusp $c \in \mathbb{Q} \cup \{\infty\}$ *of* $\overline{\Gamma(2)}$, *the stabilizer in* $\mu(\Gamma(2))$ *is infinite cyclic with a parabolic generator. The generator is the square of a generator of* $(\mu(SL_2(\mathbb{Z})))_c$. *For example, for* $c \in \{\infty, 0, 1\}$ *generators are as follows,*

c	*generator of* $(\mu(\Gamma(2))_c$
∞	$\mu^2_{(12)} = \mu(\begin{pmatrix} 1 & -2 \\ 0 & 1 \end{pmatrix})$
0	$\mu^2_{(13)} = \mu(\begin{pmatrix} 1 & 0 \\ 2 & 1 \end{pmatrix})$
1	$\mu(\begin{pmatrix} 3 & -2 \\ 2 & -1 \end{pmatrix}) = \mu^{-2}_{(12)}\mu^{-2}_{(13)}$

The cusps form three $\Gamma(2)$ *orbits, the orbits of* 0, 1 *and* ∞. *If* $c \in \mathbb{Q} \cup \{\infty\}$ *is in the orbit of* $c_0 \in \{0, 1, \infty\}$ *with* $c = \mu(B)(c_0)$ *for some* $B \in \Gamma(2)$ *then* $(\mu(\Gamma(2)))_c = \mu(B)(\mu(\Gamma(2)))_{c_0}\mu(B)^{-1}$.

(c) The lift $\kappa : \mathbb{H} \to \mathbb{C}$ *of the logarithm* $\ln : \mathbb{C} - \{0\} \dashrightarrow \mathbb{C}$ *satisfies the following.*

(i) The very fact that it is a lift with T says

$$T(\tau) = \exp(\kappa(\tau)) \quad \text{for } \tau \in \mathbb{H}.$$

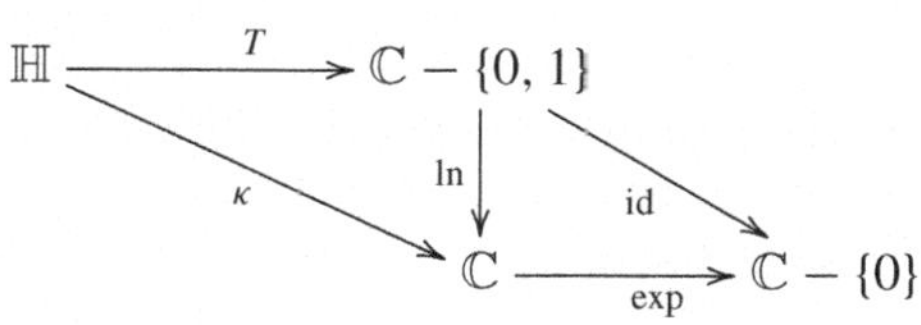

The following table shows the values under κ of some sets.

set	$1+i\mathbb{R}_{>0}=A(\infty,1)$	$A(1,0)$	$i\mathbb{R}_{>0}=A(0,\infty)$
$\kappa(set)$	$\mathbb{R}_{>0}$	$\mathbb{R}_{<0}$	$\pi i+\mathbb{R}$

(ii) Let ν be the (because of $\mu(\Gamma(2))=\langle\mu_{(12)}^2,\mu_{(13)}^2\rangle\cong G^{free,2}$ well defined) group homomorphism

$$\nu:\mu(\Gamma(2))\to(2\pi i\mathbb{Z},+)\quad\textit{with } \nu(\mu_{(12)}^2)=2\pi i,\ \nu(\mu_{(13)}^2)=-2\pi i.$$

Then

$$\kappa(\mu(B)(\tau))=\kappa(\tau)+\nu(B)\quad\textit{for } B\in\Gamma(2),\tau\in\mathbb{H}.$$

Especially

$$\kappa(\tau-2)=\kappa(\tau)+2\pi i,$$
$$\kappa\left(\frac{\tau}{2\tau+1}\right)=\kappa(\tau)-2\pi i,$$
$$\kappa\left(\frac{3\tau-2}{2\tau-1}\right)=\kappa(\tau).$$

For $B\in\Gamma(2)$ and $m\in\mathbb{Z}$

$$\kappa(\mu(B)(\tau))=\kappa(\tau)+2\pi im\quad\textit{if } \mu(B) \textit{ is conjugate in } \Gamma(2) \textit{ to } \mu_{(12)}^{2m},$$
$$\kappa(\mu(B)(\tau))=\kappa(\tau)-2\pi im\quad\textit{if } \mu(B) \textit{ is conjugate in } \Gamma(2) \textit{ to } \mu_{(13)}^{2m},$$
$$\kappa(\mu(B)(\tau))=\kappa(\tau)\quad\textit{if } \mu(B) \textit{ is conjugate in } \Gamma(2) \textit{ to } \mu\left(\begin{pmatrix}3&-2\\2&-1\end{pmatrix}\right)^m.$$

(iii) Consider a cusp $c=\mu(B)(1)$ in the $\mu(\Gamma(2))$ orbit of 1, so $B\in\Gamma(2)$. If τ moves along any geodesic in $\mathbb{H}$ to c, then $\kappa(\tau)\to\nu(B)$.

(iv) Finally for $\tau\in\mathbb{H}$

$$\kappa\left(\frac{2\tau-1}{\tau}\right)=-\kappa(\tau),$$
$$0=\pi i-\kappa(\tau-1)+\kappa(\tau)+\kappa\left(\frac{\tau-1}{\tau}\right).$$

Proof (a) and (b) are well known.

(c) (i) $T(\tau)=\exp(\kappa(\tau))$ holds by definition of a lift of the logarithm with T. The table on the values of κ follows from the table in Theorem B.1 on the values of T and from properties of the logarithm.

(ii) Because of $\mu(\Gamma(2))=\langle\mu_{(12)}^2,\mu_{(13)}^2\rangle\cong G^{free,2}$, it is sufficient to show

$$\kappa(\tau-2)=\kappa(\tau)+2\pi i \quad \text{and} \quad \kappa\left(\frac{\tau}{2\tau+1}\right)=\kappa(\tau)-2\pi i.$$

First consider a path from $z_0 \in \mathbb{R}_{>1} \subset \mathbb{C}-\{0,1\}$ to z_0 which goes once counterclockwise around 0 along a circle with center 0. The logarithm takes up the summand $2\pi i$. The lift to a path in $\mathbb{H}$ via $T:\mathbb{H}\to\mathbb{C}-\{0,1\}$, which starts at the preimage $\tau_0 \in 1+i\mathbb{R}_{>0}$ of z_0, ends at τ_0-2. Therefore $\kappa(\tau-2)=\kappa(\tau)+2\pi i$.

Now consider a path from $z_1 \in (0,1) \subset \mathbb{C}-\{0,1\}$ to z_1 which goes once clockwise around 0 along a circle with center 0 The logarithm takes up the summand $-2\pi i$. The lift to a path in $\mathbb{H}$ via $T:\mathbb{H}\to\mathbb{C}-\{0,1\}$, which starts at the preimage $\tau_1 \in A(-1,0)$ of z_1, ends at $\frac{\tau_1}{2\tau_1+1} \in A(1,0)$. Therefore $\kappa(\frac{\tau}{2\tau+1})=\kappa(\tau)-2\pi i$.

(iii) The logarithm is univalued near 1 with value $\ln(1)=0$. Therefore if $\tau\to 1$ along any geodesic in $\mathbb{H}$ then $\kappa(\tau)\to 0$. Together with $\kappa(\mu(B))(\tau)=\kappa(\tau)+\nu(B)$ for $B\in\Gamma(2)$, this shows

$$\kappa(t)\to\nu(B) \quad \text{if} \quad \tau\to\mu(B)(1).$$

(iv) Observe $\mu(\begin{pmatrix}2 & -1\\ 1 & 0\end{pmatrix}) = \mu_{(12)}^{-1}\mu_{(123)}$ and

$$\exp\left(\kappa\left(\frac{2\tau-1}{\tau}\right)+\kappa(\tau)\right)=T\left(\mu_{(12)}^{-1}\mu_{(123)}(\tau)\right)T(\tau)$$
$$=g_{(12)}^{-1}(g_{(123)}(T(\tau)))T(\tau)=g_{(23)}(T(\tau))T(\tau)=1,$$

so $\kappa(\frac{2\tau-1}{\tau})+\kappa(\tau)=2\pi i m$ for some $m\in\mathbb{Z}$. If $\tau\to 1$ along a geodesic in $\mathbb{H}$ then $\kappa(\frac{2\tau-1}{\tau})+\kappa(\tau)\to 0+0=0$, so $m=0$.

Similarly

$$\exp\left(\pi i-\kappa(\tau-1)+\kappa(\tau)+\kappa\left(\frac{\tau-1}{\tau}\right)\right)$$
$$=(-1)T(\mu_{(12)}(\tau))^{-1}T(\tau)T(\mu_{(13)}\mu_{(12)}(\tau))$$
$$=(-1)g_{(12)}(T(\tau))^{-1}T(\tau)g_{(13)}g_{(12)}(T(\tau))$$
$$=(-1)\frac{1}{1-T(\tau)}T(\tau)\frac{T(\tau)-1}{T(\tau)}=1,$$

so $\pi i-\kappa(\tau-1)+\kappa(\tau)+\kappa(\frac{\tau-1}{\tau})=2\pi i m$ for some $m\in\mathbb{Z}$. If $\tau\in 1+i\mathbb{R}_{>0}=A(\infty,1)$, then

$$\begin{aligned} \tau - 1 \in i\mathbb{R}_{>0}, \quad \text{so } \kappa(\tau - 1) &\in \pi i + \mathbb{R}, \\ \kappa(\tau) &\in \mathbb{R}_{>0}, \\ \frac{\tau - 1}{\tau} \in A(1, 0), \quad \text{so } \kappa(\frac{\tau - 1}{\tau}) &\in \mathbb{R}_{<0}, \\ \text{so } \pi i - \kappa(\tau - 1) + \kappa(\tau) + \kappa(\frac{\tau - 1}{\tau}) &\in \mathbb{R}, \end{aligned}$$

so $m = 0$. □

Appendix C
Quadratic Units Via Continued Fractions

The purpose of this appendix is to prove Lemma C.1 with two statements on the units in certain rings of algebraic integers. The proof of Lemma C.1 will be given after the proof of Theorem C.6.

A convenient and very classical tool to prove this lemma is the theory of continued fractions as best approximations of irrational numbers, applied to the case of quadratic irrationals which are algebraic integers.

Theorem C.4 below cites standard results on the continued fractions of real irrationals. It is prepared by the Definitions C.2 and C.3

Lemma C.5 provides the less well known formulas (C.2) and (C.3) for the case of a quadratic irrational. Theorem C.6 describes the unit group $\mathbb{Z}[\alpha]^*$ where α is a quadratic irrational and an algebraic integer, in terms of the continued fractions of α.

Lemma C.1 *(a) Let $x \in \mathbb{Z}_{\geq 3}$, and let $\kappa_{1/2} := \frac{x}{2} \pm \frac{1}{2}\sqrt{x^2-4}$ be the zeros of the polynomial $t^2 - xt + 1$, so $\kappa_1 + \kappa_2 = x$, $\kappa_1\kappa_2 = 1$, $\kappa_1^2 = x\kappa_1 - 1$. Then*

$$\mathbb{Z}[\kappa_1]^* = \begin{cases} \{\pm\kappa_1^l \mid l \in \mathbb{Z}\} & \text{if } x \in \mathbb{Z}_{\geq 4}, \\ \{\pm(\kappa_1 - 1)^l \mid l \in \mathbb{Z}\} & \text{if } x = 3. \end{cases}$$

κ_1 has norm 1. If $x = 3$ then $(\kappa_1 - 1)^2 = \kappa_1$, and $\kappa_1 - 1$ has norm -1.

(b) Let $x \in \mathbb{Z}_{\geq 2}$, and let $\lambda_{1/2} = x^2 - 1 \pm x\sqrt{x^2-2}$ be the zeros of the polynomial $t^2 - (2x^2-2)t + 1$, so $\lambda_1 + \lambda_2 = 2x^2 - 2$, $\lambda_1\lambda_2 = 1$, $\lambda_1^2 = (2x^2-2)\lambda_1 - 1$. Then

$$\mathbb{Z}[\sqrt{x^2-2}]^* = \begin{cases} \{\pm\lambda_1^l \mid l \in \mathbb{Z}\} & \text{if } x \geq 3, \\ \{\pm(1+\sqrt{2})^l \mid l \in \mathbb{Z}\} & \text{if } x = 2. \end{cases}$$

λ_1 has norm 1. If $x = 2$ then $(1+\sqrt{2})^2 = \lambda_1$, and $1 + \sqrt{2}$ has norm -1.

Theorem C.4 is mainly taken from several theorems in [Ai13, 1.2 and 1.3], but with part (b) from [Ca65, I 2.]. It is preceded by two definitions. According to [Bu00,

C. Hertling and K. Larabi, *Triangular Bases of Unimodular Bilinear Lattices and Induced Structures*, Lecture Notes in Mathematics 2398,
https://doi.org/10.1007/978-3-032-25997-4

5.9 Lagrange's Theorem], this part (b) is orginally due to Lagrange 1770. In fact, we will not use this part (b), but we find it enlightening.

Definition C.2 Let $\theta \in \mathbb{R} - \mathbb{Q}$ be an irrational number.

(a) Define recursively sequences $(a_n)_{n\geq 0}$, $(\theta_n)_{n\geq 0}$, $(p_n)_{n\geq -1}$, $(q_n)_{n\geq -1}$, $(r_n)_{n\geq 0}$ as follows:

$$\begin{aligned}
\theta_0 &:= \theta, \\
a_0 &:= \lfloor \theta_0 \rfloor \in \mathbb{Z}, \\
\theta_n &:= \frac{1}{\theta_{n-1} - a_{n-1}} \in \mathbb{R}_{>1} - \mathbb{Q} \quad \text{for } n \in \mathbb{N}, \\
a_n &:= \lfloor \theta_n \rfloor \in \mathbb{N} \quad \text{for } n \in \mathbb{N}, \\
(p_{-1}, p_0, q_{-1}, q_0) &:= (1, a_0, 0, 1), \\
p_n &:= a_n p_{n-1} + p_{n-2} \in \mathbb{Z} \quad \text{for } n \in \mathbb{N}, \\
q_n &:= a_n q_{n-1} + q_{n-2} \in \mathbb{N} \quad \text{for } n \in \mathbb{N}, \\
r_n &:= \frac{p_n}{q_n} \in \mathbb{Q} \quad \text{for } n \in \mathbb{Z}_{\geq 0}.
\end{aligned}$$

θ_n and a_n are defined for all $n \in \mathbb{N}$, because each θ_{n-1} is in $\mathbb{R} - \mathbb{Q}$, so $\theta_{n-1} - a_{n-1} \in (0, 1)$.

(b) Following [Ca65, Notation 2.] define

$$\|\theta\| := \min(\theta - \lfloor \theta \rfloor, \lceil \theta \rceil - \theta) \in (0, \frac{1}{2})$$

We are interested especially in the case when $\theta \in \mathbb{R} - \mathbb{Q}$ is a quadratic irrational. We recall some notations for this case.

Definition C.3 Let $\theta \in \mathbb{R} - \mathbb{Q}$ be a quadratic irrational, i.e. $\dim_{\mathbb{Q}} \mathbb{Q}[\theta] = 2$. The other root of the minimal polynomial of θ is called θ^{conj}, so $\theta + \theta^{conj} =: \widetilde{a_0} \in \mathbb{Q}$ and $-\theta\theta^{conj} =: d_0 \in \mathbb{Q}$. For any $\alpha = a + b\theta \in \mathbb{Q}[\theta]$ with $a, b \in \mathbb{Q}$ write $\alpha^{conj} := a + b\theta^{conj}$. It is the algebraic conjugate of α. The algebra homomorphism

$$\mathcal{N} : \mathbb{Q}[\theta] \to \mathbb{Q}, \quad \alpha \mapsto \alpha\alpha^{conj},$$

is the norm map. The number α is called *reduced* if $\alpha > 1$ and $\alpha^{conj} \in (-1, 0)$. Recall that α is an algebraic integer if and only if $\alpha + \alpha^{conj} \in \mathbb{Z}$ and $\mathcal{N}(\alpha) \in \mathbb{Z}$ and that in this case α is a unit in $\mathbb{Z}[\alpha]$ if and only if $\mathcal{N}(\alpha) \in \{\pm 1\}$.

Theorem C.4 *(Classical) In the situation of Definition C.2 the following holds.*

(a) [Ai13, 1.2] $a_0 \in \mathbb{Z}$, $a_n \in \mathbb{N}$ for $n \in \mathbb{N}$. For $n \in \mathbb{Z}_{\geq 0}$ the rational number r_n is

$$r_n = a_0 + \cfrac{1}{a_1 + \cfrac{1}{\ddots \cfrac{1}{a_{n-1} + \frac{1}{a_n}}}} =: [a_0, a_1, ..., a_n].$$

It is called partial quotient *or* continued fraction *or* n-th convergent *of* θ. *These numbers approximate* θ,

$$r_0 < r_2 < r_4 < \dots < \theta < \dots < r_5 < r_3 < r_1,$$
$$|\theta - r_n| < \frac{1}{q_n^2}.$$

This allows to write $\theta = [a_0, a_1, \dots]$ *as an infinite continued fraction. The numerator* p_n *and the denominator* q_n *of* r_n *are coprime,*

$$\gcd(p_n, q_n) = 1, \quad \textit{and more precisely}$$
$$p_n q_{n-1} - p_{n-1} q_n = (-1)^{n-1} \textit{ for } n \in \mathbb{Z}_{\geq 0}.$$

The denominators grow strictly from $n = 1$ *on,*

$$1 = q_0 \leq q_1 < q_2 < q_3 < \dots.$$

(b) [Ca65, I 2.] The partial quotients r_n *are in the following precise sense the only best approximations of* θ:

$$|p_n - q_n\theta| = \|q_n\theta\| \quad \textit{for } n \in \mathbb{N},$$
$$\|q_{n+1}\theta\| < \|q_n\theta\| \quad \textit{for } n \in \mathbb{N},$$
$$\|q\theta\| \geq \|q_n\theta\| \quad \textit{for } n \in \mathbb{Z}_{\geq 0} \textit{ and } q \in \mathbb{N} \textit{ with } q < q_{n+1},$$
$$|p_0 - q_0\theta| = \|q_0\theta\| > \|q_1\theta\| \quad \textit{if } q_1 > 1 (\iff a_1 > 1),$$
$$|p_0 - q_0\theta| \in (\frac{1}{2}, 1) \textit{ and } |p_0 - q_0\theta| > \|q_1\theta\| \quad \textit{if } q_1 = 1 (\iff a_1 = 1).$$

In any case

$$|p_{n+1} - q_{n+1}\theta| < |p_n - q_n\theta| \quad \textit{for } n \in \mathbb{Z}_{\geq 0}.$$

(c) [Ai13, Theorem 1.19] The partial quotients r_n *are also in the following precise sense the only best approximations of* θ: *A rational number* $\frac{p}{q}$ *with* $p \in \mathbb{Z}, q \in \mathbb{N}$ *and* $\gcd(p, q) = 1$ *satisfies*

$$|\theta - \frac{p}{q}| < \frac{1}{2q^2} \Longrightarrow (p, q) = (p_n, q_n) \quad \textit{for a suitable} \quad n \in \mathbb{Z}_{\geq 0}.$$

(d) [Ai13, Theorem 1.17 and Proposition 1.18] The continued fraction is periodic, *i.e. there exist* $k_0 \in \mathbb{Z}_{\geq 0}$ *and* $k_1 \in \mathbb{N}$ *with*

$$a_{n+k_1} = a_n \textit{ for } n \geq k_0,$$

if and only if θ is a quadratic irrational, i.e. $\dim_{\mathbb{Q}} \mathbb{Q}[\theta] = 2$. *Then one writes* $[a_0 a_1 ...] = [a_0 a_1 ... a_{k_0-1} \overline{a_{k_0} a_{k_0+1} ... a_{k_0+k_1-1}}]$. *Furthermore, then k_0 can be chosen as 0 if and only if θ is reduced. In this case the continued fraction $[a_0 a_1 ...]$ is called* purely periodic.

Lemma C.5 fixes useful additional observations for the case of a quadratic irrational θ. These observations are used in the proof of Theorem C.6. It considers an algebraic integer $\alpha \in \mathbb{R} - \mathbb{Q}$ which is a quadratic irrational. We are interested in the group $\mathbb{Z}[\alpha]^*$ of units in $\mathbb{Z}[\alpha]$. Theorem C.6 shows how to see a generator of this group (and a quarter of its elements) in the continued fractions of a certain reduced element θ in $\mathbb{Z}[\alpha]$. Theorem C.6 is not new. For example, [Bu00, Theorem 8.13] gives its main part. But the proof here is more elegant than what we found in the literature.

Lemma C.5 *Let $\theta \in \mathbb{R} - \mathbb{Q}$ be a quadratic irrational which is reduced. Let $[\overline{a_0 a_1 ... a_{k-1}}]$ be its purely periodic continued fraction of some minimal length $k \in \mathbb{N}$. We consider the objects in Definition C.2 for this θ. Then*

$$\theta_{n+k} = \theta_n \quad \text{for } n \in \mathbb{Z}_{\geq 0}. \tag{C.1}$$

θ_m is reduced for $m \in \{0, 1, ..., k-1\}$, and its purely periodic continued fraction is $[\overline{a_m ... a_{k-1} a_0 ... a_{m-1}}]$. Write

$$\begin{aligned} \widetilde{a_0} &:= \theta + \theta^{conj} \in \mathbb{Q}_{>0}, \qquad d_0 := -\mathcal{N}(\theta) = -\theta\theta^{conj} \in \mathbb{Q}_{>0}, \\ \beta &:= p_{k-1} - q_{k-1}\theta \in \mathbb{Q}[\theta] - \mathbb{Q}. \end{aligned}$$

Then for $n \in \mathbb{Z}_{\geq -1}$

$$p_{n+k} - q_{n+k}\theta = \beta \cdot (p_n - q_n\theta) \tag{C.2}$$

and for $n \in \mathbb{Z}_{\geq 0}$

$$\theta_n = \frac{-p_{n-2}p_{n-1} + p_{n-2}q_{n-1}\widetilde{a_0} + q_{n-2}q_{n-1}d_0 + (-1)^n\theta}{\mathcal{N}(p_{n-1} - q_{n-1}\theta)}. \tag{C.3}$$

Proof The natural generalization of the notation $[a_0, a_1, ..., a_m]$ to numbers $a_0 \in \mathbb{R}, a_1, ..., a_m \in \mathbb{R}_{>0}$ gives for $n \in \mathbb{Z}_{\geq 0}$

$$\theta = [a_0, a_1, ..., a_{n-1}, \theta_n] = \frac{\theta_n p_{n-1} + p_{n-2}}{\theta_n q_{n-1} + q_{n-2}},$$

see Proposition 1.9 in [Ai13]. One concludes that the continued fraction of θ_n is purely periodic, that $\theta_n = \theta_m$ if $n = kl + m$ with $l \in \mathbb{Z}_{\geq 0}$ and $m \in \{0, 1, ..., k-1\}$, and that its continued fraction is $[\overline{a_m ... a_{k-1} a_0 ... a_{m-1}}]$. Therefore θ_n is reduced. Recall $p_{n-1}q_{n-2} - p_{n-2}q_{n-1} = (-1)^n$. Inverting the equation above gives

$$\begin{aligned}\theta_n &= \frac{\theta q_{n-2} - p_{n-2}}{\theta(-q_{n-1}) + p_{n-1}} \\ &= \frac{(-p_{n-2} + q_{n-2}\vartheta)(p_{n-1} - q_{n-1}\theta^{conj})}{\mathcal{N}(p_{n-1} - q_{n-1}\theta)} \\ &= \frac{-p_{n-2}p_{n-1} + p_{n-2}q_{n-1}\widetilde{a_0} + q_{n-2}q_{n-1}d_0 + (-1)^n\theta}{\mathcal{N}(p_{n-1} - q_{n-1}\theta)}.\end{aligned}$$

The formula $\theta = \theta_k = \frac{\theta q_{k-2} - p_{k-2}}{\theta(-q_{k-1}) + p_{k-1}}$ shows

$$(1, -\theta)\begin{pmatrix} p_{k-1} & p_{k-2} \\ q_{k-1} & q_{k-2} \end{pmatrix} = (p_{k-1} - q_{k-1}\theta)(1, -\theta) = \beta(1, -\theta).$$

The inductive definition of p_n and q_n shows

$$\begin{pmatrix} p_n & p_{n-1} \\ q_n & q_{n-1} \end{pmatrix} = \begin{pmatrix} a_0 & 1 \\ 1 & 0 \end{pmatrix}\begin{pmatrix} a_1 & 1 \\ 1 & 0 \end{pmatrix}\cdots\begin{pmatrix} a_n & 1 \\ 1 & 0 \end{pmatrix}.$$

With the periodicity $a_n + k = a_n$ we obtain

$$\begin{aligned}(1, -\theta)\begin{pmatrix} p_{n+k} & p_{n-1+k} \\ q_{n+k} & q_{n-1+k} \end{pmatrix} &= (1, -\theta)\begin{pmatrix} p_{k-1} & p_{k-2} \\ q_{k-1} & q_{k-2} \end{pmatrix}\begin{pmatrix} p_n & p_{n-1} \\ q_n & q_{n-1} \end{pmatrix} \\ &= \beta(1, -\theta)\begin{pmatrix} p_n & p_{n-1} \\ q_n & q_{n-1} \end{pmatrix}.\end{aligned}$$

This gives formula (C.2). □

Theorem C.6 *Let $\alpha \in \mathbb{R} - \mathbb{Q}$ be a quadratic irrational and an algebraic integer.*

(a) There are a unique sign $\varepsilon_\alpha \in \{\pm 1\}$ and a unique number $n_\alpha \in \mathbb{Z}$ such that $\theta := \varepsilon_\alpha \alpha + n_\alpha$ is reduced. Then $\mathbb{Z}[\alpha] = \mathbb{Z}[\theta]$, and any reduced element $\widetilde{\theta} \in \mathbb{Z}[\alpha]$ with $\mathbb{Z}[\alpha] = \mathbb{Z}[\widetilde{\theta}]$ satisfies $\widetilde{\theta} = \theta$. We consider the objects in Definition C.2 for this θ. We define

$$\begin{aligned}&\widetilde{a_0} := \theta + \theta^{conj} \in \mathbb{N}, \qquad d_0 := -\mathcal{N}(\theta) = -\theta\theta^{conj} \in \mathbb{N}, \\ &\beta := p_{k-1} - q_{k-1}\theta \in \mathbb{Z}[\theta] - \mathbb{Z}.\end{aligned}$$

as in Lemma C.5. Then $a_0 = \widetilde{a_0}$ and $d_0 \in \{1, 2, ..., a_0\}$.

(b) Then β is a unit and generates together with -1 the unit group $\mathbb{Z}[\alpha]^*$, the l-th power of β is $\beta^l = p_{lk-1} - q_{lk-1}\vartheta$ for $l \in \mathbb{Z}_{\geq 0}$, and

$$\begin{aligned}&\{\pm\beta^l \mid l \in \mathbb{Z}\} = \mathbb{Z}[\alpha]^*, \\ &\{\beta^l \mid l \in \mathbb{Z}_{\geq 0}\} = \mathbb{Z}[\alpha]^* \cap \{p_{n-1} - q_{n-1}\theta \mid n \in \mathbb{Z}_{\geq 0}\}.\end{aligned}$$

The element β is uniquely characterized by the following properties:
(i) -1 and β generate the unit group $\mathbb{Z}[\alpha]^*$,
(ii) $|\beta| < 1$,
(iii) $\beta = p - q\theta$ with $p \in \mathbb{Z}, q \in \mathbb{N}$ (namely $p = p_{k-1}, q = q_{k-1}$).

Proof (a) Choose $\varepsilon_\alpha \in \{\pm 1\}$ such that $\varepsilon_\alpha(\alpha - \alpha^{conj}) > 0$. Then choose $n_\alpha \in \mathbb{Z}$ such that $\varepsilon_\alpha \alpha^{conj} + n_\alpha \in (-1, 0)$. Define $\theta := \varepsilon_\alpha \alpha + n_\alpha$. Then $\theta^{conj} = \varepsilon_\alpha \alpha^{conj} + n_\alpha \in (-1, 0)$. Also $\theta > \theta^{conj}$ and $\theta\theta^{conj} \in \mathbb{Z} - \{0\}$. This shows $\theta > 1$, so θ is reduced. Also $\mathbb{Z}[\alpha] = \mathbb{Z}[\theta]$ is clear.

Any reduced element $\widetilde{\theta} \in \mathbb{Z}[\alpha]$ with $\mathbb{Z}[\alpha] = \mathbb{Z}[\theta]$ has the shape $\widetilde{\theta} = \widetilde{\varepsilon_\alpha}\alpha + \widetilde{n_\alpha}$ with $\widetilde{\varepsilon_\alpha} \in \{\pm 1\}$ and $\widetilde{n_\alpha} \in \mathbb{Z}$. The sign $\widetilde{\varepsilon_\alpha}$ is because of $\widetilde{\theta} > 1 > 0 > \widetilde{\theta}^{conj}$ the unique sign with $\widetilde{\varepsilon_\alpha}(\alpha - \alpha^{conj}) > 0$, so $\widetilde{\varepsilon_\alpha} = \varepsilon_\alpha$. Now $\widetilde{n_\alpha}$ is the unique integer with $\varepsilon_\alpha \alpha^{conj} + n_\alpha \in (-1, 0)$, so $\widetilde{n_\alpha} = n_\alpha$. Therefore $\widetilde{\theta} = \theta$.

We have $a_0 = \lfloor \theta \rfloor = \theta + \theta^{conj} = \widetilde{a_0}$ and $d_0 = -\theta\theta^{conj} \le a_0$, both because $\theta^{conj} \in (-1, 0)$.

(b) We apply Lemma C.5. It tells us which of the elements $p_n - q_n\theta$ for $n \in \mathbb{Z}_{\ge 0}$ are units, in the following way.

Consider $n \in \mathbb{Z}_{\ge 0}$ and write $n = lk + m$ with $l \in \mathbb{Z}_{\ge 0}$ and $m \in \{0, 1, ..., k-1\}$. Recall that θ_n is reduced, that $\theta_n = \theta_m$ because of formula (C.1) and that $\theta_m = \theta$ only for $m = 0$, because for $m \in \{1, ..., k-1\}$ the purely periodic continued fractions of θ and θ_m differ. Recall also from formula (C.2) that

$$p_{n-1} - q_{n-1}\theta = \beta^l (p_{m-1} - q_{m-1}\theta).$$

If for some $n \in \mathbb{Z}_{\ge 0}$ $\mathcal{N}(p_{n-1} - q_{n-1}\theta) \in \{\pm 1\}$, then by formula (C.3) θ_n satisfies $\mathbb{Z}[\alpha] = \mathbb{Z}[\theta] = \mathbb{Z}[\theta_n]$. The uniqueness of θ in part (a) implies that then $\theta_n = \theta$, so $m = 0$. Therefore for $n \in \mathbb{Z}_{\ge 0} - k\mathbb{Z}_{\ge 0}$, $\mathcal{N}(p_{n-1} - q_{n-1}\theta) \notin \{\pm 1\}$, so then $p_{n-1} - q_{n-1}\theta$ is not a unit.

On the other hand, if $n = kl$, so $m = 0$, then $\theta_n = \theta$, and formula (C.3) tells $\mathcal{N}(p_{n-1} - q_{n-1}\theta) = (-1)^n$, so $p_{n-1} - q_{n-1}\theta$ is a unit. In fact, formula (C.2) tells $p_{n-1} - q_{n-1}\theta = \beta^l$. We see

$$\{p_{n-1} - q_{n-1}\theta \mid n \in \mathbb{Z}_{\ge 0}\} \cap \mathbb{Z}[\theta]^* = \{\beta^l \mid l \in \mathbb{Z}_{\ge 0}\}. \tag{C.4}$$

It remains to see that -1 and β generate $\mathbb{Z}[\theta]^*$.

By Dirichlet's unit theorem [BSh73, Chap. 2 4.3 Theorem 5], the set $\mathbb{Z}[\alpha]^*$ is as a group isomorphic to $\{\pm 1\} \times \mathbb{Z}$. It has two generators $\pm\widetilde{\beta}$ with $|\pm\widetilde{\beta}| < 1$. They are the unique elements in $\mathbb{Z}[\alpha]^*$ with maximal absolute value < 1. One of them has the shape $p - q\theta$ with $q \in \mathbb{N}$. This is called $\widetilde{\beta}$. Then also $p \in \mathbb{N}$, because $|\widetilde{\beta}| = |p - q\theta| < 1$ and $q\theta > 1$.

1st case, $\theta \in (1, 2)$: Then $a_0 = d_0 = 1$, and $\theta = \frac{1+\sqrt{5}}{2}$ is the golden section with $\theta^2 = \theta + 1$. This case is well known. Here the continued fraction of θ is purely periodic with period $\overline{1}$ of length one, because $a_0 = 1$ and

$$\theta_1 = (\theta_0 - a_0)^{-1} = \theta = \theta_0.$$

Here $\beta = 1 - \theta = -\theta^{-1} = \theta^{conj}$. It is well known that

$$\mathbb{Z}[\alpha]^* = \mathbb{Z}[\theta]^* = \{\pm\theta^l \mid l \in \mathbb{Z}\} = \{\pm\beta^l \mid l \in \mathbb{Z}\}$$

2nd case, $\theta > 2$: $\widetilde{\beta} = p - q\theta$ is a unit, so $\pm 1 = \mathcal{N}(p - q\theta)$. Also $|\widetilde{\beta}| < 1$ and $\theta > 2$ imply $p \geq 2q$. Therefore

$$\begin{aligned} |\frac{p}{q} - \theta| &= \frac{1}{q(p - q\theta^{conj})} = \frac{1}{q(p + q|\theta^{conj}|)} \\ &< \frac{1}{q(2q+0)} = \frac{1}{2q^2}. \end{aligned}$$

By Theorem C.4 (c) $n \in \mathbb{Z}_{\geq 0}$ with $(p, q) = (p_n, q_n)$ exists. By (C.4) $\widetilde{\beta}$ is a power of β, so $\widetilde{\beta} = \beta$. □

The parts (a) and (b) in the following proof of Lemma C.1 serve also as examples for Theorem C.6.

Proof of Lemma C.1**:** (a) Here

$$\begin{aligned} x \geq 3 \Rightarrow 2x > 5 &\Rightarrow x^2 - 4 > x^2 - 2x + 1 \\ &\Rightarrow \sqrt{x^2-4} \in (x-1, x), \\ \Rightarrow \theta = \kappa_1 - 1 = \frac{x-2}{2} &+ \frac{1}{2}\sqrt{x^2-4} \in (x - \frac{3}{2}, x-1) \\ \text{and} \quad \theta^{conj} = \kappa_2 - 1 = \frac{x-2}{2} &- \frac{1}{2}\sqrt{x^2-4} \in (-1, -\frac{1}{2}). \end{aligned}$$

Observe

$$\theta + \theta^{conj} = x - 2, \quad \theta\theta^{conj} = -x + 2.$$

Therefore

$$\begin{aligned} \theta_0 &= \theta, \quad a_0 = \lfloor\theta_0\rfloor = x - 2, \\ \theta_1 &= (\theta_0 - a_0)^{-1} = \frac{\theta^{conj} - (x-2)}{(\theta - (x-2))(\theta^{conj} - (x-2))} \\ &= \frac{-\theta}{-x+2} = \frac{\theta}{x-2}, \\ a_1 &= \lfloor\theta_1\rfloor = 1, \\ \theta_2 &= (\theta_1 - a_1)^{-1} = \frac{x-2}{\theta - (x-2)} = \frac{(x-2)(\theta^{conj} - (x-2))}{-x+2} = \theta, \\ \theta &= [\overline{x-2, 1}]. \end{aligned}$$

The continued fraction of θ is purely periodic with period $\overline{x-2, 1}$ of length 2 if $x \geq 4$ and purely periodic with period $\overline{1}$ if $x = 3$. The norm of $p - q\theta$ for $p, q \in \mathbb{Z}$ is

$$\mathcal{N}(p - q\theta) := (p - q\theta)(p - q\theta^{conj}) = p^2 - q(p+q)(x-2) \in \mathbb{Z}.$$

It is ± 1 if and only if $p - q\theta$ is a unit.

n	0	1	2	3
a_n	$x-2$	1	$x-2$	1
(p_n, q_n)	$(x-2, 1)$	$(x-1, 1)$	$(x^2-2x, x-1)$	(x^2-x-1, x)
$\mathcal{N}(p_n - q_n\theta)$	$-x+2$	1	$-x+2$	1

If $x = 3$ then $\beta = p_0 - q_0\theta = 1 - \theta$ in the notation of Lemma C.5, so $\mathbb{Z}[\theta]^*$ is generated by (-1) and β or $-\beta^{-1} = \theta = \kappa_1 - 1$, so

$$\mathbb{Z}[\theta]^* = \{\pm(1-\theta)^l \mid l \in \mathbb{Z}\} = \{\pm(\kappa_1 - 1)^l \mid l \in \mathbb{Z}\}.$$

This is also consistent with the 1st case in the proof of part (b) of Theorem C.6.

If $x \geq 4$ then $\beta = p_1 - q_1\theta = x - 1 - \theta$ in the notation of Corollary C.4, so $\mathbb{Z}[\theta]^*$ is generated by (-1) and β or $x - 1 - \theta^{conj} = \kappa_1$, so

$$\mathbb{Z}[\theta]^* = \{\pm(x-1-\theta)^l \mid l \in \mathbb{Z}\} = \{\pm\kappa_1^l \mid l \in \mathbb{Z}\}.$$

This proves part (a) of Lemma C.1.

(b) The case $x = 2$ is treated separately and first. This case is well known. Then $\theta = 1 + \sqrt{2} \in (2, 3)$, $\theta^{conj} = 1 - \sqrt{2} \in (-1, 0)$, so $a_0 = 2$. The continued fraction of θ is purely periodic with period $\overline{2}$ of length one, because $a_0 = 2$ and

$$\theta_1 = (\theta_0 - a_0)^{-1} = (\sqrt{2} - 1)^{-1} = \theta_0.$$

The element

$$p_0 - q_0\theta = 2 - \theta = 1 - \sqrt{2} = \theta^{conj}$$

is a unit. This and Theorem C.6 (b) show

$$\mathbb{Z}[\alpha]^* = \mathbb{Z}[\theta]^* = \{\pm\theta^l \mid l \in \mathbb{Z}\} = \{\pm(1+\sqrt{2})^l \mid l \in \mathbb{Z}\}.$$

Now we treat the cases $x \geq 3$. Here

$$
\begin{aligned}
x \geq 3 \Rightarrow x^2 - 2 > x^2 - x + \frac{1}{4} \Rightarrow \sqrt{x^2 - 2} > x - \frac{1}{2}, \\
\Rightarrow \theta = (x-1) + \sqrt{x^2 - 2} \in (2x - \frac{3}{2}, 2x - 1) \\
\text{and} \quad \theta^{conj} = (x-1) - \sqrt{x^2 - 2} \in (-1, -\frac{1}{2}).
\end{aligned}
$$

Observe

$$\theta + \theta^{conj} = 2x - 2, \quad \theta\theta^{conj} = -2x + 3.$$

Therefore

$$
\begin{aligned}
&\theta_0 = \theta, \quad a_0 = \lfloor \theta_0 \rfloor = 2x - 2, \\
&\theta_1 = (\theta_0 - a_0)^{-1} = \ldots = \frac{\theta}{2x-3} \in (1, 2), \quad a_1 = 1, \\
&\theta_2 = (\theta_1 - a_1)^{-1} = \ldots = \frac{\theta - 1}{2} \in (x-2, x-1), \quad a_2 = x - 2, \\
&\theta_3 = (\theta_2 - a_2)^{-1} = \ldots = \frac{\theta - 1}{2x - 3} \in (1, 2), \quad a_3 = 1, \\
&\theta_4 = (\theta_3 - a_3)^{-1} = \ldots = \theta = \theta_0, \\
&\theta = [\overline{2x - 2, 1, x - 2, 1}].
\end{aligned}
$$

The continued fraction of θ is purely periodic with period $\overline{2x - 2, 1, x - 2, 1}$ of length four. The norm of $p - q\theta$ is

$$\mathcal{N}(p - q\theta) = (p - q\theta)(p - q\theta^{conj}) = p^2 + q^2 - q(p+q)(2x-2).$$

It is ± 1 if and only if $p - q\theta$ is a unit.

n	0	1	2	3
a_n	$2x - 2$	1	$x - 2$	1
(p_n, q_n)	$(2x-2, 1)$	$(2x-1, 1)$	$(2x^2 - 3x, x - 1)$	$(2x^2 - x - 1, x)$
$\mathcal{N}(p_n - q_n\theta)$	$-2x + 3$	2	$-2x + 3$	1

We conclude with Theorem C.6 (and with the notation of Lemma C.5) that

$$\beta = p_3 - q_3\theta = (2x^2 - x - 1) - x\theta = (x^2 - 1) - x\sqrt{x^2 - 2} = \lambda_2$$

is together with (-1) a generator of $\mathbb{Z}[\sqrt{x^2 - 2}]^* = \mathbb{Z}[\theta]^*$. Therefore also λ_1 together with (-1) is a generator of $\mathbb{Z}[\sqrt{x^2 - 2}]^*$. This proves part (b) of Lemma C.1. □

Remarks C.7 In the situation of Theorem C.6, Satz 9.5.2 in [Ko97] tells that the unit group $\mathbb{Z}[\theta]^*$ is generated by -1 and $q_{k-2} + q_{k-1}\theta$. This is consistent with Theorem C.6 because of the following. Here

$$
\begin{aligned}
\theta &= \frac{\theta_k p_{k-1} + p_{k-2}}{\theta_k q_{k-1} + q_{k-2}} = \frac{\theta p_{k-1} + p_{k-2}}{\theta q_{k-1} + q_{k-2}},\\
\text{so} \quad 0 &= q_{k-1}\theta^2 - (p_{k-1} - q_{k-2})\theta - p_{k-2},\\
\text{but also} \quad 0 &= \theta^2 - \widetilde{a_0}\theta - d_0,\\
\text{so} \quad a_0 &= \widetilde{a_0} = \frac{p_{k-1} - q_{k-2}}{q_{k-1}}, \quad d_0 = \frac{p_{k-2}}{q_{k-1}},\\
p_{k-1} - q_{k-1}\theta^{conj} &= p_{k-1} - q_{k-1}(a_0 - \theta) = q_{k-2} + q_{k-1}\theta.
\end{aligned}
$$

Appendix D
Powers of Quadratic Units

The following definition and lemma treat powers of units of norm 1 in the rings of integers of quadratic number fields. Though these powers appear explicitly only in Lemma D.2(c). Lemma D.2 will be used in the proof of Theorem 5.18.

Definition D.1 (a) Define the polynomials $b_l(a) \in \mathbb{Z}[a]$ for $l \in \mathbb{Z}_{\geq 0}$ by the following recursion.

$$b_0 := 0, \quad b_1 := 1, \quad b_l := ab_{l-1} - b_{l-2} \quad \text{for } l \in \mathbb{Z}_{\geq 2}. \tag{D.1}$$

(b) Define for $l \in \mathbb{Z}_{\geq 0}$ the polynomial $r_l \in \mathbb{Z}[a]$ and for $l \in \mathbb{N}$ the rational functions $q_{0,l}, q_{1,l}, q_{2,l} \in \mathbb{Q}(t)$,

$$\begin{aligned} r_0 &:= 0, \\ r_l &:= -ab_l + 2b_{l-1} + 2 \quad \text{for } l \in \mathbb{N}, \\ q_{0,l} &:= \frac{b_l - b_{l-1}}{b_l}, \\ q_{1,l} &:= \frac{b_l - b_{l-1} - 1}{r_l b_l}, \\ q_{2,l} &:= q_{0,l} - 2q_{1,l}. \end{aligned}$$

(c) A notation: For two polynomials $f_1, f_2 \in \mathbb{Z}[a]$, $(f_1, f_2)_{\mathbb{Z}[a]} := \mathbb{Z}[a]f_1 + \mathbb{Z}[a]f_2 \subset \mathbb{Z}[a]$ denotes the ideal generated by f_1 and f_2.

Remark: If $(f_1, f_2)_{\mathbb{Z}[a]} = \mathbb{Z}[a]$ then for any integer $c \in \mathbb{Z}$ $\gcd(f_1(c), f_2(c)) = 1$.

(d) For $a \in \mathbb{Z}_{\leq -3} \cup \mathbb{Z}_{\geq 3}$ define $\kappa_a := \frac{a}{2} + \frac{1}{2}\sqrt{a^2 - 4}$ and $\kappa_a^{conj} := \frac{a}{2} - \frac{1}{2}\sqrt{a^2 - 4}$ as the zeros of the polynomial $t^2 - at + 1$, so that $\kappa_a + \kappa_a^{conj} = a$, $\kappa_a \kappa_a^{conj} = 1$, $\kappa_a^2 = a\kappa_a - 1$. They are algebraic integers and units with norm 1.

C. Hertling and K. Larabi, *Triangular Bases of Unimodular Bilinear Lattices and Induced Structures*, Lecture Notes in Mathematics 2398,
https://doi.org/10.1007/978-3-032-25997-4

The following table gives the first twelve of the polynomials $b_l(a)$. The software Maxima [Maxima22] claims that the factors in the products are irreducible polynomials as polynomials in $\mathbb{Q}[a]$. We will not use this claim.

$$
\begin{aligned}
b_0 &= 0\\
b_1 &= 1\\
b_2 &= a\\
b_3 &= (a-1)(a+1)\\
b_4 &= a(a^2-2)\\
b_5 &= (a^2-a-1)(a^2+a-1)\\
b_6 &= (a-1)a(a+1)(a^2-3)\\
b_7 &= (a^3-a^2-2a+1)(a^3+a^2-2a-1)\\
b_8 &= a(a^2-2)(a^4-4a^2+2)\\
b_9 &= (a-1)(a+1)(a^3-3a-1)(a^3-3a+1)\\
b_{10} &= a(a^2-a-1)(a^2+a-1)(a^4-5a^2+5)\\
b_{11} &= (a^5-a^4-4a^3+3a^2+3a-1)(a^5+a^4-4a^3-3a^2+3a+1)
\end{aligned}
$$

Lemma D.2 *(a) For any $l \in \mathbb{N}$*

$$1 = b_l^2 - ab_lb_{l-1} + b_{l-1}^2 = b_l^2 - b_{l+1}b_{l-1}, \tag{D.2}$$

$$(b_{l-1}, b_l)_{\mathbb{Z}[a]} = (b_l - b_{l-1}, b_l)_{\mathbb{Z}[a]} = \mathbb{Z}[a], \tag{D.3}$$

$$r_l = \begin{cases} (2-a)(b_{(l+1)/2} + b_{(l-1)/2})^2 & \text{for } l \text{ odd},\\ (2-a)(a+2)b_{l/2}^2 & \text{for } l \text{ even},\\ & (\text{also } l = 0) \end{cases} \tag{D.4}$$

$$(r_{l-1}/(2-a), r_l/(2-a))_{\mathbb{Z}[a]} = \mathbb{Z}[a], \tag{D.5}$$

and

$$q_{2,l} = 1 - \frac{r_{l-1}/(2-a)}{r_l/(2-a)}. \tag{D.6}$$

(b) For $a \in \mathbb{Z}_{\leq -3}$

$$
\begin{aligned}
b_l(a) \in (-1)^{l-1}\mathbb{N} \quad &\text{for} \quad l \geq 1,\\
b_1(a) = 1, \quad b_2(a) = a, \quad |b_2(a) + b_1(a)| &= |a| - 1,\\
|b_l(a)| > 2|b_{l-1}(a)| \geq |b_{l-1}(a)| + 1 \quad &\text{for} \quad l \geq 2,\\
|b_{l+1}(a) + b_l(a)| > |b_l(a) + b_{l-1}(a)| \quad &\text{for} \quad l \geq 1.
\end{aligned}
$$

For $a \in \mathbb{Z}_{\geq 3}$

$$\begin{aligned} b_l(a) &> 0 \quad \textit{for} \quad l \geq 1, \\ b_1(a) = 1, \quad b_2(a) = a, \quad b_2(a) + b_1(a) &= a + 1, \\ b_l(a) &> 2b_{l-1}(a) \quad \textit{for} \quad l \geq 1, \\ b_{l+1}(a) + b_l(a) &> 2(b_l(a) + b_{l-1}(a)) \quad \textit{for} \quad l \geq 1. \end{aligned}$$

(c) Consider $a \in \mathbb{Z}_{\leq -3} \cup \mathbb{Z}_{\geq 3}$ *and* $l \in \mathbb{N}$. *Then*

$$\kappa_a^l = b_l(a)\kappa_a - b_{l-1}(a), \tag{D.7}$$

$$\kappa_a = (1 - q_{0,l}(a)) + (q_{0,l}(a) - r_l(a)q_{1,l}(a))\kappa_a^l, \tag{D.8}$$

$$\kappa_a^l = \frac{2 - r_l(a)}{2} + \frac{1}{2}\sqrt{r_l(a)(r_l(a) - 4)}, \tag{D.9}$$

so κ_a^l *is a zero of the polynomial* $t^2 - (2 - r_l(a))t + 1$.

Proof (a) The recursive definition (D.1) of b_l shows immediately the equality of the middle and right term in (D.2), it shows

$$b_l^2 - ab_l b_{l-1} + b_{l-1}^2 = b_{l-1}^2 - ab_{l-1}b_{l-2} + b_{l-2}^2,$$

and it shows $b_1^2 - ab_1 b_0 + b_0^2 = 1$. This proves (D.2). It implies (D.3).

The sequence $(r_l)_{l \in \mathbb{N}}$ satisfies the recursion

$$r_0 = 0, \quad r_1 = 2 - a, \quad r_l = ar_{l-1} - r_{l-2} + 2(2 - a) \quad \text{for } l \geq 2.$$

For $l = 2$ one verifies this immediately. For $l \geq 3$ it follows inductively with (D.1),

$$\begin{aligned} r_l &= -a(ab_{l-1} - b_{l-2}) + 2(ab_{l-2} - b_{l-3}) + 2 \\ &= a(-ab_{l-1} + 2b_{l-2} + 2) - (-ab_{l-2} + 2b_{l-3} + 2) + 2(2 - a) \\ &= ar_{l-1} - r_{l-2} + 2(2 - a). \end{aligned}$$

For $l = 1$ and $l = 0$ (D.4) is obvious. For odd $l = 2k + 1 \geq 3$ as well as even $l = 2k \geq 2$, one verifies (D.4) inductively with this recursion and with (D.2), for odd $l = 2k + 1 \geq 3$:

$$\begin{aligned} &(2 - a)(b_{k+1} + b_k)^2 - ar_{2k} + r_{2k-1} - 2(2 - a) \\ &= (2 - a)[((ab_k - b_{k-1}) + b_k)^2] \\ &\quad -(2 - a)[a(a + 2)b_k^2 - (b_k + b_{k-1})^2 + 2] \\ &= (2 - a)[2b_k^2 - 2ab_k b_{k-1} + 2b_{k-1}^2 - 2] \\ &= 0 \quad \text{(with (D.2))}, \end{aligned}$$

for even $l = 2k \geq 2$:

$$\begin{aligned}
&(2-a)(a+2)b_k^2 - ar_{2k-1} + r_{2k-2} - 2(2-a) \\
&= (2-a)[(a+2)b_k^2] \\
&\quad -(2-a)[a(b_k+b_{k-1})^2 - (a+2)b_{k-1}^2 + 2] \\
&= (2-a)[2b_k^2 - 2ab_kb_{k-1} + 2b_{k-1}^2 - 2] \\
&= 0 \quad \text{(with (D.2))}.
\end{aligned}$$

(D.5) claims for $k \geq 0$

$$\begin{aligned}
&((b_{k+1}+b_k)^2, (a+2)b_k^2)_{\mathbb{Z}[a]} = \mathbb{Z}[a] \\
\text{and}\quad &((a+2)b_{k+1}^2, (b_{k+1}+b_k)^2)_{\mathbb{Z}[a]} = \mathbb{Z}[a].
\end{aligned}$$

The following **claim** is basic: For $f_1, f_2, f_3 \in \mathbb{Z}[a]$

$$(f_1, f_3)_{\mathbb{Z}[a]} = (f_2, f_3)_{\mathbb{Z}[a]} = \mathbb{Z}[a] \quad \Rightarrow \quad (f_1f_2, f_3)_{\mathbb{Z}[a]} = \mathbb{Z}[a].$$

To see this claim consider $1 = \alpha_1 f_1 + \alpha_2 f_3$, $1 = \beta_1 f_2 + \beta_2 f_3$. Then

$$\begin{aligned}
1 &= (\alpha_1 f_1 + \alpha_2 f_3)(\beta_1 f_2 + \beta_2 f_3) \\
&= \alpha_1\beta_1 f_1 f_2 + \alpha_2\beta_2 f_3^3 + \alpha_1\beta_2 f_1 f_3 + \alpha_2\beta_1 f_2 f_3.
\end{aligned}$$

The claim and (D.3) show that for (D.5) it is sufficient to prove

$$(a+2, b_{k+1}+b_k)_{\mathbb{Z}[a]} = \mathbb{Z}[a].$$

This follows inductively in k with

$$b_{k+1} + b_k = (a+2)b_k - (b_k + b_{k-1}) \quad \text{and} \quad b_1 + b_0 = 1.$$

Finally, we calculate $q_{2,l}$:

$$\begin{aligned}
q_{2,l} &= (r_lb_l)^{-1}(r_l(b_l - b_{l-1}) - 2(b_l - b_{l-1} - 1)) \\
&= 1 + (r_lb_l)^{-1}(-(-ab_l + 2b_{l-1} + 2)b_{l-1}) - 2b_l + 2b_{l-1} + 2) \\
&= 1 + (r_lb_l)^{-1}(b_l(ab_{l-1} - 2) - 2(b_{l-1}^2 - 1)) \\
&= \begin{cases} 1 + (r_l)^{-1}((ab_{l-1} - 2) - 2b_{l-2}) & \text{(with (D.2) for } l \geq 2, \\ 1 & \text{for } l = 1 \end{cases} \\
&= 1 - r_l^{-1} r_{l-1}.
\end{aligned}$$

(b) All inequalities and signs follow inductively with (D.1).

(c) (D.7) is true for $l = 1$. It follows inductively in l with the following calculation, which uses $\kappa_a^2 = a\kappa_a - 1$.

$$\begin{aligned}\kappa_a^{l+1} &= \kappa_a(b_l(a)\kappa_a - b_{l-1}(a))\\ &= b_l(a)(a\kappa_a - 1) - b_{l-1}(a)\kappa_a\\ &= (ab_l(a) - b_{l-1}(a))\kappa_a - b_l(a)\\ &= b_{l+1}(a)\kappa_a - b_l(a).\end{aligned}$$

The right hand side of (D.8) is

$$\frac{b_{l-1}(a)}{b_l(a)} + \frac{1}{b_l(a)}\kappa_a^l$$

which is κ_a by inverting (D.7). Writing $\kappa_a = \frac{a}{2} + \frac{1}{2}\sqrt{a^2-4}$ gives for κ_a^l

$$\kappa_a^l = \frac{ab_l(a) - 2b_{l-1}(a)}{2} + \frac{b_l(a)}{2}\sqrt{a^2-4}.$$

One verifies that this equals the right hand side of (D.9). □

References

[AC73] N. A'Campo, Le nombre de Lefschetz d'une monodromie. Indagationes Math. **35**, 113–118 (1973)

[AC75-1] N. A'Campo, Le groupe de monodromie du déploiement des singularités isolées de courbes planes I. Math. Ann. **213**, 1–32 (1975)

[AC75-2] N. A'Campo, Le groupe de monodromie du déploiement des singularités isolées de courbes planes II. Proceedings of the International Congress of Mathematicians, Vancouver **I**, 395–404 (1974)

[Ai13] M. Aigner, *Markov's Theorem and 100 Years of the Uniqueness Conjecture* (Springer, 2013)

[AT20] D. Aramaki, A. Takahashi, Maximally-graded matrix factorizations for an invertible polynomial of chain type. Adv. Math. **373**, 107320, p. 23 (2020)

[Ar68] V.I. Arnold, On braids of algebraic functions and the cohomology of the "swallow tails". Russian Math. Surveys **23**(4), 247–248 (1968)

[Ar73-1] V.I. Arnold, Remarks on the method of stationary phase and on the Coxeter numbers. Russian Math. Surveys **28**(5), 19–48 (1973)

[Ar73-2] V.I. Arnold, A classification of the unimodal critical points of functions. Funct. Anal. Appl. **7**, 230–231 (1973)

[AGLV98] V.I. Arnold, V.V. Goryunov, O.V. Lyashko, V.A. *Vasil'ev, Singularity Theory I* (Springer, 1998)

[AGV85] V.I. Arnold, S.M. Gusein-Zade, A.N. Varchenko, *Singularities of Differentiable Maps*, vol. I (Birkhäuser, Boston, 1985)

[AGV88] V.I. Arnold, S.M. Gusein-Zade, A.N. Varchenko, *Singularities of Differentiable Maps*, vol. II (Birkhäuser, Boston, 1988)

[Ar25] E. Artin, Theorie der Zöpfe. Abh. Math. Sem. Univ. Hamburg **4**, 47–72 (1925)

[Ar47] E. Artin, Theory of braids. Annals of Mathematics **48**, 101–126 (1947)

[BK96] L. Balke, R. Kaenders, On a certain type of Coxeter-Dynkin diagrams of plane curve singularities. Topology **35**(1), 39–54 (1996)

[BH19] S. Balnojan, C. Hertling, Real Seifert forms and polarizing forms of Steenbrink mixed Hodge structures. Bull. Braz. Math. Soc., New Ser. **50**(1), 233–274 (2019)

[BH20] S. Balnojan, C. Hertling, Conjectures on spectral numbers for upper triangular matrices and for singularities. Math. Phys. Anal. Geom. **23**(5), p. 49 (2020). https://doi.org/10.1007/s11040-019-9327-3

C. Hertling and K. Larabi, *Triangular Bases of Unimodular Bilinear Lattices and Induced Structures*, Lecture Notes in Mathematics 2398,
https://doi.org/10.1007/978-3-032-25997-4

[BH25] S. Balnojan, C. Hertling, Characterization of distinguished matrices of isolated hypersurface singularities through their spectral numbers, p 67. Preprint arXiv:2508.15497v2

[BBGKK19] B. Baumeister, K.-U. Bux, F. Götze, D. Kielak, H. Krause, *Non-crossing partitions, in European Mathematical Society Series of Congress Reports* (EMS Publishing House, Zürich, 2019), pp. 235–274

[BDSW14] B. Baumeister, M. Dyer, C. Stump, P. Wegener, A note on the transitive Hurwitz action on decompositions of parabolic Coxeter elements. Proc. Amer. Math. Soc. Ser. B **1**, 149–154 (2014)

[BWY21] B. Baumeister, P. Wegener, S. Yahiatene, Extended Weyl groups and Hurwitz transitivity and weighted projective lines II: the wild case, p. 19. Preprint arXiv:2104.07075v2

[Be83] A.F. Beardon, *The Geometry of Discrete Groups*. Graduate Texts in Mathematics, vol. 91 (Springer, 1983)

[Bi75] J.S. Birman, *Braids, Links and Mapping Class Groups*. Annals of Mathematics Studies, vol. 82 (1975)

[BH73] J.S. Birman, H.M. Hilden, *On Isotopies of Homeomorphisms of Riemann Surfaces*. Annals of Mathematics **97**, 424–439 (1973)

[BB05] A. Björner, F. Brenti, *Combinatorics of Coxeter Groups*. Graduate Texts in Mathematics, vol. 231 (Springer, New York, 2005)

[Bo90] A.I. Bondal, Representation of associative algebras and coherent sheaves. Math. USSR Izvestiya **34**(1), 23–42 (1990)

[BP94] A.I. Bondal, A.E. Polishchuk, Homological properties of associative algebras: the method of helices. Russian Acad. Sci. Izv. Math. **42**(2), 219–260 (1994)

[BO02] A. Bondal, D. Orlov, Derived categories of coherent sheaves, in *Proceedings of the International Congress of Mathematicians*, vol. II, (Beijing, 2002), pp. 47–56

[Bo04] A.I. Bondal, A symplectic groupoid of triangular bilinear forms and the braid group. Izvestiya: Math **68**(4), 659–708 (2004)

[BSh73] Z.I. Borevich, I.R. Shafarevich, *Number Theory* (Academic Press Inc, 1973),

[Br70] E. Brieskorn, Die Monodromie der isolierten Singularitäten von Hyperflächen. Manuscripta Math. **2**, 103–160 (1970)

[Br83] E. Brieskorn, Milnor Lattices and Dynkin Diagrams, in *Singularities*, ed. by P.Orlik. Proceedings of Symposia in Pure Mathematics, vol. 40, part 1 (1983), pp. 153–165

[Bu00] E.B. Burger, Exploring the Number Jungle. Student Mathematical Library, vol. 8 (American Mathematical Society, 2000)

[Ca65] J.W. S. Cassels, *An Introduction to Diophantine Approximation* (Cambridge University Press, 1965),

[CV93] S. Cecotti, C. Vafa, On classification of $N = 2$ supersymmetric theories. Commun. Math. Phys. **158**, 569–644 (1993)

[CHS25] W. Chang, F. Haiden, S. Schroll, Braid group actions on branched coverings and full exceptional sequences. Adv. Math. **472**, 110284, p. 24 (2025)

[Ch81] S.V. Chmutov, The monodromy groups of singularities of functions of two variables. Funct. Anal. Applications **15**(1), 61–66 (1981)

[Ch82] S.V. Chmutov, Monodromy groups of critical points of functions. Invent. Math. **67**, 123–131 (1982)

[Ch83] S.V. Chmutov, Monodromy groups of critical points of functions II. Invent. Math. **73**, 491–510 (1983)

[CDG24] G. Cotti, B. Dubrovin, D. Guzzetti, *Helix Structures in Quantum Cohomology of Fano Varieties*. Lecture Notes in Mathematics, vol. 2356 (Springer, 2024)

[De74] P. Deligne, Letter to Looijenga on March 9, 1974, Reprinted in the diploma thesis of P. Kluitmann, pp. 101–111. Also available on http://homepage.rub.de/christian.stump/Deligne_Looijenga_Letter_09-03-1974.pdf

[DW05] H. Derksen, J. Weyman, Quiver representations. Notices AMS **52**(2), 200–206 (2005)

[DW17] H. Derksen, J. Weyman, *An Introduction to Quiver Representations*. Graduate Studies in Mathematics, vol. 184 (AMS, 2017)

[dTVdB16] L. de Thanhoffer de Volcsey, M. van den Bergh, On an analogue of the Markov equation for exceptional collections of length 4, p. 20. Preprint arXiv:1607.04246v1

[Do83] I. Dolgachev, On the link space of a Gorenstein quasihomogeneous surface singularity. Math. Ann. **265**, 529–540 (1983)

[DM94] P. Du Bois, F. Michel, The integral Seifert form does not determine the topology of plane curve singularities. J. Algebraic Geometry **3**, 1–38 (1994)

[Du92] B. Dubrovin, Integrable systems in topological field theory. Nucl. Phys. B **379**, 627–689 (1992)

[Du96] B. Dubrovin, Geometry of 2D topological field theories, in *Integrable Systems and Quantum Groups. Montecatini*, Terme 1993, ed. by M. Francoviglia, S. Greno. Lecture Notes in Mathematics, vol. 1620 (Springer, 1996), pp. 120–348

[Du98] B. Dubrovin, Geometry and analytic theory of Frobenius manifolds, in *Proceedings of the International Congress of Mathematicians*, vol. II, (Berlin, 1998), pp. 315–326

[Du99] B. Dubrovin, Painlevé transcendents in two-dimensional topological field theory, in *Painlevé Property*, One Century Later, ed. by R. Conte. Cargèse 1996 (Springer, 1999), pp. 287–412

[Eb83] W. Ebeling, Milnor lattices and geometric bases of some special singularities, in *Noeuds, tresses et singularités, Monographie No 31 de l'enseignement Mathématique* (Genève, 1983), pp. 129–146

[Eb84] W. Ebeling, An arithmetic characterization of the symmetric monodromy groups of singularities. Invent. Math. **77**, 85–99 (1984)

[Eb87] W. Ebeling, *The Monodromy Groups of Isolated Singularities of Complete Intersections*. Lecture Notes in Mathematics, vol. 1293 (1987)

[Eb01] W. Ebeling, Funktionentheorie, *Differentialtopologie und Singularitäten* (Vieweg, 2001)

[Eb18] W. Ebeling, A note on distinguished bases of singularities. Topol. Appl. **234**, 259–268 (2018)

[Eb20] Ebeling, W., Distinguished bases and monodromy of complex hypersurface singularities, in *Handbook of Geometry and Topology of Singularities* I, ed. by J.L.C. Molina, L.D. Trâng, J. Seade (Springer, 2020), pp. 441–490

[Fo51] L.R. Ford, *Automorphic Functions* (Chelsea Publishing Company, New York, 1951)

[Fo77] O. Forster, *Riemannsche Flächen* (Springer, Berlin, Heidelberg, New York, 1977)

[Fr13] G. Frobenius, Über die *Markoffschen Zahlen* (Preuss. Akad. Wiss. Sitzungsber., 1913), pp. 458–487 or *Gesammelte Abhandlungen III* (Springer, 1968), pp. 598–627

[Ga73] A.M. Gabrielov, Intersection matrices for certain singularities. Funct. Anal. Appl. **7**, 182–193 (1973)

[Ga74-1] A.M. Gabrielov, Bifurcations, Dynkin diagrams and modality of isolated singularities. Funct. Anal. Appl. **8**, 94–98 (1974)

[Ga74-2] A.M. Gabrielov, Dynkin diagrams of unimodal singularities. Funct. Anal. Appl. **8**, 192–196 (1974)

[Ga79] A.M. Gabrielov, Polar curves and intersection matrices of singularities. Invent. Math. **54**, 15–22 (1979)

[GGI16] S. Galkin, V. Golyshev, H. Iritani, Gamma classes and quantum cohomology of Fano manifolds: gamma conjectures. Duke Math. J. **165**(11), 2005–2077 (2016)

[GH17] F. Gauß, C. Hertling, μ-constant monodromy groups and Torelli results for marked singularities, for the unimodal and some bimodal singularities, in *Singularities and Computer Algebra, Festschrift for Gert-Martin Greuel on the Occasion of his 70th Birthday*, ed. by W. Decker, G. Pfister, M. Schulze (Springer International Publishing, 2017), pp. 109–146

[GH18] F. Gauß, C. Hertling, μ-constant monodromy groups and Torelli results for the quadrangle singularities and the bimodal series. J. Singul. **18**, 119–215 (2018)

[Gi98] A.B. Givental, A mirror theorem for toric complete intersections, in Topological Field Theory, *Primitive Forms and Related Topics* (Kyoto, 1996). Progr. in Math. 160, (Birkhäuser, 1998), pp. 141–175

[Go94-1] A.L. Gorodentsev, Helix theory and nonsymmetrical bilinear forms, in *Algebraic Geometry and its Applications* (Yaroslavl', 1992), Aspects Math., E25, Friedr (Friedr. Vieweg, Braunschweig, 1994), pp. 47–59

[Go94-2] A.L. Gorodentsev, Non-symmetric orthogonal geometry of Grothendieck rings of coherent sheaves on projective spaces, p. 32. Preprint arXiv:alg-geom/9409005v1

[Gr61] A. Grothendieck, Séminaire de géométrie algébrique du Bois Marie 1960–1961, revêtements étales et groupe fondamental (SGA 1)

[GL12] M.A. Guest, C.-S. Lin, Some tt^* structures and their integral Stokes data. Commun. Number Theory Phys. **6**(4), 785–803 (2012)

[Gu74-1] S.M. Gusein-Zade, Intersection matrices for certain singularities of functions of two variables. Funct. Anal. Appl. **8**, 10–13 (1974)

[Gu74-2] S.M. Gusein-Zade, Dynkin diagrams of the functions of two variables. Funct. Anal. Appl. **8**, 295–300 (1974)

[Gu99] D. Guzzetti, Stokes matrices and monodromy of the quantum cohomology of projective spaces. Comm. Math. Phys. **207**(2), 341–383 (1999)

[Ha01] A. Hatcher, *Algebraic Topology* (Copyright A, Hatcher, 2001)

[He92] C. Hertling, Analytische Invarianten bei den unimodularen und bimodularen Hyperflächensingularitäten. Doctoral thesis. Bonner Math. Schriften **250**, (1992), p. 188

[HM99] C. Hertling, Y. Manin, Weak Frobenius manifolds. Int. Math. Res. Not. **1999**(*6*), 277–286 (1999)

[He02] C. Hertling, *Frobenius Manifolds and Moduli Spaces for Singularities*. Cambridge Tracts in Mathematics, vol. 151 (Cambridge University Press, 2002)

[He03] C. Hertling, tt^* geometry, Frobenius manifolds, their connections, and the construction for singularities. J. Reine Angew. Math. **555**, 77–161 (2003)

[He11] C. Hertling, μ-constant monodromy groups and marked singularities. Ann. Inst. Fourier, Grenoble 61(7), 2643–2680 (2011)

[HS11] C. Hertling, C. Sabbah, Examples of non-commutative Hodge structures. J. Inst. Math. Jussieu **10**(3), 635–674 (2011)

[He20] C. Hertling, Automorphisms with eigenvalues in S^1 of a $\mathbb{Z}$-lattice with cyclic finite monodromy. Acta Arith **192**(1), 1–30 (2020)

[HR21] C. Hertling, C. Roucairol, Distinguished bases and Stokes regions for the simple and the simple elliptic singularities, in *Moduli Spaces and Locally Symmetric Spaces*. ed. by L.-Z. Ji, S.-T. Yau, Surveys of Modern Mathematics 16, (Higher Education Press and International Press, Beijing-Boston, 2021), pp.39–106

[HM22-1] C. Hertling, M. Mase, The integral monodromy of cycle type singularities. J. Singul. **24**, 268–298 (2022)

[HM22-2] C. Hertling, M. Mase, The integral monodromy of isolated quasihomogeneous singularities. Algebra and Number Theory **16**(4), 955–1024 (2022)

[HM22-3] C. Hertling, M. Mase, The combinatorics of weight systems and characteristic polynomials of isolated quasihomogeneous singularities. J. Algebr. Comb. (2022), published online at https://doi.org/10.1007/s10801-022-01138-x

[HK16] A. Hubery, H. Krause, A categorification of non-crossing partitions. J. Eur. Math. Soc. **18**, 2273–2313 (2016)

[Hu90] J.E. Humphreys, *Reflection Groups and Coxeter Groups*. Cambridge Studies in Advanced Mathematics, vol. 29 (Cambridge University Press, 1990)

[IS10] K. Igusa, R. Schiffler, Exceptional sequences and clusters. J. Algebra **323**, 2183–2202 (2010)

[Ir09] H. Iritani, An integral structure in quantum cohomology and mirror symmetry for toric orbifolds. Adv. Math. **222**(3), 1016–1079 (2009)

[Ja83] W.A.M. Janssen, Skew-symmetric vanishing lattices and their monodromy groups. Math. Ann. **266**, 115–133 (1983)

[Ja85] W.A.M. Janssen, Skew-symmetric vanishing lattices and their monodromy groups. II. Math. Ann. **272**, 17–22 (1985)

[Ja86] P. Jaworski, Distribution of critical values of miniversal deformations of parabolic singularities. Invent. Math. **86**, 19–33 (1986)

[Ja88] P. Jaworski, Decomposition of parabolic singularities. Bull. Sci. Math., II Sér **112**, 143–176 (1988)

[Ka90] V.G. Kac, *Infinite-Dimensional Lie Algebras*, 3rd edn. (Cambridge University Press, 1990)

[Ka75] S.I. Kaliman, Holomorphic universal covering spaces of polynomials without multiple roots. Funct. Anal. Appl. **9**(1), 67–68 (1975)

[Ka93] S.I. Kaliman, The holomorphic universal covers of spaces of polynomials without multiple roots. Selecta Mathematica formerly Sovietica **12**, 395–405 (1993)

[KT08] C. Kassel, V. Turaev, *Braids Groups*. Graduate Texts in Mathematics, vol. 247 (Springer, 2008)

[Ki16] A. Kirillov, in Quiver Representations and Quiver Varieties. Graduate Studies in Mathematics, vol. 174 (AMS, 2016)

[Kl83] P. Kluitmann, Geometrische Basen des Milnorgitters einer einfach elliptischen Singularität. Diploma thesis (Bonn, 1983), p. 116

[Kl87] P. Kluitmann, Ausgezeichnete Basen erweiterter affiner Wurzelgitter. Doctoral thesis. Bonner Math. Schriften **185**, (1987), p. 142

[Ko97] H. Koch, Zahlentheorie. Algebraische Zahlen und Funktionen, (1997)

[Kr90] B. Krüger, Automorphe Mengen und die Artinschen Zopfgruppen. Doctoral thesis. Bonner Math. Schriften **207**, (1990), p. 111

[Ku14] A. Kuznetsov, Semiorthogonal decompositions in algebraic geometry, in *Proceedings of the International Congress of Mathematicians*, vol. II, (Seoul, 2014), pp. 635–660

[Ku17] A. Kuznetsov, Exceptional collections in surface-like categories. Sbornik: Math **208**(9), 1368–1398 (2017)

[Ku23] A. Kuznetsov, Semiorthogonal decompositions in families, in *ICM-International Congress of Mathematicians*, vol. 2 (EMS Press, 2023), pp. 1154–1200

[La09] K. Lamotke, Riemannsche Flächen, (Springer, 2009),

[La73] F. Lazzeri, A theorem on the monodromy of isolated singularities. Singularités a Cargèse, Astérisque **7–8**, 269–275 (1973)

[Le73] L.D. Tráng, Une application d'un théoréme d'A'Campo a l'equisingularité (1973). Preprint Centre de Math. de l'Ecole Polytechnique, Paris

[Le66] J. Lehner, *A Short Course in Automorphic Functions* (Holt, Rinehart and Winston, New York, 1966)

[LSh18] P. Leviant, E. Shustin, Morsifications of real plane curve singularities. J. Sing. **18**, 307–328 (2018)

[Lo74] E. Looijenga, The complement of the bifurcation variety of a simple singularity. Invent. Math. **23**, 105–116 (1974)

[Lo84] E.J.N. Looijenga, *Isolated Singular Points on Complete Intersections*. London Mathematical Society Lecture Note Series, vol. 77 (Cambridge University Press, 1984)

[Ly79] O.V. Lyashko, The geometry of bifurcation diagrams. Russian Math. Surv. **34**(3), 209–210 (1979)

[Ma83] B. Malgrange, La classification des connexions irrégulières à une variable, in *Mathematique et physique (Paris, 1979–1982), Progress in Mathematics*, vol. 37 (Birkhäuser, 1983), pp. 381–399

[Ma99] Y. Manin, *Frobenius Manifolds, Quantum Cohomology and Moduli Spaces*, vol. 47 (A.M.S. Colloquium Publications, 1999),

[Ma71] B. Maskit, On Poincaré's theorem for fundamental polygons. Adv. Math. **7**, 219–230 (1971)

[Maxima22] One homepage of the computer algebra system Maxima is this: https://maxima.sourceforge.io/, last modification 2022

[Mi19] T. Milanov, The period map for quantum cohomology of $\mathbb{P}^2$. Adv. Math. **351**, 804–869 (2019)
[Mi69] J. Milnor, On isometries of inner product spaces. Invent. Math. **8**, 83–97 (1969)
[MO70] J. Milnor, P. Orlik, Isolated singularities defined by weighted homogeneous polynomials. Topology **9**, 385–393 (1970)
[Mo69] L.J. Mordell, *Diophantine Equations* (Academic Press, New York, London, 1969)
[Ne98] A. Nemethi, The real Seifert form and the spectral pairs of isolated hypersurface singularities. Comp. Math. **98**, 33–41 (1995)
[NST25] A. Nakago, Y. Shiraishi, A. Takahashi, Frobenius manifold for the nodal quiver, p. 49. Preprint arXiv:2509.12722v1
[NT25] A. Nakago, A. Takahashi, A simple algebraic proof of the non-transitivity of the braid group action on full exceptional sequences, p. 9. Preprint arXiv:2512.03554v1
[OW71] P. Orlik, P. Wagreich, Isolated singularities of algebraic surfaces with $\mathbb{C}^*$ action. Annals of Mathematics **93**(2), 205–228 (1971)
[Or72] P. Orlik, On the Homology of Weighted Homogeneous Manifolds, in Proceedings of the second conference on compact transformation groups, ed. by H.T. Ku, L.N. Mann, J.L. Sicks, J.C. Su. Lecturer Notes in Mathematics **298**, (Springer, 1972), pp. 260–269
[OR77] P. Orlik, R.C. Randell, The monodromy of weighted homogeneous singularities. Invent. Math. **39**, 199–211 (1977)
[Ph83] F. Pham, Vanishing Homologies and the *n* Variable Saddlepoint Method, in Singularities, ed. by P. Orlik. Proceedings of Symposia in Pure Mathematics, vol. 40, part 2 (1983), pp. 319–333
[Ph85] F. Pham, La descente des cols par les onglets de Lefschetz, avec vues sur Gauss-Manin. Systèmes différentiels et singularités. Asterisques **130**, 11–47 (1985)
[Po82] H. Poincaré, Théorie des groupes fuchsiens. Acta Math. **1**, 1–62 (1882)
[RS15] T. Reichelt, C. Sevenheck, Logarithmic Frobenius manifolds, hypergeometric systems and quantum D-modules. J. Algebraic Geom. **24**, 201–281 (2015)
[Ru90] A.N. Rudakov et al., *Helices and vector bundles: Seminaire Rudakov*. London Mathematical Society Lecture Note Series, vol. 148 (Cambridge University Press, 1990)
[Sa02] C. Sabbah, *Déformations isomonodromiques et variétés de Frobenius*. Savoirs Actuels (EDP Sciences, Les Ulis, Mathématiques, 2002). English translation: *Isomonodromic Deformations and Frobenius Manifolds*. Universitext (Springer and EDP Sciences, 2007)
[Sa74] K. Saito, Einfach-elliptische Singularitäten. Invent. Math. **23**, 289–325 (1974)
[Sa82] K. Saito, A characterization of the intersection form of a Milnor's fiber for a function with an isolated critical point. Proc. Japan Acad. **58**, 79–81 (1982)
[SS20] F. Sanda, Y. Shamoto, An analogue of Dubrovin's conjecture. Ann. Inst. Fourier **70**, 621–682 (2020)
[ST71] M. Sebastiani, R. Thom, Un resultat sur la monodromie. Invent. Math. **13**, 90–96 (1971)
[Sp66] E.H. Spanier, *Algebraic Topology* (Springer, 1966)
[TZ23] A. Takahashi, H. Zhang, Number of full exceptional collections modulo spherical twists for elliptic orbifolds, p. 15. Preprint arXiv:2311:11110v1.pdf
[Tr13] M. Trivković, *Algebraic Theory of Quadratic Numbers* (Springer, 2013),
[vS09] D. van Straten, Personal Communication (2009)
[Va23] U. Varolgunes, Seifert form of chain-type invertible singularities. Kyoto J. Math. **63**(1), 155–193 (2023)
[Vi71] È.B. Vinberg, Discrete linear groups generated by reflections. Math. USSR Izvestija **5**(5), 1083–1119 (1971)
[Wa99] C.T.C. Wall, Notes on trimodal singularities. Manuscripta Math. **100**(2), 131–157 (1999)
[Wa80] B. Wajnryb, On the monodromy group of plane curve singularities. Math. Ann. **246**, 141–154 (1980)
[Yu90] J. Yu, Kombinatorische Geometrie der Stokesregionen. Doctoral thesis. Bonner Math. Schriften **212** (1990), p. 114

Index

Symbols

C. Hertling and K. Larabi, *Triangular Bases of Unimodular Bilinear Lattices and Induced Structures*, Lecture Notes in Mathematics 2398,
https://doi.org/10.1007/978-3-032-25997-4

The manufacturer's authorised representative in the EU is Springer Nature Customer Service Centre GmbH, Europaplatz 3, 69115 Heidelberg, Germany. If you have any concerns regarding our products, please contact ProductSafety@springernature.com

Printed and bound by CPI Group (UK) Ltd, Croydon, CR0 4YY
25/09/2026
02222848-0008